The Vitamins
Fourth Edition

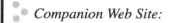

:• *Companion Web Site:*

http://www.elsevierdirect.com/companions/ 9780123819802

The Vitamins: Fundamental Aspects in Nutrition and Health
Gerald F. Combs, Jr.

Resources:

• All figures from the book available as both Power Point slides and .jpeg files

ELSEVIER

ACADEMIC
PRESS

The Vitamins
Fourth Edition

Gerald F. Combs, Jr
Professor Emeritus
Cornell University
Ithaca, NY

AMSTERDAM • BOSTON • HEIDELBERG • LONDON • NEW YORK • OXFORD • PARIS
SAN DIEGO • SAN FRANCISCO • SINGAPORE • SYDNEY • TOKYO
Academic Press is an imprint of Elsevier

Academic Press is an imprint of Elsevier
32 Jamestown Road, London NW1 7BY, UK
225 Wyman Street, Waltham, MA 02451, USA
525 B Street, Suite 1800, San Diego, CA 92101-4495, USA

First edition 1990
Second edition 1998
Third edition 2008
Fourth edition 2012

Notice
No responsibility is assumed by the publisher for any injury and/or damage to persons or property as a
matter of products liability, negligence or otherwise, or from any use or operation of any methods, products,
instructions or ideas contained in the material herein. Because of rapid advances in the medical sciences,
in particular, independent verification of diagnoses and drug dosages should be made

British Library Cataloguing-in-Publication Data
A catalogue record for this book is available from the British Library

Library of Congress Cataloging-in-Publication Data
A catalog record for this book is available from the Library of Congress

ISBN: 978-0-12-810244-2

For information on all Academic Press publications
visit our website at elsevierdirect.com

Typeset by MPS Limited, Chennai, India
www.macmillansolutions.com

Printed and bound in United States of America

12 13 14 15 16   10 9 8 7 6 5 4 3 2 1

to Barbara and Sylvia

Contents

Part II
Considering the Individual Vitamins

Contents

Part III
Using Current Knowledge of the Vitamins

Writing a book of this nature involves more than putting one's understanding on the page. For me, it involved going over again those points I thought I understood and paying special attention those in which my understanding was incomplete or to which pertinent new information had emerged. It involved searching for concordance among related findings and consensus among experts. It involved weighing the strength of evidence and considering the limitations of systematic reviews, meta-analyses and individual original reports. It involved finding clear ways to present complex information without over-striding its truth.

The vitamins occupy a central role in the field of Nutrition. Their discoveries, as factors that prevent specific diseases, marked the emergence of the field. Today, they are also important for their roles in the support of health in ways that often lack the specificity that facilitated their discoveries.

Understanding these latter roles has grown as a result of methodological trends. Improved instrumental analytical sensitivity has facilitated metabolic modeling. Application of molecular technologies has revealed polymorphisms in vitamin transporters, receptors, and enzymes that will, ultimately, allow quantitative needs for vitamins to be indexed to individual metabolic characteristics. Randomized clinical trials have produced databases from which conclusions of increased confidence can be drawn, including application of meta-analytical techniques to identify central findings.

In writing the fourth edition of *The Vitamins*, I was mindful of comments and suggestions from users of the third edition. Accordingly, I decided to maintain emphases on the roles of vitamins in human health, and to increase documentation to the key scientific literature through expanded sets of footnotes. I am most grateful for the professional assistance that I received from editors Ms Sue Armitage, Ms Caroline Johnson, Ms Carrie Bolger, and Ms Nancy Maragioglio of Academic Press.

I enjoyed writing this fourth edition of *The Vitamins*. I hope you will find it useful.

G. F. Combs, Jr
Grand Forks, North Dakota
July 2011

I have received several valuable criticisms concerning previous editions of *The Vitamins*. A couple have suggested that the text should cite more of the primary scientific literature. One suggested the inclusion of specific analytical methods. While I considered these suggestions in preparing this edition, I admit to not having responded to either to the degree that I suspect may have been intended.

In my 30 years of teaching in the academy, I became convinced that few students, if any, look to a text book, particularly one on a fairly broad topic, as a portal to the core scientific literature. Neither do I, as a scientist, expect such of the reference texts on my shelf. To get into the literature, I turn first to recent reviews, and then to PubMed and other search engines. So, with this in mind, I decided to maintain *The Vitamins* as a reference text, citing the key literature mostly as footnotes to the general conclusions they support.

As for methods, I resisted the temptation to include specific procedures. Because analytical methods, in most cases, evolve and improve with technology, I elected to address only the underlying characteristics of vitamins that can be exploited by technologies. I did, however, include a sampling of analytical protocols (Appendix B), which I expect will soon become out of date.

I did, however, make other changes that I believe enhance the usefulness of the book, particularly as a reference text. Most reflect new understanding in the field. For example, many "extra-vitamin A" roles of carotenoids are emerging, and genomics approaches are leading to new ways of understanding differences between individuals in vitamin needs and functions. The distribution of allelic variants of vitamin-dependent enzymes may explain anomalies previously considered simply as part of "biologic variability." This approach is likely to lead to individualized food and nutrition guidance.

G. F. Combs, Jr
Grand Forks, North Dakota
September, 2005

Since writing the first edition of *The Vitamins*, I have had a chance to reflect on that work and on the comments I have received from Cornell students and from various other instructors and health professionals who have used it at other institutions. This, of course, has been an appropriately humbling experience that has focused my thinking on ways to improve the book. In writing the second edition I have tried to make those improvements, but I have not changed the general format, which I have found to facilitate using the book as a classroom text.

Those familiar with the first edition will find the revision to be expanded and more detailed in presentation in several ways. The most important changes involve updating to include new information that has emerged recently; particularly, important findings in the areas of molecular biology and clinical medicine. This has included the addition of key research results, many in tabular form, and the expansion of the footnote system to include citations to important research papers. The reference lists following each chapter have been expanded, but these citations include mainly key research reviews, as in the first edition. While each chapter has been expanded in these ways, the most significant changes were made in the chapters on vitamin A (5), vitamin D (6), vitamin E (7), vitamin C (8) and folate (16), which areas have experienced the most rapid recent developments. Some smaller changes have been made in the structures of the vitamin chapters (5–18): each has been given an internal table of contents; an overview of the general significance of the vitamin in nutrition and health has been included; the topics of deficiency and toxicity are discussed under separate headings.

Perhaps the most striking aspect of the second edition will be its professional layout. That reflects the contributions of the editors and staff at Academic Press. They have made of this revision a book that is not only more handsome but also, I believe, easier to use than was the first edition, which I laid out on my own computer. I am grateful to Charlotte Brabants and Chuck Crumley, who handled *The Vitamins*, for their professionalism and for their patience.

I am also grateful to Dr Barbara Combs for her suggestions for the design and use of a teaching text in a discussion-based learning environment, and for her understanding in giving up so many weekends and evenings to my work on this project.

The Vitamins remains a book intended for use by nutrition instructors, graduate students in nutrition, dietetics, food science and medicine, clinicians, biomedical scientists and other health professionals. It can be used as a teaching text or as a desk reference. It is my hope that it will be used – that it will become one of those highly annotated, slightly tattered, note-stuffed volumes that can be found on many bookshelves.

G. F. Combs, Jr
Ithaca, New York
August, 1997

I have found it to be true that one learns best what one has to teach. And, because I have had no formal training either in teaching or in the field of education in general, it was not for several years of my own teaching that I began to realize that the good teacher must understand more than the subject matter of his or her course. In my case, that realization developed, over a few years, with the recognition that individuals learn in different ways, and that the process of learning itself is as relevant to my teaching as the material I present. This enlightenment has been for me invaluable because it has led me to the field of educational psychology from which I have gained at least some of the insights of the good teacher. In fact, it led me to write this book.

In exploring that field, I came across two books that have influenced me greatly: *A Theory of Education*[1] by another Cornell professor, Joe Novak, and *Learning How to Learn*[2] by Professor Novak and his colleague Bob Gowin. I highly recommend their work to any scientific "expert" in the position of teaching within the area of his or her expertise. From those books and conversations with Prof. Novak, I have come to understand that people think (and, therefore, learn) in terms of *concepts* – not facts. Therefore, for the past few years I have experimented in offering my course at Cornell University, "The Vitamins," in ways that are more concept-centered than I (or others, for that matter) have used previously. While I regard this experiment as an ongoing activity, it has already resulted in my shifting away from the traditional lecture format to one based on open classroom discussions aimed at involving the students, each of whom, I have found, brings a valuable personal perspective to discussions. I have found this to be particularly true for discussions concerning the vitamins; while it is certainly possible in modern societies to be misinformed about nutrition, it is virtually impossible to be truly naive. In other words, every person brings to the study of the vitamins some relevant conceptual framework, and it is thus the task of the teacher to build upon that framework by adding new concepts, establishing new linkages, and modifying existing ones where appropriate.

It quickly became clear to me that my own notes, indeed, all other available reference texts on the subject of the vitamins, were insufficient to support a concept-centered approach to the subject. Thus, I undertook to write a new type of textbook on the vitamins, one that would be maximally valuable in this kind of teaching. In so doing, I tried to focus on the key concepts and to make the book itself useful in a practical sense. Because I find myself writing in virtually any book that I really use, I gave this text margins wide enough for the reader to do the same. Because I have found the technical vocabularies of many scientific fields to present formidable barriers to learning, I have listed what I regard as the most important technical terms at the beginning of each chapter and have used each in context. Because I intend this to be an accurate synopsis of present understanding but not a definitive reference to the original scientific literature, I have cited only current major reviews that I find useful to the student. Because I have found the discussion of real-world cases to enhance learning of the subject, I have included case reports that can be used as classroom exercises or student assignments. I have designed the text for use as background reading for a one-semester upper-level college course within a nutrition-related curriculum. In fact, I have used draft versions in my course at Cornell as a means of refining it for this purpose.

While *The Vitamins* was intended primarily for use in teaching, I recognize that it will also be useful as a desk reference for nutritionists, dietitians, and many physicians, veterinarians, and other health professionals. Indeed, I have been gratified by the comments I have received from colleagues to that effect.

It is my hope that *The Vitamins* will be read, re-read, written in, and thought over. It seems to me that a field as immensely fascinating as the vitamins demands nothing less.

G.F. Combs, Jr
Ithaca, New York
August, 1991

---

1. Cornell University Press, Ithaca, NY, 1977, 324 pp.
2. Cambridge University Press, Ithaca, NY, 1984, 199 pp.

The Vitamins is intended as a teaching text for an upper-level college course within a nutrition or health-related curriculum; however, it will also be useful as a desk reference or as a workbook for self-paced study of the vitamins. It has several features that are designed to enhance its usefulness to students as well as instructors. Here is how I suggest using it.

TO THE STUDENT

When you use this text, make sure to have by your side a notebook, pencil (not pen – you may want to make changes in the notes you take), and medical dictionary. Then, before reading each chapter, take a few moments to go over the *Anchoring Concepts* and *Learning Objectives* listed on the chapter title page. *Anchoring Concepts* are the ideas fundamental to the subject matter of the chapter; they are the concepts to which the new ones presented in the chapter will be related. The *Anchoring Concepts* identified in the first several chapters should already be very familiar to you; if they are not, then it will be necessary for you to do some background reading or discussion until you feel comfortable in your understanding of these basic ideas. You will find that most chapters are designed to build upon the understanding gained through previous chapters; in most cases, the *Anchoring Concepts* of a chapter relate to the *Learning Objectives* of previous chapters. Pay attention to the *Learning Objectives*; they are the key elements of understanding that the chapter is intended to support. Keeping the *Learning Objectives* in mind as you go through each chapter will help you maintain focus on the key concepts. Next, read through the *Vocabulary* list and *mark* any terms that are unfamiliar or about which you feel unsure. Then, make a list of *your own questions* about the topic of the chapter.

As you read through the text, look for items related to your questions and for the terms that are unfamiliar to you. You will be able to find each in bold-face type, and you should be able to get a good feel for their meanings from the contexts of their uses. If this is not sufficient for any particular term, then look it up in your medical dictionary. Don't wait to do this. Cultivate the habit of being bothered by not understanding something – this will help you enormously in years to come.

As you proceed through the text, note what information the layout is designed to convey. First, note that the major sections of each chapter are indicated with a bold heading above a bar, and that the wide left margin contains key words and phrases that relate to the major topic of the text at that point. These are features that are designed to help you *scan* for particular information. Also note that the footnoted information is largely supplementary but not essential to the understanding of the key concepts presented. Therefore, the text may be read at two levels: at the basic level, one should be able to ignore the footnotes and still get the key concepts; at the more detailed level, one should be able to pick up more of the background information from the footnotes. Refer back frequently to your own list of questions and "target" vocabulary words; when you find an answer or can make a deduction, make a note. Don't be reluctant to write in the book, particularly to put a concept into your own words, or to note something you find important or don't fully understand.

When you have completed a chapter, take some time to list what you see as the key points – those that you would cover in a formal presentation. Then, skim back over the chapter.

You'll find that Chapters 5–17 are each followed by a *Case Study* comprised of one or more clinical case reports abstracted from the medical literature. For each case, use the associated questions to focus your thinking on the features that relate to vitamin functions. As you do so, try to ignore the obvious connection with the subject of the chapter; put yourself in the position of the attending physician who was called upon to diagnose the problem without prior knowledge that it involved any particular nutrient, much less a certain vitamin. You may find the additional case studies in Appendix B similarly interesting.

Take some time and go through the *Study Questions and Exercises* at the end of each chapter. These, too, are designed to direct your thinking back to the key concepts of each chapter, and to facilitate integration of those concepts with those you already have. To this end, you are asked in this section of several chapters to prepare a *concept map* of the subject matter. Many people find the *concept map* to be a powerful learning tool; therefore, if you have had no previous experience with this device, then it will be well worth your while to consult *Learning How to Learn*.[1]

When you have done all of this for a chapter, then deal with your questions. Discuss them with fellow students, or look them up. To assist you in the latter, a short reading list is included at the end of each chapter. With the exception of Chapter 2, which lists papers of landmark significance to the discovery of the vitamins, the reading lists consist of key reviews in prominent scientific journals. Thus, while primary research reports are not cited in the text, you should be able to trace research papers on topics of specific interest through the reviews that are listed.

After you have followed all of these steps, *re-read the chapter*. You will find this last step to be extraordinarily useful in gaining a command of the material.

Last, but certainly not least, have *fun* with this fascinating aspect of the field of nutrition!

TO THE INSTRUCTOR

I developed this format and presentation for my teaching a course called "The Vitamins" for over some 29 years at Cornell University. To that end, some of my experiences in using *The Vitamins* as a text for my course may be of interest to you.

I have found that *every* student comes to the study of the vitamins with *some* background knowledge of the subject, although those backgrounds are generally incomplete, frequently with substantial areas of no information and misinformation. This is true for upper-level nutrition majors and for students from other fields, the difference being largely one of magnitude. This is also true for instructors, most of whom come to the field with specific expertise that relates to only a subset of the subject matter. I demonstrate this simple fact in the first class by raising my index finger into the air (best done with a bit of dramatic flair) and saying "vitamin A." I hold that pose for 15–20 seconds, and then ask "What came to mind when I said 'vitamin A'?" Without fail, someone will say "vision" or "carrots," and some older graduate student may add "toxic"; most of the answers, by far, will relate to the clinical symptoms of vitamin A deficiency and the sources of vitamin A in diets. I catch each answer by dashing it on to a large Post-It® note that I stick haphazardly to the blackboard or wall behind me. If I hear something complex or a cluster of concepts, I make sure to question the contributor until I hear one or more individual concepts which I then record. This approach never fails to stimulate further answers, and it is common that a group of 15–20 students will generate a list of twice that number of concepts

before the momentum fades. Having used sticky notes, it is easy to move these around into clusters and thus to use this activity to construct a *concept map* of "Vitamin A" based solely on the knowledge that the students, collectively, brought into the room. I use this exercise to demonstrate what I regard to be an empowering idea: that, having at least *some* background on the subject and being motivated (by any of a number of reasons) to learn more, *every* learner brings to the study of the vitamins a unique perspective which may not be readily apparent.

I am convinced that meaningful learning is served when both instructor and students come to understand each others' various perspectives. This has two benefits in teaching the vitamins. First, it is in the instructor's interest to know the students' ideas and levels of understanding concerning issues of vitamin need, vitamin function, etc., such that these can be built upon and modified as may be appropriate. Second, I have found that many upper-level students have interesting experiences (through personal or family histories, their own research, information from other courses, etc.) that can be valuable contributions to classroom discussions. These experiences are assets that can reduce the temptation to fall back on the "instructor knows all" notion, which we all know to be false. To identify student perspectives, I have found it useful to assign on the first class period, for submission at the second class, a written autobiographical sketch. I distribute one I wrote for this purpose, and I ask each student to write "as much or as little" as he or she cares to, recognizing that I will distribute copies of whatever is submitted to each student in the class. The biographical sketches that I see range from a few sentences that reveal little of a personal nature, to longer ones that provide many good insights about their authors; I have found *every one* to help me get to know my students personally and to get a better idea of their understandings of the vitamins and of their expectations of my course. The exercise serves the students in a similar manner, thus promoting a group dynamic that facilitates classroom discussions.

I have used *The Vitamins* as the text from which I make regular reading assignments, usually a chapter at a time, for preparation for each class, which I generally conduct in an open discussion format. Long ago, I found it difficult, if not impossible, to cover in a traditional lecture format all of the information about the vitamins I deemed important for a nutritionist to know. Thus, I put that information into this text, which has allowed me to shift more of the responsibility for learning to the student to glean from assigned reading. As a result, I can use class time to assist the student by providing discussions of issues of particular interest or concern. Often, this means that certain points were not clear upon reading or that the reading itself stimulated questions not specifically addressed in the text. Usually, these questions are nicely handled by eliciting the

---

1. Novak, J. D., and Gowin, D. B. (1984). *Learning How to Learn.* Cambridge, University Press, New York, NY, 199 pp.

views and understandings of other students and by my giving supplementary information.

With this approach, my class preparation involves the collation of research data that will supplement the discussion in the text, and the identification of questions that I can use to initiate discussions. In developing my questions, I have found it useful to prepare my own concept maps of the subject matter and to ask rather simple questions about the linkages between concepts, e.g.: *"How does the mode of enteric absorption of the tocopherols relate to what we know about its physiochemical properties?"* If you are unfamiliar with concept mapping, then I strongly recommend your consulting *Learning How to Learn* and experimenting with the technique to determine whether it can assist you in your teaching.

I have found it useful to give weekly written assignments for which I use the ***Study Questions and Exercises*** or ***Case Studies***. In my experience, regular assignments keep students focused on the topic and prevent them from letting the course slide until exam time. More importantly, I believe there to be learning associated with the thought that necessarily goes into these written assignments. In order to support that learning, I make a point of going over each assignment briefly at the beginning of the class at which it is due, and of returning it by the *next* class with my written comments on *each* paper. You will find that the *Case Studies* I have included are abstracted from actual clinical reports; however, I have presented them without some of the pertinent clinical findings (e.g., responses to treatments) that were originally reported, in order to make of them learning exercises. I have found that students do well on these assignments, and that they particularly enjoy the *Case Studies*. For that reason, I have included in the revised edition additional case studies in Appendix C; I encourage you to use them in class discussions.

I evaluate student performance on the basis of class participation, weekly written assignments, a review of a recent research paper, and either one or two examinations (i.e., either a final, or a final plus a mid-term). In order to allow each student to pursue a topic of specific individual interest, I ask them to review a research paper published within the last year, using the style of *Nutrition Reviews*. I evaluate each review for its critical analysis, as well as on the importance of the paper that was selected, which I ask them to discuss. This assignment has also been generally well received. Because many students are inexperienced in research and thus feel uncomfortable in criticizing it, I have found it helpful to conduct in advance of the assignment a discussion dealing with the general principles of experimental design and statistical inference. Because I have adopted a concept-oriented teaching style, I long ago abandoned the use of short-answer questions (e.g., *"Name the species that require dietary sources of vitamin C"*) on examinations. Instead, I use brief case descriptions and actual experimental data and ask for diagnostic strategies, development of hypotheses, design of means of hypothesis testing, interpretation of results, etc. Many students may prefer the more traditional short-answer test; however, I have found that such inertia can be overcome by using examples in class discussions or homework assignments.

The Vitamins has been of great value in enhancing my teaching of the course by that name at Cornell. Thus, it is my sincere wish that it will assist you similarly in your teaching. I have been helped very much by the comments on the previous editions, which I have received both from my students and from instructors and others who have used this book. Please let me know how it meets your needs.

G.F. Combs, Jr

Perspectives on the Vitamins in Nutrition

Chapter 1

What is a Vitamin?

Chapter Outline

Anchoring Concepts

1. Certain factors, called *nutrients*, are necessary for normal physiological function of animals, including humans. Some nutrients cannot be synthesized adequately by the host, and must therefore be obtained from the external chemical environment; these are referred to as *dietary essential nutrients*.
2. *Diseases* involving physiological dysfunction, often accompanied by morphological changes, can result from insufficient intakes of dietary essential nutrients.

Imagination is more important than knowledge.

A. Einstein

Learning Objectives

1. To understand the classic meaning of the term *vitamin* as it is used in the field of nutrition.
2. To understand that the term *vitamin* describes both a concept of fundamental importance in nutrition as well as any member of a rather heterogeneous array of nutrients, any one of which may not fully satisfy the classic definition.
3. To understand that some compounds are vitamins for one species and not another, and that some are vitamins only under specific dietary or environmental conditions.
4. To understand the concepts of a *vitamer* and a *provitamin*.

VOCABULARY

Vitamer
Vitamin
Provitamin

1. THINKING ABOUT VITAMINS

Among the nutrients required for the many physiologic functions essential to life are the vitamins. Unlike other classes of nutrients, the vitamins do not serve structural functions, nor does their catabolism provide significant energy. Instead, their various uses each tend to be highly specific, and, for that reason, the vitamins are required in only small amounts in the diet. The common food forms of most vitamins require some metabolic activation to their functional forms.

Although the vitamins share these general characteristics, they show few close chemical or functional similarities, their categorization as vitamins being strictly empirical. Consider also that, whereas several vitamins function as enzyme cofactors (vitamins A, K, and C, thiamin, niacin, riboflavin, vitamin B_6, biotin, pantothenic acid, folate, and vitamin B_{12}), not all enzyme cofactors are vitamins.[1] Some vitamins function as biological antioxidants (vitamins E and C), and several function as cofactors in metabolic oxidation–reduction reactions (vitamins E, K, and C, niacin, riboflavin, and pantothenic acid). Two vitamins (vitamins A and D) function as hormones; one of them (vitamin A) also serves as a photoreceptive cofactor in vision.

1. Other enzyme cofactors are biosynthesized, e.g., heme, coenzyme Q, and lipoic acid.

The Vitamins. DOI: 10.1016/B978-0-12-381980-2.00001-3

2. VITAMIN: A REVOLUTIONARY CONCEPT

Everyday Word or Revolutionary Idea?

The term *vitamin*, today a common word in everyday language, was born of a revolution in thinking about the interrelationships of diet and health that occurred at the beginning of the twentieth century. That revolution involved the growing realization of two phenomena that are now so well understood they are taken for granted even by the layperson:

1. Diets are sources of many important nutrients.
2. Low intakes of specific nutrients can cause certain diseases.

In today's world each of these concepts may seem self-evident, but in a world still responding to and greatly influenced by the important discoveries in microbiology made in the nineteenth century, each represented a major departure from contemporaneous thinking in the area of health. Nineteenth-century physiologists perceived foods and diets as being sources of only four types of nutrients: *protein, fat, carbohydrate, ash,*[2] and *water*. After all, these accounted for very nearly 100% of the mass of most foods. With this view, it is understandable that, at the turn of the century, experimental findings that now can be seen as indicating the presence of hitherto unrecognized nutrients were interpreted instead as substantiating the presence of natural antidotes to unidentified disease-causing microbes.

Important discoveries in science have ways of directing, even entrapping, one's view of the world; resisting this tendency depends on critical and constantly questioning minds. That such minds were involved in early nutrition research is evidenced by the spirited debates and frequent polemics that ensued over discoveries of apparently beneficial new dietary factors. Still, the systematic development of what emerged as nutritional science depended on a new intellectual construct for interpreting such experimental observations.

Vitamin or Vitamine?

The elucidation of the nature of what was later to be called *thiamin* occasioned the proposition of just such a new construct in physiology.[3] Aware of the impact of what was a departure from prevailing thought, its author, the Polish biochemist Casimir Funk, chose to generalize from his findings on the chemical nature of that "vital amine"

to suggest the term *vitamine* as a generic descriptor for many such *accessory factors* associated with diets. That the factors soon to be elucidated comprised a somewhat chemically heterogeneous group, not all of which were nitrogenous, does not diminish the importance of the introduction of what was first presented as the *vitamine theory*, later to become a key concept in nutrition: the vitamin.

The term *vitamin* has been defined in various ways. While the very concept of a vitamin was crucial to progress in understanding nutrition, the actual definition of a vitamin has evolved in consequence of that understanding.

3. AN OPERATING DEFINITION OF A VITAMIN

For the purposes of the study of this aspect of nutrition, a vitamin is defined as follows. A vitamin…

- is an *organic compound* distinct from fats, carbohydrates, and proteins
- is a *natural component of foods* in which it is usually present in minute amounts
- is essential, also usually in minute amounts, for *normal physiological function* (i.e., maintenance, growth, development, and/or production)
- causes, by its absence or underutilization, a *specific deficiency syndrome*
- is *not synthesized by the host* in amounts adequate to meet normal physiological needs.

This definition will be useful in the study of vitamins, as it effectively distinguishes this class of nutrients from others (e.g., proteins and amino acids, essential fatty acids, and minerals) and indicates the needs in various normal physiological functions. It also points out the specificity of deficiency syndromes by which the vitamins were discovered. Further, it places the vitamins in that portion of the chemical environment on which animals (including humans) must depend for survival, thus distinguishing vitamins from hormones.

Some Caveats

It will quickly become clear, however, that, despite its usefulness, this operating definition has serious limitations, notably with respect to the last clause, for many species can indeed synthesize at least some of the vitamins. Four examples illustrate this point:

Vitamin C: Most animal species have the ability to synthesize ascorbic acid. Only those few that lack the enzyme L-gulonolactone oxidase (e.g., the guinea pig, humans) cannot; only for them can ascorbic acid properly be called vitamin C.

---

2. The residue from combustion, i.e., minerals.
3. This is a clear example of what T. H. Kuhn called a "scientific revolution" (Kuhn, T. H. [1968] *The Structure of Scientific Revolutions*. University of Chicago Press, Chicago, IL), i.e., the discarding of an old paradigm with the invention of a new one.

TABLE 1.1 The Vitamins: Their Vitamers, Provitamins, and Functions

Group	Vitamers	Provitamins	Physiological functions
Vitamin A	Retinol	β-Carotene	Visual pigments; epithelial cell differentiation
	Retinal	Cryptoxanthin	
	Retinoic acid		
Vitamin D	Cholecalciferol (D₃)		Calcium homeostasis; bone metabolism
	Ergocalciferol (D₂)		
Vitamin E	α-Tocopherol		Membrane antioxidant
	γ-Tocopherol		
Vitamin K	Phylloquinones (K₁)		Blood clotting; calcium metabolism
	Menaquinones (K₂)		
	Menadione (K₃)		
Vitamin C	Ascorbic acid		Reductant in hydroxylations in the formation of collagen and carnitine, and in the metabolism of drugs and steroids
	Dehydroascorbic acid		
Vitamin B₁	Thiamin		Coenzyme for decarboxylations of 2-keto acids (e.g., pyruvate) and transketolations
Vitamin B₂	Riboflavin		Coenzyme in redox reactions of fatty acids and the tricarboxylic acid (TCA) cycle
Niacin	Nicotinic acid		Coenzyme for several dehydrogenases
	Nicotinamide		
Vitamin B₆	Pyridoxol		Coenzyme in amino acid metabolism
	Pyridoxal		
	Pyridoxamine		
Folic acid	Folic acid		Coenzyme in single-carbon metabolism
	Polyglutamyl folacins		
Biotin	Biotin		Coenzyme for carboxylations
Pantothenic acid	Pantothenic acid		Coenzyme in fatty acid metabolism
Vitamin B₁₂	Cobalamin		Coenzyme in the metabolism of propionate, amino acids, and single-carbon units

Vitamin D: Individuals exposed to modest amounts of sunlight can produce cholecalciferol, which functions as a hormone. Only individuals without sufficient exposure to ultraviolet light (e.g., livestock raised in indoor confinement, people spending most of their days indoors) require dietary sources of vitamin D.

Choline: Most animal species have the metabolic capacity to synthesize choline; however, some (e.g., the chick, the rat) may not be able to employ that capacity if they are fed insufficient amounts of methyl-donor compounds. In addition, some (e.g., the chick) do not develop that capacity fully until they are several weeks of age. Thus, for the young chick and for individuals of other species fed diets providing limited methyl groups, choline is a vitamin.

Niacin: All animal species can synthesize nicotinic acid mononucleotide (NMN) from the amino acid tryptophan. Only those for which this metabolic conversion is particularly inefficient (e.g., the cat, fishes) and others fed low dietary levels of tryptophan require a dietary source of *niacin*.

With these counterexamples in mind, the definition of a vitamin can be understood as having specific connotations for animal species, stage of development, diet or nutritional status, and physical environmental conditions.[4]

The "vitamin caveat:"

- Some compounds are vitamins for one species and not another.
- Some compounds are vitamins only under specific dietary or environmental conditions.

4. THE RECOGNIZED VITAMINS

Thirteen substances or groups of substances are now generally recognized as vitamins (see Table 1.1); others have been proposed.[5] In some cases, the familiar name is actually the generic descriptor for a family of chemically related compounds having qualitatively comparable metabolic activities. For example, the term *vitamin E* refers to those analogs of tocol or tocotrienol[6] that are active in preventing such syndromes as fetal resorption in the rat and myopathies in the chick. In these cases, the members of the same vitamin family are called *vitamers*. Some carotenoids can be metabolized to yield the metabolically active form of vitamin A; such a precursor of an actual vitamin is called a *provitamin*.

Study Questions and Exercises

1. What are the key features that define a vitamin?
2. What are the fundamental differences between vitamins and other classes of nutrients … between vitamins and hormones?
3. Using key words and phrases, list briefly what you know about each of the recognized vitamins.

---

4. For this reason, it is correct to talk about vitamin C for the nutrition of humans but ascorbic acid for the nutrition of livestock.

5. These include such factors as inositol, carnitine, bioflavonoids, pangamic acid, and laetrile, for some of which there is evidence of vitamin-like activity (see Chapter 19).

6. Tocol is 3,4-dihydro-2-methyl-2-(4,8,12-trimethyltridecyl)-6-chromanol; tocotrienol is the analog with double bonds at the 3′, 7′, and 11′ positions on the phytol side chain (*see* Chapter 7).

Discovery of the Vitamins

Chapter Outline

Anchoring Concepts

1. A scientific theory is a plausible explanation for a set of observed phenomena; because theories cannot be tested directly, their acceptance relies on a preponderance of supporting evidence.
2. A scientific hypothesis is a tentative supposition that is assumed for the purposes of argument or testing, and is thus used in the generation of evidence by which theories can be evaluated.
3. An empirical approach to understanding the world involves the generation of theories strictly by observation, whereas an experimental approach involves the undertaking of operations (experiments) to test the truthfulness of hypotheses.
4. Physiology is that branch of biology dealing with the processes, activities, and phenomena of life and living organisms, and biochemistry deals with the molecular bases for such phenomena. The field of nutrition, derived from both of these disciplines, deals with the processes by which animals or plants take in and utilize food substances.

When science is recognized as a framework of evolving concepts and contingent methods for gaining new knowledge, we see the very human character of science, for it is creative individuals operating from the totality of their experiences who enlarge and modify the conceptual framework of science.

J. D. Novak

Learning Objectives

1. To understand the nature of the process of discovery in the field of nutrition.
2. To understand the impact of the vitamine theory, as an intellectual construct, on that process of discovery.
3. To recognize the major forces in the emergence of nutrition as a science.
4. To understand that the discoveries of the vitamins proceeded along indirect lines, most often through the seemingly unrelated efforts of many people.
5. To know the key events in the discovery of each of the vitamins.
6. To become familiar with the basic terminology of the vitamins and their associated deficiency disorders.

VOCABULARY

Accessory factor
Anemia
Animal model
Animal protein factor
Ascorbic acid
β-Carotene
Beriberi
Biotin
Black tongue disease
Cholecalciferol
Choline
Dermatitis
Ergocalciferol
Fat-soluble A

The Vitamins. DOI: 10.1016/B978-0-12-381980-2.00002-5

Filtrate factor
Flavin
Folic acid
Germ theory
Hemorrhage
Lactoflavin
Niacin
Night blindness
Ovoflavin
Pantothenic acid
Pellagra
Polyneuritis
Prothrombin
Provitamin
Purified diet
Pyridoxine
Retinen
Riboflavin
Rickets
Scurvy
Thiamin
Vitamin A
Vitamin B
Vitamin B complex
Vitamin B_{12}
Vitamin B_2
Vitamin B_6
Vitamin C
Vitamin D
Vitamin E
Vitamin K
Vitamine
Water-soluble B
Xerophthalmia

1. THE EMERGENCE OF NUTRITION AS A SCIENCE

In the span of only five decades commencing at the very end of the nineteenth century, the vitamins were discovered. Their discoveries were the result of the activities of hundreds of people that can be viewed retrospectively as having followed discrete branches of intellectual growth. Those branches radiated from ideas originally derived inductively from observations in the natural world, each starting from the recognition of a relationship between diet and health. Subsequently, branches were pruned through repeated analysis and deduction – a process that both produced and proceeded from the fundamental approaches used in experimental nutrition today. Once pruned, the limb of discovery may appear straight to the naive observer. Scientific discovery, however, does not occur that way; rather, it tends to follow a zig-zag course, with many participants contributing many branches. In fact, the contemporaneous view of each participant may be that of

a thicket of tangled hypotheses and facts. The seemingly straightforward appearance of the emergent limb of discovery is but an illusion achieved by discarding the dead branches of false starts and unsupported hypotheses, each of which can be instructive about the process of scientific discovery.

With the discovery of the vitamins, therefore, nutrition moved from a largely observational activity to one that relied increasingly on hypothesis testing through experimentation; it moved from empiricism to science. Both the process of scientific discovery and the course of the development of nutrition as a scientific discipline are perhaps best illustrated by the history of the discovery of the vitamins.

2. THE PROCESS OF DISCOVERY IN NUTRITIONAL SCIENCE

Empiricism and Experiment

History shows that the process of scientific discovery starts with the synthesis of general ideas about the natural world from observations of particulars in it – i.e., an *empirical phase*. In the discovery of the vitamins, this initial phase was characterized by the recognition of associations between diet and human diseases, namely night blindness, scurvy, beriberi, rickets, and pellagra, each of which was long prevalent in various societies. The next phase in the process of discovery involved the use of these generalizations to generate hypotheses that could be tested experimentally – i.e., the *experimental phase*. In the discovery of the vitamins, this phase necessitated the development of two key tools of modern experimental nutrition: the **animal model** and the **purified diet**. The availability of both of these tools proved to be necessary for the discovery of each vitamin; in cases where an animal model was late to be developed (e.g., for pellagra), the elucidation of the identity of the vitamin was substantially delayed.

3. THE EMPIRICAL PHASE OF VITAMIN DISCOVERY

The major barrier to entering the empirical phase of nutritional inquiry proved to be the security provided by prescientific attitudes about foods that persisted through the nineteenth century. Many societies had observed that human populations in markedly contrasting parts of the world tended to experience similar health standards despite the fact that they subsisted on very different diets. These observations were taken by nineteenth-century physiologists to indicate that health was not particularly affected by the kinds of foods consumed. Foods were thought important as sources of the only nutrients known at the time: *protein, available energy*, and *ash*. While the "chemical revolution," led by the French scientist Antoine Lavoisier, started probing the elemental components and metabolic fates of these nutrients, the widely read ideas of the German chemist Justus von Liebig[1] resulted in protein being regarded as the only real essential nutrient, supporting both tissue growth and repair as well as energy production. In the middle part of the century, attention was drawn further from potential relationships of diet and health by the major discoveries of Pasteur, Liebig, Koch, and others in microbiology. For the first time, several diseases, first anthrax and then others, could be understood in terms of a microbial etiology. By the end of the century, germ theory, which proved to be of immense value in medicine, directed hypotheses for the etiologies of most diseases. The impact of this understanding as a barrier to entering the inductive phase of nutritional discovery is illustrated by the case of the Dutch physician Christian Eijkman, who found a water-soluble factor from rice bran to prevent a beriberi-like disease in chickens (now known to be the vitamin thiamin) and concluded that he had found a "pharmacological antidote" against the beriberi "microbe" presumed to be present in rice.

Diseases Linked to Diet

Nevertheless, while they appeared to have little effect on the prevailing views concerning the etiology of human disease, by the late 1800s several empirical associations had been made between diet and disease.

Diseases empirically associated with diet were:

Scurvy
Beriberi
Rickets
Pellagra
Night blindness.

Scurvy

For several centuries it has been known that **scurvy**, the disease involving apathy, weakness, sore gums, painful joints, and multiple hemorrhages, could be prevented by including in the diet green vegetables or fruits. Descriptions of cases in such sources as the Eber papyrus (*ca.* 1150 BC) and writings of Hippocrates (*ca.* 420 BC) are often cited to indicate that scurvy was prevalent in those ancient populations. Indeed, signs of the disease are said to have been found in the skeletal remains of primitive humans. Scurvy was common in northern Europe during the Middle Ages, a time when local agriculture provided few sources of vitamin C that lasted through the winter. In northern Europe, it was treated by eating cresses and spruce leaves. Scurvy was very highly prevalent among seamen, particularly those on ocean voyages to Asia during which they subsisted for months at a time on dried and salted foods. The Portuguese explorer Vasco da Gama reported losing more than 60% of his crew of 160 sailors in his voyage around the Cape of Good Hope in 1498. In 1535–1536, the French explorer Jacques Cartier reported that signs of scurvy were shown by all but three of his crew of 103 men (25 of whom died) during his second Newfoundland expedition. In 1595–1597, the first Dutch East Indies fleet lost two-thirds of its seamen due to scurvy. In 1593, the British admiral Richard Hawkins wrote that, during his career, he had seen some 10,000 seamen die of the disease.

The link between scurvy and preserved foods was long evident to seafarers. The first report of a cure for the disease appears to have been Cartier's description of the rapidly successful treatment of his crew with an infusion of the bark of Arborvitae (*Thuja occidentalis*) prepared by the indigenous Hurons of Newfoundland. By 1601, the consumption of berries, vegetables, scurvy-grass (*Cochlearis officianalis*, which contains as much ascorbic acid as orange juice), citrus fruits or juices was recognized as being effective in preventing the disease. In that year, the English privateer Sir James Lancaster introduced regular issues of lemon juice (three spoonfuls each morning) on one of his found ships, finding significantly less scurvy among treated sailors. Nevertheless, the prestigious London College of Physicians viewed scurvy as a "putrid" disease in which affected tissues became alkaline, and stated that other acids could be as effective as lemon juice in treating the disease. Accordingly, in the mid-1600s British ship's surgeons were supplied with vitriol (sulfuric acid).

Against this background, in 1747 a British naval surgeon, James Lind, conducted what has been cited as the first controlled clinical trial to compare various therapies recommended for scurvy in British sailors at sea. Lind's report, published 6 years later, described 12 sailors with scurvy whom he assigned in pairs to 2-week regimens including either lemons and oranges, dilute sulfuric acid, vinegar or other putative remedies. His results were clear: the pair treated with lemons and oranges recovered almost completely within 6 days; whereas no other treatment showed improvement. In 1753 he published his now-classic *Treatise on Scurvy*, which had great impact on the medical thought of the time, as it detailed past work on the subject (most of which was anecdotal) and also presented the results of his

---

1. In his widely read book *Animal Chemistry, or Organic Chemistry in its Application to Physiology and Pathology*, Liebig argued that the energy needed for the contraction of muscles, in which he was able to find no carbohydrate or fat, must come only from the breakdown of protein. Protein, therefore, was the only true nutrient.

experiments. Lind believed that citrus contained "a saponaceous, attenuating and resolving virtue [detergent action]" that helped free skin perspiration that had become clogged by sea air; however, his results were taken as establishing the value of fresh fruits in treating the disease. Accordingly, by the 1790s the British Navy had made it a regular practice to issue daily rations of lemon juice to all seamen – a measure that gave rise to the term "limey"[2] as a slang expression for a British seaman. In the early part of the nineteenth century, there remained no doubt of a dietary cause and cure of scurvy; even so, it would be more than a century before its etiology and metabolic basis would be elucidated. Outbreaks of scurvy continued in cases of food shortages: in British prisons, during the California gold rush, among troops in the Crimean War, among prisoners in the American Civil War, among citizens during the Siege of Paris in 1871, and among polar explorers in the early twentieth century.

Beriberi

It is said that signs consistent with **beriberi** (e.g., initial weakness and loss of feeling in the legs leading to heart failure, breathlessness and, in some cases, edema) are described in ancient Chinese herbals (*ca.* 2600 BC). Certainly, beriberi has been a historic disease prevalent in many Asian populations subsisting on diets in which polished (i.e., "white" or dehulled) rice is the major food. For example, in the 1860s, the Japanese navy experienced the disease affecting 30–40% of its seamen. Interesting clinical experiments conducted in the 1870s with sailors by Dr Kanehiro Takaki, a British trained surgeon who later became Director General of the Japanese Naval Medical Service, first noted an association between beriberi and diet: Japanese sailors were issued lower protein diets than their counterparts in European navies which had not seen the disease. Takaki conducted an uncontrolled study at sea in which he modified sailors' rations to increase protein intake by including more meat, condensed milk, bread, and vegetables at the expense of rice. This cut both the incidence and severity of beriberi dramatically, which he interpreted as confirmation of the disease being caused by insufficient dietary protein. The adoption of Takaki's dietary recommendations by the Japanese navy was effective – eliminating the disease as a shipboard problem by 1880 – despite the fact that his conclusion, reasonable in the light of contemporaneous knowledge, later proved to be incorrect.

Rickets

Rickets, the disease of growing bones, which manifests itself in children as deformations of the long bones (e.g.,

bowed legs, knock knees, curvatures of the upper and/or lower arms), swollen joints, and/or enlarged heads, is generally associated with the urbanization and industrialization of human societies. Its appearance on a wide scale was more recent and more restricted geographically than that of either scurvy or beriberi. The first written account of the disease is believed to be that of Daniel Whistler, who wrote on the subject in his medical thesis at Oxford University in 1645. A complete description of the disease was published shortly thereafter (in 1650) by the Cambridge professor Francis Glisson, so it is clear that by the middle of the seventeenth century rickets had become a public health problem in England. However, rickets appears not to have affected earlier societies, at least not on such a scale. Studies in the late 1800s by the English physician T. A. Palm showed that the mummified remains of Egyptian dead bore no signs of the disease. By the latter part of the century, the incidence of rickets among children in London exceeded one-third; by the turn of the century, estimates of prevalence were as high as 80% and rickets had become known as the "English disease." Noting the absence of rickets in southern Europe, Palm in 1890 was the first to point out that rickets was prevalent only where there is relatively little sunlight (e.g., in the northern latitudes). He suggested that sunlight exposure prevented rickets, but others held that the disease had other causes – e.g., heredity or syphilis. Through the turn of the century, much of the Western medical community remained either unaware or skeptical of a food remedy that had long been popular among the peoples of the Baltic and North Sea coasts, and that had been used to treat adult rickets in the Manchester Infirmary by 1848: cod liver oil. Not until the 1920s would the confusion over the etiology of rickets become clear.

Pellagra

Pellagra, the disease characterized by lesions of the skin and mouth, and by gastrointestinal and mental disturbances, also became prevalent in human societies fairly recently. There appears to have been no record of the disease, even in folk traditions, before the eighteenth century. Its first documented description, in 1735, was that of the Spanish physician Gaspar Casal, whose observations were disseminated by the French physician François Thiery, whom he met some years later after having been appointed as physician to the court of King Philip V. In 1755, Thiery published a brief account of Casal's observations in the *Journal de Vandermonde*; this became the first published report on the disease. Casal's own description was included in his book on the epidemic and endemic diseases of northern Spain, *Historia Natural y Medico de el Principado de Asturias*, which was published in 1762, i.e., 3 years after his death. Casal regarded the disease, popularly called *mal de la rosa*,

---

2. It is a curious fact that lemons were often called *limes*, a source of confusion to many writers on this topic.

as a peculiar form of leprosy. He associated it with poverty, and with the consumption of spoiled corn (maize).

In 1771, a similar dermatological disorder was described by the Italian physician Francesco Frapolli. In his work *Animadversiones in Morbum Volgo Pelagrum*, he reported the disease to be prevalent in northern Italy. In that region corn, recently introduced from America, had become a popular crop, displacing rye as the major grain. The local name for the disease was *pelagra*, meaning rough skin. There is some evidence that it had been seen as early as 1740. At any rate, by 1784 the prevalence of *pelagra* (now spelled *pellagra*) in that area was so great that a hospital was established in Legano for its treatment. Success in the treatment of pellagra appears to have been attributed to factors other than diet – e.g., rest, fresh air, water, sunshine. Nevertheless, the disease continued to be associated with poverty and the consumption of corn-based diets.

Following the finding of pellagra in Italy, the disease was reported in France by Hameau in 1829. It was not until 1845 that the French physician Roussel associated pellagra with Casal's *mal de la rosa*, and proposed that these diseases, including a similar disease called *flemma salada*,[3] were related or identical. To substantiate his hypothesis, Roussel spent 7 months of 1847 in the area where Casal had worked in northern Spain[4] investigating *mal de la rosa* cases; on his return, he presented to the French Academy of Medicine evidence in support of his conclusion. Subsequently, pellagra, as it had come to be called, was reported in Romania by Theodari in 1858, and in Egypt by Pruner Bey in 1874. It was a curiosity, not to be explained for years, that pellagra was never endemic in the Yucatán Peninsula, where the cultivation of corn originated; the disease was not reported there until 1896.

It is not known how long pellagra had been endemic in the United States; however, it became common early in the twentieth century. In 1912, J. W. Babcock examined the records of the state hospital of South Carolina and concluded that the disease had occurred there as early as 1828. It is generally believed that pellagra also appeared during or after the American Civil War (1861–1865), in association with food shortages in the southern states. It is clear from George Searcy's 1907 report to the American Medical Association that the disease was endemic at least in Alabama. By 1909 it had been identified in more than 20 states, several of

which had impaneled Pellagra Commissions, and a national conference on the disease was held in South Carolina.

Since it first appeared, pellagra was associated with poverty and with the dependence on corn as the major staple food. Ideas arose early on that it was caused by a toxin associated with spoiled corn, yet by the turn of the century other hypotheses were also popular. These included the suggestion of an infectious agent with, perhaps, an insect vector.

Night Blindness

Night blindness, the inability to see under low levels of light, was one of the first recorded medical conditions. Writings of Ancient Greek, Roman, and Arab physicians show that animal liver was known to be effective in both the prevention and cure of the disease. The Eber papyrus (*ca.* 1150 BC) described its treatment by the squeezing of liquid from a lamb's liver (now known to be a good source of vitamin A in well-nourished animals) directly into the eyes of the affected patient. The use of liver for the prevention of night blindness became a part of the folk cultures of most seafaring communities. In the 1860s, Hubbenet and Bitot noted the presence of small, foamy white spots on the outer aspects of the conjunctiva of patients with night blindness – lesions that have become known as "Bitot's spots." Corneal ulceration, now known to be a related condition resulting in permanent blindness, was recognized in the eighteenth and nineteenth centuries in association with protein-energy malnutrition as well as such diseases as meningitis, tuberculosis, and typhoid fever. In Russia, it occurred during long Lenten fasts. In the 1880s, cod liver oil was found to be effective in curing both night blindness and early corneal lesions; by the end of the century, cod liver oil, meat, and milk were used routinely in Europe to treat both conditions. It was not until the early 1900s, however, that the dietary nature of night blindness, and the corneal lesions that typically ensued, was understood – not until the "active lipid" was investigated, i.e., the factor in cod liver oil that supported growth and prevented night blindness and xerophthalmia in the rat.

Ideas Prevalent by 1900

Thus, by the beginning of the twentieth century, four different diseases had been linked with certain types of diet. Further, by 1900 it was apparent that at least two, and possibly three, could be cured by changes in diet (see Table 2.1).

Other diseases, in addition to those listed in Table 2.1, had been known since ancient times to respond to what would now be called diet therapy. Unfortunately, much of this knowledge was overlooked, and its significance was not fully appreciated by a medical community galvanized by the new germ theory of disease. Alternative theories for the etiologies of these diseases were popular. Thus, as the

---

3. Literally meaning "salty phlegm," *flemma salada* involved gastrointestinal signs, delirium, and a form of dementia. It did not, however, occur in areas where maize was the major staple food; this, and disagreement over the similarities of symptoms, caused Roussel's proposal of a relationship between these diseases to be challenged by his colleague Arnault Costallat. From Costallat's letters describing *flemma salada* in Spain in 1861, it is apparent that he considered it to be a form of acrodynia, then thought to be due to ergot poisoning.

4. Casal practiced in the town of Oviedo in the Asturias of northern Spain.

TABLE 2.1 Diet–Disease Relationships Recognized by 1900

Disease	Associated Diet	Recognized Prevention
Scurvy	Salted (preserved) foods	Fresh fruits, vegetables
Beriberi	Polished rice-based	Meats, vegetables
Rickets	Few "good" fats	Eggs, cod liver oil
Pellagra	Corn-based	None
Night blindness	None	Cod liver oil

twentieth century began, it was widely held that scurvy, beriberi, and rickets were each caused by a bacterium or bacterial toxin rather than by the simple absence of something required for normal health. Some held that rickets might also be due to hypothyroidism, while others thought it to be brought on by lack of exercise or excessive production of lactic acid. These theories died hard, and had lingering deaths. In explanation of the lack of interest in the clues presented by the diet–disease associations outlined above, Harris (1955) mused:

Perhaps the reason is that it seems easier for the human mind to believe that ill is caused by some positive evil agency, rather than by any mere absence of any beneficial property.

Limitations of Empiricism

In actuality, the process of discovery of the vitamins had moved about as far as it could in its empirical phase. Further advances in understanding the etiologies of these diseases would require the rigorous testing of the various hypotheses – i.e., entrance into the deductive phase of nutritional discovery. That movement, however, required tools for productive scientific experimentation – tools that had not been available previously.

4. THE EXPERIMENTAL PHASE OF VITAMIN DISCOVERY

In a world where one cannot examine all possible cases (i.e., use strictly inductive reasoning), natural truths can be learned only by inference from premises already known to be true (i.e., through deduction). Both the inductive and deductive approaches may be linked; that is, probable conclusions derived from observation may be used as hypotheses for testing deductively in the process of scientific experimentation.

Requirements of Nutrition Research

In order for scientific experimentation to yield informative results, it must be both *repeatable* and *relevant*. The value of the first point, **repeatability**, should be self-evident. Inasmuch as natural truths are held to be constant, non-repeatable results cannot be construed to reveal them. The value of the second point, **relevance**, becomes increasingly important when it is infeasible to test a hypothesis in its real-world context. In such circumstances, it becomes necessary to employ a representation of the context of ultimate interest – a construct known in science as a **model**. Models are born of practical necessity, but they must be developed carefully in order to serve as analogs of situations that cannot be studied directly.

Defined Diets Provided Repeatability

Repeatability in nutrition experimentation became possible with the use of *diets of defined composition*. The most useful type of defined diet that emerged in nutrition research was the **purified diet**. Diets of this type were formulated using highly refined ingredients (e.g., isolated proteins, refined sugars and starches, refined fats) for which the chemical composition could be reasonably well known. It was the use of defined diets that facilitated experimental nutrition; such diets could be prepared over and over by the same or other investigators to yield comparable results. Results obtained through the use of defined diets were repeatable, and therefore predictable.

Appropriate Animal Models Provided Relevance

Relevance in nutrition research became possible with the identification of **animal models**[5] appropriate to diseases of interest in human medicine, or to physiological processes of interest in human medicine or animal production. The first of these was discovered quite by chance by keen observers studying human disease. Ultimately, the use of animal models would lead to the discovery of each of the vitamins, as well as to the elucidation of the nutritional

---

5. In nutrition and other biomedical research, an animal model consists of the experimental production in a conveniently managed animal species of biochemical and/or clinical changes that are comparable to those occurring in another species of primary interest but that may be infeasible, unethical, or uneconomical to study directly. Animal models are, frequently, species with small body weights (e.g., rodents, chicks, rabbits); however, they may also be larger species (e.g., monkeys, sheep), depending on the target problem and species they are selected to represent. In any case, background information on the biology and husbandry should be available. The selection and/or development of an animal model should be based primarily on representation of the biological problem of interest without undue consideration of the practicalities of cost and availability.

roles and metabolic functions of each of the approximately 40 nutrients. The careful use of appropriate animal models made possible studies that would otherwise be infeasible or unthinkable in human subjects or in other animal species of interest.

Major Forces in the Emergence of Nutritional Science

- Recognition that certain diseases were related to diet
- Development of appropriate animal models
- Use of defined diets.

An Animal Model for Beriberi

The analytical phase of vitamin discovery, indeed modern nutrition research itself, was entered with the finding of an animal model for beriberi in the 1890s. In 1886, Dutch authorities sent a commission led by Cornelius Pekalharing to their East Indian colony (now Indonesia) to find the cause of beriberi, which had become such a problem among Dutch soldiers and sailors as to interrupt military operations in Atjeh, Sumatra. Pekalharing took an army surgeon stationed in Batavia (now Jakarta), Christian Eijkman, whom he had met when each was on study leave (Pekalharing from his faculty post at the University of Utrecht, and Eijkman as a medical graduate from the University of Amsterdam) in the laboratory of the great bacteriologist, Robert Koch. The team, unaware of Takaki's work, expected to find a bacterium as the cause, and was therefore disappointed, after 8 months of searching, to uncover no such evidence. They concluded:

Beriberi has been attributed to an insufficient nourishment and to misery: but the destruction of the peripheral nervous system on such a large scale is not caused by hunger or grief. The true cause must be something coming from the outside, but is it a poison or an infection?

However, looking for a poison, they observed, would be very difficult, whereas they had techniques for looking for a microorganism that had been successful for other diseases. Thus, they tried to culture organisms from blood smears from patients, and to create the disease in monkeys, rabbits, and dogs by inoculations of blood, saliva, and tissues from patients and cadavers. When single injections produced no effects, they used multiple injection regimens. Despite the development of abscesses at the point of some injections, it appeared that multiple inoculations could produce some nerve degeneration in rabbits and dogs. Pekalharing concluded that beriberi was indeed an infectious disease, but an unusual one requiring repeated reinfection of the host. Before returning to Holland,

Pekalharing persuaded the Dutch military to allow Eijkman to continue working on the beriberi problem.

The facilities used by the Commission at the Military Hospital Batavia became a new Laboratory for Bacteriology and Pathology of the colonial government, and Eijkman was named as director, with one assistant. His efforts in 1888 to infect rabbits and monkeys with Pekalharing's micrococcus were altogether unsuccessful, causing him to posit that beriberi must require a long time before the appearance of signs. The following year, he started using chickens as his animal model. Later in the year, he noted that many, regardless of whether they had been inoculated, lost weight and started walking with a staggering gait. Some developed difficulty standing, and died. Eijkman noted on autopsy no abnormalities of the heart, brain or spinal cord, but microscopic degeneration of the peripheral nerves, particularly in the legs. The latter were signs he had observed in people dying of beriberi. He was unable, though, to culture any consistent type of bacteria from the blood of affected animals. It would have been easy for Eijkman to dismiss the thought that this avian disease, which he called "polyneuritis," might be related to beriberi.

Serendipity or a Keen Eye?

After persisting in his flock for some 5 months, the disease suddenly disappeared. Eijkman reviewed his records and found that in June, shortly before the chickens had started to show paralysis, a change in their diet had been occasioned by failure of a shipment of feed-grade brown (unpolished) rice to arrive. His assistant had used, instead, white (polished) rice from the hospital kitchen. It turned out that this extravagance had been discovered a few months earlier by a new hospital superintendent, who had ordered it stopped. When Eijkman again fed the chickens brown rice, he found affected animals recovered completely within days.

With this clue, Eijkman immediately turned to the chicken as the animal model for his studies. He found chicks showed signs of **polyneuritis** within days of being fed polished rice, and that their signs disappeared even more quickly if they were then fed unpolished rice. It was clear that there was something associated with rice polishings that protected chickens from the disease. After discussing these results, Eijkman's colleague Adolphe Verdeman, the physician inspector of prisons in the colony, surveyed the use of polished and unpolished rice and the incidence of beriberi among inmates. His results (summarized in Table 2.2), later to be confirmed in similar epidemiological investigations by other groups, showed the advantage enjoyed by prisoners eating unpolished rice: they were much less likely to contract beriberi.

This kind of information, in conjunction with his experimental findings with chickens, allowed Eijkman to

TABLE 2.2 Beriberi in Javanese Prisons ca. 1890

Diet	Population	Cases	Prevalence (cases/10,000 people)
Polished rice	150,266	4200	279.5
Partially polished rice	35,082	85	24.2
Unpolished rice	96,530	86	8.9

investigate, by means of bioassay, the beriberi-protective factor apparently associated with rice husks.

Antiberiberi Factor Is Announced

Eijkman used this animal model in a series of investigations in 1890–1897, and found that the antipolyneuritis factor could be extracted from rice hulls with water or alcohol, that it was dialyzable, but that it was rather easily destroyed with moist heat. He concluded that the water-soluble factor was a "pharmacological antidote" to the "beriberi microbe," which, although still not identified, he thought to be present in the rice kernel proper. Apparently, Gerrit Grijns, who continued Eijkman's work in Batavia when the latter went back to Holland suffering from malaria in 1896, came to interpret these findings somewhat differently. Grijns went on to show that polyneuritis could be prevented by including mung bean (*Vigna radiata*) in the diet; this led to mung beans being found effective in treating beriberi. In 1901, Grijns suggested, for the first time, that beriberi-producing diets "lacked a certain substance of importance in the metabolism of the central nervous system." Subsequently, Eijkman came to share Grijn's view; in 1906, the two investigators published a now-classic paper in which they wrote:

There is present in rice polishings a substance different from protein, and salts, which is indispensable to health and the lack of which causes nutritional polyneuritis.

5. THE VITAMINE THEORY

Defined Diets Revealed Needs for Accessory Factors

The announcement of the antiberiberi factor constituted the first recognition of the concept of the vitamin, although the term itself was yet to be coined. At the time of Eijkman's studies, but a world removed and wholly separate, others were finding that animals would not survive when fed "synthetic" or "artificial" diets formulated with purified fats, proteins, carbohydrates, and salts – i.e.,

containing all of the nutrients then known to be constituents of natural foods. Such a finding was first reported by the Swiss physiologist Nicholai Lunin, in 1888, who found that the addition of milk to a synthetic diet supported the survival of mice. Lunin concluded:

A natural food such as milk must, therefore, contain besides these known principal ingredients small quantities of other and unknown substances essential to life.

Lunnin's finding was soon confirmed by several other investigators. By 1912, Rhömann in Germany, Socin in Switzerland, Pekalharing in The Netherlands, and Hopkins in Great Britain had each demonstrated that the addition of milk to purified diets corrected the impairments in growth and survival that were otherwise produced in laboratory rodents. The German physiologist Stepp took another experimental approach. He found it possible to extract, from bread and milk, factors required for animal growth. Although Pekalharing's 1905 comments, in Dutch, lay unnoticed by many investigators (most of whom did not read that language), his conclusions about what Hopkins had called the *accessory factor* in milk alluded to the modern concept of a vitamin:

If this substance is absent, the organism loses the power properly to assimilate the well known principal parts of food, the appetite is lost and with apparent abundance the animals die of want. Undoubtedly this substance not only occurs in milk but in all sorts of foodstuffs, both of vegetable and animal origin.

Perhaps the most important of the early studies with defined diets were those of the Cambridge biochemist Frederick Gowland Hopkins.[6] His studies demonstrated that the growth-promoting activities of accessory factors were independent of appetite, and that such factors prepared from milk or yeast were biologically active in very small amounts.

Two Lines of Inquiry

Therefore, by 1912, two independently developed lines of inquiry had revealed that foods contained beneficial factor(s) in addition to the nutrients known at the time. That these factor(s) were present and active in minute amounts was apparent from the fact that almost all of the mass of food was composed of the known nutrients.

6. Sir Frederick Gowland Hopkins, Professor of Biochemistry at Cambridge University, is known for his pioneering work in biochemistry, which involved not only classic work on accessory growth factors (for which he shared, with Christian Eijkman, the 1929 Nobel Prize in Medicine), but also the discoveries of glutathione and tryptophan.

Two Lines of Inquiry Leading to the Discovery of the Vitamins

- The study of substances that prevent deficiency diseases
- The study of accessory factors required by animals fed purified diets.

Comments by Hopkins in 1906 indicate that he saw connections between the accessory factors and the deficiency diseases. On the subject of the accessory growth factors in foods, he wrote:

No animal can live on a mixture of pure protein, fat and carbohydrate, and even when the necessary inorganic material is carefully supplied the animal still cannot flourish. The animal is adjusted to live either on plant tissues or the tissues of other animals, and these contain countless substances other than protein, carbohydrates and fats. In diseases such as rickets, and particularly scurvy, we have had for years knowledge of a dietetic factor; but though we know how to benefit these conditions empirically, the real errors in the diet are to this day quite obscure . . . They are, however, certainly of the kind which comprises these minimal qualitative factors that I am considering.

Hopkins demonstrated the presence of a factor(s) in milk that stimulated the growth of animals fed diets containing all of the then-known nutrients (see Fig. 2.1).

The Lines Converge

The discovery by Eijkman and Grijns had stimulated efforts by investigators in several countries to isolate the antiberiberi factor in rice husks. Umetaro Suzuki, of Imperial University Agricultural College in Tokyo, succeeded in preparing a concentrated extract from rice bran for the treatment of polyneuritis and beriberi. He called the active fraction "oryzanin," but could not achieve its purification in crystalline form. Casimir Funk, a Polish-born chemist schooled in Switzerland, Paris, and Berlin, and then working at the Lister Institute in London, concluded from the various conditions in which it could be extracted and then precipitated that the antipolyneuritis factor in rice husks was an organic base and, therefore, nitrogenous in nature. When he appeared to have isolated the factor, Funk coined a new word for it, with the specific intent of promoting the new concept in nutrition to which Hopkins had alluded. Having evidence that the factor was an organic base, and therefore an *amine*, Funk chose the term **vitamine**[7] because it was clearly *vital*, i.e., pertaining to life.

Funk's Theory

In 1912, Funk published his landmark paper presenting the **vitamine theory**; in it he proposed, in what some have referred to as a leap of faith, four different vitamines. That the concept was not a new one, and that not all of these factors later proved to be amines (hence, the change to **vitamin**[8]), are far less important than the focus the newly coined term gave to the diet–health relationship. Funk was not unaware of the importance of the term itself; he wrote:[9]

I must admit that when I chose the name "vitamine" I was well aware that these substances might later prove not all to be of an amine nature. However, it was necessary for me to use a name that would sound well and serve as a "catch-word."

Funk's Vitamines

- Antiberiberi vitamine
- Antirickets vitamine
- Antiscurvy vitamine
- Antipellagra vitamine.

Impact of the New Concept

The vitamine theory opened new possibilities in nutrition research by providing a new intellectual construct for interpreting observations of the natural world. No longer was the elucidation of the etiologies of diseases to be constrained by the germ theory. Thus, Funk's greatest contribution involves not the data generated in his laboratory, but rather the theory produced from his thoughtful review of information already in the medical literature of the time. This fact caused Harris (1955) to write:

The interpreter may be as useful to science as the discoverer. I refer here to any man[10] who is able to take a broad view of what has already been done by others, to collect evidence and discern through it all some common connecting link.

The real impact of Funk's theory was to provide a new concept for interpreting diet-related phenomena. As the educational psychologist Novak[11] observed more recently:

As our conceptual and emotional frameworks change, we see different things in the same material.

---

7. Harris (1955) reported that the word *vitamine* was suggested to Funk by his friend, Dr Max Nierenstein, Reader in Biochemistry at the University of Bristol.

8. The dropping of the *e* from *vitamine* is said to have been the suggestion of J. C. Drummond.
9. Funk, C. (1912). The etiology of the deficiency diseases. *J. State Med.* 20, 341–368.
10. Harris's word choice reveals him as a product of his times. Because it is clear that the process of intellectual discovery to which Harris refers does not recognize gender, it is more appropriate to read this word as *person*.
11. Novak, J. D. (1977). *A Theory of Education.* Cornell University Press, Ithaca, NY.

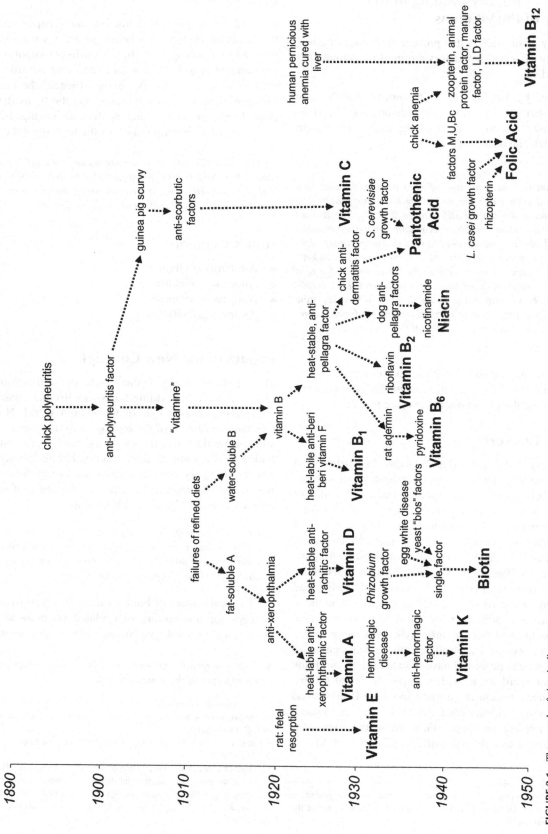

FIGURE 2.1   The cascade of vitamin discovery.

Still, it was not clear by 1912 whether the accessory factors were the same as the vitamines. In fact, until 1915 there was considerable debate concerning whether the growth factor for the rat was a single or multiple entity (it was already clear that there was more than one vitamine). Some investigators were able to demonstrate it in yeast and not butter; others found it in butter and not yeast. Some showed it to be identical with the antipolyneuritis factor; others showed that it was clearly different.

There is More Than One Accessory Factor

The debate was resolved by the landmark studies of the American investigator Elmer McCollum and his volunteer assistant Marguerite Davis at the University of Wisconsin in 1913–1915. Using diets based on casein and lactose, they demonstrated that at least two different additional growth factors were required to support normal growth of the rat. One factor could be extracted with ether from egg or butterfat (but not olive or cottonseed oils) but was non-saponifiable; it appeared to be the same factor shown earlier by the German physiologist Wilhelm Stepp, and by Thomas Osborne and Lafayette Mendel at Yale University in the same year, to be required to sustain growth of the rat. The second factor was extractable with water, and prevented polyneuritis in chickens and pigeons. McCollum called these factors **fat-soluble A** and **water-soluble B**, respectively (Table 2.3).

Accessory Factors Prevent Disease

Subsequent studies conducted by McCollum's group showed that the ocular disorders (i.e., **xerophthalmia**[12]) that developed in rats, dogs, and chicks fed fat-free diets could be prevented by feeding them cod liver oil, butter or preparations of fat-soluble A, which then became known as the *antixerophthalmic factor*. Shortly, it was found that the so-called water-soluble B material was not only required for normal growth of the rat, but also prevented polyneuritis in the chick. Therefore, it was clear that water-soluble B was identical to or at least contained Funk's antiberiberi vitamine; hence, it became known as *vitamine B*.

Accessory Factors Are the Same as Vitamines

With these discoveries, it became apparent that the biological activities of the accessory factors and the vitamines were likely to be due to the same compounds. The concept of a vitamine was thus generalized to include non-nitrogenous compounds, and the antipolyneuritis vitamine became **vitamin B**.

---

12. Xerophthalmia, from the Greek *xeros* ("dry") and *ophthalmos* ("eye"), involves dryness of the eyeball owing to atrophy of the periocular glands, hyperkeratosis of the conjunctiva, and, ultimately, inflammation and edema of the cornea, which leads to infection, ulceration, and blindness.

TABLE 2.3 McCollum's Rat Growth Factors

Factor	Found in:	Not found in:
Fat-soluble A	Milk fat, egg yolk	Lard, olive oil
Water-soluble B	Wheat, milk, egg yolk	Polished rice

Elucidation of the Vitamines

So it was, through the agencies of several factors, a useful new intellectual construct, the use of defined diets, and the availability of appropriate animal models, that nutrition emerged as a scientific discipline. By 1915, thinking about diet and health had been forever changed, and it was clear that the earlier notions about the required nutrients had been incomplete. Therefore, it should not be surprising to find, by the 1920s, mounting interest in the many questions generated by what had become sound nutritional research. That interest and the further research activity it engendered resulted, over the brief span of only five decades, in the development of a fundamental understanding of the identities and functions of about 40 nutrients, one-third of which are considered vitamins.

Crooked Paths to Discovery

The paths leading to the discovery of the vitamins wandered from Java with the findings of Eijkman in the 1890s, to England with Funk's theory in 1912, to the United States with the recognition of fat-soluble A and water-soluble B in 1915. By that time the paths had already branched, and for the next four decades they would branch again and again as scientists from many laboratories and many nations would pursue many unexplained responses to diet among many types of animal model. Some of these pursuits appeared to fail; however, in aggregate, all laid the groundwork of understanding on which the discoveries of those factors now recognized to be vitamins were based. When viewed in retrospect, the path to that recognition may seem deceptively straight – but it most definitely was not. The way was branched and crooked; in many cases, progress was made by several different investigators traveling in apparently different directions. The following recounts the highlights of the exciting search for the elucidation of the vitamins.

6. ELUCIDATION OF THE VITAMINS

New Animal Model Reveals New Vitamin: "C"

Eijkman's report of polyneuritis in the chicken and an animal model for beriberi stimulated researchers Axel Holst and Theodor Frölich at the University of Christiana in Oslo, who were interested in *shipboard beriberi*, a common problem among Norwegian seamen. Working with pigeons, they found a beriberi diet to produce the polyneuritis described by Eijkman; however, they considered that condition very different from the disease of sailors. In 1907, they attempted to produce the disease in another experimental animal species: the common Victorian household pet, the guinea pig. Contrary to their expectations, they failed to produce, by feeding that species a cereal-based diet, anything resembling beriberi; instead, they observed the familiar signs of scurvy. Eijkman's work suggested to them that, like beriberi, scurvy too might be due to a dietary deficiency. Having discovered, quite by chance, one of the few possible animal species in which scurvy could be produced,[13] Holst and Frölich had produced something of tremendous value – an animal model of scurvy[14] – showing that lesions could be prevented by feeding apples, (unboiled) cabbage, potatoes, and lemon juice.

This finding led Henriette Chick and Ruth Skelton of the Lister Institute, in the second decade of the twentieth century, to use the guinea pig to develop a bioassay for the determination of the antiscorbutic activity in foods, and S. S. Zilva and colleagues (also at the Lister Institute) to isolate from lemons the crude factor that had come to be known as **vitamin C**. It was soon found that vitamin C could reduce the dye 2,6-dichloroindophenol, but the reducing activity determined with that reagent did not always correlate with the antiscorbutic activity determined by bioassay. Subsequently, it was found that the vitamin was reversibly oxidized, but that both the reduced and oxidized forms had antiscorbutic activity.

In 1932, Albert Szent-Györgi, a Hungarian scientist working in Hopkins' laboratory at Cambridge University, and Glen King at the University of Pittsburgh established that the antiscorbutic factor was identical with the reductant *hexuronic acid*,[15] now called **ascorbic acid**.

Szent-Györgi had isolated it in crystalline form from adrenal cortex, while King had isolated it from cabbage and citrus juice.[16] After Szent-Györgi returned to Hungary to take a professorship, he was joined by an American-born Hungarian, J. Svirbely, who had been working in King's laboratory. Szent-Györgi had isolated *ca*. 500 grams of crystalline hexuronic acid from peppers, and then 25 g of the vitamin from adrenal glands, making samples available to other laboratories. On April 1, 1932, King and Waugh reported that their crystals protected guinea pigs from scurvy; two weeks later, Svirbely and Szent-Györgi reported virtually the same results. The following year, the chemical structure of ascorbic acid was elucidated by the groups of Haworth in Birmingham and Karrer in Zurich, both of which also achieved its synthesis.

Fat-Soluble A: Actually Two Factors

Pursuing the characterization of fat-soluble A, by 1919 McCollum's group[17] and others had found that, in addition to supporting growth for the rat, the factor also prevented xerophthalmia and night blindness in that species. In 1920, Drummond called the active lipid **vitamin A**.[18] This factor was present in cod liver oil, which at the turn of the century had been shown to prevent both xerophthalmia and night blindness – which Bitot, some 40 years earlier, had concluded had the same underlying cause.

Vitamin A Prevents Rickets?

Undoubtedly influenced by the recent recognition of vitamin A, Edward Mellanby, who had worked with Hopkins, undertook to produce a dietary model of rickets. For this he used puppies, which the Scottish physician Findley found developed rickets if kept indoors.[19] Mellanby fed a low-fat diet based on oatmeal with limited milk intake to puppies that he kept indoors; the puppies developed the marked skeletal deformities characteristic of rickets. When he found that these deformities could be prevented by feeding cod liver oil or butterfat without allowing the puppies outdoors, he concluded that rickets, too, was caused

---

13. Their finding was, indeed, fortuitous, as vitamin C is now known to be an essential dietary nutrient only for the guinea pig, primates, fishes, some fruit-eating bats, and some passiform birds. Had they used the rat, the mouse or the chick in their study, vitamin C might have remained unrecognized for perhaps quite a while.

14. In fact, scorbutic signs had been observed in the guinea pig more than a decade earlier, when a US Department of Agriculture pathologist noted in an annual report: "When guinea pigs are fed with cereals (bran and oats mixed), without any grass, clover or succulent vegetables, such as cabbage, a peculiar disease, chiefly recognizable by subcutaneous extravasation of blood, carries them off in four to eight weeks." That this observation was not published for a wider scientific audience meant that it failed to influence the elucidation of the etiology of scurvy.

15. It is said that when Szent-Györgi first isolated the compound, he was at a loss for a name for it. Knowing it to be a sugar, but otherwise ignorant of its identity, he proposed the name *ignose*, which was disqualified by an editor who did not appreciate the humor of the Hungarian chemist. Ultimately, the names *ascorbic acid* and *vitamin C*, by which several groups had come to refer to the antiscorbutic factor, were adopted.

16. The reports of both groups (King and Waugh, 1932; Svirbely and Szent-Györgi, 1932) appeared within 2 weeks of one another. In fact, Svirbely had recently joined Szent-Györgi's group, having come from King's laboratory. In 1937, King and Szent-Györgi shared the Nobel Prize for their work in the isolation and identification of vitamin C.

17. In 1917, McCollum moved to the newly established School of Public Health at Johns Hopkins University.

18. In 1920, J. C. Drummond proposed the use of the names *vitamin A* and *vitamin B* for McCollum's factors, and the use of the letters C, D, etc., for any vitamins subsequently to be discovered.

19. Exposing infants to sunlight has been a traditional practice in many cultures, and had been a folk treatment for rickets in northern Europe.

by a deficiency of vitamin A, which McCollum had discovered in those materials.

New Vitamin: "D"

McCollum, however, suspected that the antirachitic factor present in cod liver oil was different from vitamin A. Having moved to the Johns Hopkins University in Baltimore, he conducted an experiment in which he subjected cod liver oil to aeration and heating (100°C for 14 hours), after which he tested its antixerophthalmic and antirachitic activities with rat and chick bioassays, respectively. He found that heating had destroyed the antixerophthalmic (vitamin A) activity, but that cod liver oil had retained antirachitic activity. McCollum called the heat-stable factor **vitamin D**.

β-Carotene, a Provitamin

At about the same time (1919), Steenbock in Wisconsin pointed out that the vitamin A activities of plant materials seemed to correlate with their contents of yellow pigments. He suggested that the plant pigment *carotene* was responsible for the vitamin A activity of such materials. Yet the vitamin A activity in organic extracts of liver was colorless. Therefore, Steenbock suggested that carotene could not be vitamin A, but that it may be converted metabolically to the actual vitamin. This hypothesis was not substantiated until 1929, when von Euler and Karrer in Stockholm demonstrated growth responses to carotene in rats fed vitamin A-deficient diets. Further, Moore in England demonstrated, in the rat, a dose–response relationship between dietary β-**carotene** and hepatic vitamin A concentration. This showed that β-carotene is, indeed, a **provitamin**.

Vitamin A Linked to Vision

In the early 1930s, the first indications of the molecular mechanism of the visual process were produced by George Wald, of Harvard University but working in Germany at the time, who isolated the chromophore **retinen** from bleached retinas.[20] A decade later, Morton in Liverpool found that the chromophore was the aldehyde form of vitamin A – i.e., *retinaldehyde*. Just after Wald's discovery, Karrer's group in Zurich elucidated the structures of both β-carotene and vitamin A. In 1937, Holmes and Corbett succeeded in crystallizing vitamin A from fish liver. In 1942, Baxter and Robeson crystallized *retinol* and several

of its esters; in 1947, they crystallized the 13-*cis*-isomer. Isler's group in Basel achieved the synthesis of retinol in the same year, and that of β-carotene 3 years later.

The Nature of Vitamin D

McCollum's discovery of the antirachitic factor he called vitamin D in cod liver oil, which was made possible through the use of animal models, was actually a *rediscovery*, as that material had been long recognized as an effective medicine for rickets in children. Still, the nature of the disease was the subject of considerable debate, particularly after 1919, when Huldschinsky, a physician in Vienna, demonstrated the efficacy of ultraviolet light in healing rickets. This confusion was clarified by the findings in 1923 of Goldblatt and Soames, who demonstrated that when livers from rachitic rats were irradiated with ultraviolet light, they could cure rickets when fed to rachitic, non-irradiated rats. The next year, Steenbock's group demonstrated the prevention of rickets in rats by ultraviolet irradiation of either the animals themselves *or* their food. Further, the light-produced antirachitic factor was associated with the fat-soluble portion of the diet.[21]

Vitamers D

The ability to produce vitamin D (which could be bioassayed using both rat and chick animal models) by irradiating lipids led to the finding that large quantities of the vitamin could be produced by irradiating plant sterols. This led Askew's and Windaus's groups, in the early 1930s, to the isolation and identification of the vitamin produced by irradiation of *ergosterol*. Steenbock's group, however, found that, while the rachitic chick responded appropriately to irradiated products of cod liver oil or the animal sterol *cholesterol*, that animal did *not* respond to the vitamin D so produced from ergosterol. On the basis of this apparent lack of equivalence, Wadell suggested in 1934 that the irradiated products of ergosterol and cholesterol were different. Subsequently, Windaus's group synthesized 7-dehydrocholesterol and isolated a vitamin D-active product of its irradiation. In 1936 they reported its structure, showing it to be a side-chain isomer of the form of the vitamin produced from plant sterols. Thus, two forms of vitamin D were found: **ergocalciferol** (from ergosterol), which was called vitamin D_2;[22] and

---

20. For this and other discoveries of the basic chemical and physiological processes in vision, George Wald was awarded, with Haldan K. Hartline (of the United States) and R. Grant (of Sweden), the Nobel Prize in Chemistry in 1967.

21. This discovery, i.e., that by ultraviolet irradiation it was possible to induce vitamin D activity in such foods as milk, bread, meats, and butter, led to the widespread use of this practice, which has resulted in the virtual eradication of rickets as a public health problem.
22. Windaus's group had earlier isolated a form of the vitamin he had called vitamin D_1, which had turned out to be an irradiation-breakdown product, *lumisterol*.

cholecalciferol (from cholesterol), which was called vitamin D_3. While it was clear that the vitamers D had important metabolic roles in calcification, insights concerning the molecular mechanisms of the vitamin would not come until the 1960s. Then, it became apparent that neither vitamer was metabolically active *per se*; each is converted *in vivo* to a host of metabolites that participate in a system of calcium homeostasis that continues to be of great interest to the biomedical community.

Multiple Identities of Water-Soluble B

By the 1920s, it was apparent that the antipolyneuritis factor, called water-soluble B and present in such materials as yeasts, was not a single substance. This was demonstrated by the finding that fresh yeast could prevent both beriberi and pellagra. However, the antipolyneuritis factor in yeast was unstable to heat, while such treatment did not alter the efficacy of yeast to prevent dermititic lesions (i.e., **dermatitis**) in rodents. This caused Goldberger to suggest that the so-called vitamin B was actually at least *two* vitamins: the antipolyneuritis vitamin and a new antipellagra vitamin.

In 1926, the heat-labile antipolyneuritis/beriberi factor was first crystallized by Jansen and Donath, working in the Eijkman Institute (which replaced Eijkman's simple facilities) in Batavia. They called the factor *aneurin*. Their work was facilitated by the use of the small rice bird (*Munia maja*) as an animal model in which they developed a rapid bioassay for antipolyneuritic activity.[23] Six years later, Windaus's group isolated the factor from yeast, perhaps the richest source of it. In the same year (1932) the chemical structure was determined by R. R. Williams, who named it **thiamin** – i.e., the vitamin containing sulfur (*thios*, in Greek). Noting that deficient subjects showed high blood levels of pyruvate and lactate after exercise, in 1936 Rudolph Peters of Oxford University used, for the first time, the term "biochemical lesion" to describe the effects of the dietary deficiency. Shortly thereafter, methods of synthesis were achieved by several groups, including those of Williams, Andersag and Westphal, and Todd. In 1937, thiamin diphosphate (thiamin pyrophosphate) was isolated by Lohmann and Schuster, who showed it to be identical to the *cocarboxylase* that had been isolated earlier by Auhagen. That many research groups were actively engaged in the research on the antipolyneuritis/beriberi

factor is evidence of intense international interest due to the widespread prevalence of beriberi.

The characterization of thiamin clarified the distinction of the antiberiberi factor from the antipellagra activity. The latter was not found in maize (corn), which contained appreciable amounts of thiamin. Goldberger called the two substances the "A-N (antineuritic) factor" and the "P-P (pellagra-preventive) factor." Others called these factors vitamins F (for Funk) and G (for Goldberger), respectively, but these terms did not last.[24] By the mid-1920s the terms *vitamin B_1* and *vitamin B_2* had been rather widely adapted for these factors, respectively; this practice was codified in 1927 by the Accessory Food Factors Committee of the British Medical Research Council.

Vitamin B_2: A Complex of Several Factors

That the thermostable second nutritional factor in yeast, which by that time was called vitamin B_2, was not a single substance was not immediately recognized, giving rise to considerable confusion and delay in the elucidation of its chemical identity (identities). It should be noted that efforts to fractionate the heat-stable factor were guided almost exclusively by bioassays with experimental animal models. Yet, different species yielded discrepant responses to preparations of the factor. When such variation in responses among species was finally appreciated, it became clear that vitamin B_2 actually included *several* heat-stable factors. Vitamin B_2, as then defined, was indeed a complex.

Components of the Vitamin B_2 Complex

- The P-P factor (preventing pellagra in humans and pellagra-like diseases in dogs, monkeys, and pigs)
- A growth factor for the rat
- A pellagra-preventing factor for the rat
- An antidermatitis factor for the chick.

Vitamin B_2 Complex Yields Riboflavin

The first substance in the vitamin B_2 complex to be elucidated was the heat-stable, water-soluble rat growth factor, which was isolated by Kuhn, György, and Wagner-Jauregg at the Kaiser Wilhelm Institute in 1933. Those investigators found that thiamin-free extracts of autoclaved yeast, liver or rice bran prevented the growth failure of rats fed a thiamin-supplemented diet. Further, they noted that a yellow-green fluorescence in each extract promoted rat growth, and that the intensity of fluorescence was proportional to the

---

23. The animals, which consumed only 2 grams of feed daily, showed a high (98+%) incidence of polyneuritis within 9 to 13 days if fed white polished rice. The delay of onset of signs gave them a useful bioassay of antipolyneuritic activity suitable for use with small amounts of test materials. This point is not trivial, inasmuch as there is only about a teaspoon of thiamin in a ton of rice bran. The bioassay of Jansen and Donath was sufficiently responsive for $10\,\mu g$ of active material to be curative.

24. In fact, the name *vitamin F* was later used, with some debate as to the appropriateness of the term, to describe essential fatty acids. The name *vitamin G* has been dropped completely.

effect on growth. This observation enabled them to develop a rapid chemical assay that, in conjunction with their bio-assay, they exploited to isolate the factor from egg white in 1933. They called it **ovoflavin**. The same group then isolated, by the same procedure, a yellow-green fluorescent growth-promoting compound from whey (which they called **lactoflavin**). This procedure involved the adsorption of the active factor on fuller's earth,[25] from which it could be eluted with base.[26] At the same time, Ellinger and Koschara, at the University of Düsseldorf, isolated similar substances from liver, kidney, muscle, and yeast, and Booher in the United States isolated the factor from whey. These water-soluble growth factors became designated as **flavins**.[27] By 1934, Kuhn's group had determined the structure of the so-called flavins. These substances were thus found to be identical; because each contained a ribose-like (ribotyl) moiety attached to an isoalloxazine nucleus, the term **riboflavin** was adopted. Riboflavin was synthesized by Kuhn's group (then at the University of Heidelberg) and by Karrer's group at Zurich in 1935. As the first component of the vitamin B_2 complex, it is also referred to as vitamin B_2; however, that should not be confused with the earlier designation of the pellagra-preventive (P-P) factor.

Vitamin B_2 Complex Yields Niacin

Progress in the identification of the P-P factor was retarded by two factors: the pervasive influence of the germ theory of disease, and the lack of an animal model. The former made acceptance of evidence suggesting a nutritional origin of the disease a long and difficult undertaking. The latter precluded the rigorous testing of hypotheses for the etiology of the disease in a timely and highly controlled manner. These challenges were met by Joseph Goldberger, a US Public Health Service bacteriologist who, in 1914, was put in charge of the Service's pellagra program.

Pellagra: An Infectious Disease?

Goldberger's first study[28] is now a classic. He studied a Jackson, Mississippi, orphanage in which pellagra was endemic. He noted that whereas the disease was prevalent among the inmates, it was absent among the staff, including the nurses and physicians who cared for patients; this suggested to him that pellagra was not an infectious disease. Noting that the food available to the professional staff was much different from that served to the inmates (the former included meat and milk not available to the inmates), Goldberger suspected that an unbalanced diet was responsible for the disease. He secured funds to supply meat and milk to inmates for a 2-year period of study. The results were dramatic: pellagra soon disappeared, and no new cases were reported for the duration of the study. However, when funds expired at the end of the study and the institution was forced to return to its former meal program, pellagra reappeared. While the evidence from this uncontrolled experiment galvanized Goldberger's conviction that pellagra was a dietary disease, it was not sufficient to affect a medical community that thought the disease likely to be an infection.

Over the course of two decades, Goldberger undertook to elucidate the dietary basis of pellagra. Among his efforts to demonstrate that the disease was not infectious was the exposure, by ingestion and injection, of himself, his wife, and 14 volunteers to urine, feces, and biological fluids from *pellagrins*.[29] He also experimented with 12 male prisoners who volunteered to consume a diet (based on corn and other cereals, but containing no meat or dairy products) that he thought might produce pellagra: within 5 months half of the subjects had developed dermatitis on the scrotum, and some also showed lesions on their hands.[30] The negative results of these radical experiments, plus the finding that therapy with oral supplements of the amino acids cysteine and tryptophan was effective in controlling the disease, led, by the early 1920s, to the establishment of a dietary origin of pellagra. Further progress was hindered by the lack of an appropriate animal model. Although pellagra-like diseases had been identified in several species, most proved not to be useful as biological assays (indeed, most of these later proved to be manifestations of deficiencies of other vitamins of the B_2 complex and to be wholly unrelated to pellagra in humans).

The identification of a useful animal model for pellagra came from Goldberger's discovery in 1922 that maintaining dogs on diets essentially the same as those associated with human pellagra resulted in the animals developing a necrotic degeneration of the tongue called **black tongue disease**. This animal model for the disease led to the final solution of the problem.

---

25. Floridin, a nonplastic variety of kaolin containing an aluminum magnesium silicate. The material is useful as a decolorizing medium. Its name comes from an ancient process of cleaning or *fulling* wool, in which a slurry of earth or clay was used to remove oil and particulate dirt.
26. By this procedure, the albumen from 10,000 eggs yielded *ca.* 30 milligrams of riboflavin.
27. Initially, the term *flavin* was used with a prefix that indicated the source material; for example, ovoflavin, hepatoflavin, and lactoflavin designated the substances isolated from egg white, liver, and milk, respectively.
28. See the list of papers of key historical significance, in the section Recommended Reading at the end of this chapter.

---

29. People suffering from pellagra.
30. Goldberger conducted this study with the approval of prison authorities. As compensation for participation, volunteers were offered release at the end of the 6-month experimental period, which option each exercised immediately upon the conclusion of the study. For that reason, Goldberger was unable to demonstrate to a doubting medical community that the unbalanced diet had, indeed, produced pellagra.

Impact of an Animal Model for Pellagra

This finding made possible experimentation that would lead rather quickly to an understanding of the etiology to the disease. Goldberger's group soon found that yeast, wheat germ, and liver would prevent canine black tongue, and produce dramatic recoveries in pellagra patients. By the early 1930s, it was established that the human pellagra and canine black tongue curative factor was heat-stable and could be separated from the other B_2 complex components by filtration through fuller's earth, which adsorbed only the latter. Thus, the P-P factor became known as the *filtrate factor*. In 1937 Elvehjem isolated *nicotinamide* from liver extracts that had high antiblack-tongue activity, and showed that nicotinamide and *nicotinic acid* each cured canine black tongue. Both compounds are now called **niacin**. In the same year, several groups went on to show the curative effect of nicotinic acid against human pellagra.

It is ironic that the antipellagra factor was already well known to chemists of the time. Some 70 years earlier, the German chemist Huber had prepared nicotinic acid by the oxidation of nicotine with nitric acid. Funk had isolated the compound from yeast and rice bran in his search for the antiberiberi factor; however, because it had no effect on beriberi, nicotinic acid remained, for two decades, an entity with unappreciated biological importance. This view changed in the mid-1930s, when Warburg and Christian isolated nicotinamide from the hydrogen-transporting coenzymes I and II,[31] giving the first clue to its importance in metabolism. Within a year, Elvehjem had discovered its nutritional significance.

B_2 Complex Yields Pyridoxine

During the course of their work leading to the successful isolation of riboflavin, Kuhn and colleagues noticed an anomalous relationship between the growth-promoting and fluorescence activities of their extracts: the correlation of the two activities diminished at high levels of the former. Further, the addition of non-fluorescent extracts was necessary for the growth-promoting activity of riboflavin. They interpreted these findings as evidence for a second component of the heat-stable complex – one that was removed during the purification of riboflavin. These factors were also known to prevent dermatoses in the rat, an activity called *adermin*; however, the lack of a critical assay that could differentiate between the various components of the B_2 complex led to considerable confusion.

In 1934, György proffered a definition of what he called *vitamin B_6 activity*[32] as the factor that prevented what had formerly been called *acrodynia* or *rat pellagra*, which was a symmetrical florid dermatitis spreading over the limbs and trunk, with redness and swelling of the paws and ears. His definition effectively distinguished these signs from those produced by riboflavin deficiency, which involves lesions on the head and chest, and inflammation of the eyelids and nostrils. The focus provided by György's definition strengthened the use of the rat in the bioassay of vitamin B_6 activity by clarifying its end-point. Within 2 years, partial purification of **vitamin B_6** had been achieved by his group; and in 1938 (only 4 years after the recognition of the vitamin), the isolation of vitamin B_6 in crystalline form was achieved by five research groups. The chemical structure of the substance was quickly elucidated as 3-hydroxy-4,5-bis-(hydroxymethyl)-2-methylpyridine. In 1939, Folkers achieved the synthesis of this compound, which György called **pyridoxine**.

B_2 Complex Yields Pantothenic Acid

In the course of studying the growth factor called vitamin B_2, Norris and Ringrose at Cornell described, in 1930, a pellagra-like syndrome of the chick. The lesions could be prevented with aqueous extracts of yeast or liver, then recognized to contain the B_2 complex. In studies of B_2 complex-related growth factors for chicks and rats, Jukes and colleagues at Berkeley found positive responses to a thermostable factor that, unlike pyridoxine, was not adsorbed by fuller's earth from an acid solution. They referred to it as their *filtrate factor*.

At the same time, and quite independently, the University of Texas microbiologist R. J. Williams was pursuing studies of the essential nutrients for *Saccharomyces cerevisiae* and other yeasts. His group found a potent growth factor that they could isolate from a wide variety of plant and animal tissues.[33] They called it **pantothenic acid**, meaning "found everywhere," and also referred to the substance as *vitamin B_3*. Later in the decade, Snell's group found that several lactic and propionic acid bacteria require a growth factor that had the same properties. Jukes recognized that his filtrate factor, Norris's chick antidermatitis factor, and the unknown factors required by yeasts and bacteria were identical. He showed that both his filtrate factor and pantothenic acid obtained from Williams could prevent dermatitis in the chick. Pantothenic acid was isolated and its chemical structure was determined by

---

31. Nicotinamide adenine dinucleotide (NAD) and nicotinamide adenine dinucleotide phosphate (NADP), respectively.
32. György defined vitamin B_6 activity as "that part of the vitamin B-complex responsible for the cure of a specific dermatitis developed by rats on a vitamin-free diet supplemented with vitamin B_1, and lactoflavin."

33. The first isolation of pantothenic acid employed 250 kilograms of sheep liver. The autolysate was treated with fuller's earth; the factor was adsorbed to Norite and eluted with ammonia. Brucine salts were formed and were extracted with chloroform–water, after which the brucine salt of pantothenic acid was converted to the calcium salt. The yield was 3.0 grams of material with *ca.* 40% purity.

TABLE 2.4 Factors Leading to the Discovery of Pantothenic Acid

Factor	Bioassay
Filtrate factor	Chick growth
Chick antidermatitis factor	Prevention of skin lesions and poor feather development in chicks
Pantothenic acid	Growth of *S. cerevisiae* and other yeasts

Williams's group in 1939. The chemical synthesis of the vitamin was achieved by Folkers the following year.

The various factors leading to the discovery of pantothenic acid are presented in Table 2.4.

A Fat-Soluble, Anti-Sterility Factor: Vitamin E

Interest in the nutritional properties of lipids was stimulated by the resolution of fat-soluble A into vitamins A and D by the early 1920s. Several groups found that supplementation with the newly discovered vitamins A, C, and D and thiamin markedly improved the performance of animals fed purified diets containing adequate amounts of protein, carbohydrate, and known required minerals. However, H. M. Evans and Katherine Bishop, at the University of California, observed that rats fed such supplemented diets seldom reproduced normally. They found that fertility was abnormally low in both males (which showed testicular degeneration) and females (which showed impaired placental function and failed to carry their fetuses to term).[34] Dystrophy of skeletal and smooth muscles (i.e., those of the uterus) was also noted. In 1922, these investigators reported that the addition of small amounts of yeast or fresh lettuce to the purified diet would restore fertility to females, and prevent infertility in animals of both sexes. They designated the unknown fertility factor as *factor X*. Using the prevention of *gestation resorption* as the bioassay, Evans and Bishop found factor X activity in such unrelated materials as dried alfalfa, wheatgerm, oats, meats, and milk fat, from which it was extractable with organic solvents. They distinguished the new fat-soluble factor from the known fat-soluble vitamins by showing that single droplets of wheatgerm oil administered daily completely prevented gestation resorption, whereas cod liver oil, known to be a rich source of vitamins A and D, failed to do so.[35] In 1924, Sure, at the

University of Arkansas, confirmed this work, concluding that the fat-soluble factor was a new vitamin, which he called **vitamin E**.

A Classic Touch in Coining Tocopherol

Soon, Evans was able to prepare a potent concentrate of vitamin E from the unsaponifiable lipids of wheatgerm oil; others prepared similar vitamin E-active concentrates from lettuce lipids. By the early 1930s, Olcott and Mattill at the University of Iowa had found that such preparations, which prevented the gestation resorption syndrome in rats, also had chemical antioxidant properties that could be assayed *in vitro*.[36] In 1936, Evans isolated from unsaponifiable wheatgerm lipids allophanic acid esters of three alcohols, one of which had very high biological vitamin E activity. Two years later, Fernholz showed that the latter alcohol had a phytyl side chain and a hydroquinone moiety, and proposed the chemical structure of the new vitamin. Evans coined the term *tocopherol*, which he derived from the Greek words *tokos* ("childbirth") and *pherein* ("to bear");[37] he used the suffix *-ol* to indicate that the factor is an alcohol. He also named the three alcohols α-, β-, and γ-tocopherol. In 1938, synthesis of the most active vitamer, α-tocopherol, was achieved by the groups of Karrer, Smith, and Bergel. A decade later another vitamer, δ-tocopherol, was isolated from soybean oil; not until 1959 were the *tocotrienols* described.[38]

Antihemorrhagic Factor: Vitamin K

In the 1920s, Henrik Dam, at the University of Copenhagen, undertook studies to determine whether cholesterol was an essential dietary lipid. In 1929, Dam reported that chicks fed diets consisting of food that had been extracted with non-polar solvents to remove sterols developed subdural, subcutaneous or intramuscular **hemorrhages**, **anemia**, and abnormally long blood-clotting times. A similar syndrome in chicks fed ether-extracted fish meal was reported by McFarlane's group, which at the time was attempting to determine the chick's requirements for vitamins A and D. They found that non-extracted fish meal completely prevented the clotting defect. Holst and Holbrook found that cabbage prevented the syndrome,

---

34. The vitamin E-deficient rat carries her fetuses quite well until a fairly late stage of pregnancy, at which time they die and are resorbed by her. This syndrome is distinctive, and is called *gestation resorption.*

35. In fact, Evans and Bishop found that cod liver oil actually increased the severity of the gestation resorption syndrome, a phenomenon now understood on the basis of the antagonistic actions of high concentrations of the fat-soluble vitamins.

36. Although the potencies of the vitamin preparations in the *in vivo* (rat gestation resorption) and *in vitro* (antioxidant) assays were not always well correlated.

37. Evans wrote in 1962 that he was assisted in the coining of the name for vitamin E by George M. Calhoun, Professor of Greek and a colleague at the University of California. It was Calhoun who suggested the Greek roots of this now-familiar name.

38. The tocotrienols differ from the tocopherols only by the presence of three conjugated double bonds in their phytyl side chains.

which they took as evidence of an involvement of vitamin C. By the mid-1930s Dam had shown that the clotting defect was also prevented by a fat-soluble factor present in green leaves and certain vegetables, and distinct from vitamins A, C, D, and E. He named the fat-soluble factor **vitamin K**.[39]

At that time, Herman Almquist and Robert Stokstad, at the University of California, found that the hemorrhagic disease of chicks fed a diet based on ether-extracted fish meal and brewers' yeast, polished rice, cod liver oil, and essential minerals was prevented by a factor present in ether extracts of alfalfa, and that was also produced during microbial spoilage of fish meal and wheat bran. Dam's colleague, Schønheyder, discovered the reason for prolonged blood-clotting times of vitamin K-deficient animals. He found that the clotting defect did not involve a deficiency of tissue thrombokinase or plasma fibrinogen, or an accumulation of plasma anticoagulants; he also determined that affected chicks showed relatively poor thrombin responses to exogenous thromboplastin. The latter observation suggested inadequate amounts of the clotting factor **prothrombin**, a factor already known to be important in the prevention of hemorrhages.

In 1936, Dam partially purified chick plasma prothrombin and showed its concentration to be depressed in vitamin K-deficient chicks. It would be several decades before this finding was fully understood.[40] Nevertheless, the clotting defect in the chick model served as a useful bioassay tool. When chicks were fed foodstuffs containing the new vitamin, their prothrombin values were normalized; hence, clotting time was returned to normal and the hemorrhagic disease was cured. The productive use of this bioassay led to the elucidation of the vitamin and its functions.

Vitamers K

Vitamin K was first isolated from alfalfa by Dam in collaboration with Paul Karrer at the University of Zurich

in 1939. They showed the active substance, which was a yellow oil, to be a quinone. The structure of this form of the vitamin (called *vitamin K₁*) was elucidated by Doisy's group at the University of St Louis, and by Karrer's, Almquist's, and Feiser's groups in the same year. Soon, Doisy's group isolated a second form of the vitamin from putrified fish meal; this vitamer (called *vitamin K₂*) was crystalline. Subsequent studies showed this vitamer to differ from vitamin K₁ by having an unsaturated isoprenoid side chain at the 3-position of the naphthoquinone ring; in addition, putrified fish meal was found to contain several vitamin K₂-like substances with polyprenyl groups of differing chain lengths. Syntheses of vitamins K₂ were later achieved by Isler's and Folker's groups. A strictly synthetic analog of vitamers K₁ and K₂, consisting of the methylated head group alone (i.e., 2-methyl-1,4-naphthoquinone), was shown by Ansbacher and Fernholz to have high antihemorrhagic activity in the chick bioassay. It is therefore referred to as *vitamin K₃*.

Bios Yields Biotin

During the 1930s, independent studies of a yeast growth factor (called *bios IIb*[41]), a growth- and respiration-promoting factor for *Rhizobium trifolii* (called *coenzyme R*), and a factor that protected the rat against hair loss and skin lesions induced by raw egg white feeding (called *vitamin H*[42]) converged in an unexpected way. Kögl's group isolated the yeast growth factor from egg yolk and named it **biotin**. In 1940, György, du Vigneaud, and colleagues showed that vitamin H prepared from liver was remarkably similar to Kögl's egg yolk biotin. Owing to some reported differences in physical characteristics between the two factors, the egg yolk and liver substances were called, for a time,

39. Dam cited the fact that the next letter of the alphabet that had not previously been used to designate a known or proposed vitamin-like activity was also the first letter in the German or Danish phrase *koagulation facktor*, and was thus a most appropriate designator for the antihemorrhagic vitamin. The phrase was soon shortened to *K factor* and, hence, *vitamin K*.

40. It should be remembered that, at the time of this work, the biochemical mechanisms involved in clotting were incompletely understood. Of the many proteins now known to be involved in the process, only prothrombin and fibrinogen had been definitely characterized. It would not be until the early 1950s that the remainder of the now-classic clotting factors would be clearly demonstrated and that, of these, factors VII, IX, and X would be shown to be dependent on vitamin K. While these early studies effectively established that vitamin K deficiency results in impaired prothrombin activity, that finding would be interpreted as indicative of a vitamin K-dependent activation of the protein to its functional form.

41. *Bios IIb* was one of three essential factors for yeasts that had been identified by Wilders at the turn of the century in response to the great controversy that raged between Pasteur and Liebig. In 1860, Pasteur had declared that yeast could be grown in solutions containing only water, sugar, yeast ash (i.e., minerals), and ammonium tartrate; he noted, however, the growth-promoting activities of *albuminoid materials* in such cultures. Liebig challenged the possibility of growing yeast in the absence of such materials. Although Pasteur's position was dominant through the close of the century, Wilders presented evidence that proved that cultivation of yeast actually did require the presence of a little wort, yeast water, peptone or beef extract. (Wilders showed that an inoculum the size of a bacteriological loopful, which lacked sufficient amounts of these factors, was unsuccessful, whereas an inoculum the size of a pea grew successfully.) Wilders used the term *bios* to describe the new activity required for yeast growth. For three decades, investigators undertook to characterize Wilders's bios factors. By the mid-1920s, three factors had been identified: *bios I*, which was later identified as meso-inositol; *bios IIa*, which was replaced by pantothenic acid in some strains and by β-alanine plus leucine in others; and *bios IIb*, which was identified as biotin.

42. György used the designation *H* after the German word *haut*, meaning "skin."

TABLE 2.5 Factors Leading to the Discovery of Biotin

Factor	Bioassay
Bios IIb	Yeast growth
Coenzyme R	*Rhizobium trifolii* growth
Vitamin H	Prevention of hair loss and skin lesions in rats fed raw egg white

α-biotin and β-biotin,[43] respectively. These differences were later found to be incorrect, and the chemical structure of biotin was elucidated in 1942 by du Vigneaud's group at Cornell Medical College;[44] its complete synthesis was achieved by Folkers in the following year.

A summary of the factors leading to the discovery of biotin is presented in Table 2.5.

Anti-Anemia Factors

The last discoveries that led to the elucidation of new vitamins involved findings of anemias of dietary origin. The first of these was reported in 1931 by Lucy Wills's group as a *tropical macrocytic anemia*[45] observed in women in Bombay, India, which was often a complication of pregnancy. They found that the anemia could be treated effectively by supplementing the women's diet with an extract of autolyzed yeast.[46] Wills and associates found that a macrocytic anemia could be produced in monkeys by feeding them food similar to that consumed by the women in Bombay. Further, the monkey anemia could be cured by oral administration of yeast or liver extract, or by parenteral administration of extract of liver; these

treatments also cured human patients. The anti-anemia activity in these materials thus became known as the *Wills factor*.

Vitamin M?

Elucidation of the Wills factor involved the convergence of several lines of research, some of which appeared to be unrelated. The first of these came in 1935 from the studies of Day and colleagues at the University of Arkansas Medical School, who undertook to produce riboflavin deficiency in monkeys. They fed their animals a cooked diet consisting of polished rice, wheat, washed casein, cod liver oil, a mixture of salts, and an orange; quite unexpectedly, they found them to develop anemia, leukopenia,[47] ulceration of the gums, diarrhea, and increased susceptibility to bacillary dysentery. They found that the syndrome did not respond to thiamin, riboflavin or nicotinic acid; however, it could be prevented by feeding daily 10 grams of brewers' yeast or 2 grams of a dried hog liver–stomach preparation. Day named the protective factor in brewers' yeast *vitamin M* (for monkey).

Factors U and R, and Vitamin B$_c$

In the late 1930s, three groups (Robert Stokstad's at the University of California, Leo Norris's at Cornell, and Albert Hogan's at the University of Missouri) reported syndromes characterized by anemia in chicks fed highly purified diets. The anemias were found to respond to dietary supplements of yeast, alfalfa, and wheat bran. Stokstad and Manning called this unknown factor *factor U*; Baurenfeind and Norris called it *factor R*. Shortly thereafter, Hogan and Parrott discovered an anti-anemic substance in liver extracts; they called it *vitamin B$_c$*.[48] At the time (1939), it was not clear to what extent these factors may have been related.

Yeast Growth Related to Anemia?

At the same time, the microbiologists Snell and Peterson, who were studying the bios factors required by yeasts, reported the existence of an unidentified water-soluble factor that was necessary for the growth of *Lactobacillus casei*. This factor was present in liver and yeast, from which it could be prepared by adsorption to and then elution from Norit (a carbon-based filtering agent); for a while they called it the *yeast Norit factor*, but it quickly became known as the *L. casei factor*. Hutchings and colleagues

---

43. The substances derived from each source were reported as having different melting points and optical rotations, giving rise to these designations. Subsequent studies, however, have been clear in showing that such differences are not correct, nor do these substances show different biological activities in microbiological systems. Thus, the distinguishing of biotin on the basis of source is no longer valid.

44. du Vigneaud was to receive a Nobel Prize in Medicine for his work on the metabolism of methionine and methyl groups.

45. A *macrocytic anemia* is one in which the number of circulating erythrocytes is below normal, but the mean size of those present is greater than normal (normal range, 82–92 μm^3). Macrocytic anemias occur in such syndromes as pernicious anemia, sprue, celiac disease, and macrocytic anemia of pregnancy. Wills's studies of the macrocytic anemia in her monkey model revealed megaloblastic arrest (i.e., failure of the large, nucleated, embryonic erythrocyte precursor cell type to mature) in the erythropoietic tissues of the bone marrow, and a marked reticulocytosis (i.e., the presence of young red blood cells in numbers greater than normal [usually<1%], occurring during active blood regeneration); both signs were eliminated coincidentally on the administration of extracts of yeast or liver.

46. Wills's yeast extract was not particularly potent, as they needed to administer 4 grams of it two to four times daily to cure the anemia.

47. Leukopenia refers to any situation in which the total number of leukocytes (i.e., white blood cells) in the circulating blood is less than normal, which is generally *ca.* 5000 per mm^3.

48. Hogan and Parrott used the subscript *c* to designate this factor as one required by the chick.

at the University of Wisconsin further purified the factor from liver and found it to stimulate chick growth; this suggested a possible identity of the bacterial and chick factors. The factor from liver was found to stimulate the growth of both *L. helveticus* and *Streptococcus fecalis* R.,[49] whereas the yeast-derived factor was twice as potent for *L. helveticus* as it was for *S. fecalis*. Thus, it became popular to refer to these as the "liver *L. casei* factor" and the "yeast (or fermentation) *L. casei* factor."

Snell's group found that many green leafy materials were potent sources of something with the microbiological effects of the *Norit eluate factor* – i.e., extracts promoted the growth of both *S. fecalis* and *L. casei*. They named the factor, by virtue of its sources, **folic acid**. In 1943, a fermentation product was isolated that stimulated the growth of *S. fecalis* but not *L. casei*; this was called the *SLR factor* and, later, *rhizopterin*.

Who's on First?

It was far from clear in the early 1940s whether any of these factors were at all related, as folic acid appeared to be active for both microorganisms and animals, whereas concentrates of vitamin M, factors R and U, and vitamin B_c appeared to be effective only for animals. Clues to solving the puzzle came from the studies of Mims and associates at the University of Arkansas Medical School, who showed that incubation of vitamin M concentrates in the presence of rat liver enzymes caused a marked increase in the folic acid activity (i.e., assayed using *S. casei* and *S. lactis* R.) of the preparation. Subsequent work showed such "activation" enzymes to be present in both hog kidney and chick pancreas. Charkey, of the Cornell group, found that incubation of their factor R preparations with rat or chick liver enzymes produced large increases in their folic acid potencies for microorganisms. These studies indicated for the first time that at least some of these various substances may be related.

Derivatives of Pteroylglutamic Acid

The real key to solving what was clearly the most complicated puzzle in the discovery of the vitamins came in 1943 with the isolation of pteroylglutamic acid from liver by Stokstad's group at the Lederle Laboratories of American Cyanamid, Inc., and by Piffner's group at Parke-Davis, Inc. Stokstad's group achieved the synthesis of the compound in 1946. Soon it was found that pteroylmonoglutamic acid was indeed the substance that had been variously identified in liver as factor U, vitamin M, vitamin B_c, and the liver *L. casei* factor. The yeast *L. casei* factor was found to be the diglutamyl derivative (pteroyldiglutamic acid) and the liver-derived vitamin B_c was the hexaglutamyl derivative

TABLE 2.6 Factors Leading to the Discovery of Folic Acid

Factor	Bioassay
Wills' factor	Cure of anemia in humans
Vitamin M	Prevention of anemia in monkeys
Vitamin B_c	Prevention of anemia in chicks
Factor R	Prevention of anemia in chicks
Factor U	Prevention of anemia in chicks
yeast Norit factor	Growth of *L. casei*
L. casei factor	Growth of *L. casei*
SLR factor	Growth of *Rhizobium* species
Rhizopterin	Growth of *Rhizobium* species
Folic acid	Growth of *S. fecalis* and *L. casei*

(pteroylhexaglutamic acid). Others of these factors (the *SLR factor*) were subsequently found to be single-carbon metabolites of pteroylglutamic acid. These various compounds thus became known generically as **folic acid**.

A summary of the factors leading to the discovery of folic acid is presented in Table 2.6.

Antipernicious Anemia Factor

The second nutritional anemia that was found to involve a vitamin deficiency was the fatal condition of human patients that was first described by J. S. Combe in 1822, and became known as *pernicious anemia*.[50] The first real breakthrough toward understanding the etiology of pernicious anemia did not come until 1926, when Minot and Murphy found that lightly cooked liver, which the prominent hematologist G. H. Whipple had found to accelerate the regeneration of blood in dogs made anemic by exsanguination, was highly effective as therapy for the disease.[51,52] This indicated that liver contained a factor necessary for hemoglobin synthesis.

Intrinsic and Extrinsic Factors

Soon, studies of the antipernicious anemia factor in liver revealed that its enteric absorption depended on yet

---

49. *Streptococcus fecalis* was then called *S. lactis* R.

50. Pernicious anemia is also called *Addison's anemia* after T. Addison, who described it in great detail in 1949, and *Biemer's anemia*, after A. Biemier, who reported the disease in Zurich in 1872 and coined the term *pernicious anemia*.

51. Minot and Murphy treated 45 pernicious anemia patients with 120–240 grams of lightly cooked liver per day. The patients' mean erythrocyte count increased from 1.47×10^6 per milliliter before treatment to 3.4×10^6 per milliliter and 4.6×10^6 per milliliter after 1 and 2 months of treatment, respectively.

52. Whipple, Minot, and Murphy shared the 1934 Nobel Prize in Medicine for the discovery of whole liver therapy for pernicious anemia.

another factor in the gastric juice, which W. B. Castle, in 1928, called the *intrinsic factor*, to distinguish it from the *extrinsic factor* in liver. Biochemists then commenced a long endeavor to isolate the antipernicious anemia factor from liver. The isolation of the factor was necessarily slow and arduous for the reason that the only bioassay available was the hematopoietic response of human pernicious anemia patients, which was frequently not available. No animal model had been found, and a bioassay could not be replaced by a chemical reaction or physical method because, as is now known, this most potent vitamin is active at exceedingly low concentrations. Therefore, it was most important to the elucidation of the antipernicious anemia factor when, in 1947, Mary Shorb of the University of Maryland found that it was also required for the growth of *Lactobacillus lactis* Dorner.[53] With Shorb's microbiological assay, isolation of the factor, by that time named **vitamin B₁₂** by the Merck group, proceeded rapidly.

Animal Protein Factors

At about the same time, animal growth responses to factors associated with animal proteins or manure were reported as American animal nutritionists sought to eliminate expensive and scarce animal by-products from the diets of livestock. Norris's group at Cornell attributed responses of this time to an **animal protein factor**; the factor in liver necessary for rat growth was called *factor X* by Cary and *zoopherin*[54] by Zucker and Zucker. It soon became evident that these factors were probably identical. Stokstad's group found the factor in manure and isolated an organism from poultry manure that would synthesize a factor that was effective both in promoting chick growth and in treating pernicious anemia. That the antipernicious anemia factor was produced microbiologically was important, in that it led to an economical means of industrial production of vitamin B₁₂.

Vitamin B₁₂ Isolated

By the late 1940s, Combs[55] and Norris, using chick growth as their bioassay procedure, were fairly close to the isolation of vitamin B₁₂. However, in 1948, Folkers at Merck, using the *Lactobacillus lactis* Dorner assay, succeeded in first isolating the antipernicious anemia factor in crystalline form. This achievement was accomplished in the same year by Lester Smith's group at the Glaxo Laboratories in England (who found their pink crystals to contain cobalt),

TABLE 2.7 Factors Leading to the Discovery of Vitamin B₁₂

Factor	Bioassay
Extrinsic factor	Cure of anemia in humans
LLD factor	Growth of *L. lactis* Dorner
Vitamin B₁₂	Growth of *L. lactis* Dorner
Animal protein factor	Growth of chicks
Factor X	Growth of rats
Zoopherin	Growth of rats

assaying their material on pernicious anemia patients in relapse.[56] The elucidation of the complex chemical structure of vitamin B₁₂ was finally achieved in 1955 by Dorothy Hodgkin's group at Oxford with the use of X-ray crystallography. In the early 1960s several groups accomplished the partial synthesis of the vitamin; it was not until 1970 that the *de novo* synthesis of vitamin B₁₂ was finally achieved by Woodward and Eschenmoser.

A summary of the factors leading to the discovery of vitamin B₁₂ is presented in Table 2.7.

Vitamins Discovered in Only Five Decades

Beginning with the concept of a vitamin, which emerged with Eijkman's proposal of an antipolyneuritis factor in 1906, the elucidation of the vitamins continued through the isolation of vitamin B₁₂ in potent form in 1948 (see Table 2.8). Thus, the identification of the presently recognized vitamins was achieved within a period of only 42 years! For some vitamins (e.g., pyridoxine) for which convenient animal models were available, discoveries came rapidly; for others (e.g., niacin, vitamin B₁₂) for which animal models were late to be found, the pace of scientific progress was much slower (Fig. 2.1). These paths of discovery were marked by nearly a dozen Nobel Prizes (Table 2.9).

7. VITAMIN TERMINOLOGY

The terminology of the vitamins can be as daunting as that of any other scientific field. Many vitamins carry alphabetic or alphanumeric designations, yet the sequence of such designations has an arbitrary appearance by virtue of its many gaps

---

53. For a time, this was referred to as the *LLD factor*.
54. The term *zoopherin* carries the connotation: "to carry on an animal species."
55. Characterization of the animal protein factor was the subject of the author's father's doctoral thesis in Norris's laboratory at Cornell in the late 1940s.

56. Friedrich (1988) has pointed out that it should be no surprise that the first isolations of vitamin B₁₂ were accomplished in industrial laboratories, because the task required industrial-scale facilities to handle the enormous amounts of starting material that were needed. For example, the Merck group used a ton of liver to obtain 20 milligrams of crystalline material.

TABLE 2.8 Timelines for the Discoveries of the Vitamins

Vitamin	Proposed	Isolated	Structure Determined	Synthesis Achieved
Thiamin	1906	1926	1932	1933
Vitamin C	1907	1926	1932	1933
Vitamin A	1915	1937	1942	1947
Vitamin D	1919	1932	1932 (D_2)	1932
			1936 (D_3)	1936
Vitamin E	1922	1936	1938	1938
Niacin	1926	1937	1937	1867[a]
Vitamin B_{12}	1926	1948	1955	1970
Biotin	1926	1939	1942	1943
Vitamin K	1929	1939	1939	1940
Pantothenic acid	1931	1939	1939	1940
Folate	1931	1939	1943	1946
Riboflavin	1933	1933	1934	1935
Vitamin B_6	1934	1936	1938	1939

[a]Much of the chemistry of nicotinic acid was known before its nutritional roles were recognized.

and inconsistent application to all of the vitamins. This situation notwithstanding, the logic underlying the terminology of the vitamins becomes apparent when it is viewed in terms of the history of vitamin discovery. The familiar designations in use today are, in most cases, the surviving terms coined by earlier researchers on the paths to vitamin discovery. Thus, because McCollum and Davis used the letters A and B to distinguish the lipid-soluble antixerophthalmic factor from the water-soluble antineuritic and growth activity that was subsequently found to consist of several vitamins, such chemically and physiologically unrelated substances as thiamin, riboflavin, pyridoxine, and cobalamins (in fact, all water-soluble vitamins except ascorbic acid, which was designated before the vitamin B complex was partitioned) are all called B vitamins. In the case of folic acid, certainly the name survived its competitors by virtue of its relatively attractive sound (e.g., versus *rhizopterin*). Therefore, the accepted designations for the vitamins, in most cases, have relevance only to the history and chronology of their discovery, and not to their chemical or metabolic similarities. The discovery of the vitamins left a path littered with designations of "vitamins," "factors," and other terms, most of which have been discarded (see Appendix A for a complete listing).

TABLE 2.9 Nobel Prizes Awarded for Research on Vitamins

Year Awarded	Recipients	Discovery
Prizes in Medicine and Physiology		
1929	Christian Eijkman and Frederick G. Hopkins	Discovery of the antineuritic vitamin; discovery of the growth-stimulating vitamins
1934	George H. Whipple, George R. Minot, and William P. Murphy	Discoveries concerning liver therapy against pernicious anemia
1937	Albert von Szent-Györgi and Charles G. King	Discoveries in connection with the biological combustion, with especial reference to vitamin C, and the catalysis of fumaric acid
1943	Henrik Dam and Edward A. Doisy	Discovery of vitamin K; discovery of the chemical nature of vitamin K
1953	Fritz A. Lipmann	Discovery of coenzyme A and its importance in intermediary metabolism
1955	Hugo Theorell	Discoveries relating to the nature and mode of action of oxidizing enzymes
1964	Feordor Lynen and Konrad Bloch	Discoveries concerning the mechanism and regulation of cholesterol and fatty acid metabolism
Prizes in Chemistry		
1928	Adolf Windaus	Studies on the constitution of the sterols and their connection with the vitamins
1937	Paul Karrer and Walter N. Haworth	Researches into the constitution of carbohydrates and vitamin C
		Researches into the constitution of carotenoids, flavins, and vitamins A and B
1938	Richard Kuhn	Work on carotenoids and vitamins
1967	George Wald, H. K. Hartline, and R. Grant	Discoveries of the basic chemical and physiological processes in vision

TABLE 2.10 Quasi-vitamins

Substance	Biological Activity
Choline	Component of the neurotransmitter acetylcholine and the membrane structural component phosphatidylcholine; essential for normal growth and bone development in young poultry; can spare methionine in many animal species, and thus can be essential in diets that provide limited methyl groups
p-Aminobenzoic acid	Essential growth factor for several microbes, in which it functions as a provitamin of folic acid; reported to reverse diet- or hydroquinone-induced achromotrichia in rats, and to ameliorate rickettsial infections
myo-Inositol	Component of phosphatidylinositol; prevents diet-induced lipodystrophies due to impaired lipid transport in gerbils and rats; essential for some microbes, gerbils, and certain fishes
Bioflavonoids	Reported to reduce capillary fragility, and inhibit in vitro aldolase reductase (has a role in diabetic cataracts) and o-methyltransferase (inactivates epinephrine and norepinephrine)
Ubiquinones	Group includes a component of the mitochondrial respiratory chain; are antioxidants and can spare vitamin E in preventing anemia in monkeys, and in maintaining sperm motility in birds
Lipoic acid	Cofactor in oxidative decarboxylation of α-keto acids; essential for growth of several microbes, but inconsistent effects on animal growth
Carnitine	Essential for transport of fatty acyl CoA from cytoplasm to mitochondria for β oxidation; synthesized by most species except some insects, which require a dietary source for growth
Pyrroloquinoline quinone	Component of certain bacterial and mammalian metallo-oxidoreductases; deprivation impairs growth, causes skin lesions in mice

TABLE 2.11 Factors _Not_ Considered Vitamins

Substance	Purported Biological Activity
Laetrile	A cyanogenic glycoside with unsubstantiated claims of anti-tumorigenicity
Gerovital	Unsubstantiated anti-aging elixir
Orotic acid	Normal metabolic intermediate of pyrimidine biosynthesis with hypocholesterolemic activity
Pangamic acid	Ill-defined substance(s), originally derived from apricot pits, with unsubstantiated claims for a variety of health benefits

8. OTHER FACTORS SOMETIMES CALLED VITAMINS

Several other factors have, at various times or under certain conditions, been called vitamins. Many remain today only as historic markers of once incompletely explained phenomena, now better understood. Today, some factors would appear to satisfy, for at least some species, the operating definition of a vitamin; although in practice that term is restricted to those factors required by higher organisms.[57] Therefore, these can be referred to as *quasi-vitamins* (see Table 2.10). The biological activities of the quasi-vitamins are addressed in Chapter 18. At various times, of course, other factors have been represented as vitamins; however, no solid evidence has been sustained to support such claims. These factors (see Table 2.11) are not to be confused with the vitamins.

9. THE MODERN HISTORY OF THE VITAMINS

While the first half of the twentieth century was an exciting period of vitamin discovery, the subsequent history has been one of addressing the huge amount of additional information needed in order to use the vitamins to improve human and animal health, and to optimize the efficiency of producing food animals. Does this mean that all dietary factors that satisfy the definition of a vitamin have been discovered?[58] Perhaps; but will such a complacent position best serve the science of nutrition? Or is it better to remain open to reinterpreting the notion of vitamins in light of emerging knowledge of the metabolic roles of other bioactive factors in foods?

---

57. Organic growth-promoting substances required only by microorganisms are frequently called *nutrilites*.

58. When I was an assistant professor (in the mid-1970s), I discussed with my Dad how unlikely it was that a graduate student could be asked to undertake thesis research on "unidentified growth factors" (UGFs). I asked him why he, as a graduate student in the late 1940s, had thought it profitable to do such work himself. He pointed out that "Every UGF had proven to be an essential nutrient, so we had confidence that that 'animal protein factor' would also." He was correct, of course; his work in the Cornell laboratory of Prof. Leo Norris contributed directly to the recognition of vitamin B_{12}.

TABLE 2.12 Foci of Current Vitamin Research

Area	Research Foci
Analytical and physical chemistry	Determining chemical and biological potencies and stabilities (to storage, processing, and cooking) of the vitamins, their various vitamers and chemical derivatives; developing analytical methods for measuring vitamin contents of food
Biochemistry and molecular biology	Elucidating the molecular mechanisms of vitamin action, including roles in gene expression; elucidating pathways of vitamin metabolism; determining the interactions with other nutrients and/or factors (e.g., disease, oxidative stress) that affect vitamin functions and needs
Nutritional surveillance and epidemiology	Determining vitamin intakes and status of populations and at-risk subgroups; elucidating relationships of vitamin intake/status and disease risks
Nutrition and dietetics	Determining quantitative requirements for vitamins
Medicine	Determining roles of vitamins in etiology and/or management of chronic (e.g., cancer, heart disease), congenital (e.g., neural tube defects), and infectious diseases; determining vitamin needs over the life cycle
Agriculture and international development adequacy	Developing smallholder farming/gardening systems and other food-based approaches that support nutritional requirements with respect to vitamins and other nutrients; genetic enhancement of foods with increased contents of vitamins and other micronutrients
Food technology	Developing food processing techniques that retain vitamins in food; developing methods for the effective vitamin fortification of foods

Current recent research interest in the vitamins has centered on certain foci (see Table 2.12). This information, much of which is still emerging today, will be the subject of the following chapters.

Study Questions and Exercises

1. How did the *vitamin theory* influence the interpretation of findings concerning diet and health associations?
2. For each vitamin, list the key empirical observations that led to its initial recognition.
3. In what general ways were animal models employed in the discovery of the vitamins? What ethical issues must be addressed in this type of research?
4. Which vitamins were discovered as results of efforts to use chemically defined diets for raising animals? How would you go about developing such a diet?
5. Which vitamins were discovered primarily through human experimentation? What ethical issues must be addressed in this type of research?
6. Prepare a concept map illustrating the interrelationships of the various prevalent ideas and the many goals, approaches, and outcomes that resulted in the discovery of the vitamins.

RECOMMENDED READING

General History of the Vitamins

Baron, J.H., 2009. Sailor's scurvy before and after James Lind – a reassessment. Nutr. Rev. 67, 315.

Carpenter, K.J., 1986. The History of Scurvy and Vitamin C. Cambridge University Press, Cambridge, MA. (288 pp.)

Carpenter, K.J., 2000. Beriberi, White Rice and Vitamin B: A Disease, a Cause, and a Cure. University of California Press, Los Angeles, CA. (282 pp.)

Carpenter, K.J., 2003. A short history of nutritional science: Part 1 (1785–1885). J. Nutr. 133, 638.

Carpenter, K.J., 2003. A short history of nutritional science: Part 2 (1885–1912). J. Nutr. 133, 975.

Carpenter, K.J., 2003. A short history of nutritional science: Part 3 (1912–1944). J. Nutr. 133, 3023.

Carpenter, K.J., 2003. A short history of nutritional science: Part 4 (1945–1985). J. Nutr. 133, 3331.

Györgi, P., 1954. Early experiences with riboflavin – a retrospect. Nutr. Rev. 12, 97.

Harris, L.J., 1955. Vitamins in Theory and Practice. Cambridge University Press, Cambridge, MA. pp. 1–39.

Lepkovsky, S., 1954. Early experiences with pyridoxine – a retrospect. Nutr. Rev. 12, 257.

Olson, J.A., 1994. Vitamins: The tortuous path from needs to fantasies. J. Nutr. 124, 1771S–1776S.

Roe, D.A., 1973. A Plague of Corn: The Social History of Pellagra. Cornell University Press, Ithaca, NY.

Sebrell Jr., W.H., 1981. History of pellagra. Fed. Proc. 40, 1520.

Sommer, A., 2008. Vitamin A deficiency and clinical disease: An historical overview. J. Nutr. 138, 1835.

Wald, G., 1968. Molecular basis of visual excitation. Science 162, 230–239.

Wolf, G., 2001. The discovery of the visual function of vitamin A. J. Nutr. 131, 1647–1650.

Papers of Key Historical Significance

Vitamin concept	Funk, C., 1912. J. State Med. 20, 341.
History of vitamin discovery	Olson, J.A., 1994. J. Nutr. 124, 1771S.
Vitamin A	McCollum, E.V., Davis, M., 1913. J. Biol. Chem. 15, 167.
	Osborne, T.B., Mendel, L.B., 1917. J. Biol. Chem. 31, 149.
	Wald, G., 1933. Vitamin A in the retina. Nature (London) 132, 316.
	Steenbock, H., 1919. Science 50, 352.
Vitamin D	McCollum, E.V., Simmonds, N., Pitz, W., 1916. J. Biol. Chem. 27, 33.
Vitamin E	Evans, H.M., Bishop, K.S., 1922. Science 56, 650.
	Olcott, H.S., Mattill, H.A., 1931. J. Biol. Chem. 93, 65–70.
Vitamin K	Dam, H., 1929. Biochem. Z. 215, 475.
Ascorbic acid	Holst, A., Frolich, T., 1907. J. Hyg. (Camb.) 7, 634.
	King, C.G., Waugh, W.A., 1932. Science 75, 357.
	Svirbely, J.L., Szent-Györgi, A., 1932. Nature (London) 129, 576.
	Baron, J.H., 2010. Nutr. Rev. 67, 315.
Thiamin	Eijkman, C., 1890. Med. J. Dutch East Indies 30, 295.
	Eijkman, C., 1896. Med. J. Dutch East Indies 36, 214.
Riboflavin	Kuhn, P., Györgi, P., Wagner-Juregg, T., 1933. Ber. 66, 317.
Niacin	Elvehjem, C., Madden, R., Strong, F., Wolley, D., 1937. J. Am. Chem. Soc. 59, 1767.
	Goldberger, J., 1922. J. Am. Med. Assoc. 78, 1676.
	Warburg, O., Christian, W., 1936. Biochem. Z. 43, 287.
Vitamin B_6	Györgi, P., 1934. Nature (London) 133, 498.
Biotin	Kögl, F., Tonnis, B., 1936. Z. Physiol. Chem. 242, 43.
Pantothenic acid	Norris, L.C., Ringrose, A.T., 1930. Science 71, 643.
	Williams, R.J., Lyman, C., Goodyear, G., Truesdail, J., Holaday, D., 1933. J. Am. Chem. Soc. 55, 2912.
Folate	Jukes, T.H., 1939. J. Am. Chem. Soc. 61, 975.
	Mimms, V., Totter, J.R., Day, P.L., 1944. J. Biol. Chem. 155, 401.
Vitamin B_{12}	Castle, W.B., 1929. Am. J. Med. Sci. 178, 748.
	Minot, G.R., Murphy, W.P., 1926. J. Am. Med. Assoc. 87, 470.
	Wills, L., Contab, M.A., Lond, B.S., 1931. Br. Med. J. 1, 1059.
	Shorb, M.S., 1948. Science 107, 398.

Properties of Vitamins

Chapter Outline

Anchoring Concepts

1. The chemical composition and structure of a substance determine both its physical properties and chemical reactivity.
2. The physicochemical properties of a substance determine the ways in which it acts and is acted on in biological systems.
3. Substances tend to be partitioned between hydrophilic regions (plasma, cytosol, and mitochondrial matrix space) and hydrophobic regions (membranes, bulk lipid droplets) of biological systems on the basis of their relative solubilities; overcoming such partitioning requires actions of agents (micelles, binding or transport proteins) that serve to alter their effective solubilities.
4. Isomers and analogs of a given substance may not have equivalent biological activities.

La vie est une fonction chimique.

A. L. Lavoisier

Learning Objectives

1. To understand that the term *vitamin* refers to a family of compounds, i.e., structural analogs, with qualitatively similar biological activities but often with different quantitative potencies.
2. To become familiar with the chemical structures and physical properties of vitamins.
3. To understand the relationship between the physico-chemical properties of vitamins and their stabilities, and how these properties affect their means of enteric absorption, transport, and tissue storage.
4. To become familiar with the general nature of vitamin metabolism.

VOCABULARY

Adenosylcobalamin
Ascorbic acid
β-Carotene
β-Ionone nucleus
Binding proteins
Bioavailability
Biopotency
Biotin
Carotenoid
Cholecalciferol
6-Chromanol nucleus
Chylomicrons
Cobalamin
Coenzyme A
Corrin nucleus
Cyanocobalamin
Dehydroascorbic acid

The Vitamins. DOI: 10.1016/B978-0-12-381980-2.00003-7

Ergocalciferol
FAD
FMN
Folacin
Folic acid
HDL
Isoalloxazine nucleus
LDL
Lipoproteins
Menadione
Menaquinone
Methylcobalamin
Micelle
NAD(H)
NADP(H)
Naphthoquinone nucleus
Niacin
Nicotinamide
Nicotinic acid
Pantothenic acid
Pyridine nucleus
Phylloquinone
Portomicron
Pteridine
Pteroylglutamic acid
Pyridine nucleus
Pyridoxal
Pyridoxamine
Pyridoxine
Pyridoxol
Pyrimidine ring
Retinal
Retinoic acid
Retinoid
Retinol
Riboflavin
Steroid
Tetrahydrofolic acid
Tetrahydrothiophene (thiophane) nucleus
Thiamin
Thiamin pyrophosphate
Thiazole ring
Tocol
Tocopherol
Tocotrienol
Ureido nucleus
Vitamin A
Vitamin B_2
Vitamin B_6
Vitamin B_{12}
Vitamin C
Vitamin D
Vitamin D_2
Vitamin D_3
Vitamin E
Vitamin K
Vitamin K_1
Vitamin K_2
Vitamin K_3
VLDL

1. CHEMICAL AND PHYSICAL PROPERTIES OF THE VITAMINS

Classifying the Vitamins According to Their Solubilities

The vitamins are organic, low molecular weight substances that have key roles in metabolism. Few of the vitamins are single substances; almost all are families of chemically related substances, i.e., *vitamers*, sharing qualitatively (but not necessarily quantitatively) biological activities. Thus, the vitamers comprising a vitamin family may vary in biopotency, and the common vitamin name is actually a generic descriptor for all of the relevant vitamers. Otherwise, vitamin families are chemically heterogeneous; therefore, it is convenient to consider their physical properties (Table 3.1), which offer an empirical means of classifying the vitamins broadly.

The vitamins are frequently described according to their solubilities, that is, as being either fat-soluble or water-soluble.[1] Is interesting to note that this way of classifying the vitamins recapitulates the history of their discovery, calling to mind McCollum's "fat-soluble A" and "water-soluble B." The water-soluble vitamins tend to have one or more **polar** or ionizable groups (carboxyl, keto, hydroxyl, amino or phosphate), whereas the fat-soluble vitamins have predominantly aromatic and aliphatic characters.

Fat-soluble vitamins – appreciably soluble in non-polar solvents:

Vitamin A	Vitamin D
Vitamin E	Vitamin K

---

1. The concept of solubility refers to the interactions of solutes and solvents; a material is said to be soluble if it can disperse on a molecular level within a solvent. Solvents such as water, which can either donate or accept electrons, are said to be *polar*; whereas solvents (e.g., many organic solvents) incapable of such interactions are called *non-polar*. Compounds that themselves are polar or that have charged or ionic character are soluble in polar solvents such as water (the compounds are *hydrophilic*), but are insoluble in non-polar organic solvents (*lipophobic*). Molecules that do not contain polar or ionizable groups tend to be insoluble in water (*hydrophobic*) but soluble in non-polar organic solvents (*lipophilic*). Some large molecules (e.g., phospholipids, fatty acids, bile salts) that have local areas of charge or ionic bond density, as well as other areas without charged groups, exhibit both polar and non-polar characteristics. Such molecules, called *amphipaths*, have both hydrophilic and lipophilic internal regions; they tend to align along the interfaces of mixed polar/non-polar phases. Amphipathic molecules are important in facilitating the dispersion of hydrophobic substances in aqueous environments; they do this by surrounding those substances, forming the submicroscopic structure called the mixed **micelle**.

TABLE 3.1 Physical Properties of the Vitamins

Vitamin	Vitamer	MW	Solubility Organic[a]	Solubility H$_2$O	Absorption Maximum (nm)	Molar Absorptivity ε	Absorptivity A1%1cm	Fluorescence Excitation (nm)	Fluorescence Emission (nm)	Melting Point (°C)	Color/Form
Vitamin A	all-*trans*-Retinol	286.4	+	–	325	52,300	1,845	325	470	62–64	Yellow/crystal
	11-*cis*-Retinol	286.4	+	–	319	34,900	1,220				
	13-*cis*-Retinol	286.4	+	–	328	48,300	1,189				
	Retinal	284.4	+	–	373		1,548			61–64	Orange/crystal
	all-*trans*-Retinoic acid	300.4	+	sl[b]	350	45,300	1,510			180–182	Yellow/crystal
	13-*cis*-Retinoic acid	300.4	+	sl[b]	354	39,800	1,325			180–182	Yellow/crystal
	all-*trans*-Retinyl acetate	312.0	+	sl[b]	326		1,550			57–58	Yellow/crystal
	all-*trans*-Retinyl palmitate	508.0	+	sl[b]	325–328		975			28–29	Yellow/crystal
Provitamin A	β-Carotene	536.9	+	–	453	2,592	139			183	Purple/crystal
	α-carotene	536.9	+	–	444	2,800				187	Purple/crystal
Vitamin D	Vitamin D$_2$	396.7	+	–	264	18,300	459	No fluorescence		115–118	White/crystal
	Vitamin D$_3$	384.6	+	–	265	19,400	462	No fluorescence		84–85	White/crystal
	25(OH) vitamin D$_3$	400.7	+	–	265	18,000	449	No fluorescence			
	1α,25(OH)$_2$ vitamin D$_3$	416.6	+	–	264	19,000	418	No fluorescence			
Vitamin E	α-Tocopherol	430.7	+	–	292	3,265	75.8	295	320	2.5	Yellow/oil
	β-Tocopherol	416.7	+	–	296	3,725	89.4	297	322		Yellow/oil
	γ-Tocopherol	416.7	+	–	298	3,809	91.4	297	322	–2.4	Yellow/oil
	δ-Tocopherol	402.7	+	–	298		91–92	297	322		Yellow/oil
	α-Tocopheryl acetate	+472.8	+	–	286	1,891–2,080	40–44	290	323		Yellow/oil
	α-Tocopheryl succinate	530.8	+	–	268	2,044	38.5	285	310		
	α-Tocotrienol	424.7	+	–	292	3,652	86.0				

(Continued)

TABLE 3.1 (Continued)

Vitamin	Vitamer	MW	Solubility Organic[a]	Solubility H₂O	Absorption Maximum (nm)	Molar Absorptivity ε	Absorptivity A1%1cm	Fluorescence Excitation (nm)	Fluorescence Emission (nm)	Melting Point (°C)	Color/Form
	β-Tocotrienol	410.6	+	−	296	3,540	86.2	290	323		Yellow/oil
	γ-Tocotrienol	410.6	+	−	298	3,737	91.0	290	324		Yellow/oil
	δ-Tocotrienol	396.6	+	−	298	3,403	85.8	292	324		Yellow/oil
Vitamin K	Vitamin K₁	450.7	+	−	242	17,900	396	No fluorescence			Yellow/oil
					248	18,900	419				
					260	17,300	383				
					269	17,400	387				
					325	3,100	68				
	Vitamin K₂(20)	444.7	+	−	248	19,500	439	No fluorescence		35	Yellow/crystals
	Vitamin K₂(30)	580.0	+	−	243	17,600	304	No fluorescence		50	Yellow/crystals
					248	18,600	320				
					261	16,800	290				
					270	16,900	292				
	Vitamin K₂(35)	649.2	+	−	243	18,000	278	No fluorescence		54	Yellow/crystal
	Vitamin K₃	172.2	+	−	248	19,100	195			105–107	Yellow/crystal
					261	17,300	266				
					270	30,300	467				
					325–328	3,100	48				
Vitamin C	Ascorbic acid	176.1	−	+	245	12,200	695	No fluorescence			White/crystal
	Calcium ascorbate	390.3	−	+						190–192	White/crystal
	Sodium ascorbate	198.1	−	xs[e]						218[c]	White/crystal
	Ascorbyl palmitate	414.5	−	+							White/crystal

		MW			λ (nm)	ε	E1%	Fluor. ex (nm)	Fluor. em (nm)	MP (°C)	Color/form
Thiamin	Thiamin disulfide	562.7	–	sl[b]				No fluorescence		177	Yellow/crystal
	Thiamin HCl	337.3	–	xs[e]						246–250	White/crystal
	Thiamin mononitrate	344.3	–	+						196–200[c]	White/crystal
	Thiamin monophosphate	344.3	–								
	Thiamin pyrophosphate	424.3	–							220–222[c]	
	Thiamin triphosphate	504.3	–							228–232[c]	
Riboflavin[d]	Riboflavin	376.4	–	+	260, 375, 450	27,700, 10,600, 12,200	736, 282, 324	360, 465	521	278	Orange-yellow/crystal
	Riboflavin-5′-phosphate	456.4	260	27,100	594	594	282, 324, 594	440–500	530		Orange-yellow/crystal
	FAD	785.6	260, 375, 450	37,000, 9,300, 11,300	471, 118, 144		440–500	530			
Niacin	Nicotinic acid	123.1	–	+	260	2,800	227	No fluorescence		237	White/crystal
	Nicotinamide	122.1	–	xs[e]	261	5,800	478	No fluorescence		128–131	White/crystal
Vitamin B6	Pyridoxal HCl	203.6	–	+	390, 318	200, 8,128	9.8, 399	330[f], 310	382	165[c]	White/crystal
	Pyridoxine	169.2			254, 324	3,891, 7,244	23, 428	320	380	160	
	Pyridoxol HCl	205.6	–	+	253	3,700	180	332	400	206–208	White/crystal
	Pyridoxamine di-HCl	241.1			290, 292, 325, 253, 328	8,400, 7,720, 7,100, 4,571, 7,763	408, 375, 345, 190, 322	320, 337	370[g], 400[h]	226–227	

(Continued)

TABLE 3.1 (Continued)

Vitamin	Vitamer	MW	Solubility Organic[a]	Solubility H₂O	Absorption Maximum (nm)	Molar Absorptivity ε	Absorptivity A1%1cm	Fluorescence Excitation (nm)	Fluorescence Emission (nm)	Melting Point (°C)	Color/Form
	Pyridoxal 5'-phosphate	247.1			330	2,500	101	365	423[h]		
					388	4,900	198	360 330	430[g] 410[f]		
Biotin	d-Biotin	244.3	−	+	204	(Very weak)		No fluorescence		232–233	Colorless/crystal
Pantothenic acid	Pantothenic acid	219.2	−	xs[e]	204	(Very weak)		No fluorescence			Clear/oil
	Calcium pantothenate	467.5	−	−	no chromophore			No fluorescence		195–196[c]	White/crystal
	D-pantothenol	205.3		sl[b]	no chromophore			No fluorescence			Clear/oil
Folate	Folic acid	441.1			282 350	27,000 7,000	612 159	363	450–460[g]		
	Tetrahydrofolate FH₄	445.4			297	27,000	606	305–310	360[h]		
	10-Formyl FH₄	473.5			288	18,200	384	313	360[f]		
	5-Formyl FH₄	473.5			287	31,500	665	314	365[f]		
	5-Methyl FH₄	459.5			290	32,000	697				
	5-Formimino FH₄	472.5			285	35,400	749	308	360[f]		
	5,10-Methenyl FH₄	456.4			352	25,000	548	370	470[h]		
	5,10-Methylene FH₄	457.5			294	32,000	700				
Vitamin B₁₂	Cyanocobalamin	1355.4	−	xs[e]	278 261 551	8,700 27,600 8,700	115 204 64	No fluorescence			Dark red/crystal
	Hydroxylcobalamin (B₁₂ₐ)	1346.4	−	+	279 325 359 516 537	19,000 11,400 20,600 8,900 9,500	141 85 153 66 71	No fluorescence			Dark red/crystal

Compound	MW			λ (nm)	ε	E	Color/form
Aquacobalamin (B12b)	1347.0	−	+	274	20,600	153	Red/crystal
				317	6,100	45	
				351	26,500	197	
				499	8,100	60	
Nitrocobalamin (B12c)	1374.6	−	+	352	21,000	153	Red/crystal
				357	19,100	139	
				528	8,400	60	
				535	8,700	63	
Methylcobalamin	1344.4	−	+	266	19,900	148	Red/crystal
				342	14,400	107	
				522	9,400	70	
Adenosylcobamide	1579.6	−	+	288	18,100	115	Yellow-orange/crystal
				288	18,100	115	
				340	12,300	78	
				375	10,900	60	
				522	8,000	51	

[a] In organic solvents, fats, and oils.
[b] sl, Slightly soluble.
[c] Decomposes at this temperature.
[d] Fluoresces.
[e] xs, Freely soluble.
[f] Neutral pH.
[g] Alkaline pH.
[h] Acidic pH.

Water-soluble vitamins – appreciably soluble in polar solvents:

Thiamin	Folate	Vitamin B$_6$
Niacin	Vitamin C	Pantothenic acid
Biotin	Riboflavin	Vitamin B$_{12}$

The fat-soluble vitamins have some traits in common, in that each is composed either entirely or primarily of five-carbon **isoprenoid** units (i.e., related to *isoprene*, 2-methyl-1,3-butadiene) derived initially from acetyl-CoA in those plant and animal species capable of their biosynthesis. In contrast, the water-soluble vitamins have, in general, few similarities of structure. The routes of their biosyntheses in capable species do not share as many common pathways.

Vitamin Nomenclature

The nomenclature of the vitamins is in many cases rather complicated, reflecting both the terminology that evolved non-systematically during the course of their discovery, as well as more recent efforts to standardize the vocabulary of the field. Current standards for vitamin nomenclature policy were established by the International Union of Nutritional Sciences in 1978.[2] This policy distinguishes between generic descriptors used to describe families of compounds having vitamin activity (e.g., *vitamin D*) and to modify such terms as *activity* and *deficiency*, and trivial names used to identify specific compounds (e.g., *ergocalciferol*). These recommendations have been adopted by the Commission on Nomenclature of the International Union of Pure and Applied Chemists, the International Union of Biochemists, and the Committee on Nomenclature of the American Institute of Nutrition. The latter organization publishes the policy every few years.[3]

2. VITAMIN A

Essential features of the chemical structure:

2. Anonymous (1978). *Nutr. Abstr. Rev.* 48, 831–835.
3. Anonymous (1990). *J. Nutr.* 120, 12–19.

1. Substituted β-ionone nucleus [4-(2,6,6-trimethyl-2-cyclohexen-1-yl)-3-buten-2-one]
2. Side chain composed of three isoprenoid units joined head to tail at the 6-position of the β-ionone nucleus
3. Conjugated double-bond system among the side-chain and 5,6-nucleus carbon atoms.

Chemical structures of the vitamin A group:

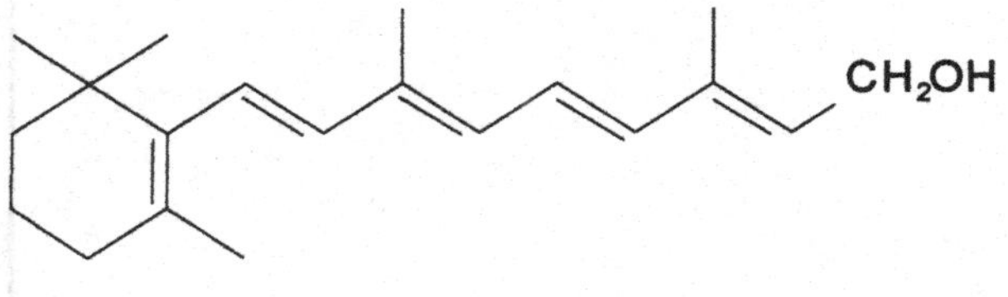

all-*trans*-Retinol

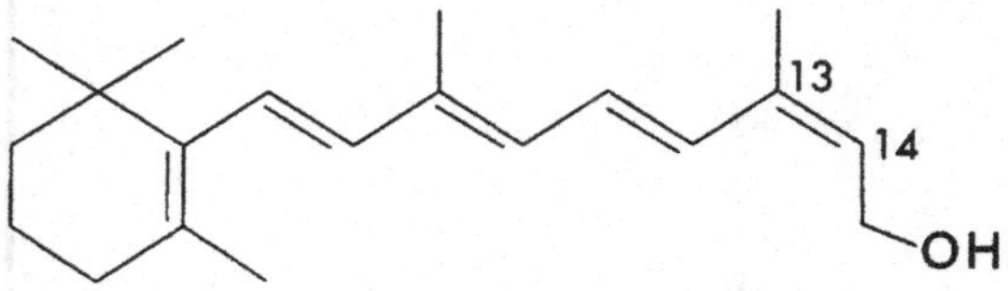

13-*cis*-Retinol

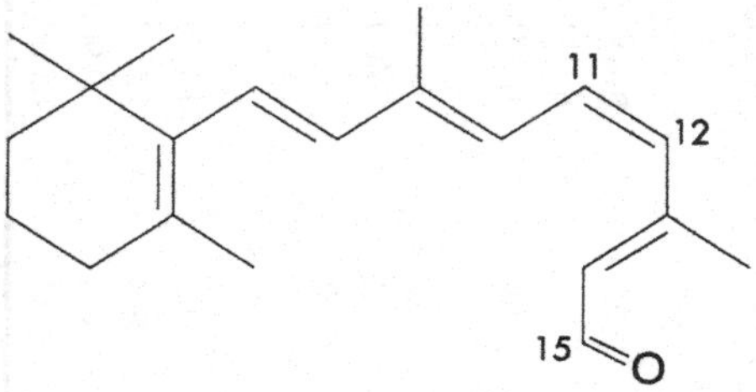

11-*cis*-Retinal

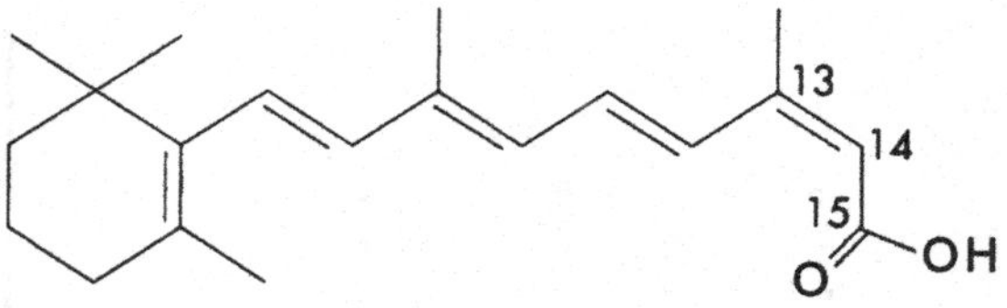

13-*cis*-Retinoic acid

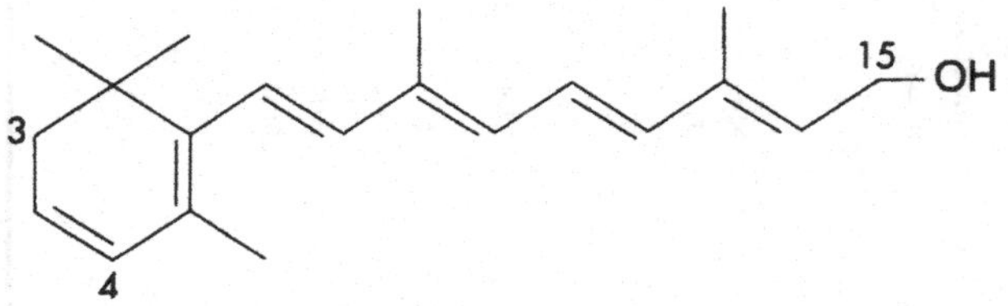

all-*trans*-3-Dehydroretinol (sometimes called *vitamin A$_2$*)

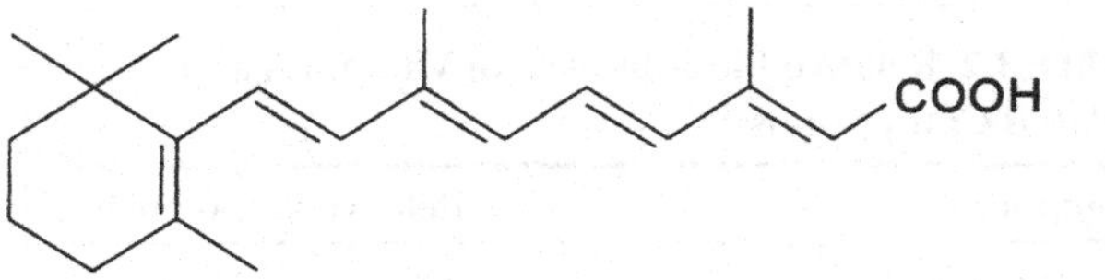

all-*trans*-Retinoic acid

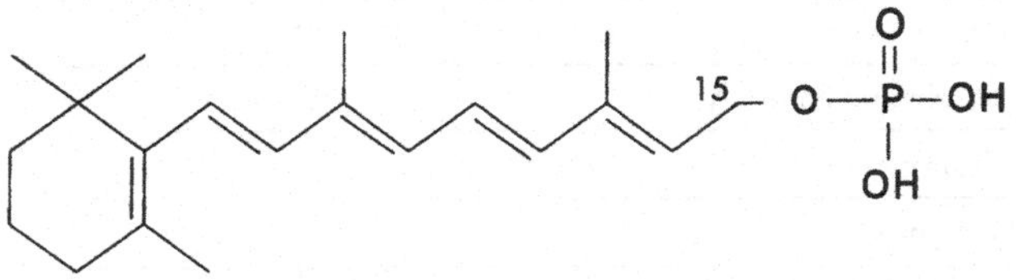

all-*trans*-Retinyl phosphate

Chemical structures of provitamins A:

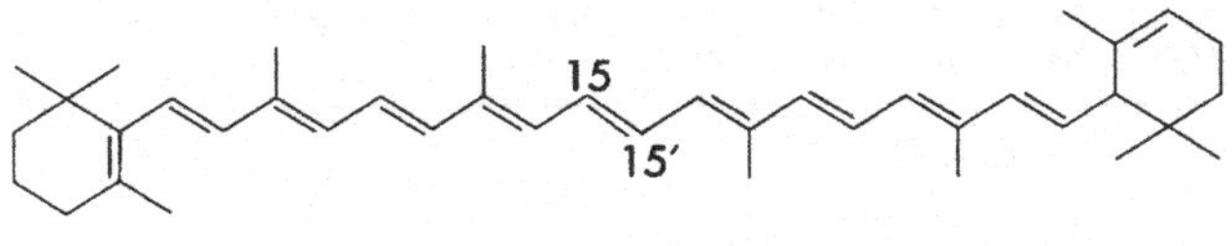

α-Carotene

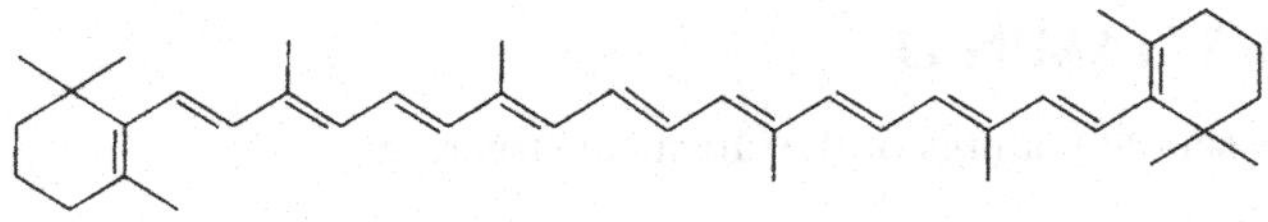

β-Carotene

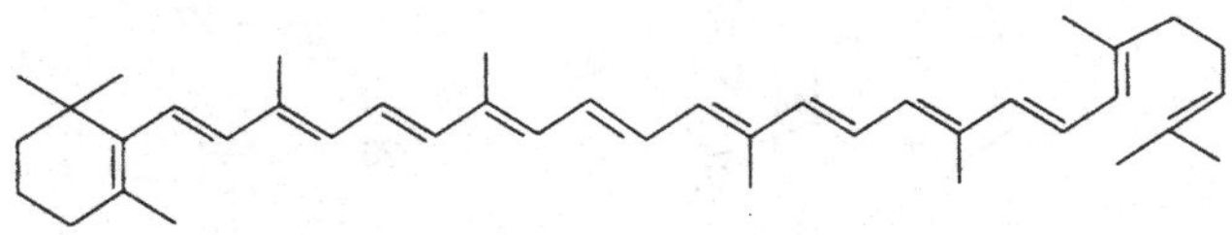

γ-Carotene

Vitamin A Nomenclature

Vitamin A is the generic descriptor for compounds with the qualitative biological activity of retinol. These compounds are formally derived from a monocyclic parent compound containing five carbon–carbon double bonds and a functional group at the terminus of the acyclic portion. Owing to their close structural similarities to retinol, they are called **retinoids**.

The Vitamin A-active retinoids occur in nature in three forms:

The alcohol ... **retinol**

The aldehyde ... **retinal** (also *retinaldehyde*)
The acid ... **retinoic acid.**

All three basic forms are found in two variants: with the β-**ionone nucleus** (vitamin A_1) or with the dehydrogenated β-ionone nucleus (vitamin A_2). However, because the former is both quantitatively and qualitatively more important as a source of vitamin A activity, the term *vitamin A* is usually taken to mean vitamin A_1. Some compounds of the class of polyisoprenoid plant pigments called **carotenoids**, owing to their relation to the carotenes, yield retinoids on metabolism, and thus also have vitamin A activity; these are called *provitamin A carotenoids*, and include β-**carotene**, which is actually a tail-conjoined retinoid dimer.

Chemistry of Vitamin A

Vitamin A is insoluble in water, but soluble in ethanol, and freely soluble in organic solvents including fats and oils. Of the 16 stereoisomers of vitamin A made possible by the four side chain double bonds, most of the potential *cis* isomers are sterically hindered; thus, only a few isomers are known. In solution, retinoids and carotenoids can undergo slow conversion by light, heat, and iodine through *cis–trans* isomerism of the side chain double bonds (e.g., in aqueous solution, all-*trans*-retinol spontaneously isomerizes to an equilibrium mixture containing one-third *cis* forms).

Contrary to what might be expected by their larger number of double bonds, carotenoids in both plants and animals occur almost exclusively in the all-*trans* form. These conjugated polyene systems absorb light, and, in the case of the carotenoids, appear to quench free radicals weakly. For the retinoids, the functional group at position 15 determines specific chemical reactivity. Thus, retinol can be oxidized to retinal and retinoic acid or esterified with organic acids; retinal can be oxidized to retinoic acid or reduced to retinol; and retinoic acid can be esterified with organic alcohols. Retinol and retinal each undergo color reactions with such reagents as antimony trichloride, trifluoroacetic acid, and trichloroacetic acids, which were formerly used as the basis of their chemical analyses by the Carr-Price reaction.

Most forms of vitamin A are crystallizable, but have low melting points (e.g., retinol, 62–64°C; retinal, 65°C). Both retinoids and carotenoids have strong absorption spectra. Vitamin A and the provitamin A carotenoids are very sensitive to oxygen in air, especially in the presence of light and heat; therefore, isolation of these compounds requires the exclusion of air (e.g., sparging with an inert gas) and the presence of a protective antioxidant (e.g., α-tocopherol). The esterified retinoids and carotenoids in native plant matrices are fairly stable.

Vitamin A Biopotency

Of the estimated 600 carotenoids in nature, only about 50 have been found to have provitamin A activity – i.e., those that can be cleaved metabolically to yield at least one molecule of retinol. Five or six of these are common in foods. While the chemical properties of each determine its **biopotency** (Table 3.2), dietary and physiological factors can affect the physiological utilization of each, referred to as its **bioavailability**.

To accommodate differences in biopotency, the reporting of vitamin A activity from its various forms in foods requires some means of standardization. Two systems are used for this purpose:

1. For foods, retinol equivalents (RE):

    1 (RE) =1 µg all-*trans*-retinol
    =2 µg all-*trans*-β-carotene in dietary supplements
    =12 µg all-*trans*-β-carotene in foods
    =24 (12–26) µg other provitamin A carotenoids (α-carotene, β-cryptoxanin) in foods

2. For pharmaceutical applications, international units (IU):

    1 IU =0.3 µg all-*trans*-retinol
    =0.344 µg all-*trans*-retinyl acetate
    =0.55 µg all-*trans*-retinyl palmitate.

In the calculation of RE values, corrections are made for the conversion efficiency of carotenoids to retinol. It is assumed that the retinol intermediate is completely absorbed (i.e., has an absorption efficiency that is 100%). Although 1 mole of β-carotene can, in theory, be converted (by cleavage of the C–15=C–15′ bond) to 2 moles of retinal, the physiological efficiency of this process appears to be much less; until recently, this has been assumed to be about 50%. When considered with a presumed 33% efficiency of intestinal absorption, this yielded a factor of one-sixth to calculate RE values from the β-carotene and, assuming an additional 50% discount, one of one-twelfth for other provitamin A carotenoids. Therefore, older recommendations used equivalency ratios of 6:1 and 12:1 in setting RE values for β-carotene in supplements and foods, respectively. However, recent work has shown purified β-carotene in oils and nutritional supplements to be utilized much more efficiently, at about half that of retinol, but that β-carotene in fruits and dark green leafy vegetables is utilized much less well. While the relevant data are sparse,[4] in 2001 the IOM[5] revised its estimates of

TABLE 3.2 Relative Biopotencies of Vitamin A and Related Compounds

Compound	Relative Biopotency[a]
all-*trans*-Retinol	100
all-*trans*-Retinal	100
cis-Retinol isomers	23–75
Retinyl esters	10–100
3-Dehydrovitamin A	30
β-Carotene	50
α-Carotene	26
γ-Carotene	21
Cryptoxanthin	28
Zeaxanthin	0

[a] *Most relative biopotencies were determined by liver storage bioassays with chicks and/or rats. In the case of 3-dehydrovitamin A, biopotency was assessed using liver storage by fish. In each case, the responses were standardized to that of all-trans-retinol.*

the vitamin A biopotency of carotenoids to the figures presented in Table 3.2.

3. VITAMIN D

Essential features of the chemical structure:

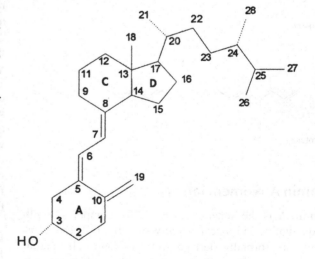

1. Side chain-substituted, open-ring **steroid**[6]
2. *cis*-Triene structure
3. Open positions on carbon atoms at positions 1 (ring) and 25 (side chain).

4. In fact, the 2001 estimate was based on a single report (Sauberlich, H.E., Hodges, R.E., Wallace, D.L., *et al.* [1974] *Vitamins Hormones* 32, 251–275) of the efficacy of β-carotene in correcting impaired dark adaptation in two volunteers.
5. Institute of Medicine, US National Academy of Sciences.

6. Steroids are four-ringed compounds related to the sterols, which serve as hormones and bile acids.

Chemical structures of the vitamin D group:

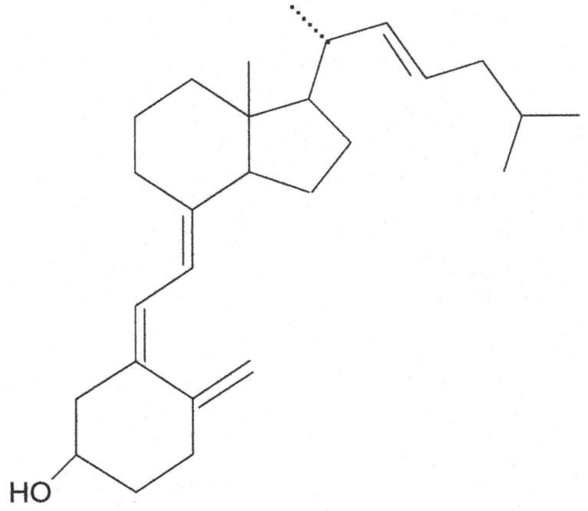

Vitamin D$_2$ (ergocalciferol)

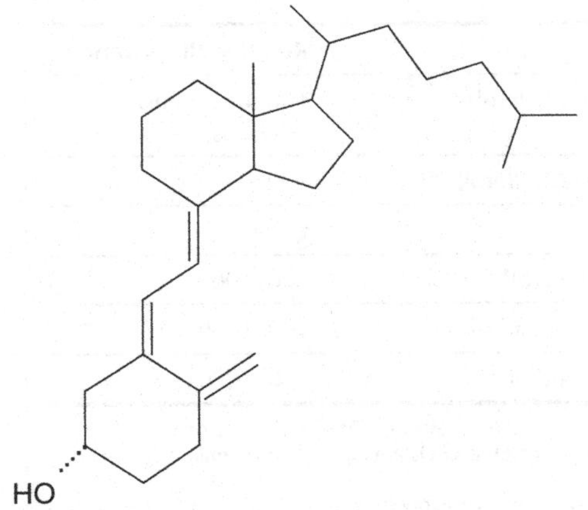

Vitamin D$_3$ (cholecalciferol)

Vitamin D Nomenclature

Vitamin D is the generic descriptor for all steroids qualitatively exhibiting the biological activity of **cholecalciferol**. These compounds contain the intact A, C, and D steroid rings,[7] being ultimately derived *in vivo* by photolysis of the B ring of 7-dehydrocholesterol. That process frees the A ring from the rigid structure of the C and D rings, yielding conformational mobility in which the A ring undergoes rapid interconversion between two chair configurations.

Vitamin D-active compounds also have either of two types of isoprenoid side chains attached to the steroid nucleus at C-17 of the D ring. One side chain contains nine carbons and a single double bond. Vitamin D-active compounds with that structure are derivatives of **ergocalciferol**, which is also called **vitamin D$_2$**. This vitamer can be produced synthetically by the photolysis of plant sterols. The other type of side chain consists of eight carbons and contains no double bonds. Vitamin D-active compounds with that structure are derivatives of **cholecalciferol**, also called **vitamin D$_3$**, which is produced metabolically through a natural process of photolysis of 7-dehydrocholesterol on the surface of skin exposed to ultraviolet irradiation, e.g., sunlight. The metabolically active forms of vitamin D are ring- (at C-1) and side chain-hydroxylated derivatives of vitamins D$_2$ and D$_3$.

Vitamin D Chemistry

Unlike the ring-intact steroids, vitamin D-active compounds tend to exist in extended conformations (shown left) due to the 180° rotation of the A ring about the 6,7 single bond (in solution, the stretched and closed conformations are probably in a state of equilibrium favoring the former).

---

7. Steroids contain a polycyclic hydrocarbon cyclopentanaperhydrophenanthrene nucleus consisting of three six-carbon rings (referred to as the A, B, and C rings) and a five-carbon ring (the D ring).

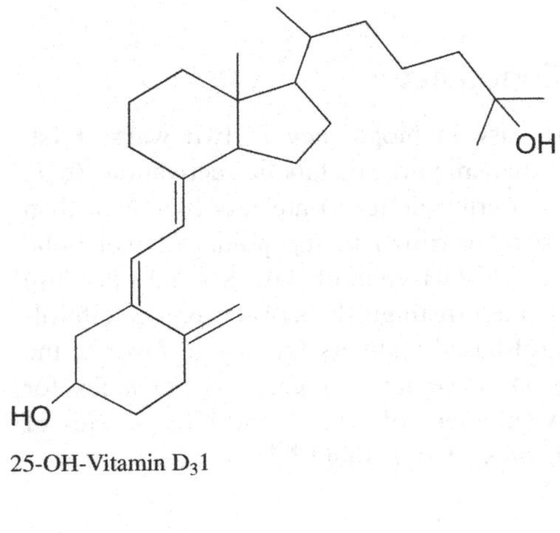

25-OH-Vitamin D$_3$1

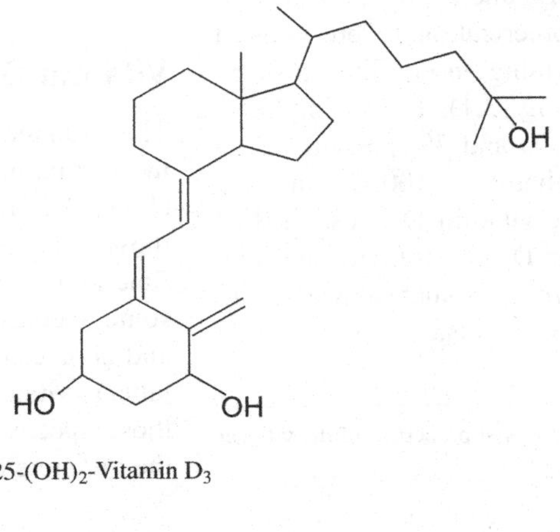

25-(OH)$_2$-Vitamin D$_3$

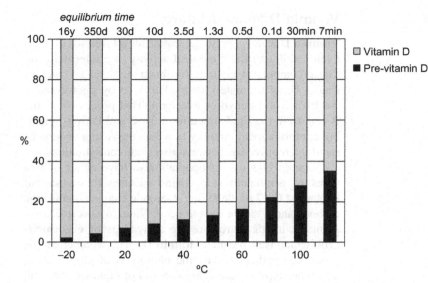

FIGURE 3.1   Thermal isomerizaton of vitamin D.

The hydroxyl group on C-3 is thus in the β position (i.e., above the plane of the A ring) in the closed forms, and in the α position (i.e., below the plane of the A ring) in the stretched forms. Rotation about the 5,6 double bond can also occur by the action of light or iodine to interconvert the biologically active 5,6-*cis* compounds to 5,6-*trans* compounds, which show little or no vitamin D activity.

Vitamins D_2 and D_3 are white to yellowish powders that are insoluble in water, moderately soluble in fats, oils, and ethanol, and freely soluble in acetone, ether, and petroleum ether. Each shows strong ultraviolet (UV) absorption, with a maximum at 264 nm. Vitamin D is sensitive to oxygen, light, and iodine. Heating or mild acidity can convert it to the 5,6-*trans* and other inactive forms. Whereas the vitamin is stable in dry form, in organic solvents and most plant oils (owing to the presence of α-tocopherol, which serves as a protective antioxidant), its thermal and photolability can result in losses during such procedures as saponification with refluxing. Therefore, it is often necessary to use inert gas environments, light-tight sealed containers, and protective antioxidants in isolating the vitamin. In solution, both vitamins D_2 and D_3 undergo reversible, temperature-dependent isomerization to pre-vitamin D (*see* Chapter 6, Fig. 6.1), forming an equilibrium mixture with the parent vitamin (Fig. 3.1). For example, at 20°C a mixture of 93% vitamin D and 7% pre-vitamin D is established within 30 days, whereas at 100°C a mixture of 72% vitamin D and 28% pre-vitamin D is established within only 30 minutes.[8] Vitamin D_3 is extremely sensitive to photodegradation of the 5,6-*trans* isomer, as well as to B-ring analogs referred to as suprasterols.

---

8. This designation connotes the light-rotational properties of the vitamer and comes from the Latin *rectus*, meaning right.

TABLE 3.3 Relative Biopotencies of Vitamin D-Active Compounds

Compound	Relative Biopotency[a]
Vitamin D_2 (ergocalciferol)	100 (mammals)[b]
	10 (birds)[c]
Vitamin D_3 (cholecalciferol)	100
Dihydrotachysterol[d]	5–10
25-OH-Cholecalciferol[e]	200–500
1,25-(OH)$_2$-Cholecalciferol[e]	500–1000
1α-OH-Cholecalciferol[f]	500–1000

[a] *Results of bioassays of rickets prevention in chicks and/or rats.*
[b] *Biopotencies of vitamins D_2 and D_3 are equivalent for mammalian species.*
[c] *Biopotency of vitamin D_2 is very low for chicks, which cannot use this vitamer effectively.*
[d] *A sterol generated by the irradiation of ergosterol.*
[e] *Normal metabolite of vitamin D_3; the analogous metabolite of vitamin D_2 is also formed and is comparably active in non-avian species.*
[f] *A synthetic analog.*

Vitamin D Biopotency

The vitamers D vary in biopotency in two ways. First, those vitamers that require metabolic activation (e.g., cholecalciferol and ergocalciferol) are less biopotent than those that are more proximal to the points of metabolic functioning (e.g., 25-OH-vitamin D). Second, because some species (avians) distinguish between ergocalciferol- and cholecalciferol-based vitamers (greatly in favor of the latter), vitamers D_3 have much greater biopotencies for those species. A summary of the relative biopotencies of the vitamers D is presented in Table 3.3.

4. VITAMIN E

Essential features of the chemical structure:

1. Side chain derivative of a methylated **6-chromanol nucleus** (3,4-dihydro-2*H*-1-benzopyran-6-ol)
2. Side chain consists of three isoprenoid units joined head to tail
3. Free hydroxyl or ester linkage on C-6 of the chromanol nucleus.

Chemical structures of the vitamin E group:

The tocopherols

The tocotrienols

Vitamer	R_1	R_2	R_3
α-Tocopherol/α-tocotrienol	CH$_3$	CH$_3$	CH$_3$
β-Tocopherol/β-tocotrienol	CH$_3$	H	CH$_3$
γ-Tocopherol/γ-tocotrienol	H	CH$_3$	CH$_3$
δ-Tocopherol/δ-tocotrienol	H	H	CH$_3$
Tocol/tocotrienol	H	H	H

Vitamin E Nomenclature

Vitamin E is the generic descriptor for all tocol and tocotrienol derivatives that exhibit qualitatively the biological activity of α-tocopherol. These compounds are isoprenoid side chain derivatives of 6-chromanol. The term **tocol** is the trivial designation for the derivative with a side chain consisting of three fully saturated isopentyl units; **tocopherol** denotes generically the mono-, di-, and trimethyl tocols irrespective of biological activity. **Tocotrienol** is the trivial designation of the 6-chromanol derivative with a similar side

chain containing three double bonds. Individual tocopherols and tocotrienols are named according to the position and number of methyl groups on their chromanol nuclei.

Because the tocopherol side chain contains two chiral center carbons (C-4′, C-8′) in addition to the one at the point of its attachment to the nucleus (C-2), eight stereoisomers are possible. However, only one stereoisomer occurs naturally: the R^9,R,R-form. The chemical synthesis of vitamin E produces mixtures of other stereoisomers, depending on the starting materials. For example, through the early 1970s the commercial synthesis of vitamin E used, as the source of the side chain, isophytol isolated from natural sources (which has the *R*-configuration at both the 4- and 8-carbons); tocopherols so produced were racemic at only the C-2 position. Such a mixture of 2RS^{10}-α-tocopherol was then called *dl*-α-tocopherol; its acetate ester was the form of commerce, and was adopted as the international standard on which the biological activities of other forms of the vitamin are still based. In recent years, however, the commercial synthesis of vitamin E has turned away from using isophytol in favor of a fully synthetic side chain. Therefore, synthetic preparations of vitamin E presently available are mixtures of all eight possible stereoisomers, i.e., 2RS,4′RS,8′RS compounds, which are designated more precisely with the prefix all-*rac*-. The acetate esters of vitamin E are used in medicine and animal feeding, whereas the unesterified (i.e., free alcohol) forms are used as antioxidants in foods and pharmaceuticals. Other forms (e.g., α-tocopheryl hydrogensuccinate, α-tocopheryl polyethylene glycol-succinate) are also used in multivitamin preparations.

Vitamin E Chemistry

The tocopherols are light yellow oils at room temperature. They are insoluble in water, but are readily soluble in nonpolar solvents. Being monoethers of a hydroquinone with a phenolic hydrogen (in the hydroxyl group at position C-6 in the chromanol nucleus), with the ability to accommodate an unpaired electron within the resonance structure of the ring (undergoing transition to a semistable chromanoxyl radical before being converted to tocopheryl quinone), they are good quenchers of free radicals, and thus serve as antioxidants. They are, however, easily oxidized, and can be destroyed by peroxides, ozone, and permanganate in a process catalyzed by light and accelerated by polyunsaturated fatty acids and metal salts. They are very resistant to acids and (only under anaerobic conditions) to bases. Tocopheryl esters, by virtue of the blocking of the C-6 hydroxyl group,

9. This designation connotes the light-rotational properties of the vitamer and comes from the Latin *sinister*, meaning left.
10. From Mulder, F.J., de Vries, E. J., and Borsje, B. (1971) *J. Assoc. Off. Anal. Chem.* 54: 1168–1174.

TABLE 3.4 Relative Biopotencies of Vitamin E-Active Compounds

Trivial Designation	Systematic Name	Biopotency (IU/mg)[a]
R,R,R-α-Tocopherol[b]	2R,4'R,8'R-5,7,8-Trimethyltocol	1.49
R,R,R-α-Tocopheryl acetate	2R,4'R,8'R-5,7,8-Trimethyltocyl acetate	1.36
all-rac-α-Tocopherol[c]	2RS,4'RS,8'RS-5,7,8-Trimethyltocol	1.1
all-rac-α-Tocopheryl acetate	2RS,4'RS,8'RS-5,7,8-Trimethyltocylacetate	1.0
R,R,R-β-Tocopherol	2R,4'R,8'R-5,8-Dimethyltocol	0.12
R,R,R-γ-Tocopherol	2R,4'R,8'R-5,7-Dimethyltocol	0.05
R-α-Tocotrienol	trans-2R-5,7,8-Trimethyltocotrienol	0.32
R-β-Tocotrienol	trans-2R-5,8-Dimethyltocotrienol	0.05
R-γ-Tocotrienol	trans-2R-5,7-Dimethyltocotrienol	—

[a] International units per milligram of material, based chiefly on rat gestation–resorption bioassay data.
[b] Formerly called d-α-tocopherol.
[c] Formerly called dl-α-tocopherol; this form remains the international standard despite the fact that it has not been produced commercially for several years.

are very stable in air, and are therefore the forms of choice as food/feed supplements. Because tocopherol is liberated by the saponification of its esters, extraction and isolation of vitamin E call for the use of protective antioxidants (e.g., propyl gallate, ascorbic acid), metal chelators, inert gas environments, and subdued light. The UV absorption spectra of tocopherols and their acetates in ethanol have maxima of 280–300 nm (α-tocopherol, 292 nm); however, their extinction coefficients are not great (70–91).[11] Because their fluorescence is significant (excitation, 294 nm; emission, 330 nm), particularly in polar solvents (e.g., diethyl ether or alcohols), this property has analytical utility.

Vitamin E Biopotency

The vitamers E vary in biopotency (Table 3.4), mainly according to the positions and numbers of their nucleus methyl groups, the most biopotent being the trimethylated (α) form.

11. UV absorption can, nevertheless, be useful for checking the quality of α-tocopherol standards: solutions containing 90% α-tocopherol show ratios of the absorption at the A_{min} (255 nm) and A_{max} (292 nm) of <0.18 (Balz, M. K., Schulte, E., and Thier, H. P. [1996] Z. Lebensm. U.-Forsch 202, 80–81.

5. VITAMIN K

Essential features of the chemical structure:

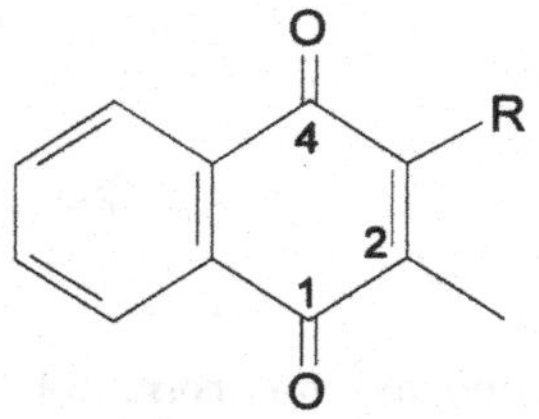

1. Derivative of 2-methyl-1,4-naphthoquinone
2. Ring structure has a hydrophobic constituent at the 3-position and can be alkylated with an isoprenoid side chain.

Chemical structures of the vitamin K group:

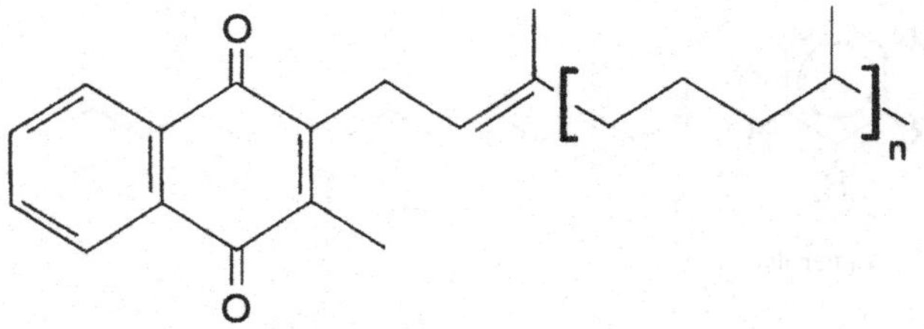

The phylloquinones

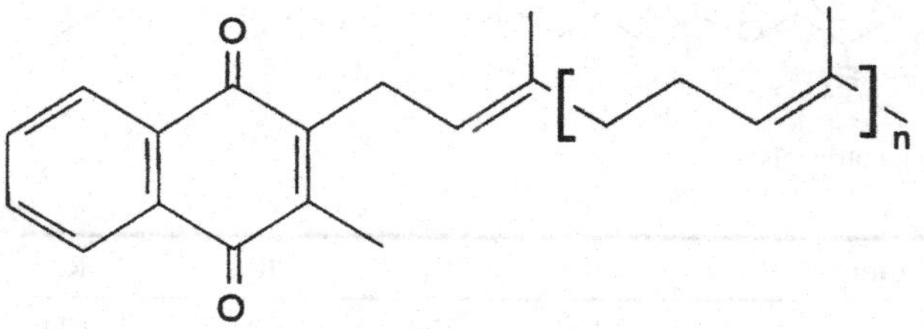

The menaquinones

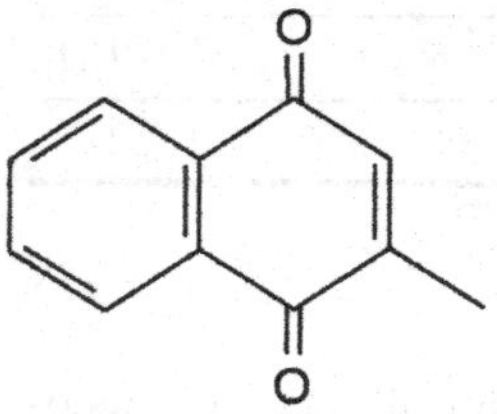

Menadione

Vitamin K Nomenclature

Vitamin K is the generic descriptor for 2-methyl-1,4-naphthoquinone and all of its derivatives exhibiting qualitatively the biological (antihemorrhagic) activity of phylloquinone. Naturally occurring forms of the vitamin

TABLE 3.5 Systems of Vitamin K Nomenclature

Chemical Name	IUPAC[a] System[b]	IUNS[c] System	Traditional
2-Methyl-3-phytyl-1,4-naphthoquinone	Phylloquinone (K)	Phytylmenaquinone (PMQ)	K_1
2-Methyl-3-multiprenyl-1,4-naphthoquinone (n)	Menaquinone-n (MK-n)	Prenylmenaquinone-n (MQ-n)	$K_{2(n)}$
2-Methyl-1,4-naphthoquinone	Menadione	Menaquinone	K_3

[a] International Union of Pure and Applied Chemists.
[b] Preferred system.
[c] International Union of Nutritional Sciences.

have an unsaturated isoprenoid side chain at C-3 of the **naphthoquinone nucleus**; the type and number of iso-prene units (not carbon atoms) form the basis of the char-acterization of the side chain, and hence the designation of the vitamer. The **phylloquinone** group includes forms with phytyl side chains and side chains that are further alkylated, thus consisting of several saturated isoprenoid units. The vitamers of this group have only one double bond in their side chains, i.e., on the proximal isoprene unit. These vitamers are synthesized by green plants. They are properly referred to as phylloquinones,[12] and are abbreviated as *K*. The **menaquinone** group also includes vitamers with side chains consisting of variable numbers of isoprenoid units; however, each isoprene unit has a double bond. These vitamers are synthesized by bacteria. They are abbreviated as *MK*[13] and were formerly referred to as **vitamin K$_2$**. For each of these groups of vitamers, a numeric system is used to indicate side chain length – e.g., the abbreviations K-n and MK-n are used for the phyllo-quinones and menaquinones, respectively, to indicate spe-cific vitamers with side chains consisting of n isoprenoid units (Table 3.5). The compound 2-methyl-1,4-naphtho-quinone (i.e., without a side chain) is called **menadione**.[14] It does not exist naturally, but has biological activity by virtue of the fact that animals can alkylate it to produce such metabolites as MK-4. Menadione is the compound of commerce; it is made in several forms (e.g., menadione sodium bisulfite complex, menadione dimethylpyrimidinol bisulfite).

Vitamin K Chemistry

Phylloquinone (K$_1$) is a yellow oil at room temperature, but the other vitamers K are yellow crystals. The vitamers K, MK, and most forms of menadione are insoluble in water, slightly soluble in ethanol, and readily soluble in ether, chloroform, fats, and oils. The vitamers K are sen-sitive to light and alkali, but are relatively stable to heat and oxidizing environments. Their oxidation proceeds to produce the 2,3-epoxide form. Being naphthoquinones, they can be reduced to the corresponding naphthohydro-quinones (e.g., with sodium hydrogen sulfite), which can be reoxidized with mild oxidizing agents. The vitamers K show the characteristic UV spectra of the naphthoqui-nones, i.e., their oxidized forms have four strong absorp-tion bands in the 240- to 270-nm range. The reduced (hydroquinone) forms show losses of the band near 270 nm and increases of the band around 245 nm. Extinction decreases with increasing side chain length.

Vitamin K Biopotency

The biopotency of vitamers K (Table 3.6) depends on both the nature and length of their isoprenoid side chains. In general, the menaquinones (K-2) tend to have greater bio-potencies than the corresponding phylloquinone analogs, and members of each series with four or five isoprenoid side chains are the most biopotent. The reported biopo-tencies of the menadiones tend to be variable; this may be due, at least in part, to varying stabilities of the prepara-tions tested, as well as to whether the vitamin K antagonist sulfaquinoxaline was used in the assay diet.

6. VITAMIN C

Essential features of the chemical structure:

1. 1.6-carbon lactone
2. 2.2,3-endiol structure.

---

12. These vitamers were formerly called the *phytylmenaquinones*, or **vita-min K$_1$**; the latter term is still encountered.
13. Formerly referred to as the *prenylmenaquinones*.
14. Formerly referred to as **vitamin K$_3$**.

TABLE 3.6 Relative Biopotencies of Vitamin K-Active Compounds

Compound	Biopotency[a] (%)
Phylloquinones (Formerly K₁)	
K-1[b]	5
K-2	10
K-3	30
K-4	100
K-5	80
K-6	50
Menaquinones (Formerly K₂)	
MK-2[b]	15
MK-3	40
MK-4	100
MK-5	120
MK-6	100
MK-7	70
Forms of menadione (Formerly K₃)	
Menadione	40–150
Menadione sodium bisulfite complex	50–150
Menadione dimethylpyrimidinol bisulfite	100–160

[a] Relative biopotency is based on chick prothrombin/clotting time bioassays using phylloquinone (K-1) as the standard.
[b] For both the phylloquinones (K) and menaquinones (MK), the number of side chain isoprenoid units (each containing five carbons) is indicated after the hyphens.

Chemical structure of vitamin C:

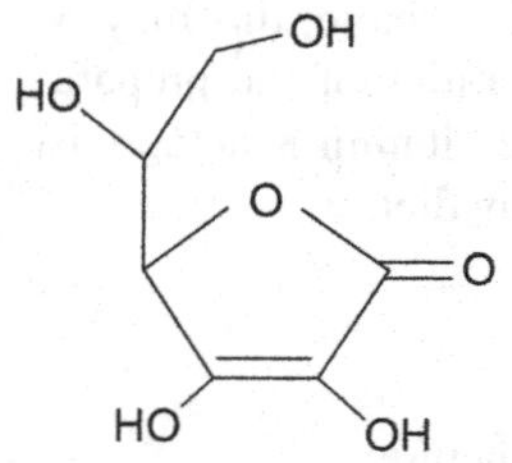

Ascorbic acid

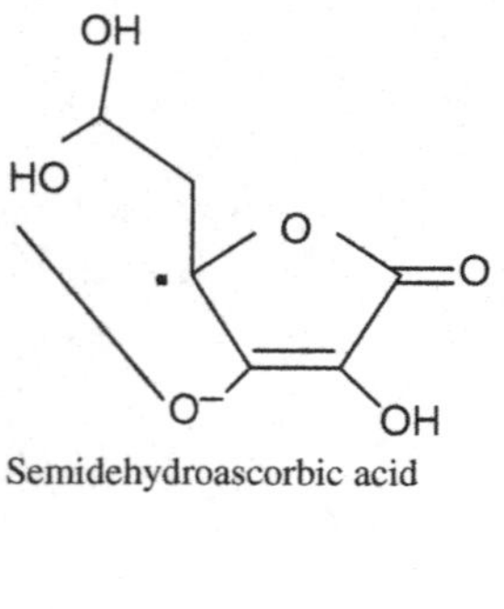

Semidehydroascorbic acid

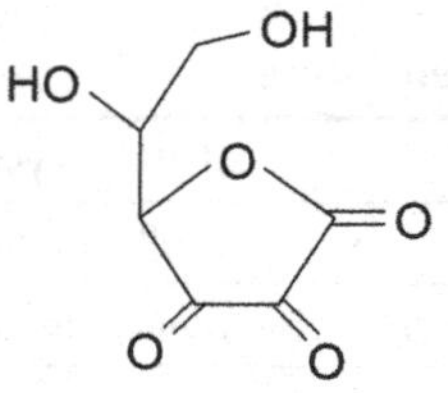

Dehydroascorbic acid

Vitamin C Nomenclature

Vitamin C is the generic descriptor for all compounds exhibiting qualitatively the biological activity of ascorbic acid. The terms L-*ascorbic acid* and **ascorbic acid** are both trivial designators for the compound 2,3-didehydro-L-threo-hexano-1,4-lactone, which was formerly known as *hexuronic acid*. The oxidized form of this compound is called L-*dehydroascorbic acid* or **dehydroascorbic acid**.

Vitamin C Chemistry

Ascorbic acid is a dibasic acid (with pK_a values of 4.1 and 11.8), because both enolic hydroxyl groups can dissociate. It forms salts, the most important of which are the sodium and calcium salts, the aqueous solutions of which are strongly acidic. A strong reducing agent, ascorbic acid is oxidized under mild conditions to dehydroascorbic acid via the radical intermediate semidehydroascorbic acid (sometimes called *monodehydroascorbic acid*). The semiquinoid ascorbic acid radical is a strong acid ($pK_a = -0.45$); after the loss of a proton, it becomes a radical anion that, owing to resonance stabilization, is relatively inert but disproportionates to ascorbic acid and dehydroascorbic acid. Thus, the three forms (ascorbic acid, semidehydroascorbic acid, and dehydroascorbic acid) compose a reversible redox system. Thus, it is an effective quencher of free radicals such as singlet oxygen ($^1O_2$). It reduces ferric (Fe^{3+}) to ferrous (Fe^{2+}) iron (and other metals analogously), and the superoxide radical (O_2^-) to H_2O_2 and is oxidized to monodehydroascorbic acid in the process. Ascorbic acid complexes with disulfides (e.g., oxidized glutathione, cystine), but does not reduce those disulfide bonds.

Dehydroascorbic acid is not ionized in environments of weakly acidic or neutral pH; therefore, it is relatively hydrophobic and is better able to penetrate membranes than is ascorbic acid. In aqueous solution, dehydroascorbic acid is unstable and is degraded by hydrolytic ring opening to yield 2,3-dioxo-L-gulonic acid. Dehydroascorbic acid reacts with several amino acids to form brown colored products, a reaction contributing to the spoilage of food.

TABLE 3.7 Relative Biopotencies of Vitamin C-Active Substances

Compound	Relative Biopotency (%)
Ascorbic acid	100
Ascorbyl-5,6-diacetate	100
Ascorbyl-6-palmitate	100
6-Deoxy-6-chloro-L-ascorbic acid	70–98
Dehydroascorbic acid	80
6-Deoxyascorbic acid	33
Ascorbic acid 2-sulfate	±[a]
Isoascorbic acid	5
L-Glucoascorbic acid	3

[a] This form is active in fishes, which have an intestinal sulfohydrase that liberates ascorbic acid; it is inactive in guinea pigs, rhesus monkeys, and humans, which lack the enzyme.

Chemical structure of thiamin:

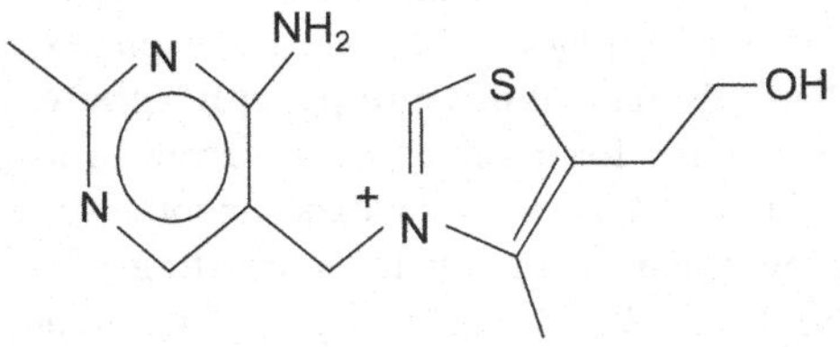

Thiamin (free base)

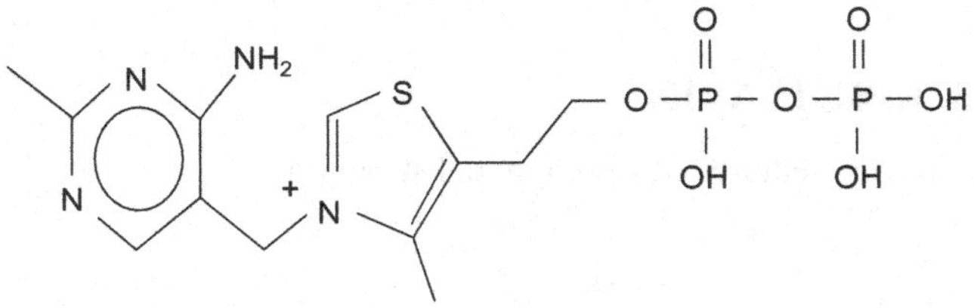

Thiamin pyrophosphate

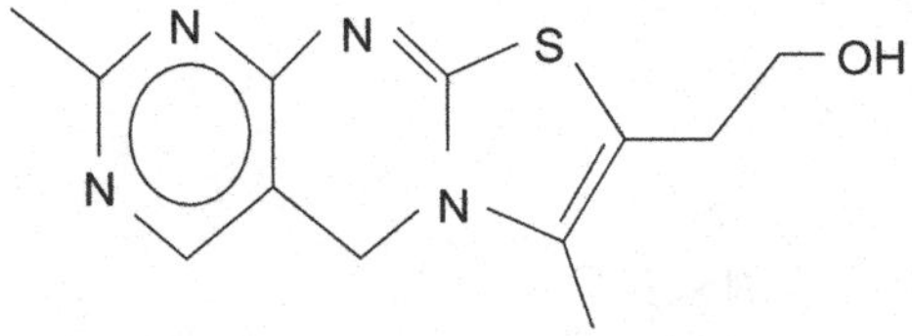

Thiochrome

Vitamin C Biopotency

Several synthetic analogs of ascorbic acid have been made. Some (e.g., 6-deoxy-L-ascorbic acid) have biological activity, whereas others (e.g., D-isoascorbic acid and L-glucoascorbic acid) have little or no activity. Several esters of ascorbic acid are converted to the vitamin *in vivo*, and thus have good biological activity (e.g., ascorbyl-5,6-diacetate, ascorbyl-6-palmitate, 6-deoxy-6-chloro-L-ascorbic acid; see Table 3.7). Esters of the C-2 position show variable vitamin C activity among different species.

7. THIAMIN

Essential features of the chemical structure:

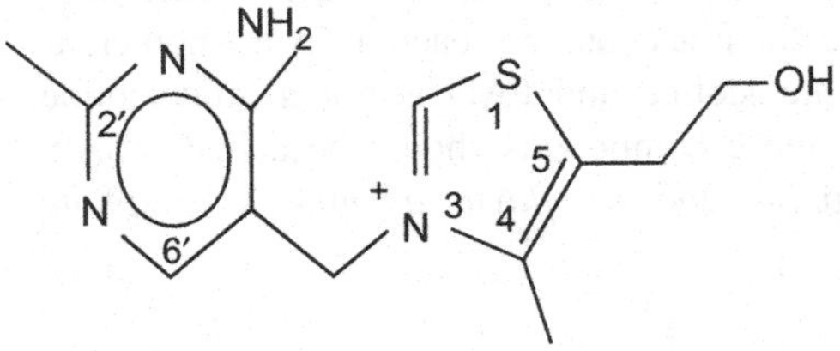

1. Conjoined pyrimidine and thiazole rings
2. Thiazole ring contains a quaternary nitrogen, an open C-2, and a phosphorylatable alkyl group on C-5
3. Amino group on C-4 of the pyrimidine ring.

Thiamin Nomenclature

The term **thiamin** is the trivial designation of the compound 3-[(4-amino-2-methyl-5-pyrimidinyl)methyl]-5-(2-hydroxyethyl)-4-methylthiazolium, formerly known as *vitamin B₁, aneurine*, and *thiamine*.

Thiamin Chemistry

Free thiamin is unstable because of its quaternary nitrogen; in water it is cleaved to the thiol form. For this reason, the hydrochloride and mononitrate forms are used in commerce. Thiamin hydrochloride (actually, thiamin chloride hydrochloride) is a colorless crystal that is very soluble in water (1 g/ml, thus making it a very suitable form for parenteral administration), soluble in methanol and glycerol, but practically insoluble in acetone, ether, chloroform, and benzene. The protonated salt has two positive charges: one associated with the **pyrimidine ring** and one associated with the **thiazole ring**. The mononitrate form is more stable than the hydrochloride form, but it is less soluble in water (27 mg/ml). It is used in food/feed supplementation and in dry pharmaceutical preparations.

Free thiamin is easily oxidized to thiamin disulfide and other derivatives including thiochrome, a yellow biologically inactive product with strong blue fluorescence that can be used for the quantitative determination of thiamin. The thiazole hydroxyethyl group can be phosphorylated *in vivo* to form thiamin mono-, di-, and triphosphates. Thiamin diphosphate, also called **thiamin pyrophosphate**, is the metabolically active form sometimes referred to as *cocarboxylase*. Thiamin antagonists of experimental significance include pyrithiamin (the analog consisting of a pyridine moiety replacing the thiazole ring) and oxythiamin (the analog consisting of a hydroxyl group replacing the C-4 amino group on the pyrimidine ring).

8. RIBOFLAVIN

Essential features of the chemical structure:

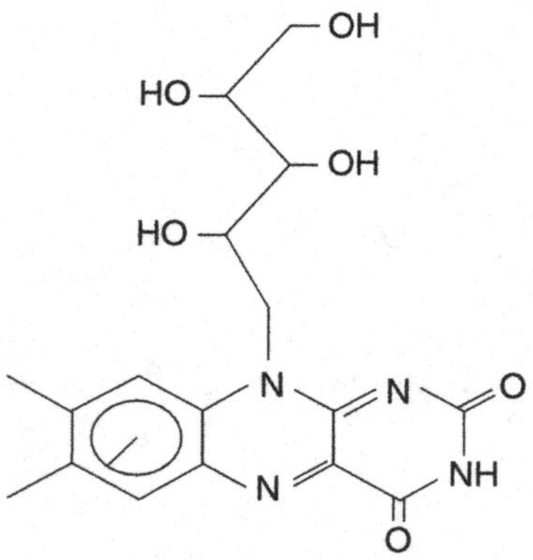

1. Substituted isoalloxazine nucleus
2. D-Ribityl side chain
3. Reducible nitrogen atoms in nucleus.

Chemical structures of riboflavin and its nucleotide forms:

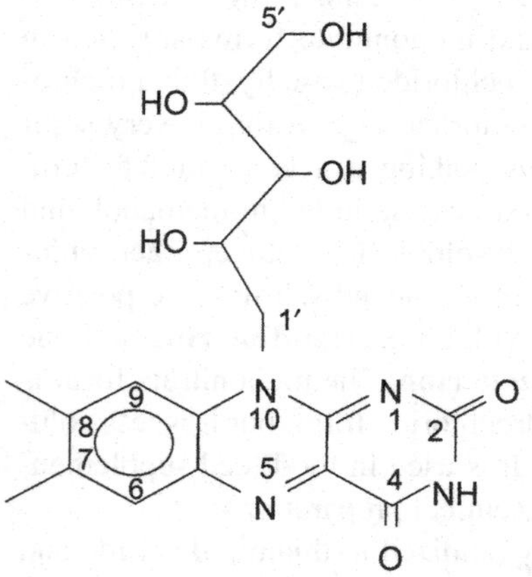

Riboflavin

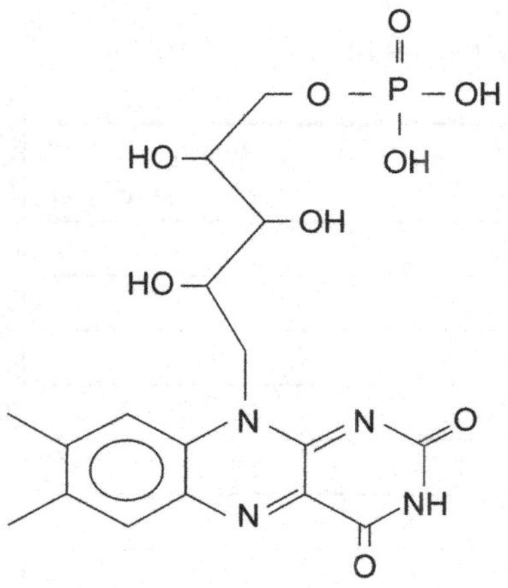

Flavin mononucleotide (FMN)

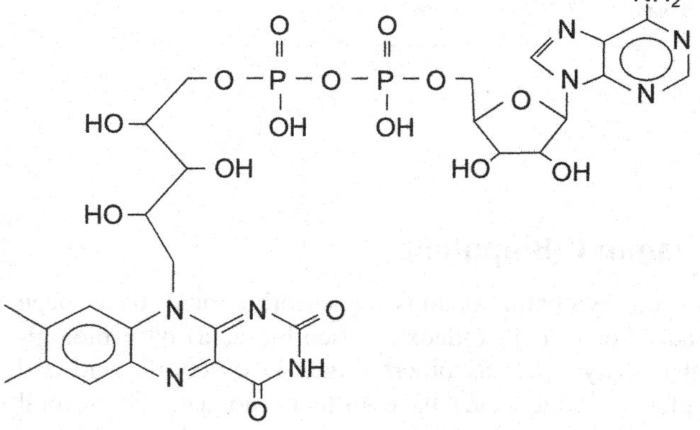

Flavin adenine dinucleotide (FAD)

Riboflavin Nomenclature

Riboflavin is the trivial designation of the compound 7,8-dimethyl-10-(1′-D-ribityl)isoalloxazine, formerly known as **vitamin B$_2$**, vitamin G, lactoflavine or riboflavine. The metabolically active forms are commonly called **flavin mononucleotide** (**FMN**) and **flavin adenine dinucleotide** (**FAD**). Despite their acceptance, each is a misnomer, as FMN is not a nucleotide and FAD is not a dinucleotide. More properly, these compounds should be called *riboflavin monophosphate* and *riboflavin adenine diphosphate*, respectively.

Riboflavin Chemistry

Riboflavin is a yellow tricyclic molecule that is usually phosphorylated (to FMN and FAD) in biological systems. In FAD, the isoalloxazine and adenine nuclear systems are arranged one above the other and are nearly coplanar.

The flavins are light-sensitive, undergoing photochemical degradation of the ribityl side chain, which results in the formation of such breakdown products as lumiflavin and lumichrome. Therefore, the handling of riboflavin must be done in the dark or under subdued red light.

Riboflavin is moderately soluble in water (10–13 mg/dl) and ethanol, but insoluble in ether, chloroform, and acetone. It is soluble but unstable under alkaline conditions. Because riboflavin cannot be extracted with the usual organic solvents, it is extracted with chloroform as lumiflavin after photochemical cleavage of the ribityl side chain. Flavins show two absorption bands, at ~370 nm and ~450 nm, with fluorescence emitting at 520 nm.

The catalytic functions of riboflavin are carried out primarily at positions N-1, N-5, and C-4 of the **isoalloxazine nucleus**. In addition, the methyl group at C-8 participates in covalent bonding with enzyme proteins. The flavin coenzymes are highly versatile redox cofactors because they can participate in either one- or two-electron redox reactions, thus serving as switching sites between obligate two-electron donors (e.g., NAD(H), succinate) and obligate one-electron acceptors (e.g., iron–sulfur proteins, heme proteins). They serve this function by undergoing reduction through a two-step sequence involving a radical anion intermediate. Because the latter can also react with molecular oxygen, flavins can also serve as cofactors in the two-electron reduction of O_2 to H_2O, and in the reductive four-electron activation and cleavage of O_2 in the monooxygenase reactions. In these redox reactions, riboflavin undergoes changes in its molecular shape, i.e., from a planar oxidized form to a folded reduced form. Differences in the affinities of the associated apoprotein for each shape affect the redox potential of the bound flavin.

Riboflavin antagonists include analogs of the isoalloxazine ring (e.g., diethylriboflavin, dichlororiboflavin) and the ribityl side chain (e.g., D-araboflavin, D-galactoflavin, 7-ethylriboflavin).

9. NIACIN

Essential features of the chemical structure:

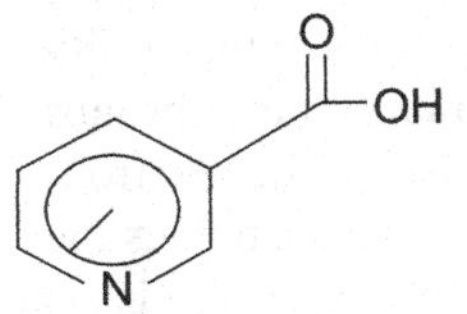

1. **Pyridine nucleus** substituted with a β-carboxylic acid or a corresponding amine
2. Pyridine nitrogen must be able to undergo reversible oxidation/reduction (i.e., quaternary pyridinium ion to/from tertiary amine)

3. Pyridine carbons adjacent to the nuclear nitrogen atom must be open.

Chemical structures of niacin:

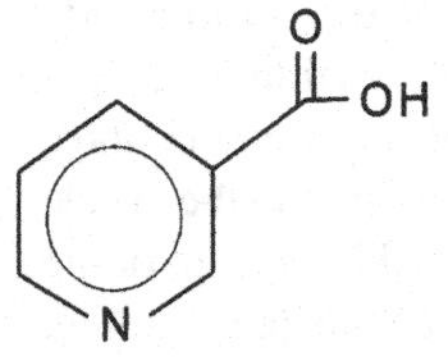

Nicotinic acid

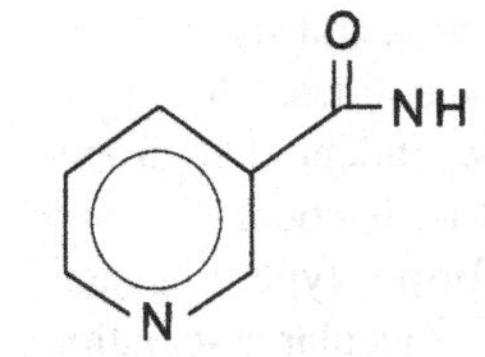

Nicotinamide

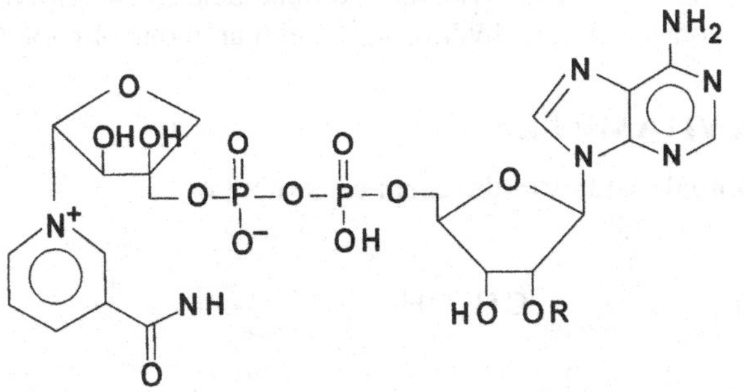

Nicotinamide adenine dinucleotide (NAD^+), R = H
Nicotinamide adenine dinucleotide phosphate ($NADP^+$), R = PO_3H_2

Niacin Nomenclature

Niacin is the generic descriptor for pyridine 3-carboxylic acid and derivatives exhibiting qualitatively the biological activity of nicotinamide.[15]

Niacin Chemistry

Nicotinic acid and **nicotinamide** are colorless crystalline substances. Each is insoluble or only sparingly soluble in organic solvents. Nicotinic acid is slightly soluble in water and ethanol; nicotinamide is very soluble in water and moderately soluble in ethanol. The two compounds have similar absorption spectra in water, with an absorption maximum at ~262 nm.

---

15. This compound is sometimes referred to as *niacinamide*, which, because its use would suggest that nicotinic acid should be called niacin, invites confusion and is not recommended.

Nicotinic acid is amphoteric, and forms salts with acids as well as bases. Its carboxyl group can form esters and anhydrides, and can be reduced. Both nicotinic acid and nicotinamide are very stable in dry form, but in solution nicotinamide is hydrolyzed by acids and bases to yield nicotinic acid.

The coenzyme forms of niacin are the pyridine nucleotides, **NAD(H)** and **NADP(H)**. In each of these compounds, the electron-withdrawing effect of the N-1 atom and the amide group of the oxidized pyridine nucleus enables the pyridine C-4 atom to react with many nucleophilic agents (e.g., sulfite, cyanide, and hydride ions). It is the reaction with hydride ions (H^-) that is the basis of the enzymatic hydrogen transfer by the pyridine nucleotides; the reaction involves the transfer of two electrons in a single step.[16] The hydride transfer of nonenzymatic reactions of the pyridine nucleotides, plus those catalyzed by the pyridine nucleotide-dependent dehydrogenases, is stereospecific with respect to both coenzyme and substrate. At least for reactions of the former type, this stereospecificity results from a specific intramolecular association between the adenine residue and the pyridine nucleus.

Several substituted pyridines are antagonists of niacin in biological systems: pyridine-3-sulfonic acid, 3-acetylpyridine, isonicotinic acid hydrazine,[17] and 6-aminonicotinamide.

10. VITAMIN B₆

Essential features of the chemical structure:

1. Derivative of 3-hydroxy-2-methyl-5-hydroxypyridine
2. Phosphorylatable 5-hydroxymethyl group
3. Substituent at ring carbon para to the pyridine nitrogen must be metabolizable to an aldehyde.

Chemical structures of vitamin B₆:

General structure of vitamin B₆

R=CH₂OH	pyridoxine
R=CHO	pyridoxal
R=COOH	pyridoxic acid
R=CH₂NH₂	pyridoxamine

Pyridoxal phosphate

Pyridoxamine phosphate

Vitamin B₆ Nomenclature

Vitamin B₆ is the generic descriptor for all 3-hydroxy-2-methylpyridine derivatives exhibiting the biological activity of pyridoxine in rats. The term **pyridoxine** is the trivial designation of one vitamin B₆-active compound, i.e., 3-hydroxy-4,5-bis(hydroxymethyl)-2-methylpyridine, which was formerly called *adermin* or **pyridoxol**. The biologically active analogs of pyridoxine are the aldehyde **pyridoxal** and the amine **pyridoxamine**.

Vitamin B₆ Chemistry

Vitamers B₆ are colorless crystals at room temperature. Each is very soluble in water, weakly soluble in ethanol, and either insoluble or sparingly soluble in chloroform. Each is fairly stable in dry form and in solution.

Pyridoxine is oxidized *in vivo* and under mild oxidizing conditions *in vitro* to yield pyridoxal. The prominent feature of the chemical reactivity of pyridoxal is the ability of its aldehyde group to react with primary amino groups (e.g., of amino acids) to form Schiff bases.[18] The electron-withdrawing effect of the resulting Schiff base labilizes the other bonds on the bound carbon, thus serving as the basis of the catalytic roles of pyridoxal and pyridoxamine.

---

16. It has been argued that the enzyme-catalyzed oxidation of NADH occurs in two steps with the intermediate formation of the NAD radical. Such a radical has been demonstrated, but it spontaneously dimerizes to an enzymatically inactive form, (NAD)₂, thus making it unlikely that such a mechanism plays a significant role in the redox functions of the pyridine nucleotides.

17. Also called *isoniazid*, this compound (4-pyridinecarboxylic acid hydrazide) is used as an antituberculous and antiactinomycotic agent.

---

18. This reaction can occur with tris(hydroxymethyl)amino methane (i.e., Tris) and may, thus, affect results of biochemical studies of vitamin B₆ in which this common buffering agent is employed.

11. BIOTIN

Essential features of the chemical structure:

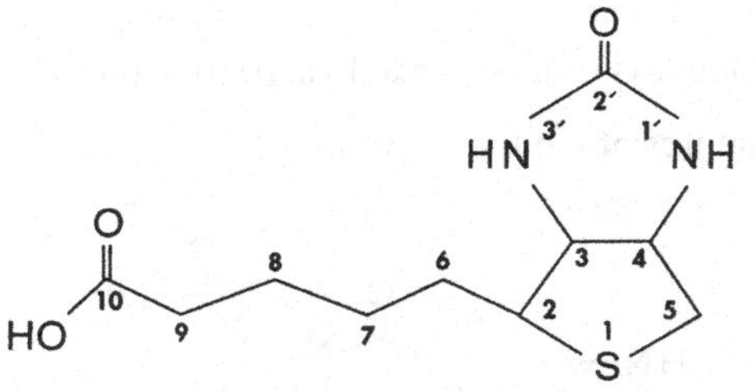

1. Conjoined ureido and tetrahydrothiophene nuclei
2. Ureido 3′ nitrogen is sterically hindered, preventing substitution
3. Ureido 1′ nitrogen is poorly nucleophilic.

Chemical structure of biotin:

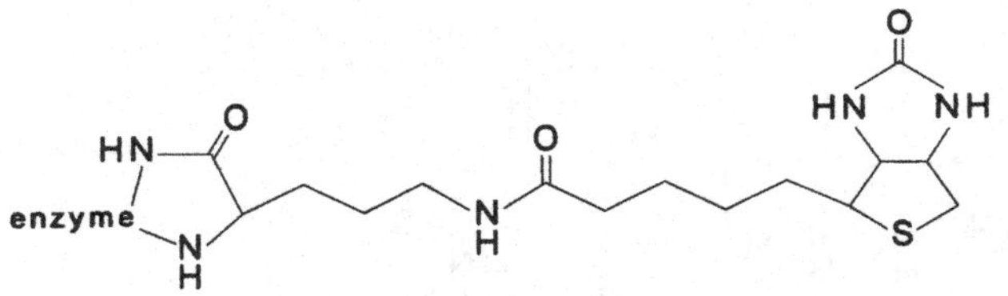

Enzyme-bound biotin (i.e., *biocytin*)

Biotin Nomenclature

Biotin is the trivial designation of the compound *cis*-hexahydro-2-oxo-1*H*-thieno[3,4-*d*]imidazole-4-pentanoic acid, formerly known as *vitamin H* or *coenzyme R*.

Biotin Chemistry

Biotin is a white crystalline substance that, in dry form, is fairly stable to air, heat, and light. In solution, however, it is sensitive to degradation under strongly acidic or basic conditions. Its structure consists of a planar **ureido nucleus** and a folded **tetrahydrothiophene (thiophane) nucleus**, which results in a boat configuration with a plane of symmetry passing through the S-1, C-2′, and O positions in such a way as to elevate the sulfur atom above the plane of the four carbons. The molecule has three asymmetric centers; however, of the eight possible stereoisomers, only the (+)-isomer (called *d*-biotin) has biological activity. Biotin is covalently bound to its enzymes by an amide bond to the ε-amino group of a lysine residue and C-2 of the thiophane nucleus. This bond is flexible, allowing the coenzyme to move between the active centers of some enzymes. The biotin molecule is activated by polarization of the O and N-1′ atoms of the ureido nucleus. This leads to increased nucleophilicity at N-1′, which promotes the formation of a covalent bond between the electrophilic

carbonyl phosphate formed from bicarbonate and ATP, and allows biotin to serve as a transport agent for CO_2.

12. PANTOTHENIC ACID

Essential features of the chemical structure:

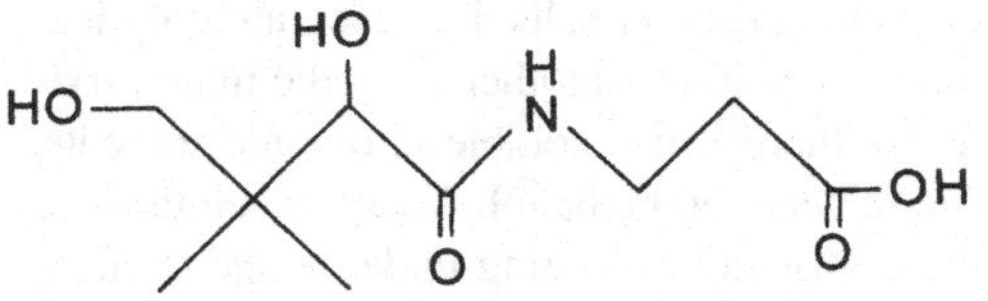

1. Formal derivative of pantoic acid and alanine
2. Optically active.

Chemical structure of pantothenic acid:

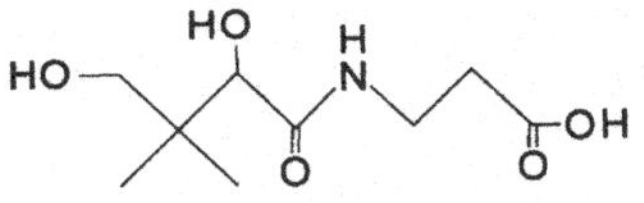

Pantothenic acid

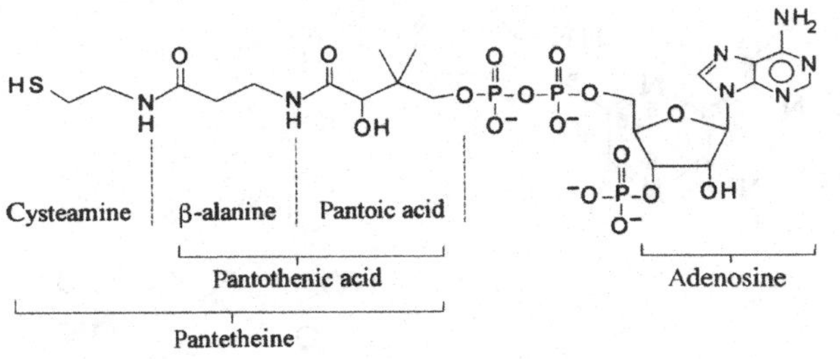

Coenzyme A (showing its constituent parts)

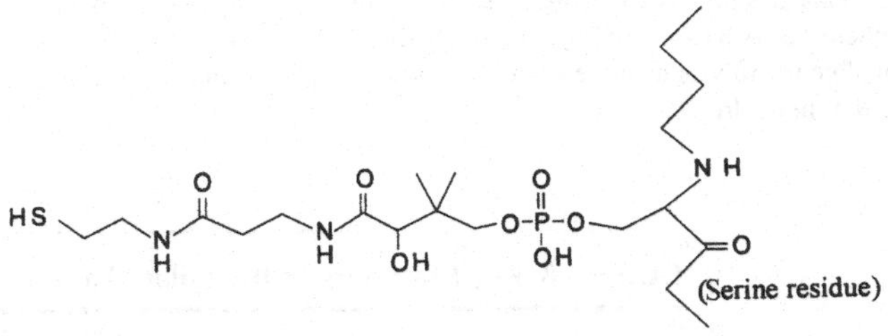

Acyl-carrier protein

Pantothenic Acid Nomenclature

Pantothenic acid is the trivial designation for the compound dihydroxy-β,β-dimethylbutyryl-β-alanine, which was formerly known as *pantoyl-β-alanine*. It has two metabolically active forms: **coenzyme A**, in which the vitamin is linked via a phosphodiester group with adenosine-3′5′-diphosphate; and acyl-carrier protein, in which it is linked via a phosphodiester to a serinyl residue of the protein.

Pantothenic Acid Chemistry

Pantothenic acid is composed of β-alanine joined to 2,4-dihydroxy-3,3-dimethylbutyric acid via an amide linkage. The molecule has an asymmetric center, and only the *R*-enantiomer, usually called D-(+)*pantothenic acid*, is biologically active and occurs naturally. Pantothenic acid is a yellow, viscous oil. Its calcium and other salts, however, are colorless crystalline substances; calcium pantothenate is the main product of commerce. Neither form is soluble in organic solvents, but each is soluble in water and ethanol. Aqueous solutions of pantothenic acid are unstable to heating under acidic or alkaline conditions, resulting in the hydrolytic cleavage of the molecule (to yield β-alanine and 2,4-dihydroxy-3,3-dimethylbutyrate).[19] The analog panthenol (in which the carboxyl group is replaced by an hydroxymethyl group) is fairly stable in solution. In dry form, the salts are stable to air and light; but they (particularly sodium pantothenate) are hygroscopic.

13. FOLATE

Essential features of the chemical structure:

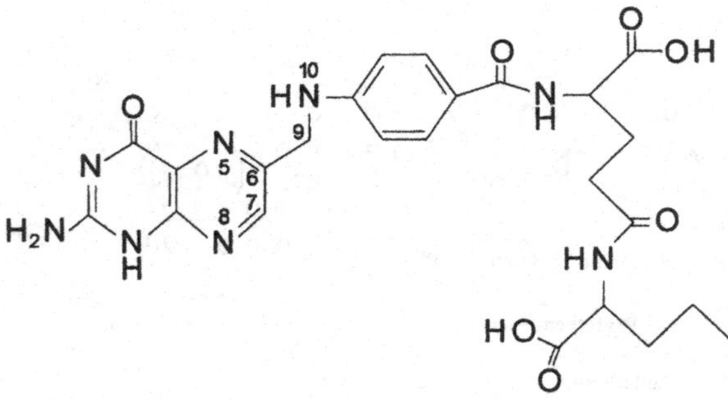

1. Pteridine derivative
2. Variable degree of hydrogenation of pteridine nucleus
3. Single-carbon units can bind nitrogens at position 5 and/or 10
4. One or more glutamyl residues linked via peptide bonds.

Chemical structures of the folate group:

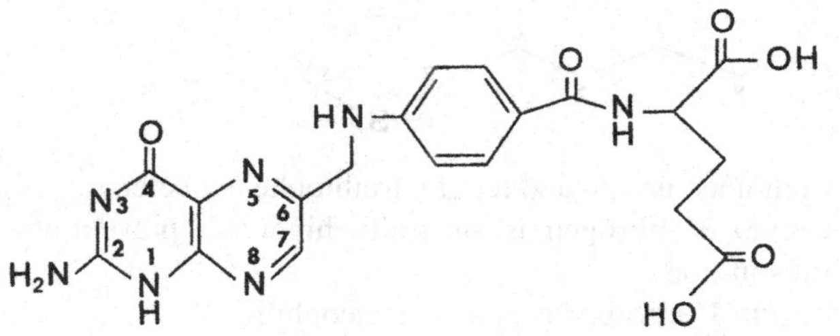

Pteroylglutamic acid

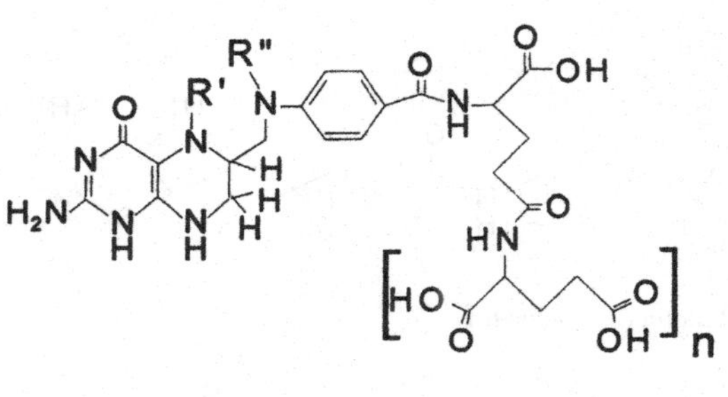

Tetrahydrofolic acid and its derivatives

Key members of the folate family are listed in Table 3.8.

Folate Nomenclature

Folate is the generic descriptor for **folic acid** (pteroyl-monoglutamic acid) and related compounds exhibiting the biological activity of folic acid. The terms *folacin*, *folic acids*, and *folates* are used only as general terms for this group of heterocyclic compounds based on the

---

19. This reaction is often used in the chemical determination of pantothenic acid by quantifying colorimetrically the β-alanine released on alkaline hydrolysis using reagents such as 1,2-naphthoquinone 4-sulfonic acid or ninhydrin.

TABLE 3.8 Key Members of the Folate Family

Vitamer	Abbreviation	R' (at N-5)	R (at N-10)
Tetrahydrofolic acid	FH_4	H	H
5-Methyltetrahydrofolic acid	$5\text{-}CH_3\text{-}FH_4$	CH_3	H
5,10-Methenyltetrahydrofolic acid	$5,10\text{-}CH^+\text{-}FH_4$	$-CH^+\text{-}$(bridge)	
5,10-Methylenetetrahydrofolic acid	$5,10\text{-}CH_2=FH_4$	$-CH_2=$(bridge)	
5-Formyltetrahydrofolic acid	$5\text{-}HCO\text{-}FH_4$	HCO	H
10-Formyltetrahydrofolic acid	$10\text{-}HCO\text{-}FH_4$	H	HCO
5-Forminintetrahydrofolic acid	$5\text{-}HCNH\text{-}FH_4$	HCNH	H

N-[(6-pteridinyl)methyl]-p-aminobenzoic acid skeleton conjugated with one or more L-glutamic acid residues. Folates can consist of a mono- or polyglutamyl conjugate; these are named for the number of glutamyl residues (n), using such notations as PteGlu$_n$.[20] The reduced compound tetrahydropteroylglutamic acid is called **tetrahydrofolic acid**; its single-carbon derivatives are named according to the specific carbon moiety bound.

Folate Chemistry

The folates include a large number of chemically related species, each differing with respect to the various substituents possible at three sites on the **pteroylglutamic acid** basic structure. Each is a formal derivative of **pteridine**.[21] With 3 known reduction states of the **pyrazine nucleus**, 6 different single-carbon substituents on N-5 and/or N-10, and as many as 8 glutamyl residues on the benzene ring, more than 170 different folates are theoretically possible.[22] Not all of these occur in nature, but it has been estimated that as many as 100 different forms are found in animals. The compound called *folic acid*, i.e., pteroylmonoglutamic acid, is probably not present in living cells, being rather an artifact of isolation of the vitamin. The folates from most natural sources usually have a single carbon unit at N-5 and/or N-10; these forms participate in the metabolism of the *single-carbon pool*. The single-carbon units that may be transported and stored by folates can vary in oxidation state from the methyl (e.g., 5-CH$_3$-FH$_4$) to the formyl (e.g., 5-HCO-FH$_4$, 10-HCO-FH$_4$). Intracellular folates contain poly-γ-glutamyl chains usually of 2 to 8 glutamyl residues, sometimes extending to 12 in bacteria. Tissues contain enzymes called *conjugases* that hydrolytically remove glutamyl residues to release the monoglutamyl form, i.e., folic acid. While the actual biochemical role of the polyglutamyl side chain is not presently clear, it appears that the folylpolyglutamates are the actual coenzyme forms active intracellularly, and that the monoglutamates, which can pass through membranes, are transport forms.

The folates have an asymmetric center at C-6. This introduces stereospecificity in the orientation of hydrogen atoms on reduction of the pteridine system; that is, they add to carbons 6 and 7 in positions below the plane of the pyrazine ring. The UV absorption spectra of the folates are characterized by the independent contributions of the pterin and 4-aminobenzoyl moieties; most have absorption maxima in the region of 280–300 nm.

Folic acid (pteroylmonoglutamic acid) is an orange-yellow crystalline substance that is soluble in water but insoluble in ethanol or less polar organic solvents. It is unstable to light, to acidic or alkaline conditions, to reducing agents, and, except in dry form, to heat. It is reduced *in vivo* enzymatically (or *in vitro* with a reductant such as dithionite) first to 7,8-dihydrofolic acid (FH$_2$) and then to FH$_4$; both of these compounds are unstable in aerobic environments and must be protected by the presence of an antioxidant (e.g., ascorbic acid, 2-mercaptoethanol).

Two derivatives of folic acid, each having an amino group in the place of the hydroxyl at C-4, are folate antagonists of biomedical use: *aminopterin* (4-aminofolic acid) and *methotrexate* (4-amino-N^{10}-methylfolic acid). Aminopterin is used as a rodenticide; methotrexate is an antineoplastic agent.

14. VITAMIN B$_{12}$

Essential features of the chemical structure:

1. Cobalt (Co)-centered corrin nucleus
2. Cobalt α position (below the plane of the corrin ring as shown) may be open or occupied by a side chain heterocyclic nitrogen, or solvent
3. Cobalt β position (above the plane of the corrin ring as shown) may be occupied by a hydroxo, aqua, methyl, 5-deoxyadenosyl, CN$^-$, Cl$^-$, Br$^-$, nitro, sulfito or sulfato group.[23]

Chemical structures of vitamin B$_{12}$:

Cyanocobalamin

---

20. Although they are still frequently used, the abbreviations using PteGlu to indicate pteroylglutamic acid are not suggested by current IUPAC-IUNS recommendations for vitamin nomenclature.
21. More specifically, the folates are pterins, namely 2-amino-4-hydroxypteridines. The pteridines are yellow compounds first isolated from butterfly wings, for which they were named (i.e., *pteron* is the Greek word meaning "wing"); many are folate antagonists.
22. This estimate is low, as bacteria are known to have as many as 12 residues in their polyglutamyl chains.

23. Only the first four liganded forms of vitamin B$_{12}$ are found in biological systems.

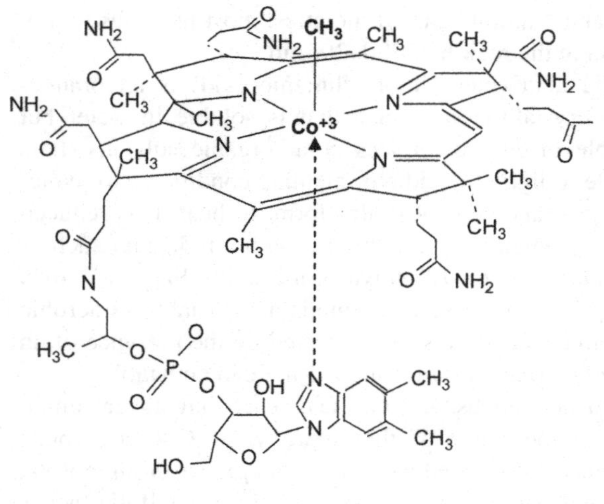

Methylcobalamin

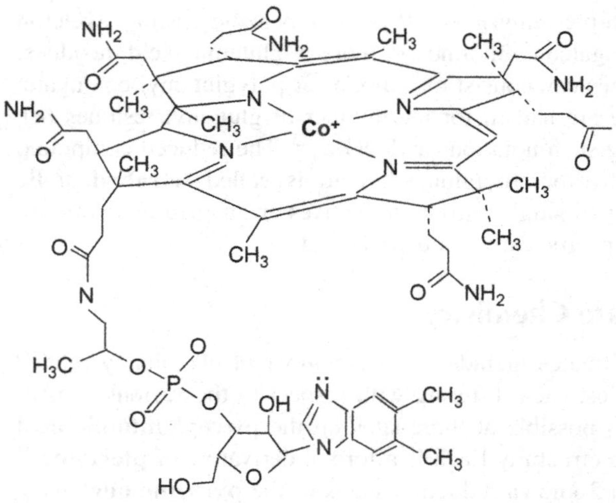

Cob(I)alamin

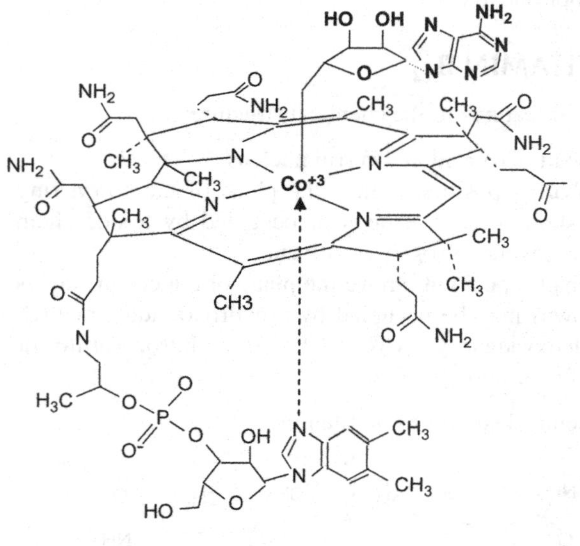

5′-Deoxyadenosylcobalamin

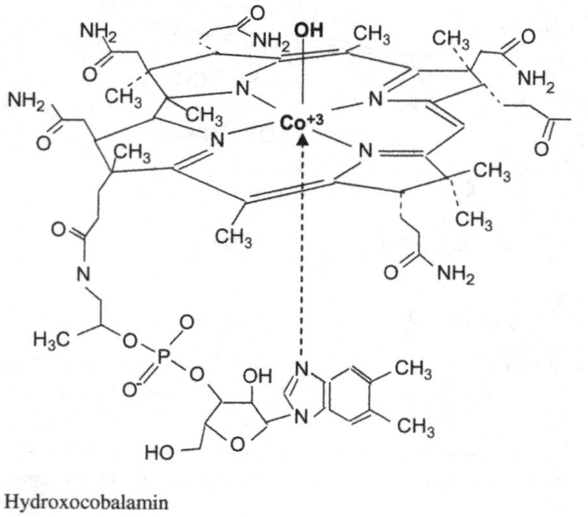

Hydroxocobalamin

Vitamin B$_{12}$ Nomenclature

Vitamin B$_{12}$ is the generic descriptor for all corrinoids (i.e., compounds containing the **corrin nucleus**) exhibiting the qualitative biological activity of **cyanocobalamin**. Cyanocobalamin is the trivial designation of the vitamin B$_{12}$-active corrinoid (also called **cobalamin**) with a cyano ligand (CN$^-$) at the β position of the cobalt atom. The analogs containing methyl-, 5′-deoxyadenosyl-, hydroxo- (OH) groups at that position are called **methylcobalamin**, **adenosylcobalamin**, and *hydroxocobalamin* (formerly vitamin B$_{12b}$), respectively. Those, as well as a form with an unliganded, reduced cobalt center, *cob(I)alamin*, and are found intracellularly. Other analogs with vitamin B$_{12}$ activity include *aquacobalamin* (formerly vitamin B$_{12a}$) and *nitritocobalamin* (formerly vitamin B$_{12c}$), and contain aqua- (H$_2$O) and nitrite groups, respectively.

Vitamin B$_{12}$ Chemistry

Vitamin B$_{12}$ is an octahedral cobalt complex consisting of a porphyrin-like, cobalt-centered macroring (called a *corrin ring* or *nucleus*), a nucleotide, and a second cobalt-bound group (e.g., CH$_3$, H$_2$O, CN$^-$). The corrin nucleus consists of four reduced pyrrole nuclei linked by three methylene bridges and one direct bond. The triply ionized cobalt atom (i.e., Co^{3+}) can form up to six coordinate bonds, and is tightly bound to the four pyrrole nitrogen atoms. It can also bond a nucleotide and a small ligand below and above, respectively, the plane of the ring system. The cobalt atom is removed *in vitro* only with difficulty, resulting in loss of biological activity.

The corrinoids are red, red–orange or yellow crystalline substances that show intense absorption spectra above 300 nm owing to the π–π transitions of the corrin nucleus. They are soluble in water, and are fairly stable to heat but decompose at temperatures above ~210°C without melting.

Vitamin B_{12} reacts with ascorbic acid, resulting in the reduction and subsequent degradation of the former, which releases its cobalt atom as the free ion. Cobalamins with relatively strongly bound ligands (e.g., cyano-, methyl-, and adenosylcobalamin) are less reactive, and are therefore more stable in the presence of ascorbic acid. The cobalamins are unstable to light. Cyanocobalamin undergoes a photoreplacement of the CN^- ligand with water; the organocobalamins (methyl- and adenosylcobalamin) undergo photoreduction of the cobalt–carbon bond, resulting in the loss of the ligand and the reduction of the corrin cobalt. The vitamin can bind to proteins in the vitamin B_{12} enzymes through an imidazole nitrogen of a histidyl residue on the protein which serves as the ligand to the lower axial position of the cobalt atom instead of the dimethylbenzimidazole grouping.

15. GENERAL PROPERTIES OF THE VITAMINS

Multiple Forms of Vitamins

Few of the vitamins are biologically active without metabolic conversion to another species and/or binding to a specific protein. Thus, any consideration of the vitamins in nutrition involves, for each vitamin group, a number of vitamers and metabolites; some of these are important in the practical sense for food and diet supplementation (Table 3.9), whereas others are important in the physiological sense as they participate in metabolism.

Vitamin Stability

For the use of vitamins as food/feed additives, in diet supplements, and as pharmaceuticals, stability is a prime concern. In general, the fat-soluble vitamins, vitamin C, thiamin, riboflavin, and biotin are poorly stable to

TABLE 3.9 The Most Important Forms of the Vitamins

Vitamin	Representative	Metabolically Active Forms	Important Dietary Forms
Vitamin A	Retinol	Retinol Retinal Retinoic acid	Retinyl palmitate and acetate, provitamins (β-carotene, other carotenoids)
Vitamin D	Cholecalciferol	25-OH-Cholecalciferol 1,25-$(OH)_2$-Cholecalciferol	Cholecalciferol, ergocalciferol
Vitamin E	α-Tocopherol	α-, β-, γ-, δ-Tocopherols	R,R,R-α-Tocopherol; all-rac-α-tocopheryl acetate
Vitamin K	Phylloquinone	Phylloquinones (K) Menaquinones (MK)	K, MK, menadione, menadione sodium bisulfite complex
Vitamin C	Ascorbic acid	Ascorbic acid Dehydroascorbic acid	L-Ascorbic acid, sodium ascorbate
Thiamin	Thiamin	Thiamin pyrophosphate	Thiamin; thiamin pyrophosphate, disulfide, HCl, mononitrate
Riboflavin	Riboflavin	FMN, FAD	FMN, FAD, flavoproteins, riboflavin
Niacin	Nicotinamide	NAD, NADP	NAD, NADP, nicotinamide, nicotinic acid
Vitamin B_6	Pyridoxine	Pyridoxal 5'-phosphate Pyridoxamine 5'-phosphate	Pyridoxal HCl, pyridoxal- and pyridoxamine-5'-phosphates
Biotin	d-Biotin	d-Biotin	Biocytin, d-biotin
Pantothenic acid	Pantothenic acid	Coenzyme A	Calcium pantothenate, coenzyme A, acyl-CoAs
Folate	Pteroylglutamic acid	Pteroylpolyglutamates	Pteroyl poly- and monoglutamates
Vitamin B_{12}	Cyanocobalamin	Methylcobalamin 5'-Deoxyadenosylcobalamin	Cyano-, aqua-, hydroxo-, methyl-, and 5'-deoxyadenosylcobalamins

oxidation. They must be protected from heat, oxygen, metal ions (particularly Fe^{2+} and Cu^{2+}), polyunsaturated lipids undergoing peroxidation, and ultraviolet light; antioxidants are frequently used in their formulations. For vitamins A and E, the more stable esterified forms are used for these purposes. Because of the instabilities of their naturally occurring vitamers, the amounts of the fat-soluble vitamins in natural foods and feedstuffs are highly variable, being greatly affected by the conditions of food production and processing. Niacin, vitamin B_6, pantothenic acid, folate, and vitamin B_{12} tend to be more stable under most practical conditions (Table 3.10). Some

TABLE 3.10 Stabilities of the Vitamins

Vitamin	Vitamer	UV	Heat[a]	O₂	Acid	Alkali	Metals[b]	To Enhance Stability
Vitamin A	Retinol	+		+	+		+	Keep in the dark, exclude O₂, use antioxidants
	Retinal			+	+		+	Exclude O₂, use antioxidants
	Retinoic acid							Good stability
	Dehydroretinol			+				Exclude O₂, use antioxidants
	Retinyl esters							Good stability
	β-Carotene	+		+				Keep in the dark, exclude O₂, use antioxidants
Vitamin D	D₂	+	+[c]	+	+		+	Keep cool, in the dark, exclude O₂, use antioxidants
	D₃	+	+[c]	+	+	+	+	Keep cool, in the dark, exclude O₂, use antioxidants
Vitamin E	Tocopherols		+	+	+	+	+	Keep cool, at neutral pH
	Tocopheryl esters				+	+		Good stability
Vitamin K	K	+		+		+	+	Avoid reductants[c], work in subdued light
	MK	+		+		+	+	Avoid reductants[c], work in subdued light
	Menadione	+				+	+	Avoid reductants[c], work in subdued light
Vitamin C	Ascorbic acid			+[b]		+	+	Exclude O₂, at neutral pH
Thiamin	Disulfide form		+	+	+	+	+	Keep at neutral pH[d]
	Hydrochloride[e]		+	+	+	+	+	Exclude O₂, at neutral pH[d]
Riboflavin	Riboflavin	+[f]	+			+	+	Keep in the dark, at pH 1.5–4[d]
Niacin	Nicotinic acid							Good stability
	Nicotinamide							Good stability
Vitamin B₆	Pyridoxal	+	+					Keep cool, work in subdued light
	Pyridoxol HCl			+		+		Good stability
Biotin	Biotin			+		+		Exclude O₂, at pH 4–9, use antioxidants, work in subdued light
Pantothenic acid	Free acid[g]	+		+		+		Cool, neutral pH
	Calcium salt[e]		+					Exclude O₂, at pH 5–7
Folate	FH₄	+	+	+	+[h]		+	Good stability[d]
Vitamin B₁₂	Cyano-B₁₂	+			+[i]		+[j]	Good stability[c] at pH 4–7

[a] That is, 100°C.
[b] In solution with Fe^{3+} and Cu^{2+}.
[c] isomerization to the pre-vitamin form may be unavoidable, but tachysterol can also be formed under acid conditions in samples exposed to light.
[d] Unstable to reducing agents.
[e] Slightly hygroscopic.
[f] Especially in alkaline solution.
[g] Very hygroscopic.
[h] pH <5.
[i] pH <3.
[j] pH >9.

vitamins can undergo degradation by reacting to factors in foods during sample storage and/or preparation: ascorbic acid with plant ascorbic acid oxidase; thiamin with sulfites or with plant or microbial thiaminases; folates with nitrites; pantothenic acid with microbial pantothenases.

Vitamin Analysis

A variety of methods is available for the quantitative determination of the vitamins (Table 3.11). Because many vitamers are bound to proteins or other factors in biological specimens and foods, their extraction necessitates disruption of those complexes and separation from interfering substances. This must be done in ways that are both quantitative and accommodate the intrinsic characteristics of each vitamer. Accordingly, conditions of sample extraction and clean-up must stabilize the vitamin(s) of interest in order to yield accurate results. Chromatographic separations have proven useful for determining vitamins A, D, E, K, C, thiamin, riboflavin, niacin, and vitamin B_6. They depend on separation by phase-partitioning (liquid–liquid,[24] or gas–liquid[25]) of vitamers for specificity, ascertained by comparison to authentic standards, and a suitable means of detection (e.g., ultraviolet–visible absorption, fluorescence, electrochemical reactivity) for sensitivity. Microbiological assays are available for thiamin, riboflavin, niacin, vitamin B_6, pantothenic acid, biotin, folate, and vitamin B_{12}. These methods are based on the absolute requirement of certain microorganisms for particular vitamins for multiplication, which can be measured turbidimetrially or by the evolution of CO_2 from substrate provided in the growth media. Some forms of vitamins A, E, and C can be measured by chemical colorimetric reactions; however, only the dye reduction methods for ascorbic acid have appropriate specificity and reliability to be recommended.[26] Competitive protein-binding assays have been developed for biotin, folate, and vitamin B_{12}.[27] A radioimmunoassay and an enzyme-linked immunoabsorbent assay have been developed for pantothenic acid.

16. PHYSIOLOGICAL UTILIZATION OF THE VITAMINS

Vitamin Bioavailability

Because not all vitamins in foods are completely utilized by the body, a measurement of the gross amounts of vitamins in foods (yielded by analysis) is insufficient to understand the actual nutritional value of those vitamin sources. This is due to several factors (Table 3.12): differing losses, dietary effects, physiological effects, and health status, as well as differing biopotencies. The actual rate and extent to which a vitamin is absorbed and utilized at the cellular level are referred to as its *bioavailability*. This concept differs from that of biopotency, which refers only to properties intrinsic to particular vitamers. Bioavailability refers to the integrated effects of those and other physiological and dietary factors. Some authors have used the term *bioefficacy* with a similar connotation.

Bioavailability

The concept of bioavailability, as applied to the vitamins and other essential nutrients, refers to the portion of ingested nutrient that is absorbed, retained, and metabolized through normal pathways in a form or forms that can be utilized for normal physiologic functions.

Vitamin Absorption

The means by which the vitamins are absorbed are determined by their chemical and associated physical properties. The fat-soluble vitamins (and hydrophobic substances such as carotenoids and cholesterol), which are not soluble in the aqueous environment of the alimentary canal, are associated with and dissolved in other lipid materials. In the upper portion of the gastrointestinal tract, they are dissolved in the bulk lipid phases of the emulsions that are formed of dietary fats[28] by the mechanical actions of mastication and gastric churning. Emulsion oil droplets, however, are generally too large (e.g., 1,000 Å) to gain the intimate proximity to the absorptive surfaces of the small intestine that is necessary to facilitate the diffusion of these substances into the hydrophobic environment of the brush border membranes of intestinal mucosal cells.

---

24. High-performance liquid chromatography, HPLC.
25. Gas-liquid chromatography, GLC.
26. *Vitamin A*: The Carr-Price method, which uses the time-sensitive production of a blue complex of retinol and antimony trichloride, is no longer recommended due to its lack of specificity and negative bias. *Vitamin E*: The Fe^{2+}-dependent reduction of a fat-soluble dye such as bathophenanthroline by vitamin E is not recommended due to its lack of specificity, although many interfering substances can be partitioned into aqueous solvents during sample preparation. *Vitamin C*: The reaction of ascorbic acid with the dye 2,4-dinitrophenolindolphenol remains a useful method due to the fact that most interfering substances can be partitioned into organic solvents during sample preparation.
27. Biotin and avidin; folate and folate-binding protein; vitamin B_{12} and R proteins.

---

28. While it follows that the fat-soluble vitamins cannot be well absorbed from low-fat diets, the minimum amount of fat required is not clear. One study (Roodenburg, A. J., Leenen, R., van het Hof, K. H., Weststrate, J. A., and Tilburg, L. B. [2000] *Am. J. Clin. Nutr.* 71, 1187–1193) found that 3 g fat per meal was sufficient for optimal absorption of some provitamin A carotenoids (α- and β-carotenes), but another study (Brown, M. J., Ferruzzi, M. G., Nguyen, M. L., Cooper, D. A., Eldridge, A.L., Schwartz, S. J., and White, W. S. [2004] *Am. J. Clin. Nutr.* 80, 396–403) found that 29.5 g fat per meal was insufficient.

TABLE 3.11 Methods of Vitamin Analysis

Vitamin	Sample Preparation	Instrumental Analysis		Microbiological Assay
		Analyte Separation	Analyte Detection	
Vitamin A	Direct solvent extraction; alkaline HPLC[a] hydrolysis[b], extraction into organic solvents	UV absorption		
Vitamin D	Alkaline hydrolysis with extraction into organic solvents	HPLC	UV absorption	
		LC-IT-TOF[c]	Molecular mass spectrometry	
Vitamin E	Alkaline hydrolysis with extraction into organic solvents	HPLC	Fluorescence, UV absorption	
Vitamin K	Direct solvent extraction; supercritical fluid extraction[d]; enzymatic hydrolysis	HPLC	UV absorption	
Vitamin C	Acid hydrolysis	HPLC, IEC[e], MECC[f]	UV absorption	
Thiamin	Acid hydrolysis; enzymatic hydrolysis[g]	IEC, GLC[h], HPLC	Fluorescence, UV absorption, FID[j]	*Lactobaccillus viridescens* (12706)[i]
Riboflavin	Acid hydrolysis	HPLC, MECC	Fluorescence, UV absorption	*Lactobaccillus casei* subsp. *rhamnsus* (7469) *Enterococcus fecalis* (10100)
Niacin	Alkaline hydrolysis	IEC, HPLC GLC, MECC	UV absorption, FID	*Lactobaccillus plantarum* (8014)[k] *Leuconostoc mesenteroides* subsp. *mesenteroides* (9135)
Vitamin B$_6$	Acid hydrolysis	HPLC, IEC, GLC, MECC	Fluorescence, UV absorption, FID	*Saccharomyces carlbergensis* (9080) *Kloectra apiculata* (8714)
Pantothenic Acid	Alkaline hydrolysis; enzymatic hydrolysis	GLC	FID	*Lactobaccillus plantarum*[l]
Folate	Enzymatic hydrolysis[m]	IEC		*Lactobaccillus casei* subsp. *rhamnsus* (7469) [n] *Enterococcus hirae* (8043)[o]
Biotin	Acid hydrolysis; enzymatic hydrolysis[p]			*Lactobaccillus plantarum* (8014)
Vitamin B$_{12}$	Direct solvent extraction	MECC	UV absorption	*Lactobaccilus delbrueckii*, subsp. *lactis* (4797)

[a] High-performance liquid chromatography.
[b] Saponification.
[c] Liquid chromatography-ion trap, time-of-flight mass spectrometry.
[d] Supercritical fluids are gases held above their critical temperature and critical pressure, which confers solvating properties similar to organic solvents with very low viscosities and very high diffusivities.
[e] Ion-exchange chromatography.
[f] Micellar electrokinetic capillary electrophoresis.
[g] Thiaminase or other phosphatase.
[h] Gas-liquid chromatography.
[i] Flame ionization detection (used with GLC separation).
[j] American Type Culture Collection number.
[k] Responds to nicotinic acid only.
[l] Responds to free vitamer only.
[m] Responds to all vitamers; yields "total" folate activity.
[n] Folyl conjugase.
[o] Papain.

TABLE 3.12 Several Factors Determining Vitamin Bioavailability

Factor	Description
	Extrinsic factors
Differing biopotencies	Different vitamers can have inherent differences in biopotencies, e.g., ergocalciferol is markedly less biopotent for the chick in comparison with cholecalciferol
Losses	Some vitamins in foods show significant losses during storage, processing, and/or cooking, e.g., the vitamin C content of potatoes can drop by one-third within 1 month of storage
Dietary effects	The compositions of meals and diets can affect the absorption of some vitamins by affecting intestinal transit time and/or the enteric formation of mixed micelles, e.g., vitamin A and provitamin A carotenoids are absorbed very poorly from very low-fat diets
	Intrinsic factors
Physiological effects	Age-related differences in gastrointestinal function can affect the absorption and postabsorptive utilization of certain vitamins, e.g., the absorption of vitamin B_{12} is reduced in many older persons who experience loss of gastric parietal cell function
Health status	Some illnesses can affect the absorption and postabsorptive utilization of certain vitamins, e.g., folate absorption is impaired in patients with sprue

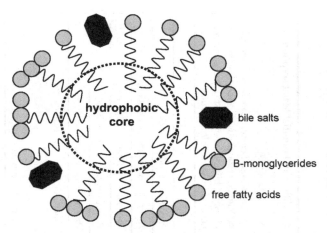

FIGURE 3.2 Mixed micelles form in the intestinal lumen by the spontaneous association of the products of triglyceride digestion, β-monoglycerides and free fatty acids, and bile salts. Their hydrophobic cores provide environments for the fat-soluble vitamins (A, D, E, and K) and other lipophilic dietary components. The absorption across the intestinal microvillar surface is facilitated by their very small size (10–50 Å diameter).

However, *lipase*, which is present in the intestinal lumen, having been synthesized in and exported from the pancreas via the pancreatic duct, binds to the surface of emulsion oil droplets, where it catalyzes the hydrolytic removal of the α- and α′-fatty acids from triglycerides, which make up the bulk of the lipid material in these large particles. The products of this process (i.e., free fatty acids and β-monoglycerides) have strong polar regions or charged groups, and thus will dissolve to some extent monomerically in this aqueous environment. However, they also have long-chain hydrocarbon non-polar regions; therefore, when certain concentrations (*critical micellar concentrations*) are achieved, these species and bile salts, which have similar properties, combine spontaneously to form small particles called *mixed micelles* (Fig. 3.2). Mixed micelles thus contain free fatty acids, β-monoglycerides, and bile salts in which the non-polar regions of each are associated interiorly, and the polar or charged regions of each are oriented externally and are associated with the aqueous phase. The core of the mixed micelle is hydrophobic, and thus serves to solubilize the fat-soluble vitamins and other non-polar lipid substances. Because they are small (10–50 Å

in diameter), mixed micelles can gain close proximity to microvillar surfaces of intestinal mucosa, thus facilitating the diffusion of their contents into and across those membranes. Because the enteric absorption of the fat-soluble vitamins depends on micellar dispersion, it is impaired under conditions of lipid malabsorption or very low dietary fat (<10 g/day). Only one fat-soluble vitamer (phylloquinone) is known to be absorbed by active transport.

The water-soluble vitamins, which are soluble in the polar environment of the intestinal lumen, can be taken up by the absorptive surface of the gut more directly. Some (vitamin C, thiamin, niacin, vitamin B_6, biotin, pantothenic acid, folate, vitamin B_{12}) are absorbed as the result of passive diffusion; others are absorbed via specific carriers as a means of overcoming concentration gradients unfavorable to simple diffusion. Several (vitamin C, vitamin B_{12}, thiamin, niacin, folate) are absorbed via carrier-dependent mechanisms at low doses,[29] and by simple diffusion (albeit at lower efficiency) at high doses.

The absorption of at least three water-soluble vitamins (vitamin C, riboflavin, and vitamin B_6) appears to be regulated in part by the dietary supply of the vitamin in a feedback manner. Thus, it has been questioned whether high doses of one vitamin/vitamer may antagonize the absorption of related vitamins. There is some evidence for such mutual antagonisms among the fat-soluble vitamins, as well as in the case of α-tocopherol, a high intake of which antagonizes the utilization of the related γ-vitamer.

A summary of modes of enteric absorption of the vitamins is presented in Table 3.13.

---

29. These processes show apparent K_m values in the range of 0.1–300 μmol/l.

TABLE 3.13 Enteric Absorption of the Vitamins[a]

Vitamer	Digestion	Site	Enterocytic Metabolism	Efficiency (%)	Conditions of Potential Malabsorption
Micelle-Dependent Diffusion					
Retinol	—	D, J	Esterification	80–90	Pancreatic insufficiency (pancreatitis, selenium deficiency, cystic fibrosis, cancer), β-carotene cleavage, biliary atresia, obstructive jaundice, celiac disease, very low-fat diet
Retinyl esters	De-esterified	D, J	Re-esterification		
Vitamins D	—	D, J	Esterification	50–60	Pancreatic or biliary insufficiency
Tocopherols	—	D, J	—	~50	Pancreatic or biliary insufficiency
Tocopherol esters	De-esterified[b]	D, J	—	20–80	Pancreatic or biliary insufficiency
MKs	—	D, J	—	10–70	Pancreatic or biliary insufficiency
Menadione	—	D, J	—	10–70	Pancreatic or biliary insufficiency
Active Transport					
Phylloquinone	—	D, J	—	~80	Pancreatic or biliary insufficiency
Ascorbic acid	—	I	—	70–80	D-Isoascorbic acid
Thiamin	—	D	Phosphorylation		Pyrithiamin, excess ethanol
Thiamin di-P	Dephosphorylation[b]	D	Phosphorylation		Pyrithiamin, excess ethanol
Riboflavin	—	J	Phosphorylation		
FMN, FAD	Hydrolysis[b]	J	Phosphorylation		
Flavoproteins	Hydrolysis[b]	J	Phosphorylation		
Folylmono-glu	—	J	Glutamation		Celiac sprue
Folylpoly-glu	Hydrolysis[b]	J	Glutamation		Celiac sprue
Vitamin B$_{12}$	Hydrolysis[b]	I	Adenosinylation, methylation	>90	Intrinsic factor deficiency (pernicious anemia)
Facilitated Diffusion[c]					
Nicotinic acid	—	J		>90[d]	
Nicotinamide	—	J		~100[d]	
Niacytin	Hydrolysis[b]	J			
NAD(P)	Hydrolysis[b]	J			

Compound	Site	Percent absorbed	Metabolism during absorption	Released in absorbable form	Inhibitors / comments
Biotin	J			—	Biotinidase deficiency, consumption of raw egg white (avidin)
Biocytin	J			Hydrolysis[b]	Biotinidase deficiency, consumption of raw egg white (avidin)
Pantothenate				–	
Coenzyme A	J			Hydrolysis[b]	
Simple Diffusion					
Ascorbic acid[e]	D, J, I	<50	—		
Thiamin[e,f]	J		Phosphorylation		
Nicotinic acid	J		—		
Nicotinamide	J		—		
Pyridoxol	J		Phosphorylation		
Pyridoxal	J		Phosphorylation		
Pyridoxamine	J		Phosphorylation		
Biotin	D, J	>95	—		Consumption of raw egg white (avidin)
Pantothenate	J		—		
Folylmono-glu[e]	J		Glutamation		
Vitamin B$_{12}$	D, J	~1	Adenosinylation, methylation		

a Abbreviations: D, duodenum; J, jejunum; I, ileum; thiamin di-P, thiamin diphosphate; folylmono-glu, folylmonoglutamate; folylpoly-glu, folylpolyglutamate.
b Yields vitamin in absorbable form.
c Na$^+$-dependent saturable processes.
d Estimate may include contribution of simple diffusion.
e Simple diffusion important only at high doses.
f Symport with Na$^+$.

TABLE 3.14 Postabsorptive Transport of the Vitamins in the Body

Vehicle	Vitamin	Form Transported	Vehicle	Vitamin	Form Transported
Lipoprotein Bound			**Free in Plasma**		
Chylomicrons[a]	Vitamin A	Retinyl esters	—	Vitamin C	Ascorbic acid
	Vitamin A	β-Carotene	—	Thiamin	Free thiamin
	Vitamin D	Vitamin D[b]	—	Thiamin	Thiamin pyrophosphate
	Vitamin E	Tocopherols	—	Riboflavin	Flavin mononucleotide
	Vitamin K	K, MK, menadione	—	Pantothenic acid	Pantothenic acid
VLDL[c]/HDL[d]	Vitamin E	Tocopherols	—	Biotin	Free biotin
	Vitamin K	Mainly MK-4	—	Niacin	Nicotinic acid
Associated Non-Specifically with Proteins			—	Niacin	Nicotinamide
Albumin	Riboflavin	Free riboflavin	—	Folate	Pteroylmonoglutamates[f]
	Riboflavin	Flavin mononucleotide	**Bound to Specific Intracellular Binding Proteins**		
	Vitamin B_6	Pyridoxal	Cellular RBP (CRBP)	Vitamin A	all-*trans*-Retinol
	Vitamin B_6	Pyridoxal phosphate	Cellular RBP, type II (CRBPII)	Vitamin A	all-*trans*-Retinol
Immunoglobulins[e]	Riboflavin	Free riboflavin	Interstitial RBP (IRBP)	Vitamin A	all-*trans*-Retinol
Bound to Specific Binding Proteins			Cellular retinal BP (CRALBP)	Vitamin A	all-*trans*-Retinal
Retinol BP (RBP)	Vitamin A	all-*trans*-Retinol	Cellular retinoic acid BP (CRABP)	Vitamin A	all-*trans*-Retinoic acid
Transcalciferin (vitamin D BP)	Vitamin D	D_2; D_3; 25-OH-D; 1,25-$(OH)_2$-D; 24,25-$(OH)_2$-D	Vitamin D receptor	Vitamin D	1,25-$(OH)_2$-D
Thiamin BP	Thiamin	Free thiamin	Vitamin E BP	Vitamin E	Tocopherols
Riboflavin BP	Riboflavin	Riboflavin	Flavoproteins	Riboflavin	Flavin mononucleotide
Biotinidase	Biotin	Free Biotin		Riboflavin	Flavin adenine dinucleotide
Folate BP	Folate	Folate	Transcobalamin I	Vitamin B_{12}	Vitamin B_{12}
Transcobalamin II	Vitamin B_{12}	Methylcobalamin			
Transcobalamin III	Vitamin B_{12}	Vitamin B_{12}			
Carried in Erythrocytes					
Erythrocyte membranes	Vitamin E	Tocopherols			
Erythrocytes	Vitamin B_6	Pyridoxal phosphate			
	Pantothenic acid	Coenzyme A			

[a] In mammals, *lipids are absorbed into the lymphatic circulation, where they are transported to the liver and other tissues as large lipoprotein particles called chylomicra (singular, chylomicron); in birds, reptiles, and fishes, lipids are absorbed directly into the hepatic portal circulation and the analogous lipoprotein particle is called a* **portomicron**.
[a] *Representation of vitamin D without a subscript is meant to refer to both major forms of the vitamin: ergocalciferol (D_2) and cholecalciferol (D_3).*
[b] *VLDL, Very low-density lipoprotein.*
[c] *HDL, High-density lipoprotein.*
[d] *For example, IgG, IgM, and IgA.*
[e] *Especially 5-CH_3-tetrahydrofolic acid.*

Vitamin Transport

The mechanisms of postabsorptive transport of the vitamins also vary according to their particular physical and chemical properties (Table 3.14). Again, therefore, the problem of solubility in the aqueous environments of the blood plasma and lymph is a major determinant of ways in which the vitamins are transported from the site of absorption (the small intestine) to the liver and peripheral organs. The fat-soluble vitamins, because they are insoluble in these transport environments, depend on carriers that are soluble there. These vitamins, therefore, are associated with the lipid-rich **chylomicrons**[30] that are elaborated in intestinal mucosal cells, largely of re-esterified triglycerides from free fatty acids and β-monoglycerides that have just been absorbed. As the lipids in these particles are transferred to other **lipoproteins**[31] in the liver, some of the fat-soluble vitamins (vitamins E and K) are also transferred to those carriers. Others (vitamins A and D) are transported from the liver to peripheral tissues by specific **binding proteins** of hepatic origin (*see* Table 3.14). Some of the water-soluble vitamins are transported by protein carriers in the plasma, and therefore are not found free in solution. Some (riboflavin, vitamin B_6) are carried via weak, non-specific binding to albumin, and may thus be displaced by other substances (e.g., ethanol) that also bind to that protein. Others are tightly associated with certain immunoglobulins (riboflavin) or bind to specific proteins involved in their transport (riboflavin, vitamins A, D, E, and B_{12}). Several vitamins (e.g., vitamin C, thiamin, niacin, riboflavin, pantothenic acid, biotin, folate) are transported in free solution in the plasma.

Tissue Distribution of the Vitamins

The retention and distribution of the vitamins among the various tissues also vary according to their general physical and chemical properties (Table 3.15). In general, the fat-soluble vitamins are well retained; they tend to be stored in association with tissue lipids. For that reason, lipid-rich tissues such as adipose and liver frequently have appreciable stores of the fat-soluble vitamins. Storage of these vitamins means that animals may be able to accommodate widely variable intakes without consequence by mobilizing their tissue stores in times of low dietary intakes.

TABLE 3.15 Tissue Distribution of the Vitamins

Vitamin	Predominant Storage Form(s)	Depot(s)
Vitamin A	Retinyl esters (e.g., palmitate)	Liver
Vitamin D	D_3; 25-OH-D	Plasma, adipose, muscle
Vitamin E	α-Tocopherol	Adipose, adrenal, testes, platelets, other tissues
Vitamin K[a]	K: K-4, MK-4 MK: MK-4 Menadione: MK-4	Liver All tissues All tissues
Vitamin C	Ascorbic acid	Adrenals, leukocytes
Thiamin	Thiamin pyrophosphate[b]	Heart, kidney, brain, muscle
Riboflavin	Flavin adenine dinucleotide[b]	Liver, kidney, heart
Vitamin B_6	Pyridoxal phosphate[b]	Liver[c], kidney[c], heart[c]
Vitamin B_{12}	Methylcobalamin	Liver[d], kidney[c], heart[c], spleen[c], brain[c]
Niacin	No appreciable storage	—
Biotin	No appreciable storage[b]	—
Pantothenic acid	No appreciable storage	—
Folate	No appreciable storage	—

[a] *The predominant form of the vitamin is shown for each major form of dietary vitamin K consumed.*
[b] *The amounts in the body are composed of the enzyme-bound coenzyme.*
[c] *Small amounts of the vitamin are found in these tissues.*
[d] *Predominant depot.*

30. Chylomicrons are the largest (*ca.* 1 μm in diameter) and the lightest of the blood lipids. They consist mainly of triglyceride with smaller amounts of cholesterol, phospholipid, protein, and the fat-soluble vitamins. They are normally synthesized in the intestinal mucosal cells and serve to transport lipids to tissues. In mammals, these particles are secreted into the lymphatic drainage of the small intestine (hence their name). However, in birds, fishes, and reptiles they are secreted directly into the renal portal circulation; therefore, in these species they are referred to as *portomicrons*. In either case, they are cleared from the plasma by the liver, and their lipid contents are either deposited in hepatic stores (e.g., vitamin A) or released back into the plasma bound to more dense particles called *lipoproteins*.

31. As the name would imply, a *lipoprotein* is a lipid–protein combination with the solubility characteristics of a protein (i.e., soluble in the aqueous environment of the blood plasma), and hence involved in lipid transport. Four classes of lipoproteins, each defined empirically on the basis of density, are found in the plasma: chylomicrons/portomicrons, high-density lipoproteins (**HDLs**), low-density lipoproteins (**LDLs**), and very low-density lipoproteins (**VLDLs**). The latter three classes are also known by names derived from the method of electrophoretic separation, i.e., α-, β-, and pre-β-lipoproteins, respectively.

In contrast, the water-soluble vitamins tend to be excreted rapidly and not retained well. Few of this group are stored to any appreciable extent. The notable exception is vitamin B_{12}, which, under normal circumstances, can accumulate in the liver in amounts adequate to satisfy the nutritional needs of the host for periods of years.

17. METABOLISM OF THE VITAMINS

Some Vitamins Have Limited Biosynthesis

By definition, the vitamins as a group of nutrients are obligate factors in the diet (i.e., the chemical environment) of an organism. Nevertheless, some vitamins do not quite fit that general definition by virtue of being biosynthesized regularly by certain species or by being biosynthesized under certain circumstances by other species (see

---

TABLE 3.16 Vitamins That Can Be Biosynthesized

Vitamin	Precursor	Route	Conditions Increasing Dietary Need
Niacin	Tryptophan	Conversion to NMN via picolinic acid	Low 3-OH-anthranilic acid oxidase activity High picolinic acid carboxylase activity Low dietary tryptophan High dietary leucine[a]
Vitamin D_3	7-Dehydrocholesterol	UV photolysis	Insufficient sunlight/UV exposure
Vitamin C[b]	Glucose	Gulonic acid pathway	L-Gulonolactone oxidase deficiency

[a] The role of leucine as an effector of the conversion of tryptophan to niacin is controversial (see Chapter 12).
[b] Humans and other higher primates, guinea pigs, the Indian fruit bat, and a few other species are capable of vitamin C biosynthesis.

the "vitamin caveat" in Chapter 1). The biosynthesis of such vitamins (Table 3.16) thus depends on the availability, either from dietary or metabolic sources, of appropriate precursors (e.g., adequate free tryptophan is required for niacin production; the presence of 7-dehydrocholesterol in the surface layers of the skin is required for its conversion to vitamin D_3; flux through the gulonic acid pathway is needed to produce ascorbic acid), as well as the presence of the appropriate metabolic and/or chemical catalytic activities (e.g., the several enzymes involved in the tryptophan–niacin conversion; exposure to UV light for the photolysis of 7-dehydrocholesterol to produce vitamin D_3; the several enzymes of the gulonic acid pathway, including L-gulonolactone oxidase, for the formation of ascorbic acid).

Most Vitamins Require Metabolic Activation

Only a few vitamins are directly metabolically active: vitamin E, some vitamers K (e.g., MK-4), and vitamin C. The others require metabolic activation or linkage to a cofunctional species (e.g., an enzyme) (Table 3.17). The transformation of dietary forms of the vitamins into their respective, metabolically active forms may involve substantive modification of a vitamin's chemical structure and/or its combination with another species. Thus, factors that affect the metabolic (i.e., enzymatic) activation of vitamins to their functional species can have profound influences on their nutritional efficacy.

Vitamin Binding to Proteins

Some vitamins, even some requiring metabolic activation, are biologically active only when bound to a specific protein (Table 3.18). In most such cases, this happens when the vitamin serves as the prosthetic group of an enzyme, remaining bound to the enzyme protein during catalysis.[32] In other cases, this involves a vitamer binding to a specific nuclear receptor to elicit transcriptional modulation of one or more protein products (Table 3.19).

Vitamin Excretion

In general, the fat-soluble vitamins, which tend to be retained in hydrophobic environments, are excreted with the feces via the enterohepatic circulation[33] (Table 3.20). Exceptions include vitamins A and E, which to some extent have water-soluble metabolites (e.g., short-chain derivatives of retinoic acid; and the so-called *Simon's metabolites, carboxylchromanol metabolites*, of vitamin E), and menadione, which can be metabolized to a polar salt; these vitamin metabolites are excreted in the urine. In contrast, the water-soluble vitamins are generally excreted in the urine, both in intact forms (riboflavin, pantothenic acid) and as water-soluble metabolites (vitamin C, thiamin, niacin, riboflavin, pyridoxine, biotin, folate, vitamin B_{12}).

---

32. Vitamins of this type are properly called *coenzymes*; those that participate in enzymatic catalysis but are not firmly bound to enzyme protein during the reaction are, more properly, *cosubstrates*. This distinction, however, does not address the mechanism, but only the tightness, of binding. For example, the associations of NAD and NADP with certain oxidoreductases are weaker than those of FMN and FAD with the flavoprotein oxidoreductases. Therefore, the term *coenzyme* has come to be used to describe enzyme cofactors of both types.

33. These substances are discharged from the liver with the bile; the amounts that are not subsequently reabsorbed are eliminated with the feces.

TABLE 3.17 Vitamins That Must Be Activated Metabolically[a]

Vitamin	Active Form(s)	Activation Step	Condition(s) Increasing Need
Vitamin A	Retinol	Retinal reductase Retinol hydrolase	Protein insufficiency
	11-*cis*-Retinol 11-*cis*-Retinal	Retinyl isomerase Alcohol dehydrogenase	Zinc insufficiency
Vitamin D	1,25-$(OH)_2$-D	Vitamin D 25-hydroxylase 25-OH-D 1-hydroxylase	Hepatic failure Renal failure, lead exposure, estrogen deficiency, anticonvulsant drug therapy
Vitamin K	All forms	Dealkylation of Ks, MKs Alkylation of Ks, MKs, menadione	Hepatic failure
Thiamin	Thiamin-diP	Phosphorylation	High carbohydrate intake
Riboflavin	FMN, FAD	Phosphorylation Adenosylation	
Vitamin B_6	Pyridoxal-P	Phosphorylation Oxidation	High protein intake
Niacin	NAD(H), NADP(H)	Amidation (nicotinic acid)	Low tryptophan intake
Pantothenic acid	Coenzyme A	Phosphorylation Decarboxylation ATP condensation Peptide bond formation	
	ACP	Phosphorylation Peptide bond formation	
Folate	C_1-FH_4	Reduction Addition of C_1	
Vitamin B_{12}	Methyl-B_{12}	Cobalamin methylation CH_3 group insufficiency	Folate deficiency
	5'-Deoxyadenosyl-B_{12}	Adenosylation	

[a]Abbreviations: Thiamin-diP, thiamin pyrophosphate; ACP, acyl carrier protein; C_1-FH_4, tetrahydrofolic acid.

TABLE 3.18 Vitamins That Must Be Linked to Enzymes and Other Proteins

Vitamin	Form(s) Linked
Biotin	Biotin
Vitamin B_{12}	Methylcobalamin, adenosylcobalamin
Vitamin A	11-*cis*-Retinal
Thiamin	Thiamin pyrophosphate
Riboflavin	FMN, FAD
Niacin	NAD, NADP
Vitamin B_6	Pyridoxal phosphate
Pantothenic acid	Acyl carrier protein
Folate	Tetrahydrofolic acid (FH_4)

TABLE 3.19 Vitamins That Have Nuclear Receptor Proteins

Vitamin	Form(s) Linked	Receptor
Vitamin A	all-*trans*-Retinoic acid	Retinoic Acid Receptors (RARs)
	9-*cis*-Retinoid acid	Retinoid X receptors (RXRs)
Vitamin D	1,25-$(OH)_2$-vitamin D_3	Vitamin D receptor (VDR)

TABLE 3.20 Excretory Forms of the Vitamins

Vitamin	Urinary Form(s)	Fecal Form(s)
Vitamin A	Retinoic acid Acidic short-chain forms	Retinoyl glucuronides Intact-chain products
Vitamin D		$25,26\text{-}(OH)_2\text{-}D$ $25\text{-}(OH)_2\text{-}D\text{-}23,26\text{-lactone}$
Vitamin E	Some carboxylchromanol metabolites	Tocopheryl quinone Tocopheronic acid and its lactone
Vitamin K (K, MK, and menadione)		Vitamin K-2,3-epoxide $2\text{-}CH_3\text{-}3(5'\text{-Carboxy-}3'\text{-}CH_3\text{-}2'\text{-pentenyl})\text{-}1,4\text{-}$naphthoquinone $2\text{-}CH_3\text{-}3(3'\text{-Carboxy-}3'\text{-methylpropyl})\text{-}1,4\text{-}$napthoquinone Other unidentified metabolites
Menadione	Menadiol phosphate, menadiol sulfate	Menadiol glucuronide
Vitamin C[a]	Ascorbate-2-sulfate Oxalic acid 2,3-Diketogulonic acid	
Thiamin	Thiamin, thiamin disulfide Thiamin pyrophosphate, thiochrome 2-Methyl-4-amino-5-pyrimidine carboxylic acid 4-Methyl-thiazole-5-acetic acid 2-Methyl-4-amino-5-hydroxymethyl pyrimidine 5-(2-Hydroxyethyl)-4-methylthiazole 3-(2'-Methyl-4-amino-5'-pyrimidinylmethyl)-4-methylthiazole-5-acetic acid 2-Methyl-4-amino-5-formylaminomethylpyrimidine Several other minor metabolites	
Riboflavin	Riboflavin, 7- and 8-hydroxmethylriboflavins, 8β-sulfonylriboflavin, riboflavinyl peptide ester, 10-hydroxyethylflavin, lumiflavin, 10-formyl-methylflavin, 10-carboxymethylflavin, lumichrome	
Niacin	N^1-Methylnicotinamide Nicotinuric acid Nicotinamide-N^1-oxide N^1-Methylnicotinamide-N^1-oxide N^1-Methyl-4-pyridone-3-carboxamide N^1-Methyl-2-pyridone-5-carboxamide	
Vitamin B_6	Pyridoxol, pyridoxal, pyridoxamine, and phosphates 4-Pyridoxic acid and its lactone 5-Pyridoxic acid	
Biotin	Biotin; *bis-nor*-biotin Biotin *d*- and *l*-sulfoxide	
Pantothenic acid	Pantothenic acid	
Folate	Pteroylglutamic acid 5-Methyl-pteroylglutamic acid 10-HCO-FH_4, pteridine Acetamidobenzoylglutamic acid	Intact folates
Vitamin B_{12}	Cobalamin	Cobalamin

[a] Substantial amounts are also oxidized to CO_2 and are excreted across the lungs.

TABLE 3.21 Metabolic Functions of the Vitamins

Vitamin	Activities
Antioxidants	
Vitamin E	Protects polyunsaturated membrane phospholipids and other substances from oxidative damage via conversion of tocopherol to tocopheroxyl radical and, then, to tocopheryl quinone
Vitamin C	Protects cytosolic substances from oxidative damage
Hormones	
Vitamin A	Signals coordinate metabolic responses of several tissues
Vitamin D	Signals coordinate metabolism important in calcium homeostasis
H^+/e^- Donors/Acceptors (Cofactors)	
Vitamin K	Converts the epoxide form in the carboxylation of peptide glutamyl residues
Vitamin C	Oxidizes dehydroascorbic acid in hydroxylation reactions
Niacin	Interconverts $NAD^+/NAD(H)$ and $NADP^+/NADP(H)$ couples in several dehydrogenase reactions
Riboflavin	Interconverts $FMN/FMNH/FMNH_2$ and $FAD/FADH/FADH_2$ systems in several oxidases
Pantothenic acid	Oxidizes coenzyme A in the synthesis/oxidation of fatty acids
Coenzymes	
Vitamin A	Rhodopsin conformational change following light-induced bleaching
Vitamin K	Vitamin K-dependent peptide-glutamyl carboxylase
Vitamin C	Cytochrome *P*-450-dependent oxidations (drug and cholesterol metabolism, steroid hydroxylations)
Thiamin	Cofactor of α-keto acid decarboxylases and transketolase
Niacin	$NAD(H)/NADP(H)$ used by more than 30 dehydrogenases in the metabolism of carbohydrates (e.g., glucose-6-phosphate dehydrogenase), lipids (e.g., α-glycerol-phosphate dehydrogenase), protein (e.g., glutamate dehydrogenase); Krebs cycle, rhodopsin synthesis (alcohol dehydrogenase)
Riboflavin	FMN: L-Amino-acid oxidase, lactate dehydrogenase, pyridoxine (pyridoxamine); 5'-phosphate oxidase FAD: D-Amino-acid and glucose oxidases, succinic and acetyl-CoA dehydrogenases; glutathione, vitamin K, and cytochrome reductases
Vitamin B_6	Metabolism of amino acids (aminotransferases, deaminases, decarboxylases, desulfhydratases), porphyrins (δ-aminolevulinic acid synthase), glycogen (glycogen phosphorylase), and epinephrine (tyrosine decarboxylase)
Biotin	Carboxylations (pyruvate, acetyl-CoA, propionyl-CoA, 3-methylcrotonyl-CoA carboxylases) and transcarboxylations (methylmalonyl-CoA carboxymethyltransferase)
Pantothenic acid	Fatty acid synthesis/oxidation
Folate	Single-carbon metabolism (serine–glycine conversion, histidine degradation, purine synthesis, methyl group synthesis)
Vitamin B_{12}	Methylmalonyl-CoA mutase, N^5-CH_3-FH_4:homocysteine methyltransferase

18. METABOLIC FUNCTIONS OF THE VITAMINS

Vitamins Serve Five Basic Functions

The 13 families of nutritionally important substances called *vitamins* comprise two to three times that number of practically important vitamers, and function in metabolism in five general and not mutually exclusive ways. The type of metabolic function of any particular vitamer or vitamin family, of course, is dependent on its tissue/cellular distribution and its chemical reactivity, both of which are direct or indirect functions of its chemical structure. For example, the antioxidative function of vitamin E reflects the ability of that vitamin to form semistable radical intermediates; its lipophilicity allows vitamin E to discharge this antioxidant function within the hydrophobic regions of biomembranes, thus protecting polyunsaturated membrane phospholipids. Similarly, the redox function of riboflavin is due to its ability to undergo reversible reduction/oxidation involving a radical anion intermediate. These functions (summarized in Table 3.21), and the fundamental aspects of their significance in nutrition and health, are the subjects of Chapters 5–17.

Vitamins function as:

- Antioxidants – antioxidants can inhibit the process of oxidation, which frequently produces free radicals that can start chain reactions that damage lipids and proteins, and affect cellular function. Antioxidants interrupt such processes by being oxidized themselves, and thus removing free radicals.
- Gene transcription elements – factors that affect the first step in gene expression, the process of "transcription" by which a complementary RNA copy of a DNA sequence is made.
- H^+/e^- donors/acceptors – factors that can undergo changes in oxidation state in metabolism by being oxidized (losing electrons to an acceptor) or reduced (accepting electrons to a donor acceptor) in metabolism.
- Hormones – metabolites released by cells or glands in one part of the body that affect cell function in another part of the body.
- Coenzymes – metabolites that link to enzymes and are required for enzyme activity.

Study Questions and Exercises

1. Prepare a concept map of the relationships between the chemical structures, the physical properties, and the modes of absorption, transport and tissue distributions of the vitamins.

2. For each vitamin, identify the key feature(s) of its chemical structure. How is/are this/these feature(s) related to the stability and/or biologic activity of the vitamin?

3. Discuss the general differences between the fat-soluble and water-soluble vitamins, and the implications of those differences in diet formulation and meal preparation.

4. Which vitamins would you suspect might be in shortest supply in the diets of livestock? in your own diet? Explain your answer in terms of the physicochemical properties of the vitamins.

5. Which vitamins would you expect to be stored well in the body? Which would you expect to be unstable in foods or feeds?

6. What factors would you expect to influence the absorption of specific vitamins?

RECOMMENDED READING

Vitamin Nomenclature

Anonymous, 1987. Nomenclature policy: Generic descriptors and trivial names for vitamins and related compounds. J. Nutr. 120, 12–19.

Vitamin Chemistry and Analysis

Ball, G.F.M., 1998. Bioavailability and Analysis of Vitamins in Foods. Chapman & Hall, New York, NY, 569 pp.

Eitenmiller, R.R., Landen Jr, W.O., 1999. Vitamin Analysis for the Health and Food Sciences. CRC Press, New York, NY, 518 pp.

Zempleni, J., Rucker, R.B., McCormick, D.B., Suttie, J.W. (Eds.),, 2007. Handbook of Vitamins, fourth ed. CRC Press, New York, NY, 593 pp.

Vitamin Deficiency

Chapter Outline

Anchoring Concepts

1. A *disease* is an interruption or perversion of function of any of the organs with characteristic signs and/or symptoms caused by specific biochemical and morphological changes.
2. *Deficient intakes* of essential nutrients can cause disease.

These diseases ... were considered for years either as intoxication by food or as infectious diseases, and twenty years of experimental work were necessary to show that diseases occur which are caused by a deficiency of some essential substance in the food.

C. Funk

Learning Objectives

1. To understand the concept of *vitamin deficiency*.
2. To understand that deficient intakes of vitamins lead to *sequences of lesions* involving changes starting at the biochemical level, progressing to affect cellular and tissue function and, ultimately, resulting in morphological changes.
3. To appreciate the range of possible morphological changes in organ systems that can be caused by vitamin deficiencies.
4. To get an overview of *specific clinical signs and symptoms* of deficiencies of each vitamin in animals, including humans, as background for further study of the vitamins.
5. To appreciate the relationships of clinical manifestations of vitamin deficiencies and lesions in the biochemical functions of those vitamins.

VOCABULARY

Achlorhydria
Achromatrichia
Acrodynia
Age pigments
Alopecia
Anemia
Anorexia
Arteriosclerosis
Ataxia
Beriberi
Bradycardia
Brown bowel disease
Brown fat disease
Cage layer fatigue
Capillary fragility
Cardiomyopathy
Cataract
Cervical paralysis
Cheilosis
Chondrodystrophy
Cirrhosis
Clinical signs
Clubbed down
Convulsion
Cornification
Curled toe paralysis
Dermatitis
Desquamation
Dystrophy
Edema
Encephalomalacia
Encephalopathy
Exudative diathesis
Fatty liver and kidney syndrome
Geographical tongue

The Vitamins. DOI: 10.1016/B978-0-12-381980-2.00004-9

Glossitis
Hyperkeratosis
Hypovitaminosis
Inflammation
Keratomalacia
Leukopenia
Lipofuscin(osis)
Malabsorption
Mulberry heart disease
Myopathy
Necrosis
Nephritis
Neuropathy
Night blindness
Nyctalopia
Nystagmus
Opisthotonos
Osteomalacia
Osteoporosis
Pellagra
Perosis
Photophobia
Polyneuritis
Retrolental fibroplasia
Rickets
Scurvy
Steatitis
Stomatitis
Symptom
Vitamin deficiency
Wernicke–Korsakoff syndrome
White muscle disease
Xerophthalmia
Xerosis

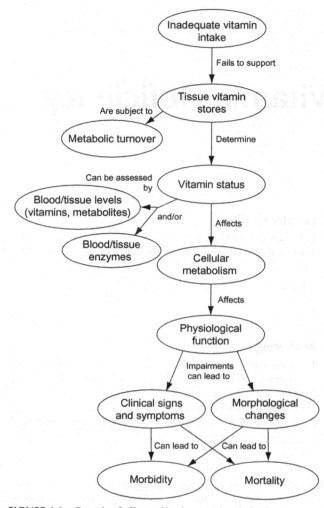

FIGURE 4.1   Cascade of effects of inadequate vitamin intake.

1. THE CONCEPT OF VITAMIN DEFICIENCY

What Is Meant by the Term *Vitamin Deficiency*?

Because the gross functional and morphological changes caused by deprivation of the vitamins were the source of their discovery as important nutrients, these signs have become the focus of attention for many with interests in human and/or veterinary health. Indeed, freedom from clinical diseases caused by insufficient vitamin nutriture has generally been used as the main criterion by which vitamin requirements have been defined. The expression **vitamin deficiency** therefore simply refers to the basic condition of **hypovitaminosis**. Vitamin deficiency is distinct from but underlies the various biochemical changes, physiological and/or functional impairments, or other overt disease signs by which the need for a vitamin is defined.

A vitamin deficiency is … the shortage of supply of a vitamin relative to its needs by a particular organism.

Vitamin Deficiencies Involve Cascades of Progressive Changes

The diseases associated with low intakes of particular vitamins typically represent clinical manifestations of a progressive sequence of lesions that result from biochemical perturbations (e.g., diminished enzyme activity due to lack of a coenzyme or cosubstrate; membrane dysfunction due to lack of a stabilizing factor) that lead first to cellular and subsequently to tissue and organ dysfunction. Thus, the clinical signs of a vitamin deficiency are actually the end result of a chain of events that starts with the diminution in cells and tissues of the metabolically active form of the vitamin (*see* Fig. 4.1).

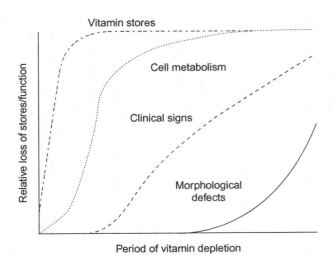

FIGURE 4.2 Over time, thiamin-deprived human subjects experience four progressive stages of deficiency, starting with tissue depletion and ending with morphological changes.

Stages of Vitamin Deficiency

Marginal deficiency:

Stage I	Depletion of vitamin stores, which leads to ...
Stage II	Cellular metabolic changes, which lead to ...

Observable deficiency:

Stage III	Clinical defects, which ultimately produce ...
Stage IV	Morphological changes

Marks[1] illustrated this point with the results of a study of thiamin depletion in human volunteers (Fig. 4.2). When the subjects were fed a thiamin-free diet, no changes of any type were detected for 5–10 days, after which the first signs of decreased saturation of erythrocyte transketolase with its essential cofactor, thiamin pyrophosphate (TPP), were noted. Not until nearly 200 days of depletion – i.e., long after tissue thiamin levels and transketolase–TPP saturation had declined – were classic **clinical signs**[2] of thiamin deficiency (**anorexia**, weight loss, malaise, insomnia, hyperirritability) detected.

Marginal deficiencies of vitamins in which the impacts of poor vitamin status are not readily observed without chemical or biochemical testing are often referred to as *subclinical deficiencies* for that reason. They involve depleted reserves or localized abnormalities, but without

1. Marks, J. (1968). *The Vitamins in Health and Disease: A Modern Reappraisal*. Churchill, London, UK.
2. A symptom is a change, whereas a clinical sign is a change detectable by a physician.

the presence of overt functional or morphological defects. The traditional perspective has been that the absence of overt, clinical manifestations of deficiency constitutes good nutrition; this perspective ignores the importance of preventing the early functional impairments that can progress to overt clinical signs. Therefore, the modern view of nutritional adequacy must focus on the maintenance of normal metabolism and, in several cases, body reserves as criteria of adequate vitamin status.[3]

2. CLINICAL MANIFESTATIONS OF VITAMIN DEFICIENCIES

Many Organ Systems Can Be Affected by Vitamin Deficiencies

Every organ system of the body can be the target of a vitamin deficiency. Some vitamin deficiencies affect certain organs preferentially (e.g., vitamin D deficiency chiefly affects calcified tissues); others affect several or many organs in various ways. Because the diagnosis of a vitamin deficiency involves its differentiation from other potential causes of similar clinical signs, it is useful to consider the various morphologic lesions caused by vitamin deficiencies from an organ system perspective. After all, anatomical and/or functional changes in organs are the initial presentations of deficiencies of each of the vitamins. This point is show in Table 4.1, which shows the organ systems affected by vitamin deficiencies.

3. VITAMIN DEFICIENCY DISEASES: MANIFESTATIONS OF BIOCHEMICAL LESIONS

Relationships Between Biochemical Lesions and Clinical Diseases of Vitamin Deficiencies

The clinical signs and symptoms that characterize the vitamin deficiency diseases are manifestations of impairments (i.e., *lesions*) in biochemical function that result from insufficient vitamin supply. This is a fundamental concept in understanding the roles of the vitamins in nutrition and health. Hypovitaminosis of sufficient magnitude and duration is causally related to the morphological and/or functional changes associated with the latter stages of vitamin deficiency. Although the validity of this concept is readily apparent in the abstract, documentary evidence for it in the case of each of the vitamin deficiency diseases, that is,

3. By such criteria, marginal vitamin status in the United States appears to be quite prevalent, even though the prevalence of clinically significant vitamin deficiencies appears to be very low. Estimates of marginal status with respect to one or more vitamins have been as high as 15% of adolescents, 12% of persons 65 years of age and older, and 20% of dieters.

TABLE 4.1 Organ Systems Affected by Vitamin Deficiencies in Humans and Other Animals

Organ Systems	Vitamin A	Vitamin D	Vitamin E	Vitamin K	Ascorbic Acid	Thiamin	Riboflavin	Niacin	Pyridoxine	Biotin	Pantothenic Acid	Folate	Vitamin B₁₂
General													
Appetite	+	+	+		+[a]	+	+	+	+	+			
Growth	+	+	+	+	+[a]	+	+	+	+	+	+	+	+
Integument													
Skin	+						+	+					
Hair, nails, feathers	+						+			+			
Musculature													
Skeletal muscles			+										
Heart						+							
Gizzard			+										
Vascular system													
Vessels			+	+	+[a]				+				
Blood cells			+				+					+	+
Clotting system				+									
Gastrointestinal tract													
Stomach						+		+					
Mouth							+						
Tongue								+					
Small intestine				+		+							
Colon							+	+					
Skeletal system													
Bone	+	+			+[a]		+						
Teeth	+	+			+[a]								
Vital organs													
Liver			+			+							
Kidney	+						+		+				
Thymus											+		

	1	2	3	4	5	6	7	8	9
Adrenals		+			+				
Pancreas			+						
Adipose							+		
Ocular system									
Eye					+	+	+		
Reproductive system									
Vagina									+
Uterus							+		
Ovary					+				+
Egg								+	
Testes									+
Fetus					+				+
Nervous system									
General		+	+	+	+	+		+	+
Spinal cord						+			+
Brain							+		
Peripheral nerves	+						+		
Psychological, emotional				+		+			

aOnly humans, higher primates, the guinea pig, and some birds are affected by ascorbic acid deprivation.

direct cause–effect linkages of specific biochemical lesions and clinical changes, is, for many vitamins, not complete.

Vitamin A offers a case in point of this fact. While the role of vitamin A in preventing nyctalopia (night blindness) is clear from presently available knowledge of the essentiality of retinal as the prosthetic group of rhodopsin and several other photosensitive visual receptors in the retina, the amount of vitamin A in the retina, and thus available for visual function, is only about 1% of the total amount of vitamin A in the body. Further, it is clear from the clinical signs of vitamin A deficiency that the vitamin has other essential functions unrelated to vision, especially some relating to the integrity and differentiation of epithelial cells. However, although evidence indicates that vitamin A is involved in the metabolism of mucopolysaccharides and other essential intermediates, present knowledge cannot adequately explain the mechanism(s) of action of vitamin A in supporting growth, in maintaining epithelia, and so on. It has been said that 99% of our information about the mode of action of vitamin A concerns only 1% of the vitamin A in the body.

The ongoing search for a more complete understanding of the mechanisms of vitamin action is therefore largely based on the study of biochemical correlates of changes in physiological function or morphology effected by changes in vitamin status. Most of this knowledge has come from direct experimentation, mostly with animal models. Also edifying in this regard has been information acquired from observations of individuals with a variety of rare, naturally occurring, hereditary anomalies involving vitamin-dependent enzymes, and transport proteins. Most of the documented inborn metabolic errors have involved specific mutations manifest as either a loss or aberration in single factors in vitamin metabolism – a highly targeted situation not readily produced experimentally.[4]

Tables 4.2 and 4.3 summarize in general terms current information concerning the interrelationships of the biochemical functions of the vitamins and the clinical manifestations of their deficiencies or anomalous metabolism.

4. THE MANY CAUSES OF VITAMIN DEFICIENCIES

Primary and Secondary Causes of Vitamin Deficiencies

The balance of vitamin supply and need for a particular individual at a given point in time is called *vitamin status*. Reductions in vitamin status can be produced either by reductions in effective vitamin supply or by increases in effective vitamin need. Vitamin deficiency occurs when vitamin status is reduced to the point of having metabolic

impact (i.e., stage II); if not corrected, continued reductions in vitamin status lead inevitably to the observable stages of vitamin deficiency (stages III and IV), at which serious clinical and morphological changes can become manifest. When these changes occur as a result of the failure to ingest a vitamin in sufficient amounts to meet physiological needs, the condition is called a *primary deficiency*. When these changes come about as a result of the failure to absorb or otherwise utilize a vitamin owing to an environmental condition or physiological state, and not to insufficient consumption of the vitamin, the condition is called a *secondary deficiency*.

The two fundamental ways in which vitamin deficiencies can be caused are:

- *Primary deficiencies* … involve failures to ingest a vitamin in sufficient amounts to meet physiological needs.
- *Secondary deficiencies* … involve failures to absorb or otherwise utilize a vitamin post-absorptively.

Causes of Vitamin Deficiencies in Humans

Many of the ways in which vitamin deficiencies can come about are interrelated. For example, poverty is often accompanied by gross ignorance of what constitutes a nutritionally adequate diet. People living alone, and especially the elderly and others with chronic disease, tend to consume foods that require little preparation but that may not provide good nutrition. Despite these potential causes of vitamin deficiency, in most of the technologically developed parts of the world the general level of nutrition is high. In those areas, relative few persons can be expected to show signs of vitamin deficiency; those that do present such signs will most frequently be found to have a potentiating condition that affects either their consumption of food or their utilization of nutrients. In the developing parts of the world, however, famine is still the largest single cause of general malnutrition today. People affected by famine show signs of multiple nutrient deficiencies, including lack of energy, protein, vitamins, and minerals.

Potential Causes of Vitamin Deficiencies in Humans

Primary deficiencies have psychosocial and technological causes:

- Poor food habits
- Poverty (i.e., low food-purchasing power)
- Ignorance (i.e., lack of sound nutrition information)
- Lack of total food (e.g., crop failure)
- Lack of vitamin-rich foods (e.g., consumption of highly refined foods)
- Vitamin destruction (e.g., during storage, processing, and/or cooking)

---

4. However, it is theoretically possible to produce transgenic animal models with similar metabolic anomalies.

- Anorexia (e.g., homebound elderly, infirm, dental problems)
- Food taboos and fads (e.g., fasting, avoidance of certain foods)
- Apathy (lack of incentive to prepare adequate meals).

Secondary deficiencies have biological causes:

- Poor digestion (e.g., **achlorhydria** – absence of stomach acid)

- **Malabsorption** (impaired intestinal absorption of nutrients; e.g., as a result of diarrhea, intestinal infection, parasites, pancreatitis)
- Impaired metabolic utilization (e.g., certain drug therapies)
- Increased metabolic need (e.g., pregnancy, lactation, rapid growth, infection, nutrient imbalance)
- Increased vitamin excretion (e.g., diuresis, lactation, excessive sweating).

TABLE 4.2 The Underlying Biological Functions of the Vitamins

Vitamin	Active Form(s)	Deficiency Disorders	Important Biological Functions or Reactions
Vitamin A	Retinol, retinal, retinoic acid	Night blindness, xerophthalmia, keratomalacia	Photosensitive retinal pigment
			Regulation of epithelial cell differentiation
		Impaired growth	Regulation of gene transcription
Vitamin D	1,25-$(OH)_2$-D	Rickets, osteomalacia	Promotion of intestinal calcium absorption, mobilization of calcium from bone, stimulation of renal calcium, resorption, regulation of PTH secretion, possible function in muscle
Vitamin E	α-Tocopherol	Nerve, muscle degeneration	Antioxidant protector for membranes
Vitamin K	K, MK	Impaired blood clotting	Cosubstrate for γ-carboxylation of glutamyl residues of several clotting factors and other calcium-binding proteins
Vitamin C	Ascorbic acid, dehydroascorbic acid	Scurvy	Cosubstrate for hydroxylations in collagen synthesis, drug and steroid metabolism
Thiamin	Thiamin pyrophosphate	Beriberi, polyneuritis, Wernicke–Korsakoff syndrome	Coenzyme for oxidative decarboxylation of 2-keto acids (e.g., pyruvate and 2-keto-glutarate); coenzyme for pyruvate decarboxylase and transketolase
Riboflavin	FMN, FAD	Dermatitis	Coenzymes for numerous flavoproteins that catalyze redox reactions in fatty acid synthesis/degradation, TCA cycle
Niacin	NAD(H), NADP(H)	Pellagra	Cosubstrates for hydrogen transfer catalyzed by many dehydrogenases, e.g., TCA cycle respiratory chain
Pyridoxine	Pyridoxal-5'-phosphate	Signs vary with species	Coenzyme for metabolism of amino acids, e.g., side chain, decarboxylation, transamination, racemization
Folate	Polyglutamyl tetrahydrofolates	Megaloblastic anemia	Coenzyme for transfer of single-carbon units, e.g., formyl and hydroxymethyl groups in purine synthesis
Biotin	1'-N-Carboxybiotin	Dermatitis	Coenzyme for carboxylations, e.g., acetyl-CoA/malonyl-CoA conversion
Pantothenic acid	Coenzyme A	Signs vary with species	Cosubstrate for activation/transfer of acyl groups to form esters, amides, citrate, triglycerides, etc.
	Acyl carrier protein		Coenzyme for fatty acid biosynthesis
Vitamin B_{12}	5'-Deoxyadenosyl-B_{12}	Megaloblastic anemia	Coenzyme for conversion of methylmalonyl-CoA to succinyl-CoA
	Methyl-B_{12}	Impaired growth	Methyl group transfer from 5-CH_3-FH_4 to homocysteine in methionine synthesis

Abbreviations: PTH, parathyroid hormone; TCA, tricarboxylic acid cycle; CoA, coenzyme A; 5-CH_3-FH_4, 5-methyltetrahydrofolic acid.

TABLE 4.3 Vitamin-Responsive Inborn Metabolic Lesions

Curative Vitamin	Missing Protein or Metabolic Step Affected	Clinical Condition
Vitamin A	Apolipoprotein B	Abetalipoproteinemia; low tissue levels of retinoids
Vitamin D	Receptor	Unresponsive to $1,25(OH)_2$-D; osteomalacia
Vitamin E	Apolipoprotein B	Abetalipoproteinemia; low tissue levels of tocopherols
Thiamin	Branched-chain 2-oxoacid dehydrogenase	Maple syrup urine disease
	Pyruvate metabolism	Lactic acidemia; neurological anomalies
Riboflavin	Methemoglobin reductase	Methemoglobinemia
	Electron transfer flavoprotein	Multiple lack of acyl-CoA dehydrogenations, excretion of acyl-CoA metabolites, i.e., metabolic acidosis
Niacin	Abnormal neurotransmission	Psychiatric disorders, tryptophan malabsorption, abnormal tryptophan metabolism
Pyridoxine	Cystathionine β-synthase	Homocysteinuria
	Cystathionine γ-lyase	Cystathioninuria; neurological disorders
	Kynureninase	Xanthurenic aciduria
Folate	Enteric absorption	Megaloblastic anemia, mental disorders
	Methylene-FH_4-reductase	Homocysteinuria, neurological disorders
	Glutamate formiminotransferase	Urinary excretion of FIGLU[a]
	Homocysteine/methionine conversion	Schizophrenia
	Tetrahydrobiopterin-phenylalanine hydrolase	Mental retardation, PKU[b]
	Dihydrobiopteridine reductase	PKU, severe neurological disorders
	Tetrahydrobiopterin formation	PKU, severe neurological disorders
Biotin	Biotinidase	Alopecia, skin rash, cramps, acidemia, developmental disorders, excess urinary biotin and biocytin
	Propionyl-CoA carboxylase	Propionic academia
	3-Methylcrotonyl-CoA carboxylase	3-Methylcrontonylglycinuria
	Pyruvate carboxylase	*Leigh disease*, accumulation of lactate and pyruvate
	Acetyl-CoA carboxylase	Severe brain damage
	Holocarboxylase synthase	Lack of multiple carboxylase activities, urinary excretion of metabolites
Vitamin B_{12}	Intrinsic factor	Juvenile pernicious anemia
	Enteric absorption	
	Transcobalamin	Megaloblastic anemia, growth impairment
	Methylmalonyl-CoA mutase	Methylmalonic acidemia

[a]*FIGLU, Formiminoglutamic acid.*
[b]*PKU, Phenylketonuria.*

High-Risk Groups for Vitamin Deficiencies

> Pregnant women
> Infants and young children
> Elderly people
> Poor (food-insecure) people
> People with intestinal parasites or infections
> Dieters
> Smokers.

Causes of Vitamin Deficiencies in Animals

Many of the same primary and secondary causal factors that produce vitamin deficiencies in humans can also produce vitamin deficiencies in animals. In livestock, however, most of the very serious cases of vitamin deficiency in animals are due to human errors involving improper or careless animal husbandry.

Potential Causes of Vitamin Deficiencies in Livestock and Other Managed Animal Species

Primary deficiencies have physical causes:

- Improperly formulated diet (i.e., error in vitamin premix formulation)
- Feed mixing error (e.g., omission of vitamin from vitamin premix)
- Vitamin losses (e.g., during pelleting, extrusion, and/or storage)
- Poor access to feed (e.g., competition for limited feeder space, improper feeder placement, breakdown of feed delivery system).

Secondary deficiencies have biological and social causes:

- Poor feed intake (e.g., poor palatability, heat stress)
- Other deficiencies (e.g., deficiencies of protein and/or zinc can impair retinol transport)
- Poor digestion
- Malabsorption (e.g., diarrhea, parasites, intestinal infection due to poor hygiene)
- Impaired post-absorptive utilization (e.g., certain drug therapies)
- Increased metabolic demand (e.g., infection, low environmental temperature, egg/milk production, rapid growth, pregnancy, lactation).

Intervention is Most Effective in Early Stages of Vitamin Deficiency

That vitamin deficiencies[5] are not simply specific morphological events but, rather, cascades of biochemical,

physiological, and anatomical changes is the key point in both the assessment of vitamin status and the effective treatment of hypovitaminoses. Because the early biochemical and metabolic effects of specific vitamin deficiencies are almost always readily reversed by therapy with the appropriate vitamin, in contrast to the later functional and anatomical changes which may be permanent, intervention to correct hypovitaminoses is most effective when the condition is detected in its early and less severe stages. In this respect, the management of vitamin deficiencies is no different from that of other diseases – treatment is generally most effective when given at the stage of cellular biochemical abnormality, rather than waiting for the appearance of the ultimate clinical signs.[6] For this reason, the early detection of insufficient vitamin status using biochemical indicators has been, and will continue to be, a very important activity in the clinical assessment of vitamin status.

Table 4.4 can be used to diagnose vitamin deficiency diseases based on a three-step analysis:

1. Identify vitamin deficiencies possibly involved based on mapping of signs/symptoms to those reported in the scientific literature
2. Use the appropriate clinical biochemical test(s) to exclude possibilities
3. Determine the actual deficiency(ies) involved based on responses to treatment.

Vitamin deficiency disorders are often treated using a medical/pharmacological approach involving a two-step analysis: (1) diagnosis of the deficiency; and (2) treatment with an appropriate form of the relevant vitamin (*see* Fig. 4.3). This approach offers the advantages of speed and efficacy, which are often significant in the context of treating subjects in need. However, they do not address the multiple, underlying causes of deficiency; these must be addressed for the sustainable prevention of vitamin deficiency, especially in populations. This goal calls for considering the deficiency in the context of the social and biological systems in which it occurs, to the end of identifying root-causes – i.e., the underlying, contributing conditions, one or more of which will likely to be amenable to change. An example of a root-cause analysis of xerophthalmia due to vitamin A deficiency is shown in Fig. 4.4.

---

5. This discussion, of course, is pertinent to any class of nutrients.

6. Marks (1968) makes this point clearly with the example of diabetes, which should be treated once hypoglycemia is detected, thus reducing the danger of diabetic arteriosclerosis and retinopathy.

TABLE 4.4 Diagnosis of Vitamin Deficiencies in Humans and Other Animals

Signs and Symptoms by Organ System and Organ

System	Organ	Signs/Symptoms	Vitamin Deficiencies Possibly Involved	Humans Affected	Other Species Affected
General	General weakness		Vitamin A		Cat
			Vitamin D	+	
			Ascorbic acid	+	
			Thiamine	+	Rat
			Riboflavin		Pig, dog, fox
			Niacin	+	Chick
			Pyridoxine		Rat, chick
			Pantothenic acid	+	Chick
	Reduced appetite		Vitamin A		Rat, chick, mouse, pig, calf
			Vitamin D		Rat, chick, mouse, pig, calf
			Vitamin K		Rat, chick, mouse, pig, calf
			Ascorbic acid		Guinea pig
			Thiamine[a]	+	Rat, chick, mouse, pig, calf
			Riboflavin		Rat, chick, mouse, pig, calf
			Niacin	+	Rat, chick, mouse, pig, calf
			Pyridoxine		Rat, chick, mouse, pig, calf
			Biotin		Rat, chick, mouse, pig, calf
	Growth retardation		Ascorbic acid		Guinea pig
			Other Vitamins		Rat, mouse, chick, dog, calf
Integument					
Skin	Dermis	Scaly dermatitis[b]	Vitamin A		Cattle
			Riboflavin		Pig

Criteria for deficient/low status (humans)

Vitamin	Criteria
Vitamin A	Plasma retinol <10 µg/dl (<5 mos., >17 yrs); <20 µg/dl (5 mos.-17 yrs)
Vitamin D	Plasma 25(OH)D3 <3 ng/ml
Vitamin E	Plasma a-tocopherol <3.5 µg/ml
Vitamin K	Clotting time >10 min
Ascorbic acid	Plasma ascorbic acid <2 µg/ml; WBC ascorbic acid <8 µg/ml
Thiamin	RBC transketolase TPP-stimulation >25%; urine thiamin <40 µg/24 hr
Riboflavin	RBC GSH reductase FAD-stimulation >40%; urine riboflavin <40 µg/24 hr
Niacin	Urine N'-methyl-2-pyridone-5-carboxamide <1 µg/24 hr
Pyridoxine	Plasma pyridoxal phosphate <60 nM
Biotin	Blood biotin <0.4 ng/ml; urine biotin <10 µg/24 hr
Pantothenic acid	Plasma pantothenic acid <6 µg/dl; urine pantothenic acid <1 mg/24 hrs
Folate	Plasma folates <3 ng/ml; RBC folates <140 ng/ml
Vitamin B12	Plasma Vitamin B12 <100 pg/ml

Symptom	Vitamin		Species	
	Pyridoxine		Rat (acrodynia[c])	+
	Biotin		Rat, mouse, hamster, cat, mink, fox	+
	Pantothenic acid		Rat	+
Cracking dermatitis	Niacin	+ (Pellagra)	Chick	+
	Biotin		Pig, poultry, monkey	+
	Pantothenic acid		Chick (feet)	+
Desquamation[d]	Riboflavin	+	Monkey, rat, chick, dog	+
	Niacin	+	Chick	+
	Pyridoxine	+		+
	Biotin		Rat	+
Hyperkeratosis[e]	Riboflavin		Rat	+
	Niacin	+ (Pellagra)	Chick	+
	Biotin		Rat, mouse, hamster	+
Hyperpigmentation	Niacin	+ (Pellagra)		+
Photosensitization	Niacin	+ (Pellagra)		+
Rough	Vitamin A		Cattle, poultry (feathers)	+
Hair, nails, feathers	Biotin		Poultry (feathers)	
Achromatrichia[f]	Biotin		Rat, rabbit, cat, mink, fox, monkey	+
	Pantothenic acid		Rat	+
Alopecia[g]	Riboflavin		Rat, pig, calf	+
	Niacin		Rat, pig	+
	Biotin		Rat, mouse, hamster, rabbit, pig, chick, cat, mink, fox: (spectacle eye)	+
"Blood"-caked whiskers[h]	Pantothenic acid		Rat	+
	Pantothenic acid		Rat	+
Impaired growth	Biotin		Poultry	+
	Folate		Poultry	+

(Continued)

TABLE 4.4 (Continued)

Signs and Symptoms by Organ System and Organ

System	Organ	Signs/Symptoms	Vitamin Deficiencies Possibly Involved	Humans Affected	Other Species Affected	Vitamin A	Vitamin D	Vitamin E	Vitamin K	Ascorbic acid	Thiamin	Riboflavin	Niacin	Pyridoxine	Biotin	Pantothenic acid	Folate	Vitamin B₁₂
Musculature																		
Skeletal muscles		Myopathy^i	Vitamin E		Rat, guinea pig, pig, rabbit, chick, duck, calf, horse, goat, salmon, mink, catfish (white muscle dis.); lamb (stiff lamb dis.)			+										
			Ascorbic acid	+ (Scurvy)	Guinea pig					+								
			Thiamin	+ (Beriberi)	Rat						+							
			Pantothenic acid		Pig											+		
Heart	Rhythm	Bradycardia	Thiamin	+ (Beriberi)	Rat						+							
	Muscle	Cardiomyopathy^j	Thiamin	+ (Beriberi)	Rat						+							
			Vitamin E		Pig (mulberry heart dis.); guinea pig, rabbit, rat, dog, calf, lamb, goat			+										
Gizzard^k		Myopathy	Vitamin E		Turkey poults, ducklings			+										
Vascular system																		
Vessels	General	Arteriosclerosis^l	Pyridoxine		Monkey									+				
	Capillary	Edema^m	Vitamin E		Chick (exudative diathesis), pig (visceral edema)			+										
			Thiamin	+ (Beriberi)							+							
		Hemorrhage	Vitamin K		Poultry				+									
			Ascorbic acid	+ (Scurvy)	Guinea pig, monkey					+								
Blood cells	Erythrocyte	Hemolytic anemia^n	Vitamin E	+	Pig, monkey			+										
		Hemorrhagic anemia^o	Vitamin K		Rat, chick				+									

Organ/system	Cell	Sign	Vitamin	Humans	Animals
		Normocytic hypochromic anemia^p	Riboflavin	+	Monkey, baboon
		Megaloblastic anemia^q	Folate	+	Rat, chick
		Megaloblastic anemia	Vitamin B12	+	Rat, chick
		Fragility	Vitamin E	+	Rat, pig, monkey
	Leukocyte	Leukopenia^r	Riboflavin	+	Rat, guinea pig
			Folate	+	Rat, guinea pig
	Platelet	Thrombocytosis^s	Vitamin E	+	Rat
		Excess aggregation	Vitamin E	+	Rat
Clotting system		Prolonged clotting time	Vitamin K	+	Rat, chick, pig, calf
Gastrointestinal tract					
Stomach	Epithelium	Achlorhydria^t	Niacin	+ (Pellagra)	
		Gastric distress	Thiamin	+ (Beriberi)	
Mouth		Stomatitis^u	Riboflavin	+	Calf
			Niacin	+ (Pellagra)	
			Biotin		Chick
			Pantothenic acid		Chick
		Cheliosis^v	Riboflavin	+	
Tongue		Glossitis^w	Niacin	+ (Pellagra)	
			Riboflavin	+	Rat
			Niacin	+ (Pellagra)	
Small intestine	Mucosa	Inflammation	Thiamin	+	Rat
			Riboflavin	+	Chick, dog, pig
			Niacin	+	Chick
		Ulcer	Thiamin	+	Rat
	Enterocyte	Lipofuscinosis^x	Vitamin E	+	Dog (brown bowel disease)

(Continued)

TABLE 4.4 (Continued)

Signs and Symptoms by Organ System

System	Organ	Signs/Symptoms	Vitamin Deficiencies Possibly Involved	Humans Affected	Other Species Affected	Vitamin A	Vitamin D	Vitamin E	Vitamin K	Ascorbic acid	Thiamin	Riboflavin	Niacin	Pyridoxine	Biotin	Pantothenic acid	Folate	Vitamin B₁₂
		Hemorrhage	Vitamin K		Poultry				+									
			Thiamin		Rat						+							
			Niacin		Dog								+					
	Colon	Diarrhea	Riboflavin		Chick, dog, pig, calf							+						
			Niacin	+ (Pellagra)	Dog, pig, poultry								+					
			Vitamin B₁₂		Young pigs													+
		Constipation	Niacin	+ (Pellagra)									+					
Skeletal system																		
Bone	Periosteum	Excessive growth	Vitamin A		Pig, dog, calf, horse, sheep	+												
	Epiphyses	Undermineralization malformations	Vitamin D	+ (Osteomalacia^y: rickets in children)	Chick, dog, calf (rickets)		+											
			Ascorbic acid	Children						+								
	Cortical bone	Demineralization, increased fractures	Vitamin D	+ (Osteomalacia, osteoporosis^z)	Laying hen (caged layer fatigue)		+											
		Chondrodystrophy^aa	Niacin		Chick, poult (perosis)								+					
			Biotin		Chick, poult (perosis)										+			
		Congenital deformities	Riboflavin		Rat							+						
			Pyridoxine		Rat									+				
Teeth	Dentin	Caries	Vitamin D	Children			+											
			Pyridoxine	+										+				

Vital organs

Liver	Hepatocyte	Necrosis	Vitamin E	+	Pig (hepatosis dietetica); rat, mouse
		Steatosis[bb]	Thiamin	+	Rat
			Biotin	+	Chick (fatty liver and kidney syndrome)
			Pantothenic acid	+	Chick, dog
		Cirrhosis[cc]	Choline	+	Rat, dog, monkey
Kidney	Nephron	Nephritis[dd]	Vitamin A	+	
		Steatosis	Riboflavin	+	Pig
			Biotin	+	Chick
		Hemorrhagic necrosis[ee]	Choline		Rat, mouse, pig, rabbit, calf
		Calculi	Pyridoxine	+	Rat
Thymus	Thymocyte	Necrosis	Pantothenic acid	+	Chick, dog
Adrenals		Hypertrophy	Pantothenic acid	+	Pig
		Hemorrhage	Riboflavin	+	Pig
		Lymphoid necrosis	Riboflavin	+	Baboon
			Pantothenic acid	+	Rat
Pancreas	Eyelets	Insulin insufficiency	Pyrodoxine	+	Rat
Adipose	Adipocyte	Lipofuscinosis	Vitamin E	+	Rat, mouse, hamster, cat, pig, mink (brown fat disease)

Ocular system

Eye	Eyeball	Nystagmus[ff]	Thiamin	+	(Wernicke-Korsakoff synd.)
	Retina	Photophobia	Riboflavin	+	
		Nyctalopia[gg]	Vitamin A	+	Rat, pig, cat, sheep
		Pigmented retinopathy	Vitamin E	+	Rat, dog, cat, monkey
	Cornea	Xerophthalmia[hh]	Vitamin A	+	Rat, calf
		Keratomalacia[ii]	Vitamin A	+	Rat, calf
	Lens	Cataract[jj]	Vitamin E	+	Rabbit, turkey embryo

(Continued)

TABLE 4.4 (Continued)

Signs and Symptoms by Organ System and Organ

System	Organ	Signs/Symptoms	Vitamin Deficiencies Possibly Involved	Humans Affected	Other Species Affected	Vitamin A	Vitamin D	Vitamin E	Vitamin K	Ascorbic acid	Thiamin	Riboflavin	Niacin	Pyridoxine	Biotin	Pantothenic acid	Folate	Vitamin B12
Reproductive system																		
Vagina	Epithelium	Cornification[kk]	Vitamin A	+	Rat	+												
Uterus	Epithelium	Lipofuscinosis	Vitamin E		Rat			+										
Ovary	????	Degeneration	Vitamin A		Poultry	+												
		Estrus	Anestrus[ll]	Riboflavin		Rat							+					
			Low egg production	Vitamin A		Poultry	+											
				Riboflavin		Poultry							+					
				Pyridoxine		Poultry									+				
Egg	Shell	Thinning	Vitamin D		Poultry		+											
Testes	Germinal epithelium	Degeneration	Vitamin A		Rat, bull, cat	+												
			Vitamin E		Rat, rooster, dog, pig, guinea pig, hamster, rabbit, monkey			+										
Fetus		Developmental abnormalities	Riboflavin		Rat, chick (clubbed down)							+						
			Folate		Chick (parrot beak)												+	
		Death	Vitamin A		Poultry	+												
			Vitamin E		Rat			+										
			Riboflavin		Chick							+						
			Folate		Chick												+	
			Vitamin B12		Poultry													+
Nervous system																		
General		Ataxia[mm]	Vitamin A		Chick, pig, calf, sheep	+												
			Vitamin E	+	Chick			+										

Region	Sign	Vitamin	Humans	Rat, mouse, chick, pig, rabbit, calf, monkey
		Thiamin	+ (Wernicke-Korsakoff synd.[nn])	
	Tremors	Riboflavin		Rat, pig
		Niacin	+ (Pellagra)	
	Tetany	Vitamin D	+ (Rickets)	Chick, pig
	Abnormal gait	Pantothenic acid		Pig, dog (goose-stepping)
	Seizures	Pyridoxine	Infants	Rat
	Paralysis	Riboflavin		Chick (curled toe paralysis); rat
		Pyridoxine		Chick
		Pantothenic acid		Poult (cervical paralysis); chick
	Irritability	Thiamin	+ (Beriberi)	
		Niacin	+ (Pellagra)	
		Pyridoxine	+	Rat
		Vitamin B12		Pig
Spinal cord / Cerebrospinal fluid	Excess pressure	Vitamin A		Chick, pig, calf
Brain	Opithotonus[oo]	Thiamin		Chick, pigeon (star gazing)
	Encephalopathy[pp]	Vitamin E		Chick (encephalomalacia[qq])
		Thiamin	+ (Wernicke-Korsakoff synd.)	
Peripheral nerves	Neuropathy	Vitamin E		Rat, dog, duck, monkey
		Thiamin	+ (Beriberi)	Chick (polyneuritis[rr])
		Riboflavin	+	
		Niacin		Rat, pig
		Pyridoxine		Rat
		Pantothenic acid	+ (Burning feet)	
		Vitamin B12	+	

(Continued)

TABLE 4.4 (Continued)

System	Organ	Signs/Symptoms	Vitamin Deficiencies Possibly Involved	Humans Affected	Other Species Affected	Vitamin A	Vitamin D	Vitamin E	Vitamin K	Ascorbic acid	Thiamin	Riboflavin	Niacin	Pyridoxine	Biotin	Pantothenic acid	Folate	Vitamin B12
		Signs and Symptoms by Organ System and Organ																
Psychological, emotional																		
		Depression	Thiamin	+ (Beriberi)							+							
			Niacin	+ (Pellagra)									+					
		Anxiety	Thiamin	+ (Beriberi)							+							
		Dizziness	Niacin	+ (Pellagra)									+					
		Irritability	Thiamin	+ (Beriberi)							+							
			Niacin	+ (Pellagra)									+					
			Pyridoxine	+										+				
		Dementia	Niacin	+ (Pellagra)									+					
		Psychosis	Thiamin	+ (Wernicke-Korsakoff synd.)							+							

a Severe initiation of rapid onset.
b Inflammation of the skin.
c Swelling and **necrosis** (i.e., tissue and/or organ death) of the paws, tips of the ears and nose, and lips.
d Shedding of skin.
e Thickening of the stratum corneum.
f Loss of normal pigment from hair or feathers.
g Loss of hair or feathers.
h Whiskers accumulate porphyrins shed in tears.
i General term for disease of muscle.
j General term for disease of the heart muscle.
k Muscular portion of the forestomach of birds.
l General term for hardening, i.e., loss of elasticity, of medium or large arteries.
m Abnormal fluid retention.
n Abnormally low erythrocyte count due to their fragility, rupture, and clearance.
o Abnormally low erythrocyte count due to hemorrhage.
p Abnormally low hemoglobin content in otherwise normal erythrocytes.
q Abnormally low erythrocyte count due to impaired DNA synthesis, with erythroblast growth without division, i.e., forming macrocytes.
r Abnormally low white blood cell count.
s Abnormally high platelet count.
t Lack of gastric acid production due to dysfunction of gastric parietal cells.
u Inflammation of the oral mucosa (soft tissues of the mouth).
v Angular stomatitis, i.e., inflammatory lesions (cracks, fissures) at the labial commissure (corners of the mouth).
w Inflammation of the tongue.
x Accumulation of lipid oxidation products.

^yDemineralization leading to softening of bones.

^zProgressive demineralization leading to thinning of bones, as in rickets in children.

^{aa}Disorders of cartilaginous components of growing ends of bones.

^{bb}Abnormal intracellular retention of lipids.

^{cc}Chronic liver disease involving replacement of normal tissue with fibrosis and presence of regenerative nodules.

^{dd}Inflammation of nephrons.

^{ee}Cell death.

^{ff}Involuntary eye movement.

^{gg}Night blindness, i.e., difficulty seeing in low light.

^{hh}Failure to produce tears due to dysfunction of lacrimal glands, resulting in dryness and thickening of the conjunctiva and cornea and leading to ulceration and blindness.

ⁱⁱDrying and clouding of the cornea due to xerophthalmia.

^{jj}Clouding of the lens.

^{kk}Formation of an epidermal barrier in stratified squamous epithelial tissue by increased expression of keratin proteins.

^{ll}Cessation of female ovulatory cycle.

^{mm}Gross lack of muscular coordination.

ⁿⁿConfusion, ataxia, nystagmus, and double vision due to damage in thalamus and hypothalamus (Wernicke's encephalopathy) progressing to loss of memory, confabulation, and hallucination (Korsokoff syndrome).

^{oo}State of severe hyperextension and spasm of the axial muscles along the spinal column, causing an individual's head, neck, and spinal column to assume an arching position.

^{pp}Term for global brain disease.

^{qq}Degenerative disease of brain involving function: blindness, ataxia, circling, and terminal coma.

^{rr}Neuropathy affecting multiple peripheral nerves.

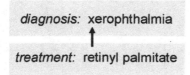

diagnosis: xerophthalmia

treatment: retinyl palmitate

FIGURE 4.3   The medical/pharmacological approach to addressing vitamin deficiencies focuses on *treatment*.

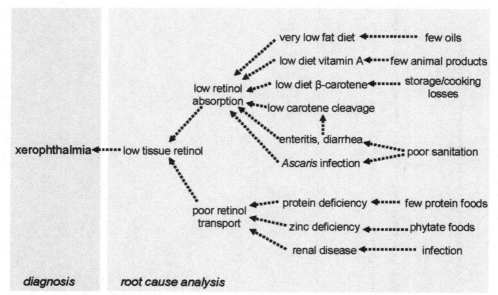

diagnosis   *root cause analysis*

FIGURE 4.4   A fish bone-type, root-cause analysis uses a systems approach in considering nutritional deficiencies; it seeks to identify the underlying contributing factors and focus on *prevention*. This is illustrated in the case of xerophthalmia due to vitamin A deficiency.

Study Questions and Exercises

1. For a major organ system, discuss the means by which vitamin deficiencies may affect its function.
2. List the clinical signs that have special diagnostic value (i.e., are specifically associated with insufficient status with respect to certain vitamins) for specific vitamin deficiencies.
3. For a fat-soluble and a water-soluble vitamin, discuss the relationships between tissue distribution of the vitamin and organ site specificity of the clinical signs of its deficiency.
4. List the animal species and deficiency diseases that, because they show specificity for certain vitamins, might be particularly useful in vitamin metabolism research.
5. Develop a decision tree for determining whether lesions of a particular organ system may be due to insufficient intakes of one or more vitamins

RECOMMENDED READING

Marks, J., 1968. The Vitamins in Health and Disease. Churchill, London, UK.

Zempleni, J., Rucker, R.B., McCormick, D.B., Suttie, J.W. (Eds.), 2007. Handbook of Vitamins (fourth ed.). CRC Press, New York, NY (593 pp.).

Considering the Individual Vitamins

Vitamin A

Anchoring Concepts

1. Vitamin A is the generic descriptor for compounds with the qualitative biological activity of *retinol*, i.e., retinoids and some (provitamin A) carotenoids.
2. Vitamin A-active substances are *hydrophobic*, and thus are insoluble in aqueous environments (intestinal lumen, plasma, interstitial fluid, cytosol). *Accordingly, vitamin A-active substances are absorbed by micelle-dependent diffusion.*
3. Vitamin A was discovered by its ability to prevent *xerophthalmia*.

3. To become familiar with the various carriers involved in the extra- and intracellular transport of vitamin A.
4. To understand the metabolic conversions involved in the activation and degradation of vitamin A in its absorption, transport and storage, cellular function, and excretion.
5. To understand current knowledge of the biochemical mechanisms of action of vitamin A and their relationships to vitamin A-deficiency diseases.
6. To understand the physiologic implications of high doses of vitamin A.

In ... four [southeast Asian] countries alone at least 500,000 preschool age children every year develop active xerophthalmia involving the cornea. About half of this number will be blind and a very high proportion, probably in excess of 60%, will die. The annual prevalence for these same countries of non-corneal xerophthalmia is many times higher, probably on the order of 5 million ... There can be no doubt about the claim that vitamin A deficiency is the most common cause of blindness in children and one of the most prevalent and serious of all nutritional deficiency diseases.

D. S. McLaren

Learning Objectives

1. To understand the nature of the various sources of vitamin A in foods.
2. To understand the means of vitamin A absorption from the small intestine.

VOCABULARY

Abetalipoproteinemia
Acyl-CoA:retinol acyltransferase (ARAT)
Alcohol dehydrogenase
all-*trans*-Retinal
all-*trans*-Retinoic acid
Apo-RBP4
β-Carotene
β-Carotene 15,15′-oxygenase
β-Cryptoxanthin
Bleaching
Canthoxanthin
Carotenodermia
Carotenoid
cGMP phosphodiesterase
Chylomicron

The Vitamins. DOI: 10.1016/B978-0-12-381980-2.00005-0

11-*cis*-Retinal
9-*cis*-Retinoic acid
11-*cis*-Retinoic acid
Conjunctival impression cytology
CRABP (cellular retinoic acid-binding protein)
CRABP(II) (cellular retinoic acid-binding protein type II)
CRALBP (cellular retinal-binding protein)
CRBP (cellular retinol-binding protein)
CRBP(II) (cellular retinol-binding protein type II)
3,4-Didehydroretinol
Glycoproteins
High-density lipoproteins (HDLs)
Holo-RBP4
Hyperkeratosis
Iodopsins
IRBP (interphotoreceptor retinol-binding protein)
Keratomalacia
Lecithin–retinol acyltransferase (LRAT)
Low-density lipoproteins (LDLs)
Lycopene
Melanopsin
Metarhodopsin II
Modified relative dose–response (MRDR) test
MRDR (modified relative dose–response) test
Night blindness
Nyctalopia
Opsins
Pancreatic nonspecific lipase
Peroxisome-proliferator activation receptor (PPAR)
Protein-calorie malnutrition
Relative dose–response (RDR) test
Retinal
Retinal isomerase
Retinal oxidase
Retinal reductase
Retinoic acid
Retinoic acid receptors (RARs)
Retinoic acid response elements (RAREs)
Retinoid X receptors (RXRs)
Retinoids
Retinol
Retinol dehydrogenases
Retinol equivalents (RE)
Retinol phosphorylase
Retinyl acetate
Retinyl ester hydrolase
Retinyl palmitate
Retinyl phosphate
Retinyl stearate
Retinyl β-glucuronide
Rhodopsin
Thyroid hormone (T_3)
Transducin
Transgenic
Transthyretin
Very low-density lipoproteins (VLDLs)
Xerophthalmia
Xerosis
Zeaxanthin

1. THE SIGNIFICANCE OF VITAMIN A

Vitamin A is a nutrient of global importance. Shortages in its consumption are estimated to affect an estimated 140 million children worldwide, some 90% of whom live in Southeast Asia and Africa. In recent years, substantial progress has been made in reducing the magnitude of this problem. In 1994, nearly 14 million preschool children (three-quarters from south Asia) were estimated to have clinical eye disease (*xerophthalmia*) due to vitamin A deficiency. By 2005, that prevalence had declined; yet some 5.2 million children remain affected by night blindness. If untreated, two-thirds of those children die within months of going blind owing to their increased susceptibility to infections enhanced by the deficiency. Vitamin A deficiency remains the single most important cause of childhood blindness in developing countries.

At the same time the prevalence of subclinical deficiency (serum retinol levels <0.7 μmol/l) has increased. Of the world's preschool children, 33% (estimates range from 75 to 254 million) appear to be growing up with insufficient vitamin A (Table 5.1). More than 19 million pregnant women in developing countries are also vitamin A-deficient; a third are affected by night blindness.

Subclinical vitamin A deficiency is also associated with increased child mortality, having public health significance in at least 122 countries in Africa, southern and Southeast Asia, and some parts of Latin America and the western Pacific. High rates of morbidity and mortality have long been associated with vitamin A deficiency; recent intervention trials have indicated that providing vitamin A can reduce child mortality by about 25%, and birth-related maternal mortality by 40%. Vitamin A deficiency in these areas does not necessarily imply insufficient national or regional supplies of food vitamin A, as vitamin A deficiency can also be caused by insufficient dietary intakes of protein, fats, and oils. Still, most studies show that children with histories of xerophthalmia consume fewer dark green leafy vegetables than their counterparts without such histories.

2. SOURCES OF VITAMIN A

Dietary Sources of Vitamin A

Vitamin A exists in natural products in many different forms. It exists as preformed **retinoids**, which are stored in animal tissues, and as provitamin A **carotenoids**, which are synthesized as pigments by many plants and are found in green, orange, and yellow plant tissues. In milk, meat, and eggs, vitamin A exists in several forms, mainly as long-chain fatty acid esters of **retinol**, the predominant one being *retinyl palmitate*. The carotenoids are present in both plant and animal food products; in animal products, their occurrence results from dietary exposure. Carotenoid

TABLE 5.1 Global Prevalence of Vitamin A Deficiency Among Preschool Children and Pregnant Women

Region	Children (0–5 years)		Pregnant Women	
	Night Blindness % (millions)	Low Serum Retinol[a] % (millions)	Night Blindness % (millions)	Low Serum Retinol[a] % (millions)
Africa	2.0 (2.55)	44.4 (56.4)	9.8 (3.02)	13.5 (4.18)
Americas	0.6 (0.36)	15.6 (8.68)	4.4 (0.50)	2.0 (0.23)
Southeast Asia	0.5 (1.01)	49.9 (91.5)	9.9 (3.84)	17.3 (6.69)
Europe	0.8 (0.24)	19.7 (5.81)	3.5 (0.22)	11.6 (0.72)
Eastern Mediterranean	1.2 (0.77)	20.4 (13.2)	7.2 (1.09)	16.1 (2.42)
Western Pacific (including China)	0.2 (0.26)	12.9 (14.3)	4.8 (1.09)	21.5 (4.90)
Global	**0.9 (5.17)**	**33.3 (190)**	**7.8 (9.75)**	**15.3 (19.1)**

[a]*Serum retinol <0.7 μmol/l.*
Source: WHO Global Database on Vitamin A Deficiency (2009). Global Prevalence of Vitamin A Deficiency in Populations at Risk 1995–2005. WHO, Geneva, 55 pp.

pigments are widespread among diverse animal species, with more than 500 different compounds estimated. About 60 of these have *provitamin A activity*, i.e., those that can be cleaved by animals to yield at least one molecule of retinol. In practice, however, only five or six of these pro-vitamins A are commonly encountered in foods.

Therefore, actual vitamin A intakes depend on the patterns of consumption of vitamin A-bearing animal food products and provitamin A-bearing fruits and vegetables (Table 5.2), the relative contributions of which are influenced by food availability and personal food habits. In the diets of American omnivores, vitamin A needs are satisfied by animal products and fortified foods; however, in the diets of vegetarians and of many poor families in developing countries, the only real sources of vitamin A are plant foods – particularly vegetables.

The bioconversion of carotenoids to vitamin A has been found experimentally to vary considerably (10–90%). Variation has also been apparent in practical circumstances, as not all interventions with vegetables have produced improvements in vitamin A status (i.e., serum retinol concentrations) of deficient individuals. While studies from different parts of the world have shown positive impacts on the vitamin A status of provitamin A-containing vegetables and fruits, others have failed to show such effects. For example, local green herbs cooked and consumed by Indonesian women were ineffective in raising serum retinol levels.[1,2]

1. de Pee, S., West, C. E., Muhilal, X., *et al.* (1995). Lack of improvement in vitamin A status with increased consumption of dark-green leafy vegetables. *Lancet* 346, 75–81.
2. If this phenomenon is found to be generally true, then there may be limits to the contributions of horticultural approaches (e.g., "home-gardening" programs) to solving problems of vitamin A deficiency.

Expressing the Vitamin A Activities in Foods

Owing to the fact that vitamin A exists in foods in many different preformed retinoids and provitamin A carotenoids, the reporting of vitamin A activity in foods requires some means of standardization. Two systems are used for this purpose: **retinol equivalents** (**RE**)[3] for food applications, and **international units** (*IU*) for pharmaceutical applications.

1 RE	=1 μg all-*trans*-retinol
	=2 μg all-*trans*-β-carotene in dietary supplements
	=12 μg all-*trans*-β-carotene in foods
	=24 (12–26) μg other provitamin A carotenoids (α-carotene, β-cryptoxanin) in foods[4]
1 IU	=0.3 μg all-*trans*-retinol
	=0.344 μg all-*trans*-retinyl acetate
	=0.55 μg all-*trans*-retinyl palmitate

3. These equivalencies were established by the Food and Nutrition Board in 2001. USDA database for standard reference lists both IU and RE; FAO tables list μg of retinol and β-carotene. INCAP (Instituto de Nutrición de Centroamérica y Panamá) tables list vitamin A as μg retinol; those values are not the same as REs, as a factor of one-half was used to convert β-carotene to retinol.
4. The data of van Lieshout *et al.* (*Am. J. Clin. Nutr.* [2002] 77, 12–28) suggest a ratio of 2.6 based on stable isotope (or circulating retinol) dilution studies in more than a hundred Indonesian children. Other isotope dilution studies show the conversion of β-carotene to retinol by humans to be quite variable; for example, Wang *et al.* (*Br. J. Nutr.* [2004] 91, 121–131) found individual equivalency ratios to vary from 2:1 to 12:1, and Haskell *et al.* (*Am. J. Clin. Nutr.* [2004] 80, 705–714) reported equivalency ratios from 6:1 to 13:1. West and colleagues (*Am. J. Clin. Nutr.* [1998] 68, 1058–1067; *J. Nutr.* [2002] 132, S2920–2926) suggested that in developing countries this conversion ratio may be as great as 21:1.

TABLE 5.2 Sources of Vitamin A in Foods

Food	Percentage Distribution of Vitamin A Activity		
	Retinol	β-Carotene	Non-β-Carotenoids
Animal Foods			
Red meat	90	10	
Poultry meat	90	10	
Fish and shellfish	90	10	
Eggs	90	10	
Milk, milk products	70	30	
Fats and oils	90	10	
Plant Foods			
Maize, yellow		40	60
Legumes and seeds		50	50
Green vegetables		75	25
Yellow vegetables[a]		85	15
Pale sweet potatoes		50	50
Yellow fruits[b]		85	15
Other fruits		75	25
Red palm oil		65	35
Other vegetable oils		50	50

[a]For example, carrots and deep-orange sweet potatoes.
[b]For example, apricots.
Source: Leung, W. and Flores, M. (1980). *Food Composition Table for Use in Latin America*. Institute of Nutrition of Central America and Panama, Guatemala City, Guatemala; and Interdepartmental Committee on Nutrition for National Defense, Washington, DC.

TABLE 5.3 Vitamin A Activities of Foods

Food	Vitamin A Activity	
	IU/100 g	RE (μg)/100 g
Animal Products		
Red meats	—	—
Beef liver	10,503	35,346
Poultry meat	41	140
Mackerel	130	434
Herring	28	94
Egg	552	1,839
Swiss cheese	253	856
Butter	754	3,058
Plant Products		
Corn	—	—
Peas	15	149
Beans	171	171
Chick peas	7	67
Lentils	4	39
Soybeans	2	24
Green peppers	—	420
Red peppers	—	21,600
Carrots	—	11,000
Peach	—	1,330
Pumpkin	—	1,600
Yellow squash	—	—
Orange sweet potatoes	—	—
Margarine[a]	993	3,307

[a]Fortified with vitamin A.
Source: USDA (1978). USDA Science and Education Series, Handbook 8.

These retinol equivalencies represent a 2001 revision by the Food and Nutrition Board of the Institute of Medicine from the Board's previous values for provitamin A carotenoids. However, the Board's estimated equivalency ratio for purified β-carotene in oils and nutritional supplements, 2:1, is weakly supported.[5] It is clear from isotope dilution studies that β-carotene from plant foods is utilized with much poorer efficiency than has been previously thought. This appears to be especially the case in resource-poor countries in which children rely almost entirely on the conversion of β-carotene from fruits and

vegetables for their vitamin A. In such circumstances, several factors may reduce that bioconversion efficiency: low fat intakes, intestinal roundworms, recurrent diarrhea, tropical enteropathy, and other factors that affect the absorptive function of the intestinal epithelium and intestinal transit time. Solomons and Bulux (1993) have suggested that these factors may reduce by more than half the expected bioconversion of plant carotenoids.

Foods Rich in Vitamin A

Several foods contain vitamin A activity (Table 5.3); however, relatively few are rich dietary sources, those being

5. The 2001 estimate was based on a single report (Sauberlich, H. E., Hodges, R. E., Wallace, D. L., *et al.* [1974]. Vitamin A metabolism and requirements in the human studied with the use of labeled retinol. *Vitamins Hormones* 32, 251–275) of the efficacy of β-carotene in correcting impaired dark adaptation in two volunteers in a six-person study.

green and yellow vegetables,[6] liver, oily fishes, and vitamin A-fortified products such as margarine. It should be noted that, for vitamin A and other vitamins that are susceptible to breakdown during storage and cooking, values given in food composition tables are probably high estimates of amounts actually encountered in practical circumstances.

3. ABSORPTION OF VITAMIN A

Absorption of Retinoids

Most of the preformed vitamin A in the diet is in the form of **retinyl esters**. Retinyl esters are hydrolyzed in the lumen of the small intestine to yield retinol; this step is catalyzed by hydrolases produced by the pancreas and situated on the mucosal brush border[7] or intrinsic to the brush border itself. The retinyl esters, as well as the carotenoids, are hydrophobic, and thus depend on micellar solubilization for their dispersion in the aqueous environment of the small intestinal lumen. For this reason, vitamin A is poorly utilized from low-fat diets. The micellar solubilization of vitamin A facilitates access of soluble hydrolytic enzymes to their substrates (i.e., the retinyl esters), and provides a means for the subsequent presentation of retinol to the mucosal surface across which free retinol and intact β-carotene diffuse passively into the mucosal epithelial cells. The overall absorption of retinol from retinyl esters appears to be fairly high (e.g., about 75%); this process appears to be minimally affected by the level and type of dietary fat, although the absorption is appreciably less efficient at very high vitamin A doses.

Studies have shown that vitamin A can also be absorbed via non-lymphatic pathways. Rats with ligated thoracic ducts retain the ability to deposit retinyl esters in their livers. That such animals fed retinyl esters show greater concentrations of retinol in their portal blood than in their aortic blood suggests that, in mammals, the portal system may be an important alternative route of vitamin A absorption when the normal lymphatic pathway is blocked. This phenomenon corresponds to the route of vitamin A absorption in birds, fishes, and reptiles, which, lacking lymphatic drainage of the intestine, rely strictly on portal absorption. That retinyl esters can also be absorbed by other epithelial cells is evidenced by the fact that the use of vitamin A-containing toothpaste or inhalation of vitamin A-containing aerosols can increase plasma retinol levels.

Absorption of Carotenoids

The major sources of vitamin A activity for most populations are the provitamin A carotenoids. Their utilization involves three steps:

1. *Release from food matrices.* A major factor limiting the utilization of carotenoids from food sources is their release from physical food matrices. Carotenoids can occur in cytosolic crystalline complexes or in chromoplasts and chloroplasts, where they can be associated with proteins, polysaccharides, fibers, and phenolic compounds. Many carotenoid complexes are resistant to digestion without heat treatment.

2. *Micellar solubilization in the intestinal lumen.* The enteric absorption of carotenoids depends on their solubilization in mixed lipid micelles, the formation of which requires the consumption and digestion of lipids (see "Vitamin Absorption" in Chapter 3). Absorption, particularly of the less polar carotenoids, can be impaired by the presence of undigested lipids or sucrose polyesters in the intestinal lumen. Gastric acidity may also be an affector, as patients with pharmaceutically obliterated gastric acid production showed reduced blood responses to test doses of β-carotene.[8]

3. *Uptake by the intestinal mucosa.* Release of carotenoids from micelles is thought to involve the diffusion directly through the plasma membranes of the enterocytes. The process appears to be impaired by soluble dietary fiber and, likely, other factors that interfere with the contact of the micelle with the mucosal brush border. Limited evidence suggests that carotenoids may be mutually competitive during absorption. For example, high doses of **canthoxanthin** or **lycopene** have been shown to reduce the absorption of β-carotene. For these reasons, it has been technically difficult to estimate quantitatively the efficiency of intact carotenoid absorption. Available evidence indicates that, for β-carotene, it tends to be highly variable (10–90%). While β-carotene can cross the mucosal epithelial cell intact, most is metabolized within the cell.

Carotenoid Metabolism Linked to Absorption

Carotenoid absorption typically results in the accumulation in enterocytes of more all-*trans*-β-carotene than 9-*cis*-β-carotene. This suggests enterocytic capacity for *cis-trans* isomerization, which is also indicated by the fact that humans given 9-*cis*-β-carotene show detectable levels of

---

6. It is estimated that carotene from vegetables contributes two-thirds of dietary vitamin A worldwide, and more than 80% in developing countries.
7. One of these activities appears to be the same enzyme that catalyzes the intralumenal hydrolysis of cholesteryl esters; it is a relatively non-specific carboxylic ester hydrolase. It has been given various names in the literature, the most common being pancreatic non-specific lipase and *cholesteryl esterase*.

---

8. This finding has implications for millions of people, as atrophic gastritis and hypochlorhydria are common conditions, particularly among older people.

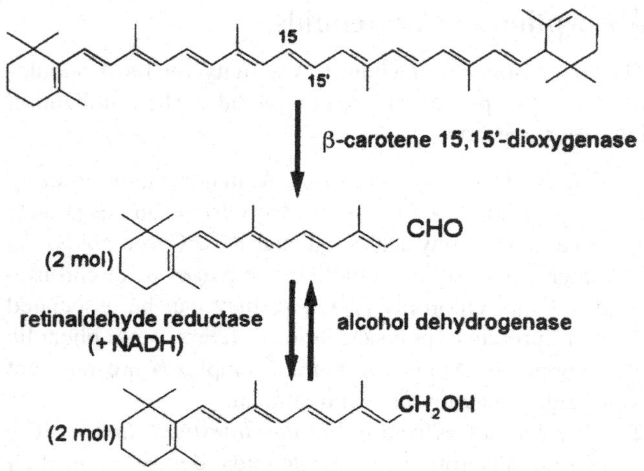

FIGURE 5.1 Bioconversion of provitamins A to retinal.

9-*trans*-retinol in their plasma. The capacity for isomerization would serve to limit the distribution of 9-*cis*-retinoids to tissues, and render both isomers of β-carotene capable of being metabolized to **retinal**, thus serving as effective provitamins A.

Carotenoids[9] are provitamins A to the extent that they are converted to retinal. This metabolism is catalyzed by carotene oxygenases[10] (Fig. 5.1). Most of this bioconversion occurs via the central cleavage of the polyene moiety by a predominantly cytosolic enzyme, **β-carotene 15,15′-oxygenase (BCO1)**, found in the intestinal mucosa, liver, and corpus luteum. The enzyme requires iron as a cofactor; studies have shown that its activity responds to dietary iron intake as well as to factors affecting iron utilization (e.g., copper, fructose). It contains ferrous iron (Fe^{2+}) linked to a histidinyl residue at the axis of a seven-bladed, β-propeller-chain fold covered by a dome structure formed by six large loops in the protein. Upon binding within that structure, the three consecutive *trans* double bonds of the carotenoid are isomerized to a *cis–trans–cis* conformation, leading to the oxygen-cleavage of the central *trans* bond. The expression of BCO1 is repressed by the intestinal transcription factor ISX, which is induced by retinoic acid. ISX also appears to repress the expression of a receptor (scavenger receptor B type 1, SR-B1) thought to facilitate the intestinal absorption of lipids including β-carotene. By this mechanism, both the absorption of β-carotene as well as its cleavage to produce retinal are reduced under conditions of vitamin A-adequacy.

β-Carotene oxygenase, sometimes also called *carotene cleavage enzyme*, cleaves β-carotene into two molecules

of retinal. However, the enzyme is not highly specific for β-carotene, cleaving other carotenoids, which have provitamin A activities to the extent that they can also yield retinal. Apo-carotenals yield retinal; epoxy-carotenoids are not metabolized. The reaction requires molecular oxygen, which reacts with the two central carbons (C-15 and C-15′), followed by cleavage of the C–C bond. It is inhibited by sulfhydryl group inhibitors and by chelators of ferrous iron (Fe^{2+}). The enzyme has been found in a wide variety of animal species;[11] enzyme activities were found to be greatest in herbivores (e.g., guinea pig, rabbit), intermediate in omnivores (e.g., chicken, tortoise, fish), and absent in the only carnivore studied (cat). The enzyme activity is enhanced by the consumption of triglycerides,[12] suggesting that its regulation involves long-chain fatty acids. It is diminished by high intakes of β-carotene and protein deprivation, is induced by vitamin A deficiency, and can be inhibited by quercetin and other flavonols. The symmetric, central cleavage of β-carotene is highly variable between individuals (35–90% of consumed β-carotene). In the bovine corpus luteum, which also contains a high amount of β-carotene, BCO1 activity has been shown to vary with the estrous cycle, showing a maximum on the day of ovulation. Studies with the rat indicate that the activity is stimulated by vitamin A deprivation, and reduced by dietary protein restriction.

Low BCO1 activities are associated with the absorption of intact carotenoids; this phenomenon is responsible for the yellow-colored adipose tissue, caused by the deposition of absorbed carotenoids, in cattle. Thus, at low doses β-carotene is essentially quantitatively converted to vitamin A by rodents, pigs, and chicks; cats, in contrast, cannot perform the conversion, and therefore β-carotene cannot support their vitamin A needs.

The asymmetric cleavage of carotenoids also occurs by a second intestinal mucosal enzyme, β-carotene oxygenase 2 (BCO2), although this appears to be a quantitatively minor pathway. This enzyme cleaves the carotene-9′,10′-bond to form apo-10′-β-carotenal (Fig. 5.2), which can apparently be chain-shortened directly to yield retinal or, first, oxidized to the corresponding apo-carotenoic acids and, then, chain-shortened to yield **retinoic acid**.[13] It has been suggested that BCO2 functions in the metabolism of

9. Fewer than 10% of naturally occurring carotenoids are provitamins A.
10. More than 100 enzymes in this group are known. Two occur in animals: β-carotene-15,15′-oxygenase and β-carotene-9′,10′-oxygenase.

11. The β-carotene 15,15′-dioxygenase has also been identified in *Halobacterium halobium* and related halobacteria, which use retinal, coupled with an opsin-like protein, to form bacteriorhodopsin, an energy-generating light-dependent proton pump.
12. Triglycerides also increase CRBP(II) levels.
13. Studies of these processes are complicated by the inherent instability of carotenoids under aerobic conditions; many of the products thought to be produced enzymatically are the same as those known to be produced by autoxidation.

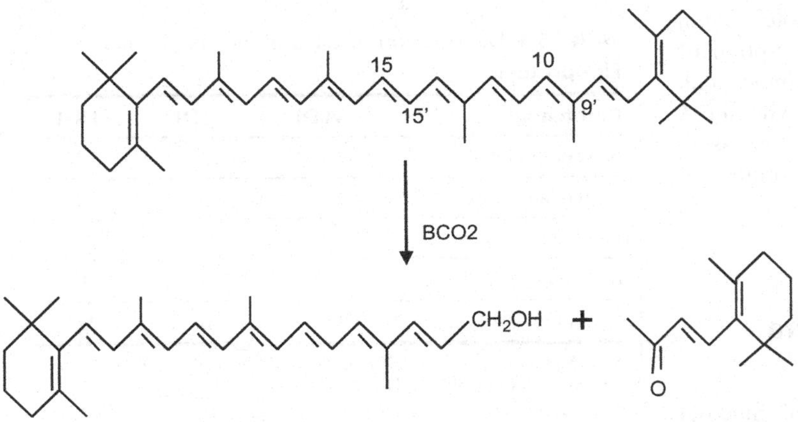

FIGURE 5.2 The asymmetric (or eccentric) cleavage of β-carotene by the β-carotene oxygenase 2 (BCO2) yields apo-10′-β-carotenal and β-ionone.

β-Carotene —dioxygenase→ retinal —reductase→ retinol —ARAT→ retinyl esters

β-Carotene —dioxygenase→ CRBP(II)–retinal —reductase→ CRBP(II)–retinol —LRAT→ retinyl esters

FIGURE 5.3 Intestinal metabolism of vitamin A.

acylic, non-provitamin A carotenoids such as lycopene, which accumulates when the enzyme is not expressed. Intestinal enzymes can cleave 9-*cis*-β-carotene (which comprises 8–20% of the β-carotene in fruits and vegetables, but seems less well utilized than the all-*trans* isomer) to 9-*cis*-retinal, which in turn appears to be oxidized to 9-*cis*-retinoic acid.

The turnover of carotenoids in the body occurs via first-order mechanisms that differ for individual carotenoids. For example, in humans the biological half-life of β-carotene has been determined to be 37 days, whereas those of other carotenoids vary from 26 days (lycopene) to 76 days (lutein).

Mucosal Metabolism of Retinol

Retinol, formed either from the hydrolysis of dietary retinyl esters or from the reduction of retinal cleaved from β-carotene,[14] is absorbed by facilitated diffusion via a specific transporter.[15] Retinal produced by the central cleavage of β-carotene is reduced in the intestinal mucosa to retinol (Fig. 5.1) by **retinaldehyde reductase**, which is also found in the liver and the eye. The reduction requires a reduced pyridine nucleotide (NADH/NADPH) as a cofactor, and has an apparent K_m of 20 mmol/l. It can also be catalyzed by a **short-chain alcohol dehydrogenase/aldehyde reductase**, and there is some debate concerning whether the two activities reside on the same enzyme.

Retinol is quickly re-esterified with long-chain fatty acids in the intestinal mucosa, whereupon retinyl esters are transported to the liver (i.e., 80–90% of a retinol dose[16]). The composition of lymph retinyl esters is remarkably independent of the fatty acid composition of the most recent meal. **Retinyl palmitate** typically comprises about half of the total esters, with **retinyl stearate** comprising about a quarter, and retinyl oleate and retinyl linoleate being present in small amounts. Two pathways for the enzymatic re-esterification of retinol have been identified in the microsomal fraction of the intestinal muscosa. A low-affinity route (Fig. 5.3, top) involves uncomplexed retinol; it is catalyzed by **acyl-CoA:retinol acyltransferase (ARAT)**. A high-affinity route (Fig. 5.3, bottom) involves retinol complexed with a specific binding protein, **cellular retinol-binding protein type II [CRBP(II)[17]]**; it is catalyzed by **lecithin–retinol acyltransferase (LRAT)**. The expression of LRAT mRNA is induced by retinoic acid and downregulated by vitamin A depletion. It has been suggested that LRAT serves to esterify low doses of retinol, whereas ARAT serves to esterify excess retinol, when CRBP(II) becomes saturated. The identification of

14. It has been estimated that humans convert 35–71% of absorbed β-carotene to retinyl esters.
15. This transporter was discovered by Ong (1994), who found it to transport all-*trans*-retinol and 3-dehydroretinol. They found other retinoids to be taken up by enterocytes by passive diffusion.
16. Humans fed radiolabeled β-carotene showed the ability to absorb some of the unchanged compound directly in the lymph, with only 60–70% of the label appearing in the retinyl ester fraction.
17. CRBP(II) is a low molecular mass (15.6kDa) protein that constitutes about 1% of the total soluble protein of the rat enterocyte.

a retinoic acid-responsive element in the promoter region of the CRBP(II) gene suggests that the transcription of that gene may be positively regulated by retinoic acid, leading to increased CRBP(II) levels at high vitamin A doses. Experiments have shown that CRBP (II) expression is enhanced under conditions of stimulated absorption of fats, especially unsaturated fatty acids.

4. TRANSPORT OF VITAMIN A

Retinyl Esters Conveyed by Chylomicra in Lymph

Retinyl esters are secreted from the intestinal mucosal cells in the hydrophobic cores of chylomicron particles, by which absorbed vitamin A is transported to the liver through the lymphatic circulation, ultimately entering the plasma[18] compartment through the thoracic duct. Carnivorous species in general, and the dog in particular, typically show high plasma levels. Retinyl esters are almost quantitatively retained in the extrahepatic processing of chylomicra to their remnants; therefore, chylomicron remnants are richer in vitamin A than are chylomicra. Retinyl and cholesteryl esters can undergo exchange reactions between lipoproteins including chylomicra in rabbit and human plasma by virtue of a **cholesteryl ester transfer protein** peculiar to those species.[19] Although this kind of lipid transfer is probably physiologically important in those species, the demonstrable transfer involving chylomicra is unlikely to be a normal physiological process.

Transport of Carotenoids

Carotenoids appear to be transported across the intestinal mucosum by a facilitated process similar to that of cholesterol. They are not metabolized in the epithelium, but are transported from that organ by chylomicra via the lymphatic circulation to the liver, where they are transferred to lipoproteins. It is thought that strongly non-polar species such as β-carotene and lycopene are dissolved in the chylomicron

TABLE 5.4 Distribution of Carotenoids in Human Lipoproteins

Carotenoid	VLDL	LDL	HDL
Zeaxanthin/lutein (%)	16	31	53
Cryptoxanthin (%)	19	42	39
Lycopene (%)	10	73	17
α-Carotene (%)	16	58	26
β-Carotene (%)	11	67	22

Source: Reddy, P. P., Clevidance, B. A., Berlin, E., Taylor, P. R., Bieri, J. G., and Smith, J. C. (1989). FASEB J. 3, A955.

core, whereas species with polar functional groups may exist at least partially at the surface of the particle. Such differences in spatial distribution would be expected to affect transfer to lipoproteins during circulation and tissue uptake. Indeed, the distribution of carotenoids among the lipoprotein classes reflects their various physical characteristics, with the hydrocarbon carotenoids being transported primarily in **low-density lipoproteins (LDLs)**, and the more polar carotenoids being transported in a more evenly distributed manner among LDLs and **high-density lipoproteins (HDLs)** (Table 5.4). It is thought that small amounts of the non-polar carotenoids are transferred from chylomicron cores to HDLs during the lipolysis of the triglycerides carried by the former particles; however, because HDL transports only a small fraction of plasma β-carotene, the carrying capacity of the latter particles for hydrocarbon carotenoids would appear to be small. Therefore, it is thought that β-carotene is retained by the chylomicron remnants to be internalized by the liver for subsequent secretion in very low-density lipoproteins (VLDLs).

Impact of Abetalipoproteinemia

The absorption of vitamin A (as well as the other fat-soluble vitamins) is a particular problem in patients with **abetalipoproteinemia**, a rare autosomal recessive disorder characterized by general lipid malabsorption, acanthosis (diffuse epidermal hyperplasia), and hypocholesterolemia. These patients lack apo B, and consequently cannot synthesize any of the apo B-containing lipoproteins (i.e., LDLs, VLDLs, and chylomicra). Having no chylomicra, they show hypolipidemia and low plasma vitamin A levels. However, when given oral vitamin A supplementation, their plasma levels are normal. Although the basis of this response is not clear, it has been suggested that these patients can transport retinol from the absorptive cells via their remaining lipoprotein (HDLs), possibly by the portal circulation.

---

18. On entering the plasma, chylomicra acquire apolipoproteins C and E from the plasma high-density lipoproteins (HDLs). Acquisition of one of these (apo C-II) activates lipoprotein lipase at the surface of extrahepatic capillary endothelia, which hydrolyzes the core triglycerides, thus causing them to shrink and transfer surface components (e.g., apo A-I, apo A-II, some phospholipid) to HDLs and fatty acids to serum albumin, and lose apo A-IV and fatty acids to the plasma or other tissues. These processes leave a smaller particle called a chylomicron remnant, which is depleted of triglyceride but relatively enriched in cholesteryl esters, phospholipids, and proteins (including apo B and apo E). Chylomicron remnants are almost entirely removed from the circulation by the liver, which takes them up by a rapid, high-affinity receptor-mediated process stimulated by apo E.

19. This protein has not been found in several other mammalian species examined.

Vitamin A Uptake by the Liver

Most of the recently absorbed vitamin A is taken up by the liver from chylomicron remnants. Because those particles are cleared by the liver rapidly, vitamin A (mostly as retinyl esters with smaller amounts of β-carotene) circulates in chylomicra for only a short time. Chylomicron clearance occurs by receptor-mediated[20] endocytosis into hepatic parenchymal cells[21] within which the remnants are degraded by lysosomal enzymes.

Vitamin A is Stored in the Liver

After being taken up by the liver, retinyl esters are hydrolyzed to yield retinol in parenchymal cells from which it is transferred by a retinol-binding protein to *stellate cells*,[22] which also contain appreciable amounts of triglycerides, phospholipids, free fatty acids, and cholesterol. There, it is re-esterified and stored in droplets (some that are membrane-bound and appear to be derived from lysosomes and other, larger ones not associated with membranes). It is likely that the re-esterification of retinol proceeds by a reaction similar to that of the intestinal microsomal acyl-CoA:retinol acyltransferase (ARAT). The liver thus serves as the primary storage depot for vitamin A, normally containing 50–80% of the total amount of the vitamin in the body.[23] Most of this (80–90%) is stored in stellate cells, which account for only about 2% of total liver volume. The balance is stored in parenchymal cells. These are the only types of hepatocytes that contain retinyl ester hydrolase activities. Almost all (about 95%) of hepatic vitamin A occurs as long-chain retinyl esters, the predominant one being retinyl palmitate. Kinetic studies of vitamin A turnover indicate the presence, in both liver and extrahepatic tissues, of two effective pools (i.e., fast- and slow-turnover pools) of the vitamin. Of rat liver retinoids, 98% were in the slow-turnover pool (retinyl esters of stellate cells), with the balance corresponding to the retinyl esters of parenchymal cells.

In addition to retinol ester hydrolases and ARAT, stellate cells contain two other retinoid-related proteins: **cellular retinoid-binding protein (CRBP)** and **cellular retinoic acid-binding protein (CRABP).** The storage or retinyl esters appears not to depend on the expression of CRBP, as **transgenic** mice that overexpressed CRBP in several organs have not shown elevated vitamin A stores in those organs. The metabolism of vitamin A by hepatic cytosolic retinal dehydrogenase increases with increasing hepatic retinyl ester stores.

Mobilization of Vitamin A from the Liver

Vitamin A is mobilized as retinol from the liver by hydrolysis of hepatic retinyl esters. This mobilization accounts for about 55% of the retinol discharged to the plasma, the balance coming from recycling from extrahepatic tissues. The **retinyl ester hydrolase** involved in this process remains poorly characterized; it shows extreme variation between individuals.[24] The activity of this enzyme is known to be low in protein-deficient animals, and has been found to be inhibited, at least *in vitro*, by vitamins E and K.[25]

Retinol Transported Protein-Bound in the Plasma

Once mobilized from liver stores, retinol is transported to peripheral tissues by means of a specific carrier protein, **plasma retinol-binding protein (RBP4).** Human RBP4 consists of a single polypeptide chain of 182 amino acid residues, with a molecular mass of 21 kDa. Like several other of the retinoid binding proteins,[26] it is classified as a member of the lipocalin family of lipid-binding proteins. These are composed of an eight-stranded, antiparallel β-sheet that is folded inwards to form a hydrogen-bonded β-barrel that comprises the ligand-binding domain, the entrance of which is flanked by a single loop scaffold. Within this domain, a single molecule of all-*trans*-retinol is completely encapsulated, being stabilized by hydrophobic interactions of the β-ionone ring and the isoprenoid chain with the amino acids lining the interior of the barrel structure. This structure protects the vitamin from oxidation during transport. RBP4 is synthesized as a 24 kDa *pre-RBP4* by parenchymal cells, which also convert it to RBP4 by the cotranslational removal of a 3.5 kDa polypeptide.[27]

---

20. Chylomicron remnants are recognized by high-affinity receptors for their apo E moiety.
21. The parenchymal cell is the predominant cell type of the liver, comprising in the rat more than 90% of the volume of that organ.
22. These are also known as pericytes, fat-storing cells, interstitial cells, lipocytes, Ito cells or vitamin A-storing cells.
23. Mean hepatic stores have been reported in the range of 171–723 μg/g in children and 0–320 μg/g in adults (Panel on Micronutrients, Food and Nutrition Board [2002] *Dietary Reference Intakes for Vitamin A, Vitamin K, Arsenic, Boron, Chromium, Copper, Iodine, Iron, Manganese, Molybdenum, Nickel, Silicon, Vanadium and Zinc*. Washington, DC: National Academy Press, p. 95).

24. In the rat, hepatic retinyl ester hydrolase activities can vary by 50-fold among individual rats and by 60-fold among different sections of the same liver.
25. Each vitamin has been shown to act as a competitive inhibitor of the hydrolase. This effect may explain the observation of impaired hepatic vitamin A mobilization (i.e., increased total hepatic vitamin A and hepatic retinyl esters with decreased hepatic retinol) of animals fed very high levels of vitamin E.
26. The cellular retinal and retinoic acid binding proteins; see section on "Other Vitamin A-Binding Proteins Involved in Vitamin A Transport" and Table 5.6, both later in this chapter.

TABLE 5.5 Percentage Distribution of Vitamin A in Sera of Fasted Humans

Fraction	Retinol (%)	Retinyl Palmitate (%)
VLDL	6	71
LDL	8	29
HDL	9	—
RBP4	77	—
Total:	100	100[a]

[a]This represents only about 5% of the total circulating vitamin A.

This protein product (**apo-RBP4**) is secreted in a 1:1 complex with all-*trans*-retinol (**holo-RBP4**). Stellate cells also contain low amounts of RBP4; however, it is not clear whether they synthesize it or whether their apo-RBP4 derives from parenchymal cells to mobilize retinol to the circulation. According to the latter view, stellate cells may be important in the control of retinol storage and mobilization, a complex process that is thought to involve retinoid-regulated expression of cellular retinoid-binding proteins.

The secretion of RBP4 from the liver is regulated in part by estrogen levels,[28] and by vitamin A status (i.e., liver vitamin A stores) protein, and zinc status; deficiencies of each markedly reduce RBP4 secretion, and thus reduce circulating levels of retinol (Table 5.5). In cases of protein-energy malnutrition, RBP4 levels (and thus serum retinol levels) can be decreased by as much as 50%. Except in the postprandial state, virtually all plasma vitamin A is bound to RBP4. In the plasma, almost all RBP4 forms a 1:1 complex with **transthyretin**[29] (TTR, a tetrameric, 55kDa protein that strongly binds four thyroxine molecules). The formation of the RBP4–transthyretin complex (RBP4–TTR) appears to reduce the glomerular filtration of RBP4 and, thus, its renal catabolism.[30] The kidney appears to be the only site of catabolism of RBP4, which turns over rapidly.[31,32] Under normal conditions, the turnover of holo-RBP4 is rapid – 11–16 hours in humans.

Computer modeling studies indicate that more than half of hepatically released holo-RBP4 comes from apo-RBP4 recycled from RBP4–TTR complexes. Apo-RBP4 is not secreted from the liver. Vitamin A-deficient animals continue to synthesize apo-RBP4, but the absence of retinol inhibits its secretion (a small amount of denatured apo-RBP4 is always found in the plasma). Owing to this hepatic accumulation of apo-RBP4, vitamin A-deficient individuals may show a transient overshooting of normal plasma RBP4 levels on vitamin A realimentation.

Other factors can alter the synthesis of RBP4 to reduce the amount of the carrier available for binding retinol, and secretion into the plasma. This occurs in response to dietary deficiencies of protein (e.g., **protein-energy malnutrition**) and/or zinc, which reduce the hepatic synthesis of apo-RBP4. Because RBP4 is a negative acute phase reactant, subclinical infections or inflammation can also decrease circulating retinol levels. Thus, low serum retinol levels in malnourished individuals may not be strictly indicative of a dietary vitamin A deficiency. Also, because vitamin A deprivation leads to reductions in plasma RBP4 only after the depletion of hepatic retinyl ester stores (i.e., reduced retinol availability), the use of plasma RBP4/retinol as a parameter of nutritional vitamin A status can yield false-negative results in cases of vitamin A deprivation of short duration.

A two-point mutation in the human RBP4 gene has been shown to result in markedly impaired circulating retinol levels; surprisingly, this is associated with no signs other than night blindness.[33] This suggests that other pathways are also important in supplying cells with retinol, presumably via retinyl esters and/or β-carotene, and/or with retinoic acid. That the systemic functions of vitamin A can be discharged by retinoic acid, which is ineffective in supporting vision, indicates that the metabolic role of RBP4 is to deliver retinol to the pigment epithelium as a direct requirement of visual function, and to other cells as a precursor to retinoic acid. Retinoic acid is not transported by RBP4, but it is normally present in the plasma, albeit at very low concentrations (1–3 ng/ml), tightly bound to albumin. It is presumed that the cellular uptake of retinoic acid from serum albumin is very efficient.

27. Retinol-binding proteins isolated from several species, including rat, chick, dog, rabbit, cow, monkey, and human, have been found to have similar sizes and binding properties.

28. Seasonally breeding animals show three-fold higher plasma RBP4 levels in the estrous compared with the anestrous phase. Women using oral contraceptive steroids frequently show plasma RBP4 levels that are greater than normal.

29. Previously called prealbumin.

30. Studies have shown holo-RBP4 bound to transthyretin to have a half-life in human adult males of 11–16h; that of free RBP4 was only 3.5h. These half-lives increase (i.e., turnover decreases) under conditions of severe protein-calorie malnutrition.

31. For this reason, patients with chronic renal disease show greatly elevated plasma levels of both RBP4 (which shows a half-life 10- to 15-fold that of normal) and retinol, while concentrations of transthyretin remain normal.

32. Turnover studies of RBP4 and retinol in the liver and plasma suggest that some retinol does indeed recirculate to the liver; however the mechanism of such recycling, perhaps involving transfer to lipoproteins, is unknown.

33. Biesalski, H. K., Frank, J., Beck, S. C., *et al.* (1999). *Am. J. Clin. Nutr.* 69, 931–936.

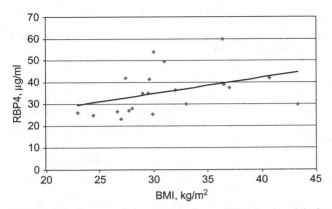

FIGURE 5.4 Inverse relationship of plasma RBP4 level and body mass index (BMI). Data from Graham, T. E., Yang, Q., Blűher, M. *et al.* (2006). *New Eng. J. Med.* 354, 2552–2563.

RPB4 is also expressed in adipose and other tissues, although liver is the predominant source of the protein in circulation. Appreciable amounts of vitamin A are stored in adipocytes; 15–20% of the total body store, more than half as retinyl esters. That visceral fat expresses more RBP4 than subcutaneous fat makes serum apo-RBP4 a candidate biomarker for visceral adiposity. Unlike other tissues, which take up retinol from RBP4, adipocytes take up retinyl esters from chylomicra. Studies with the rat have shown that the mobilization of vitamin A from adipocytes also differs from that process in other cells. A cAMP-sensitive, hormone-dependent lipase converts adipocyte retinyl esters to retinal in a manner analogous to the liberation of free fatty acids from adipocyte triglyceride depots.

RBP4 secreted from adipocytes has been found to be elevated in overweight/obese individuals, although much of this appears not to be bound to retinol. That elevated apo-RBP4 levels may be related to the development of type 2 diabetes was suggested by the finding of increased gluconeogenic capacity in mice treated with recombinant RPB4. Studies with humans have found the plasma RPB4 level to be positively correlated with body mass index (BMI, kg/m^2) (Fig. 5.4), the degree of insulin resistance and impaired glucose tolerance in subjects with obesity or family histories of type 2 diabetes.[34] In contrast, plasma RBP4 levels have been found to be reduced by treatments that reduce insulin resistance, such as exercise training, or gastric bypass surgery, and positively correlated with the expression of p85[35] in adipose tissue. This evidence has made adipose RBP4 synthesis of interest as a target for antidiabetic therapies.

Other Vitamin A-Binding Proteins Involved in Vitamin A Transport

The transport, storage, and metabolism of the retinoids involve their binding to several other binding proteins, four of which (**cellular retinol-binding protein types I and II, CRBPs (I) and (II), and cellular retinoic acid-binding protein (CRABPs) types (I) and (II)**; see Table 5.6) have the same general tertiary structure as the lipocalins, a class of low molecular weight proteins whose members bind hydrophobic ligands (e.g., fatty acids, cholesterol, biliverdin). Proteins of this family have multi-stranded β-sheets that are folded to yield a deep hydrophobic pocket suitable for binding appropriate hydrophobic ligands. Each of the retinoid-binding proteins has approximately 135 amino acid residues with pair-wise sequence homologies of 40–74%. Each has been highly conserved (91–96% sequence homology among the human, rat, mouse, pig, and chick proteins). Their genes also show similarities: each contains four exons and three introns, the latter being positioned identically. Therefore, it appears that these proteins share a common ancestral gene. The very close separation of the genes for CRBPs (I) and (II) (only 3 centimorgans) suggests that this pair resulted from the duplication of one of these two genes. They do, however, show very different tissue distributions: CRBP(I) is expressed in most fetal and adult tissues, particularly those of the liver, kidney, lung, choroid plexus, and pigment epithelium; CRBP(II) is expressed only in mature enterocytes in the villi of the mucosal epithelium (especially the jejunum), and in the fetal and neonatal liver.

Other retinoid-binding proteins, which are larger and not members of this family, have been characterized: **cellular retinal-binding protein (CRALBP)** and **interphotoreceptor retinol-binding protein (IRBP)**. These are classified in the group of intracellular lipid-binding proteins that include the fatty acid-binding proteins. Like the lipocalins, they also bind their lipophilic ligands with an antiparallel, β-barrel structure. However, unlike those other proteins, they bind vitamin A in the reverse orientation – with its polar group buried internally and the β-ionone ring close to the surface. The IRBPs are unusual in that they can bind three retinol molecules (all other retinoid-binding proteins bind only a single ligand molecule) as well as two long-chain fatty acid molecules.

34. Graham, T. E., Yang, Q., Blűher, M., *et al.* (2006). *New Eng. J. Med.* 354, 2552–2563; Broch, M., Vendrell, J., Ricart, W., *et al.* (2007) *Diabetes Care* 30, 1802–1806; Chavez, A. O., Coletta, D. K., Kamath, S., *et al.* (2008). *Am. J. Physiol. Endocrinol. Metab.* 296, E768–E764; Chavez, A. O., Coletta, D. K., Kamath, S., *et al.* (2009) *Am. J. Physiol. Endocrinol. Metab.* 296, E768–E764; Kelly, K. R., Kashyap, S. R., O'Leary, V. B., *et al.* (2009). *Obesity* 18, 663–666.

35. p85 is a regulatory subunit of a protein involved in insulin signaling, phosphoinositol 3-kinase (PI3K). PI3K is activated by phosphorylation by insulin receptor substrate-1.

TABLE 5.6 Vitamin A-Binding Proteins

Binding Protein		Molecular Weight			
	Abbreviation	kDa	Ligand	nM	Location
Retinol-BP	RBP4	21	all-*trans*-Retinol	20	Plasma
Cellular retinol-BP, type I	CRBP (I)	15.7	all-*trans*-Retinol	<10	Cytosol of most tissues except heart, adrenal, and ileum
			all-*trans*-Retinal	50	
Cellular retinol-BP, type II	CRBP(II)	15.6	all-*trans*-Retinol	—	Cytosol of enterocytes, fetal, and neonatal liver
			all-*trans*-Retinal	90	
Cellular retinal-BP	CRALBP	36	11-*cis*-Retinol		Cytosol of retina
			11-*cis*-Retinal	15	
Cellular retinoic acid-BP, type I	CRABP(I)	15.5	all-*trans*-Retinoic acid	0.1	Cytosol of most tissues except liver, jejunum, ileum
Cellular retinoic acid-BP, type II	CRABP(II)	15	all-*trans*-Retinoic acid	0.1	Cytosol of embryonic limb bud
Epididymal retinoic acid-BP	—	18.5	all-*trans*-Retinoic acid		Lumen of epididymis
Uterine retinol-BP	—	22	all-*trans*-Retinol		Cytosol of uterus (sow)
Interphotoreceptor retinol-BP	IRBP	140	all-*trans*-Retinol and	50–100	Interphotoreceptor space
			11-*cis*-Retinol	50–100	
Retinol pigment epithelium proteinRPE65	65	65	all-trans-retinyl esters		Retinal pigment epithelium
Nuclear retinoic acid receptor-α	RARα	50	all-*trans*-Retinoic acid		Nuclei of most tissues except adult liver
Nuclear retinoic acid receptor-α	RARβ	50	all-*trans*-Retinoic acid		Nuclei of most tissues except adult liver
Nuclear retinoic acid receptor-γ	RARγ	50	all-*trans*-Retinoic acid		Nuclei of most tissues except adult liver
Nuclear retinoid X receptor-γ	RARγ	50	all-*trans*-Retinoic acid		Nuclei of most tissues except adult liver
Rhodopsin	—	41	11-*cis*-Retinal		Cytosol of retina
Melanopsin	—		11-*cis*-Retinal		Cytosol of retina, brain, skin[a]

[a]*In* Xenopus *sp.*

It is clear that tissue levels of the mRNAs for the CRBPs are influenced by nutritional vitamin A status. Both CRBP(I) protein and mRNA are reduced by deprivation of the vitamin; however, CRBP(II) protein and mRNA levels are increased by vitamin A deficiency. The CRBP gene appears to be inducible by retinoic acid, and a number of **retinoic acid response elements** have been identified in both the CRBP(I) and CRBP(II) promotors. Apparently discrepant results have been obtained regarding the retinoic acid inducibility of the CRBP(II) gene. Whether other transcription factors also bind to those elements remains to be learned, yet it appears that these genes are responsive to other hormones, including glucocorticoids and 1,25-dihydroxyvitamin D_3, which have been shown to have negative effects on CRBP(I) and CRBP(II), respectively.

Cellular Uptake of Retinol

Due to their hydrophobic character the plasma membranes do not present a barrier to retinol uptake, and thus retinol can enter target cells by non-specific partitioning into the plasma membrane from holo-RBP4. Nevertheless, most of the vitamin appears to enter cells through specific holo-RBP4–TTR receptor-mediated mechanisms.[36] The retinol ligand of holo-RBP4–TTR is taken into cells via binding of the complex to a specific cell surface receptor recently identified as a transmembrane protein STRA6. STRA6 is expressed in all tissues *except* liver, which clears more vitamin A from chylomicrons than do extrahepatic tissues (notable exceptions being the mammary gland and bone marrow). Release of retinol to the target cell increases the negative charge of the resulting apo-RBP4; this reduces its affinity for TTR, which is subsequently lost. The residual apo-RBP4 can then be filtered by the kidney, where it is degraded. Thus, plasma apo-RBP4 levels are normally low, but can be elevated (by about 50%) under conditions of acute renal failure. Studies have shown that apo-RBP4 can be recycled to the holo form; injections of apo-RBP4 into rats produced marked (70–164%) elevations in serum retinol levels. It is thought, therefore, that circulating apo-RBP4 may be a positive feedback signal from peripheral tissues for the hepatic release of retinol, the extent of which response is dependent on the size of hepatic vitamin A stores.

On entry into the target cell, retinol combines with CRBP. Because CRBP(I) is present at high levels in cells that synthesize and secrete RBP4, it has been suggested that it may interact at specific sites to effect the transfer of retinol to RBP4 for release to the general circulation. The synthesis and/or the retinol-binding affinity of CRBP(I) may be affected by thyroid and growth hormones; both hormones promote the cellular uptake of retinol. Cells of many tissues also contain other retinoid-binding proteins that are thought to be involved in the transport of the hydrophobic retinoids within and between cells, and to effect presentation of their retinoid ligands to enzymes and to the nuclear receptors. These include CRABPs (I) and (II), and CRALBP,[37] which bind the dominant, hormonally active form retinoic acid, and several nuclear **retinoic acid receptors** (the **RAR** and **RXR** proteins).[38] Other vitamin A-binding proteins, with narrow tissue distributions, have been identified: IRBP in the retina, retinoic acid-binding proteins of the lumen of the epididymis, and a retinol-binding protein in the uterus.

Transport Roles of Vitamin A-Binding Proteins

The intracellular vitamin A-binding proteins appear to be important in the cellular uptake and the intracellular and transcellular transport of various vitamin A metabolites. Both CRBP and CRABP have been shown to serve as carriers of their ligands (retinol and retinoic acid, respectively) from the cytoplasm into the nucleus, where the latter are transferred to the chromatin such that the binding proteins are released, possibly to return to the cytoplasm.

CRBP also appears to have more specialized transport functions in certain tissues. In the liver, CRBP concentrations increase with increasing retinyl ester contents, suggesting that the binding protein may function in the transport of retinol from parenchymal cells into the stellate cells, which store retinyl esters. Specific and rich localization of CRBP has been identified in endothelial cells of the brain microvasculature, in cuboidal cells of the choroid plexus, in the Sertoli cells of the testis, and in the pigment epithelium of the retina. Because these tight-junctioned cells also have surface receptors for the plasma holo-RBP4–TTR complex, it is thought that their abundant CRBP(I) concentrations are involved in the transport of retinal across the blood–brain, blood–testis, and retinal blood–pigment epithelium barriers. Studies with mice have shown that CRBP(I) is necessary for the hepatic uptake of retinol: CRBP(I)-null individuals exhausted their hepatic retinyl ester stores even when they were fed vitamin A.

CRBP(II) appears to be restricted largely to the enterocytes of the small intestine (particularly, the jejunum). Its abundance in mature enterocytes (where it comprises 1% of the total soluble protein) as well as the absence of CRBP(I) in these cells suggest that CRBP(II) is involved in enteric absorption of vitamin A, presumably by transporting it across the cell. Both CRBP(II), as well as a high-capacity esterase that esterifies CRBP(II)-bound retinol, have been identified in hepatic parenchymal cells of fetal and newborn rats. After birth, CRBP(II) appears to be replaced by CRBP(I), such that mature animals show none of the former binding protein in that organ. The presence of CRBP(II) in fetal liver corresponds to the increased concentration of retinyl palmitate in that organ at birth.

Some retinoid-binding proteins found extracellularly are believed to serve similar transport functions. These include two low molecular weight retinoic acid-binding proteins, generally related to RBP4, that are secreted into the lumen of the epididymis, where they are thought to participate in the delivery of all-*trans*-retinoic acid to sperm in that organ, which also contains a particularly abundant supply of CRBP(I).[39] Other retinol-binding proteins are synthesized in the uterine endometrium and secreted into the uterus; these show some sequence homology with RBP4, but are slightly larger. They are thought to be involved in the transport of retinol to the fetus.

---

36. The existence of a membrane-associated retinol transporter is still controversial; that retinol can move spontaneously between the two layers of artificial phospholipid bilayers would argue against the need for a specific transporter.
37. CRABP is related to the hepatic tocopherol binding protein.
38. See the section "Vitamin A Regulation of Gene Transcription" for a discussion of the retinoid receptors.

---

39. The initial segment of the epididymis contains the greatest concentration of CRBP found in any tissue.

The IRBP of the interphotoreceptor space is thought to function in the transport of retinol between the pigment epithelium and photoreceptor cells. It is synthesized by the latter, in which its mRNA has been detected. Unlike the other retinoid-binding proteins, the IRBP is a large (140kDa) glycoprotein; it can bind 6 moles of long-chain fatty acid in addition to 2 moles of retinol. It has been suggested that its relatively low affinity for retinol, in comparison with the other retinol-binding proteins, facilitates rapid, high-volume transport of that ligand along a series of IRBPs. That IRBP is involved in the visual process, perhaps by delivering the chromophore, is indicated by the finding that its binding specificity shifts from mainly 11-*cis*-retinol to mostly all-*trans*-retinol as eyes become more completely light-adapted.[40]

The two IRBP binding sites for retinoids have been shown, using fluorescence techniques, to be quite different: a strongly hydrophobic binding pocket, and a surface site that interacts with retinol via its polar head group. The protein shows higher affinities for all-*trans*-retinol and 11-*cis*-retinal than for other retinoids. Studies have shown that docosahexanoic acid (DHA) induces a rapid and specific release of 11-*cis*-retinal from the IRBP hydrophobic site, whereas palmitic acid is without effect. This finding suggests that DHA may function in the targeting of 11-*cis*-retinal to photoreceptor cells, the DHA concentrations of which are much greater than those of pigment epithelial cells.

A carotenoid-binding protein has been characterized in rat liver.[41] This protein binds β-carotene, and is distributed predominantly in the mitochondria and lysosomes. The partially purified *carotenoprotein* is very sensitive to bright light and temperatures above 4°C.

Retinol Recycling

The majority of retinol that leaves the plasma appears to be recycled, as plasma turnover rates have been found to exceed (by more than an order of magnitude) utilization rates. Thus, kinetic studies in rats have indicated that a retinol molecule recycles via RBP4 7–13 times before its irreversible utilization. Such data indicate that, in the rat, newly released retinol circulates in the plasma for 1–3.5 hours before leaving that compartment, and that it may take a week or more to recycle to the plasma. It is estimated that some 50% of plasma turnover in the rat is to the kidneys, 20% to the liver, and 30% to other tissues. It has been suggested that retinol leaves the plasma bound to RBP4. Although the source of RBP4 for retinol recycling is not established, it is worth noting that mRNA for RBP4 has been identified in many extrahepatic tissues, including kidney.

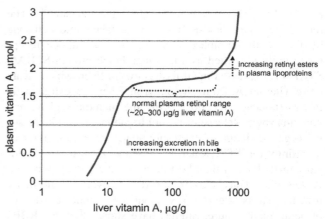

FIGURE 5.5 Regulation of plasma vitamin A levels. *After* Olson, J.A. (1984). *J. Natl Cancer Inst.* 73, 1439–1446.

Plasma Retinol Homeostasis

In healthy individuals, plasma retinol is maintained within a narrow range (40–50mg/ml in adults; typically about half that in newborn infants) in spite of widely varying intakes of vitamin/provitamin A (Fig. 5.5).[42] This control appears to be effected by several factors: regulation of CRBP(II) expression in stellate cells, regulation of enzymes that esterify retinol and hydrolyze retinyl esters, and other factors that may affect retinol release to the plasma and/or removal from it. The liver and kidneys appear to play important roles in these various processes. Indeed, renal dysfunction has been shown to increase plasma retinol levels; this has been suggested to involve a regulatory signal to the liver that alters the secretion of RBP4–retinol. Serum retinol levels can also be affected by nutrition status with respect to zinc, which is required for the hepatic synthesis of RBP4.

Plasma levels of carotenoids, in contrast, do not appear to be regulated; they reflect intake of carotenoid-rich foods. Careful studies have revealed, however, cyclic changes of up to nearly 30% in the plasma β-carotene concentrations during the menstrual cycles of women. Whether these fluctuations are physiologically meaningful[43] or whether they relate to fluctuations in plasma lipids is not clear.

Vitamin A in the Eye

Vitamin A taken up by the retinal pigment epithelium is transported to rod cells by IRBP for the discharge of the

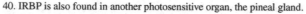

40. IRBP is also found in another photosensitive organ, the pineal gland.
41. Others have been identified in crustaceans, cyanobacteria, and carrots.

42. In contrast, plasma levels of all-*trans*-retinoic acid and 4-oxo-retinoic acid can respond to the level of ingested vitamin A.
43. It has been suggested that the oxidative enzymatic reactions involved in the synthesis of progesterone from cholesterol by the corpus luteum at mid-cycle may constitute an oxidative stress calling for protection by lipid-soluble antioxidants including β-carotene.

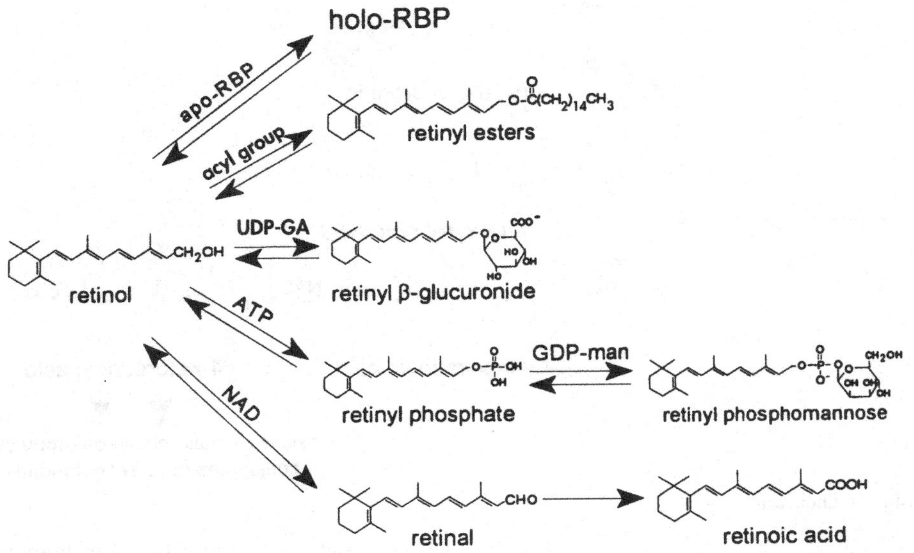

FIGURE 5.6 Metabolic fates of retinol.

visual function of the vitamin (see "Vitamin A in Vision," later in this chapter). RBP4 is also expressed in the lacrimal glands; retinol appears to reach the cornea via holo-RBP4 secreted in tears.

Milk Retinol

Retinol is transferred from mother to infant through milk. The retinol A content of milk is a function of two factors: the stage of lactation, and the vitamin A status of the mother. Human breast milk from well nourished, vitamin A-adequate mothers typically drops from *ca.* 5–7 μmol/l in colostrum, to *ca.* 3–5 μmol/l in transitional milk, to 1.4–2.6 μmol/l in mature milk. These levels are enough to meet the infant's immediate metabolic needs while also supporting the development of adequate vitamin A stores.[44] Such an infant will consume, over the first 6 months of life, nearly 60 times as much vitamin A from breast milk (*ca.* 300 μmoles) than it accumulated throughout gestation. Vitamin A-deficient mothers, however, produce breast milk that is low in the vitamin; in vitamin A-deficient areas of the world, levels average *ca.* 1 μmol/l (levels <1.05 μmol/l are considered indicative of maternal vitamin A deficiency[45]). This level appears to be sufficient to meet an infant's immediate metabolic requirements, as breastfed infants of vitamin A-deficient mothers are largely protected from xerophthalmia; however, higher levels (at least 1.75 μmol/l) are required to support adequate vitamin A stores to protect against the development of xerophthalmia during weaning. For this reason, vitamin A supplementation of mothers in vitamin A-deficient areas is regarded as a prudent public health strategy. A meta-analysis of randomized controlled trials showed that such measures have reduced infant mortality by 23% in children under 5 years of age in populations at risk of vitamin A deficiency.[46] The success of postpartum maternal supplementation depends on the prevailing breast-feeding practices, with the simultaneous promotion of optimal practices (including the feeding of colostrum) being highly effective in improving infant vitamin A status. For example, the administration of 60,000 RE to a low-vitamin A mother can produce a 29% increase in the retinol contents of her breast milk over 6 months.

5. METABOLISM OF VITAMIN A

Metabolic Fates of Retinol

The metabolism of vitamin A (Fig. 5.6) centers on the transport form, retinol, and the various routes of conversion available to it: *esterification*, *conjugation*, *oxidation*, and *isomerization*.

Esterification

As discussed above, retinol is esterified in the cells of the intestine and most other tissues via enzymes of the endoplasmic reticulum, which use acyl groups from either phosphatidylcholine (lecithin–retinol acyltransferase, LRAT) or acylated coenzyme A (acyl-CoA:retinol acyltransferase, ARAT). These systems show marked specificities for saturated fatty acids, in particular palmitic acid; thus, the most abundant product is retinyl palmitate.

44. The normal weight (*ca.* 3.2 kg) infant of a well nourished, vitamin A-adequate mother is born with hepatic vitamin A stores of *ca.* 5 μmol/l.
45. Stoltzfus, R. J. and Underwood, B. A. (1995). *Bull WHO* 73, 703–711.
46. Beaton, G. H., Mortorell, R., Aronson, K. J., *et al.* (1993). Nutrition Policy Discussion Paper No. 13, UN Administrative Committee of Coordination – Subcommittee on Nutrition, New York, NY, 120 pp.

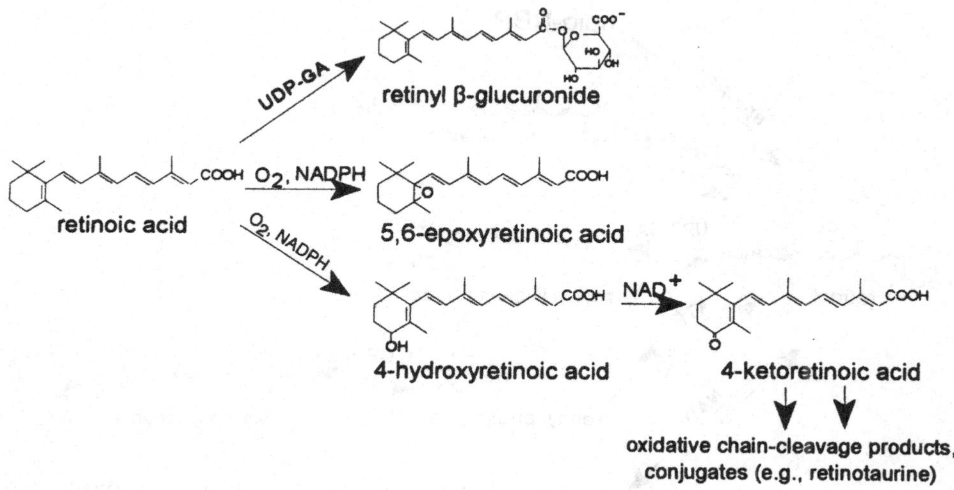

FIGURE 5.7   Catabolism of retinoic acid.

Conjugation

Retinol may also be conjugated in either of two ways. The first entails the reaction catalyzed by retinol-UDP-glucuronidase, present in the liver and probably other tissues, which yields **retinyl β-glucuronide**, a metabolite that is excreted in the bile.[47] The second path of conjugation involves ATP-dependent phosphorylation to yield **retinyl phosphate** catalyzed by **retinol phosphorylase**. That product, in the presence of guanosine diphosphomannose (GDP-man), can be converted to the glycoside retinyl phosphomannose, which can transfer its sugar moiety to glycoprotein receptors. However, because only a small amount of retinol appears to undergo phosphorylation *in vivo*, the physiological significance of this pathway is not clear.

Oxidation

Retinol can also be reversibly oxidized to retinal by NADH- or NADPH-dependent **retinol dehydrogenases** that are also dependent on zinc. These cytosolic and microsomal activities are found in many tissues, the greatest being in the testis.[48] A short-chain alcohol dehydrogenase has been described that can oxidize 9-*cis*- and 11-*cis*-retinol to the corresponding aldehyde. This activity has been identified in several tissues, including the retinal pigment epithelium, liver, mammary gland, and kidney. That 9-*cis*-retinol can be converted to 9-cis-retinoic acid is evidenced by the finding of a 9-*cis*-retinol dehydrogenase. The enzyme in both humans and mice is inhibited by 13-*cis*-retinoic acid

at levels similar to those found in human plasma, suggesting that 13-*cis*-retinoic acid may play a role in the regulation of retinoid metabolism. Retinal can be irreversibly oxidized by **retinal oxidase** to retinoic acid. Because retinoic acid is the active ligand for the nuclear retinoid receptors, it is very likely that this metabolism is tightly regulated.[49] The rate of that reaction is several-fold greater than that of retinol dehydrogenase; that, plus the fact that the rate of reduction of retinal back to retinol is also relatively great, results in retinal being present at very low concentrations in tissues. The increasing analytical sensitivity afforded by developments in high-performance liquid—liquid partition chromatography and mass spectrometry has resulted in the identification of an increasing number (approaching that of steroid hormones) of retinoic acid isomers in the plasma of various species;[50] the number of these with physiological significance is presently unclear.

Fates of Retinoic Acid

All-*trans*-retinoic acid is converted to forms that can be readily excreted (Fig. 5.7). It may be directly conjugated by glucuronidation in the intestine, liver, and possibly other tissues to retinyl β-glucuronide. Alternatively, it can be catabolized by several further oxidized products, including the 4-hydrox-, 4-oxo-, 4-dihydroxy-, 18-hydroxy-, 3-hydroxy-, and 5,6-epoxy metabolites. Several cytochrome *P*-450 enzymes have

---

47. About 30% of the retinyl β-glucuronide excreted in the bile is reabsorbed from the intestine and is recycled in an enterohepatic circulation back to the liver. For this reason, it was originally thought to be an excretory form of the vitamin. That view has changed with the findings that retinyl β-glucuronide is biologically active in supporting growth and tissue differentiation, and that it is formed in many extrahepatic tissues.

48. Male rats fed retinoic acid instead of retinol become aspermatogenic and experience testicular atrophy. It has been proposed that retinoic acid is required for spermatogenesis, but cannot cross the blood–testis barrier; this is supported by the fact that the rat testis is also rich in CRABP.

49. Two microsomal proteins that catalyze this reaction have been isolated from rat liver; one has been shown to cross-link with holo (but not apo)-CRBP in the presence of NADP.

50. In addition to all-*trans*-, 13-*cis*-, and 9-*cis*-retinoic acid, this number includes the 9,13-*dicis*-, 4-hydroxy-, 4-oxo-, 18-hydroxy-, 3,4-dihydroxy-, and 5,6-epoxy isomers, as well as such derivatives as retinotaurine.

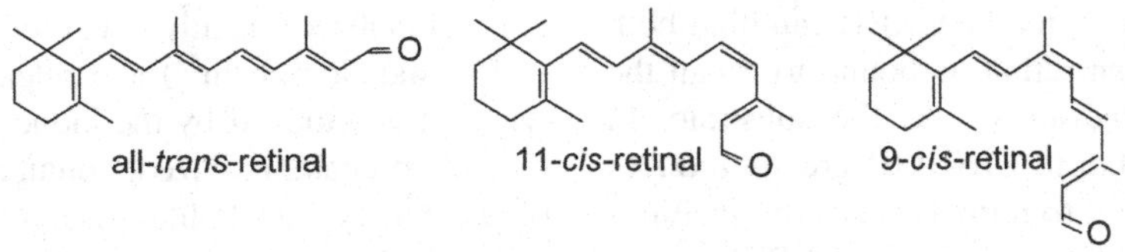

FIGURE 5.8 Structures of the all-*trans*, 11-*cis* and 9-*cis* isomers of retinal.

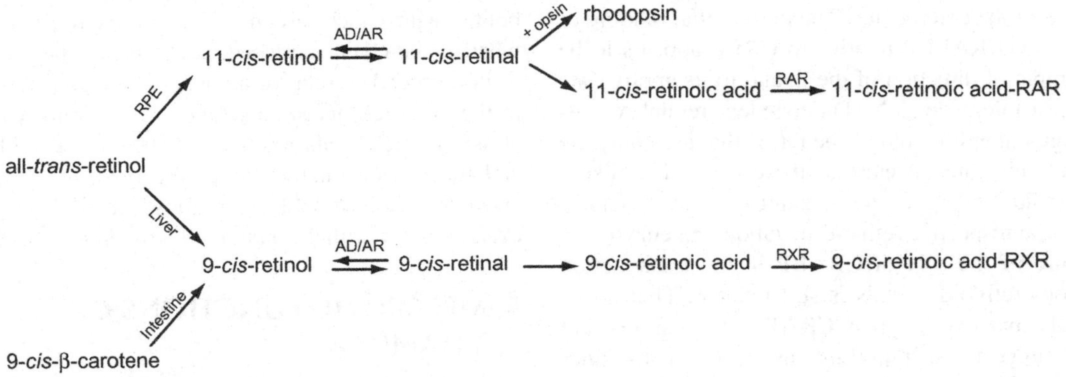

FIGURE 5.9 Roles of binding proteins in vitamin A metabolism.

been implicated in this metabolism. These include different families: CYP2, CYP3, CYP4, and CYP26. Of these, three in the CYP26 family appear to be most important. These are mono-oxygenases that convert retinoic acid to the 4-hydroxy- (CYP26A), 4-oxo- (CYP26B), and 18-hydroxy- (CYP26C) metabolites. Their pattern of expression varies according to tissue and stage of development. The expression of CYP26A is known to be upregulated by dietary vitamin A and retinoic acid, and downregulated by vitamin A depletion.

The oxidative chain-cleavage metabolites are conjugated with glucuronic acid, taurine[51] or other polar molecules; these, being more polar, can readily be excreted. Glucuronides comprise a significant portion of the retinoids excreted in the bile. That retinoyl-β-glucuronide has been found to show some vitamin A activity suggests that there may be some hydrolysis to yield retinoic acid.

It should be noted that both the production and catabolism of all-*trans*-retinoic acid involve unidirectional processes. Whereas the reduced forms of vitamin A (retinol, retinyl esters, retinal) can be converted *in vivo* to retinoic acid, the latter *cannot* be converted to any of the reduced forms.

Isomerization

Interconversion of the most common all-*trans* forms of vitamin A and various *cis* forms occurs in the eye, and is a key aspect of the visual function of the vitamin, as the conformational change caused by the isomerization alters the

binding affinity of retinal for the visual pigment protein *opsin*. In the eye, light induces the conversion of **11-*cis*-retinal** to **all-*trans*-retinal** (Fig. 5.8); the conversion back to the 11-*cis* form is catalyzed by the enzyme **retinal isomerase**, which also catalyzes the analogous isomerization (in both directions) of 11-*cis*- and all-*trans*-retinol. Some isomers (e.g., the 13-*cis* form) tend to be isomerized to the all-*trans* form more rapidly than others. The conversion of all-*trans*-retinoic acid to 9-*cis*-retinoic acid has also been demonstrated.

Role of Retinoid-Binding Proteins in Modulating Vitamin A Metabolism

In addition to serving as reserves of retinoids, the cellular retinoid-binding proteins also serve to modulate vitamin A metabolism, apparently both by holding the retinoids in ways that render them inaccessible to the oxidizing environment of the cell, and by channeling the retinoids via protein–protein interactions among its enzymes (Fig. 5.9; Table 5.7). Both CRBP and CRBP(II) function in directing the metabolism to their bound retinoid ligands by shielding them from some enzymes that would use the free retinoid substrate, and by making them accessible to other enzymes important in metabolism. For example, the esterification of retinol by LRAT occurs while the substrate is bound to CRBP or CRBP(II). The abundance of CRBP in the liver and its high affinity for retinol suggest that its presence directs the esterification of the retinoid ligand to the reaction catalyzed by LRAT, rather than that catalyzed by ARAT, which can use only free retinol. This also appears to be the

---

51. Significant amounts of retinotaurine are excreted in the bile.

case for CRBP(II). Although it, unlike CRBP, can bind both retinol and retinal, only when retinal is bound to it can the reducing enzyme **retinal reductase** use the substrate. In addition, the binding of retinol to CRBP(II) greatly reduces the reverse reaction (oxidation to retinal). Thus, by facilitating retinal formation and inhibiting its loss, CRBP(II) seems to direct the retinoid to the appropriate enzymes, which sequentially convert retinal to the esterified form in which it is exported from the enterocytes. The preferential binding of 11-*cis*-retinal by CRALBP relative to CRBP appears to be another example of direction of the ligand to its appropriate enzyme, i.e., a microsomal NAD-dependent retinal reductase in the pigment epithelium of the retina that uses only the carrier-bound substrate. In each of these cases, it is likely that the retinoid-binding protein–ligand complex interacts directly with the respective retinoid-metabolizing enzyme.

The nature of the role of CRABP in the transduction of endogenous retinoid signals is still unclear. That transgenic animals that over-express CRABP show significant pathology[52] suggests an important function. On the other hand, mice in which expression of the gene is knocked out have been found to have normal phenotypes; importantly, they show normal susceptibility to the teratogenic effects of high doses of retinoic acid.

Excretion of Vitamin A

Vitamin A is excreted in various forms in both the urine and the feces. Under normal physiological conditions, the efficiency of enteric absorption of vitamin A is high (80–95%), with 30–60% of the absorbed amount being deposited in esterified form in the liver. The balance of absorbed vitamin

A is catabolized (mainly at C-4 of the ring and at C-15 at the end of the side chain[53]) and released in the bile or plasma, where it is removed by the kidney and excreted in the urine (i.e., short-chain, oxidized, conjugated products). About 30% of the biliary metabolites (i.e., retinoyl β-glucuronides) are reabsorbed from the intestine into the enterohepatic circulation back to the liver, but most are excreted in the feces with unabsorbed dietary vitamin A. In general, vitamin A metabolites with intact carbon chains are excreted in the feces, whereas the chain-shortened, acidic metabolites are excreted in the urine. The relative amounts of vitamin A metabolites in the urine and feces thus vary with vitamin A intake (i.e., at high intakes fecal excretion may be twice that of the urine) and the hepatic vitamin A reserve (i.e., when reserves are above the low–normal level of 20μg/g, both urinary and fecal excretion vary with the amount of vitamin A in the liver).

6. METABOLIC FUNCTIONS OF VITAMIN A

Feeding provitamin A carotenoids retinyl esters, retinol, and retinal can support the maintenance of healthy epithelial cell differentiation, normal reproductive performance, and visual function (Table 5.8). Each of these forms can be metabolized to retinol, retinal or retinoic acid; however, unlike retinol and retinal, retinoic acid cannot be reduced to retinal or retinol. Feeding retinoic acid can support only the *systemic* functions of vitamin A, e.g., epithelial cell differentiation. These observations and knowledge of retinoid metabolism led to the conclusion that whereas retinal discharges the visual functions, retinoic acid (and specifically **all-*trans*-retinoic acid**) must support the systemic functions of the vitamin.

Vitamin A in Vision

The best elucidated function of vitamin A is in the visual process, where, as 11-*cis*-retinal, it serves as the

TABLE 5.7 Apparent Metabolic Functions of the Retinoid-Binding Proteins

Binding Protein	Apparent Function(s)
CRBPs (I)	Directs retinol to LRAT and oxidative enzymes; regulates retinyl ester hydrolase
CRBP(II)	Directs retinol to LRAT and oxidative enzymes
CRABPs (I) and (II)	Directs retinoic acid to catabolizing enzymes; regulates free retinoic acid concentrations
CRALBP	Regulates enzymatic reactions of the visual cycle

TABLE 5.8 Known Functional Forms of Vitamin A

Active form	Function
Retinol	Transport, reproduction (mammals)
Retinyl esters	Storage
Retinal	Vision
Retinoic acid	Epithelial differentiation, gene transcription, reproduction

52. For example, transgenic mice that expressed CRABP under the influence of lens-specific αA-crystallin promoter developed cataracts and, later, pancreatic endocrine tumors.

53. The chain-terminal carbon atoms (C-14 and C-15) can be oxidized to CO_2; retinoic acid is oxidized to CO_2 to a somewhat greater extent than retinol.

photosensitive chromophoric group of the visual pigments of rod and cone cells of the retina. Rod cells contain the pigment **rhodopsin**; cone cells contain one of three possible **iodopsins**. In each case, 11-*cis*-retinal is bound (via formation of a Schiff base) to a specific lysyl residue of the respective apo-protein (collectively referred to as **opsins**) (*see* Fig. 5.10).

The visual functions of rhodopsin and the iodopsins differ only with respect to their properties of light absorbency,[54] which are conferred by the different opsins involved. In each, photoreception is effected by the rapid, light-induced isomerization of 11-*cis*-retinal to the all-*trans* form. That product, present as a protonated Schiff base of a specific lysyl residue of the protein, produces a highly strained conformation. This results in the dissociation of the retinoid from the opsin complex. This process (**bleaching**) is a complex series of reactions, involving progression of the pigment through a series of

unstable intermediates of differing conformations[55] and, ultimately, to *N*-retinylidene opsin, which dissociates to all-*trans*-retinal and opsin (*see* Fig. 5.11).

The dissociation of all-*trans*-retinal and opsin is coupled to nervous stimulation of the vision centers of the brain. The bleaching of rhodopsin causes the closing of Na$^+$ channels in the rod outer segment, thus leading to hyperpolarization of the membrane. This change in membrane potential is transmitted as a nervous impulse along the optic neurons. This response appears to be stimulated by the reaction of an unstable "activated" form of rhodopsin, **meta-rhodopsin II**, which reacts with **transducin**, a membrane-bound G protein of the rod outer segment disks. This results in the binding of the transducin α subunit with **cGMP phosphodiesterase**, which activates the latter to catalyze the hydrolysis of cGMP to GMP. Because cGMP maintains Na$^+$ channels of the rod plasma membrane in the open state, the resulting decrease in its concentration causes a marked reduction in Na$^+$ influx. This results in hyperpolarization of the membrane and the generation of a nerve impulse through the synaptic terminal of the rod cell.

The visual process is a cyclic one, in that its constituents are regenerated. All-*trans*-retinal can be converted enzymatically in the dark back to the 11-*cis* form. After bleaching, all-*trans*-retinal is rapidly reduced to all-*trans*-retinol, in the rod outer segment. The latter is then transferred (presumably via IRBP) into the retinal pigment epithelial cells, where it is esterified (again, predominantly with palmitic acid) and stored in the bulk lipid of those cells. The regeneration of rhodopsin, which occurs in the dark-adapted eye, involves the simultaneous hydrolysis and isomerization of

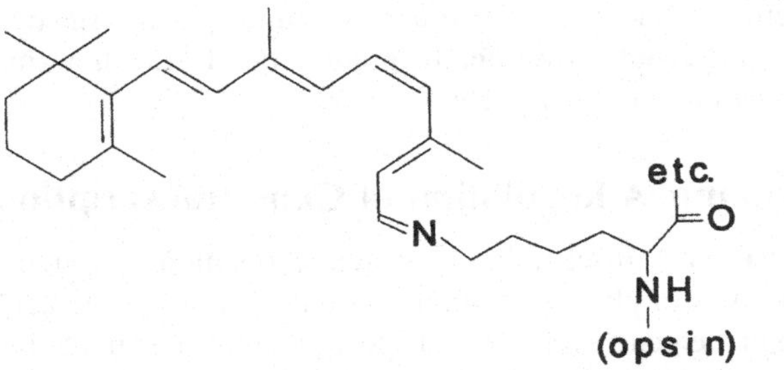

FIGURE 5.10 11-*cis*-Retinal binds to photopigment proteins via a lysyl linkage.

FIGURE 5.11 Vitamin A in the visual cycle. Rh, rhodopsin; T, transducin; PDE-I, phosphodiesterase (inactive); PDE-A, phosphodiesterase (active).

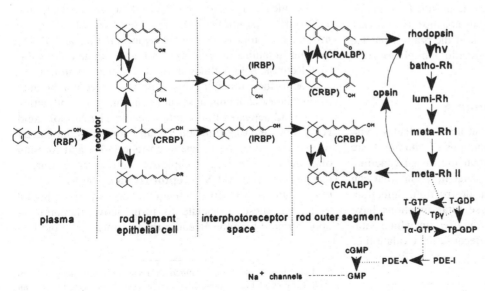

54. The absorbance maxima of the pigments from the human retina are as follows: rhodopsin (rods), 498 nm; iodopsin (blue cones), 420 nm; iodopsin (green cones), 534 nm; iodopsin (red cones), 563 nm.

55. The conformation of rhodopsin is changed to yield a transient photopigment, *bathorhodopsin*, which, in turn, is converted sequentially to *lumirhodopsin*, *metarhodopsin I*, and (by deprotonation) *metarhodopsin II*.

retinyl esters to yield 11-*cis*-retinol and then 11-*cis*-retinal, which is transferred into the rod outer segment via IRBP. Studies have revealed that a protein, RPE65, has a key role in the *trans-cis* isomerization in the retinal pigment epithelium. That protein not only binds and stabilizes all-*trans*-retinyl esters, but also extracts from the retinal pigment epithelium endoplasmic reticular membrane for delivery to the isomerizing enzyme isomerohydrolase. Its activity appears to be regulated through the addition/release of palmitic acid: palmitoylation of the protein converts the protein from a form soluble in the cytosol to a membrane-bound form, thus controlling its capacity to present its retinoid ligand to cytosolic isomerohydrolase. Nervous recovery is effected by the GTPase activity of the transducin α subunit, which, by hydrolyzing GTP to GDP, causes the reassociation of transducin subunits, and hence the loss of its activating effect on cGMP phosphodiesterase. Metarhodopsin II is also removed by phosphorylation to a form incapable of activating transducin, and by dissociation to yield opsin and all-*trans*-retinal.

The visual cycle of cones[56] differs from that of rods. Cones have much lower (100-fold) light sensitivity but faster (10-fold) recovery rates than rods. In cones, the oxidation of 11-*cis*-retinol to 11-*cis*-retinal is NADP-dependent, and the isomerization of all-*trans* to 11-*cis*-retinal occurs in a two-step process apparently in Müller cells. In rods, the oxidation step is NAD-dependent and the isomerization is a one-step process in the retinal pigment epithelium.

The vitamer 11-*cis*-retinal can also bind another rhodopsin-like protein in the eye, melanopsin. The photosensitive pigment undergoes light-induced isomerization of 11-*cis*-retinal to 11-*trans*-retinal, but, unlike rhodopsin, it does not dissociate from the prosthetic group. That melanopsin is found in the inner retina but not in rods or cones, and in the site of the circadian clock of the brain,[57] has led to the suggestion that the protein functions in photoperiod regulation.

Systemic Functions of Vitamin A

Vitamin A-deficient animals die, but not from lack of visual pigments. The extra-retinal functions of vitamin A are of greater physiological impact than the visual function. Deprivation of vitamin A impairs vital functions that can be life-threatening (e.g., corneal destruction, infection, stunted growth). Collectively, these vital functions have been referred to as the *systemic functions* in which vitamin A acts much like a hormone. Because the oxidation of

retinal to retinoic acid is irreversible, retinoic acid can support only the systemic functions. Animals fed diets containing retinoic acid as the sole source of vitamin A grow normally and appear healthy in every way except that they go blind.

Chief among the systemic functions of vitamin A is its role in the differentiation of epithelial cells. It is well documented that vitamin A-deficient individuals (humans or animals) experience replacement of normal mucus-secreting cells by cells that produce keratin, particularly in the conjunctiva and cornea of the eye, the trachea, the skin, and other ectodermal tissues. Less severe effects are also produced in tissues of mesodermal or endodermal origin. It appears that retinoids affect cell differentiation through actions analogous to those of the steroid hormones – i.e., they bind to the nuclear chromatin to signal transcriptional processes. In fact, studies have revealed that the differentiation of cultured cells can be stimulated by exposure to retinoids, and that abnormal mRNA species are produced by cells cultured in vitamin A-deficient media.[58] Further, retinoic acid has been found to stimulate, synergistically with **thyroid hormone**, the production of growth hormone in cultured pituitary cells.

Vitamin A Regulation of Gene Transcription

Vitamin A discharges its systemic functions through the abilities of all-*trans*-retinoic acid and 9-*cis*-retinoic acid to regulate gene expression at specific target sites in the body. The hormone-like regulation of transcription by retinoids is receptor-mediated. Retinoic acid binds to two members of a highly conserved superfamily of proteins that act as nuclear receptors for steroid hormones including 1,25-$(OH)_2$-vitamin D_3 and thyroid hormone (T_3).[59] These nuclear receptors have similar ligand-binding and DNA-binding domains, as well as substantial sequence homology. Retinoic acid is thought to interact with them in ways similar to their other ligands, with each receptor binding to regulatory elements upstream from the gene and acting as a ligand-activated transcription factor. The **retinoic acid receptors** (**RAR**s) function by attracting low molecular weight co-activators, releasing co-repressors, and forming obligate heterodimeric complexes with another retinoid receptor, the **retinoid X receptor** (**RXR**). The RXRs form transcriptionally inactive homotetramers when not bound to ligand; these dissociate to form active homodimers upon retinoid binding. Because the tetramers can bind two

---

56. Cones comprise only 5% of all photoreceptors in the human eye, although they are more numerous in other species. For example, cones comprise 60% of the photoreceptors in the chicken eye.
57. The suprachiasmatic nucleus.

58. Epidermal keratinocytes cultured in a vitamin A-deficient medium made keratins of higher molecular weight than those made by vitamin A-treated controls; this shift toward larger keratins was corrected by treatment with vitamin A. Different mRNA species were identified, which encoded the different proteins produced under each condition.
59. T_3, triiodothyronine.

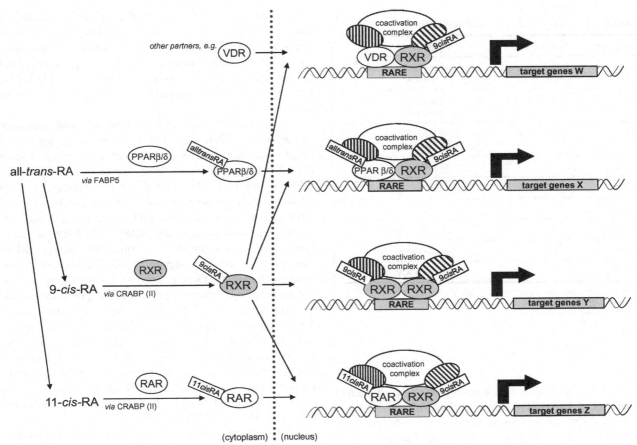

FIGURE 5.12 Participation of retinoid receptors and other nuclear receptors in integrating metabolic signaling in controlling expression of multiple genes. Retinoid activation of receptors that participate in complexes that bind to cognate DNA sequences to cause gene expression. The involvement of retinoid receptors (RARs and RXRs) with multiple binding partners allows retinoids to signal transcription of multiple pathways (*see text* for abbreviations.)

DNA recognition sequences simultaneously, the retinoid-induced shift to the liganded dimer results in a change in DNA geometry. The involvement of RXRs with multiple binding partners, including peroxisome proliferator-activated receptors (PPARs), the vitamin D receptor (VDR), and farnesoid X receptors (FXR), makes them the master regulators of multiple pathways signaling transcription in response to lipophilic nutrients and hormones (Fig. 5.12). The regulation of gene expression also involves co-activators and co-repressors that act by modifying chromatin.

All isomers of retinoic acid bind the RARs, although the greatest affinities *in vitro* are shown by all-*trans*-retinoic acid (K_d = 1–5 nmol/l). Three RAR subtypes have been identified (α, β, and γ). That only all-*trans*-retinoic acid functions as the endogenous ligand for RAR is suggested by the demonstration that growth arrest in mice lacking retinaldehyde dehydrogenase, thus incapable of producing all-*trans*-retinoic acid, can be rescued by the all-*trans*- but not the 9-*cis*- isomer.

The RARs resemble thyroid hormone receptors; the expression of RARs varies distinctly during development

(Table 5.9). The ligand-binding domains of the RARs are highly conserved (showing 75% identity in terms of amino acid residues). That $RAR\alpha_1$ and $RAR\alpha_2$ share 7 of their 11 exons suggests that they have arisen from a common ancestral RAR gene. Different promoters, however, direct the expression of each in an unusual organization involving the 5′-untranslated region of genes divided among different axons. In the case of $RAR\alpha_1$, the 5′ region is encoded in three axons: two contain most of the untranslated region; the third encodes the remainder of that region plus the first 61 amino acids that are peculiar to $RAR\alpha_1$.

Both the RARs and RXRs can bind **9-*cis*-retinoic acid**[60] *in vitro* with high affinity (K_d = 10 nmol/liter); however, it is not clear that this isomer is of physiological consequence, as this isomer has not been identified

---

60. Yet to be identified *in vivo* and originally questioned as to whether it was a physiological metabolite, 9-*cis*-retinoic acid is now known to be produced from 9-*cis*-retinol (by 9-*cis*-retinol dehydrogenase with subsequent oxidation, by cleavage of 9-*cis*-β-carotene, and from the isomerization of all-*trans*-retinol in the lung.

TABLE 5.9 Nuclear Retinoic Acid Receptors

Receptor	Isoforms[a]	Ligands
RARα	RARα$_1$, RARα$_2$	all-trans-RA[b], 9-cis-RA[c], 13-cis-RA[d]
RARβ	RARβ$_1$, RARβ$_2$, RARβ[e]	all-trans-RA, 9-cis-RA[c], 13-cis-RA[d]
RARγ	RARγ$_1$, RARγ$_2$[c]	all-trans-RA, 9-cis-RA[c], 13-cis-RA[d]
RXRα		9-cis-RA
RXRβ		9-cis-RA
RXRγ		9-cis-RA
PPARβ/δ		all-trans-RA[c]
PPARα		all-trans-RA[c]
PPARγ		all-trans-RA[c]

[a]Isoforms differ only in their N-terminal regions.
[b]Retinoic acid.
[c]Binding shown only in vitro.
[d]Binding (weak) shown only in vitro.
[e]Identified in Xenopus laevis.

in vivo. Retinaldehyde has also been found to bind weakly to RXR (and PPARγ) to inhibit activation (and is linked to inhibition of adipogenesis). Three RXR subtypes have been identified (α, β, and γ); RXRα responds to somewhat higher retinoic acid levels than other RXR isoforms. The RXRs show only weak homology with the RARs, the highest degree of homology (61%) being in their DNA-binding domains. On the basis of homologies with an insect locus, it is thought that the RXRs may have evolved as the original retinoid-signaling system. Both RARs and RXRs are found in most tissues.[61]

Recent studies have shown that all-*trans*-retinol can also bind **peroxisome proliferation-activated receptors** (**PPARs**), PPARβ/δ having the greatest affinity. The shuttling of the retinoid to PPARβ/δ is accomplished by the fatty acid binding protein 5 (FABP5), in contrast to CRABP(II) which delivers retinoids to other receptors. Therefore, the partitioning of retinoid between PPARβ/δ and the RARs and RXRs is a function of the relative amounts of FABP5 and CRABP(II) in cells. Cells in which CRABP(II) predominates express primarily through the RARs; cells with relatively high FABP5 levels express primarily through PPARβ/δ. In the presence of the co-activator SRC-1, the retinoid-activation of PPARβ/δ led to the upregulation of expression of 3-phosphoinositide-dependent kinase 1 (PDK1), an activator of the anti-apoptotic factor Akt1. Activation of PPARδ, which is downregulated in adipose tissue in obese individuals, induces expression of genes affecting lipid and glucose homeostasis, including the insulin-signaling gene *PDK1*, resulting in improved insulin action. This mechanism is thought to underlie the funding of retinoic acid-induced weight loss in obese-prone mice.[62]

Two types of high-affinity **retinoic acid response elements** have been identified in the promoter regions of target genes near the transcription start: those that recognize the RXR homodimer; and those that recognize the RXR–RAR heterodimer. The RXR–RAR heterodimer binds to retinoic acid response elements (RAREs), which consist of direct repeats of the consensus half-site sequence AGGTCA, usually separated by five nucleosides. The RXR homodimer binds to cognate retinoid X response elements (RXREs), most of which are direct repeats of AGGTCA with only one nucleoside spacing. Gene expression is effected by activating each response element present in the promoter regions of responsive genes.[63] The RXRs can also form homotetramers as well as dimers with other members of the steroid/thyroid/retinoic acid family; heterodimerization in this system has usually been found to increase the efficiency of interactions with DNA, and thus transcriptional activation. Further regulation is effected in this signaling system as the RXR–RAR heterodimer appears to repress the transcription-activating function of RXR–RXR.

In the absence of retinoic acid, the apo-receptor pair (RXR–RAR/RXR) binds to the RAREs of target genes, and RAR recruits co-repressors that mediate negative transcriptional effects by recruiting histone deacetylase complexes which modify histone proteins to induce changes in chromatin structure that reduce the accessibility of DNA to transcriptional factors. This process is reversed upon retinoic acid binding: a conformational change in the ligand-binding domain results in the release of the co-repressor and the recruitment of co-activators of the AF-2 region of the receptor. Some cofactors interact directly to enhance transcriptional activation, while others can affect the acetylation histone proteins causing the conformational opening of chromatin and the activation of transcription of the target gene. Impairments in the process can lead to carcinogenesis (*see* "Anticarcinogenesis," page 124).

The general picture is one of RXRs forming heterodimers with RARs that are activated by retinoids, and with

---

61. Greatest concentrations have been found in adrenals, hippocampus, cerebellum, hypothalamus, and testis.

62. Berry, D. C. and Noy, N. (2009). All-trans-retinoic acid represses obesity and insulin resistance by activating both peroxisome proliferation-activated receptor β/δ and retinoic acid receptor. Mol. Cell. Biol. 29, 3286–3296.

63 The RXR–RAR response elements consist of polymorphic arrangements of the nucleotide sequence motif 5′-RG(G/T)TCA-3′. These gene elements are also responsive to thyroxine, suggesting that retinoic acid and thyroid hormone may control overlapping networks of genes.

other receptors of the same superfamily – i.e., thyroid hormone receptors, the vitamin D$_3$ receptor, the peroxisome proliferator-activating receptor, and, probably, others yet unidentified. The metabolite 9-*cis*-retinoic acid, which targets RXR, causes the formation of RXR homodimers that recognize certain RAREs. However, the same ligands inhibit the formation of heterodimers of RXR and the thyroid hormone receptor (TR), which reduce the expression of thyroid hormone-responsive genes.[64] In contrast, RXR-specific ligands do not appear to affect the formation of RAR-containing heterodimers. The retinoid binding protein CRABP(II) is directly involved in facilitating the interactions of RARα−RXRα heterodimers in the formation of a gene-bound receptor complex. In this role, CRABP(II) serves both to deliver retinoic acid to its nuclear receptors, and also to act as a co-activator of the expression of retinoic acid-responsive genes. Retinoid responses appear to be restricted to a subset of retinoid-responsive genes through the action of the orphan COUP (chicken ovalbumin upstream promoter) receptors, which have been found to form homodimers that avidly bind several retinoic acid response elements and repress both RAR–RXR and RXR homodimer activities. Retinoic acid receptors are abundant in the brain: RARα is distributed throughout, whereas RXRα and RXRδ are found specifically in the striatal regions with dopaminergic neurons. These receptors in both the brain and pituitary gland function in the regulation of expression of the dopamine receptor.

Vitamin A and T$_3$ appear to play compensatory signaling roles in this system. Studies with rats have shown that deprivation of either factor impairs thyroid signaling in the brain through reduced expression of RAR and TR, as well as a neuronal protein neurogranin; this effect can be corrected by administering either vitamin A or T$_3$. Similarly, the regulation of the anterior pituitary hormone, thyroid-stimulating hormone (TSH), has been found to be dependent on the binding of both TR and RXR (which are activated by T$_3$ and 9-*cis*-retinoic acid, respectively) to the TSH gene.

These nuclear retinoic acid receptors comprise a two-component signaling system for the activation of the transcription of genes. Retinoic acid appears to act in either a paracrine or an autocrine manner; both all-*trans*-retinoic acid and 9-*cis*-retinoic acid can be synthesized within the target cell (from retinol *via* retinal) or delivered to that cell from the circulation. On intracellular transport to the nucleus (via CRABP) the ligand is thought to be transferred to the appropriate receptor (RAR, RXR), which then can bind to its respective cognate response element

to regulate the transcription of target genes. The retinoid receptors thus participate in a larger system of nuclear receptors (RARs, RXRs, PPARs, VDR, FXR, etc.) that integrates a range of metabolic signaling (from retinoids, carotenoid metabolites, and fatty acids) in controlling the expression of multiple genes and affecting many tissues.

A Coenzyme Role for Vitamin A?

A coenzyme-like role has been proposed for vitamin A. According to this hypothesis, vitamin A acts as a sugar carrier in the synthesis of **glycoproteins** (which function on the surfaces of cells to effect intercellular adhesion, aggregation, recognition, and other interactions). Indeed, it has been observed that retinol can be phosphorylated to yield retinyl phosphate, which accepts mannose from its carrier GDP-mannose (to form retinyl phosphomannose) and donates it to a membrane-resident acceptor for the production of glycoproteins. Further, vitamin A-deficient animals synthesize less glycoprotein in general (particularly in plasma, intestinal goblet cells, and corneal and trachea epithelial cells), relative to vitamin A-sufficient animals, and in addition produce abnormal glycoproteins.

Although changes in the glycan moieties of glycoproteins can certainly have great effects on cell functions, the actual physiologic significance of this sugar carrier role of vitamin A is not as clear. This is because the form that supports the systemic functions of vitamin A, retinoic acid, cannot serve as a sugar carrier because it cannot be reduced to form retinol. Further, retinyl phosphate does not accept mannose. This problem has not been resolved; however, it has been proposed that retinoic acid may actually be hydroxylated *in vivo* to a derivative that is capable of being phosphorylated and, thus, serving as a sugar carrier.

Vitamin A in Embryogenesis

Retinoids play fundamental roles as differentiating agents in morphogenesis. Deprivation of vitamin A in the Japanese quail results in the loss of normal specification of heart left–right asymmetry. That this effect is associated with the decreased expression of RARβ$_2$ in the presumptive cardiogenic mesoderm suggests that retinoids may direct the differentiation of mesoderm into the heart lineage. Studies of the regenerating amphibian limb have revealed profound effects of retinoids in providing positional information to enable cells to differentiate into the pattern of structures relevant to their appropriate spatial locations. Such observations suggested that the morphogenic role of vitamin A may involve concentration gradients of RARs/RXRs, due to differential induction of the receptor by the retinoid, which establishes positional identity. Accumulating evidence suggests a far more complex mechanism.

---

64. This has been shown for the expression of uridine-5′-diphosphate-glucuronyl transferase which is involved in the phase II metabolism of xenobiotic and endogenous substrates.

It now appears that the embryo has multiple areas with different responsiveness to retinoic acid caused by local differences in the production and binding of, and sensitivity to, retinoic acid. These differences appear to vary among tissues during development, with retinoic acid acting primarily in a paracrine manner in pluripotent cells. In general, in early development retinoic acid signals the posterior neuroectoderm, trunk mesoderm, and foregut endoderm in the organization of the trunk, while in later development it signals development of the other organs, including the eye (*see* Table 5.10).

Many proteins are known to appear during retinoic acid-induced cell differentiation. These include several that have been found to be induced by activation through RXR/RAR binding: growth hormone[65] (in cultured pituitary cells), the protein laminin (in mouse embryo cells), the respiratory chain-uncoupling protein of brown adipose tissue (suggesting a role in heat production and energy balance), the vitamin K-dependent matrix Gla protein, and the RARs. The latter finding indicates autoregulation – i.e., retinoic acid induces its own receptor. In fact, the induction of RARs appears to be differentially selective among various tissues; retinoic acid has been found to induce mainly RARα in hemopoietic cells, but RARβ in other tissues.

Vitamin A in Reproduction

Vitamin A is necessary for reproduction, but the biochemical basis of this function is not known. It is apparent, however, that this role is different from the systemic one, as maintenance of reproduction is discharged by retinol and *not* retinoic acid, at least in mammals.[66] For example, rats maintained with retinoic acid grow well and appear healthy, but lose reproductive ability – i.e., males show impaired spermatogenesis, and females abort and resorb their fetuses. Injection of retinol into the testis restores spermatogenesis, indicating that vitamin A has a direct role in that organ. It has been proposed that these effects are secondary to lesions in cellular differentiation and/or hormonal sensitivity. Several researchers have found that vitamin A-deficient dairy cows show reduced corpus luteal production of progesterone, and increased intervals between luteinizing hormone peak and ovulation. There is some evidence indicating responses of these hormonal parameters to oral treatment with β-carotene, but not preformed vitamin A.

Vitamin A in Bone Metabolism

Vitamin A has an essential role in the normal metabolism of bone. This is indicated by results of animal studies that have shown both low and high vitamin A intakes to reduce bone mineral density. The results of observational studies in humans, however, have been inconsistent in this regard, with only some showing high intakes of vitamin A reducing bone mineral density or increasing fracture risk.[67]

The metabolic role of vitamin A in bone is not clear. Retinoids are thought to be involved in regulating the phenotypic expression of bone-mobilizing cells, osteoclasts, which are reduced in vitamin A deficiency. This

TABLE 5.10 Genes Regulated by Retinoic Acid in Embryonic Development

Aspect of Development	Expression Induced	Expression Repressed
Hindbrain anterioposterior patterning	*Hoxa1, Hoxa3, Hoxb1, Hoxd4, vHnf1*	
Spinal cord motor neuron differentiation	*Pax6, Olig2*	
Early somite formation	*Cdx1*	*Fgf8*
Heart anteroposterior patterning		*Fgf8*
Forelimb differentiation		*Fgf8?*
Pancreas differentiation	*Pdx1*	
Lung differentiation	*Hoxa5*	*TGF-β1*
Anterior eye formation	*Pitx2*	
Kidney formation	*Ret*	
Meiosis induction	*Stra8*	

From Duester, G. (2008). Retinoic acid synthesis and signaling during early organogenesis. *Cell* 134, 921–931.

TABLE 5.11 Prevalence of Vitamin A Deficiency Related to Hemoglobin Status in Indonesian Preschool Children

Hemoglobin (g/dl)	Prevalence of Vitamin A Deficiency (%)[a]
<11.0	54.2
11.0–11.9	43.3
≥12.0	34.3

[a]Based on conjunctival impression cytological assessment.
From Lloyd-Puryear, M. A., Mahoney, J., Humphrey, F., *et al.* (1991). *Nutr. Rev.* 11, 1101–1110.

65. That vitamin A may be required for the expression of growth hormone in humans is suggested by observations of a correlation of plasma retinol and nocturnal growth hormone concentrations in short children.

66. In the chicken, retinoic acid supports normal spermatogenesis, but in all mammalian species examined this function is supported only by retinol or retinal.

67. Ribaya-Mercado, J. D. and Blumberg, J. B. (2007). Vitamin A: Is it a risk factor for osteoporosis and bone fracture? *Nutr. Rev.* 65, 425–438.

TABLE 5.12 Efficacy of Vitamin A Supplementation on Increasing Hemoglobin Levels in Anemic Subjects

Country	Subject Age (years)	Vitamin A Dosage	Follow-up (months)	n	Hemoglobin, g/dl	
					Baseline	Follow-up
Indonesia[a]	<6	0	5	240	11.4 ± 1.6	11.2 ± 1.5
		240 μg RE/d	5	205	11.3 ± 1.6	12.3 ± 1.6*[b]
Indonesia[c]	17–35	0	2	62	10.4 ± 0.7	10.7 ± 0.6
		3 mg RE/d	2	63	10.3 ± 0.8	11.2 ± 0.8*
Guatemala[d]	1–8	0	2	20	10.4 ± 0.7	10.7 ± 0.6
		2.4 mg RE/d	2	25	10.3 ± 0.8	11.2 ± 0.8*

*$P < 0.05$.
[a]Mahilal, P. D., Idjradinata, Y. R., and Muheerdiyantiningsih, K. D. (1988). Am. J. Clin. Nutr. 48, 1271–1276.
[b]Significantly different from baseline level, $P < 0.05$.
[c]Suharno, D., West, C. E., Muhilal, K. D., and Hautvast, J. G. (1993). Lancet 342, 1325–1328.
[d]Mejia, L. A. and Chew, F. (1988). Am. J. Clin. Nutr. 48, 595–600.

consequently unchecked function of bone-forming cells, osteoblasts, results in excessive deposition of periosteal bone and a reduction in the degradation of glycosaminoglycans. It also appears that 9-*cis*-retinoic acid can serve as an effector of the vitamin D-induced renal calcification involving the vitamin K-dependent matrix γ-carboxyglutamic acid protein.

Vitamin A in Hematopoiesis

Because chronic deprivation of vitamin A leads to anemia, a role for the vitamin in hematopoiesis has been suggested. Cross-sectional studies have shown low hemoglobin levels to be associated with the prevalence of signs of xerophthalmia in children (Table 5.11), and children with mild-to-moderate vitamin A deficiency or mild xerophthalmia to have lower circulating hemoglobin levels than non-deficient children; serum retinol level has been shown to explain 4–10% of the variation in hemoglobin level among pre-adolescent children in developing countries. In such cases, the hematological response to vitamin A deficiency is biphasic: an initial fall in both hemoglobin and erythrocyte count due to an apparent interference in hemoglobin synthesis, followed by the rise in both late in deficiency due to hemoconcentration apparently resulting from dehydration from reduced water intake and/or diarrhea.

Supplemental vitamin A has been shown to increase iron status in anemic, vitamin A-deficient animals and humans (Table 5.12). Clinical trials have shown intervention with both iron and vitamin A to be more effective in correcting anemia than intervention with iron alone.

Retinoids are involved in the differentiation of myeloid cells into neutrophils, which occurs in the bone marrow; this function appears to involve all-*trans*-retinoic acid, as RARα is the predominant retinoid receptor type found in hematopoietic cells. Vitamin A-deficient animals have been found to sequester retinol in their bone marrow.[68] That vitamin A-deficient individuals do not necessarily show neutropenia suggests that local retinol sequestration is sufficient to meet the needs of myeloid cells for growth and differentiation into neutrophils, thus mitigating against the effects of low intakes of the vitamin.

The metabolic basis of the role of vitamin A in hematopoiesis appears to involve the mobilization and transport of iron from body stores, as well as the enhancement of non-heme-iron bioavailability. The results of cross-sectional studies in developing countries have found serum iron to be positively correlated with serum retinol levels, and animal studies have shown vitamin A deprivation to cause decreases in both hematocrit and hemoglobin levels, which precede other disturbances in iron storage and absorption. Vitamin A deficiency reduces the activity of ceruloplasmin, a copper-dependent protein with ferroxidase activity which is important in the enteric absorption of iron. This effect appears to occur as the result of a post-transcriptive disruption in the activity. In addition, the results of *in vitro* studies demonstrate that all-*trans*-retinol induces the differentiation and proliferation of pluripotent hemopoietic cells. These findings suggest that the anemia of vitamin A deficiency is initiated by impairments in erythropoesis, and accelerated by subsequent impairments in iron metabolism. The presence of vitamin A or β-carotene has been found to increase the enteric absorption of iron from both inorganic and plant sources; this has been explained on the basis of the formation of complexes with iron that are soluble in the intestinal lumen, thus blocking the inhibitory effects of iron absorption of such

---

68. Twinig *et al.* (*J. Nutr.* [1996] 126, 1618) found bone marrow of vitamin A-deficient rats to contain about four times the retinol contained by that tissue from vitamin A-fed rats.

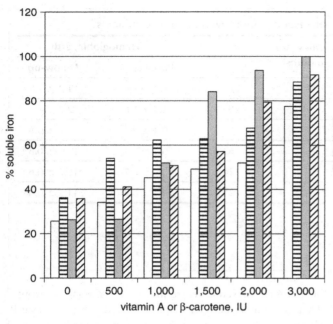

FIGURE 5.13 Effect of vitamin A and β-carotene on iron solubility (at pH 6). *From* Garcia-Casal, M. N., Layrisse, M., Solano, L. *et al.* (1998). *J. Nutr.* 128, 646–650.

☐ Vit A + ferrous fumarate
⊟ Vit A + ferrous sulfate
▨ β-carotene + ferrous fumarate
▨ β-carotene + ferrous sulfate

TABLE 5.13 Effects of Vitamin A and β-Carotene on the Bioavailability of Non-Heme Plant Iron for Humans

Iron source	Vitamin A (μmol)	β-Carotene (μmol)	Iron Absorption (%)
Rice	0	0	2.1
	1.51	0	4.6[a]
	0	0.58	6.4[a]
	0	0.95	8.8[a]
Corn	0	0	3.0
	0.61	0	6.6[a]
	0	0.67	8.5[a]
	1	1.53	6.3[a]
Wheat	0	0	3.0
	0.66	0	5.5[a]
	0	8.5	8.3[a]
	0	2.06	8.4[a]

[a]P <0.05, n = 11–20 subjects.
From Garcia-Casal, M. N., Layrisse, M., Solano, L., *et al.* (1998). *J. Nutr.* 128, 646–650.

antagonists as phytates and polyphenols (*see* Fig. 5.13; Table 5.13).

7. VITAMIN A IN HEALTH AND DISEASE

Vitamin A has been shown to play roles in a variety of functions directly related to health maintenance.

Immune Function

Vitamin A-deficient animals and humans are typically more susceptible to infection than are individuals of adequate vitamin A nutriture.[69] They show changes in lymphoid organ mass, cell distribution, histology, and lymphocyte characteristics.

Vitamin A deficiency is typically associated with malnutrition, particularly protein-energy malnutrition. This may be due to the common origins of each condition, i.e., in grossly unbalanced diets and poor hygiene, resulting in the fact that malnourished children are likely to be deficient in vitamin A and other essential nutrients. Protein deficiency also impairs the synthesis of apo-RBP4, CRBP, and other retinol-binding proteins, impairing vitamin A transport and cellular utilization. Vitamin A deficiency is known to induce or exacerbate inflammation; inflammatory changes in the colon of the vitamin A-deficient rat resemble those observed in human colitis.

Epidemiologic studies have found that low vitamin A status is frequently associated with increased incidences of morbidity and mortality. Many studies have found positive associations between mild xerophthalmia and risks of diarrhea, respiratory infection, and measles among children. A large longitudinal study of pre-school children in Indonesia revealed that the overall mortality rate in children with xerophthalmia was four to five times that of children with no ocular lesions. Children with xerophthalmia were found to have low $CD4^+$:$CD8^+$ ratios, as well as other immune abnormalities in T cell subsets (Table 5.14),

---

69. In practice, it can be difficult to ascribe such effects simply to the lack of vitamin A, as vitamin A-deficient individuals generally also have protein-calorie malnutrition which, itself, leads to impaired immune function.

TABLE 5.14 T Cell Abnormalities in Children

Measure	Without Xerophthalmia	With Xerophthalmia
CD4/CD8	1.11 ± 0.04	0.99 ± 0.05
% CD4/CD45RA (naive)	34.9 ± 1.7	29.9 ± 2.1[a]
% CD4/CD45RO (memory)	18.0 ± 1.1	17.4 ± 1.2
% CD8/CD45RA	37.3 ± 1.7	41.6 ± 2.1
% CD8/CD45RO	7.6 ± 0.6	10.2 ± 0.9[a]
Plasma retinol (μM)	0.84 ± 0.06	0.57 ± 0.04[a]

[a]$P < 0.05$.
Source: Semba, R. D., Muhilal, X., Ward, B. J., et al. (1993). Lancet 341, 5–8.

each of which was reversible on supplementation with vitamin A. It appears that vitamin A deficiency can affect resistance to infection even before it is severe enough to cause xerophthalmia, but that child mortality increases with increasing severity of the eye disease, and affected children die at nine times the rate of normal children.

Vitamin A deficiency leads to histopathological changes providing environments conducive to bacterial growth and secondary infection in loci obstructed by keratinizing debris.[70] This is supported by clinical findings, e.g., excess bacteriuria among xerophthalmic compared to non-xerophthalmic, malnourished children in Bangladesh; and negative correlation of plasma retinol level and bacteria adherent to nasopharyngeal cells of children in India. Such outcomes involve impairments in epithelial integrity (including mucosal immunity), antibody responses, and lymphocyte differentiation. Studies have shown that vitamin A deficiency can reduce child survival by 30%[71] or more.

Underlying these effects are roles of vitamin A in inducing heightened primary immune responses, enhancing memory responses, and accelerating the expansion of the mature B-lymphocyte pool. These functions appear to be discharged by retinoic acid acting at the nuclear level, i.e., by ligating RAR–RXR receptors to alter the expression of genes affecting immune function. Retinoic acid has therefore been called a "fourth signal" of the antibody

response, the others being: receptor-binding of antigen (signal 1); receptor-binding co-stimulatory/accessory factors (signal 2); and binding of "danger signals," e.g., lipopolysaccharide to toll-like receptors (signal 3). In this way, retinoic acid induces changes that "imprint" antibody-secreting cells by permanently committing them. Studies have shown this function to be particularly important in the maintenance of mucosal immunity, which depends on the retinoic acid-dependent modification of B cells by gut-associated dendritic cells within the lymphoid tissues of the small intestine. Both RAR and RXR receptors are constitutively expressed in both B and T cells. Vitamin A, which has long been known to prevent atrophy of the thymus, functions as retinoic acid in at least some sub-types of T cell, including T helper lymphocytes, the antigen-presenting cells that stimulate B cells to produce immunoglobulin A (IgA). It is known that retinoic acid is a modulator of the differentiation of T cells from interleukin-17 secreting T helper cells and towards Foxp3$^+$ T regulatory cells, and it may also play a role in determining the development of T cells towards CD^{+4} vs. CD^{+8} status.

Active infection appears to alter the utilization or, at least, the distribution of vitamin A among tissues. Plasma retinol concentrations drop during malarial attacks, chickenpox, diarrhea, measles, and respiratory disease. Ocular signs of xerophthalmia following measles outbreaks associated with declines in plasma retinol levels depend on the severity and duration of infection, and can be as great as 50%. Episodes of acute infection have been found to be associated with substantive (e.g., eight-fold) increases in the urinary excretion of retinol and RBP4. That such insults to vitamin A status can be of clinical significance is indicated by the fact that vitamin A treatment can greatly reduce the case morbidity and mortality rates in measles and respiratory diseases.

Low serum retinol levels have been found to be more common among HIV$^+$ than in non-infected individuals, and to be highly predictive of vaginal HIV-1 DNA shedding. Whether this is cause or effect is not clear, as serum retinol is known to decrease in the acute-phase response to infection. Intervention studies have found vitamin A supplements effective in improving maternal vitamin A status. Clinical trials with HIV$^+$ children have found vitamin A supplementation to reduce mortality; however, those with HIV$^+$ mothers have not shown benefits in reducing the progress of disease, mother-to-child transmission of the virus or the prevalence of infection in infants (Table 5.15). That vitamin A may affect other viral infections has been suggested by clinical findings of reduced viremia in some vitamin A-treated subjects with hepatitis C.

Stimulation of immunity and resistance to infection are thought to underlie the observed effects of vitamin A supplements in reducing risks of mortality and morbidity from some forms of diarrhea, measles, HIV infection, and

70. Such an example is *Bitot's spots*, which are patches of xerotic conjunctiva with keratin debris and bacillus growth. In Indonesia the presence of Bitot's spots is associated with intestinal worms, to the point that the former are referred to as "worm feces."
71. See meta-analyses by Fawzi, W. W., Chalmers, T. C., Herrara, M. G., and Mostelleer, F. (1993). Vitamin A supplementation and child mortality. A meta-analysis. *J. Am. Med. Assoc.* 269, 898–903; Glasziou, P. and Mackerras, D. (1993). Vitamin A supplementation in infectious diseases: A meta-analysis. *Br. Med. J.* 306, 366–370.

TABLE 5.15 Vitamin A Supplementation has Failed to Reduce Mother-to-Child HIV Transmission

Trial	Intervention Agent	RR[a] to being HIV+, by age			
		Birth	6 weeks	12 weeks	2 years
Malawi[b]	Retinol	—	0.96	—	0.84
Tanzania[c]	Retinol + β-carotene	1.60	1.22	—	1.38[d]
South Africa[e]	Retinol + β-carotene	0.85	—	0.91	—

[a]Ratio of % HIV+ children in treatment group to % HIV+ in control group.
[b]Kunwenda, N., Motti, P. G., Thaha, T. E., et al. (2002). Clin. Infect. Dis. 35, 618–624.
[c]Fawzi, W. W., Msamanga, G. I., Hunter, D., et al. (2002). AIDS 16, 1935–1944.
[d]P <0.05.
[e]Coutsoudis, A., Pillay, K., Spooner, E., et al. (1999). AIDS 13, 1517–1524.

TABLE 5.16 Efficacy of Low-Dose Vitamin A Supplements in Reducing Mortality Related to Pregnancy in Nepal

Parameter	Placebo	Vitamin A	β-Carotene
Serum Levels, Mid-Pregnancy[a], μmol/l			
Retinol	1.02 ± 0.35	1.30 ± 0.33	1.14 ± 0.39
β-Carotene	0.14 ± 0.12	0.15 ± 0.14	0.20 ± 0.17
Mortality, Deaths/100,000 Pregnancies (RR, 95% C.L.)			
During pregnancy	235 (1.0)	142 (0.60, 0.26–1.38)	111 (0.47, 0.18–1.20)
0–6 weeks post-partum	359 (1.0)	232 (0.65, 0.34–1.25)	222 (0.62, 0.31–1.23)
7–12 weeks post-partum	110 (1.0)	52 (0.47, 0.13–1.76)	28 (0.25, 0.04–1.42)

[a]The 3.5-year trial involved some 44,646 women who had more than 22,000 pregnancies, 7,200–7,700 in each treatment group. Vitamin A (retinol) or β-carotene were given in weekly dosages; post-hoc analyses revealed that only half of subjects took 80% of the intended doses (7,000IU), suggesting that the study underestimated the potential impact of vitamin A supplementation.
Source: West, K. P. Jr, Katz, J., Kharty, S. K., et al. (1999). Br. Med. J. 318, 570–574.

malaria in children. Night-blind women have a five-fold increased risk of dying from infections, and low doses of vitamin A have been found to reduce peri- and post-partum mortality in women (Table 5.16), presumably due to reduction in the severity of infections. Indeed, vitamin A supplementation has been found to reduce the incidence of uncomplicated malaria by more than 30%.

Restoration of adequate vitamin A status of deficient children can reduce morbidity rates, particularly for diarrhea and measles (Table 5.17). Meta-analyses of community-based vitamin A intervention studies indicate an average 23% (range: 6–52%)[72] reduction in pre-school mortality (Table 5.18). Vitamin A supplementation of children with active, severe, complicated measles has been shown to reduce in-hospital mortality by at least 50%. In other populations, vitamin A treatment has been shown to reduce the symptoms of diarrhea nearly as much (Table 5.19), as well as the symptoms of pneumonia and other infections substantially. Although attendant reductions in morbidity would be expected, studies have shown those effects to be variable. That not all interventions with vitamin A have reduced mortality rates of vitamin A-deficient children is not surprising, as other factors (e.g., time of initiation of breastfeeding, poverty, poor sanitation, inadequate diets) clearly contribute to the diminished survival of vitamin A-deficient children. Because enteric pathogens induce unique immune responses, not all of which may be comparably affected by the differential action of vitamin A, the efficacy of vitamin A supplementation is likely to depend on the dominant pathogens present in particular communities.

Of the newborn vitamin A supplementation trials conducted to date, vitamin A has been found to reduce mortality in trials with subjects of low vitamin A status and/or late initiation of breastfeeding, whereas benefits have not been observed in trials with subjects of relatively good vitamin A

72. Beaton, G. H., Martorell, R., L'Abbe, K. A., et al. (1992). Report to CIDA, University of Toronto; Sommer, A., West, K. P. Jr, Olson, J. A., and Ross, C. A. (1996). Vitamin A Deficiency: Health, Survival, and Vision. Oxford University Press, New York, NY, p. 33.

TABLE 5.17 Effects of Vitamin A on Morbidity of Children with Measles in South Africa

Outcome	Hospital Morbidity			6-Month Morbidity	
	Placebo	Vitamin A	Outcome	Placebo	Vitamin A
Clinical pneumonia (days)	5.7 ± 0.8	3.8 ± 0.4[a]	Weight gain (kg)	2.37 ± 0.24	2.89 ± 0.23[a]
Diarrhea duration (days)	4.5 ± 0.4	3.2 ± 0.7	Diarrheal episodes	6	3
Fever duration (days)	4.2 ± 0.5	3.5 ± 0.3	Respiratory infections	8	3[a]
Clinical recovery (<8 days), %	65	96	Pneumonia episodes	3	0[a]
Integrated morbidity score[b]	1.37 ± 0.40	0.24 ± 0.15[a]	Integrated morbidity score[b]	4.12 ± 1.13	0.60 ± 0.22

[a]$P <0.05$.
[b]Based on incidence/severity of diarrhea, upper respiratory infections, pneumonia, and laryngotracheobronchitis.
Source: Coutsoudis, A., Broughton, M., and Coovadia, H. M. (1991). *Am. J. Clin. Nutr.* 54, 890.

TABLE 5.18 Effects of Vitamin A Supplementation on Child Mortality

Trial	Observation (months)	Deaths/Total		
		Control	Vitamin A	Odds Ratio (95%CL)
Sarlahi, Nepal	12	210/14,143	152/14,487	0.70 (0.57–0.87)
Northern Sudan	18	117/14,294	123/14,446	1.04 (0.81–1.34)
Tamil Nadu, India	12	80/7,655	37/7,764	0.45 (0.31–0.67)
Aceh, Indonesia	12	130/12,209	101/12,991	0.73 (0.56–0.95)
Hyderabad, India	12	41/8,084	39/7,691	1.00 (0.64–1.55)
Jumia, Nepal	5	167/3,4111	138/3,786	0.73 (0.58–0.93)
Java, Indonesia	12	250/5,445	186/5,775	0.69 (0.57–0.84)
Bombay, India	48	32/1,644	7/1,784	0.20 (0.09–0.45)

Source: Fawzi, W. W., Chalmers, T. C., Merrara, M. G., and Mosteller, F. (1993). *J. Am. Med. Assoc.* 269, 898–903.

TABLE 5.19 Effects of Vitamin A Supplementation on Child Cause-Specific Mortality

Study Country	Relative Risk[a] of Death, by Disease		
	Measles	Diarrhea	Respiratory Disease
Indonesia[b]	0.58	0.48	0.67
Nepal[c]	0.24	0.61	1.00
Nepal[d]	0.67	0.65	0.95
Ghana[e]	0.82	0.66	1.00

[a]Ratio of deaths occurring in the vitamin A-treated group to those occurring in the untreated control group.
[b]Rahmathullah, L., Underwood, B. A., Thulasiraj, R. D., et al. (1990). N. Engl. J. Med. 323, 929–935.
[c]Reanalysis of data of West, K. P. Jr, Pokhrel, R. P., Katz, J., et al. (1991). Lancet 338, 67–71, cited in Sommer, A., West, K. P. Jr, Olson, J. A., and Ross, C. A. (1996). Vitamin A Deficiency: Health, Survival, and Vision. Oxford University Press, New York, NY, p. 41.
[d]Daulaire, N. M. P, Starbuck, E. S., Houston, R. M., et al. (1992). Br. Med. J. 304, 207–210.
[e]Ghana VAST Team (1993). Lancet 342, 7–12.

status and/or early initiation of breastfeeding.[73] It is likely that excess mortality occurs not only among xerophthalmic pre-schoolers, but also among those who are mildly to marginally deficient in vitamin A but have not developed corneal lesions. In fact, a large portion of the deaths averted by vitamin A supplementation may be in this low-vitamin A group. Sommer and colleagues have estimated that the improvement of serum retinol levels in mildly deficient, asymptomatic children (with serum retinol levels of 18–$20\,\mu g/dl$) to serum levels of $30\,\mu g/dl$ would be expected to reduce mortality by 30–50%. A recent meta-analysis showed a 62% reduction in the risk of measles in response to a two-dose regimen of vitamin A administered with measles vaccination; however, other recent negative findings for vitamin A administered to children at the time of vaccination[74] raise the question of whether vitamin A may benefit mostly children who are not vaccinated, which has too often been the case in poor countries.

Skin Health

Vitamin A has a role in the normal health of the skin. Its vitamers, as well as carotenoids, are typically found in greater concentrations in the subcutis than in the plasma (significant amounts are also found in the dermis and epidermis), indicating the uptake of retinol from plasma RBP4. Epithelial cell phenotypes are regulated by hormonal cycles and vitamin A intake; vitamin A deficiency impairs the terminal differentiation of human keratinocytes and causes the skin to be thick, dry and scaly. It also results in obstruction and enlargement of the hair follicles.[75]

Owing to their similarities to changes observed in vitamin A-deficient animals, certain dermatologic disorders of keratinization (e.g., ichthyosis, Darier's disease, pityriasis rubra pilaris) have been treated with large doses of retinol. Clinical success of such treatments generally has been variable, and the high doses of the vitamin needed for efficacy commonly produce unacceptable side effects. Greater therapeutic efficacy has been achieved with all-*trans*-retinoic

acid,[76] 13-*cis*-retinoic acid,[77] and an ethyl ester of all-*trans*-retinoic acid.[78] The most successful of these has been 13-*cis*-retinoic acid for the treatment of *acne vulgaris*, as it affects all of the major pathogenic mechanisms. It decreases sebum production, inhibits the development of blackheads, reduces bacterial numbers in both the ducts and at the surface, and reduces inflammation by inhibiting the chemotactic responses of monocytes and neutrophils.

Retinoids have also been found to produce rapid reductions in the incidence of new non-melanoma skin cancers in high-risk patients. Therefore, it has been suggested that they may produce regressions of prediagnostic malignant and/or premalignant lesions. Indeed, regressions of cutaneous metastases of malignant melanoma and cutaneous T cell lymphoma have been reported in response to retinoid therapy. Topical treatment with all-*trans*-retinoic acid has been found to protect against photoaging signs by stimulating collagen synthesis and accumulation[79] in the upper papillary dermis, and downregulating the induction by UV light of metalloproteinase expression, thereby increasing collagen replacement. The action of retinoids in psoriasis appears to involve thinning of the stratum corneum, reduced keratinocyte proliferation, and reduced inflammation.

The therapeutic value of retinoids is limited by their dose-limiting side effects (see "Vitamin A Toxicity," this chapter). Some retinoids, e.g., 13-*cis*-retinoic acid, can be teratogenic, limiting their use especially for women of child-bearing age. Therefore, alternative approaches have been of interest: use of synthetic, mono- and poly-aromatic retinoids; and use of inhibitors (imidazoles and triazoles) of the cytochrome *P*-450 dependent 4-hydroxylation of all-*trans*-retinoic acid to sustain its intracellular concentrations. Clinical evaluation of the retinoic acid metabolism inhibitor liarozol has shown it to be comparably effective as 13-*cis*-retinoic acid in treating psoriasis and ichthyosis; when used topically, the azole compound was more effective.

Obesity

Overweight/obese individuals show elevated apo-RBP4 levels (*see* Fig. 5.4), which have also been shown to increase with obesity in mice. These observations reflect increased secretion of the transport protein under conditions in which

73. Thurman, D. (2010). Newborn vitamin A dosing and neonatal mortality. *Sight Life* 1, 19–26.
74. Sudfeld, C. R., Navar, A. M., and Halsey, N. A. (2010). Effectiveness of measles vaccination and vitamin A treatment. *Intl J. Epidemiol.* 39, 148–155; Benn, C. S., Aaby, P., Nielsen, J., *et al.* (2009). Does vitamin A supplementation interact with routine vaccinations? An analysis of the Ghana vitamin A supplementation trial. *Am. J. Clin. Nutr.* 90, 626–639; Benn, C. S., Rodrigues, A., Yazdanbakhsh, M., *et al.* (2009). The effect of high-dose vitamin A supplementation administered with BCG vaccine at birth may be modified by subsequent DTP vaccination. *Vaccine* 27, 2891–2898.
75. That is, follicular hyperkeratosis; this condition can also be caused by deficiencies of niacin and vitamin A.
76. This compound, known generically as *tretinoin*, is effective in the treatment of acne vulgaris, photoaging, and actinic keratoses.
77. 13-*cis*-Retinoic acid, known generically as *isotretinoin*, is effective in the treatment of cystic acne, rosacea, gram-negative folliculitis, pyoderma faciale, hidradentis suppurativa, and cancers.
78. This compound, known generically as *etretinate*, is effective in the treatment of psoriasis, ichthyosis, Darier's disease, palmoplantar keratodermas, and pityriasis rubra pilaris.
79. Particularly collagen I, which comprises some 85% of total dermal collagen.

adipocytes downregulate the expression of GLUT4, the insulin-responsive transporter required for cellular uptake of glucose. Because adipose GLUT4 is known to be downregulated under conditions of food deprivation, it has been suggested that the obesity-stimulated secretion of apo-RPB4 may constitute a signal for restricted glucose uptake by peripheral tissues, which system would be expected to offer advantages under conditions of food scarcity.

Adipose tissue is a major storage site of carotenoids, which partition into fat. Carotenoid concentrations tend to be inversely related to percentage body fat, due to the fact that the caloric excesses that drive adiposity tend to be unrelated to the intake of carotenoid-rich foods (fruits, vegetables). Therefore, body fat would appear to be a determinant of the tissue distribution of carotenoids, including those with pro-vitamin A potential.

Drug Metabolism

That the level of vitamin A intake has been found to affect negatively the genotoxic effect of several chemical carcinogens suggests that the vitamin may play a role in the cytochrome *P*-450-related enzyme system. Indeed, several studies have shown that vitamin A deficiency can reduce hepatic cytochrome *P*-450 contents and related enzyme activities, and vitamin A supplementation has been shown to increase the activities of cytochrome *P*-450 isozymes.[80]

Antioxidant Protection

It has been suggested that actions of vitamin A in supporting the health of the skin and immune systems may involve effects on systems that provide protection against the adverse effects of prooxidants.[81] Yet it is unlikely that vitamin A itself is physiologically significant in this regard, as retinol and retinal cannot quench singlet oxygen ($^1O_2$) and have only weak capacities to scavenge free radicals. It can, however, affect tissue levels of other antioxidants; animal studies have shown that deprivation of vitamin A leads to marked increases in the concentrations of α-tocopherol in the liver and plasma, whereas high intakes of retinyl esters can enhance the bioavailability of selenium, an essential constituent of several glutathione-dependent peroxidases.

Several carotenoids, on the other hand, have been shown to have direct antioxidant activities. These include β-carotene, lycopene, and some oxycarotenoids (zeaxanthin, lutein), which can quench $^1O_2$ or free radicals in the lipid membranes into which they partition (Table 5.20). These antioxidant activities are due to their extended systems of conjugated double bonds, which are thought to delocalize the unpaired electron of a free-radical reactant.[82] At low (physiologic) partial pressures of oxygen, carotenoids can also participate in the reduction of free radicals; xanthophyll carotenoids (lutein, lycopene, and **β-cryptoxanthin**) are more effective than β-carotene and more efficient than α-tocopherol *in vitro*. Despite these differences, the carotenoids tend to be less plentiful in tissues, for which reason their contributions to physiologic antioxidant protection are likely to be less important than those of the tocopherols except, perhaps, in cases of high carotenoid intake. Cooperative antioxidant interactions between α-tocopherol and β-carotene have been observed in model systems, and it is likely that *in vivo* carotenoids may serve to protect tocopherols.

The interactions of carotenoids with radicals result in the production of oxidation products of the former and in the bleaching of these pigments;[83] the decomposed product cannot be regenerated metabolically, thus destroying its provitamin A potential. In ultraviolet (UV)-irradiated skin, lycopene is more susceptible to bleaching than is β-carotene, suggesting that it may be more important than β-carotene in antioxidant protection of dermal tissues. Supplementation with β-carotene has been found to improve antioxidant status *in vivo*. These results have included the following: reduced pentane[84] breath output in smokers; reduced plasma

---

80. These include CYP3A in rats, rabbits, and guinea pigs, and CYP2A in hamsters.

81. Although aerobic systems rely on oxygen as the terminal electron acceptor for respiration, they must also protect themselves against the deleterious effects of highly reactive oxygen metabolites that can be formed either metabolically or through the action of such physical agents as ultraviolet light or ionizing radiation. These reactive oxygen species include singlet oxygen ($^1O_2$), superoxide (O_2^-), hydroxyl radical (OH·), and nitric oxide (NO·); they can react, either directly or indirectly, with polyunsaturated membrane phospholipids (to form scission products), protein thiol groups (to form disulfide bridges), non-protein thiols (to form disulfides), and DNA (to cause base changes) to alter cellular function. They can also react with polyunsaturated fatty acid components of circulating lipoprotein complexes; such oxidative changes in low-density lipoproteins (LDLs) appear to be important in the development of atherosclerotic lesions. The systems that protect against these oxidative reactions include several reductants (e.g., tocopherols and carotenoids in membranes and lipoprotein complexes; glutathione, ascorbic acid, urate, and bilirubin in the soluble phases of cells) and antioxidant enzymes (e.g., superoxide dismutases, selenium-dependent glutathione peroxidases, catalase).

82. This mechanism differs from that of the tocopherols, which donate a hydrogen atom to the lipid free radical to produce a semistable lipid peroxide; the tocopherols in turn become semiquinone radicals. (The antioxidant function of the tocopherols is discussed in detail in Chapter 7.)

83. This is seen in the loss of pigmentation from the shanks of poultry, which sign has been used historically by poultry keepers to determine the reproductive status of their hens. Immature pullets deposit carotenoids in dermis, whereas those that are actively laying deposit the pigments in the lipids of the developing oocyte. The bleaching of the skin, therefore, is a positive sign of good laying condition.

84. *n*-Pentane is a scission product of the peroxidative degradation of ω-6 fatty acids.

TABLE 5.20 Antioxidant Abilities of Carotenoid and Other Antioxidants

Compound	ROO˙ Reduction[a]	$^1O_2$ Quenching[a]
Lycopene	—	9×10^9
β-Carotene	1.5×10^9	5×10^9
α-Tocopherol	5×10^8	8×10^7
L-Ascorbate	2×10^8	1×10^7

[a]Bimolecular rate constants (per M/s).
Source: Sies, H. and Stahl, W. (1995). *Am. J. Clin. Nutr.* 62(Suppl.), 1315S–1321S.

concentrations of malonyldialdehyde[85] in cystic fibrotics; reduced lipid peroxidation products (TBARS) in mice; reduced lethality to cultured cells of pro-oxidant drugs; and reduced acetaminophen toxicity[86] in mice. These and other non-provitamin A properties of carotenoids are discussed in greater detail in Chapter 18.

Cardiovascular Health

Epidemiologic investigations have repeatedly found inverse relationships between the level of consumption of provitamin A-containing fruits and vegetables and risks of cardiovascular disease. Indeed, plasma retinol levels have been found to be related inversely to the risk of ischemic stroke, and low plasma β-carotene concentrations are associated with increased risk of myocardial infarction.[87] Such findings have provided the bases for hypothetical actions of vitamin A or, more often, provitamin A carotenoids in chronic disease prevention. Unfortunately, many of these hypotheses have not withstood experimental challenge. Well-designed, randomized, double-blind clinical intervention trials have found supplements of β-carotene[88] or

a combination of β-carotene and/or α-tocopherol[89] to be ineffective in reducing risk of either cardiovascular disease or angina pectoris. In fact, the ATBC Cancer Prevention Trial found a slight increase in the incidence of angina associated with β-carotene use.

Anticarcinogenesis

Because vitamin A deficiency characteristically results in a failure of differentiation of epithelial cells without impairing proliferation (i.e., the *keratinizing* of epithelia), it has been reasonable to question the possible role of vitamin A in the etiology of epithelial cell tumors, i.e., carcinomas. The squamous metaplastic changes seen in vitamin A deficiency are morphologically similar to precancerous lesions induced experimentally. Indeed, patients with oral leukoplakia, a precancerous condition of the buccal mucosa, have been found to have lower serum retinol levels than healthy controls, and treatment with retinol has been found to reduce the development of new lesions and to cause remissions in the lesions of some patients. It has been proposed that retinoic acid, which in high doses can inhibit the conversion of papillomas (benign lesions) to carcinomas, can upregulate its receptors (RARs),[90] which can in turn complex with proto-oncogenes such as c-fos to prevent malignant transformation.

Studies with animal tumor models have found vitamin A deficiency to enhance susceptibility to chemical carcinogenesis, and large doses of vitamin A (i.e., supranutritional but not toxic) to inhibit carcinogenesis in some models (Table 5.21). Retinoids appear to suppress carcinogenesis and tumor growth by being able to induce apoptosis (programmed cell death) and/or terminal differentiation. The pro-apoptotic effects appear to be mediated by RARs, the target genes of which include players in the intrinsic (caspase cascade initiated by cell stress, DNA damage or deprivation of growth factor) and the extrinsic (triggered by activation of death receptor-associated caspases) pathways of apoptosis.

Studies have demonstrated the efficacy of retinoic acid in inhibiting the growth of several types of cancer cells and tumors that do not express the RARβ gene even in the presence of physiological levels of vitamin A. The mechanism of silencing of RARβ gene expression is thought to involve hypermethylation of the gene due to loss of heterozygosity of chromosome 3p24, the locus of RARβ,

85. Malonyldialdehyde is also a peroxidative scission product of polyunsaturated fatty acids. It can be detected by reaction with 3-thiobarbituric acid; it is the predominant, but not only, reactant in biological specimens. Because of the lack of absolute specificity, results are frequently expressed as total thiobarbituric-reactive substances (TBARS).
86. The microsomal metabolism of acetaminophen, like that of other prooxidant drugs metabolized by cytochrome *P*-450-related enzymes, is known to produce O_2^-. Antioxidant status has been shown to affect its acute toxicity in animal modes.
87. The Physicians' Health Study (a prospective study of 22,071 male American physicians) (Hak, A. E., Stampfer, M. J., Campos, H., *et al.* [2003] *Circulation* 108, 802–807.)
88. The Dartmouth Skin Cancer Study involved 1,188 male and 532 female Americans treated with 50 mg/day β-carotene for more than 4 years (Greenberg, E. R., Baron, J. A., Karagas, M. R., *et al.* [1996] *J. Am. Med. Assoc.* 275, 699–703.).

89. The Alpha Tocopherol and Beta Carotene (ATBC) Cancer Prevention Trial involved 29,133 Finish male smokers treated with β-carotene and α-tocopherol for nearly 5 years (Törnwall, M. E., Virtamo, J., Korhonen, P. A., *et al.* [2004] *Eur. Heart J.* 25, 1171–1178).
90. *In situ* hybridization studies have revealed RARα and RARγ in columnar and squamous epithelial cells of the skin.

TABLE 5.21 Inhibition by β-Carotene of Chemical Carcinogenesis in Rats: Reduced Hepatic γ-Glutamyltranspeptidase-Positive Foci

Treatment[a]	Foci (number/cm^2)	Focal area (% total area)
Control	37.1 ± 9.7	1.267 ± 1.121
Retinyl acetate (10 mg/kg/2 days)	34.8 ± 9.6	0.911 ± 0.901
β-Carotene (70 mg/kg/2 days)	20.1 ± 12.5[b]	0.308 ± 0.208

[a]Rats were also treated with diethylnitrosamine/2-acetylaminofluorene.
[b]$P < 0.05$.
Source: Moreno, F. S., Wu, T. S., Penteado, M. V. C., et al. (1995). Intl J. Vit. Nutr. Res. 65, 87–94.

TABLE 5.22 Results of a Meta-Analysis of Results of 57 Epidemiological Studies of β-Carotene Status and Human Cancer Risk

Design	Pooled Estimate of Risk
Cohort	1.013 (0.884, 1.16)[a]
Nested case–control	0.977 (0.864, 1.105)
Case–control	0.729 (0.640, 0.831)

[a]Mean (95% confidence limits).
Source: Musa-Velosa, K., Card, J. W., Wong, A. W. and Cooper, D. A. (2009). Influence of observational study design on the interpretation of cancer risk reduction by carotenoids.

and/or impaired expression of other factors involved in RARβ expression.[91] Loss of RAR function under vitamin A-adequate conditions is associated with a variety of different cancers, the best studied of which is acute promyelocytic leukemia (APL[92]). This cancer has been found to result from a non-random chromosomal translocation or deletion that leads to the production of a fusion of RARα gene on chromosome 17 to the promyelocytic (PML) gene on chromosome 15. When expressed, the fusion product represses translation and initiates leukemogenesis. This appears to occur through that action of the PML-RAR protein, which is a transcriptional activator of retinoic acid target genes. Studies with one specific target, the tumor suppressor gene RARβ2, have revealed that its promoter contains a high-affinity retinoic acid response element (RARE) near the transcription start site, but is inactivated by methylation.[93] It appears that the PML-RAR fusion protein can form a complex with histone deacetylase, which, in turn, becomes oncogenic by recruiting DNA methyltransferases to the promoters of RARβ2, locking them in a stably silenced chromatin state by hypermethylation. Most APL patients (80%), however, respond to treatment with very high doses of all-trans-retinoic acid, resulting in complete remission in more than half of cases. This effect appears to involve retinoic acid causing the dissociation of the PML-RAR protein–histone deacetylase complex, which converts the fusion protein into a

transcriptional activator resulting in leukemia cell differentiation. Studies with breast cancer cells[94] suggest that retinoic acid can also influence histone acetylation (due to the release of histone deactylase) of the RARβ2 gene, resulting in the inhibition of cell growth. Thus, in sensitive cells retinoic acid can reactivate RARβ2 gene expression through epigenetic means; such reactivation has been shown to suppress malignancy in lung cancer cells.

While the use of retinoic acid, which is rapidly metabolized and eliminated from the body, avoids the problem of chronic hypervitaminosis, its substantial toxicity makes it unsuitable for regular clinical use. Therefore, more than 1,500 retinoids have been synthesized and tested for potential anticarcinogenicity.[95] A number of these[96] have been found to effectively inhibit experimentally induced tumors in several organs of animals,[97] and have yielded hopeful results in clinical trials. The consensus, however, is that although retinoids currently available can delay tumorigenesis, they cannot do so at doses that are not themselves toxic.

Epidemiological investigations of vitamin A intake and human cancer have produced mostly negative results, depending on study design. Findings of significant, inverse associations of intakes/plasma level and cancer risk have come mostly from retrospective case–control studies (Table 5.22). More than 60% of these have indicated reductions in the prevalences of cancers of the lung, colon/

91. Possibilities include the orphan receptors nurr77 and COUP-TF, both of which are over-expressed in retinoic acid-resistant cells.
92. APL is characterized by a block in myeloid differentiation, resulting in the accumulation in the bone marrow of abnormal promyelocytes and in a coagulopathy involving disseminated intravascular coagulation and fibrinolysis.
93. DiCroce, L, Raker V. A., Corsaro, M., et al. (2002). Methyltransferase recruitment and DNA hypermethylation of target promoters by an oncogenic transcription factor. Science 295, 1079–1082.

94. Sirchia, S. M., Ren, M., Pili, R., et al. (2002). Endogenous reactivation of the RARβ2 tumor suppressor gene epigenetically silenced in breast cancer. Cancer Res. 62, 2455–2461.
95. These compounds are formal derivatives of retinal differing by changes in the isoprenoid side chain (including modification of the polar end and cyclization of the polyene structure), or modifications of the cyclic head group (including replacement with other ring systems).
96. For example, 13-cis-retinoic acid, N-ethylretinamide, N-(2-hydroxyethyl)-retinamide, N-(4-hydroxyphenyl)-retinamide, etretinate, N-(pivaloyloxyphenyl)-retinamide, N-(2,3-dihydroxypropyl)-retinamide.
97. For example, several studies have shown that retinoids can inhibit the initiation and promotion of mammary tumorigenesis induced in rodents by dimethylbenz(α)anthracene or N-methyl-N-nitrosourea, as well as the induction of ornithine decarboxylase, an enzyme the induction of which appears to be essential in the development of neoplasia.

TABLE 5.23 Lack of Cancer Protection by β-Carotene among Male Smokers

Cancer Site	Cancer Incidence (per 10,000 person-years)	
	No β-Carotene	β-Carotene
Lung	47.5	56.3
Prostate	13.2	16.3
Bladder	9.0	9.3
Colon–rectum	8.6	9.0
Stomach	6.6	8.3
Other	44.9	42.3

Source: Alpha-Tocopherol, Beta-Carotene Cancer Prevention Study Group (1994). *N. Engl. J. Med.* 330, 1029–1035.

TABLE 5.24 Increased Mortality among Male Smokers Taking β-Carotene

Causes of Death	Mortality Rate (per 10,000 person-years)	
	No β-Carotene	β-Carotene
Lung cancer	30.8	35.6
Other cancers	32.0	33.1
Ischemic heart disease	68.9	77.1
Hemorrhagic stroke	6.0	7.0
Ischemic stroke	6.5	8.0
Other cardiovascular disease	14.8	14.8
Injuries and accidents	19.3	20.3
Other causes	23.5	22.5

Source: Alpha-Tocopherol, Beta-Carotene Cancer Prevention Study Group. (1994). *N. Engl. J. Med.* 330, 1029–1035.

rectum, skin, and prostate, whereas such protective effects have been found in less than 20% of prospective cohort or nested case–control studies. Nevertheless, the results of such surveys have fostered the hypothesis that β-carotene may have some beneficial effect unrelated to its role as a precursor of vitamin A.

The cancer-chemopreventive potential of supplemental β-carotene/retinoids has been tested in at least five well-designed, placebo-controlled, double-blind clinical intervention trials, none of which found significant reduction in cancer in high-risk subjects. In fact, the results of two studies found β-carotene treatment harmful. The first, the Carotene and Retinol Efficiency Trial (CARET), a 12-year study involving more than 18,314 Americans, found no protection against lung or prostate cancer in men, but an increase in lung cancer in women (mostly among former smokers).[98] The second was the Alpha-Tocopherol and Beta-Carotene trial (ATBC),[99] conducted in Finland with more than 29,000 men with histories of smoking; this evaluated the health impacts of modest supplements of β-carotene (20 mg/day) and/or α-tocopherol (50 mg/day). Within 5–8 years of follow-up, results showed significantly greater total mortality (8%) and lung cancer incidence (18%) among men taking β-carotene in comparison with those not taking that supplement (Tables 5.23, 5.24).

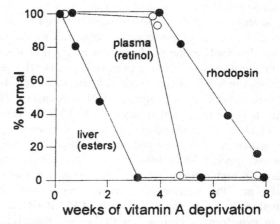

FIGURE 5.14 Hepatic vitamin A stores must be depleted before changes in circulating retinol levels or photopigment concentrations occur.

The interpretation of these findings is not straightforward, as ample evidence also indicates that β-carotene at these levels is generally safe. It is possible that β-carotene at these levels of intake can redox-cycle to act as a pro-oxidant and co-carcinogen in the lungs of smokers. It is also possible that resulting retinoid metabolites may activate RAR to produce anti-apoptotic effects (PPARβ/γ target genes include those that activate survival pathways) and/or increase cell proliferation.

8. VITAMIN A DEFICIENCY

The appreciable storage of vitamin A in the body tends to mitigate against the effects of low dietary intakes of the vitamin, as tissue stores are mobilized in response to low-vitamin A conditions. However, because there are two effective

98. The Beta-Carotene and Retinol Efficacy Trial (CARET) involved 18,314 American male and female current and ex-smokers and asbestos-exposed men supplemented with both β-carotene and retinyl palmitate for 4 years of treatment. (Omenn, G. S., Goodman, G., Thornquist, M., *et al.* [1996] Chemoprevention of lung cancer: The beta-carotene and retinol efficacy trial (CARET) in high-risk smokers and asbestos-exposed workers. *IARC Sci. Publ.* 136, 67–85).

99. Albanes, D., Hainonen, O. P., Huttenen, J. K., *et al.* (1995). Effects of alpha-tocopherol and beta-carotene supplements on cancer incidence in the Alpha-Tocopherol Beta-Carotene Cancer Prevention Study. *Am. J. Clin. Nutr.* 62, 1427S–1430S.

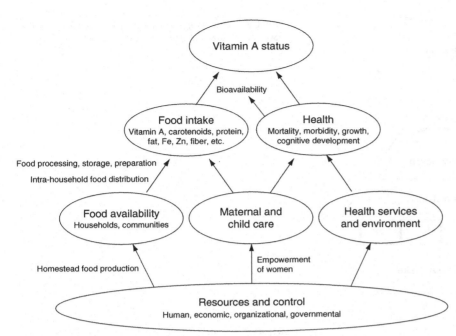

FIGURE 5.15 WHO conceptual framework of causes of vitamin A deficiency in human populations. *From UN Food Nutr. Bull.* 19, 1998.

pools of vitamin A in the body, the rate of mobilization varies between tissues according to their respective proportions of fast-turnover and slow-turnover pools (*see* Fig. 5.14). For this reason, rats showed faster losses of vitamin A from intestine than liver (71 vs. 53%) after being fed a vitamin A-deficient diet. It should be noted that, while hepatic stores are great enough to provide retinol, the plasma retinol level is only minimally affected by vitamin A deprivation (e.g., in the same experiment, it decreased by only 8%). Cellular functions of vitamin A can be expected to change only after transport of the vitamin is reduced – i.e., after vitamin A stores have dropped to such levels as to reduce plasma retinol–RBP4 concentrations. Hepatic retinyl ester stores >20 µg (>0.07 µmol) RE per gram indicate adequate vitamin A status.

Vitamin A deficiency can occur either because of a lack of both provitamins A and preformed vitamin A in diets (*primary vitamin A deficiency*), or because of failures in their physiologic utilization (*secondary vitamin A deficiency*). Primary vitamin A deficiency can occur among children and adults who consume diets composed of few servings of yellow and green vegetables and fruits, and of liver. For infants and young children, early weaning can increase the risk of primary deficiency. For livestock, it can occur with unsupplemented diets containing low amounts of yellow maize (corn) and corn gluten meal.

Secondary vitamin A deficiency can occur in several ways. One involves chronically impaired enteric absorption of lipids, such as in diseases affecting the exocrine pancreas (e.g., pancreatitis, cystic fibrosis, nutritional selenium deficiency) or bile production and release (e.g., biliary atresia,

some mycotoxicoses in livestock), or due to the consumption of diets containing very low amounts of fat.[100] Chronic exposure to oxidants can also induce vitamin A depletion; an example is benzo(α)pyrene in cigarette smoke. Nutritional deficiencies of zinc can also impair the absorption, transport, and metabolism of vitamin A, as zinc is essential for the hepatic synthesis of RBP4 and the oxidation of retinol to retinal, which is catalyzed by a zinc-dependent retinol dehydrogenase. Malnourished populations, which typically have low intakes of several essential nutrients, including vitamin A and zinc, are at risk of vitamin A deficiency. A prevalence of 25% or more of individuals with plasma retinol levels <0.70 µmol/l (<20 µg/dl) is indicative of population-wide inadequacy with respect to the vitamin.

Like most nutritional deficiencies in human populations, vitamin A deficiency is an outcome of a bio-eco-social system that fails to provide sources of the vitamin in ways that are at once accessible and utilizable. The complexity of this system has been captured in the WHO conceptual model (Fig. 5.15).

Vitamin A functions in many organs of the body. Therefore, insufficient intakes of the vitamin lead to a sequence of physiological events (Figs. 5.16 and 5.17) that ultimately are manifest in several clinical signs (Table 5.25). In fact, the only unequivocal signs of vitamin A deficiency are the ocular lesions nyctalopia and xerophthalmia. The

---

100. There are few data on which to estimate the minimum amounts of dietary fat that are needed to support the absorption of vitamin A and the other fat-soluble vitamins; in the absence of empirical data, the estimate of 5 g/day is frequently used.

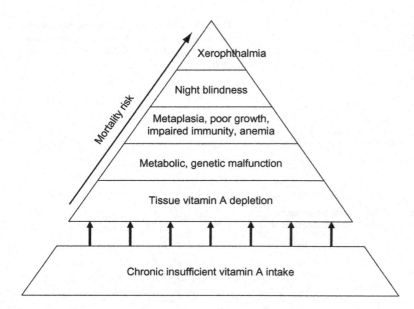

FIGURE 5.16   Progression of vitamin A deficiency. *After* West, K. P. (2002). *J. Nutr.* 132, 2857S–2866S.

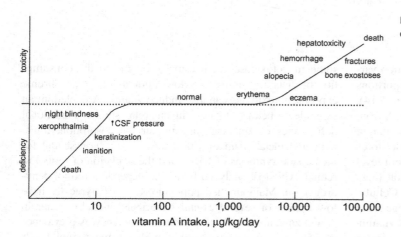

FIGURE 5.17   Both low and high intakes of vitamin A can cause clinical signs.

TABLE 5.25 Signs of Vitamin A Deficiency

Organ System	Sign
General	Loss of appetite, retarded growth, drying and keratinization of membranes, infection, death
Dermatologic	Rough scaly skin, rough hair/feathers
Muscular	Weakness
Skeletal	Periosteal overgrowth, restriction of cranial cavity and spinal cord, narrowed foramina
Vital organs	Nephritis
Nervous system	Increased cerebrospinal fluid pressure, ataxia, constricted optic nerve at foramina
Reproductive	Aspermatogenesis, vaginal cornification, fetal death and resorption
Ocular	Nyctalopia, xerophthalmia, keratomalacia, constriction of optic nerve

former disorder of dark adaptation of the retina can take a year to develop after the initiation of a vitamin A-deficient diet, but responds rapidly to vitamin A treatment. The latter disorder involves permanent morphological changes of the anterior segment of the eye that are not correctable without scarring. Early intervention is very important in cases of xerophthalmia (Table 5.26) in order to interrupt the progressive lesions in early stages before permanent blindness occurs.

Detection of Vitamin A Deficiency

Vitamin A deficiency can be detected by clinical diagnosis or assessment of biochemical or histological indicators. Clinical signs include impaired dark adaptation, nyctalopia, and ocular lesions (Figs 5.18–5.20, Table 5.27). Nyctalopia is the first functional sign of vitamin A

deficiency that can be measured. It can be diagnosed by instrumental observation in the ophthalmology clinic, but in the field it is usually necessary to obtain this information about children from their caregivers. The condition can be detected by using the papillary response to a graduated light stimulus. Examination by slit lamp can reveal

FIGURE 5.19 Keratomalacia in a vitamin A-deficient child. Courtesy of D. S. McLaren, American University of Beirut.

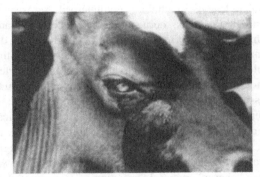

FIGURE 5.20 Vitamin A-deficient calf: blind with copious lacrimation. Courtesy of J. K. Loosli, University of Florida.

TABLE 5.26 Stages of Xerophthalmia

Stage	Signs
1. Xerosis	Dryness of conjunctiva
	Bitot's spots[a] (gray-white, foamy, greasy, "cheesy" deposits on the conjunctiva near the cornea; contain fatty degenerated epithelial cells and leukocytes)
	Ultimate extension to cornea
2. Keratomalacia	Softening of cornea
	Ultimate involvement of iris/lens
	Secondary infection

[a]There is some question about the relation of this sign, which can also occur in vitamin A-adequate individuals, to the deficiency.

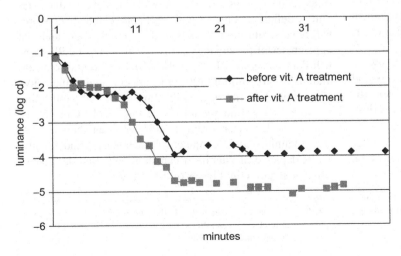

FIGURE 5.18 Dark adaptation in a vitamin A-deficient individual before and after vitamin A treatment. *After* Russell, R. M., Multack, R., Smith, V., *et al.* (1973). *Lancet* 2, 1161–1164.

TABLE 5.27 Clinical Classification of Eye Lesions Caused by Vitamin A Deficiency (From WHO)

Site Affected	Clinical Sign	Designation
Retina	Night blindness *also* nyctalopia	XN
	Fundus specs	XF
Conjunctiva	Xerosis	X1A
	Bitot's spots	X1B
Cornea	Xerosis	X2
	Ulceration/keratomalacia <one-third of surface	X3A
	Ulceration/keratomalacia = one-third of surface	X3B
	Scar	XS

fundus[101] specks and disrupted rod outer segments (with the possible involvement of similarly disrupted cones), signs suggestive of visual field alterations.

Ocular lesions, including conjunctival xerosis with/ without Bitot's spots, corneal xerosis, ulceration, and/or keratomalacia, and corneal scars, can be diagnosed by direct examination of the eye. Morphological changes in epithelial cells blotted from the conjunctival surface can be detected by histologic examination, a procedure called **conjunctival impression cytology**.[102] The presence of enlarged, flattened epithelial cells and few or no goblet cells is indicative of vitamin A deficiency. With progressing vitamin A deficiency, the conjunctival surface takes on a dry, corrugated, irregular surface, ultimately developing an overlay of white, foamy or "cheesy" material consisting of desquamated keratin and a heavy bacterial growth.[103] **Bitot's spots**, the *sine qua non* of conjunctival xerosis, are almost always bilateral oval or triangular structures first appearing temporal to the limbus and comprising a thickened, superficial layer of flattened cells usually with a keratinizing surface, a prominent granular cell layer, and acanthotic thickening with a disorganized basal cell layer but with no goblet cells.

Conjunctival impression cytology has been show to be capable of correctly identifying 82–93% of cases and 70–90% of normals; sensitivity declines and specificity increases when serum retinol or retinol relative dose

response cut-offs are also used in the definition of a case. The use of impression cytology yields estimates of vitamin A deficiency that are 5–10 times the rates of diagnosed xerophthalmia. Somer has suggested that vitamin A deficiency should be considered a public health problem when the prevalence of abnormal impression cytology reaches 20% in either women or children.

The earliest corneal sign of vitamin A deficiency is fine, fluorescein-positive, superficial punctuate keratopathy that usually begins in the inferior aspect of the cornea, particularly infernasally. These lesions can be seen using the slit lamp microsope. Studies have revealed punctuate keratopathy in 60–75% of patients with nyctalopia or vitamin A-responsive conjunctival xerosis. With progressive vitamin A deficiency the lesions become more numerous and concentrated, involving larger portions of the corneal surface. By the time most of the corneal surface is involved, the lesions are generally apparent by hand-light examination as a haziness and diminished wettability on the corneal surface. At that point the condition is called **corneal xerosis**. In addition to punctuate keratopathy, corneas affected at this level also show stromal edema, again mostly in the inferior aspect. If untreated, the condition progresses to the point of corneal ulceration, frequently characterized by the presence of a single (or, in a minority of cases, two to three) sharply defined ulcer(s) varying in depth, usually one-fourth to one-half of corneal thickness, but sometimes deep enough to effect stromal loss. This can lead to deep stromal necrosis characterized by gray-yellow, edematous and cystic lesions of varying size, from 2 mm to most of the cornea.

The vitamers A can be accurately quantified in biological specimens using high-performance liquid–liquid partition chromatography (HPLC). Serum retinol is therefore a convenient parameter of vitamin A status. The homeostatic control of circulating retinol levels makes this parameter useful only in identifying subjects with chronically low vitamin A intakes sufficient to exhaust their hepatic stores that support the synthesis of RBP4. With the caveats that vitamin A-deficient subjects with still-appreciable liver stores will have serum retinol levels in the normal range, and that deficiencies of protein and zinc, as well as hepatic illnesses, will depress serum RBP4 levels, RBP4 and retinol concentrations[104] <0.70 µmol/l (20 µg/dl) are considered indicative of vitamin A deficiency. Still, almost 20% of night-blind children have serum retinol above this level (Table 5.28), children with greater serum retinol levels can show clinical xeropthalmia and/or conjunctival metaplasia, and healthy adults depleted of vitamin A can show impaired dark adaptation at serum retinol levels of 20–30 µg/dl.

---

101. The fundus of the eye is the interior surface opposite the lens and including the retina, optic disc, macula and fovea, and posterior pole.

102. This procedure involves blotting onto a piece of filter paper cells that are then fixed, stained, and examined microscopically. Normal conjunctival cells appear in sheets of small, uniform, non-keratinized epithelial cells with abundant mucin-secreting goblet cells.

103. Most commonly the *xerosis* bacillus.

---

104. It should be remembered that RBP4 and retinol exist in the circulation in equimolar concentrations. Levels between 0.70 and 1.05 µmol/l indicate marginal vitamin A status.

TABLE 5.28 Relationship of Serum Retinol Level and Clinical Vitamin A Deficiency in Indonesian Pre-school Children

Clinical Status		n	Case Frequency, by Serum Retinol Level		
XN	X1B		<10 µg/dl	10–20 µg/dl	>20 µg/dl
+	−	174	27%	55%	18%
−	+	51	31%	57%	12%
+	+	79	38%	53%	9%
−	−	252	8%	37%	55%

From Sommer, A., Hussaini, G., Muhilal, L., *et al.* (1980). *Am. J. Clin. Nutr.* 33, 887–891.

Because of the uncertainties of interpreting serum retinol values, tests have been devised to assess the mobilizable vitamin A capacity of the liver based on serum responses to oral doses of a retinyl ester (acetate or palmitate in the **relative dose–response (RDR)** test) or the synthetic retinoid **3,4-didehydroretinol**[105] (in the **modified relative dose–response (MRDR)** test). The RDR test requires two samples of blood: a fasting sample drawn immediately before the administration of the test dose; and another sample drawn 5 hours later. Retinol is determined in each sample, and the percentage response over baseline is calculated. For the MRDR, only a single blood sample is required; it is taken 4–6 hours after oral administration of 3,4-didehydroretinol. Both retinol and 3,4-didehydroretinol are determined in the sample, and the response is taken as the molar ratio of the two retinoids. Both tests assume that the appearance of retinoid in the plasma is a function of the amount also entering from endogenous hepatic stores. Therefore, an RDR value of ≥20% or an MRDR value of 20–30% is taken as indicative of inadequate hepatic storage (<0.07 µmol/g) of the vitamin; children with such low stores are almost certain to be vitamin A deficient. Determination of serum retinoids, which is accomplished using high performance liquid chromatography (HPLC), requires relatively large sample volumes; therefore, it has been of interest to use RPB4, which can be determined in very small volumes by enzyme-linked immunoassay. Such RBP4–RDR results, however, have shown high rates of false-positives, particularly in subjects with a high BMs,[106] apparently due to the presence in serum of adipose-derived apo-RPB4 which reduces the retinol:RBP4 ratio. Therefore, the RDR and MRDR tests remain the instruments of choice as categorical indicators of vitamin A status. While neither gives quantitative measures of hepatic vitamin A stores, each can indicate changes in vitamin A status from low to adequate. The MRDR test is useful in assessing impacts of interventions in which serum retinol is not expected to change – e.g., in subjects of marginal status.

Such approaches have shown that pre-term infants generally have reduced hepatic vitamin A stores. The concentrations of vitamin A in breast milk (i.e., primarily retinyl palmitate in milk fat) drop in vitamin A deficiency, and can therefore be used to detect the deficiency in mothers.[107]

Treatment of Vitamin A Deficiency

Because vitamin A is stored in appreciable amounts in the liver, it can be administered in relatively large, infrequent doses with efficacy. In cases of clear or suspected xerophthalmia, particularly in communities in which the deficiency is prevalent, vitamin A is administered orally in large doses, followed by an additional dose the next day and a third a few weeks later (Table 5.29). Oral administration of water-miscible or oil solutions of the vitamin are as effective as water-miscible preparations administered parenterally. Water-miscible preparations are much more effective than oil solutions when administered parenterally, i.e., by intramuscular injection. Topical administration on the skin is ineffective.

Night blindness due to vitamin A deficiency responds within hours to days upon the administration of vitamin A, although full recovery of visual function may take weeks, and the fading of retinal lesions may take up to 3 months. Active Bitot's spots and the accompanying xerosis responds rapidly to vitamin A treatment; in most cases lesions regress in days and disappear in 2–3 weeks. Punctate keratopathy of the cornea also responds rapidly to vitamin A, improving within a week in response to a large oral dose.

---

105. This compound has also been called dehydroretinol and vitamin A$_2$.

106. Fujita, M., Brindle, E., Rocha, A., *et al.* (2009). Assessment of the relative dose–response test based on serum retinol-binding protein instead of serum retinol in determining vitamin A stores. *Am. J. Clin. Nutr.* 90, 217–224.

107. Breast milk retinyl palmitate concentrations typically fall in the range of 1.75–2.45 µmol/l; levels ≤1.05 µmol/l (i.e., ≤8 µg/g milk fat) are generally considered as indicative of vitamin A deficiency.

TABLE 5.29 Treatment Schedule for Xerophthalmia

Case	Time	Vitamin A Dose[a], by Age		
		<6 Months	6–12 Months	1 Year
Xerophthalmia	Day 1	15,000 RE	30,000 RE	60,000 RE
	Next day	15,000 RE	30,000 RE	60,000 RE
	2–4 weeks later	15,000 RE	30,000 RE	60,000 RE
Women of reproductive age without severe corneal lesions	Daily for 2 weeks		3,000 RE	
Subjects with complicated measles	Day 1	15,000 RE	30,000 RE	60,000 RE
	Next day	15,000 RE	30,000 RE	60,000 RE
Subjects with severe protein-energy malnutrition	Day 1	15,000 RE	30,000 RE	60,000 RE
	Every 4–6 months[b]			60,000 RE
For prevention in high-risk populations	Once	15,000 RE	30,000 RE	60,000 RE[c]
	Every 4–6 months		60,000 RE	

[a]Oil solution administered orally.
[b]Until signs of protein-energy malnutrition subside.
[c]Including post-partum mothers.
From WHO (1997) Vitamin A Supplementation: A Guide to Their Use in the Treatment and Prevention of Vitamin A Deficiency and Xerophthalmia, 2nd edn. WHO, Geneva.

While low doses of retinyl palmitate (e.g., RDA-amounts given daily or seven times RDA-amounts given weekly) have been found to be very well absorbed and effective in supporting adequate serum retinol levels; periodic dosing with large oral doses (e.g., 100,000−200,000 IU every 4–6 months) was found effective in reducing the prevalence of xerophthalmia and improving survival of children in studies conducted in Indonesia, Kenya, and South Africa. These regimens appear to be comparably effective in reducing diarrhea incidence and duration; however, the low dose may be more effective in reducing the incidence and duration of acute respiratory infections.

The successes of vitamin A supplementation trials have led to the widespread use of vitamin A capsules: UNICEF estimates that each year more than half a billion capsules are distributed to some 200 million children in 100 countries. The large-dose, medical approach has been questioned regarding how effective it is in reducing child morbidity.[108] A meta-analysis of nine randomized controlled trials found that vitamin A supplementation had no consistent effect on the incidence of diarrhea and may slightly increase the risk of respiratory tract infection,[109] and a very large trial in Ghana found vitamin A supplementation of women without effect on pregnancy-related mortality.[110] West and colleagues[111] have pointed out that in fact maternal vitamin A supplementation has reduced maternal mortality in areas of prevalent gestational night blindness and where the risk of maternal mortality is very high. Maternal interventions also offer the potential to produce health benefits, such as improved respiratory function, that become apparent in pre-adolescent children.

9. VITAMIN A TOXICITY

The hepatic storage of vitamin A tends to mitigate against the development of intoxication due to intakes in excess of physiological needs. However, persistent large overdoses (more than 1,000 times the nutritionally required amount) can exceed the capacity of the liver to store and catabolize, and will thus produce intoxication[112] (Fig. 5.17). This is

108. Latham, M. (2010). J. World Pub. Health Nutr. Assoc. 1, 12–24.
109. Grotto, I., Mimouni, M., Gdalevich, M., and Mimouni, D. (2003). J. Pediatrics 142, 297–304.

110. In 1999, West et al. (Br. Med. J. 318, 570–575) reported vitamin A supplementation to reduce pregnancy-related mortality by 44%, although the apparent effects seemed unrelated to infection/immunity; however, in 2010 Kirkwood et al. (Lancet 376, 1643–1644) found no such protection in an intervention trial with 207,781 women in Ghana.
111. West, K. P, Christian, P, Katz, J., et al. (2010). Effect of vitamin A supplementation on maternal survival. Lancet 376, 873–874.
112. It has been suggested that the hepatic damage following hepatitis B infection may be due to the toxic effects of retinoids, which accumulate in the cholestatic liver.

marked by the appearance in the plasma of high levels of retinyl esters that, because they are associated with lipoproteins rather than RBP4, are outside the normal strict control of vitamin A transport to extrahepatic tissues.

Four aspects of vitamin A metabolism tend to protect against hypervitaminosis:

1. Relatively inefficient conversion of the provitamins A in the gut
2. The unidirectional oxidation of the vitamin to a form (retinoic acid) that is rapidly catabolized and excreted
3. The relative excess capacity of CRBP(II) to bind retinol
4. Accelerated vitamin A catabolism.

Hypervitaminosis A therefore requires high exposures. Signs of hypervitaminosis A are associated with plasma retinol levels >3μmol/l and increases in serum retinyl ester levels without substantial increases in circulating holo-RBP4. In humans, signs are manifest after single large doses (>660,000 IU for adults, >330,000 IU for children) or after doses >100,000 IU/day have been taken for several months. Children experiencing hypervitaminosis A develop transient (1–2 days) signs: nausea, vomiting, signs due to increased cerebrospinal fluid pressure (headache, vertigo, blurred or double vision); and muscular incoordination. Most field studies have found that 3–9% of children given high single doses (ca. 200,000 IU) for prophylaxis show transient nausea, vomiting, headache, and general irritability; a similar percentage of younger children may show bulging fontanelles[113] subsiding within 48–96 hours. Those reacting to extremely large doses of the vitamin show drowsiness, loss of appetite, and malaise, followed by skin exfoliation, itching (circumocular), and recurrent vomiting. Several studies have found that about 30% of acne patients treated with 13-cis-retinoic acid show increased LDL/HDL ratios which, if persistent, would suggest increased risk of ischemic heart disease. Some 15% of such patients report arthralgia and myalgia.

Chronic hypervitaminosis A occurs with recurrent exposures exceeding 12,500 IU (infants) to 33,000 IU (adult). It is manifest mainly as changes in the skin and mucous membranes. Dry lips (cheilitis) are a common early sign in humans, often followed by dryness and fragility of the nasal mucosa, leading to dry eyes and conjunctivitis. Skin lesions include dryness, pruritis, erythema, scaling, peeling of the palms and soles, hair loss (alopecia), and nail fragility. Headache, nausea, and vomiting (signs of increased intracranial pressure) have also been reported.

Hypervitaminosis A can affect bone, perhaps by activating abnormal gene expression by the direct membranolytic activities of retinoids. Infants and young children can show painful periostitis[114] and, rarely, premature closure of lower limb epiphyses (manifest as "hyena disease" in calves, characterized by shortened hind limbs). It has been suggested that chronic high intakes of vitamin A may contribute to the pathogenesis of osteoporosis in adults. While most epidemiological studies have not revealed a relationship between vitamin A status and fracture risk, one study[115] found reduced bone mineral density with intakes greater than 0.6 mg RE (2,000 IU)/day. The Third National Health and Nutrition Examination Survey (NHANES III) found that half of American adults consume more than that level. An analysis of results from more than 72,000 post-menopausal women in the Nurses' Health Study[116] found that individuals reporting intakes of at least 2,000 IU/d had twice the risk of hip fracture due to mild or moderate trauma than those reporting intakes of less than 500 IU/d. Hypervitaminotic A animals frequently show bone abnormalities that apparently result from changes in impaired osteoclastic activities and enhanced osteoblastic activities, resulting in overgrowth of periosteal bone in a non-vitamin D-dependent manner. This in turn can lead to impairments in visual function by restricting the optic foramina and pinching the optic nerve, and in motor function by increasing intracranial pressure. Intracranial hypertension is a well known side effect of therapeutic doses of 13-cis-retinoic acid (for acne) and all-trans-retinoic acid (acute promyelitic leukemia). That this condition resolves more rapidly if patients undergo weight loss suggests that RBP4 or adipose origin may be involved in triggering these signs.

Therapeutic doses of all-trans-retinoic acid are generally well tolerated, although some patients experience headache, nausea, and visual changes. In rare cases, muscular stiffness and epileptiform seizures have been reported. A few cases of myocarditis have been reported in acute promyelitic leukemia patients.

Signs of hypervitaminosis A (Table 5.30) are usually reversed on cessation of exposure to the vitamin.

Embryotoxic Potential of High Levels of Vitamin A

Retinoids can be toxic to maternally exposed embryos – a fact that limits their therapeutic uses and raises concerns

---

113. i.e., the convex displacement of the infant's "soft spot" (the membranous covering of the cranial sutures) caused by fluid accumulation in the skull cavity or increased intracranial pressure.

114. Inflammation of the periosteum, the membranous tissue surrounding bone.
115. The Rancho Bernardo Study of 570 women and 388 men found a U-shaped relationship of bone mineral density and vitamin A intake, with optimal bone mineral density occurring at 2000–2800 IU (0.6–0.9 mg RE) per day (Promislow, J. H. E., Goodman-Gruen, D., Slymen, D. J., and Barret-Connor, E. [2002] J. Nutr. 129, 2246–2250).
116. Feskanich, D., Singh, V., Willet, W. C., and Colditz, G. A. (2002). J. Am. Med. Assoc. 287, 47–54.

TABLE 5.30 Signs of Hypervitaminosis A

Organ Affected	Signs
General	Muscle and joint pains, headache
Skin	Erythema, desquamation, alopecia
Mucous membranes	Cheilitis, stomatitis, conjunctivitis
Liver	Dysfunction
Skeletal	Thinning and fracture of long bones

TABLE5. 31 Teratogenicity of Vitamin A in Rodent Models

Species	Retinyl Palmitate[a]		All-*trans*-Retinoic Acid[a]
	Highest Non-Teratogenic	Lowest Teratogenic	Teratogenic
Rat[b]	30	90	6
Mouse[b]	15	50	3
Rabbit[c]	2	5	6
Hamster	—	—	7

[a]Dosage level (mg/kg/day).
[b]Exposed on gestational days 6–15.
[c]Exposed on gestational days 6–18.
Source: Kamm, J. J. (1982). *J. Am. Acad. Dermatol.* 64, 552–559.

TABLE 5.32 Teratogenic Risk of High Prenatal Exposures to Preformed Vitamin A

Retinol Intake (IU/day)	Pregnancies	Cranial–Neural Crest Defects	Total Defects
0–5,000	6,410	33 (0.51%)	86 (1.3%)
5,001–10,000	12,688	59 (0.47%)	196 (1.5%)
10,001–15,000	3,150	20 (0.63%)	42 (1.3%)
>15,000	500	9 (1.80%)	15 (3.0%)

Source: Rothman, K. J., Moore, L. L., Singer, M. R., et al. (1995). *N. Engl. J. Med.* 333, 1369–1373.

The critical period for fetal exposure to maternally derived retinoids is when organogenesis is occurring – i.e., before many women suspect they are pregnant. This is also before the development of fetal retinoid receptors as well as cellular and transport binding proteins, which serve to restrict maternal–fetal transfer of retinoids. Fetal malformations of cranial–neural crest origin have been reported in cases of oral use of all-*trans*-retinoic acid in treating *acne vulgaris* and of regular prenatal vitamin A supplements in humans. The latter have generally been linked to daily exposures at or above 20,000–25,000 IU. A retrospective epidemiologic study reported an increased risk of birth defects associated with an apparent threshold exposure of about 10,000 IU of preformed vitamin A per day;[118] however, the elevated risk of birth defects was observed in a small group of women whose average intake of the vitamin exceeded 21,000 IU/day (Table 5.32).

Carotenoid Toxicity

The toxicities of carotenoids are considered low, and circumstantial evidence suggests that β-carotene intakes of as much as 30 mg/day are without side effects other than the accumulation of the carotenoid in the skin, with consequent yellowing of the skin (**carotenodermia**). Regular, high intakes of β-carotene can lead to accumulation in fatty tissues, and thus to this condition. An intervention study with a small number of subjects showed that a daily intake of 30 mg of β-carotene from carotene-rich foods produced carotenodermia within 25–42 days of exposure;[119] the effect persisted for at least 14

about the safety of high-level vitamin A supplementation for pregnant animals and humans. This is especially true for 13-*cis*-retinoic acid, which is very effective in the treatment of acne but can cause severe disruption of cephalic neural crest cell activity that results in birth defects characterized by craniofacial, central nervous system, cardiovascular, and thymus malformations. Similar effects have been induced in animals by high doses of retinol, all-*trans*-retinoic acid or 13-*cis*-retinoic acid. Animal model studies suggest that the teratogenic effects of excess vitamin A are due to the embryonic exposure to all-*trans*-retinoic acid (Table 5.31), although those effects can be induced without substantially increasing maternal plasma concentrations of that metabolite. It has been proposed that the mechanism of teratogenic action of retinoids involves the elevated production of mRNAs for particular isoforms of RARs[117] that lead to an imbalance in heterodimers among the various RARs, RXRs, and other hormone receptors, consequently affecting the expression of genes not expressed in normal metabolism.

---

117. That teratogenic doses of all-*trans*-retinoic acid produce prolonged increases in RARα$_2$ mRNA levels supports this hypothesis.

118. Rothman *et al.* (*N. Engl. J. Med.* [1995] 33, 1369) estimated that apparent threshold, using regression techniques. That level of exposure to preformed vitamin A was associated with a birth defect risk of 1 in 57. This report has been criticized for suspected misclassification of malformations. Note that the current RDA for pregnant women is 800 μg of RE (i.e., 2700 IU) per day.

119. Carotenodermia was diagnosed only after plasma total carotenoid concentrations exceeded 4.0 mg/l.

days, and in some cases for more than 42 days, after cessation of carotene exposure. In addition to yielding retinal by symmetric cleavage, β-carotene can be cleaved asymmetrically to yield apo-carotenals and apo-carotenoic acids by a co-oxidative mechanism. Thus, it appears that under highly oxidative conditions β-carotene can yield oxidative breakdown products that can diminish retinoic acid signaling by interfering with the binding of retinoic acid to RAR. This effect has been proposed as the basis for the finding that a regular daily dose of β-carotene increased lung cancer risk among smokers (see "Anticarcinogenesis," this chapter).

10. CASE STUDIES

Instructions

Review each of the following case reports, paying special attention to the diagnostic indicators on which the respective treatments were based. Then answer the questions that follow.

Case 1

The physical examination of a 5-month-old boy with severe **marasmus**[120] showed extreme wasting, apathy, and ocular changes: in the left eye, Bitot's spots, and conjunctival and corneal xerosis; in the right eye, corneal liquefaction and keratomalacia with subsequent prolapse of the iris, extrusion of the lens, and loss of vitreous humor. The child was 65 cm tall and weighed 4.5 kg. His malnutrition had begun at cessation of breastfeeding at 4 months, after which he experienced weight loss and diarrhea.

Laboratory Results:

Parameter	Patient	Normal Range
Hb (hemoglobin)	10.7 g/dl	12–16 g/dl
HCT (hematocrit)	36 ml/dl	35–47 mg/dl
WBC (white blood cells)	15,000/μl	5,000–9,000/μl
Serum protein	5.6 g/dl	6–8 g/dl
Serum albumin	2.49 g/dl	3.5–5.5 g/dl
Plasma sodium	139 mEq/l	136–145 mEq/l
Plasma potassium	3.5 mEq/l	3.5–5.0 mEq/l
Blood glucose	70 mg/dl	60–100 mg/dl
Total bilirubin	1.1 mg/dl	<1 mg/dl
Serum retinol	5.5 μg/dl	30–60 μg/dl
Serum β-carotene	10.7 μg/dl	50–250 μg/dl
Serum vitamin E	220 μg/dl	500–1,500 μg/dl

The child had an infection, showing **otitis media**[121] and *Salmonella* septicemia,[122] which responded to antibiotic treatment in the first week. The patient was given, by nasogastric tube, an aqueous dispersion of retinyl palmitate (with a non-ionic detergent) at the rate of 3,000 μg/kg per day for 4 days. This increased his plasma retinol concentration from 5 to 35 μg/dl by the second day, at which level it was maintained for the next 12 days. The child responded to general nutritional rehabilitation with a high-protein, high-energy formula that was followed by whole milk supplemented with solid foods. He recovered, but was permanently blind in the right eye and was left with a mild corneal opacity in the left eye. He returned to his family after 10 weeks of hospitalization.

Case 2

An obese 15-year-old girl, 152 cm tall and weighing 100 kg, was admitted to the hospital for partial jejuno-ileal bypass surgery for morbid obesity. She had a past history of obsessive eating that had not been correctable by diet. Except for massive obesity, her physical examination was negative.

Initial Laboratory Results:

Parameter	Patient	Normal range
Hb	14 g/dl	12–15 g/dl
RBC	$4.5 \times 10^6/\mu l$	$4–5 \times 10^6/\mu l$
WBC	8,000/μl	5,000–9,000/μl
Serum retinol	38 μg/dl	30–60 μg/dl
Serum β-carotene	12 μg/dl	50–300 μg/dl
Serum vitamin E	580 μg/dl	500–1500 μg/dl
Serum 25-OH-D$_3$	11 ng/dl	8–40 ng/dl

The following test results were within normal ranges: serum electrolytes, calcium, phosphorus, triglycerides, cholesterol, total protein, albumin, total bilirubin, iron, TBIC (total serum iron-binding capacity), copper, zinc, folic acid, thiamin, and vitamin B$_{12}$.

The patient encountered few postoperative complications except for mild bouts of diarrhea and some fatigue. Over the next year, she lost 45 kg of body weight while ingesting a liberal diet. She reported having three to four stools daily, but denied having any objectionable diarrhea or changes in stool appearance. Two years after surgery, she noted the onset of inflammatory horny lesions above her knees and elbows, and she experienced some difficulty

---

120. Extreme emaciation or general atrophy, occurring especially in young children; it is caused by extreme undernutrition, owing primarily to lack of energy and protein.

121. Inflammation of the middle ear.
122. Presence in the blood of pathogenic, gram-negative, rod-shaped bacteria of the genus *Salmonella*.

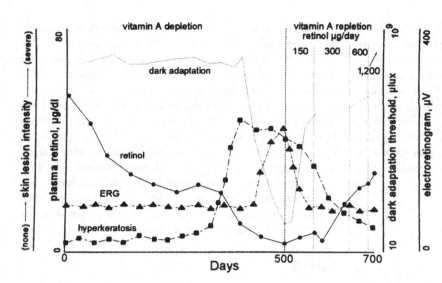

FIGURE 5.21 Subject responses to vitamin deprivation and repletion.

in seeing at dusk. The skin lesions failed to respond to topical corticosteroids and oral antihistamine therapy. Because of intensification of these signs, she sought medical help; however, the cause was not determined.

She was readmitted to the hospital, complaining of her skin disorder and night blindness. At that time, she showed evidence of mild liver dysfunction and her serum concentrations of retinol and β-carotene were 16 and 14 μg/dl, respectively. Her fecal fat was 70 g/day (normal, <7 g/day). Biopsies of the skin of her left thigh and right upper arm each showed **hyperkeratosis** and horny plugging of dilated follicles. She was treated with 15,000 μg of retinyl palmitate given orally three times daily for 6 months. By 1 month, the follicular hyperkeratosis had cleared and healed with residual pigmentation. By 2 months, the night blindness had subsided. At that time, her serum retinol concentration was 54 μg/dl, β-carotene was 7 μg/dl, α-tocopherol was 1.6 μg/ml, and urinary [^{57}Co]B$_{12}$ was 6.7% (normal, 7–8%). She has been well on a daily oral supplement of 1,500 μg of retinyl palmitate.

Case 3

A 41-year-old man was housed in a metabolic ward for 2 years during a clinical investigation of vitamin A deficiency. He weighed 77.3 kg, and was healthy by standard criteria (history, physical examination, and laboratory studies). For 505 days, he was fed a casein-based formula diet that contained <10 μg of vitamin A per day. His initial plasma retinol concentration was 58 μg/dl, and his body vitamin A pool, determined by isotope dilution, was 766 mg (10 mg/kg). At the end of 1 year, his plasma retinol had declined to 25 μg/dl and he began to show follicular hyperkeratosis (*see* Fig. 5.21). On day 300, his plasma retinol was 20 μg/dl, and he showed a mild anemia (Hb 12.6 mg/dl). Two months later, by which

time his plasma retinol had dropped to 10 μg/dl, he developed night blindness as evidenced by changes in dark adaptation and electroretinogram. When his plasma retinol reached 3 μg/dl, his body vitamin A pool was 377 mg and repletion with vitamin A was begun, with increasing doses starting with 150 μg and rising to 1,200 μg of retinol per day over a 145-day period. After receiving 150 μg of retinol per day for 82 days, his night blindness was partially repaired, but his skin keratinization remained and his plasma retinol level was only 8 μg/dl. Then, after receiving 300 μg of retinol per day for 42 days, his follicular hyperkeratosis resolved and his plasma retinol level was 20 μg/dl. At the 600 μg of retinol per day level, his plasma retinol was in the normal range and all signs of vitamin A deficiency disappeared.

Case Questions and Exercises

1. For each case, what signs/symptoms indicated vitamin A deficiency?
2. Propose hypotheses to explain why the patients of Cases 1 and 2 each responded to oral vitamin A treatment even though they had very different medical conditions. Outline tests of those hypotheses.
3. Comment on the value of serum retinol concentration for the diagnosis of nutritional vitamin A status.

Study Questions and Exercises

1. Discuss how the absorption, transport, tissue distribution, and intracellular activities of vitamin A relate to the concept of solubility.
2. Construct a flow diagram showing vitamin A, in its various forms, as it passes from ingested food, through the body where it functions in its various physiologic roles, and ultimately to its routes of elimination.

3. Construct a decision tree for the diagnosis of vitamin A deficiency in a human or animal.
4. Night blindness is particularly prevalent among alcoholics. Propose a hypothesis for the metabolic basis of this phenomenon and outline an experimental approach to test it.
5. Discuss the points of control, and intervention possibilities for each, in the WHO conceptual framework for vitamin A deficiency.

RECOMMENDED READING

ACC/SCN Consultative Group, 1994. Controlling vitamin A deficiency. Nutrition policy paper No. 14. In: Gillespie, S., Mason, J.B. (Eds.), United Nations Administrative Committee of Coordination – Subcommittee on Nutrition, New York, NY, 81 pp.

Azais-Braesco, V., Pascal, G., 2000. Vitamin A in pregnancy: Requirements and safety limits. Am. J. Clin. Nutr. 71, 1325S–1333S.

Bellemère, G., Stamatas, G.N., Bruère, V., et al., 2009. Antiaging action of retinol: From molecular to clinical. Skin Pharmacol. Physiol. 22, 200–209.

Bendich, A., Langseth, L., 1989. Safety of vitamin A. Am. J. Clin. Nutr. 49, 358–371.

Bloem, M.W., 1995. Interdependence of vitamin A and iron: An important association for programmes of anemia control. Proc. Nutr. Soc. 54, 501–508.

Campo-Paysaa, F., Marlétaz, F., Laudet, V., Schubert, M., 2008. Retinoic acid signaling in development: Tissue-specific functions and evolutionary origins. Genesis 46, 640–656.

Casetenmiller, J.J.M., West, C.E., 1998. Bioavailability and bioconversion of carotenoids. Ann. Rev. Nutr. 18, 19–38.

Chroni, E., Monastrili, A., Tsambo, D., 2010. Neuromuscular adverse effects associated with systemic retinoid dermatotherapy: Monitoring and treatment algorithm for clinicians. Drug Saf. 33, 25–34.

Congdon, N.G., West Jr, K.P., 2002. Physiological indicators of vitamin A status. J. Nutr. 132, 2889S–2894S.

Duester, G., 2008. Retinoic acid synthesis and signaling during early organogenesis. Cell 134, 921–931.

Duriancik, D.M., Lackey, D.E., Hoag, K.A., 2010. Vitamin A as a regulator of antigen presenting-molecules. J. Nutr. 140, 1395–1399.

Futoryan, T., Gilchrest, B.A., 1994. Retinoids and the skin. Nutr. Rev. 52, 299–310.

Graham, T.E., Yang, Q., Blüher, M., et al., 2010. β-Carotene is an important vitamin A source for humans. J. Nutr. 140, 2268S–2285S.

Harrison, E.H., 2005. Mechanisms of digestion and absorption of dietary vitamin A. Ann. Rev. Nutr. 25, 87–103.

Latham, M., 2010. The great vitamin A fiasco. J. World Pub. Health Nutr. Assoc. 1, 12–24.

Leitz, G., Lange, J., Rimbach, G., 2010. Molecular and dietary regulation of β,β-carotene 15,15′-monooxygenase 1 (BCMO1). Arch. Biochem. Biophys. 502, 8–16.

Li, E., Norris, A.W., 1996. Structure/function of cytosolic vitamin A-binding proteins. Annu. Rev. Nutr. 16, 205–234.

Libien, J., Blaner, W.S., 2007. Retinol and retinol-binding protein in cerebrospinal fluid: Can vitamin A take the "idiopathic" out of idiopathic intracranial hypertension? J. Neuro-Ophthalmol. 4, 253–257.

Lobo, G.P., Hessel, S., Eichinger, A., et al., 2010. ISX is a retinoic acid-sensitive gatekeeper that controls intestinal β-carotene absorption and vitamin A production. FASEB J. 24, 1656–1666.

Mason, J.B., Lotfi, M., Dalmiya, N., et al., 2001. The Micronutrient Report: Current Progress in the Control of Vitamin A, Iodine and Iron Deficiencies. The Micronutrient Initiative, Ottawa, 116 pp.

McClaren, D.S., Frigg, M., 1997. Sight and Life Manual on Vitamin A Deficiency Disorders (VADD). Task Force for Sight and Life, Basel, 138 pp.

Mora, J.R., von Andrian, U.H., 2009. Role of retinoic acid in the imprinting of gut-homing IgA-secreting cells. Semin. Immunol. 21, 28–35.

Noy, N., 2010. Between death and survival: Retinoic acid in regulation of apotosis. Ann. Rev. Nutr. 30, 201–217.

Ong, D.E., 1994. Cellular transport and metabolism of vitamin A: Roles of the cellular retinol-binding proteins. Nutr. Rev. 52, S24–S31.

Penniston, K.L., Tanumihardjo, S.A., 2006. The acute and chronic toxic effects of vitamin A. Am. J. Clin. Nutr. 83, 191–201.

Pino-Lago, K., Benson, M.J., Noelle, R.J., 2008. Retinoic acid in the immune system. Ann. NY Acad. Sci. 1143, 170–187.

Ramakrishnan, U., Darnton-Hill, I., 2002. Assessment and control of vitamin A deficiency disorders. J. Nutr. 132, 2947S–2953S.

Reifen, R., 2002. Vitamin A as an anti-inflammatory agent. Proc. Nutr. Soc. 61, 397–400.

Ribaya-Mercado, J.D., Blumberg, J.B., 2007. Vitamin A: Is it a risk factor for osteoporosis and bone fracture? Nutr. Rev. 65, 425–438.

Ross, C., Harrison, E.H., 2007. Vitamin A: Nutritional aspects of retinoids and carotenoids. In: Zempleni, J., Rucker, R.B., McCormick, D.B., Suttie, J.W. (Eds.), Handbook of Vitamins, fourth ed. CRC Press, New York, NY, pp. 1–40.

Ross, C., Chen, Q., Ma, Y., 2009. Augmentation of antibody responses by retinoic acid and costimulatory molecules. Semin Immunol. 21, 42–50.

Solomons, N.W., 1993. Plant sources of pro-vitamin A and human niture. Nutr. Rev. 51, 199–204.

Solomons, N.W., Bulux, J., 2006. Present Knowledge in Nutrition. In: Bowman, B.A. Russel, R.M. (Eds.), Vitamin A, ninth ed., vol I. ILSI Press, Washington, DC, pp. 157–183.

Sommer, A., 2008. Vitamin A deficiency and clinical disease: An historical overview. J. Nutr. 138, 1835–1839.

Sommer, A., West Jr, K.P., Olson, J.A., Ross, C.A., 1996. Vitamin A Deficiency: Health, Survival, and Vision. Oxford University Press, New York, NY. (452 pp.)

Soprano, D.R., Soprano, K.J., 1995. Retinoids as teratogens. Annu. Rev. Nutr. 15, 111–132.

Soprano, D.R., Qin, P., Soprano, K.J., 2004. Retinoic acid receptors and cancers. Annu. Rev. Nutr. 24, 201–221.

Tang, G., 2010. Bioconversion of dietary provitamin A carotenoids to vitamin A in humans. Am. J. Nutr. 91, 1468S–1473S.

Thatcher, J.E., Isoheranen, N., 2009. The role of CYP26 enzymes in retinoic acid clearance. Expert. Opin. Drug Metab. Toxicol. 5, 875–886.

Underwood, B.A., 1989. Teratogenecity of vitamin A. In: Walter, P., Brubacher, G., Stähelin, H. (Eds.), Elevated Dosages of Vitamins: Benefits and Hazards. Hans Huber, Berlin, pp. 42–55.

Underwood, B.A., 1994. Maternal vitamin A status and its importance in infancy and early childhood. Am. J. Clin. Nutr. 59, 517S–524S.

Van Poppel, G., Goldbohm, R.A., 1995. Epidemiologic evidence for β-carotene and cancer prevention. Am. J. Clin. Nutr. 62, 1393S–1402S.

Villamour, E., Fawzi, W.W., 2005. Effects of vitamin A supplementation on immune responses and correlations with clinical outcomes. Clin. Micro. Rev. 18, 446–464.

Wang, Z., Yin, S., Zhao, X., et al., 2004. β-Carotene–vitamin A equivalence in Chinese adults assessed by an isotope dilution technique. Br. J. Nutr. 91, 121–131.

West Jr, K.P., 2002. Extent of vitamin A deficiency among preschool children and women of reproductive age. J. Nutr. 132, 2857S–2866S.

Willett, W.C., Hunter, D.J., 1994. Vitamin A and cancers of the breast, large bowel and prostate: Epidemiologic evidence. Nutr. Rev. 52, S53–S59.

Wolf, G., 2004. The visual cycle of the cone photoreceptors of the retina. Nutr. Rev. 62, 283–286.

Wolf, G., 2007. Serum retinol-binding protein: A link between obesity, insulin resistance, and type 2 diabetes. Nutr. Rev. 65, 251–256.

World Health Organization. (2009). Global Prevalence of Vitamin A Deficiency in Populations at Risk 1995–2005. WHO Global Database on Vitamin A Deficiency, Geneva, World Health Organization. <www.who.int/vmnis/> (accessed 25.02.11).

Yeum, K.J., Russell, R.M., 2002. Carotenoid bioavailability and bioconversion. Ann. Rev. Nutr. 22, 483–504.

Ziouzenkova, O., Plutzky, J., 2008. Retinoid metabolism and nuclear receptor responses: New insights into coordinated regulation of the PPAR-RXR complex. FEBS Letts, 582, 32–38.

Vitamin D

Chapter Outline

Anchoring Concepts

1. Vitamin D is the generic descriptor for *steroids* exhibiting qualitatively the biological activity of cholecalciferol (i.e., vitamin D_3).
2. Most vitamers D are *hydrophobic*, and thus are insoluble in aqueous environments (e.g., plasma, interstitial fluids, cytosol).
3. Vitamin D is not required in the diets of animals or humans adequately exposed to sources of *ultraviolet light* (e.g., sunlight).
4. Deficiencies of vitamin D lead to structural lesions of *bone*.

By following the reasoning that vitamin D is not required in the diet under conditions of adequate ultraviolet irradiation of skin and that it is the precursor of a hormone, it is likely that the vitamin is not truly a vitamin but must be regarded as a pro-hormone. These arguments, however, are only semantic; the fact remains that vitamin D is taken in the diet and is an extremely potent substance which prevents a deficiency disease.

H. F. DeLuca

Learning Objectives

1. To understand the nature of the various sources of vitamin D.
2. To understand the means of endogenous production of vitamin D.
3. To understand the means of enteric absorption of vitamin D.
4. To understand the transport and metabolism involved in the activation of vitamin D to its functional forms.
5. To understand the role of vitamin D and other endocrine factors in calcium homeostasis.
6. To understand the roles of vitamin D in non-calcified tissues.
7. To understand the genomic bases of vitamin D action.
8. To understand the physiologic implications of high doses of vitamin D.

VOCABULARY

Cage layer fatigue
Calbindins
Calcidiol
Calcinosis
Calcipotriol
Calcitonin
Calcitriol
Calcitroic acid
Calcium (Ca)
Calcium-binding protein (CaBP)
Calcium-sensing receptor (CaR)
Calmodulin
Cathelicidin antimicrobial protien (CAMP)
Cholecalciferol
DBP (vitamin D-binding protein)
7-Dehydrocholesterol
Diabetes
1,25-Dihydroxyvitamin D (1,25-[OH]$_2$-vitamin D)
24,25-Dihydroxyvitamin D (24,25-[OH]$_2$-vitamin D)
Diuresis
Epiphyseal plate

The Vitamins. DOI: 10.1016/B978-0-12-381980-2.00006-2

Ergocalciferol
Ergosterol
Genu varum
25-Hydroxyvitamin D (25-OH-vitamin D)
Hypercalcemia
Hyperphosphatemia
Hypersensitivity
Hypocalcemia
Hypoparathyroidism
Hypophosphatemia
Lead (Pb)
Melanin
Milk fever
25-OH-Vitamin D 1-hydroxylase
Osteoblast
Osteochondrosis
Osteoclast
Osteomalacia
Osteon
Osteopenia
Osteoporosis
Parathyroid gland
Parathyroid hormone (PTH)
Privational osteomalacia
Privational rickets
Prolactin
Provitamin D
Pseudofracture
Pseudohypoparathyroidism
Psoriasis
Rickets
Sarcopenia
Tibial dyschondroplasia
Transcalciferin
Transcaltachia
24,25,26-Trihydroxyvitamin D (24,25,26-[OH]$_3$-vitamin D)
Varus deformity
Vitamin D binding protein (DBP)
Vitamin D-dependent rickets
Vitamin D receptors (VDRs)
Vitamin D$_2$
Vitamin D$_3$
Vitamin D 25-hydroxylase
Vitamin D-resistant rickets

1. THE SIGNIFICANCE OF VITAMIN D

Vitamin D,[1] the "sunshine vitamin," is actually a hormone produced from sterols in the body by the photolytic action of ultraviolet light on the skin; individuals who receive modest exposures to sunlight are able to produce their own vitamin D. However, this is not the case for many people, such as those who live in northern latitudes, spend most of their days indoors, and/or have darker skin, and for animals reared in controlled environments. Such individuals must obtain the nutrient from their diets; for them, vitamin D is a vitamin in the traditional sense.

Vitamin D plays an important role, along with the essential minerals calcium, phosphorus, and magnesium, in the maintenance of healthy bones and teeth – where problems appear to have existed in a variety of past populations from various parts of the world.[2] They remain contemporary issues of major public health impact; an estimated 40–90% of adults worldwide are of insufficient vitamin D status. The prevalence of low vitamin D status has been estimated at 11–84% of adults in the United States, 5–40% of adults in the United Kingdom, and 93% of youth in Canada.[3]

Rickets, the deforming and debilitating disease involving delayed or failed endochondral ossification (mineralization at the growth plates) of the long bones, remains a problem in many countries, having been reported at prevalences as great as 10% among infants exclusively fed breast milk and children with little sun exposure. Shockingly high prevalences have been reported in Yemen (27%), Ethiopia (42%), Tibet (66%), and Mongolia (70%).[4]

The global prevalence of osteomalacia in adults is not well documented. Each year Americans experience an estimated 13 million fractures,[5] including more than 350 hip fractures, and metabolic bone diseases are estimated to cost the United States some $13 billion in immediate medical care costs, with as much as $8 billion lost annually from the US economy in the form of extended treatment and lost productivity.[6] Metabolic bone diseases target

---

1. The use of the letter D without a subscript indicates either vitamin D$_2$ (ergocalciferol) or vitamin D$_3$ (cholecalciferol).

2. Archeological studies have revealed low bone mass. That these findings have not always been associated with osteoporotic fractures may suggest that earlier populations may have had greater calcium intakes than most people today (Nelson, D. A., Sauer, N. J., and Agarwal, S. C. [2004]. Evolutionary aspects of bone health. In *Nutrition and Bone Health* [Holick, M. F. and Dawson-Hughes, B., eds]. Humana Press, Totowa, NJ, pp. 3–18).
3. Egan, K. M., Signorello, L. B., Munro, H. M., *et al.* (2008). Vitamin D insufficiency among African-Americans in the southeastern United States: Implications for cancer disparities (United States). *Cancer Causes Control* 19, 527–535; Prentice, A. (2008). Vitamin D deficiency: A global perspective. *Nutr. Rev.* 66, S153–S164; Mark, S., Gray-Donald, K. Delvin, E. E., *et al.* (2008) Low vitamin D status in a representative sample of youth from Quebec, *Canada. Clin. Chem.* 54, 1283–1289; Holick, M. F. (2008). Vitamin D: A D-lightful health perspective. *Nutr. Rev.* 66, S182–S194.
4. Prentice, A. (2008). Vitamin D deficiency: A global perspective. *Nutr. Rev.* 66, S153–S164.
5. Fractures of the vertebrae, hip, forearm, leg, and ankle, in that order, are the most common, although in many cases they may be asymptomatic. In 1999, hip fractures in the United States accounted for an estimated 338,000 physician office visits. Canadian health statistics show the risk of radial fractures for men and women to be about 25 per 100,000 until the fifth decade of life, after which it increases for women linearly to more than 200 per 100,000 by the seventh to eighth decades. Men, in contrast, do not show an appreciable age-related increase in risk.
6. Only 25% of hip fracture patients can be expected to make full recoveries; 40% will require admission to nursing homes; half will be dependent on a cane or walker; and 20% will die within a year of various complications.

women, who are more susceptible to **osteoporosis** (i.e., the loss of bone leading to increased bone fragility) than are men. Starting during the fourth or fifth decade of life, both men and women lose bone mass at similar annual rates;[7] however, after the onset of menopause, the rate of bone loss in women can increase as much as 10-fold[8] owing to diminished production of estrogen, which is required along with vitamin D to maintain bone mineralization. It has been estimated that among 50-year-old Americans, half of all women and one-fifth of all men have signs of osteoporosis; accordingly, 75–80% of all hip fractures occur in women. As the US population ages, these rates can be expected to increase; that of hip fractures is expected to triple by the year 2050.

Vitamin D status affects more than bone. The vitamin also functions in the regulation of cellular development and differentiation of most cells, in the regulation of the parathyroid gland and immune system function, in the skin, in cancer prevention, and in the metabolism of foreign compounds.

2. SOURCES OF VITAMIN D

Distribution in Foods

Vitamin D, as either **ergocalciferol** (**vitamin D$_2$**) or **cholecalciferol** (**vitamin D$_3$**), is rather sparsely represented in nature; however, its provitamins are common in both plants and animals. Ergocalciferol and its precursor **ergosterol** are found in plants, fungi,[9] molds, lichens, and some invertebrates (e.g., snails and worms). In fact, some microorganisms are quite rich in ergosterol, in which it may comprise as much as 10% of the total dry matter.[10] Ergosterol does not occur naturally in higher vertebrates, but it can be present in low amounts in tissues of those species as the result of their consuming it. The actual distribution of ergocalciferol in nature is much more limited and variable than that of ergosterol (e.g., grass hays and alfalfa contain vitamin D only after they have been cut and left to dry in the sun). Whereas vitamin D$_2$ is probably present only in small amounts from natural sources, it has been a major synthetic form used in animal and human nutrition for several decades.

Cholecalciferol is widely distributed in animals, but has an extremely limited distribution in plants. In animals, tissue cholecalciferol concentrations are dependent on the vitamin D$_3$ content of the diet and/or the exposure to sunlight. Few foods, however, are rich in the vitamin. Fish liver and oils[11] are particularly rich sources of vitamin D$_3$, which occurs in those materials in free form as well as esters of long-chain fatty acid esters. Oily fishes can provide significant amounts of vitamin D by virtue of their positions in the upper levels of food chains that include lower trophic level species, including those that consume ergosterol-containing plants. However, farm-raised fish fed formulated diets may not contain significant amounts of vitamin D unless it has been added as a supplement to their rations. With a few notable exceptions, vitamin D$_3$ is not found in plants. Those exceptions include the species[12] *Solanum glaucophyllum*, *Solanum malacoxylon*, *Cestrum diurnum*, and *Trisetum flavescens*, in which the vitamin occurs as water-soluble β-glycosides of vitamin D$_3$, 25-hydroxyvitamin D$_3$ (25-OH-D$_3$), and 1,25-dihydroxyvitamin D$_3$ (1,25-[OH]$_2$-D$_3$).[13]

Most species appear not to discriminate between the vitamers D$_2$ and D$_3$. Studies in humans have demonstrated that each supports comparable plasma 25-OH-D responses, whether given individually or in mixtures.[14] Birds, however, are notably unable to utilize vitamin D$_2$, for which reason the D$_3$ vitamer is used in vitamin supplements for poultry diets.

Because most foods contain only very low amounts of vitamin D (*see* Table 6.1), it is the practice in many countries to fortify certain frequently consumed foods (e.g., baked goods, grain products, milk,[15] yogurt, cheeses,

7. Typical rates of bone loss of men and premenopausal women are in the range of 0.3–0.5%/year.
8. Women can lose as much as 20% of their total bone mass within the first 5–7 years following menopause; by age 70–80, many women have lost 30–50% of their bone mass (in contrast with 20–30% losses for men).
9. Ergosterol was named for the parasitic fungus, *ergot*, from which the sterol was first isolated. Sun-dried mushrooms can be a good source of vitamin D.
10. Provitamin D$_2$ accounts for virtually all of the sterols in *Aspergillus niger* and 80% of those in *Saccharomyces cerevisiae* (i.e., brewers' yeast).

11. Fish oils typically have vitamin D$_3$ concentrations of about 50 μg/g, but cod, tuna, and mackerel oils can contain 20 times that level. Marine mammals, which consume large amounts of such cold water fishes, accumulate vitamin D$_3$ from those sources in their livers. Therefore, peoples (e.g., Inuit, Faroe Islanders) who consume seals and whales obtain vitamin D$_3$ from those foods.
12. Consumption of these plants has been associated with calcinosis in grazing ruminants; this observation was the basis of the discovery that they contained vitamin D.
13. Both vitamin D$_3$-25-hydroxylase and 25-OH-vitamin D$_3$ 1-hydroxylase activities have been found in *S. malacoxylon*, indicating that its ability to metabolize vitamin D is similar to that of higher animals.
14. Holick, M. F., Biancuzzo, R. M., Chen, T. C., *et al.* (2008). *J. Clin. Endocrinol. Metab.* 93, 677.
15. In the late 1950s, the American Medical Association recommended that milk be fortified with 400 IU (10 μg) per quart; the US Food and Drug Administration has specified that milk should contain 400–600 IU/qt. However, a survey by Tanner *et al.* (Tanner, J. T., Smith, J., Defibaugh, P., *et al.* [1988] *J. Assoc. Off. Anal. Chem.* 71, 607–610) showed that most fortified milk products failed to contain the specified amounts of vitamin D stated. Most European countries do not fortify milk with vitamin D.

TABLE 6.1 Vitamin D Activities in Foods

Food	Vitamin D (IU[a]/100 g)
Animal Products	
Milk	
Cow	0.3–54[b]
Human	0–10
Dairy Products	
Butter	35
Cheese	12
Cream	50
Eggs	28
Fish Products	
Cod	85
Cod liver oil	10,000
Herring	330
Herring liver oil	140,000
Mackerel	120
Salmon	220–440
Sardines	1,500
Shrimp	150
Meats	
Beef	13
Pork	84
Poultry	80
Poultry skin	900
Liver	
Beef	8–40
Chicken	50–65
Pork	40
Plant Products	
Cabbage	0.2
Corn oil	9
Spinach	0.2

[a] 1 IU = 0.025 mg of vitamin D_2 or vitamin D_3.
[b] US regulations specify that milk be fortified with 400 IU of vitamin D_3 per quart (about 37 IU/100 ml).

margarines, orange juice, infant foods, and some breads). Both vitamers D_2 and D_3 are used in the fortification of foods. Some other foods may be enriched indirectly as a result of the supplementation of animal feeds with the

vitamin. In general, therefore, the chief food sources of vitamin D in Western diets are fortified milk, juices, and cereals; and fatty fish.

Vitamin D can also be obtained from nutritional supplements. For example, multivitamin supplements typically contain 400 IU vitamin D, and pharmaceutical preparations can contain as much as 50,000 IU vitamin D_2 per capsule/tablet.

Biosynthesis of Vitamin D_3

Vitamin D is formed in animals by the action of ultraviolet light in the UVB range (290–310 nm) on 7-dehydrocholesterol in the skin (see Fig. 6.1). The activation reaction depends on the absorption of UV light (optimally, 295–300 nm) by the 5,7-diene of the B ring of the sterol nucleus, causing it to open[16] and isomerize to form the energetically more stable s-trans, s-cis-previtamin D_3. This physicochemical reaction appears to convert only 5–15% of the available 7-dehydrocholesterol to vitamin D_3.[17] That efficiency is affected by the physical properties of the skin and of the environment; thus, it differs between individuals and species, and shows great variation according to time of day, season, and latitude.

The provitamin D sterol, 7-dehydrocholesterol, is both a precursor to and a product of cholesterol (via different pathways); it is synthesized in the sebaceous glands of the skin and is secreted rather uniformly onto the surface, where it is reabsorbed into the various layers of the epidermis.[18] Thus, the skin contains very high concentrations of the sterol (e.g., 200 times that of liver). The distribution of 7-dehydrocholesterol in the epidermis varies according to distance from the surface. In humans, the greatest concentrations are found in the deeper *Malpighian layer*; whereas, in the rat it is distributed more superficially in the *stratum corneum*. In consequence of such a difference in **provitamin D_3** distribution, photoproduction can occur in the rat stratum corneum but only deeper in humans. This makes the process less efficient in humans, as absorption by the stratum corneum reduces the amount of UV reaching 7-dehydrocholesterol-rich layers.

Vitamin D biosynthesis is thus deteremined by environmental exposure to UV light, which also can increase risk to skin cancers in individuals experiencing episodes of severe

16. For this reason, the vitamers D are called "secosteroids," denoting the presence of a "broken" ring by the Latin prefix *seco*, "to cut."
17. Excess irradiation does little to increase the efficiency of this activation step. Instead, it increases the photoproduction of biologically inactive forms, e.g., lumisterol-3, tachysterol-3, and 5,6-*trans*-vitamin D_3.
18. In humans, the epidermis (particularly the *stratum basale* and *stratum spinosum*) contains twice as much 7-dehydrocholesterol as the dermis.

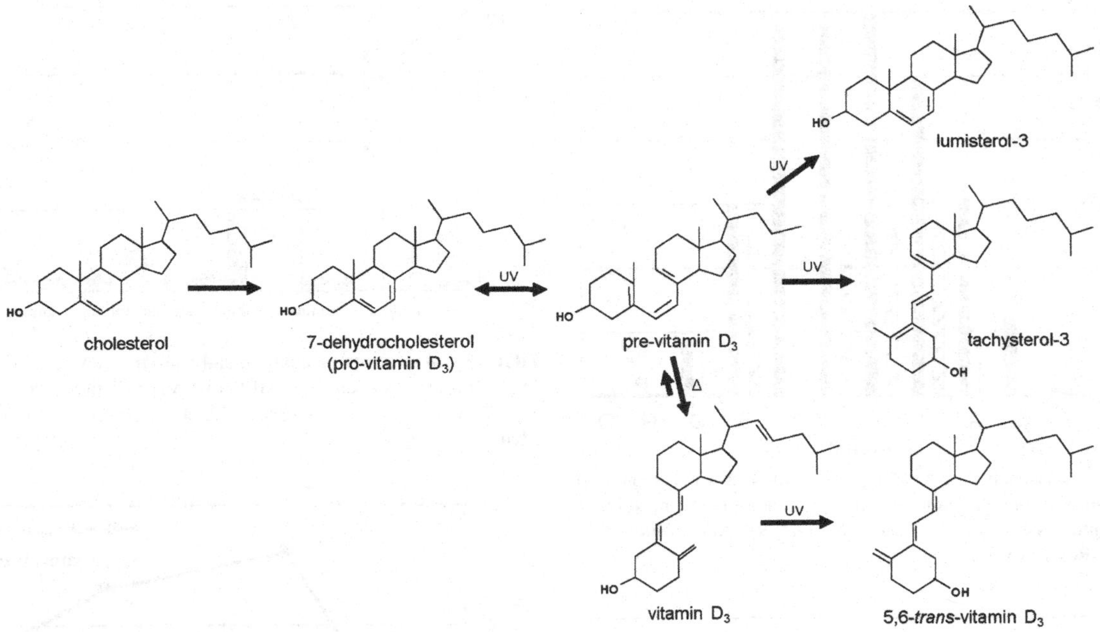

FIGURE 6.1 Biosynthesis of vitamin D_3 and its photolytic by-products.

burning.[19] The amount of sunlight required to support adequate vitamin D status is substantially less than that which increases skin cancer risk. Holick has estimated that exposure of only 6–10% of body surface to a median erythemal dose (MED)[20] of sunlight can be equivalent to consuming 600–1,000 IU of vitamin D. He recommends exposures of one-quarter of that amount of sunlight two to three times weekly to support the synthesis of physiologically relevant amounts, *ca.* 15,000 IU per week, of the vitamin.[21]

Holick's Rule*

Sun exposure of one-quarter MED over one-quarter of the body is equivalent to 1,000 IU of oral vitamin D_3.

*Holick, M. F. (2001). *Lancet* 357, 4.

Because sunlight exposure varies with latitude and season, vitamin D biosynthesis also varies according to those factors (Table 6.2). Vitamin D-producing UV irradiation varies with the zenith angle of the sun, being greatest at noon (60% occurs between 10 AM and 2 PM), reaching an

TABLE 6.2 Determinants of Vitamin D_3 Biosynthesis

Exogenous Factors Affecting Sunlight Exposure
Season
Latitude
Time spent out of doors
Use of topical sunscreens
Endogenous factors affecting UVB responsiveness
Skin thickness
Skin pigmentation

annual peak at midsummer (Fig. 6.2), and declining with the distance from the Earth's equator. In winter there is almost no UV light at latitudes 50°N/S,[22] and above 40°N/S[23] there is virtually no vitamin D biosynthesis in the skin.

The vitamin D biosynthetic capacity of skin, while great, is diminished by factors that block UV penetration into the skin. Both the thickness and 7-dehydrocholesterol content of skin decline with age, such that the vitamin D

19. UV radiation exposure is a risk factor for basal and squamous cell carcinomas, and malignant melanomas. Basal cell carcinomas comprise 70–85% of skin cancers; malignant melanomas, while far less prevalent, account for the majority of skin cancer deaths.

20. 1 MED is that amount of sunlight that causes the skin to turn slightly pink, producing detectable erythema – the sign of mild sunburn.

21. Holick, M. F. and Jenkins, M. (2003). *The UV Advantage*, ibooks, Inc., New York, NY, p. 93.

22. That is, that of Winnipeg, Frankfurt, Prague, and Kiev in the northern hemisphere, and Launceston (Tasmania) and the southern tips of Patagonia (Argentina) and Chile in the southern hemisphere. Note: the entire African and Australian contents lie north of 50°S.

23. That is, that of Denver, Philadelphia, Toledo (Spain), Ankara, and Beijing in the northern hemisphere, and San Martin de los Andes (Argentina) in the southern hemisphere. Note: the entire African and Australian continents lie north of 40°S.

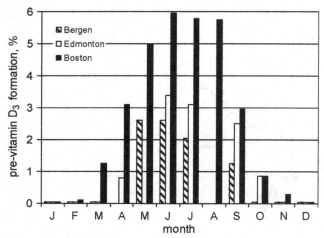

FIGURE 6.2 Seasonality of vitamin D biosynthesis (in response to 1-hour exposures to sunlight) at three different latitudes (Bergen, Norway, 60°N; Edmonton, Alberta, 52°N; Boston, MA, 42°N). After *Holick, M.F. (2008)*. Nutr. Rev. *66, S182.*

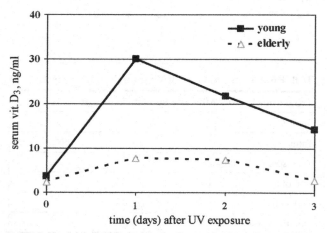

FIGURE 6.3 Circulating vitamin D3 responses of young and elderly human subjects. After *Holick, M. F., Siris, E. S., Binkley, N., et al. (1989)*. Lancet 4, *1104–1105.*

biogenic response to solar irradiation is also diminished in older people (*see* Fig. 6.3). Accordingly, an age-related decline in plasma 25-OH-D_3 concentration is typically seen in individuals without significant dietary intakes of the vitamin.[24] The epidermal pigment **melanin**[25] efficiently absorbs UVB, for which reason dark-skinned individuals require greater UV doses than light-skinned ones for comparable vitamin D biosynthesis. Compared to a person with

24. Need, A. G., Morris, H. A., Horowitz, M., *et al.* ([1993] *Am. J. Clin. Nutr.* 58, 882–885) found that the skinfold thickness of the back of the hand was significantly less for subjects in their sixties or seventies than for subjects in their forties or fifties; this corresponded to lower serum concentrations of 25-OH-D_3 in the oldest group.
25. Melanin absorbs solar radiation over the broad range of 290–700 nm. It has been proposed as an evolutionary adaptation to protect against hypervitaminosis D due to excessive sunlight exposure in tropical latitudes, which was lost by populations migrating to areas distant from the equator.

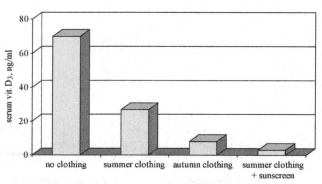

FIGURE 6.4 Effects of clothing and sunscreen on circulating vitamin D_3 responses of humans to 1 MED of UVB radiation. After *Matsuoka, L. Y., Wortsman, J., Dannenberg, M. J. et al. (1992)*. J. Clin. Endocrinol. Med. *75, 1099–1103.*

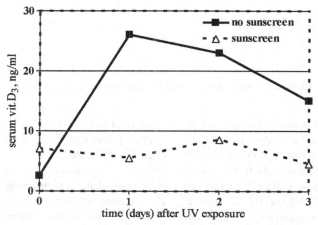

FIGURE 6.5 Effects of sunscreen (SPF8) application on circulating vitamin D_3 responses of humans to 1 MED of UVB radiation. After *Matsuoka, L. Y., Wortsman, J. , Hanifan, N., et al. (1987)*. J. Clin. Endocrinol. Med. *64, 1165–1168.*

light, type 1 skin (easily sunburned; never tan), an individual with dark, type 5 or 6 skin (seldom or never burn; always tan) can require 5–10 times as much solar exposure to produce the same amount of vitamin D_3.[26]

Physical factors that reduce the exposure of the skin to UV light also reduce the biosynthesis of vitamin D_3 (Figs 6.4, 6.5). These include factors associated with the lifestyle of humans (e.g., clothing, indoor living [glass and plexiglass absorb UV light], use of sun screens) and practical management of livestock (e.g., confined indoor housing). Properly applied topical suncreens with sun protection factors of 8 and greater have been shown to reduce cutaneous vitamin D_3 production by >95%.[27]

26. Bauer, J. M. and Freyberg, R. H. (1946). *J. Am. Med. Assoc.* 130, 1208–1215.
27. Holick, M. F. (*Am. J. Clin. Nutr.* [2004] 79, 362–371) has pointed out that because most people apply sunscreens in no more than half of recommended amounts, their use is not a practical limitation to cutaneous vitamin D_3 production.

TABLE 6.3 Contributions of Sun and Dietary Vitamin D to the Vitamin D Status of Women

Parameter	Low Sunlight Exposure		High Sunlight Exposure	
	Low Vitamin D	High Vitamin D	Low Vitamin D	High Vitamin D
Age (years)	84.9 ± 0.8	81.3 ± 1.6	83.6 ± 0.6	81.6 ± 1.5
Summer				
Vitamin D intake (mg/day)	3.58 ± 0.53	16.05 ± 1.38^a	4.08 ± 0.65	14.53 ± 1.15^a
Plasma 25-OH-D_3 (nmol/l)	44 ± 6	80 ± 8^a	57 ± 5	74 ± 6^a
Winter				
Vitamin D intake (mg/day)	3.48 ± 0.45	16.33 ± 1.33^a	4.40 ± 0.78	14.63 ± 1.43^a
Plasma 25-OH-D_3 (nmol/l)	35 ± 3	81 ± 7^a	42 ± 4	64 ± 4^a

[a]$P <0.05$.
Source: Salamone, L. M., Dallal, G. E., Zantos, D., *et al.* (1993). *Am. J. Clin. Nutr.* 58: 80–86.

TABLE 6.4 Bioactive Vitamin D Analogs

Dehydrotachysterol-2

1-OH-analogs, e.g., 1-OH-D_3

Vitamin D_2 derivatives, e.g., 1-OH-D_2; 1,25,28-$(OH)_3$-D_2; 1,24S-$(OH)_2$-D_2

Side chain derivatives

  cyclopropane ring (consisting of side chain carbons 25, 26 and 27) derivatives, e.g., calcipitriol

  20-epi and 20-methyl side chain derivatives

  analogs with one or more added carbons in the side chain or on the branching methyl groups

  unsaturated side chain derivatives, e.g., $C_{16}=C_{17}$, $C_{22}=C_{23}$

  oxa-containing (oxygen-for-carbon substitution), e.g., 22-oxa-calcitriol

  fluorinated derivatives, e.g., F_6-1,25-$(OH)_2$-D_3

That many people show seasonal changes in their serum 25-OH-D_3 levels (greatest concentrations occurring in the autumn, i.e., after a summer of relatively great solar exposure; Table 6.3) indicates that sunlight is generally more important than diet as a source of this critical nutrient. Under such conditions, vitamin D cannot be considered a vitamin at all; it is, instead, a pro-hormone produced in the skin. At the same time, several environmental and lifestyle and, for livestock, management factors can limit vitamin D_3 biosynthesis for many people and animals. This puts them in need of exogenous (dietary) vitamin D. Under such conditions, wherein endogenous synthesis is not sufficient to meet needs, vitamin D becomes a vitamin in the traditional sense.

Bioactive Vitamin D Analogs

More than 2,000 analogs have been developed for use in treating disorders of vitamin D metabolism, such as vitamin D refractory rickets, hypocalcemia and osteodystropy secondary to chronic renal disease, and some types of cancer. The goal of such efforts has been to develop drugs active in the vitamin D-dependent functions in regulating cell proliferation and growth while also having low calcemic potential, thus avoiding the adverse effects of high levels of vitamin D in causing calcinosis. Vitamin D-active analogs include pro-drugs activated metabolically, typically by the same cytochrome P-450-based enzymes used to activate vitamin D.[28] These contain structural deviations in the C and D rings of the steroid nucleus or of the side chain (Table 6.4).

3. ENTERIC ABSORPTION OF VITAMIN D

Micelle-Dependent Passive Diffusion

Vitamin D is absorbed from the small intestine by nonsaturable passive diffusion that is dependent on micellar solubilization, and, hence, the presence of bile salts. The fastest absorption appears to be in the upper portions of the small intestine (i.e., the duodenum and ileum), but, owing

28. Jones, G. (2005). In *Vitamin D*, 2nd edn (Feldman, D., Pike, J. W., and Glorieux, F. H., eds), Elsevier, New York, NY, pp. 1423–1448.

TABLE 6.5 Distribution of Vitamin D Metabolites in Human Plasma

Metabolite	Percentage Distribution			Normal Concentration
	DBP	Lipoprotein	Albumin	
Vitamin D_3	60	40	0	2–4 ng/ml
25-OH-D_3	98	2	0	15–38 ng/ml
24,25-$(OH)_2$-D_3	98	2	0	—
1,25-$(OH)_2$-D_3	62	15	23	20–40 pg/ml

to the longer transit time of food in the distal portion of the small intestine, the greatest amount of vitamin D absorption probably occurs there. Like other hydrophobic substances absorbed by micelle-dependent passive diffusion in mammals, vitamin D enters the lymphatic circulation[29] predominantly (about 90% of the total amount absorbed) in association with chylomicra, with most of the balance being associated with the α-globulin fraction.[30] The efficiency of this absorption process for vitamin D appears to be about 50%.

4. TRANSPORT OF VITAMIN D

Transfer from Chylomicra to Plasma

Almost all absorbed vitamin D is retained in non-esterified form, which is thought to be associated with the surface of chylomicrons, as a portion of the vitamin D can be transferred from chylomicra to a binding protein in the plasma, either directly or during the process of chylomicron degradation. That portion of vitamin D not transferred in the plasma is taken up with chylomicron remnants by the liver where it is transferred to the same binding protein and released to the plasma.

Vitamin D-Binding Protein

Like other sterols, vitamin D is transported in the plasma largely in association with protein. While some birds and mammals transport vitamin D in association with albumin, and fishes with cartilaginous skeletons (e.g., sharks and rays) transport it in association with plasma lipoproteins, most species[31] use a protein that has been called **transcalciferin** or, more commonly, **vitamin D-binding protein (DBP)** (Table 6.5). The DBP is in the same gene family as

albumin and α-fetoprotein. It is a glycosylated, cysteine-rich, α-globulin of 458 amino acids[32] with a molecular weight of 55–58 kDa, its size being dependent on its glycosylation state. It has three internally homogous α-helical domains and exists as multiple isoforms[33] due to differences in both the primary structure of the protein (involving the presence/absence of N-acetylneuraminic acid on a threonine residue at position 420) and the carbohydrate moiety that is added post-translationally. Three alleles are common.

The DBP binds vitamins D_2 or D_3 and their metabolites stoichiometrically,[34] with ligand-binding dependent on the *cis*-triene structure and C_3-hydroxyl grouping. In adequately nourished individuals, DBP binds some 88% of the 25-OH-D_3 in serum, binding 25-OH-D with an affinity an order of magnitude greater than that of 1,25-$(OH)_2$-D.[35] The concentration of DBP in the plasma, typically 4–8 μmol/l, greatly exceeds that of 25-OH-D_3 (*ca.* 50 nmol/l) and is remarkably constant, being unaffected by sex, age or vitamin D status. The excess binding capacity means that only 5% of DBP actually carries the vitamin. That DBP can also bind the plasma actin monomer and fatty acids suggests that the vitamin D transport protein may also have other functions in normal metabolism.[36] Its

---

29. In birds, reptiles, and fishes, vitamin D, like other lipids, is absorbed into the portal circulation via portomicra.

30. This is probably identical to the carrier, vitamin D-binding protein (DBP), in the plasma.

31. At least 140 different species in five classes have been examined.

32. In contrast, the chicken has *two* distinct DBPs (54 and 60 kDa), each of which binds vitamin D_3 and its metabolites with greater affinities than vitamin D_2 and its metabolites.

33. In humans, DBP is identical with group-specific component (G_c protein), a genetic marker of use in population studies and forensic medicine.

34. DBP also has a high affinity for actin, the reasons for which are unclear. (The formation of this complex can interfere with the assay of 25-OH-D 1-hydroxylase in kidney homogenates.) It is likely, however, that the interaction of DBP with this widely distributed cellular protein may be the basis of reports of an intracellular 25-OH-D-binding protein.

35. K_A values: for 25-OH-D_3, $5 \times 10^8 M^{-1}$; for 1,25-$(OH)_2$-D_3, $4 \times 10^7 M^{-1}$ (Haddad, J. G. [1999]. In *Vitamin D: Physiology, Molecular Biology, and Clinical Applications* [Holick, M. F. ed.], Humana Press, Totowa, NJ, p. 102.)

36. That DBP binds the actin monomer with high affinity, and that the actin–DBP complex is cleared from the circulation at three times the rate of the non-liganded DBP, suggests that DBP may function in extracellular actin scavenging.

turnover in plasma is less than half that of 25-OH-D$_3$, suggesting recycling of the ligand. DBP can be taken up endocytotically by the renal proximal tubule, being internalized from the glomerular filtrate via binding to the multi-ligand binding receptors megalin and cubilin. This system results in delivery of 25-OH-D$_3$ to the cell with catabolized DBP.

In a survey of some 80,000 individuals, DBP was absent in none. Being synthesized by the liver, however, it is depressed in patients with hepatic disease. Its synthesis is increased by trauma, during estrogen therapy or pregnancy. It does not appear to cross the placenta; fetal DBP is immunologically distinct from the maternal protein. The frequency of rare alleles of DBP polymorphisms has been found to be inversely related to circulating 25-OH-D$_3$ level, with effects comparable in magnitude to those of vitamin D intake.[37]

In addition to facilitating the peripheral distribution of vitamin D obtained from the diet, DBP functions to mobilize the vitamin produced endogenously in the skin. Indeed, vitamin D$_3$ found in the skin is bound to DBP.[38] It has been suggested that the efficiency of endogenously produced vitamin D$_3$ is greater than that given orally for the reason that the former enters the circulation strictly via DBP, whereas the latter enters as complexes of DBP as well as chylomicra. This would indicate that oral vitamin D remains longer in the liver, and is thus more quickly catabolized to excretory forms. In support of this hypothesis, it has been noted that high oral doses of vitamin D can lead to very high levels of 25-OH-D$_3$ (>400 ng/ml) associated with intoxication, whereas intensive UV irradiation can rarely produce plasma 25-OH-D$_3$ concentrations greater than one-fifth that level, and hypervitaminosis D has never been reported from excessive irradiation. The DBP protein has also been found on the surfaces of lymphocytes and macrophages, although the functional significance of such binding is not clear.

Tissue Distribution

Unlike the other fat-soluble vitamins, vitamin D is *not* stored by the liver.[39] It reaches the liver within a few hours after being absorbed across the gut or synthesized in the skin, but from the liver it is distributed relatively evenly among the various tissues, where it resides in hydrophobic compartments. Therefore, fatty tissues such as adipose show slightly greater concentrations. However, in that tissue

the vitamin is found in the bulk lipid phase, from which it is only slowly mobilized. About half of the total vitamin D in the tissues occurs as the parent vitamin D$_3$ species, with the next most abundant form, 25-OH-D$_3$, accounting for about 20% of the total. In the plasma, however, the latter metabolite predominates by several-fold.[40,41] Tissues including those of the kidneys, liver, lungs, aorta, and heart also tend to accumulate 25-OH-D$_3$.[42] It is thought that the uneven tissue distribution of vitamin D, in its various forms, relates to differences in both tissue lipid content and tissue-associated vitamin D-binding proteins, the latter fraction being the smaller of the two intracellular pools of the vitamin.

The concentrations of both 25-OH-D$_3$ and 1,25-(OH)$_2$-D$_3$ are lower in the cord sera of fetuses and newborn infants than in the sera of their mothers. That fetal 25-OH-D$_3$ levels correlate with maternal levels (and show the same seasonal variations) suggests that the metabolite crosses the placenta. Such a correlation is not apparent for 1,25-(OH)$_2$-D$_3$. The extent of transplacental movement of the latter metabolite is not known; however, the placenta appears to be able to produce it from maternally derived 25-OH-D$_3$.

5. METABOLISM OF VITAMIN D

Metabolic Activation

The metabolism of vitamin D (i.e., cholecalciferol/ergocalciferol) involves its conversions to a variety of hydroxylated products, each of which is more polar than its parent (Fig. 6.6).[43] The production of these metabolites, some of which are the actual metabolically active forms of the vitamin, explains the lag time that is commonly observed between the administration of vitamin D$_3$ and the earliest biological response. The metabolism of vitamin D therefore includes reactions that affect the metabolic activation of the ingested or endogenously produced vitamin.

25-Hydroxylation

Most of the vitamin D taken up by the liver from either DBP or lipoproteins is converted by hydroxylation of side chain carbon C-25 to yield 25-OH-D$_3$, also called

---

37. Sinotte, M., Diorio, C., Berube, S., *et al.* (2009). *Am. J. Clin. Nutr.* 89, 634–640.

38. DBP has practically no affinity for lumisterol$_3$ or tachysterol$_3$; therefore, these forms produced under conditions of excessive irradiation are not well mobilized from the skin.

39. The high concentrations of vitamin D found in the livers of some fishes are important exceptions.

40. The next most abundant form is 24,25-(OH)$_2$-vitamin D.

41. The plasma concentrations of 25-OH-D$_3$ and 24,25-(OH)$_2$-D$_3$ of free-living persons vary seasonally, showing maxima in the summer months and minima in the winter. In contrast, plasma levels of 1,25-(OH)$_2$-D$_3$ are rather constant, indicating an effective regulatory mechanism for that metabolite.

42. These organs are also prone to calcification in hypervitaminosis D.

43. In fact, it was the finding in the late 1960s of radioactive peaks migrating ahead of vitamin D (indicating greater polarity relative to vitamin D) in gel filtration of plasma from animals given radiolabeled cholecalciferol that first evidenced the conversion of vitamin D to other species, some of which have subsequently been found to be the metabolically active forms of the vitamin.

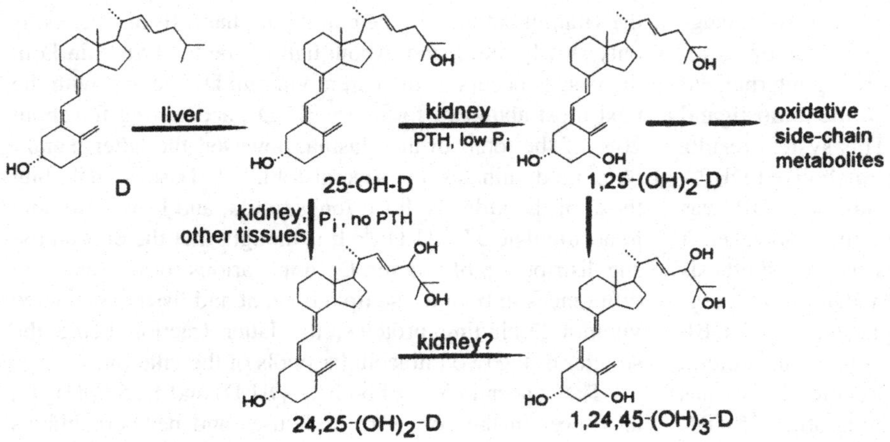

FIGURE 6.6　Metabolism of vitamin D.

liver → 25-OH-D

kidney
PTH, low P_i → 1,25-$(OH)_2$-D → — — → oxidative side-chain metabolites

D

kidney, other tissues | P_i, no PTH

24,25-$(OH)_2$-D

kidney?

1,24,45-$(OH)_3$-D

calcidiol, the major circulating form of the vitamin. This metabolism occurs in the liver in mammals, but in both liver and kidney in birds. That activity, vitamin D 25-hydroxylase, involves cytochrome *P*-450-dependent mixed-function oxygenases[44] of two types: a low-affinity, high-capacity enzyme associated with the endoplasmic reticulum[45] (CYP2R1); and a high-affinity, low-capacity enzyme located in the mitochondria (CYP27A1). The latter involves a cytochrome *P*-450 isoform that also occurs in kidney and bone, suggesting extrahepatic 25-hydroxylation of vitamin D_3. The presence of two different mechanisms of 25-hydroxylation would appear to facilitate the maintenance of adequate vitamin D status under both deficient and excessive conditions of vitamin D intake/production. 25-Hydroxyvitamin D_3 is not retained within the cell but is released to the plasma, where it accumulates by binding with DBP. At normal plasma concentrations of this metabolite, only small amounts of 25-OH-D_3 are released from this pool to enter tissues. Therefore, the circulating level of 25-OH-D_3, normally 10–40 ng/ml (25–125 nmol/l), is a good indicator of vitamin D status.

1-Hydroxylation

The initial hydroxylation product of vitamin D (25-OH-D_3) is further hydroxylated at the C-1 position of the A ring to yield 1,25-$(OH)_2$-D_3 which, being produced at a site distant from its target tissue to which it is transported in the blood, is properly classified as a hormone. This 25-OH-vitamin D 1-hydroxylase is located primarily in renal cortical mitochondria, but also in mitochondrial and microsmal fractions of at least some extrarenal tissues (e.g., bone cells, keratinocytes,

liver, placenta[46]), and has been detected in cultured cell lines (skin, bone, embryonic intestine, keratinocytes, dendritic cells, monocytes, calvarial cells). It is worth noting that keratinocytes not only produce vitamin D_3 with solar exposure, but can also metabolize it to 1,25-$(OH)_2$-D_3. Their contribution to the circulating pool of 1,25-$(OH)_2$-D_3 is substantially less than that of the kidney, but can be significant in patients with renal disease in whom reduced plasma 1,25-$(OH)_2$-D_3 levels induce increased expression of the skin 1-hydroxylase.

The 1-hydroxylase uses $NADPH_2$ as the electron donor and has three constituent proteins: ferridoxin reductase, ferridoxin, and a cytochrome *P*-450 isoform, CYT27B1. The activity is widely distributed among animal species, being found in all but 2 of 28 species surveyed, with highest activities in rachitic chicks. Despite its key role in discharging the actions of vitamin D, its tightly regulated production and relatively fast turnover[47] make 1,25-$(OH)_2$-D_3 unuseful as a biomarker of vitamin D status. Circulating levels tend to be about 40 pg/ml (100 nmol/l).

Epimerization

25-Hydroxyvitamin D appears to undergo epimerization at the C3 position[48] by a metabolic process that has not been characterized. The epimer 3-epi-25-(OH)-D_3 can account for a significant portion of the total circulating 25-(OH)-D_3,[49] leading to overestimation of vitamin

---

44. A mixed-function oxygenase uses molecular oxygen (O_2) but incorporates only one oxygen atom into the substrate; a monooxygenase.

45. In the rat, the microsomal hydroxylase is five-fold more active in males than females; in males it may involve cytochrome P4502C11, which does not occur in females.

46. Maternal levels of 25-OH-vitamin D increase in the third trimester, presumably to assist the mother in providing calcium for the mineralization of the fetal skeleton.

47. Its half-life in the serum is 4–6 hours.

48. The 3-epimer of 25-(OH)-D3 has the C3-hydroxyl moiety extending above the plane of the steroid "A" ring, whereas in the normal metabolite of that grouping it extends below that plane.

49. 3-Epi-25-(OH)-D_3 was detected in the sera of nearly a quarter of infants in a small study, with levels comprising 9–61% of total 25-(OH)-D_3 (Singh, R. J., Taylor, R. L., Reddy, G. S., *et al.* [2006] *J. Clin. Endocrinol. Metab.* 91, 3055–3061).

D status by radioimmunoassay and mass spectrometric methods that are incapable of discriminating between the epimers. 3-Epi-25-(OH)-D$_3$ can be converted by the renal 1-hydroxylase to 3-epi-1,25-(OH)$_2$-D$_3$.[50]

Catabolism

The catabolism of 1,25-(OH)$_2$-D$_3$ occurs through a cytochrome *P*-450-dependent pathway. The ultimate products of this oxidative metabolism are of no metabolic significance other than as excretory products of the vitamin: 1,24,25-(OH)$_3$-D$_3$ (also called **calcitroic acid** or **calcitriol**), 1,25,26-(OH)$_3$-D$_3$, and a 1,25-(OH)$_2$-D$_3$-23,26-lactone. These can be glucuronidated or sulfated. Most are excreted in the feces via the bile.

24-Hydroxylation

The major enzyme involved in vitamin D catabolism is the 24-hydroxylase, CYP24. CYP24 catalyzes hydroxylation at the C-24 of 25-OH-D$_3$ or 1,25-(OH)$_2$-D$_3$ to produce the di- and tri-hydroxy metabolites 24,25-(OH)$_2$-D$_3$ (also called **calcidiol**) and 1,24,25-(OH)$_3$-D$_3$, respectively. These reactions are catalyzed by the same activity, which may also catalyze further hydroxylations at the C-23 position. The 24-hydroxylase has a 10-fold greater affinity for 1,25-(OH)$_2$-D$_3$ than for 25-OH-D$_3$, but the 1,000-fold excess of the latter in the plasma suggests that the primary physiological significance of the hydroxylase is in the catabolism of excess 25-OH-D$_3$. The greatest activity of CYP24 is found in renal mitochondria. It is very strongly induced by 1,25-(OH)$_2$-D$_3$. Both calcitriol and 24,25-(OH)$_2$-D$_3$ appear to be produced under conditions of vitamin D adequacy and normal calcium homeostasis. The latter metabolite has been shown to inhibit the stimulatory effect of PTH on bone resorption by **osteoclasts**, suggesting that it may participate in local osteotropic control in bone. Calcitriol is a major biliary metabolite of the vitamin.

Other Hydroxylations

More than three dozen other metabolites of vitamin D have been identified, most (if not all) of which appear to be physiologically inactive excretory forms.[51] The first to be identified is 25,26-(OH)$_2$-D$_3$, which has a strong affinity for DBP and is detectable in the plasma of animals given large doses of vitamin D$_3$. Other 26-hydroxylated metabolites have been identified, the most abundant being the 26,23-lactone of 25-OH-D$_3$, which appears to be produced through sucessive 23- and 26-hydroxylations of 25-OH-D$_3$ mediated by extrahepatic CYP24A1. The lactone can bind the DPB. It accumulates in hypervitaminotic D animals, and is the major excretory metabolite in some species (guinea pig, opossum). These side chain hydroxylations can produce 1-OH-24,25,26,27-*tetranor*-23-carboxycholecalciferol, which accounts for nearly one-fifth of the biliary excretion of the vitamin. Subsequent chain-shortening metabolism of hydroxylated metabolites with low affinities for DBP accounts for their clearance from the circulation, and their subsequent conversion to fatty acid esters in tissues and to excretory forms including biliary glucuronides.

Regulation of Vitamin D Metabolism

Vitamin D has been found to have a physiological half-life of about 2 months. This relatively long half-life likely reflects its partial sequestration in adipose tissue, as the half-lives of its circulating (25-OH-D$_3$) and metabolically active (1,25-[OH]$_2$-D$_3$) forms have been found to be shorter, at about 15 days and 15 hours, respectively.[52] Regulation of vitamin D metabolism is effected by tight control of the activity of the 1-hydroxylase by several factors: 1,25-(OH)$_2$-D$_3$, PTH, calcitonin (CT), several other hormones,[53] as well as circulating levels of Ca^{2+} and phosphate. The 25-hydroxylase activity appears to be only poorly regulated, primarily by the hepatic concentration of vitamin D, with little or no inhibition by 25-OH-D$_3$. It is increased by inducers of cytochrome *P*-450 (phenobarbital, diphenylhydantoin),[54] and is inhibited by isoniazid.[55]

The dominant renal synthesis of 1,25-(OH)$_2$-D$_3$ is effected by the responses of PTH and CT to serum levels of Ca^{2+} and phosphate. It is increased under three conditions:

- when serum Ca^{2+} is low, the Ca receptor-mediated stimulation of the parathyroid to produce PTH which stimulates an increase in the renal 1-hydroxylase activity

50. While 3-epi-1,25-(OH)$_2$-D$_3$ binds the vitamin D receptor (VDR; see "Vitamin D$_3$ as a Steroid Hormone" in this chapter), it appears to affect the transcription of VDR-regulated gene differentially (Molenár, F., Sigüeiro, R., Sato, Y., *et al.* [2011] *PlosOne* 6, e18124).

51. It should be noted that about 95% of vitamin D excretion occurs *via* the bile; of that amount, only 2–3% of an oral or parenteral dose of vitamin D appears as vitamin D or the mono- or dihydroxy metabolites.

52. Jones, G. (2008). *Am. J. Clin. Nutr.* 88, 582S–586S.

53. For example, prolactin, estradiol, testosterone, growth hormone.

54. Antiepileptic agents such as these reduce the half-life of vitamin D apparently by enhancing its conversion to 25-OH-D and other hydroxylated products.

55. This appears to be the basis for the development of bone disease among patients on long-term isoniazid therapy.

- when serum phosphate is low (in the presence of normal serum Ca^{2+}), an unknown mechanism that appears to involve a pituitary gland hormone increases the 1-hydroxylase
- when serum levels of both Ca^{2+} and phosphate are low, both mechanisms result in the super-stimulation of the 1-hydroxylase.

In all cases, the stimulation is lost upon the return of serum Ca^{2+} to normal levels (as a result of 1,25-[OH]$_2$-D$_3$-stimulated enteric absorption and bone mobilization of Ca^{2+}). It has been suggested that some elderly people who cannot adapt to a low Ca diet by increasing enteric Ca absorption may suffer impaired PTH-dependent 1,25-(OH)$_2$-D$_3$ upregulation.

The hormones PTH and CT both stimulate the renal 1-hydroxylase, the effect of the former being rapid and mediated by cAMP, whereas the effect of the latter is relatively slow, apparently acting at the level of transcription.[56] That ovariectomy reduces the synthesis of 1,25-(OH)$_2$-D$_3$ in rat renal slices suggests that estrogen is also involved in the regulation of the 1-hydroxylase. That effect appears to be mediated via PTH, as parathyroidectomy has been found to block the stimulation of 1,25-(OH)$_2$-D$_3$ production by estrogen treatment.

The regulation of the 1-hydroxylase in extra-renal tissues appears to be completely different. In macrophages, the production of 1,25-(OH)$_2$-D$_3$ has been found to be insensitive to PTH, but to be stimulated by such immune stimuli as interferon-gamma and lipopolysaccharide.

The catabolism of 1,25-(OH)$_2$-D$_3$ is strictly regulated to prevent hypercalcemia and hyperphosphatemia. This is accomplished by PTH, serum phosphate, and factors affecting the principle catabolizing enzyme, the hepatic 24-hydroxylase (CYP24). Other enzymes also metabolize 1,25-(OH)$_2$-D$_3$, as some 40 metabolites have been identified. The 1-hydroxylase is inhibited by strontium and is feedback-inhibited by 1,25-(OH)$_2$-D$_3$.[57] Thus, when circulating levels of 1,25-(OH)$_2$-D$_3$ are high, its renal synthesis is low. The tight regulation of the 1-hydroxylase activity results in the maintenance of nearly constant plasma concentrations of 1,25-(OH)$_2$-D$_3$. 1,25-Dihydroxyvitamin D$_3$ activates its own breakdown by stimulating the transcription of the $_2$4-hydroxylase gene; this stimulation is suppressed in conditions of low serum phosphate.

---

56. It has been suggested that the function of the CT-sensitive 1-hydroxylase, which is elevated in the fetus, may be to accommodate situations of increased need for 1,25-(OH)$_2$-vitamin D.

57. The hydroxylase is also inhibited by the 25-OH-D$_3$-binding protein–actin complex. The effect of this inhibitor *in vitro* can be overcome by using large amounts of 25-OH-D$_3$ to saturate the binding protein; otherwise, it masks the 1-hydroxylase activity in kidney homogenates.

Role of Protein Binding

The DBP plays a critical role in the regulation of vitamin D metabolism by controlling the tissue distribution of vitamin D metabolites. Due to the excess of DBP over its ligands (4–8 µmol/l vs. 50 nmol/l for 25-OH-D$_3$), nearly 90% of circulating vitamin D metabolites are bound to fewer than 5% of available DBP binding sites in vitamin D-adequate individuals. Further, because of its avid binding of 25-OH-D$_3$, that metabolite accumulates in the plasma rather than other tissues, with concentrations of the free metabolite maintained at very low levels. Plasma DBP levels also correlate with those of 1,25-(OH)$_2$-D$_3$; increasing plasma levels of the latter are accommodated by increased protein binding, maintaining very low circulating concentrations of the free form of the active metabolite.

Differential Metabolism of Vitamin D$_2$

Although a minor dietary form of the vitamin, vitamin D$_2$ is metabolized in many ways analogously to vitamin D$_3$. That is, the enzymes involved in the side chain 1-, 24- and 25-hydroxylations do not discriminate between these vitamers. Vitamin D$_2$, however, appears to be metabolized to a number of mono- (24-), di- (1,24-; 24,26-) and tri-(1,25,28-) hydroxylated metabolites that are not produced from vitamin D$_3$. These, in turn, appear to be metabolized to a number of more polar tri- and tetra-hydroxylated forms, some of which (e.g., 1,25,28-[OH]$_3$-D$_2$) are biologically active. That the net effect of this differential metabolism in most species is small is indicated by the fact that, for them, vitamins D$_2$ and D$_3$ have comparable biopotencies, for which reason vitamin D$_2$ is a useful dietary/supplemental form (*see* Table 6.6).

For birds and some mammals, however, vitamin D$_2$ is much less biopotent that vitamin D$_3$. This is due to the fact that, in those species, the mono- and di-hydroxylated metabolites of vitamin D$_2$ are cleared faster than those of vitamin D$_3$. For example, in the chick, the plasma turnover rates of vitamin D$_2$, 25-OH-D$_2$, and 1,25-(OH)$_2$-D$_2$ are 1.5-, 11- and 33-fold faster than those of the respective vitamin D$_3$ analogs. These differences in turnover rates are greater than those for the binding affinities to DBP (5-, 3.6-, and 3-fold, respectively) which (for each of the chicken's *two* DBPs) are greater for vitamin D$_3$ and its metabolites than for vitamin D$_2$ and its metabolites.

6. METABOLIC FUNCTIONS OF VITAMIN D

Vitamin D$_3$ as a Steroid Hormone

At least some, if not all, of the mechanisms of action of vitamin D fit the classic model of a steroid hormone. That is, it has specific cells in target organs with specific

TABLE 6.6 Effect of Vitamin D_2 Supplementation on Plasma Vitamin D Metabolites in Humans

Metabolite	Baseline	Control	Vitamin D_2-Supplemented
D_2 (ng/ml)	0.3 ± 0.1	—	1.2 ± 0.2
D_3 (ng/ml)	1.1 ± 0.2	0.3 ± 0.2	1.0 ± 0.2
25-OH-D_2 (ng/ml)	1.7 ± 0.6	—	17.4 ± 1.6[a]
25-OH-D_3 (ng/ml)	16.9 ± 0.8	17.3 ± 2.0	19.0 ± 1.2
24,25-$(OH)_2$-D_2 (ng/ml)	—	—	0.5 ± 0.1[a]
24,25-$(OH)_2$-D_3 (ng/ml)	1.1 ± 0.4	1.0 ± 0.2	0.8 ± 0.3
1,25-$(OH)_2$-D_2 (pg/ml)	—	—	9.9 ± 0.2[a]
1,25-$(OH)_2$-D_3 (pg/ml)	47.2 ± 9.2	43.6 ± 5.2	35.4 ± 2.3

[a]$P < 0.05$.
Source: Takeuchi, A., Okano, T., and Tsugawa, N. (1989). *J. Nutr.* 119, 1639–1646.

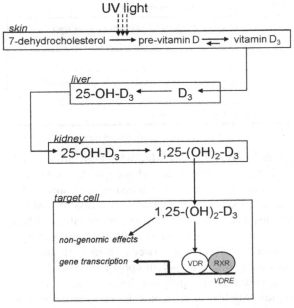

FIGURE 6.7 Metabolic activation of vitamin D (see text for abbreviations).

receptor proteins; and the receptor–ligand complex moves to the nucleus, where it binds to the chromatin at specific DNA sequences and stimulates the transcription of certain downstream genes to produce specific mRNAs that encode the synthesis of specific proteins (Fig. 6.7).

1,25-$(OH)_2$-D_3 as the Metabolically Active Form

The clearest physiological role of vitamin D is in the maintenance of calcium and phosphate homeostasis, impairment of which produces the lesions in bone called **rickets** and **osteomalacia**. In addition, other roles (e.g., induction of cell differentiation) have been proposed. In these roles, it is clear that vitamin D itself is not the functionally active form, but that it must be converted metabolically to a form(s) that exerts its biological activity. That form appears to be the di-hydroxylated metabolite, 1,25-$(OH)_2$-D_3. The earliest evidence for this function was the shorter time to response and greater molar efficacy of 1,25-$(OH)_2$-D_3 than those of any other metabolite of the vitamin.

Pathways of Vitamin D Function

Genomic Pathways – the Vitamin D Receptor

Target tissues for vitamin D contain a specific nuclear receptor, the vitamin D receptor (VDR), to which the active metabolite binds. Autoradiographic studies have shown that 1,25-$(OH)_2$-D_3 is localized in the nuclei of many cell types that contain the VDR. Vitamin D receptors have been identified in more than 30 different cell types; these include cells closely related to the maintenance of calcium homeostasis (bone, kidney, intestine), as well as immune, endocrine, hematopoietic, skin, and tumor cells (Table 6.7). Such findings indicate a wide breadth of vitamin D genomic functions.

That VDRs plays key roles in mediating the physiological function of vitamin D is evidenced by the findings of: (1) low VDR levels being associated with instestinal vitamin D resistance *in vitro*; (2) defects being manifest as hereditary hypocalcemic, vitamin D-resistant rickets; and (3) allelic variation predicting bone mineral mass/density and fracture risk.[58] In neonates VDR is absent until

---

58. In an 18-month study of elderly subjects typed by VDR allelic variant, Rizzoli and colleagues (Ferrari, S., Rizzoli, R., Chevalley, T., *et al.* [1995] *Lancet* 345, 423–424) found 78% of *BB* homozygotes to lose bone mineral density at rates >0.48%/year, whereas only 31% of *bb* homozygotes and only 41% of *Bb* heterozygotes experienced comparable losses.

TABLE 6.7 Distribution of Known Nuclear Vitamin D Receptors

Organ System	Cell Type
Bone	Osteoblasts
Connective tissue	Cartilage chrondrocytes, fibroblasts, stroma
Alimentary tract	Epithelial cells, enterocytes, colonocytes
Liver	Hepatocytes
Kidney	Epithelial (proximal and distal) cells
Heart	Atrial myoendocrine cells, heart muscle cells
Skeletal, smooth muscle	Myocytes
Cartilage	Chondrocytes
Hematolymphopoietic	Activated T and B cells, macrophages, monocytes, spleen, thymus reticular cells and lymphocytes, lymph nodes, tonsillary dendritic cells
Reproductive	Amnion, chorioallantoic membrane, epididymus, mammary gland alveolar and ductal cells, ovary, oviduct, placenta, testis Sertoli and Leydig cells, uterus, yolk sac
Skin	Epidermis, fibroblasts, hair follicles, keratinocytes, melanocytes, sebaceous glands
Nervous	Brain (hippocampus, cerebellar Purkinje and granule cells, bed nucleus, stria terminalis, amygdala central nucleus), sensory ganglia, spinal cord
Immune	Thymus, bone marrow, B cells, T cells
Other endocrine	Adrenal medulla and cortex, pancreatic β cells, pituitary, thyroid follicles and C cells, parathyroid gland
Other	Bladder, choroid plexus, lung, endothelial cells, parotid gland, adipocytes

weaning,[59] being induced, on the feeding of solid foods, by a process that apparently involves cortisol. In at least some organs, VDRs appear to be inducible by estrogen treatment; in the guinea pig (and presumably in humans, who also require vitamin C) VDRs can be reduced in number by deprivation of ascorbic acid.[60]

The VDR is a 51 kDa protein classified in the family of steroid, thyroid hormone, and retinoic acid receptor genes on the basis of its similar primary amino acid structure. The VDRs of different species vary in size (e.g., human, 48 kDa; avian, 60 kDa), but specific domains (e.g., the DNA binding domain) show >95% homology between species. Each is a sulfhydryl protein with an N-terminal recognition domain containing a cysteine-rich cluster comprising two "zinc finger" structures[61] and a C-terminal region that binds $1,25\text{-}(OH)_2\text{-}D_3$ with high affinity, utilizing the ligand's three hydroxyl groups. Multiple polymorphic variations have been idenfied in the human VDR gene that are likely to affect mRNA expression stability and patterns, and thus VDR protein concentrations. These include a *FokI* polymorphism in exon II, *BsmI* and *ApaI* allelic variants in the intron between exons VIII and IX, a *TaqI* restriction fragment polymorphism in exon IX, and a repeat mononucleotide polymorphism in the 3′ untranslated region. Studies have demonstrated contributions of these polymorphisms to interindividual variations in circulating levels of $25\text{-}OH\text{-}D_3$, enteric Ca^{2+} absorption and bone mineral density, and to risks of osteoarthritis, type 2 diabetes, autoimmune disease, fracture, and cancer.[62]

The liganded VDR is thought to be translocated from the cytosol into the nucleas via interactions with microtubules. While this mechanism is still unclear, it is thought to involve a sequence seen for other sterols: stimulation of protein kinase C (PKC), PKC-activation of guanylate cyclase, phosphorylation of microtubule-associated proteins, and association of VDR with importins. That the liganded VDR is less readily extractable from the nucleus

59. The VDR content of intestinal cells is surprisingly low, <2,000 copies/cell; VDR abundance in various tissues appears to vary in the range of 200–25,000 copies/cell.
60. Guinea pig intestinal $1,25\text{-}(OH)_2\text{-}D_3$ receptors (occupied and unoccupied) were reduced by vitamin C deprivation. Whether this finding relates to the rickets-like bone changes observed in vitamin C-deficient guinea pigs and humans is not clear.
61. "Zinc fingers" are finger-like structures folded around a zinc atom tetrahedonally coordinated through the sulfhydryls of cysteine residues in the primary structure of the protein. This motif is similar to DNA-binding motifs found in other transcriptional regulating proteins.

62. See review by Utterlinden *et al.* (2005). In *Vitamin D*, 2nd edn (Feldman, D., Pike, J. W., and Glorieux, F. H., eds). Elsevier, New York, NY, pp. 1121–1157.

than is the free receptor suggests that $1,25\text{-}(OH)_2\text{-}D_3$ binding causes a conformational change in the protein which increases its affinity for DNA.[63] After the manner of the other steroid hormone receptors, the VDR is an essential transactivator of the $1,25\text{-}(OH)_2\text{-}D_3$-dependent transcription of mRNAs for various proteins involved in calcium transport, the bone matrix, and cell cycle regulation. The VDR of intestinal epithelial cells can also bind bile acids, ultimately leading to the detoxification of those inducers through the upregulation of a cytochrome P-450 (CYP3A).

Vitamin D-Responsive Elements (VDREs)

Specific DNA promotor sequences acting as VDREs have been identified. These are similar to the responsive elements mediating the gene expression responses of thyroid hormone or retinoic acid, in that they consist of imperfect direct repeats of six-base pair half-elements.[64] Binding of the $1,25\text{-}(OH)_2\text{-}D_3$–VDR complex to VDREs involves one of the retinoid X receptors (RXRs). The preferred active species appears to be a VDR–RXR heterodimer.[65] Most cells contain both VDRs and RXRs; therefore, it may be the availability of 9-cis-retinoic acid that determines which set of genes is regulated by $1,25\text{-}(OH)_2\text{-}D_3$. Transcriptional regulation of gene expression by vitamin D acting through this sysetem is thought to involve a conformational change in VDR effected by the phosphorylation of a specific serinyl residue upon the binding of $1,25\text{-}(OH)_2\text{-}D_3$. This facilitates recruitment of co-activator proteins that induce chromatin remodeling and expose domains of the protein capable of interacting with VDREs to influence RNA polymerase-mediated transcription.

Genes Regulated by Vitamin D

Some 50 genes have been identified as being regulated by vitamin D status (Table 6.8). These include genes associated with many aspects of metabolism, including cell differentiation and proliferation, energy metabolism, hormonal signaling, mineral homeostasis, oncogenes, and chromosomal proteins, as well as vitamin D metabolism. For most of these, the regulation appears to involve $1,25\text{-}(OH)_2\text{-}D_3$-dependent modulation of mRNA levels (i.e., regulation of transcription and/or message stability). To date, $1,25\text{-}(OH)_2$-D_3-regulated transcription has been established for less than a dozen of these genes, and VDREs have been reported for only a few (e.g., calbindin$_{9K}$, integrin$_{\alpha\beta3}$, osteocalcin

and the plasma membrane Ca^{2+} pump). Evidence for post-transcriptional regulation of calbindin$_{9K}$ has been presented. From this emerging picture, it is clear that changes in vitamin D status have potential for pleiotropic actions.

The first gene product to be recognized as inducible by $1,25\text{-}(OH)_2\text{-}D_3$ was for many years called **calcium-binding protein (CaBP)**. Different forms of CaBP have subsequently been described; these are now called **calbindins**.[66] Calbindins are widespread in animal tissues, with greatest concentrations found in avian and mammalian duodenal mucosa, where they can comprise 1–3% of the total soluble protein of the cell. Two calbindins have been identified: calbindin-$_{D9k}$, occurring primarily in mammalian intestinal mucosa but also in kidney, uterus, and placenta; and calbindin-$_{D28k}$, occuring in mammalian kidney (distal convoluted tubules), pancreas (β cells), and brain, and avian intestine and kidney. Calbindin-$_{D9k}$ can bind two Ca^{2+} atoms while calbindin-$_{D28k}$ can bind four Ca^{2+} atoms. It is thought that calbindins function in the enteric absorption of calcium by facilitating the movement of calcium through the enterocytic cytosol while keeping the intracellular concentration of the free Ca^{2+} ion below hazardous levels. Calbindins are not expressed in vitamin D deficiency, but are expressed in response to $1,25(OH)_2\text{-}D_3$. That such treatment increases the expression of the protein without affecting its message suggests that vitamin D regulation of calbindin may occur at the translational level.

VDR also downregulates the expression of some genes. These include genes encoding **parathyroid hormone (PTH)** (see Table 6.8). This appears to involve binding of VDR homodimers or VDR–RXR heterodimers to a negative response element (nVDRE), which is transcriptionally active in the absence of $1,25(OH)_2\text{-}D_3$. It is thought that downregulation occurs by VDR directly binding an activator of the nVDRE.

Evidence for Non-Genomic Pathways

Several reports have shown responses to vitamin D within seconds of $1,25\text{-}(OH)_2\text{-}D_3$ treatment, suggesting signaling in ways independent of genomic responses, which typically take hours to days. The first such response to be observed was transcaltachia, the rapid transport of Ca^{2+} across the intestinal mucosa. To date, more than four dozen reports have shown such responses (Table 6.9). Studies with vitamin D analogs suggest that the receptor involved in these responses is different from the classical VDR involved in genomic responses and apparently associated with the plasma membrane and involved in regulating kinses, phosphatases, and ion channels. It has been suggested that this receptor may initiate signaling through cognate promotor elements and/or affect expression of

---

63. Two binding domains have been identified in nVDRs: a ligand-binding domain at the carboxyl-terminal end of the protein, and a DNA-binding domain at the amino-terminal end.

64. Owing to their direct repeats, these lack the dyad symmetry of the classic steroid hormone-responsive elements.

65. Interaction of these receptor proteins is thought to involve C-terminal dimerization interfaces in both.

---

66. Calbindins are members of a large family of Ca^{2+}-binding proteins having a distinctive helix–loop–helix sequence, the so-called EF "hand."

TABLE 6.8 Genes Known to Be Regulated by Vitamin D

Gene	Tissue in Which Regulation Demonstrated
Upregulated	
Aldolase subunit B	Chick kidney
Alkaline phosphatase	Chick, rat intestine
ATP synthase	Chick intestine; rat intestine
Calbindin-$D_{28\,kDa}$	Chick brain, kidney, uterus, intestine; mouse kidney; rat kidney, brain
Calbindin-$D_{9\,kDa}$	Chick kidney, skin, bone; rat intestine, skin, bone
Carbonic anhydrase	Marrow; myelomonocytes
CCAT enhancer binding protein ß	Mouse intestine, osteoblasts
Cytochrome *c* oxidase	
Subunit I	Chick intestine; rat intestine
Subunit II	Chick intestine; rat intestine
Subunit III	Chick intestine; rat intestine
Cytochrome *P*-450 isoform CYP3A	Mouse intestine
Fibronectin	MG-63, TE-85, HL-60 cells
c-Fms	HL-60 cells
c-Fos	HL-60 cells
Glyceraldehyde-3-phosphate dehydrogenase	BT-20 cells
Heat shock protein 70	Peripheral blood monocytes
Integrin$_{\alpha \nu \beta 3}$	Chick osteoclasts
Interleukin 6	U937 cells
Interleukin 1	U937 cells
Interleukin 3 receptor	MC3T3 cells
c-Ki-Ros	BALB-3T3 cells
Matrix Gla protein	UMR106-01, ROS cells
Metallothionein	Rat keratinocytes
NADH Dehydrogenase	
Subunit II	Chick intestine
Subunit III	Chick intestine
Nerve growth factor	L-929 cells
Neutrophil-activating polypeptide	HL-60 cells
c-Myc	MG-63 cells
Osteocalcin	ROS cells
Osteopontin	ROS cells
1-OH-D_3 24-hydroxylase	Rat kidney
1,25-$(OH)_2$-D_3 receptor	Mouse fibroblasts
Plasma membrane Ca^{2+} pump	Chick intestine
Prolactin	Rat pituitary cells
Protein kinase C	HL-60 cells

(Continued)

TABLE 6.8 (Continued)

Gene	Tissue in Which Regulation Demonstrated
Tumor necrosis factor α	U-937, HL-60 cells
Vascular endothelial growth factor (VEGF)	Mouse fibroblasts
Vitamin D receptor	Rat intestine, pituitary
Downregulated	
ATP synthase	Chick kidney
Calcitonin	Rat thyroid gland
CD-23	peripheral blood monocytes
Collagen, type I	Rat fetal calvaria
Cytochrome *b*	Chick kidney
Cytochrome *c* oxidase	
Subunit I	Chick kidney
Subunit II	Chick kidney
Subunit III	Chick kidney
Cytochrome *P*-450 isoform CYP2yB1	Mouse liver
Fatty acid-binding protein	Chick intestine
Ferridoxin	Chick kidney
Granulocyte-macrophage colony-stimulating factor	Human T lymphocytes
Histone H_4	HL-60 cells
Interleukin 2	Human T lymphocytes
Interferon γ	Human T lymphocytes
c-Myb	HL-60 cells
c-Myc	HL-60, U937 cells
NADH dehydrogenase subunit I	Chick kidney
25-OH-D_3 1-hydroxylase	Rat kidney
Prepro-PTH	Rat, bovine parathyroid
Protein kinase inhibitor	Chick kidney
PTH	Rat parathyroid
PTH-related protein	T lymphocytes
Transferrin receptor	PBMCs
α-Tubulin	Chick intestine

Abbreviations: PTH, parathyroid hormone; BMCs, blood mononuclear cells.

VDRE-regulated genes by post-translational modification of the classical VDR.

Roles of Vitamin D in Calcium and Phosphorus Metabolism

The most clearly elucidated and, apparently, most physiologically important function of vitamin D is in the homeostasis of Ca^{2+} and phosphate. This is effected by a multi-hormonal system involving the controlled production of 1,25-$(OH)_2$-D_3 which functions in concert with PTH and calcitonin (CT). Regulation of this system occurs at the points of intestinal absorption, bone accretion and mobilization, and renal excretion.

TABLE 6.9 Rapid Responses Reported for
$1,25\text{-}(OH)_2\text{-}D_3{}^a$

Response	Organ/Cell
Ca^{2+} transport	Intestinal mucosa, CaCo-2 cells
PKC activation	Intestinal mucosa, chondrocytes, liver, muscle
MAPK activation	Intestinal mucosa, liver, leukemia cells
Ca^{2+} signaling	Adipocytes, pancreatic β cells
Ca^{2+} channel opening	Osteoblasts
G protein activation	Intestinal mucosa
Cell differentiation	Promyelocytic NB4 cells
Src activation	Keratinocytes
Raf activation	Keratinocytes
Insulin secretion	Pancreatic β cells
Increased contraction/ relaxation	Cardiomyocytes

^{a}From *Mizwicki, M. T. and Norman, W. N. (2009). Science Signaling 2, 1–14.*

Intestinal Absorption of Ca^{2+}

Calcium is absorbed in the small intestine[67] by both transcellular and paracellular mechanisms. The former is an active, saturable process, occuring in mammals primarily in the duodenum and upper jejunum, and constitutes the most important means of absorbing calcium under conditions of low intake of the mineral; the latter is a non-saturable process occuring throughout the intestine, and is the most important means of absorbing calcium when calcium intake is high. The active metabolite, $1,25\text{-}(OH)_2\text{-}D_3$, stimulates the enteric absorption of calcium through roles in both mechanisms, although its mechanism in the paracellular process is unclear. The availability of calcium for both processes is affected by both exogenous (e.g., inhibition by food phytates or phosphate) and endogenous (e.g., gastric acid secretion) factors.

The transcellular absorption of calcium progresses in three steps, each of which is dependent on $1,25\text{-}(OH)_2\text{-}D_3$:

- *Uptake of Ca^2 from the intestinal lumen to the microvillus border.* This uptake of Ca^{2+} at the brush border involves diffusion through a Ca^{2+} channel or integral membrane transporter (CaT1) followed by channel-like flow gated by the intercellular concentration of Ca^{2+}, which affords controlled movement of Ca^{2+} down a

steep electrochemical gradient.[68] Ca^{2+} channels are expressed primarily in Ca^{2+}-transporting epithelial cells. Vitamin D treatment of cells in culture increases calcium uptake, which has been associated with increased expression of CaT1; increases expression of Ca^{2+}-binding proteins, **calbindins**,[69] which appear to function in the control of the microvillar Ca^{2+} channel; increases synthesis of phosphatidylcholine at the expense of phosphatidylethanolamine (perhaps affecting membrane fluidity); and alters intracellular distribution of the Ca^{2+}-binding protein **calmodulin**,[70] increasing its association with the brush border.

- *Translocation of Ca^{2+} across the cell to the basolateral membrane.* The movement of calcium across the enterocyte remains unclear. This appears to involve the vitamin D-dependent synthesis of calbindins, which serve as facilitators of cytosolic diffusion of Ca^{2+}. Translocation of Ca^{2+} also appears to occur within membrane-bound structures through vesicular transport.

- *Active extrusion of Ca^2 into the circulation.* This occurs against a substantial thermodynamic gradient involving a 50,000-fold differential in Ca^{2+} concentration and a positive electrical potential[71] facilitated by an Ca^{2+}-ATPase[72] in the basolateral membrane. This Ca^{2+} pump is stimulated by calmodulin and calbindin. Ca^{2+} is also extruded by a membrane Na^+/Ca^{2+} exchanger.[73]

The paracellular, diffusional process of enteric calcium absorption is less well understood, although evidence suggests that it, too, is stimulated by $1,25\text{-}(OH)_2\text{-}D_3$. This nonsaturable process occurs along the paracellular pathway that exists between adjacent enterocytes. Its permeability is regulated by the proteins comprising the tight junction

---

67. Some 70–80% of enteric calcium absorption occurs in the ileum. The colon may be responsible for 3–8% of Ca absorption.

68. Luminal concentrations of Ca^{2+} can be in the mmol/l range, whereas intracellular concentrations of the free ion are in the range of 50–100 nmol/l.

69. Calbindins have been identified in many species. The mammalian form (calbindin-D_{9k}) are 10 kDa proteins with two high-affinity Ca^{2+}-binding sites. Avian calbindin-D_{28K} is larger, 30 kDa; it is also expressed in the shell gland (uterus) of hens.

70. Calmodulin is a 17 kDa acidic protein (also of the "EF hand" family) that is expressed in many cell types and subcellular compartments. It can bind as many as four Ca^{2+} ions, which causes conformational changes and post-translational modifications that allow it also to bind to more than 100 target proteins. In this way, calmodulin serves as a major transducer of Ca^{2+}-signals in the control of cellular metabolism.

71. It is estimated that the movement of 1 mole of Ca^{2+} against this gradient requires about 9.3 kcal.

72. The CaATPase spans the membrane, with a Ca^{2+}-binding domain on the cytoplasmic side. It appears that phosphorylation-induced conformational changes in the protein allow it to form a channel-like opening through which Ca^{2+} is expelled, thus serving as a Ca "pump."

73. The efflux of 1 mole of Ca^{2+} is linked to the influx of 3 moles of Na^+, thus generating negative cytosolic electropotential.

complex at the apical region. Diffusion of Ca^{2+} through this pathway occurs when intralumenal concentrations exceed 2–6 nmol/l. Some researchers have found vitamin D to promote calcium uptake by this pathway, perhaps by 1,25-$(OH)_2$-D_3-mediated activation of protein kinase C, which is known to increase paracellular permeability.

Intestinal Phosphate Absorption

Healthy individuals absorb dietary phosphate (P_i)[74] with efficiencies of 60–65%, most absorption occurring in the duodenum and jejunum. Phosphate is absorbed by two mechanisms: a non-saturable paracellular pathway; and a saturable, energy-requiring, Na^+-dependent process on the mucosal surface that is driven by a Na^+ gradient maintained by the Na^+, K^+-ATPase on the basolateral membrane. The latter process appears to be rate-limiting to P_i absorption. Vitamin D, as 1,25-(OH)2-D3, increases net P_i uptake by increasing Na^+/P_i co-transport across the mucosal brush border, apparently through upregulation of the transporter.

Renal Resorption of Calcium and Phosphate

Vitamin D, as 1,25-$(OH)_2$-D_3, stimulates the resorption of both P_i and Ca^{2+} in the renal tubule.[75] The quantitative significance of this effect is greater for P_i, some 60% of which is reabsorbed in the proximal tubule by a Na^+/P_i co-transport mechanism analogous to those in the intestinal epithelium and in bone. The Na^+/P_i co-transporter, Npt2a, is expressed in proximal tubular cells, where the major portion of P_i is reabsorbed; its expression is also upregulated by both PTH and P_i supply.

In contrast, Ca^{2+} is reabsorbed mainly in the distal tubule, mostly (80%) by passive, vitamin D-independent, paracellular routes in the proximal tubules and ascending loop of Henle. Some 8,000 mg of Ca is filtered at the glomerulus daily,[76] 98% of which is reabsorbed in the tubules. The transcellular process resembles that of the intestine in having a Ca^{2+} channel component, cytosolic Ca^{2+}-binding proteins (calbindins -D_{9k} and -D_{28k}), and a plasma membrane Ca^{2+} ATPase, all of which are located in the distal portions of 1,25-$(OH)_2$-D_3-responsive nephrons.

Bone Mineral Turnover

Bone is the predominant target organ for vitamin D, accumulating more than one-quarter of a single dose of the vitamin within a few hours of its administration. That lesions in bone mineralization (rickets, osteomalacia) occur in vitamin D deficiency has long indicated its vital function in the metabolism of this organ.[77] The pattern of vitamin D metabolites in bone differs from that in intestine; whereas the latter contains mainly 1,25-$(OH)_2$-D_3, bone contains mainly 25-OH-D_3 (accounting for >50% of the vitamin D metabolites present, with 1,25-$(OH)_2$-D_3 comprising less than 35%). As in plasma, the level of 24,25-$(OH)_2$-D_3 in bone is fairly constant relative to that of 25-OH-D_3.

Vitamin D plays roles both in the formation (*mineralization*) and the mobilization of bone mineral (*demineralization*). Bone mineralization is epitaxial, occuring by co-deposition of Ca^{2+}, P_i (i.e., PO_4^{3-}) and hydroxyl (OH^-) ions at multiple sites on the surfaces of pre-existing crystals or on topographically similar protein/lipid surfaces. The resulting structure is one comprised of small (<200-Å) crystals with an average chemistry resembling that of hydroxyapatite, $Ca_{10}(PO_4)_6(OH)_2$. Bone mineral may also contain magnesium, sodium, potassium, fluoride, strontium, acid phosphate, and citrate. The amount of bone mineral is therefore a function of the balance of the laying down of mineral[78] by bone-forming cells called osteoblasts, and the dissolution of bone crystal[79] by bone-resorbing cells called osteoclasts. Calcium deposition in the skeleton involves the intracellular synthesis of collagen and fibrils by osteoblasts, which extrude these fibrils to form the extracellular matrix of bone, portions of which can be mineralized. Bone demineralization is directed by multinucleated osteoclasts that release proteins, and lysosomal enzymes that dissolve bone mineral and lyse its organic matrix. The accretion/mobilization of bone Ca^{2+} therefore involves the relative activities of osteoblasts and osteoclasts with the bone surface serving, in effect, as a calcium buffer. Bone growth results from the dominance of osteoblastic activity, which in the long bones is organized by the arraying of chondrocytes to affect periosteal apposition along epiphyseal growth plates (Fig. 6.8). Balanced demineralization–mineralization affects the coordinated growth ("modeling") of skeletal bone, and the "remodeling" (primarily in the endosteal area) of mature bone to replace damage and prevent senescence.

While the ultimate effects on bone involve both osteoblasts and osteoclasts, vitamin D targets only osteoblasts

---

74. P_i is an essential constituent of bone and teeth which accounts for some 85% of total body P_i. It is also important in regulating the genes, *PHEX* and *FGF23*. P_i comprises 1% of adult body weight.

75. Each day a human filters some 8 g calcium at the glomerulus, 98% of which is reabsorbed.

76. The concentration of ultrafilterable Ca in plasma, *ca.* 1.35 mmol/l (*ca.* 55% of total plasma Ca), is similar to that of the glomerular fluid.

77. Thus, the involvement of vitamin D in the metabolism of Ca and P was clear, as structural bone contains 99% of total body Ca and 85% of total body P, which in a 70-kg man is about 1,200 and 770 g, respectively.

78. Mineralization is preceded by secretion of the bone matrix, unmineralized osteoid, by chondrocytes.

79. Dissolution of bone mineral is followed by depolymerization of glycosaminoglycans and digestion of collagen and other bone matrix proteins.

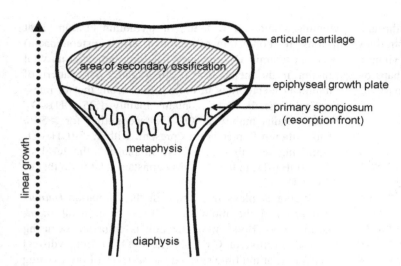

linear growth

- articular cartilage
- area of secondary ossification
- epiphyseal growth plate
- primary spongiosum (resorption front)
- metaphysis
- diaphysis

FIGURE 6.8 Linear growth of long bones occurs by deposition of bone along the epiphyseal growth plates comprised of arrays of chondrocytes apical of a highly vascularized zone of spongiform bone.

TABLE 6.10　Vitamin D-Responsive Osteoblast Genes

Collagen
Alkaline phosphatase
Osteocalcin
Osteopontin
Bone sialoprotein
Transforming growth factor-β (TGF-β)
Vascular endothelial growth factor (VEGF)
Matrix metalloproteinase-9 (MMP-9)
β3-Integrin
Receptor activity of NFκB (RANKL)
Osteopetegrin

and osteoprogenitor cells, both of which have VDRs that are stimulated by glucocorticoids. 1,25-Dihydroxyvitamin D_3 affects the expression of several osteoblast genes (Table 6.10). It is generally thought that 24,25-$(OH)_2$-D_3, which is concentrated in the epiphyseal cartilage, may also be involved in this process. While the mechanism remains unclear, it appears to involve PTH[80] and, because PTH stimulates adenylate cyclase activity, perhaps acts via cAMP. Vitamin D affects osteoclasts only by indirect means, as osteoclasts are activated by osteoblasts through the induction of the membrane-associated receptor activator of nuclear factor κB (NFκB) ligand (RANKL). Vitamin D also has a

role in the differentiation of macrophages to osteoclasts; the number of these giant multinucleated bone-degrading cells is very low in bone from vitamin D-deficient animals.

In the absence of adequate levels of 1,25-$(OH)_2$-D_3, the failure of mineralization and/or net excess of osteoclastic demineralization have structural and functional consequences to bone, as bone density is a primary determinant of bone strength. This ultimately results in the well-known clinical signs (see "Signs of Vitamin D Deficiency," this chapter). Inadequate vitamin D status has also been associated with fracture risk, particularly in older individuals (Tables 6.11, 6.12). A recent meta-analysis of 12 randomized controlled trials[81] revealed that vitamin D doses of 700–800 IU/day reduced the relative risk of hip fracture by 26% and of any non-vertebral fracture by 23%. Risk reductions were not oberved for trials that used a lower vitamin dose (400 IU/day).

In light of the clear effect of low vitamin D status in stimulating mobilization of Ca from bone, the situation of many African-Americans would seem anomalous. Their lower circulating levels of 25-OH-D_3 than White Americans (Fig. 6.9) would suggest increased risks of fracture relative to Whites. In fact, the reverse is true. Studies have shown that the prevalence of both hip fracture and osteoporosis of African-Americans is about half that of White Americans.[82] This has been explained on the basis of multiple characteristics of African-Americans that affect fracture risk.[83] Compared to Whites,

---

80. Because PTH is secreted in response to hypophosphatemia, nutritional deprivation of P_i can also lead to bone demineralization with increases in the activity of 25-OH-vitamin D 1-hydroxylase and accumulation of 1,25-$(OH)_2$-D_3 in target tissues.

81. Bischoff-Ferrari, H. A., Willett, W. C., Wong, J. B., *et al.* (2005). *J. Am. Med. Assoc.* 293, 2257–2264.
82. Barrett-Conner, E., Siris, E. S., Wehren, L. E., *et al.* (2005). *J. Bone Min. Res.* 20, 185–194.
83. Cosman, F., Nieves, J., Dempster, D., *et al.* (2007). *J. Bone Min. Res.* 22, V34–V38; Aloia, J. F. (2008). *Am. J. Clin. Nutr.* 88, 545S–550S.

TABLE 6.11 Abnormalities in Vitamin D Status of Hip Fracture Patients

Parameter	Controls	Patients
Age (years)	75.6 ± 4.2	75.9 ± 11.0
Sunshine exposure (*n*)		
Low	9	51
Intermediate	26	38
High	39	31
Dietary intakes		
Ca (mg/day)	696 ± 273	671 ± 406
vitamin D (IU/day)	114 ± 44	116 ± 63
Serum analytes		
Ca (mmol/l)	2.35 ± 0.12	2.13 ± 0.16
P_i (mmol/l)	1.09 ± 0.15	1.11 ± 0.26
Alkaline phosphatase (units)	2.1 ± 0.5	2.0 ± 0.7
Albumin (g/l)	41.9 ± 2.8	32.5 ± 4.8[a]
DBP (mg/l)	371 ± 44	315 ± 60[a]
25-OH-D_3 (nmol/l)	32.9 ± 13.6	18.5 ± 10.6[a]
24,25-$(OH)_2$-D_3 (nmol/l)	1.8	0.5[a]
1,25-$(OH)_2$-D_3 (pmol/l)	105 ± 3	179 ± 46[a]
PTH (µg Eq/l)	0.12 ± 0.05	0.11 ± 0.05

[a]*P* <0.05.
Source: Lips, P., van Ginkel, F. C., Jongen, M. H. M., *et al.* (1987). *Am. J. Clin. Nutr.* 46, 1005–1001.

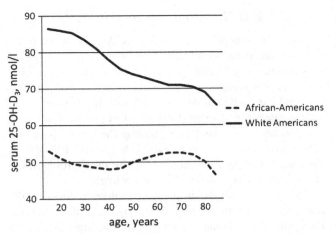

FIGURE 6.9 Serum 25-OH-D_3 levels in African-American and White Americans in the third National Health and Nutrition Examination Survey.

African-Americans tend to have greater peak bone mass, greater muscle mass, and lower bone turnover rates. They are also more likely to be obese, which gives their bone mineralization the positive stimulation of loading. While their risk of falls is not different from Whites, their shorter hip axis length protects against osteoporotic fractures caused by falls. Studies with adolescent girls have found that African-Americans have better enteric Ca absorption and renal Ca conservation that Whites. Further, African-American adults have higher levels of PTH without associated bone loss, indicating skeletal resistance to PTH. That these advantages diminish with age is evidenced by the fact that, like other groups, elderly African-Americans with elevated PTH show bone loss.

VDR genotype appears to contribute significantly to the variation observed in bone mineral density in populations, some 80% of which is thought to be due to genetic factors. Two VDR polymorphisms, Fok1 and Bsm1, have been identified as independent risk factors for stress fracture. Individuals with the B- or f-containing genotypes with respect to each polymorphism were more likely to develop fractures than individuals without those alelles.[84]

Circulating concentrations of 25-OH-D_3 have been found to be inversely correlated with the risk of periodontal disease in adults. This effect appears to be independent of those of bone mineral density. That it may involve the anti-inflammatory effects is supported by the finding that serum 25-OH-D_3 levels were inversely related to susceptibility to gingival inflammation and tooth loss in the NHANES III survey.

TABLE 6.12 Reduction of Fracture Risk with 1,25-$(OH)_2$-Vitamin D_3

Treatment	Year	Women in Study	Women With New Fractures	Number of New Fractures
1,25-$(OH)_2$-D_3	1	262	14	23
	2	236	14	22
	3	213	12	21
Ca	1	253	17	26
	2	240	30[a]	60[b]
	3	219	44[b]	69[b]

[a]*P* <0.01.
[b]*P* <0.001.
Source: Tilyard, M. W., Spears, G. F. S., Thomson, J., and Dovey, S. (1992). *N. Engl. J. Med.* 326, 357–362.

84. Chatzipapas, C., Boikos, S., Drosos, G. I., *et al.* (2009). *Horm. Metab. Res.* 41, 635–640; Diogenes, M. E. L., Bezerra, F. F., Cabello, G. M. K., *et al.* (2010). *Eur. J. Appl. Physiol.* 108, 31–38.

Calcium and Phosphate Homeostasis

Homeostatic control of Ca and P is dependent on functions of the parathyroid and thyroid glands. Each gland senses serum Ca^{2+} level[85] by a cell surface G protein-like, Ca^{2+}-sensing receptor (CaR) in chief cells of the parathyroid and parafollicular cells (C cells) of the thyroid.[86] The CaR appears sensitive to fluctuations in plasma Ca^{2+} of a few percent, responding by regulating the synthesis of two calcitropic hormones, parathyroid hormone (PTH)[87] by the parathyroid, and calcitonin (CT) by the thyroid, which elevate and lower plasma Ca^{2+}, respectively.

When Ca^{2+} levels drop, the parathyroid loses VDR, reducing its sensitivity to $1,25\text{-}(OH)_2\text{-}D_3$. It also secretes PTH into the circulation, which acts on target tissues to restore plasma Ca^{2+} by stimulating renal tubular Ca^{2+} reabsorption and renal 1-hydroxylation of $25\text{-}OH\text{-}D_3$.[88] The resulting increase in circulating $1,25\text{-}(OH)_2\text{-}D_3$ stimulates enteric Ca^{2+} absorption and osteoclastic activity. Demineralization of bone serves to mobilize Ca^{2+} and P_i from that reserve, thus maintaining the homeostasis of those minerals in the plasma.[89] Increasing plasma Ca^{2+} levels evoke signaling by CaR in thyroid C cells to increase the expression of CT, which inhibits bone resorption and, at high doses, increases urinary Ca^{2+} excretion. This system feeds back to regulate the synthesis of PTH through inhibition by $1,25\text{-}(OH)_2\text{-}D_3$ binding a negative VDRE near the promotor of the PTH gene. A similar mechanism has been proposed for the downregulation of CT by $1,25\text{-}(OH)_2\text{-}D_3$. The interplay of these hormones with Ca^{2+} and $1,25\text{-}(OH)_2\text{-}D_3$ produces fine control of circulating Ca^{2+} levels at 4–5.6 mg/dl[90] (Fig. 6.10).

Under hypercalcemic conditions, calcitonin (CT) is secreted by the thyroid. This hormone suppresses bone mobilization, and is also thought to increase the renal excretion of both Ca^{2+} and P_i. In that situation, the 25-OH-vitamin D 1-hydroxylase may be feedback-inhibited by $1,25\text{-}(OH)_2\text{-}D_3$, which may actually convert to the catalysis of the 24-hydroxylation of $25\text{-}OH\text{-}D_3$. In the case of egg-laying birds, which show relatively high circulating concentrations of $1,25\text{-}(OH)_2\text{-}D_3$, the 1-hydroxylase activity remains stimulated by the hormone prolactin.

Secondary hyperparathyroidism, characterized by elevated serum PTH concentrations, is common among elderly people. The condition can reflect some degree of renal insufficiency, with associated reduction in renal 25-OH-vitamin D_3 1-hydroxylase activity. Serum concentrations of PTH can also increase owing to privational vitamin D deficiency, in which case it is manifested by low circulating levels of $25\text{-}OH\text{-}D_3$ (Fig. 6.11). Accordingly, the PTH levels of people living in northern latitudes are highest during the winter for subjects not taking supplemental vitamin D.

Roles of Other Minerals

Vitamin D function can be affected by several other mineral elements:

- *Zinc.* Deprivation of zinc has been found to diminish the $1,25\text{-}(OH)_2\text{-}D_3$ response to low calcium intake,[91] and it has been suggested that zinc may indirectly affect renal $25\text{-}OH\text{-}D_3$ 1-hydroxylase activity.
- *Iron.* Iron deficiency has been shown to be associated with low serum concentrations of $24,25\text{-}(OH)_2\text{-}D_3$ and reduced $25\text{-}OH\text{-}D_3$ responses to supplementation with vitamin D_3. It has been suggested that iron deficiency, which is known to impair the enteric absorption of fat and vitamin A, may also impair the absorption of vitamin D.
- *Lead.* Exposure to lead appears also to impair the 1-hydroxylation of $25\text{-}OH\text{-}D_3$. Ironically, that effect increases the amount of enterically absorbed lead that can be bound by the vitamin D-responsive protein, calbindin, implicated in enteric Ca^{2+} absorption. In children, blood levels of $1,25\text{-}(OH)_2\text{-}D_3$ and calcium have been found to be inversely related to blood lead concentrations,[92] suggesting that lead may inhibit the renal 1-hydroxylation of the vitamin, perhaps constituting an adaptation to protect against lead toxicity. This is consistent with $1,25\text{-}(OH)_2\text{-}D_3$-mediated enhanced enteric lead absorption as a chief factor contributing to the elevated body lead burden generally observed in calcium deficiency (Table 6.13). Indeed, chronic ingestion of lead by calcium-deficient animals has been shown to reduce the metabolic production of $1,25\text{-}(OH)_2\text{-}D_3$. This combination of effects allows the hormone system that is impaired by chronic lead ingestion to contribute to susceptibility to lead toxicity.[93]

---

85. Normal concentrations of serum Ca^{2+} are 10,000 times greater than those within cells. Disturbance in this ion gradient to produce transient increases in intracellular Ca^{2+} concentrations that can signal a variety of cellular responses (e.g., transcriptional control, neurotransmitter release, muscular contraction) through the actions of binding proteins.

86. CaR is also expressed in renal tubules.

87. PTH is a small (9,600 Da) protein with high sequence homology across species.

88. Parathyroidectomized animals cannot mount this 1-hydroxylase response unless treated with PTH.

89. The normal ranges of these parameters in human adults are as follows: Ca, 8.5–10.6 mg/dl; P, 2.5–4.5 mg/dl.

90. This corresponds to 8–10 mg total Ca per deciliter, about half of which is present in ionized form (Ca^{2+}).

---

91. This finding may have clinical significance, as in many parts of the world children are undernourished with respect to both Zn and Ca.

92. The prevalence of lead toxicity among children is seasonal, i.e., greatest in the summer months.

93. Lead poisoning is a serious environmental health issue. In the United States alone, high blood lead levels are estimated in as many as 5 million school children and 400,000 pregnant women.

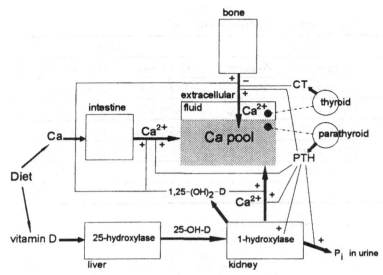

FIGURE 6.10 Calcium homeostasis.

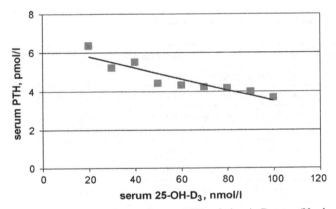

FIGURE 6.11 Relationship of serum PTH and vitamin D status (Need, A. G., Horowitz, M., Morris, H. A., and Nordin, B. C. [2000] *Am. J. Clin. Nutr.* 71, 1577–1581).

● *Boron.* An interaction of vitamin D, magnesium, and calcium has been reported with boron, an element known to be essential for plants. One laboratory has reported that the feeding of low-boron diets increased serum alkaline phosphatase activities (suggesting bone mobilization) in vitamin D-deficient chicks and rats if they were also fed deficient levels of magnesium and calcium. The mechanisms of these effects are not clear.

Vitamin D Functions in Non-Calcified Tissues

That $1,25\text{-}(OH)_2\text{-}D_3$ and nuclear VDRs occur in tissues not directly involved in Ca^{2+} homeostasis (e.g., pancreatic β cells,[94] skin Malpighian layer cells, specific brain cells,

pituitary, muscle, mammary gland, endocrine cells of the stomach, and chorioallantoic membrane surrounding chick embryos)[95] suggests that the vitamin functions in the regulation and differentiation of many cells (Table 6.14). These functions occur via VDR: at least 100 proteins (including several oncogenes) are known to be regulated by $1,25\text{-}(OH)_2\text{-}D_3$. Responses to $1,25\text{-}(OH)_2\text{-}D_3$ are observed at concentrations two to three orders of magnitude greater than circulating levels; hence it is possible that under normal circumstances they may be limited to specific sites of local production of the active metabolite.

Immune Regulation

VDRs have been identified in most immune cells, including most antigen-presenting cells (e.g., macrophages, dendritic cells) and $CD4^+$, $CD8^+$ T lymphocytes. These cells can also express the $25\text{-}OH\text{-}D_3$ 1- and -24-hydroxylases; they are thus capable of producing and catabolizing $1,25\text{-}(OH)_2\text{-}D_3$. Some (dendritic cells) can also express the vitamin D_3-25-hydoxylase. Vitamin D treatment *in vitro* has been found to enhance T cell activation, and divert immature dendritic cells from development as effector T cells toward development as regulatory ($CD4^+$, $CD25^+$) T cells. These effects involve vitamin D regulation of cytokine production – e.g., promotion of the NFκB pathway to downregulate interleukin-12 (IL-12) production; downregulation of the production of both interferon-γ (IFN-γ) and IL-4. In addition, $1,25\text{-}(OH)_2\text{-}D_3$ has been found to enhance macrophage and monocyte phagocytosis, bacterial killing, and heat shock protein production.

94. It is of interest to note that circulating insulin levels are reduced in vitamin D deficiency and respond quickly to treatment with $1,25\text{-}(OH)_2\text{-}D_3$.

95. Cytologic localization of $1,25\text{-}(OH)_2\text{-}D_3$ has been achieved by the technique of frozen section autoradiography, made possible by the availability of radiolabeled $1,25\text{-}(OH)_2\text{-}D_3$ of high specific activity.

TABLE 6.13 Vitamin D-Stimulated Uptake and Retention of Lead

Diet		Kidney			Tibia	
Vitamin D	Pb (%)	Ca (ppm)	Pb (ppm)	Ca (% ash)	Pb (ppm ash)	
—	0	69.0 ± 8.8^a	0^a	33.2 ± 0.4^a	23.4 ± 14.3^a	
1,25-(OH)$_2$-D$_3$	0	64.1 ± 1.5^a	0^a	36.0 ± 0.2^b	10.9 ± 6.7^a	
—	0.2	56.6 ± 3.8^a	4.8 ± 0.5^b	33.5 ± 0.3^a	133.1 ± 24.1^b	
1,25-(OH)$_2$-D$_3$	0.2	80.2 ± 13.4^a	$13.7 \pm 2.4^{c,d}$	$35.5 \pm 0.2^{b,c}$	335.1 ± 15.8^c	
—	0.8	62.4 ± 2.4^a	9.2 ± 0.9^c	32.8 ± 0.2^a	299.8 ± 4.8^c	
1,25-(OH)$_2$-D$_3$	0.8	90.1 ± 7.6^b	32.4 ± 7.6^d	34.7 ± 0.4^c	1008.8 ± 71.2^d	

$P < 0.05$.
Source: Fullmer, C. S. (1990). *Proc. Soc. Exp. Biol. Med.* 194, 258–264.

TABLE 6.14 Experimental Evidence for Vitamin D Functions in Non-Calcified Tissues

Putative Role	Observations
Cell differentiation	Promotion by 1,25-(OH)$_2$-D$_3$ of myeloid leukemic precursor cells to differentiate into cells resembling macrophages
Membrane structure	Alteration of the fatty acid composition of enterocytes, reducing their membrane fluidity
Mitochondrial metabolism	Decrease in isocitrate lyase and malate synthase (shown in rachitic chicks)
Muscular function	Stimulation of Ca^{2+} transport into the sarcoplasmic reticulum of cultured myeloblasts by 1,25-(OH)$_2$-D$_3$
	Easing, on treatment with vitamin D, of electrophysiological abnormalities in muscle contraction and relaxation in vitamin D-deficient humans
	Reduction, on treatment with vitamin D, of muscular weakness in humans
Pancreatic function	Stimulation of insulin production by pancreatic β cells in rats by 1,25-(OH)$_2$-D$_3$
	Impairment of insulin secretion, unrelated to the level of circulating calcium, shown in vitamin D-deficient humans
Immunity	Stimulation of immune cell functions by 1,25-(OH)$_2$-D$_3$
	Control of inflammation by 1,25-(OH)$_2$-D$_3$-dependent regulation of cytokine production
Neural function	Region-specific enhancement of choline acetyltransferase in rat brain by 1,25-(OH)$_2$-D$_3$
Skin	Inhibition of DNA synthesis in mouse epidermal cells by 1,25-(OH)$_2$-D$_3$
Parathyroid function	Inhibition of transcription of PTH gene via interaction of 1,25-(OH)$_2$-D$_3$ and DNA in parathyroid cells

Vitamin D deficiency has been associated with inflammation; studies have shown the circulating marker of inflammation, C-reactive protein, to be inversely correlated with serum concentrations of 25-OH-D$_3$, and decreased in response to vitamin D treatment.[96] However, that VDR-knockout mice show no immune abnormalities suggests that this function of the vitamin must be selective, depending on the nature of the immune challenge, or simply part of a higly redundant system.

Pancreatic Endocrine Function

For some years it has been known that vitamin D deficiency reduces the secretion of insulin (but not other hormones) by pancreatic β cells. Further, the insulin responses

96. Timms, P. M., Mannan, N., Hitman, G. A., *et al.* (2002). *Q. J. Med.* 95, 787–796.

and glucose clearance rates of free-living humans have been found to vary according to VDR type. Therefore, it is reasonable to ask whether low vitamin D status may play a role in non-insulin-dependent diabetes mellitus (type 2 diabetes, T2D), which is characterized by insulin resistance.

Skin

Vitamin D has a paracrine function in the skin. Keratinocytes express 25-$(OH)_2$-D_3 1-hydroxylase; therefore, they can not only produce vitamin D_3 with solar exposure, but also metabolize it to 1,25-$(OH)_2$-D_3. VDRs are also expressed throughout the epidermis as well as in hair follicles. Among the gene products induced by VDR activation in the skin is cathelicidin antimicrobial peptide (CAMP), which functions both in the direct killing of pathogens as well as a host response involving cytokine release, inflammation, and cellular immune response. Mutations of VDR occur in patients with hereditary vitamin D-resistant rickets. They show alopecia, the basis of which is unclear as the condition is not caused by vitamin D deficiency *per se* or by loss of 25-OH-D_3 1-hydroxylase activity.

Muscular Function

That vitamin D plays an important role in muscle is evidenced by the muscle weakness that is typical of vitamin D-deficient subjects, the presence of VDR in myocytes, and the lack of muscle development observed in VDR-knockout mice. Vitamin D, 1,25-$(OH)_2$-D_3, has been shown to be essential for the homeostatic control of intracellular Ca^{2+}, thus affecting both contractility and myogenesis. The former effect involves the transcriptional regulation of various Ca^{2+} binding proteins, including calbindin-D_{9K}, as well as enzymes related to the synthesis of phosphatidylcholine. Myocytes treated with 1,25-$(OH)_2$-D_3 respond by activating protein kinase C and transporting Ca^{2+} in the the sarcoplasmic reticulum, an effect necessary for muscle contraction. They also show upregulation of tyrosine phosphorylation of the MAP kinase cascade, stimulating proliferation and growth. Rapid responses of skeletal muscle have been observed for 1,25-$(OH)_2$-D_3: Ca^{2+} uptake through voltage-dependent channels, and Ca^{2+} release by intracellular stores. It has been suggested, based on their rapidity, that these may represent non-genomic mechanisms of vitamin D action; however, they may indicate only the short time required for VDR to enter the nucleus upon binding its ligand. Thus, it is not surprising that muscular weakness, hypotonia, and atrophy are seen in rickets, or that patients with ostomalacia frequently show myopathy. The latter affects primarily the type II muscle fibers that are the first to be recruited to avoid falling, resembling the sarcopenia[97] associated with aging.

Brain Development

A role of vitamin D in brain development and function was first indicated by the finding of 25-OH-D_3 in cerebrospinal fluid, and of nVDR and 25-OH-D_3 1-hydroxylase activity in brain tissue. Maternal deprivation of vitamin D affects rat pups, which show alterations in brain morphology, stem cell proliferation, gene expression, and expression of neurotrophic factors.[98] Some of these effects persist to adulthood, at which time abnormalities in gene expression and behavior (sensitivity to psychosis-inducing drugs) are observed. On the basis of such findings, it has been suggested that vitamin D may act as a neurosteroid with direct effects on brain development, including reduction of risk of neuropsychiatric disorders.

7. VITAMIN D IN HEALTH AND DISEASE

Studies have shown inverse associations of vitamin D intake, regardless of sun exposure, and the incidences of several diseases.

Autoimmune Diseases

These disorders are prevalent in northern latitudes. VDR polymorphisms have been associated with increased risks for at least some of these. That VDR-knockout mice do not show inflammatory responses to experimentally induced asthma suggests that 1,25-$(OH)2$-D_3 may be immunosuppressive, inhibiting signs/symptoms of T helper 1 (Th1) cell-driven autoimmune disease.

Rheumatoid Arthritis (RA)

VDR is expressed by articular chondrocytes in osteoarthritic cartilage, which also express matrix metalloproteinases (MMPs) not found in normal cartilage. Such observations, and the finding that vitamin D treatment prevented experimentally induced RA and prostaglandin E_2 production in animal models, suggest a role of the vitamin in this immune-mediated disease characterized by articular inflammation leading to disability. One study noted an inverse relationship of vitamin D intake, particularly from supplements, and risk of developing RA.[99]

97. Loss of muscle.
98. Levenson, C. W. and Figueirôa, S. M. (2008). *Nutr. Rev.* 66, 726–729; Eleyes *et al.* (2009). *Psychoneuroendocrinology* 345, S247–S257.
99. Merlino, L. A., Curtis, J., Mikuls, T. R., *et al.* (2004). *Arthritis Rheumatol.* 50, 72–77.

Multiple Sclerosis

It has been long recognized that multiple sclerosis (MS), an autoimmune disease characterized by immune attacks on the myelin sheaths of nerves, is more prevalent in northern, temperate parts of the world than in the tropics. In fact, the prevalence has been found to be strongly inversely related to the numbers of hours of annual or winter sunlight. Two studies have found the use of vitamin D supplements to reduce the risk of developing MS by as much as 40%.[100] The only study to date with MS subjects found vitamin D supplementation to increase circulating levels of the anti-inflammatory cytokine transforming growth factor β1 (TGF-β1), suggesting potential for alleviating symptoms.

Insulin-Dependent Diabetes (Type 1 Diabetes, T1D)

It has been suggested that vitamin D may play a role in reducing the risk of T1D, which results from the T-cell dependent destruction of insulin-producing pancreatic β cells by cytokines and free radicals from inflammatory infiltrates. It has been suggested that vitamin D may be beneficial in the treatment of T1D by inhibiting the production of IL-12 and suppressing the activity of IL-12-dependent Th1 cells in activating cytotoxic CD8$^+$ lymphocytes and macrophages.

Epidemological studies have found the incidence of T1D to be positively associated with latitude and negatively associated with hours of sunlight. Two prospective trials have been conducted to evaluate the efficacy of vitamin D in preventing T1D: one found positive effects of vitamin D doses of 50 μg/day;[101] the other detected no benefits using lower doses (<10 μg/d).[102] A large, multi-center case–control study[103] and a cohort study[104] have found vitamin D supplementation in infancy reduced T1D risk in later life. High doses of 1,25-(OH)$_2$-D$_3$ have been shown to arrest diabetes in the non-obese diabetic mouse.[105]

Inflammatory Skin Diseases

The finding that 1,25-(OH)$_2$-D$_3$ can inhibit proliferation and induce terminal differentiation of cultured keratinocytes stimulated the study of its potential value in the treatment of proliferative skin disorders. These effects appear to be mediated by VDRs, which are expressed throughout the epidermis as well as in hair follicles and skin immune cells,[106] and involve increases in intracellular free Ca^{2+} associated with increased phosphoinositide (inositol triphosphate and diacylglycerol) levels. Studies have shown that 1,25-(OH)$_2$-D$_3$ can decrease keratinocyte sensitivity to epidermal growth factor (EGF) receptor-mediated growth factors, increase the transcription of transforming growth factor β$_1$ (TGF-β$_1$), and regulate a cytokine cascade[107] involved in the accumulation of leukocytes during skin inflammation. Clinical studies have shown that both oral and topical applications of appropriate doses of either 1,25-(OH)$_2$-D$_3$ or the synthetic analog 1α,25-(OH)$_2$-D$_3$[108] can be safe and effective in the management of psoriasis,[109] a Th1-mediated, hyperproliferative autoimmune disease that is chronically inflammatory in nature. Psoriasis has been shown to involve dysfunctional expression of the VDR-induced gene product cathelicidin antimicrobial peptide (CAMP), a neutrophil-specific granule protein that normally functions in pathogen killing and stiumlating protective host responses (cytokine release, inflammation, cellular immune response).[110] In psoriatics, the cathelicidin peptide converts otherwise inert self-DNA and -RNA into autoimmune stimuli. Decreased levels of cathelicidin have been observed in atropic dermatitis, and abnormal processing of the cathelicidin peptide has been found to be involved in the inflammatory and vascular responses in rosacea. Because the use of 1,25-(OH)$_2$-D$_3$ carries risks of hypercalcemia and hypercalciuria, there is interest in developing treatment regimens for such diseases involving the application of high doses of the vitamin in a safe and effective manner.

Other Autoimmune Diseases

Vitamin D has been effective in reducing signs in animal models of Th1-induced *inflammatory bowel disease*,

100. Munger, K. L., Zhang, S. M., O'Reilly, E., *et al.* (2004). *Neurology* 62, 60–65; Munger, K. L., Levin, L. I., Hollis, B. W., *et al.* (2006). *J. Am. Med. Assoc.* 296, 2832–2838.
101. Hyppönen, E., Läärä, E., Reunanen, A., *et al.* (2001). *Lancet* 358, 1500–1503.
102. Stene, L. C., Ulriksen, J.., Magnus, P., and Joner, G. (2003). *Am. J. Clin. Nutr.* 78, 1128–1134.
103. EURODIAB Substudy 2 Study Group (1999). *Diabetologia* 42, 51–54.
104. Hyppönen, E., Läärä, E., Reunanen, A., *et al.* (2001). *Lancet* 358, 1500–1503.
105. Mathieu and colleagues reported a 28% reduction in the autoimmune NOD mouse (Mathieu, C., Waer, M., Laureys, J., *et al.* [1994] *Diabetologia* 37, 552–558), but no effect in the BB rat model (Mathieu *et al.* [1997]. In *Vitamin D and Diabetes* [Feldman, D., Glorieux, F. and Pike, J., eds]. Academic Press, San Diego, CA, pp. 1183–1196).

106. VDRs are expressed by the majority of Langerhans cells, macrophages, and T lymphocytes in skin.
107. 1,25-(OH)$_2$-D$_3$ has been shown to inhibit the expression of mRNA for the neutrophil-activating peptide interleukin 8 (IL-8) in keratinocytes, dermal fibroblasts, and monocytes; these cells produce IL-8 in the skin of psoriatic but not nonpsoriatic patients.
108. This analog is also called calcipotriol.
109. Results of one clinical series showed that topical application of 1,25-(OH)$_2$-vitamin D$_3$ caused complete clearing of lesions in 60% of patients, with an additional 30% of patients showing significant decreases in scale, plaque thickness, and erythema.
110. Dombrowski, H. *et al.* (2010). *Arch. Dermatol. Res.* 302, 410–408; White, J. H. (2010). *J. Steroid Biochem. Mol. Biol.* 121, 234–238.

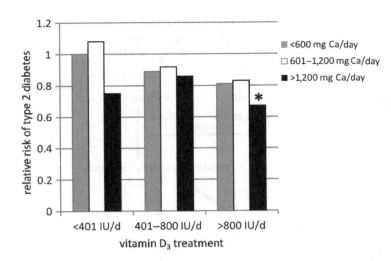

FIGURE 6.12 Reduction of type 2 diabetes risk associated with vitamin D and calcium supplement use by 83,779 women in the Nurses' Health Study (*indicates significant difference [*P* <0.05] from the low vitamin D, low Ca group). After *Pittas, A. G., Dawson-Hughes, B., Li, T.,* et al. *(2006).* Diabetes Care *29, 650–656.*

but it is not know whether it may be useful in treating ulcerative colitis or Crohn's disease in human patients.[111] Vitamin D treatment has been found to reduce metabolic and dermatologic lesions and prolong survival in animal models of the T-cell dependent autoimmune disorder *systemic lupus erythematosus*. One study also found vitamin D treatment to exacerbate the disease in one SLE-prone mouse strain.

Graft Rejection

Vitamin D treatment has been found to reduce rejection of allografts of heart, kidney, liver, pancreatic islets, small intestine, and skin in animals given high, non-hypercalcemic doses in both the presence and the absence of the immunosuppressive drug cyclosporin A. These effects involve the suppression of antibody-presenting cells (particularly dendritic cells and T cells that play major roles in immunological graft rejection) and enhancing numbers of su ppressive T (CD4$^+$, CD25$^+$) cells.

Non-Insulin-Dependent Diabetes (Type 2 Diabetes, T2D)

Protection against T2D by vitamin D may involve the vitamin protecting pancreatic β cells from inflammatory damage. Subclinical, low-intensity, chronic inflammation has been associated with insulin resistance, which has been found to be inversely related to serum 25-OH-D$_3$ concentrations over a wide range.[112] The results of the Third National Health and Nutrition Examination Survey (NHANES III) showed serum level of 25-OH-D$_3$ to be inversely associated with diabetes risk in a multi-ethnic sample of over 6,000 adults.[113] Swedish researchers have found T2D incidence to be highest during the winter months, when circulating 25-OH-D$_3$ levels are lowest.[114] A 20-year follow-up of the Nurses' Health Study cohort found T2D risk to be one-third less for women reporting the use of vitamin D and Ca supplements (Fig. 6.12). While T2D risk was inversely associated with both vitamin D and Ca intakes, most of these effects were attributed to the use of supplements of these nutrients.

The hypothesis that improving vitamin D status could improve glycemic control was recently tested in a randomized clinical conducted in Iran, a country with high reported prevalences of low vitamin D status, metabolic syndrome, and T2D. This showed that vitamin D supplementation improved glycemic control in T2D patients, as indicated by reductions in fasting glucose level and insulin resistance (Fig. 6.13).[115]

Two VDR polymorphisms, *BsmI* and *ApaI*, allelic variants in the intron between exons VIII and IX, have been associated with increased fasting glucose levels and elevated T2D risk.[116]

---

111. These recurring inflammatory diseases affect the terminal ileum and colon.

112. Chiu, K. C., Chu, A., Go, V., and Saad, M. F. (2004). *Am. J. Clin. Nutr.* 79, 820–825.

113. Scragg, R., Sowers, M., and Bell, C. (2004). *Diabetes Care* 27, 2813–2818.

114. Berger, B., Stenstrom, G., and Sundkist, G. (1999). *Diabetes Care* 22, 773–777.

115. Insulin resistance was measured by HOMA-IR, the homeostasis model assessment of insulin resistance, i.e., the products of the fasting serum concentrations of insulin and glucose.

116. Hitman, G. A., Mannan, N., McDermott, M. F., *et al.* (1998). *Diabetes* 47, 688–690; Oh, J.-Y. and Barrett-Connor, E. (2002). *Metabolism* 51, 356–359; Ortlepp, J. R., Metrikat, J., Albrecht, M., *et al.* (2003). *Diabet. Med.* 20, 451–454.

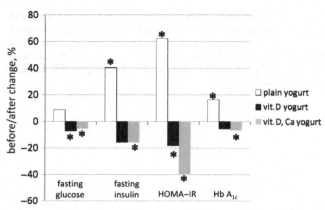

FIGURE 6.13 Effects of 12-week, yogurt-based supplementation with vitamin D (500 IU/day) and Ca (150 mg/day) on parameters of glycemic control in type 2 diabetes patients (fasting serum glucose >126 mg/dl) adults (n = 30 per treatment group); *indicates significant difference (P <0.05) from the respective baseline value (Nikooyeh, B., Neyestani, T. R., Farvid, M., *et al.* [2011] *Am. J. Clin. Nutr.* 93, 764771).

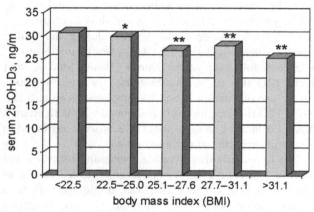

FIGURE 6.14 Inverse relationship of vitamin D status and body mass index apparent in NHANES III subjects. Comparisons with lowest quintile: *P <0.05, **P <0.01. After *Black, P. N. and Scragg, R. (2005). Clin. Invest.* 128, 3792–3798.

Effects of Obesity

The efficacy of vitamin D in preventing T2D may depend on an individual's adiposity. Studies[117] have shown that serum 25-OH-D_3 levels are inversely correlated with body mass index[118] (Fig. 6.14), body fat mass, and insulin resistance, and directly associated with weight loss resultant of caloric restriction. The relationship of vitamin D status and T2D risk appears to be greatest among overweight/obese

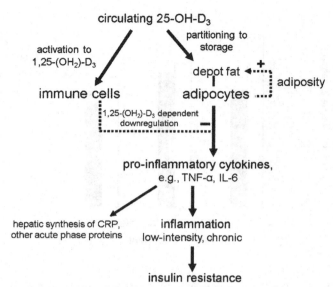

FIGURE 6.15 Hypothetical relationships of vitamin D status and adiposity in affecting insulin resistance. *Abbreviations*: TNF-α, tumor necrosis factor α; IL-6, interleukin-6; CRP, C-reactive protein.

individuals.[119] At least one single nucleotide polymorphism has been associated with increased waist circumference.[120] It has been suggested that the exacerbation of the T2D risk of low vitamin D status by adiposity may be due to the sequestration of vitamin by partitioning into bulk lipid depots in adipose tissue. According to this hypothesis (Fig. 6.15), adipocytes would be involved in two opposing ways: responding to 1,25-(OH)$_2$-D_3 in the regulation of cytokine production, and sequestering 25-OH-D_3.

It has been proposed that vitamin D deficiency may also cause obesity.[121] According to this hypothesis, the circulating 25-OH-D_3 level evolved to serve as a UVB-sensitive photoreceptor. In species with low dietary vitamin D intakes, 25-OH-D_3 levels decline with shortening of daylength during winter, to stimulate accumulation of fat and induce a winter metabolism – i.e., the metabolic syndrome. This would imply that vitamin D supplements may be effective in preventing obesity. This is supported by findings that 1,25-(OH)$_2$-D_3 can induce apoptosis in adipocytes, an effect that inolves VDR binding of peroxisome proliferator-activated receptor-γ (PPARγ), thus reducing the availability of the ligand to the retinoid X receptor (RXR). In addition, 1,25-(OH)$_2$-D_3 has been shown to suppress adipose fat deposition by stimulating the expression of the steroid-metabolizing enzyme 11β-hydroxysteroid hydroxylase, thus promoting glucocorticoid production. While a few clinical

117. Arunabh, S., Pollack, S., Yeh, J., and Aloia, J. F. (2003). *J. Clin. Endocrinol. Metab.* 88, 157–161; Parikh, J., Edelman, G. I., Uwaifo, R. J., *et al.* (2004). *J. Clin. Endocrinol. Metab.* 89, 1196–1199.

118. BMI = weight (kg)/height (m²) or [weight (lb)/height (in²)] × 705.

119. Isaia, G., Giorgino, R., and Adami, S., *et al.* (2001). *Diabetes* 24, 1496–1503.

120. Ochs-Balcom, H. M., Chennamaneni, R., Millen, A. E., *et al.* (2011). *Am. J. Clin. Nutr.* 93, 5–10.

121. Foss, Y. J. (2009). *Med. Hyp.* 72, 314–321.

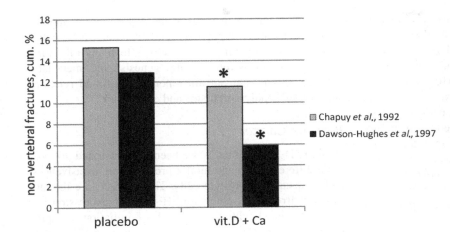

FIGURE 6.16 Prevention of fractures by vitamin D supplementation in two cohorts of older adults (Chapuy *et al.*, 2,790 women, 84 ± 6 years, receiving 1,200 mg Ca and 800 IU vitamin D daily; Dawson-Hughes *et al.*, 389 men and 213 women, 72 ± 5 years, receiving 500 mg Ca and 700 IU vitamin D daily). * Indicates significant difference from respective placebo level, P <0.05. (Chapuy, M. C., Arlot, M. E., Duboeuf, F., *et al.* [1992] *New England J. Med.* 327, 1637–1642; Dawson-Hughes, B., Harris, S. S., Krall, E. A., *et al.* [1997] *New England J. Med.* 337, 670–676).

trials have found weight loss to improve 25-OH-D_3 levels and vitamin D_3 supplements to improve insulin sensitivity, such effects have not been observed in most trials.

Risk of Falling

It has been suggested that insufficient vitamin D status may increase the risk to bone fracture by affecting strength, balance, and gait. Falls have been shown to be the most frequent causes of fractures (causing >90% of hip fractures) and are known to increase with age, particularly after 70 years. Their prevalence has also been shown to vary with both season (being greater in the winter months) and latitute (being greater at northern latitudes). Several observational studies with older adults have found significant positive associations of plasma levels of 25-OH-D_3 and/or $1,25\text{-(OH)}_2\text{-D}_3$ and factors related to risk of falling: postural balance and strength (measured by such parameters as leg extension power, quadriceps strength, arm muscle strength, handgrip strength, ability to climb stairs, and physical activity).[122]

Two meta-analyses of the eight published randomized controlled trials showed that supplementation with 700–1000 IU/day reduced by 14% and 19% the risk of falling by older subjects,[123] although individual studies have reported reductions by nearly 50%. The greatest benefits were observed when vitamin D was given with supplemental Ca (Fig. 6.16).

Cardiovascular Health

Epidemiological studies have shown that low circulating 25-OH-D_3 levels, as well as factors known to affect vitamin D status (latitude, altitude, season), are associated with the prevalence of coronary risk factors and cardiovascular disease (CVD) mortality. Vitamin D is known to suppress several mechanisms of CVD pathogenesis: proliferation of vascular smooth muscle, vascular calcification, production of pro-inflammatory cytokines, and regulation of the renin–angiotensin system. Therefore, it has been suggested that low vitamin D status may be a risk factor for CVD. Two intervention trials have been conducted to test this hypothesis; they found small (<10%) reductions in systolic and/or diastolic blood pressure in hypertensive patients treated with vitamin D. A large, prospective, general population study found no effects.

Anticarcinogenesis

That vitamin D may be protective against colon cancer was proposed nearly 75 years ago, based on epidemiologic associations with sunlight exposure. Several studies in Europe and the United States have found positive associations between latitude and risks of cancers of the prostate, colon, and breast. Residents of the northern US have nearly a two-fold higher risk of total cancer mortality than those of the southern states. Participants in the Nurses' Health Study showed a significant inverse linear association between plasma 25-OH-D_3 and colon cancer risk.[124] Studies have shown low serum $1,25\text{-(OH)}_2\text{-D}_3$ levels to be associated with increased risk to cancers of the prostate and breast (Fig. 6.17). Risk factors for cancers of the prostate, colon, and breast are related to vitamin D status: dark skin color, northern latitude residence, and increasing age.

122. Mowé, M., Haug, E., and Bohmer, T. (1999). *J. Am. Geriatr. Soc.* 47, 220–226; Bischoff, H. A., Stahelin, H. B., Urscheler, N., *et al.* (1999). *Arch. Phys. Med. Rehabil.* 80, 54–58; Annweiller, C., Schott-Petelaz, A. M., Berrut, G., *et al.* (2009). *J. Am. Geriatr. Soc.* 57, 368–369; Annweiller, C., Montero-Odasso, M., Schot, A. M., *et al.* (2009). *J. Nutr. Health Aging* 13, 90–95.
123. Bischoff-Ferrari, H. A., Dawson-Hughes, B., Staehelin, H. B., *et al.* (2009). *Br. Med. J.* 339, 3692–3603; Kalyani, R. R., Stein, B., Valiyil, R., *et al.* (2010). *J. Am. Geriatr. Soc.* 58, 1299–1310.

124. Feskanich, D., Ma, J., Fuchs, C.S., *et al.* (2004). *Cancer Epidemiol. Biomarkers* 13, 1502–1508.

The highest risk for each is seen among densely pigmented individuals who also tend to have lower circulating levels of 25-OH-D$_3$. A meta-analysis found carriers of the VDR *Bsm1 B* allele to have 6–7% less total cancer.[125] A 4-year clinical trial found that a high dose (1,000 IU/day) of vitamin D$_3$ significantly increased the protective effect of supplemental Ca against all-cause cancer (Fig. 6.18).

Numerous studies have demonstrated anticarcinogenic effects of 1,25-(OH)$_2$-D$_3$ in more than a dozen animal models (Table 6.15). These have shown vitamin D to inhibit cancer cell growth, angiogenesis, and metastasis. These effects appear to involve VDRs, which have been identified in many tumors and malignant cell types. VDR activation is known to regulate the expression of apoptosis genes to induce apoptosis, and to induce differentiation into normally functioning cells, effectively inhibiting proliferation, invasiveness, angiogenesis, and metastatic potential. Some, but not all, reports have found relationships of VDR polymorphisms and cancer risk.

Prostate Cancer

Vitamin D metabolites have been demonstrated to promote differentiation and inhibit proliferation, invasiveness, and metastasis in prostate cells. These effects are thought to result from VDR activation. High VDR expression has been associated with reduced risk of initiation and progression of prostate cancer in younger men.[126] Studies with rat and human prostates have shown VDR activation by vitamin D to upregulate the expression of the androgen receptor, and androgens, in turn, to upregulate the VDR, suggesting that antiproliferative action of vitamin D in that tissue may be androgen-dependent. A meta-analysis found prostate cancer risk is less for carriers of the VDR *Bsm1 B* allele than for those without that allele.[127]

Prostate cells contain a 25-OH-D$_3$ 1-hydroxylase activity; however, that activity appears to be suppressed in individuals with clinically detected prostate tumors. It is not clear whether this reflects a diminution of expression in the diseased state.

Colorectal Cancer

Mortality from colorectal cancer is significantly higher in the northern and northeastern United States than in

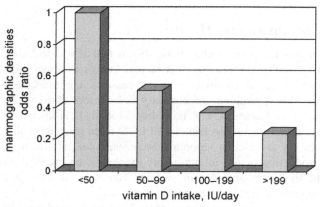

FIGURE 6.17 Inverse relationship of frequency of mammographic breast density and vitamin D status in a cohort of 543 women aged 40–60 years undergoing screening mammographies (Berube, S., Diorio, C., Verhoek-Oftedahl, W., and Brisson, J. [2004] *Cancer Epidemiol. Biomarkers Prev.* 13, 1466 1472).

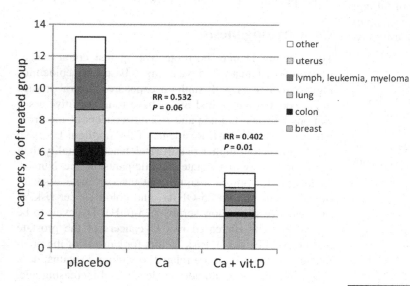

FIGURE 6.18 Cancer incidence in 1,180 women randomized to treatment with a placebo, Ca (1,400 or 1,500 mg per day) or Ca$^+$ vitamin D (1,400 or 1,500 mg Ca and 1,000 IU per day) for 4 years. After *Lappe, J. M. , Travers- Gustafson, D., Davies, K. M., et al. (2007).* Am. J. Clin. Nutr. *85, 1586–1591.*

125. Raimondi, S., Johansson, H., Maisonneuve, P., and Gandini, S. (2009). *Carcinogenesis* 30, 1170–1180.

126. Ahonen, M. H., Tenkanen, L., Teppo, L., *et al.* (2000). *Cancer Causes Control* 11, 847–852.
127. Berndt, S. I., Dodson, J. L. Huang, W. Y. *et al.* (2006). *J. Urol.* 175, 1613.

the southwest, Hawaii, and Florida. Observational studies have been inconsistent in linking this phenomenon to vitamin D status; however, a meta-analysis has shown significant reduction in colorectal risk associated with serum 25-0H-D$_3$ levels >20 ng/ml or intakes of at least 1,000–2,000 IU/day.[128]

In vitro studies have demonstated that 1,25-(OH)$_2$-D$_3$ can both attenuate the growth of rapidly dividing colonic tumor cells[129] and reverse colonocytes from a malignant to a normal phenotype. These effects appear to require high doses, and depend on activating VDR to induce apoptosis. High VDR expression has been associated with a more favorable prognosis for colorectal cancer patients.[130] VDR can also bind the secondary bile acid lithocholic acid, a potent enteric carcinogen. Lithocholic acid has been found to activate VDR to upregulate the expression of CYP3A, the cytochrome *P*-450 isoform that detoxifies bile acids.[131] This activity would appear to be the basis of vitamin D protection against colorectal carcinogenesis, which is stimulated by bile acids and xenobiotic compounds that are metabolized by CYP3A. This suggests that VDR may serve as a bile-acid sensor.

Colonocytes have a high 25-OH-D$_3$ 1-hydroxylase capacity, converting 5–10% of 25-OH-D$_3$ to the active metabolite. The colonic 1-hydroxylase activity (CYP27B1) has been found to be increased in animals by feeding genistein that also markedly reduces the activity of the 24-hydroxylase (CYP24). This suggest that this phytoestrogenic soy isoflavone may be effective in upregulating cellular production of 1,25-(OH)$_2$-D$_3$, providing a plausible hypothesis for the relatively lower risks to colon (and breast and prostate) among groups with relatively high consumption of soy foods.

Breast Cancer

Physiological concentrations of 25-OH-D$_3$ have been shown to inhibit the growth of mammary cancer cells, which can also produce 1,25-(OH)$_2$-D$_3$. These effects appear to occur via the VDR, which is expressed in normal and tumorous mammary tissues. The liganded receptor appears to function to arrest growth, induce differentiation, induce apoptosis, and oppose estrogen-driven proliferation by inducing the expression of factors[132] involved in the

TABLE 6.15 Animal Tumor Models in which Anticarcinogenic Activity of 1,25-(OH)$_2$-D$_3$ has been Demonstrated

Species	Site	Mode of Induction
Mouse	Skin	Oral treatment: dimethybenzanthracine, phorbol ester
Mouse (athymic)	Adenocarcinoma	Implantation: CAC-8 cells
	Kaposi sarcoma	Implantation: KS YH-1 cells
	Melanoma	Implantation: human melanoma cells
	Osteosarcoma	Implantaton: LNCaP cells
	Retinoblastoma	Implantation: malignant cells
Mouse, APCmin	Colon	(Spontaneous)
Rat	Mammary	Oral treatment: N-methylnitrosourea, dimethylhydrazine
	Colon	Implantation: human colon cancer cells
		Oral treatment: dimethylhydrazine
	Leydig	Implantation: Leydig tumor cells
	Prostate	Implantation: Dunning LyLu cells
	Walker carcinoma	Implantation: Walker carcinoma cells

128. Gorham, E. D., Garland, C. F., Garland, F. C., *et al.* (2007). *Am. J. Prev. Med.* 32, 210.

129. Caco-2 cells.

130. Evans, S. R. T., Nolla, J., Hanfelt, J., *et al.* (1998). *Clin. Cancer Res.* 4, 1591–1595.

131. Makishima, M., Lu, T. T., Xie, W., *et al.* (2002). *Science* 296:1313–1316.

132. For example, cyclins C and D1, Kip1, WAF1, c-Fos, C-Myc, c-JUN, and members of the TBF-β family

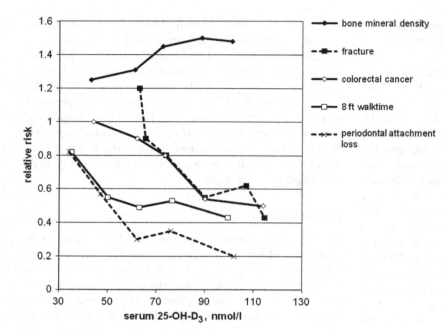

FIGURE 6.19   Relationship of vitamin D status and multiple health risks. After *Bischoff-Ferrari, H. A., Giovannucci, E., Willett, W. C., et al.* (2006). Am. J. Clin. Nutr. *84, 18–28.*

regulation of cell proliferation. A pooled analysis of two published observational studies showed a negative association of vitamin D status and breast cancer risk, with individuals with serum 25-OH-D$_3$ levels averaging 52 ng/ml having half the risk of those with serum 25-OH-D$_3$ levels of <13 ng/ml, and about 60% of the risk of those with serum levels around 20 ng/ml.[133] A meta-analysis found breast cancer risk is less for individuals of the VDR *Fok1 FF* genotype than for those of the ff genotype.[134]

Other Cancers

That vitamin D may have a role in protecting against leukemia is suggested by the finding that 1,25-(OH)$_2$-D$_3$ treatment can suppress cell division and induce differentiation in human leukemic cells. This effect involves downregulating the expression of the proto-oncogene c-*myc*. Case–control studies have provide strong evidence of relationships of vitamin D status and risk to non-Hodgkin lymphoma. A meta-analysis found individuals of the VDR *Fok1 FF* genotype to have lower risk of skin cancer (including malignant melanoma) than those of the ff genotype.[135] Evidence is inconsistent for ovarian cancer and lacking for other cancers.

Healthful Vitamin D Status

The most informative indicator of vitamin D status is the concentration of 25-OH-D$_3$ in serum/plasma.[136] Maximal PTH levels are seen at 25-OH-D$_3$ levels of *ca.* 100 nmol/l; levels less than 45–65 nmol/l are associated with secondary hyperparathyroidism. The recent report by the Institute of Medicine considered 50 nmol/l adequate for bone health in adults and children,[137] although an expert concensus considered the level of 70–80 nmol/l as optimal.[138] An analysis of multiple endpoints related to bone and dental health, lower extremity function, risks of falls, fractures, and cancer found the optimal serum 25-OH-D$_3$ level to be 90–100 nmol/l (Fig. 6.19). Such levels require regular daily vitamin D intakes of greater than 1000 IU (40 μg).[139] The maintenance of such serum 25-OH-D$_3$ levels requires the use of solar radiation, vitamin D supplements, and/or vitamin D-fortified foods. Table 6.16 presents this author's recommended criteria.

Genetic polymorphisms have been found to affect circulating levels of 25-OH-D$_3$, in some cases by as much as can deprivation of vitamin D. These include variants at three loci near genes encoding 7-dehydrocholesterol

133. Garland, C. F., Gorham, E. D., Mohr, S. B., *et al.* (2007). *J. Steroid Biochem. Mol. Biol.* 103, 708.

134. Tang, C., Chen, N., Wu, M., *et al.* (2009). *Breast Cancer Res. Treat.* 30, 1170.

135. Mocellin, S. and Nitti, D. (2008). *Cancer* 113, 2398.

136. It is more suitable than 1,25-(OH)$_2$-D$_3$ because the latter can be produced locally by many tissues and is not reduced in individuals with moderately low circulating levels of 25-OH-D$_3$.

137. Institute of Medicine (2011). *Dietary Reference Intakes: Calcium, Vitamin D.* National Academy Press, Washington, DC, 1115 pp.

138. Dawson-Hughes, B., Heaney, R. P., Holick, M. F., *et al.* (2005). *Osteopor. Intl* 16, 713–716.

139. An analysis by Bischoff-Ferrari and colleagues (Bischoff-Ferrari, H. A., Giovannucci, E., Willett, W. C., *et al.* [2006] *Am. J. Clin. Nutr.* 84, 18–28) indicated that vitamin D intakes of at least 1,000 μg/day are required to bring 50% of healthy American adults above the serum 25-OH-D$_3$ level of 75 nmol/l. Current recommended intakes for vitamin D are 200 and 600 IU/day in younger and older adults, respectively.

TABLE 6.16 Recommended Criteria of Vitamin D Status

25-OH-D$_3$ level		Status
nmol/l	ng/ml	
<50	<20	Deficient
50–75	20–29	Insufficient
75–375	≥30	Adequate
>250	>100	Excess (risk of hypervitaminosis)

reductase, DBP, and a cytochrome *P*-450 isoform CYP2R1.[140] Two of these are involved in the synthesis of the vitamin precursor and the transport of the dominant metabolite. The third, CYP2R1, has been suggested as having vitamin D$_3$ 25-hydroxylase activity. Serum 25-OH-D$_3$ measurement may easily overestimate vitamin D intakes of individuals with such polymorphisms, who may also be more susceptible to adverse effects of vitamin D deprivation.

Serum 25-OH-D$_3$ levels are more frequently low in pregnant women compared to non-pregant women. A study in Boston found that 76% of new mothers (and 81% of their newborns) had serum 25-OH-D$_3$ levels <20 ng/ml.[141] Vitamin D status has been inversely associated with risks of pre-eclampsia, gestational diabetes, bacterial vaginosis, and primary cesarean section (perhaps related to pelvic deformities and/or weakened muscular function during labor). Circulating levels of 1,25-(OH)$_2$-D$_3$ increase by two- to three-fold in the first trimester of pregancy, apparently due to the expression of 25-OH-D$_3$ 1-hydroxylase activity and suppression of 24-hydroxylase activity in placental and decidual cells, which also express VDR. It has been suggested that 1,25-(OH)$_2$-D$_3$ functions in the placenta to suppress Th1-dependent immunity to facilitate immune tolerance to implantation and successful fetal maintenance.

Serum 25-OH-D$_3$ levels have also been found to be inversely associated with risks of acute respiratory infection[142] and symptoms of asthma.

Recent studies indicate a decline in the vitamin D status of Americans in recent decades. Comparisons of serum 25-OH-D$_3$ levels measured in serum of participants in the National Health and Examination Surveys (NHANES) conducted in 1988–1994 and 2000–2004 showed a significant 7.1-nmol/l (*ca.* 9%) decline among men over that period.[143] It is not clear whether this decline may be related to other changes known to have occurred over that period – namely, increased prevalence of overweight/obesity, decreased prevalence of milk consumption, increased use of topical sunscreens.

8. VITAMIN D DEFICIENCY

Causes of Vitamin D Deficiency

Vitamin D deficiency can result from inadequate irradiation of the skin, from insufficient intake from the diet or from impairments in the metabolic activation (hydroxylations) of the vitamin. Although sunlight can provide the means of biosynthesis of vitamin D$_3$, it is a well-documented fact that many people, particularly those in extreme latitudes during the winter months, do not receive sufficient solar irradiation to support adequate vitamin D status. Even people in sunnier climates may not produce adequate vitamin D if their lifestyles or health status keep them indoors, or if such factors as air pollution or clothing reduce their exposure to UV light. Most people therefore show strong seasonal fluctuations in plasma 25-OH-D$_3$ concentration; for some, this can be associated with considerable periods of suboptimal vitamin D status if not corrected by an adequate dietary source of the vitamin. Until the practice of vitamin D fortification of foods became widespread, at least in technologically developed countries, it was difficult to obtain adequate vitamin D from the diet, as most foods contain only minuscule amounts.[144]

Vitamin D deficiency can have privational and non-privational causes.

- **Privational causes** involve inadequate vitamin D supply. They include:
  - inadequate exposure to sunlight, and
  - insufficient consumption of food sources of vitamin D.
- **Non-privational causes** relate to impairments in the absorption, metabolism or nuclear binding of the vitamin. They include:

  - diseases of the gastrointestinal tract (e.g., small bowel disease, gastrectomy, pancreatitis), involving malabsorption of the vitamin from the diet
  - diseases of the liver (biliary cirrhosis, hepatitis), involving reduced activities of the 25-hydroxylase
  - diseases of the kidney (e.g., nephritis, renal failure), involving reduced activities of the 1-hydroxylase,

140. Sinotte M., Diorio, C., Berube, S., *et al.* (2009). *Am. J. Clin. Nutr.* 89, 634–640; Wang, T. J., Zhang, F., Richards, J. B., *et al.* (2010). *Lancet* 376, 180–188.

141. Lee, J. M., Smith, J. R., Philipp, B. L., *et al.* (2007). *Clin. Pediatr.* 46, 42–44.

142. It has been suggested that the winter lows in circulating 20-OH-D$_3$ levels in the northern latitudes may be the "seasonal stimulus" for epidemic influenza.

143. Looker, A. C., Pfeiffer, C. M., Lacher, D. A., *et al.* (2008). *Am. J. Clin. Nutr.* 88, 1519–1527.

144. Eggs are the notable exception. Even cows' milk and human milk contain only very small amounts of vitamin D.

the major source of $1,25\text{-}(OH)_2\text{-}D_3$,[145] or of $25\text{-}OH\text{-}D_3$ as in individuals with nephrotic syndrome,[146] who lose $25\text{-}OH\text{-}D_3$ along with its globulin-binding protein into the urine

- exposure to certain drugs (e.g., the anticonvulsives phenobarbital and diphenylhydantoin) which induce the catabolism of $25\text{-}OH\text{-}D_3$ and $1,25\text{-}(OH)_2\text{-}D_3$, reduce circulating levels of the former, and reduce elevated PTH levels[147]
- impaired parathyroid function resulting in **hypoparathyroidism** (reduced production of PTH), which impairs the ability to respond to hypocalcemia[148] by increasing the conversion of $25\text{-}OH\text{-}D_3$ to $1,25\text{-}(OH)_2\text{-}D_3$
- genetic mutations resulting in impaired expression of the renal $25\text{-}(OH)\text{-}D_3$ 1-hydroxylase in the condition referred to as **vitamin D-dependent rickets type I**, which can be managed using low doses of $1,25\text{-}(OH)_2\text{-}D_3$ or $1\alpha\text{-}OH\text{-}D_3$
- expression of a non-functional VDR and impairing the transcription of vitamin D-regulated genes involved in Ca and phosphorus homeostasis in the condition referred to as **vitamin D-dependent rickets type II**, the management of which requires relatively high doses of $1,25\text{-}(OH)_2\text{-}D_3$ or $1\alpha\text{-}OH\text{-}D_3$
- resistance of PTH target cells, resulting in **pseudohypoparathyroidism** and involving hypocalcemia without compensating renal retention or bone mobilization of Ca despite normal PTH secretion; the condition responds to low doses of $1,25\text{-}(OH)_2\text{-}D_3$ or $1\alpha\text{-}OH\text{-}D_3$[149]
- **vitamin D-resistance**[150] involving impaired phosphate transport in the intestine and reabsorption in the proximal renal tubules, hypersensitivity to PTH, and impaired 1-hydroxylation of $25\text{-}OH\text{-}D_3$; the condition responds to phosphate plus either high-dose vitamin D_3 (25,000–50,000 IU/day) or low doses of $1,25\text{-}(OH)_2\text{-}D_3$ or $1\alpha\text{-}OH\text{-}D_3$.

It is also likely that genetic polymorphisms in DBP and other proteins involved in vitamin D metabolism/function that result in suppressed circulating levels of $25\text{-}OH\text{-}D_3$ may render individuals at risk of vitamin D deprivation.

Signs of Vitamin D Deficiency

General Signs

Frank deficiency of vitamin D affects several systems, most prominently skeletal and neuromuscular (Table 6.17).

Vitamin D Deficiency in Humans

Rickets

Rickets first appears in 6- to 24-month old children, but can manifest at any time until the closure of the bones' epiphyseal growth plates. It is characterized by impaired mineralization of the growing bones with accompanying bone pain, muscular tenderness, and hypocalcemic tetany (Table 6.17). Tooth eruption may be delayed, the fontanelle may close late, and knees and wrists may appear swollen. Affected children develop deformations of their softened, weight-bearing bones, particularly those of the rib cage (Fig. 6.20) legs and arms; hence the characteristic leg signs, *bow-leg*,[151] *knock knee*,[152] and *sabre tibia*, which occur in nearly half of cases (Fig. 6.21). Radiography reveals enlarged epiphyseal growth plates resulting from their failure to mineralize and continue growth. Rickets is most frequently associated with low dietary intakes of calcium, as in the lack of access to or avoidance of milk products.[153]

Osteomalacia

Osteomalacia occurs in older children and adults with formed bones whose epiphyseal closure has rendered that region of the bone unaffected by vitamin D deficiency. The signs and symptoms of osteomalacia are more generalized than those of rickets; e.g., muscular weakness and bone tenderness and pain, particularly in the spine, shoulder,

145. Chronic kidney disease leading to bone disease is called *renal osteodystrophy*. It is a frequent complication in renal dialysis patients, in which its severity varies directly with the reduction in glomerular filtration rate. It is more common among children than adults, presumably owing to the greater sensitivity of growing bone to the deprivation of vitamin D, phosphate, and PTH that occurs in renal disease.

146. A clinical condition, involving renal tubular degeneration, characterized by edema, albuminuria, hypoalbuminemia, and usually hypercholesterolemia.

147. Vitamin D supplements (up to 4,000 IU/day) are recommended to prevent rickets in children on long-term anticonvulsant therapy.

148. Affected individuals show continued hypocalcemia leading to hyperphosphatemia. These conditions typically respond to treatment with $1,25\text{-}(OH)_2\text{-}D_3$ or high levels of vitamin D_3.

149. 0.25–3 mg/day.

150. This is also called *hypophosphatemic rickets/osteomalacia* and *phosphate diabetes*.

151. *Genu varum.*

152. *Genu valgum.*

153. Despite the notion that rickets has been eliminated (as a result of vitamin D fortification of dairy products), the facts show the disease to have re-emerged. In the past decade, rickets has been reported in some 22 countries. Cases in Africa and South Asia appear to be cause primarily by deficiencies of calcium, which some have suggested may increase the catabolism of vitamin D. Other cases, however, appear to be due to insufficient vitamin D. These include most of the recent published cases in the US, 83% of which were described as African-American or Black, and 96% of which were breastfed, with only 5% vitamin D supplementation during breast feeding.

TABLE 6.17 Signs of Vitamin D Deficiency

Organ system	Rickets	Osteomalacia	Osteoporosis
General	Occurs in young childen; loss of apetite, retarded growth	Occurs in older children and adults	Most common in post-menopausal women and older men
Bone/teeth	Failed mineralization, deformation, swollen joints, paint, tenderness, delayed tooth eruption	Demineralized formed bone (e.g., spine, shoulder, ribs, pelvis), fractures (wrist, pelvis), bone pain, tenderness	Loss of trabecular bone with retained structure, high fracture incidence
Skin	Not affected	Not affected	Not affected
Muscle	Weakness, myotonia, pain	Weakness, sarcopenia, pain	Not affected
Vital organs	Not affected	Not affected	Not affected
Nervous	Tetany, ataxia	Not affected	Not affected
Reproductive	Birds: thin eggshell	Low sperm motility and number	Not affected
Ocular	Not affected	Not affected	

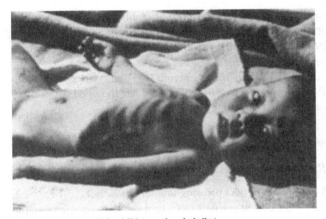

FIGURE 6.20 Rachitic child (note beaded ribs).

ribs or pelvis (Table 6.17). Lesions involve the failure to mineralize bone matrix, which continues to be synthesized by functional osteoblasts; therefore, the condition is characterized by an increase in the ratio of non-mineralized bone to mineralized bone. Radiographic examination reveals abnormally low bone density (osteopenia) and the presence of pseudofractures, especially in the spine, femur, and humerus. Patients with osteomalacia are at increased risk of fractures of all types, but particularly those of the wrist and pelvis.

Osteoporosis

Although it is sometimes confused with osteomalacia, osteoporosis is a very different disease, being characterized by decreased bone mass with retention of normal histological appearance (Table 6.17). Its etiology (loss of trabecular bone with retention of bone structure) is not fully understood; it is considered a multifactorial disease associated with aging and involving impaired vitamin D metabolism and/or function associated with low or decreasing estrogen levels. The disease is the most common bone disease of postmenopausal women, and also occurs in older men[154] (e.g., non-ambulatory geriatrics, postmenopausal women) and in people receiving chronic steroid therapy, which groups show high incidences of fractures, especially of the vertebrae, hip, distal radius, and proximal femur.[155]

In women, osteoporosis is characterized by rapid loss of bone (e.g., 0.5–1.5%/year) in the first 5–7 years after menopause.[156] The increased skeletal fragility observed in osteoporosis appears not to be due solely to reductions in bone mass, but also involves changes in skeletal architecture and bone remodeling (e.g., losses of trabecular connectivity, as well as inefficient and incomplete micro-damage repair). Affected individuals show abnormally low circulating levels of $1,25\text{-}(OH)_2\text{-}D_3$, suggesting that

154. Osteoporosis is estimated to affect 25 million Americans, costing the US economy some $13–18 billion per year. In women, bone loss generally begins in the third and fourth decades and accelerates after menopause; in men, bone loss begins about a decade later.

155. Osteoporotic fractures appear to involve different syndromes. Type I osteoporosis is characterized by distal radial and vertebral fractures, and occurs primarily in women ranging in age from 50 to 65 years; it is probably due to postmenopausal decreases in the amount of calcified bone at the fracture site. Type II osteoporosis occurs primarily among individuals over 70 years, and is characterized by fractures of the hip, proximal humerus, and pelvis, where there has been loss of both cortical and trabecular bone.

156. Therefore, the primary determinant of fracture risk from postmenopausal or senile osteoporosis in older people is the mass of bone each had accumulated during growth and early adulthood. This includes cortical bone, which continues to be accreted after closure of the epiphyses until about the middle of the fourth decade.

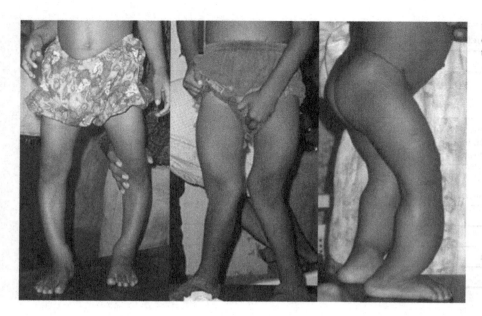

FIGURE 6.21 Characteristic leg signs of rickets: bow-leg, knock knee, and sabre tibia.

estrogen loss may impair the renal 1-hydroxylation step – i.e., that the disease may involve a bihormonal deficiency. Studies of the use of various vitamers D in the treatment of osteoporotic patients, most of which have involved low numbers of subjects, have produced inconsistent results. Results of the Nurses' Health Study showed that adequate vitamin D intake ($\geq 12.5\,\mu g$/day) was associated with a 37% reduction in risk of osteoporotic hip fracture;[157] a meta-analysis of randomized intervention trials showed that $1,25$-$(OH)_2$-D_3 treatment at doses of 0.5–$1\,\mu g$/day decreased vertebral and at least some non-vertebral (e.g., forearm) fractures in postmenopausal women.[158]

Musculoskeletal Pain

Deep pain is common among rickets and osteoporosis patients. Some reports have indicated persistent, non-specific musculoskeletal pain among asymptomatic adults with low circulating levels of 25-OH-D_3; however, a systematic review of published data[159] found no convincing evidence of either low vitamin D status or latitude being associated with chronic pain prevalence in non-cases. Similarly, well controlled intervention trials have largely been negative.

Vitamin D Deficiency in Renal Patients

Low circulating levels of 25-OH-D_3 are frequently observed in patients with chronic renal disease and those with nephrotic syndrome and normal renal function. Some studies have found treatment with vitamin D analogs to reduce proteinuria in patients with chronic renal disease. A meta-analysis of 16 clinical trials[160] concluded that such treatments are effective in increasing serum Ca^{2+} and decreasing serum PTH, but ineffective in reducing either the need for dialysis or survival.

Vitamin D Deficiency in Animals

Rickets

Vitamin D-deficient growing animals show rickets (Figs 6.22, 6.23). Species at greatest risk are those that experience rapid early growth, such as the chick. Rachitic signs are similar in all affected species: impaired mineralization of the growing bones with structural deformation in weight-bearing bones.

Osteoporosis

Older vitamin D-deficient animals show the undermineralization of bones that characterizes osteoporosis. This can be a practical problem in the high-producing laying hen,[161]

157. Feskanich, D., Willett, W. C., Colditz, G. A., *et al.* (2003). *Am. J. Clin. Nutr.* 77, 504–511.

158. Papadimitropoulos, E., Wells, G., Shea, B., *et al.* (2002). *Endocrine Rev.* 23, 560–569.

159. Straube, Andrew Moore, R., Derry, S., *et al.* (2009). *Pain* 141, 10–13.

160. Palmer, S. C., McGregor, D. O., Craig, J. C., *et al.* (2009). *Cochrane Database Syst. Rev.* CD008175.

161. In well-managed flocks it is not uncommon for a hen to lay more than 300 eggs in a year, with 40 of these laid during the first 40 days after commencing egg-laying. As each eggshell contains about 2 g Ca and the hen is able to absorb only 1.8–1.9 g Ca from the diet each day, she experiences a Ca debt of 0.1–0.2 g/day during that period. She accommodates this by mobilizing medullary bone; but, as her total skeleton contains only about 35 g Ca, chronic demineralization at that rate without either decreasing the rate of egg production or increasing the efficiency of calcium absorption leads to osteoporosis characterized by fractures of the ribs and long bones.

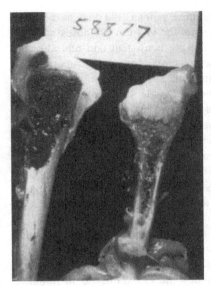

FIGURE 6.22 Tibiae of normal (*left*) and rachitic (*right*) chicks.

FIGURE 6.23 Rachitic puppy.

in which it is called **cage layer fatigue.** The condiction is associated with reductions in egg production, feed intake, efficiency of feed utilzation, and survival.

Tibial Dyschondroplasia

There appear to be other situations of impaired renal 1-hydroxylation of 25-OH-D$_3$, thus limiting the physiological function of the vitamin. One is the failure of bone mineralization seen in rapidly growing, heavy-bodied chickens and turkeys, called tibial dyschondroplasia. The disorder is similar to the condition called osteochondrosis in rapidly growing pigs and horses; it is characterized by the failure of vascularization of the proximal metaphyses of the tibiotarsus and tarsalmetatarsus. It occurs spontaneously, but can be produced in animals made acidotic, that condition reducing the conversion of 1-hydroxylation of 25-OH-D$_3$. Both the incidence and severity of tibial dyschondroplasia can be reduced by treatment with

1,25-(OH)$_2$-D$_3$ or 1α-OH-D$_3$, but not by higher levels of vitamin D$_3$ alone. That lesions in genetically susceptible poultry lines cannot be completely prevented by treatment with vitamin D metabolites suggests that tibial dyschondroplasia may involve a functional impairment in VDRs.

Milk Fever

High-producing dairy cows can become hypocalcemic at the onset of lactation when they have been fed calcium-rich diets before calving. The condition, called milk fever, occurs when plasma calcium levels decrease to less than about 5.0 mg/dl; it is characterized by tetany and coma, which can be fatal. Milk fever results from the inability of the postparturant cow to withstand massive lactational calcium losses by absorbing dietary calcium and mobilizing bone at rates sufficient to support plasma calcium at normal levels. It can be prevented by preparing the pregnant cow for upregulated bone mobilization and enteric calcium absorption. In field practice, this is done by feeding her a relatively low-calcium diet (100 g/day); parenteral treatment with 1,25-(OH)$_2$-D$_3$ is also effective.

9. VITAMIN D TOXICITY

Excessive intakes of vitamin D are associated with increases in circulating levels of 25-OH-D$_3$; this is especially true for vitamin D$_3$, exposure to high levels of which produces higher serum levels of the 25-OH metabolite than do comparable intakes of vitamin D$_2$.[162] The 25-OH metabolite is believed to be the critical metabolite in vitamin D intoxication. At high levels[163] it appears to compete successfully for VDR binding, thus bypassing the regulation of the 25-OH-D$_3$ 1-hydroxylase to induce transcriptional responses normally signaled only by 1,25-(OH)$_2$-D$_3$. Therefore, risk to hypervitaminosis D is increased under conditions such as chronic inflammation,[164] in which the normal feedback regulation of the renal 25-OH-D$_3$ 1-hydroxylase is compromised.

Hypervitaminosis D involves increased enteric absorption and bone resorption of calcium, producing hypercalcemia, with attendant decreases in serum PTH and glomerular filtration rate and, ultimately, loss of calcium homeostasis. The mobilization of bone also results in increased serum concentrations of zinc from that reserve. Vitamin D-intoxicated individuals show a variety of signs (Table 6.18), including anorexia, vomiting, headache, drowsiness, diarrhea, and polyuria. With chronically elevated serum calcium and phosphorus levels, the ultimate result is **calcinosis** – i.e., the deposition of calcium and phosphate

---

162. Vitamin D$_3$ is 10–20 times more toxic than vitamin D$_2$.
163. That is, 100 times normal physiological requirements.
164. Other conditions include tuberculosis and sarcoidosis.

TABLE 6.18 Signs of Hypervitaminosis D

Anorexia
Gastrointestinal distress, nausea, vomiting
Headache
Weakness, lameness
Polyuria, polydypsia
Nervousness
Hypercalcemia
Calcinosis

in soft tissues, especially heart and kidney, but also the vascular and respiratory systems and practically all other tissues.[165] It is not known whether calcinosis involves specific tissue lesions induced by high levels of vitamin D metabolites or whether it is simply a consequence of the induced hypercalcemia. Thus, the risk of hypervitaminosis D is dependent not only on exposure to vitamin D, but also on concomitant intakes of calcium and phosphorus. Calcinosis in grazing livestock has been traced to the consumption of water-soluble glycosides of $1,25\text{-}(OH)_2\text{-}D_3$ present in some plants. These appear to be deglycosylated to yield $1,25\text{-}(OH)_2\text{-}D_3$, which is 100 times more toxic than the dominant circulating metabolite $25\text{-}OH\text{-}D_3$.

A 7-year study of some 36,000 postmenopausal women found a 17% increase in the diagnosis of renal stones in subjects given a daily supplement of 1,000 mg Ca plus 400 IU vitamin D_3 compared to those given a placebo.[166] However, this effect is more likely to be related to total Ca intake, which was estimated to be some 2,000 mg/day, as that vitamin D dose would be expected to increase serum $25\text{-}OH\text{-}D_3$ levels by only ca. 7 nmol/l to levels below those (>600 nmol/l) found to produce hypercalcemia. Other clinical intervention trials have used similar vitamin D_3 doses without adverse effects, although a few (of several) that have used higher doses of vitamin D_3 (e.g., 2,000 IU) in combination with Ca (≥500 mg) have reported calciuria.

The recent report of the IOM[167] suggested that serum $25\text{-}OH\text{-}D_3$ levels greater than 50–74.9 nmol/l may be associated with increased risk to all-cause mortality, but a careful examination of the primary data does not provide much confidence in that conclusion. A systematic review

of the clinical trial literature pointed out that no adverse effects have been reported for vitamin D_3 doses as high as 10,000 IU/day, and that no consistent and reproducible effects, including hypercalcemia, have been reported for doses five times that amount.[168] There are no documented cases of hypervitaminosis D due to excessive sunlight exposure. A few cases of hypervitaminosis D, characterized by elevated serum concentrations of $25\text{-}OH\text{-}D_3$ and vitamin D_3, and hypercalcemia, have been documented among consumers of milk that, through processing errors, was sporadically fortified with very high levels of the vitamin.[169] Vitamin D_3 has been found safe for pregnant and lactating women and their children at oral doses of 100,000 IU/d.[170]

Vitamin D hypersensitivity has been proposed as the basis for Williams–Beuren syndrome, a rare (1:47,000) condition of hypercalcemia and calcium hyperabsorption in humans. The syndrome is manifested in infancy; it is characterized by failure to thrive, with mental handicap and long-term morbidity. Patients have been found to have normal circulating levels of $25\text{-}OH\text{-}D_3$, but they appear to have exaggerated responses to oral doses of vitamin D_3 and one report presented elevated serum levels of $1,25\text{-}(OH)_2\text{-}D_3$ in patients.

The availability of synthetic $1\alpha\text{-}OH\text{-}D_3$ in recent years has meant that it can be used at very low doses to treat vitamin D-dependent or -resistant osteopathies.[171] This has reduced the risks of hypervitaminosis that attend the use of the massive doses of vitamin D_3 that are needed to provide effective therapy in such cases.

10. CASE STUDIES

Instructions

Review each of the following case reports, paying special attention to the diagnostic indicators on which the

165. The condition idiopathic infantile hypercalcemia, formerly thought to be due to hypervitaminosis D, appears to be a multifactorial disease with genetic as well as dietary components.

166. Jackson, M. D., Lacroix, A. Z., Gass, M., et al. (2006). N. Engl J. Med. 354, 669–683.

167. Institute of Medicine (2011). *Dietary Reference Intakes: Calcium, Vitamin D.* National Academy Press, Washington, DC, 1115 pp.

168. Hathcock, J. N., Shao, A., Vieth, R., and Heaney, R. (2007). *Am. J. Clin. Nutr.* 85, 6–18.

169. Eight cases were described (Jacobus, C. H., Holick, M. F., Shao, Q., et al. [1992] N. Engl. J. Med. 326, 1173–1177); each consumed a local dairy's milk, samples of which were highly variable in vitamin D content (some samples contained as much as 245,840-IU of vitamin D_3 per liter). United States federal regulations stipulate that milk is to contain 400-IU/qt "within limits of good manufacturing practice;" however, a small survey concluded that milk and infant formula preparations rarely contain the amounts of vitamin D stated on the label, owing to both under- and overfortification.

170. Goodenday, L. S. and Gordon, G. S. (1971). *Ann. Intern. Med.* 75, 807–808.

171. These diseases include hypoparathyroidism, genetic or acquired hypophosphatemic osteomalacias, renal osteodystrophy, vitamin D-dependent rickets, and osteomalacia associated with liver disease and enteric malabsorption.

respective treatments were based. Then answer the questions that follow.

Case 1

When the patient was first evaluated at the National Institutes of Health, he was a thin, short, bowlegged, 20-year-old male. His height at that time was 159 cm (below the 1st percentile) and he weighed 52 kg. In addition to his dwarfism, he showed a **varus deformity**[172] of both knees, and he walked with a waddling gait. Radiographs showed diffusely decreased bone density, subperiosteal resorption, and a **pseudofracture**[173] of the left ischiopubic ramus.[174]

Laboratory Results:

Parameter	Patient	Normal range
Serum calcium	8.0 mg/dl	8.5–10.5 mg/dl
Serum phosphorus	2.2 mg/dl	3.5–4.5 mg/dl
Serum alkaline phosphatase	152 U/ml	<77 U/ml
Urine chromatography	Generalized aminoaciduria	

The patient's history revealed that he had been a normal, full-term infant weighing 3.2 kg. He had been breast-fed and had been given supplementary vitamin D. At 20 months, however, he failed to walk unsupported and was diagnosed as having active rickets, as revealed by **genu varum**, irregular cupped metaphyses, and widened growth plates,[175] with reductions of both calcium and phosphorus in his blood. The rickets did not respond to oral doses of ergocalciferol (normally effective in treating nutritional rickets), but healing was observed radiographically after intramuscular administration of 1,500,000 IU (37.5 mg) of vitamin D_2 weekly for 5 months. The patient continued to receive vitamin D in the form of cod liver oil, approximately 5,000–20,000 units/day. At 4 years of age, corrective surgery was performed for deformities of the tibias and femurs. At age 14, the patient's height was in the 15th percentile. Additional surgery was performed, after which vitamin D therapy was stopped; over the next 2 years, weakness and severe bone pain became evident. At age 19, bilateral femoral osteotomies[176] were performed again.

As an outpatient at the NIH Clinical Center, the patient received oral ergocalciferol, 50,000 IU daily, for the next 6 years and experienced remission of pain and weakness, and normalization of serum calcium and phosphorus levels. His height reached 161 cm (63.3 in.), i.e., still below the 1st percentile. At 27 years of age, his radiographs showed improved density of the skeletal cortices and healing of the pseudofractures, but the patient still showed the clinical stigmata[177] of rickets.

Laboratory Results:

Parameter	Patient	Normal range
Serum PTH	0.31 ng/ml	<0.22 ng/ml
Urine cAMP	6 nmol/dl	2.3 ± 1.2 nmol/dl
^{47}Ca absorption	19%	33–43%
Plasma 25-OH-D_3	25 ng/ml	10–40 ng/ml
Plasma 1,25-$(OH)_2$-D_3	213 pg/ml	20–60 pg/ml
Plasma 24,25-$(OH)_2$-D_3	1.0 ng/ml	0.8–3 ng/ml

Two hundred micrograms of 25-OH-D_3 was then given orally daily for 2 weeks. Calcium retention improved, urinary cAMP fell, and plasma phosphorus and calcium rose, each to the normal level. Vitamin D_3 maintenance doses (about 40,000 IU, i.e., 1 mg/day) were given periodically to prevent recurrent osteomalacia.

Case 2

This was a sister of the patient described in Case 1. She was first evaluated at the NIH when she was 18 years old. She was a thin female dwarf (147 cm tall, below the 1st percentile) weighing 44.8 kg. She walked with a waddling gait and had mild bilateral varus deformities of the knees. *Chvostek's sign*[178] was present bilaterally. Analyses of her serum showed 7.0 mg/dl of calcium and 3.0 mg/dl of phosphorus, and alkaline phosphatase at 110 U/ml. Skeletal radiographs showed delayed ossification of several epiphyses and a pseudofracture in the left tibia. Her plasma 25-OH-D was 44 ng/ml, 1,25-$(OH)_2$-D_3 was 280 pg/ml, and 24,25-$(OH)_2$-D_3 was 2.5 ng/ml.

Her history showed that she had been a normal, full-term infant who weighed 3.8 kg at birth. At 5 months of age, she showed radiographic features of rickets. During infancy and childhood, she received vitamin D as cod liver oil, in doses of 2,000–10,000 IU/day. She began to

---

172. That is, bow legs.
173. That is, new bone detected radiographically as thickening of the periosteum at the site of an injury to the bone.
174. That is, a narrow process of the pelvis.
175. Failure of mineralization of the growing ends of long bones.
176. Surgical correction of bone shape.

---

177. Abnormalities.
178. That is, facial spasm (as in tetany), induced by a slight tap over the facial nerve.

walk at 9 months, and developed slight varus deformity of both legs. Her rate of growth was at the 5th percentile until the vitamin D was discontinued when she was 11 years old. Within 3 years, her height fell below the 1st percentile. From ages 15 to 16, the bowing of her legs progressed moderately. When she was 18 years old, at the time of her first admission to the NIH Clinical Center, she was treated with $200\,\mu g$ of 25-OH-D_3 per day for 2 weeks. During this time, her calcium retention improved, and her serum calcium and phosphorus increased. Studies showed that $500\,\mu g$ of vitamin D_3 per day was required to maintain her plasma calcium in the normal range. At this dose, her 25-OH-D_3 was 141 ng/ml, 1,25-$(OH)_2$-D_3 was 640 pg/ml, and 24,25-$(OH)_2$-D_3 was 3.6 ng/ml (above normal). When she was 24 years old, i.e., 6 years after her first admission to the center, she was readmitted for studies of the effectiveness of oral 1,25-$(OH)_2$-D_3 with a supplement of 800 mg of calcium per day. Serum calcium remained below normal on doses of 2–10 μg of 1,25-$(OH)_2$-D_3 per day. Only when the dose was increased to 14–17 μg of 1,25-$(OH)_2$-D_3 per day did her plasma calcium reach the normal range. Parathyroid hormone remained elevated at 0.40 ng/ml. At these high doses of 1,25-$(OH)_2$-D_3, her plasma 25-OH-D_3 was 26 ng/ml, and her 1,25-$(OH)_2$-D_3 was 400 pg/ml. While on 1,25-$(OH)_2$-D_3, her osteomalacia improved, and serum calcium and phosphorus entered normal ranges.

Case Questions and Exercises

1. What are the common clinical features (physical and biochemical observations, response to treatment, etc.) of these two cases?
2. What can you infer about the nature of vitamin D metabolism in these siblings?
3. Propose a hypothesis to explain these cases of vitamin D-resistant rickets. How might you test this hypothesis?

Study Questions and Exercises

1. Construct a flow diagram showing the metabolism of vitamin D to its physiologically active and excretory forms.
2. Construct a "decision tree" for the diagnosis of vitamin D deficiency in a human or animal. How can deficiencies of vitamin D and calcium be distinguished?
3. How does the concept of solubility relate to vitamin D utilization? What features of the chemical structure of vitamin D relate to its utilization?
4. Relate the concept of organ function to the concept of vitamin D utilization/status.
5. Discuss the concept of homeostasis, using vitamin D as an example.

RECOMMENDED READING

Aloia, J.F., 2008. African Americans, 25-hydroxyvitamin D, and osteoporosis: A paradox. Am. J. Clin. Nutr. 88, 545S–550S.

Angelo, G., Wood, R.J., 2002. Novel intercellular proteins associated with cellular vitamin D action. Nutr. Rev. 60, 209–214.

Annweiler, C., Montero-Odasso, M., Schott, A.M., et al. 2010. Fall prevention and vitamin D in the elderly: An overview of the key role of the non-bone effects. J. NeuroEngin. Rehab. 7, 50–63.

Baeke, F., Takiishi, T., Korf, H., et al. 2010. Vitamin D: Modulator of the immune system. Curr. Opin. Pharmacol. 10, 482–496.

Bikke, D.D., 2010. Vitamin D and the skin. J. Bone Miner. Metab. 28, 117–130.

Bikke, D.D., 2010. Vitamin D: Newly discovered actions require consideration of physiologic requirements. Trends Endocrinol. Metab. 21, 375–384.

Bischoff-Ferrari, H.A., Giovannucci, E., Willett, W.C., et al. 2006. Estimation of optimal serum concentrations of 25-hydroxyvitamin D for multiple outcomes. Am. J. Clin. Nutr. 84, 18–28.

Bonner, F., 2003. Mechanisms of intestinal calcium absorption. J. Cell. Biochem. 88, 387–393.

Borradale, D., Kimlin, M., 2009. Vitamin D in health and disease: An insight into traditional functions and new roles for the "sunshine vitamin." Nutr. Res. Rev 22, 118–136.

Bouillon, R., Van Cromphaut, S., Carmeliet, G., 2003. Intestinal calcium absorption: Molecular vitamin D mediate mechanisms. J. Cell. Biochem. 88, 332–339.

Campbell, F.C., Xu, H., El-Tanani, M., et al. 2010. The yin and yang of vitamin D receptor (VDR) signaling in neoplastic progression: Operational networks and tissue-specific growth control. Biochem. Pharmacol. 79, 1–9.

Christakos, S., Barletta, F., Huening, M., et al. 2003. Vitamin D target proteins: Function and regulation. J. Cell. Biochem. 88, 238–244.

Crannery, A., Weiler, H.A., O'Donnel, S., Puil, L., 2008. Summary of evidence-based review of vitamin D efficacy and safety in relation to bone health. Am. J. Clin. Nutr. 88, 513S–519S.

Dawson-Hughes, B., 2004. Calcium and vitamin D for bone health in adults. In: Holick, M.F., Dawson-Hughes, B. (Eds.), Nutrition and Bone Health. Humana Press, Totowa, NJ, pp. 197–210.

DeLuca, H.F., 2008. Evolution of our understanding of vitamin D. Nutr. Rev. 66, S73–S87.

Dombrowski, Y., Peric, M., Koglin, S., et al. 2010. Control of cutaneous antimicrobial peptides by vitamin D3. Arch. Dermatol. Res. 302, 401–408.

Dror, D.K., Allen, L.H., 2010. Vitamin D inadequacy in pregancy: Biology, outcomes, and interventions. Nutr. Rev. 68, 465–477.

Eyles, D.W., Feron, F., Cui, X., et al. 2009. Psychoneuroendocrinology 345, S247–S257.

Feldman, D., Pike, J.W., Glorieux, F.H. (Eds.), 2006. Vitamin D, second ed. vols I and II. Elsevier, New York, NY.

Fleet, J.C., 2004. Rapid, membrane-initiated actions of 1,25-dihydroxyvitamin D: What are they and what do they mean? J. Nutr. 134, 3215–3218.

Flores, M., 2005. A role of vitamin D in low-intensity chronic inflammation and insulin resistance in type 2 diabetes? Nutr. Res. Rev. 18, 175–182.

Garland, C.F., Garland, F.C., Gorham, E.D., et al. 2006. The role of vitamin D in cancer prevention. Am. J. Pub. Health 96, 252–261.

Grant, W.B., Holick, M.F., 2005. Benefits and requirements of vitamin D for optimal health: A review. Alt. Med. Rev. 10, 94–111.

Hamilton, B., 2010. Vitamin D and human skeletal muscle. Scand. J. Med. Sci. Sports 20, 182–190.

Harris, D.M., Go, V.L., 2004. Vitamin D and colon carcinogenesis. J. Nutr. 134, 3463S–3471S.

Hathcock, J.N., Shao, A., Vieth, R., Heaney, R., 2007. Risk assessment for vitamin D. Am. J. Clin. Nutr. 85, 6–18.

Hayes, D.P., 2010. Vitamin D and aging. Biogerontology 11, 1–16.

Heaney, R.P., 2003. Long-latency deficiency disease: Insights into calcium and vitamin D. Am. J. Clin. Nutr. 78, 912–919.

Hoenderop, J.G.J., Nilius, B., Bindels, R.J.M., 2002. Molecular mechanism of active Ca^{2+} reabsorption in the distal nephron. Ann. Rev. Physiol. 64, 529–549.

Holick, M.F., 2006. Vitamin D: Its role in cancer prevention and treatment. Progr. Biophys. Mol. Biol. 92, 49–59.

Holick, M.F., 2008. Vitamin D: A D-lightful health perspective. Nutr. Rev. 66, S182–S194.

Holick, M.F., 2008. The vitamin D deficiency pandemic and consequences for nonskeletal health: Mechanism of action. Mol. Asp. Med. 29, 361–368.

Holick, M.F., 2009. Vitamin D: Extraskeletal health. Endocrinol. Metabol. Clin. North Am. 39, 381–400.

Holick, M.F. (Ed.), 2010. Vitamin D: Physiology, Molecular Biology, and Clinical Applications, second ed. Humana Press, New York, NY 1155 pp.

Institute of Medicine, 2011. Dietary Reference Intakes: Calcium Vitamin D. National Academy Press, Washington, DC. 1115 pp.

Jones, G., 2008. Pharmacokinetics of vitamin D toxicity. Am. J. Clin. Nutr. 88, 582S–586S.

Kamen, D.L., Tangpricha, V., 2010. Vitamin D and molecular actions on the immune system: Modulation of innate and autoimmunity. J. Mol. Med. 88, 441–450.

Lee, J.H., O'Keefe, J.H., Bell, D., et al. 2008. Vitamin D deficiency: An important, common, and easily treatable cardiovascular risk factor? J. Am. Coll. Cardiol. 52, 1949–1956.

Levine, A., Li, Y.C., 2005. Vitamin D and its analogues: Do they protect against cardiovascular disease in patients with kidney disease? Kidney Intl 68, 1973–1981.

Lou, Y.R., Qiao, S., Talonpoika, R., et al. 2004. The role of vitamin D3 metabolism in prostate cancer. J. Steroid Biochem. Mol. Biol 92, 317–325.

Mathieu, C., Gysemans, C., Giulietti, A., Bouillon, R., 2005. Vitamin D and diabetes. Diabetologia 48, 1247–1257.

Maxwell, C.S., Wood, R.J., 2011. Update on vitamin D and type 2 diabetes. Nutr. Rev. 69, 291–295.

Maxwell, J.D., 1994. Seasonal variation in vitamin D. Proc. Nutr. Soc. 53, 533–543.

Mizwicki, M.T., Norman, A.W., 2009. The vitamin D sterol–vitamin D receptor ensemble model offers unique insights into both genomic and rapid-response vitamin D signaling. Sci. Signal. 2, 1–14.

Montaro-Odasso, M., Duque, G., 2005. Vitamin D in the aging musculoskeletal system: An authentic strength preserving hormone. Mol. Aspects Med. 26, 203–219.

Nagpal, S., Na, S., Rathnachalam, R., 2005. Noncalcemic actions of vitamin D receptor ligands. Endocrinol. Rev. 26, 662–687.

Norman, A.W., 2008. A vitamin D nutritional cornucopia: New insights concerning serum 25-hydroxyvitamin D status of the US population. Am. J. Clin. Nutr. 88, 1455–1456.

Norman, A.W., Bouillon, R., 2010. Vitamin D nutritional policy needs a vision for the future. Exp. Biol Med. 235, 1034–1045.

Norman, A.W., Henry, H.L., 2006. Vitamin D. In: Bowman, B.A., Russell, R.M. (Eds.), Present Knowledge in Nutrition, ninth ed. vol. I. ILSI Press, Washington, DC, pp. 198–210.

Norman, A.W., Henry, H.L., 2007. Vitamin D. In: Zempleni, J., Rucker, R.B., McCormick, D.B., Suttie, J.W. (Eds.), Handbook of Vitamins, fourth ed. CRC Press, New York, NY, pp. 41–109.

Omdahl, J.L., Morris, H.A., May, B.K., 2002. Hydroxylase enzymes of the vitamin D pathway: Expression, function and regulation. Ann. Rev. Nutr. 22, 139–166.

Peterlik, M., Boonen, S., Cross, H.S., Lamberg-Allardt, C., 2009. Vitamin D and calcium insufficiency-related chronic diseases: An emerging world-wide public health problem. Intl. J. Environ. Res. Public Health 6, 2585–2607.

Posner, G., 2002. Low-calcemic vitamin D analogs (deltanoids) for human cancer prevention. J. Nutr. 132, 3802S–3803S.

Prentice, A., 2008. Vitamin D deficiency: A global perspective. Nutr. Rev. 66, S153–S164.

Rojas-Rivera, J., De La Piedra, C., Ramos, A., et al. 2010. The expanding spectrum of biological actions of vitamin D. Nephrol. Dial. Transplant. 25, 2850–2865.

Samuel, S., Sitrin, M.D., 2008. Vitamin D's role in cell proliferatioin and differentiation. Nutr. Rev. 66, S116–S124.

Schwartz, G.G., 2005. Vitamin D and the epidemiology of prostate cancer. Semin. Dial. 18, 276–289.

Sergeev, I.N., 2005. Calcium signaling in cancer and vitamin D. J. Steroid Biochem. Mol. Biol. 97, 145–151.

Shaffer, P.L., Gewith, D.T., 2004. Vitamin D receptor–DNA interactions. Vitam. Horm. 68, 257–273.

Solomon, A.J., Whitham, R.H., 2010. Multiple sclerosis and vitamin D: A review and recommendations. Curr. Neurol. Neurosci. Rep. 10, 389–396.

Standing Committee on the Scientific Evaluation of Dietary Reference Intakes, 1997. Dietary Reference Intakes for Calcium, Phophorus, Magnesium, Vitamin D and Fluoride. National Academy Press, Washington, DC. pp. 250–287.

Straube, S., Moore, R.A., Derry, S., McQuay, H.J., 2009. Vitamin D and chronic pain. Pain 141, 10–13.

Teegarden, D., Donkin, S.S., 2009. Vitamin D: Emerging new roles in insulin sensitivity. Nutr. Res. Rev. 22, 82–92.

van Etten, E., Stoffels, K., Gysemans, C., et al. 2008. Regulation of vitamin D homeostasis: Implications for the immune system. Nutr. Rev. 66, S125–S134.

Veith, R., 1999. Vitamin D supplementation, 25-hydroxyvitamin D concentrations, and safety. Am. J. Clin. Nutr. 69, 842–856.

Wagner, C.L., Greer, F.R., 2008. Prevention of rickets and vitamin D deficiency in infants, children, and adolescents. Am. Acad. Ped. 122, 1142–1152.

Wang, S., 2009. Epidemiology of vitamin D in health and disease. Nutr. Res. Rev. 22, 188–203.

Wargovich, M.J., 1991. Calcium, vitamin D and the prevention of gastrointestinal cancer. In: Moon, T.E., Micozzi, M.S. (Eds.), Nutrition and Cancer Prevention. Marcel Dekker, New York, NY, pp. 291–304.

Wasserman, R.H., 2004. Vitamin D and the dual processes of intestinal calcium absorption. J. Nutr. 134, 3137–3139.

Wharton, B., Bishop, N., 2003. Rickets. Lancet 362, 1389–1400.

Whitfield, G.K., Hsieh, J.C., Jurutka, P.W., et al. 1995. Genomic actions of 1,25-dihydroxyvitamin D3. J. Nutr. 125, 1690S–1694S.

Willett, A.M., 2005. Vitamin E status and its relationship with parathyroid hormone and bone mineral status in older adults. Proc. Nutr. Soc. 64, 193–203.

Ylikomi, T., Iaaksi, I., Lou, Y.R., et al. 2002. Antiproliferative action of vitamin D. Vitam. Horm. 64, 357–406.

Zemel, M.B., Sun, X., 2008. Calcitriol and energy metabolism. Nutr. Rev. 66, S139–S146.

Zhang, R., Naughton, D.P., 2010. Vitamin D in health and disease: Current perspectives. Nutr. J. 9, 65–78.

Zittermann, A., 2003. Vitamin D in preventive medicine: Are we ignoring the evidence? Br. J. Nutr. 89, 552–572.

Zittermann, A., Schleithoff, S.S., Koerfer, R., 2005. Putting cardiovascular disease and vitamin D insufficiency into perspective. Br. J. Nutr. 94, 483–492.

Vitamin E

Chapter Outline

Anchoring Concepts

1. Vitamin E is the generic descriptor for all tocopherol and tocotrienol derivatives exhibiting qualitatively the biological activity of α-tocopherol.
2. The vitamers E are hydrophobic and, thus, are insoluble in aqueous environments (e.g., plasma, interstitial fluids, cytosol).
3. By virtue of the phenolic hydrogen on the C-6 ring hydroxyl group, the vitamers E have antioxidant activities *in vitro*.
4. Deficiencies of vitamin E have a wide variety of clinical manifestations in different species.

Learning Objectives

1. To understand the various sources of vitamin E.
2. To understand the means of enteric absorption, transport and cellular uptake of vitamin E.
3. To understand the metabolic functions of vitamers E.
4. To understand the interrelationships of vitamin E and other nutrients.
5. To understand the physiologic implications of high doses of vitamin E.

VOCABULARY

Abetalipoproteinemia
Antioxidant
Apolipoprotein E (apoE)
Ataxia
Ataxia with vitamin E deficiency (AVED)
α-Carboxyethylhydroxychroman (α-CEHC)
5′-α-Carboxymethylbutylhydroxychroman (5′-α-CMBHC)
Catalase
Chylomicra
Conjugated diene
Cysteine
Cytochrome *P*-450
Encephalomalacia
Ethane
Exudative diathesis
Familial isolated vitamin E (FIVE) deficiency

Vitamin E is a focal point for two broad topics, namely, biological antioxidants and lipid peroxidation damage. Vitamin E is related by its reactions to other biological antioxidants and reducing compounds (that) stabilize polyunsaturated lipids and minimize lipid peroxidation damage. In vivo lipid peroxidation has been identified as a basic deteriorative reaction in cellular mechanisms of aging processes, in some phases of atherosclerosis, in chlorinated hydrocarbon hepatotoxicity, in ethanol-induced liver injury and in oxygen toxicity. These processes may be a universal disease of which the chemical deteriorative effects might be slowed by use of increased amounts of antioxidants.

A. L. Tappel

The Vitamins. DOI: 10.1016/B978-0-12-381980-2.00007-4

Foam cells
Free radicals
Free-radical theory of aging
Glutathione (GSH)
Glutathione peroxidases
Glutathione reductase
Hemolysis
Hemolytic anemia
High-density lipoproteins (HDLs)
Hydroperoxide
Hydrogen peroxide (H_2O_2)
Hydroxyl radical (HO·)
Intraventricular hemorrhage
Ischemia-reperfusion injury
Lipid peroxidation
Lipofuscin
Lipoprotein lipase
Liver necrosis
Low-density lipoproteins (LDLs)
Malonyldialdehyde
Mitochondrial hormesis
Mulberry heart disease
Myopathy
5-Nitro-tocopherol
Oxidative stress
Oxidized LDLs
Pentane
Peroxide
Peroxide tone
Peroxyl radical (ROO·)
Phospholipid transfer protein (PLTP)
Polyunsaturated fatty acids (PUFAs)
Pro-oxidant
Reactive oxygen species (ROS)
Resorption-gestation syndrome
Respiratory burst
Scavenger receptors
Selenium
Simon's metabolites
Steatorrhea
Superoxide dismutases
Superoxide radical, O_2^-
α-Tocopherol
α-Tocopherol transfer protein (α-TTP)
β-Tocopherol
δ-Tocopherol
γ-Tocopherol
Tocopherol-associated proteins (TAPs)
α-Tocopheronic acid
α-Tocopheronolactone
α-Tocopheroxyl radical
α-Tocopheryl hydroquinone
α-Tocopheryl phosphate
α-Tocopheryl polyethyleneglycol-succinate
α-Tocopheryl quinone
α-Tocopheryl succinate
Tocotrienols
Very low-density lipoprotein (VLDL)
White muscle disease

1. THE SIGNIFICANCE OF VITAMIN E

Vitamin E has a fundamental role in the normal metabolism of all cells. Therefore, its deficiency can affect several different organ systems. Its function is related to those of several other nutrients and endogenous factors that, collectively, comprise a multicomponent system that provides protection against the potentially damaging effects of reactive species of oxygen formed during metabolism or that are encountered in the environment. Both the need for vitamin E and the manifestations of its deficiency can be affected by such nutrients as **selenium** and vitamin C, and by exposure to such **pro-oxidant** factors as **polyunsaturated fatty acids (PUFAs)**, air pollution, and ultraviolet (UV) light. Recent evidence indicates that vitamin E may also have non-antioxidant functions in regulating gene expression and cell signaling.

Unlike other vitamins, vitamin E is not only essentially non-toxic, but also appears to be beneficial at dose levels appreciably greater than those required to prevent clinical signs of deficiency. Most notably, supranutritional levels of the vitamin have been useful in reducing the oxidation of **low-density lipoproteins (LDLs)**, and thus reducing the risk of atherosclerosis. Although vitamin E is present in most plants, only plant oils are rich sources, and most people consume less than recommended levels.[1] Its low regular intake, the ubiquitous and complex nature of its biological function, its demonstrated safety, and its apparent usefulness in combating a variety of **oxidative stress** disorders have generated enormous interest in this vitamin among the basic and clinical science communities as well as the lay public.

2. SOURCES OF VITAMIN E

Distribution in Foods

Vitamin E is synthesized only by photosynthetic organisms – plants, algae, and some cyanobacteria – where it is thought to function as a protective antioxidant in germination and cold adaptation. All higher plants appear to contain **α-tocopherol** in their leaves and other green parts. Because α-tocopherol is contained mainly in the chloroplasts of plant cells (whereas the β-, γ-, and δ-vitamers are usually found outside of these particles), green plants tend to contain more vitamin E than yellow plants. The richest food sources are plant oils. Wheatgerm, sunflower, and safflower oils are rich sources of α-(*RRR*)-tocopherol, whereas corn and soybean oils contain mostly γ-(*RRR*)-tocopherol. Some plant tissues, notably bran and germ

1. Maras, J. E., Bermudez, O. I., Qiao, N., *et al.* (2004). *J. Am. Diet. Assoc.* 104, 567–575.

TABLE 7.1 Relative Biopotencies of Tocopherols and Tocotrienols by Different Bioassays

Compound	Fetal Resorption (Rat)	Hemolysis[a] (Rat)	Myopathy Prevention (Chick)	Myopathy Cure (Rat)
α-Tocopherol	100	100	100	100
β-Tocopherol	25–40	15–27	12	—
γ-Tocopherol	1–11	3–20	5	11
δ-Tocopherol	1	0.3–2	—	—
α-Tocotrienol	28	17–25	—	28
β-Tocotrienol	5	1–5	—	—

[a]Disruption of erythrocyte membranes causing cell lysis.

γ-tocopherols) and tri-(α-tocopherol) methyl derivatives differ in vitamin E activity, the epimeric configuration at the 2-position being important in determining biological activity. Therefore, the use of an international standard facilitated the referencing of these various sources of vitamin E activity, which presumably relates to differences in their absorption, transport, retention, and/or metabolism. Although the original preparation "d,l-α-tocopheryl acetate"[5] that served as the international standard has not been extant for more than 30 years, R,R,R-α-tocopherol is now used as the international standard[6] (see Chapter 3). This system distinguishes only the methylated analogs, and not the particular diastereoisomers possible for each.

Some of the vitamers E common in foods (β- and **γ-tocopherol**, the **tocotrienols**) have little biological activity. The most biopotent vitamer, i.e., the vitamer of greatest interest in nutrition, is α-tocopherol, which occurs naturally as the *RRR* stereoisomer [(*RRR*)-α-tocopherol].

fractions,[2] can also contain tocotrienols, often in esterified form – unlike the tocopherols, which exist only as free alcohols. Animal tissues tend to contain low amounts of α-tocopherol, the highest levels occurring in fatty tissues. These levels vary according to the dietary intake of the vitamin.[3] Because vitamin E occurs naturally in fats and oils, reductions in fat intake can be expected also to reduce vitamin E intake. An amphipathic metabolite, **α-tocopheryl phosphate**,[4] has also been idenfied at trace levels in foods and animal tissues.

Synthetic Forms

Synthetic preparations of vitamin E are mixtures of all eight diastereoisomers, i.e., 2*RS*,4′*RS*,8′*RS*-vitamers designated more commonly with the prefix all-*rac*- in both unesterified (all-*rac*-α-tocopherol) and esterified (all-*rac*-a-tocopheryl acetate) forms. Other forms used commercially include **all-*rac*-α-tocopheryl succinate** and **all-*rac*-α-tocopheryl polyethylene glycol-succinate**.

Expressing Vitamin E Activity

Vitamin E activity is shown by several side chain isomers and methylated analogs of tocopherol and tocotrienol (Table 7.1). These mono- (δ-tocopherol), di-, (β- and

Dietary Sources of Vitamin E

The important sources of vitamin E in human diets and animal feeds are vegetable oils and, to lesser extents, seeds and cereal grains. The dominant dietary form (70% of tocopherols in American diets) is γ-tocopherol (Tables 7.2, 7.3). Wheatgerm oil is the richest natural source, containing 0.9–1.3 mg of α-tocopherol per gram, i.e., about 60% of its total tocopherols. The seeds and grains from which these oils are derived also contain appreciable amounts of vitamin E. Plants also synthesize tocotrienols. The richest food sources are rice-bran oil, in which tocotrienols comprise most of the vitamers E, and palm oil, in which tocotrienols comprise 70% of total vitamers E. Cereals contain small amounts of tocotrienols. Accordingly, cereals in general and wheatgerm in particular are good sources of the vitamin. Foods that are formulated with vegetable oils (e.g., margarine, baked products) tend to vary greatly in vitamin E content owing to differences in the types of oils used, and to the thermal stabilities of the vitamers E present.[7] α-Tocopherol is used in dietary supplements.[8] Regardless of the form consumed, α-tocopherol is the main form found in tissues.

The processing of foods and feedstuffs can remove substantial amounts of vitamin E. Vitamin E losses can

2. Palm oil and rice bran have high concentrations of tocotrienols; other natural sources include coconut oil, cocoa butter, soybeans, barley, and wheat germ.
3. Muscle from beef fed high levels of vitamin E (e.g., 1300 IU/day) before slaughter can yield vitamin E in excess of 16 nmol/g; this level is effective in reducing *post-mortem* oxidation reactions, thus delaying the onset of meat discoloration due to hemoglobin oxidation and the development of oxidative rancidity.
4. Water-soluble and resistant to both acid and alkaline hydrolysis, this metabolite has been missed by traditional methods of vitamin E analysis.

5. At the time, the standard was called *d,l*-α-tocopheryl acetate; now it would be called (2*RS*)-α-tocopheryl acetate. Because of uncertainty about the proportions of the two diastereoisomers in that mixture, once the supply was exhausted it was impossible to replace it.
6. 1 mg α-tocopherol equivalent = 1.49 IU.
7. Tocotrienols tend to be less stable to high temperatures than tocopherols; therefore, baking tends to destroy them selectively.
8. Water-dispersible formulations have been developed for treating lipid-malabsorbing patients.

TABLE 7.2 Vitamin E in Fats and Oils

Item	Tocopherols (%)			Tocotrienols (%)			
	α	γ	δ	α	β	γ	δ
Animal Fats							
Lard	>90	<5		<5			
Butter	>90	<10					
Tallow	>90	<10					
Plant Oils							
Soybean	4–18	58–69					
Cotton	51–67	33–49					
Maize	11–24	76–89					
Coconut	14–67		<17	<14	<3	<53	<17
Peanut	48–61	39–52					
Palm	28–50		<9	16–19	4	34–39	<9
Safflower	80–94	6–20					
Olive	65–85					15–35	

Source: Chow, C. K. (1985). *World Rev. Nutr. Diet.* 45, 133.

TABLE 7.3 Vitamin E in Grains and Oil Seeds

Item	Tocopherols (%)				Tocotrienols (%)	
	α	β	γ	δ	α	γ
Grains						
Maize	6–15		29–55		5–10	34–77
Oats	4–8	<1			10–22	
Milo	4–7		14–17		<1	
Barley	8–10	1–2	3–4		23–28	3
Wheat	8–12	4–6			2–3	
Oil Seeds						
Soybean	1–3		3–33	2–6		Trace
Cotton seed	1–18		5–18			1–2

Source: Cort, W. M., Vicente, T. S., Woysek, E. H., and Williams, B. D. (1983). *J. Agric. Food Chem.* 31, 1330.

occur as a result of exposure to peroxidizing lipids formed during the development of oxidative rancidity of fats, and to other oxidizing conditions such as drying in the presence of sunlight and air, the addition of organic acids,[9] irradiation, and canning. Milling and refining can reduce vitamin E content by removal of tocopherol-rich bran and germ fractions as well as through the use of bleaching agents (e.g., hypochlorous acid) to improve the baking characteristics of the flour. Some foods (e.g., milk and milk products) also show marked seasonal fluctuations in vitamin E content related to variations in vitamin E intake of the host (e.g., vitamin E intake is greatest when fresh forage is consumed). The many potential sources of

---

9. The addition of 1% propionic acid (as an antifungal agent) to fresh grain can destroy up to 90% of its vitamin E.

vitamin E loss mean that the vitamin E contents of foods and feedstuffs vary considerably.[10]

3. ABSORPTION OF VITAMIN E

The primary site of absorption appears to be the medial small intestine. Esterified forms of the vitamin E are hydrolyzed, probably by a duodenal mucosal *esterase*; the predominant forms absorbed are free alcohols. Most studies have shown no appreciable differences in the efficiency of absorption of the acetate ester and free alcohol forms, nor differences in the absorption of the various tocopherol and tocotrienol vitamers. It is clear, however, that regardless of the form absorbed, higher intakes lead to higher rates of absolute absorption, but lower absorption efficiencies (that is, fractional absorption). At nutritionally important intakes, variable (generally, 20–70%[11]) absorption efficiencies have been reported, with a large portion of ingested vitamin E appearing in the feces.

Micelle-Dependent Diffusion

It has been thought that vitamin E, like other hydrophobic substances, is absorbed by non-saturable passive diffusion dependent on micellar solubilization and, hence, the presence of bile salts and pancreatic juice. It is clear that the enteric absorption of vitamin E is dependent on the adequate absorption of lipids; the process requires the presence of fat in the lumen of the gut, as well as the secretion of pancreatic esterases for the release of free fatty acids from dietary triglycerides and of bile acids for the formation of mixed micelles and esterases for the hydrolytic cleavage of tocopheryl esters when those forms are consumed. Individuals unable to produce pancreatic juice or bile (e.g., patients with biliary obstruction, cholestatic liver disease, pancreatitis, cystic fibrosis) can be expected to show impaired absorption of vitamin E, as well as other fat-soluble nutrients that are dependent on micelle-facilitated diffusion for their uptake. The micelle-dependent absorption of vitamin E would imply a need for dietary fat to facilitate the process, but the interaction of tocopherols with PUFAs in the intestinal lumen can result in absorption being stimulated by intragastric medium-chain triglycerides and inhibited by linoleic acid. The need for lipid would explain reports of vitamin E in dietary supplements not being well absorbed unless taken with a meal.[12] Studies with radiolabeled α-tocopherol have shown its enteric absorption in

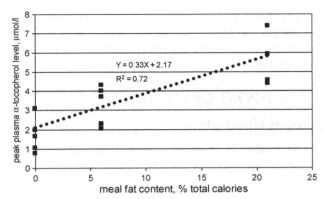

FIGURE 7.1 Effect of the fat-level of a meal on the absorption of deuterium-labeled α-tocopherol from that meal by healthy adults. After *Bruno, R. S., Leonard, S. W., Park, S. I.,* et al. *(2006). Am. J. Clin. Nutr. 83, 299.*

humans to be impaired by dietary fat levels less than *ca.* 10% (i.e., 21% of total calories) (Fig. 7.1);[13] however, absorption of that vitamer by the rat was not impaired by feeding a diet containing only 7 grams of fat per kilogram (i.e., 1.6% of total calories). It has been suggested that children can adequately absorb the fat-soluble vitamins with fat intakes as low as 5 grams per day.

Role of Mucosal Receptors

Evidence has been presented for receptor-mediated uptake of α-tocopherol involving the scavenger receptor class B type I (SR-BI).[14] The inhibition of α-tocopherol enteric absorption by carotenoids, green tea catechins, and γ-tocopherol may involve competitive binding to these receptors.

Uptake into Lymphatic Circulation

Absorbed vitamin E, like other hydrophobic substances, enters the lymphatic circulation[15] in association with the triglyceride-rich chylomicra. Studies with radiolabeled compounds have shown the preferential lymphatic uptake of α-tocotrienol compared with γ- and δ-tocotrienols and α-tocopherol. The kinetics of vitamin E absorption are biphasic, reflecting the initial uptake of the vitamin by existing chylomicra, followed by a lag phase due to the assembly of new chylomicra and intestinal VLDLs. Within the enterocytes, vitamin E combines with other lipids and

---

10. For example, refining losses in edible plant oils are typically 10–40%, but can sometimes be much greater.
11. The enteric absorption of γ-tocopherol appears to be only 85% of that of α-tocopherol.
12. Leonard, S. W., Good, C. K., Gugger, E. T., and Traber, M. G. (2004). *Am. J. Clin. Nutr.* 79, 86.

13. Bruno, R. S., Leonard, S. W., Park, S. I., *et al.* (2006). *Am. J. Clin. Nutr.* 83, 299.
14. Mice lacking SR-BI show marked reductions in the amounts of α-tocopherol in plasma (particularly, in the HDL fraction) and tissues (Mardones, P., Strobel, P., Miranda, S., *et al.* [2002] *J. Nutr.* 132, 443).
15. That is, the portal circulation in birds, fishes, and reptiles.

apolipoproteins to form chylomicra and very low-density lipoproteins (VLDLs). Chylomicra are released into the lymphatics in mammals, or into the portal circulation in birds and reptiles.

4. TRANSPORT OF VITAMIN E

Roles of Lipoproteins

Plasma vitamin E levels respond to increases in dietary vitamin E intake, but do so with diminishing increases –

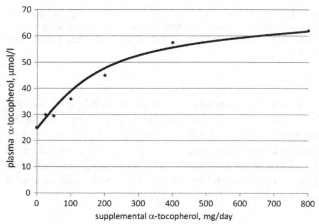

FIGURE 7.2 Plasma response to supplemental α-tocopherol. After *Princen, H. M., van Duyvenvoorde, W., Buytenhek, R., et al. (1995).* Arterioscler. Throm. Vasc. Biol. *15, 325.*

particularly as intakes exceed 200 mg/day (Fig. 7.2). This is due to the increasing uptake of the vitamin by the liver, where α-tocopherol is selectively incorporated into nascent VLDLs secreted by the liver. Unlike vitamins A and D, vitamin E does not have a specific carrier protein in the plasma. Instead, it is rapidly transferred from chylomicra to plasma lipoproteins, to which it binds nonspecifically. The metabolism of circulating chylomicra can result in tocopherols being transferred directly to tissues by partitioning into their plasma membranes, or indirectly by transfer to and between circulating lipoproteins (Fig. 7.3).

Although the majority of the triglyceride-rich VLDL remnants are returned to the liver, some are converted by lipoprotein lipase to LDLs; it appears that, during this process, vitamin E also transfers spontaneously to apolipoprotein B-containing lipoproteins, including the **very low-density lipoproteins (VLDLs), low-density lipoproteins (LDLs), and high-density lipoproteins (HDLs)**. Therefore, plasma tocopherols are distributed among these three lipoprotein classes, with the more abundant LDL and HDL classes comprising the major carriers of vitamin E. Nevertheless, as each class of lipoproteins derives its tocopherols ultimately from chylomicra, α-tocopherol transport by the latter is the major source of interindividual variation in response to ingested vitamin E.

These transport processes can be disrupted under dyslipidemic conditions. Patients with hypercholesterolemia and/or hypertriglyceridemia show reduced plasma uptake

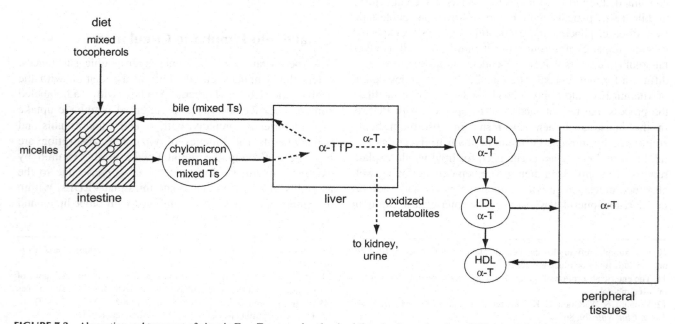

FIGURE 7.3 Absorption and transport of vitamin E. α-T, α-tocopherol: mixed Ts, mixed tocopherols; α-TTP, α-tocopherol transfer protein.

of newly absorbed vitamin E.[16] Transport can also be disrupted by impairments in the expression of apolipoprotein B; patients with apobetalipoproteinemia become vitamin E-deficient due to very low rates of uptake regardless of dietary vitamin E status. The co-transport of vitamin E with other polyunsaturated lipids ensures protection of the latter from free-radical attack, and circulating tocopherol levels tend to correlate with those of total lipids and cholesterol.[17] That postprandial levels of tocopherols exceed those of tocotrienols reflects the more rapid metabolic degradation of the latter.

Tocopherol exchanges rapidly between the lipoproteins, mediated by the **phospholipid transfer protein (PLTP)**,[18] and between lipoproteins and erythrocytes (about one-quarter of total erythrocyte vitamin E turns over every hour); thus, the concentrations of vitamin E and of erythrocytes in plasma are highly correlated[19] (the latter containing 15–25% of total vitamin E in the plasma). As vitamin E is membrane-protective, plasma tocopherol levels are inversely related to susceptibility to oxidative hemolysis. This relationship makes the plasma α-tocopherol level useful as a parameter of vitamin E status; in healthy people, values ≥ 0.5 mg/dl are associated with protection against hemolysis and are taken to indicate nutritional adequacy (Table 7.4). Maternal tocopherol levels increase during pregnancy, but fetal levels remain low,[20] suggesting a barrier to transplacental movement of the vitamin. Because **apolipoprotein E (apoE)** affects the hepatic binding and catabolism of several classes of lipoproteins, it has been suggested that it can also affect the metabolism of tocopherols (Table 7.5).

Cellular Uptake

The process of cellular uptake of vitamin E appears to occur by the established mechanisms of lipid transfer between lipoproteins and cells.

TABLE 7.4 Serum α-Tocopherol Concentrations in Humans

Group	α-Tocopherol (mg/dl)	Group	α-Tocopherol (mg/dl)
Healthy adults	0.85 ± 0.03	Infants	
Postpartum mothers	1.33 ± 0.40	Full term	0.22 ± 0.10
Children, 2–12 years	0.72 ± 0.02	Premature at delivery	0.23 ± 0.10
Cystic fibrotics, 1–19 years	0.15 ± 0.15	Premature at 1 month	0.13 ± 0.05
		2 months, bottle-fed	0.33 ± 0.15
Biliary atresiacs, 3–15 months	0.10 ± 0.10	2 months, breastfed	0.71 ± 0.25
		5 months	0.42 ± 0.20
		2 years	0.58 ± 0.20

Source: Gordon, H. H., Nitowsky, H. M., Tildon, J. T., and Levin, S. (1958). *Pediatrics* 21, 673.

TABLE 7.5 Relationship of apoE Genotype to Plasma Tocopherol Levels in Free-Living Children

apoE Genotype	n	Plasma α-Tocopherol (μmol/l)	Plasma γ-Tocopherol (μmol/L)
E2/2	6	26.5^a (23.8–29.2)	3.10^a (2.27–4.22)
E3/2	89	20.8^b (20.1–21.5)	1.90^b (1.75–2.07)
E3/3	660	21.3^b (21.1–21.6)	$2.06^{a,b}$ (2.00–2.12)
E4/3	150	21.4^b (20.9–21.9)	$2.05^{a,b}$ (1.92–2.18)
E4/2	8	$21.7^{a,b}$ (19.4–23.9)	$1.81^{a,b}$ (1.39–2.36)
E4/4	13	19.0^b (17.2–20.8)	$1.84^{a,b}$ (1.49–2.27)

Source: Ortega, H. Castilla, P., Gomez-Coronado, D., *et al.* (2005). *Am. J. Clin. Nutr.* 81, 624.
P <0.05; means with common superscripts are not significantly different.

Lipid Transfer Proteins and Lipases

Uptake of α-tocopherol from the amphipathic lipoprotein outer layer is mediated in a directional way by PLTP[21] and by way of lipoprotein lipase-mediated exchange of α-tocopherol from chylomicra and lipoprotein remodeling. This lipase is thought to be involved in the transport of α-tocopherol across the blood–brain barrier and into the central nervous system.[22]

16. Hall, W. L. Jeanes, Y. M., and Lodge, J. K. (2005). *J. Nutr.* 135, 58.
17. Therefore, high plasma vitamin E levels occur in hyperlipidemic conditions (hypothyroidism, diabetes, hypercholesterolemia), whereas low plasma vitamin E levels occur in conditions involving low plasma lipids (abetalipoproteinemia, protein malnutrition, cystic fibrosis).
18. Rats, horses, and chicks transport 70–80% of plasma α-tocopherol with HDLs, 18–22% with LDLs, and <8% with VLDLs. Human females, too, transport α-tocopherol preferentially with HDLs; but males transfer most (65%) with LDLs, only 24% with HDLs, and 8% with VLDLs.
19. Patients with abetalipoproteinemia are notable exceptions. They may show normal erythrocyte tocopherol concentrations even though their serum tocopherols levels are undetectable.
20. Serum tocopherol concentrations in infants are about 25% of those of their mothers. They increase to adequate levels within a few weeks after birth, except in infants with impaired abilities to utilize lipids (e.g., premature infants, biliary atresiacs); they show very low circulating levels of vitamin E.

21. Mice lacking PLTP show high plasma levels of α-tocopherol in apo-B- containing lipoproteins (Jiang, X.-C., Lagrost, L., and Tall, A. F. [2002] *J. Biol. Chem.* 277, 31850).
22. Mice lacking lipoprotein lipase show low brain α-tocopherol levels (although no associated pathologies have been reported) (Goti, D., Balazs, Z., Panzenboeck, U., *et al.* [2002] *J. Biol. Chem.* 277, 28537).

Receptor-Mediated Endocytosis of Lipoproteins

Evidence has been presented that the binding of lipoproteins to specific cell surface receptors must occur to allow the vitamin E to enter cells either by diffusion and/or by bulk entrance of lipoprotein-bound lipids. The inability of LDL receptor-deficient cells to take up LDL-bound vitamin E at normal rates suggests the involvement of those receptors. Such deficiencies do not necessarily reduce tissue tocopherol levels, although studies in animal models indicate apolipoprotein E genotype to be a determinant of both circulating and tissue tocopherol levels.[23] Evidence has been presented for receptor-mediated uptake of α-tocopherol without uptake of the apolipoprotein, in the manner described for the cellular uptake of cholesterol from HDLs. This appears to involve the scavenger receptor SR-BI serving as a cell surface receptor for α-tocopherol uptake.[24] SR-BI knockout animals show reduced tissue levels and biliary excretion of α-tocopherol, and increased plasma HDL-α-tocopherol levels. A related scavenger receptor, cluster differentiation 36 (CD36, also called fatty acid translocase), has been implicated in vitamin E uptake and metabolism; single nucleotide polymorphisms in CD36 have been associated with differences in plasma α-tocopherol concentration.[25]

α-Tocopherol Transfer Protein

While all vitamers E are taken up by the liver, only α-tocopherol is released into the circulation. This is due to the function of a specific tocopherol-binding protein, the α-tocopherol transfer protein (α-TTP), which was originally described in rat-liver cytosol where it was found to facilitate the transfer of α-tocopherol between microsomes and mitochondria. The 32 kDa protein binds with high affinity α-tocopherol; it has been identified in liver, brain, spleen, lung, kidney, uterus, and placenta. It appears to have been highly conserved: the rat and human liver proteins show 94% sequence homology, as well as some homology to the interphotoreceptor retinol-binding protein (IRBP), cellular retinal-binding protein (CRALBP), and a phospholipid transfer protein.

The α-TTP consists of an N-terminal helical domain and a C-terminal domain, which contains a fold that forms a binding pocket for α-tocopherol. In that pocket binds the ligand and four water molecules, two of which

are hydrogen-bonded to the hydroxyl group on the chroman ring. Binding affinity for vitamers E is determined by the degree of methylation of the chroman ring, which determines the extent of van der Waals contacts with the pocket. The relative affinities are: α-tocopherol > β-tocopherol > γ-tocopherol (affinity half that of the β-vitamer) > δ-tocopherol (affinity 5% that of the α-vitamer). The α-TTP has very low affinities for tocotrienols (11% of those for α-tocopherol), the unsaturated side chains of which lack the capability of the tocopherol phytyl side chains to bend within the ligand-binding pocket. Accordingly, α-TTP binds *RRR*-α-tocopherol more avidly than it does the *SRR* stereoisomer.[26]

The liganded α-TTP acts as a chaperone for α-tocopherol, taking up the vitamer from endocytic vesicles and moving it through the cytoplasm to transport vesicles that travel to the plasma membrane such that the vitamer is ultimately secreted complexed to lipoprotein particles in the circulation.[27] In the liver, the uptake phase of this process is thought to involve the scavenger receptor SR-BI. The discharge phase appears to involve the transporter ABCA1, which interacts with α-TTP to release the ligand from its binding pocket.

The selectivity of α-TTP for α-tocopherol contributes to the differences in tissue retention and biopotency of these vitamers,[28] explaining the fact that while γ-tocopherol is the dominant dietary form of vitamin E, α-tocopherol constitutes 90% of body vitamin E burden. Animal models that do not express α-TTP absorb α-tocopherol normally, showing normal levels in chylomicra, but fail to release the vitamer from the liver, showing very low VLDL-associated α-tocopherol.

The human α-TTP gene has been localized on chromosome 8. Allelic varients in the gene have been associated with differences in circulating α-tocopherol levels,[29] and cases of more serious genetic defects in the α-TTP gene have been described. This includes a group of American patients with sporadic or familial vitamin E deficiency;[30] they show poor incorporation of (*RRR*)-α-tocopherol into their VLDLs, and an inability to discriminate between the

23. apoE4 mice show lower tissue α-tocopherol levels than apoE3 mice (Huebbe, P., Lodge, J. K., and Ribach, G. (2010). *Mol. Nutr. Food Res.* 54, 623).
24. Mice lacking SR-BI show marked reductions in the amounts of a-tocopherol in plasma (particularly, in the HDL fraction) and tissues (Mardones, P., Strobel, P., Miranda, S., *et al.* [2002] *J. Nutr.* 132, 443).
25. Lecompte, S., de Edelenyi, F.S., Goumide, L., *et al.* (2011). *Am. J. Clin. Nutr.* 93, 644–651.
26. The preferential incorporation of the *RRR*-α-isomer into milk by the lactating sow (Lauridson, C., Engel, H., Jensen, S. K., *et al.* [2002] *J. Nutr.* 132, 1258) suggests the presence of α-TTP in the mammary gland.
27. Qian, J., Morley, S., Wilson, K., *et al.* (2005). *J. Lipid Res.* 46, 2072; Qian, J. Altkinson, J., and Manor, D. (2006). *Biochem.* 45, 8236.
28. Neither LDL receptor nor lipoprotein lipase mechanisms of vitamin E uptake by cells discriminate between these stereoisomers; yet α-tocopherol predominates in plasma owing to its preferential incorporation into nascent VLDLs, while the form often predominating in foods, γ-tocopherol, is left behind only to be more rapidly excreted.
29. Wright, M. E., Peters, U., Gunter, M. J. *et al.* (2011). *Cancer Res.* 69, 1429.
30. That is, **familial isolated vitamin E-deficiency**, also referred to as *FIVE deficiency*.

RRR and *SRR* vitamers. These patients have exceedingly low circulating tocopherol concentrations unless maintained on high-level vitamin E supplements (e.g., 1 g/day); if untreated, they experience progressive peripheral neuropathy (characterized by pathology of the large-caliber axons of sensory neurons) and **ataxia**. Chromosome 8 defects have also been identified among the members of a number of inbred Tunisian families who show low serum tocopherol levels and ataxia responsive to high-level vitamin E supplements.[31] This defect involves the deletion of the terminal 10% of the α-TTP peptide chain. A third group of subjects with isolated vitamin E deficiency, identified in Japan, have a point mutation[32] that results in their α-TTP having only 11% of the transfer activity of the wild-type protein. Individuals heterozygous for this trait show no clinical signs, but have circulating tocopherol levels 25% lower than those of normal subjects. Deletion of the α-TTP gene in the mouse resulted in the accumulation of dietary α-tocopherol in the liver at the expense of α-tocopherol in peripheral tissues.[33]

The expresson of α-TTP occurs predominantly in the liver, apparently in response to vitamin E. Studies with the rat have shown that the α-TTP messenger RNA is increased in response to treatment with either α- or γ-tocopherol.[34] As α-TTP binds only the former, this finding suggests roles of these vitamers in regulating gene expression.

Tocopherol-Associated Proteins (TAPs)

Other proteins with vitamin E-binding capacities have been identified. Three such TAPs[35] have been described. The binding specificities of the TAPs are wide; they can bind the four tocopherols, and tocotrienols, as well as such non-vitamin E substances as squalene, phosphatidylinositol, and phosphatidylcholine. TAPs have been found in most tissues, with greatest concentrations in liver, brain, and prostate. The TAPs are thought to be involved in the intracellular transport of vitamers E to mitochondria and other subcellular compartments. Allelic variants in the gene for human TAP1 (*SEC*14L2) have been associated with differences in circulating α-tocopherol levels.

Tocopherol can also bind interphotoreceptor retinol binding protein (IRBP, see Chapter 5 section "Transport of Vitamin A"), apparently in the same site as retinol, as the

TABLE 7.6 Concentrations of α-Tocopherol in Human Tissues

Tissue	α-Tocopherol	
	µg/g Tissue	µg/g Lipid
Plasma	9.5	1.4
Erythrocytes	2.3	0.5
Platelets	30	1.3
Adipose	150	0.2
Kidney	7	0.3
Liver	13	0.3
Muscle	19	0.4
Ovary	11	0.6
Uterus	9	0.7
Testis	40	1.0
Heart	20	0.7
Adrenal	132	0.7
Hypophysis	40	1.2

Source: Machlin, L. J. (1984). *Handbook of Vitamins*, Marcel Dekker, New York, NY, p. 99.

latter readily displaces it. This raises the possibility that some of the retinoid-binding proteins may also function in the intracellular transport of vitamin E.

Tissue Tocopherols

The tocopherol contents of other tissues tend to be related exponentially to vitamin E intake; unlike most other vitamins, they show no deposition or saturation thresholds. Thus, the tocopherol contents of tissues vary considerably (Table 7.6), and are not consistently related to the amounts or types of lipids present. Neural tissues exhibit very efficient retention, i.e., very low apparent turnover rates, of the vitamin.[36]

Kinetic studies indicate that such tissues have two pools of the vitamin: a *labile*, rapidly turning over pool; and a *fixed*, slowly turning over pool. The labile pools predominate in such tissues as plasma and liver, as the tocopherol contents of those tissues are depleted rapidly under conditions of vitamin E deprivation. Non-*RRR*-α-tocopherols are quickly removed from the plasma. In humans, *RRR*-α-tocopherol remains in plasma nearly four times longer than *SRR*-α-tocopherol (apparent half-life:

31. Previously called *Friedreich's ataxia*, this condition is now called **ataxia with vitamin E-deficiency (AVED)**.
32. This is a missense mutation that inserts histidine in place of glutamine at position 101.
33. Leonard, S. W., Terasawa, Y., Farese, R. V. Jr., and Traber, M. G. (2002). *Am. J. Clin. Nutr.* 75, 555–560.
34. Fechner, H. (1998). *Biochem. J.* 331, 577–581.
35. In humans: hTAP1, hTAP2 and hTAP3.

36. For example, weanling rats from vitamin E-adequate dams do not show neurologic signs of vitamin E deficiency for as long as 7 weeks when fed a vitamin E-free diet.

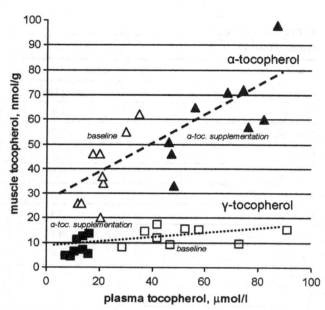

FIGURE 7.4 Correlation of α-tocopherol and γ-tocopherol contents of muscle (*m. gastrocnemius*) and plasma, respectively. Note opposite effects of supplemental α-tocopherol (800 IU/d for 30 days): increased α-tocopherol and reduced γ-tocopherol. After *Meydani, M., Fielding, R. A., Cannon, J. G.* et al. *(1997).* Nutr. Biochem. *8, 74.*

13 hours vs. 48 hours), and three times longer than *RRR*-γ-tocopherol.[37]

The preferential uptake of α-tocopherol results in that vitamer predominating in tissues. Increased intake of α-tocopherol results in increased α-tocopherol levels but displacement of non-α-vitamers in tissues (Fig. 7.4).

Membrane Vitamin E

In most non-adipose cells, vitamin E (mostly as α-tocopherol) is localized almost exclusively in membranes. The highest concentrations are found in the Golgi membranes and lysosomes, where the ratio of vitamin E:phospholipids is approximately 1:65. Other subcellular membranes contain less vitamin E by an order of magnitude. Some years ago, it was proposed that the vitamin may actually reside in intimate contact with PUFAs by virtue of their complementary three-dimensional structures (Fig. 7.5). Fluorescence techniques have revealed that vitamin E partitions into membranes where its weak surface-active properties orient it at the interface between the aqueous phase and hydrophobic domain, with its phenoxy group being hydrogen-bonded to the carbonyl group of the fatty acid ester in the phospholipid bilayer. Dynamic

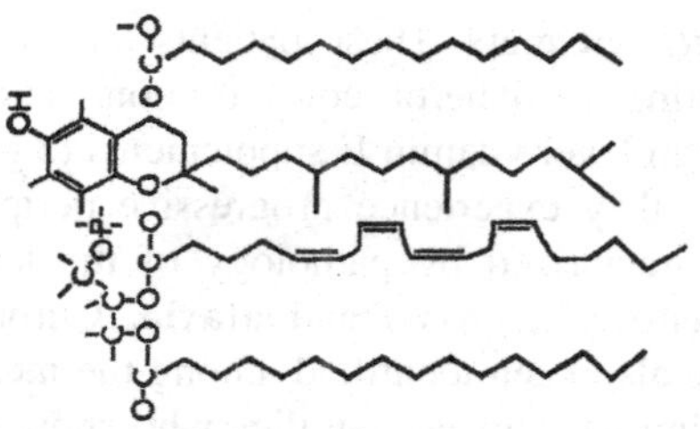

FIGURE 7.5 Proposed interdigitation of tocopherols and polyunsaturated fatty acids in biological membranes.

spectroscopic studies have shown that α-tocopherol (and α-tocotrienol) so oriented can rotate about its long axis perpendicular to the plane of the membrane, and can diffuse laterally among leaflets of the phospholipid bilayer.

While vitamin E is the major antioxidant in membranes, it has long been puzzling how it could function effectively compared to the relatively enormous quantities of polyunsaturated lipids also present in membranes. This may be due to the vitamin clustering in membrane locations of greatest need, which α-tocopherol has been found to do by forming complexes with membrane lysolipids, particularly choline lysophosphatides.[38] Such complex formation results in vitamin E being non-randomly distributed in the phospholipid membrane bilayer, instead being associated with structures analogous to "lipid rafts" – i.e., highly disordered, PUFA-rich microdomains depleted of cholesterol and sphingomeylin. Because membrane hydrolytic products are known to have destabilizing effects on membranes, the formation of such complexes of α-tocopherol (but not other isomers) has been shown to reduce membrane fluidity, and thus to stablize brush border membranes. It has been proposed that, by affecting protein–lipid and/or protein–protein interactions, vitamin E can affect embedded signal transduction pathways.

Adipose Tissue

Some 90% of vitamin E in the body is contained in adipose tissue (Table 7.6), where it resides predominantly in the bulk lipid phase. This constitutes a fixed pool from which the vitamin is slowly mobilized (Fig. 7.6), thus having long-term physiological significance. After a change in α-tocopherol intake, adipose tissue tocopherols may not reach a new steady state for 2 or more years. The amounts of vitamin E in adipose tissue can be nearly normal, even in animals showing clinical signs of vitamin E deficiency. That adipose comprises a sink for vitamin E is indicated by the fact that circulating tocopherols are inversely related to body mass index (BMI), and people on

37. Traber. M. G., Ramakrishnan, R., and Kayden, H. J. (1994). *Proc. Natl Acad. Sci.* 91, 10005; Leonard, S. W., Paterson, E., Atkinson, J. E., et al. (2005). *Free Radical Biol. Med.* 38, 857.

38. A lysophosphatide results from the partial hydrolysis of a phospholipid (e.g., phosphatidylcholine) which removes one of the fatty acid moieties. Such hydrolysis is catalyzed by phospholipase A$_2$.

weight-loss programs do not lose vitamin E from their adipose tissues. However, circulating tocopherol levels have been found to rise significantly (10–20%) during intensive exercise, and it has been suggested that vitamin E may be mobilized from its fixed pools by way of the lipolysis induced under such conditions.

5. METABOLISM OF VITAMIN E

Most α-tocopherol is transported to the tissues without metabolic transformation; subsequent metabolism involves head group and side chain oxidation.[39] The selective accumulation of *RRR*-α-tocopherol in tissues is the result of discrimination among the various tocols consumed. This is effected in two ways: selective retention of α-tocopherol via α-TTP binding; and selective metabolism of non-α vitamers.

Oxidation of the Chroman Group

The chromanol hydroxyl group renders tocopherols and tocotrienols capable of undergoing both one- and two-electron oxidations. α-Tocopherol is thus converted to α-tocopheryl quinone and (5,6- or 2,3-)-epoxy-α-tocopherylquinones, respectively (Fig. 7.7), thus enabling

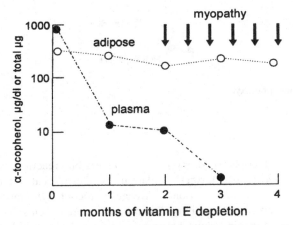

FIGURE 7.6 Retention of α-tocopherol in guinea-pig adipose tissue during vitamin E depletion. Source: *Machlin, L. J., Keating, J., Nelson, J., et al. (1979). J. Nutr. 109, 105.*

---

39. Excretion of the non-metabolized α-tocopherol occurs only at high doses (e.g., >50 mg), which apparently exceed the binding capacity of α-TTP.

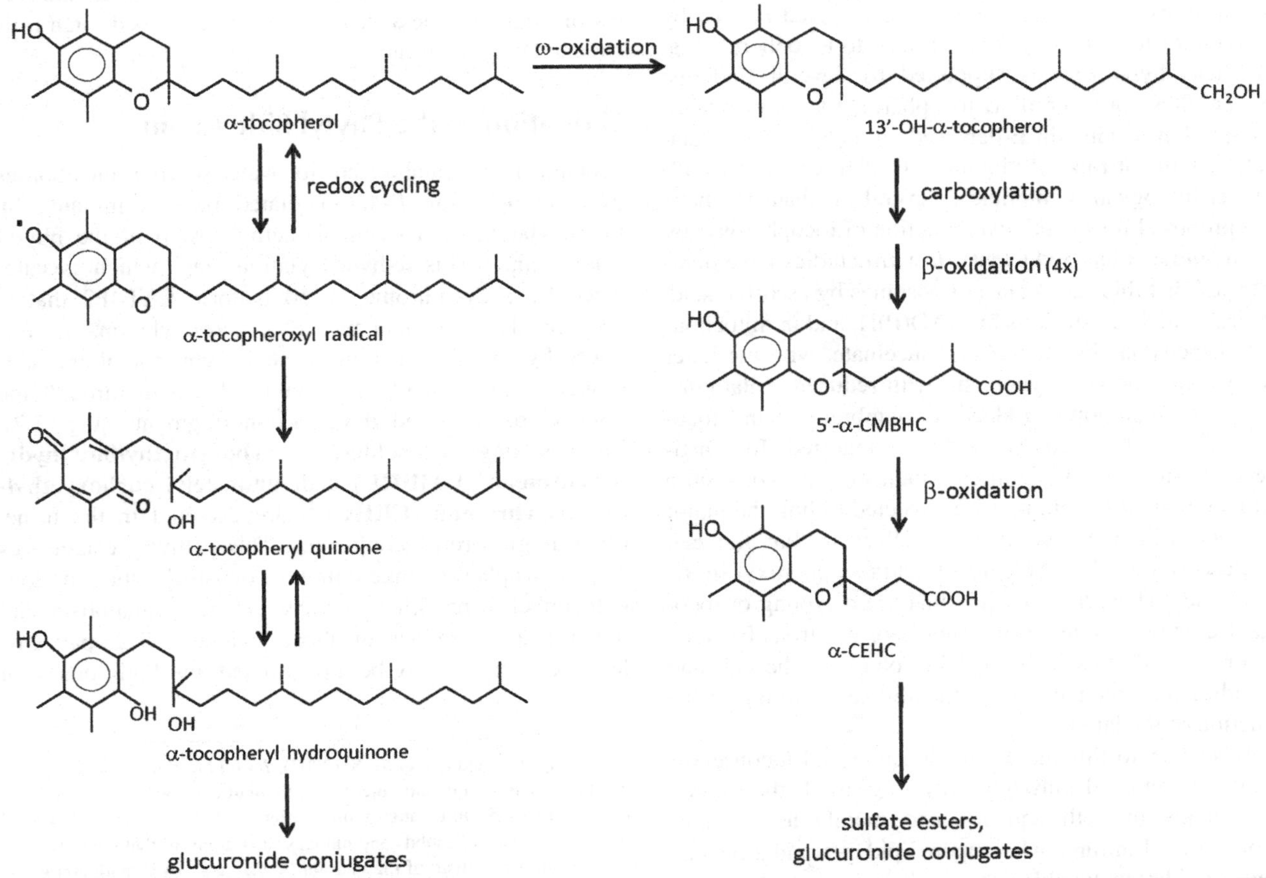

FIGURE 7.7 Vitamin E metabolism (shown with α-tocopherol). *Abbreviations*: 5′-α-CMBHC, 5′-α-carboxymethylbutylhydroxychroman; α-CEHC, α-carboxyethylhydroxychroman.

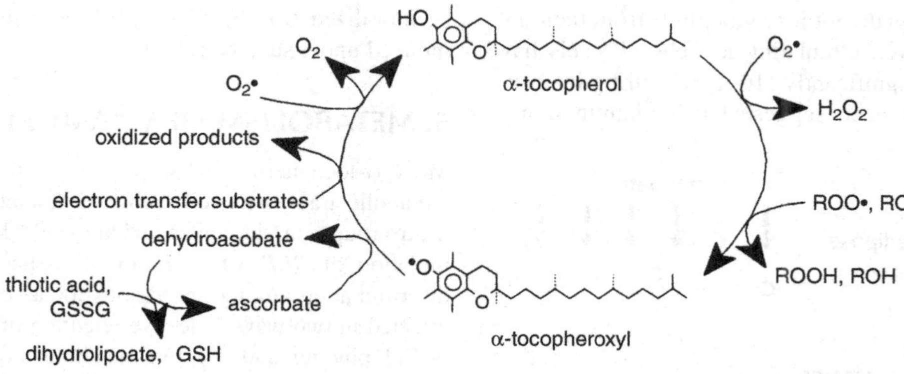

FIGURE 7.8   The vitamin E redox cycle.

it to scavenge free radicals such as peroxynitrite and lipid peroxyl radicals.

Redox Cycling

Oxidation of the chromanol ring is the basis of the *in vivo* antioxidant function of the vitamin. It involves oxidation primarily to tocopherylquinone, which proceeds through the semistable tocopheroxyl radical intermediate. A significant portion of vitamin E may be recycled *in vivo* by reduction of tocopheroxyl radical back to tocopherol (Fig. 7.8). This hypothesis is supported by several findings: the very low turnover of α-tocopherol, the slow rate of its depletion in vitamin E-deprived animals, and the relatively low molar ratio of vitamin E to PUFA (about 1:850) in most biological membranes. Several mechanisms have been proposed for the *in vivo* reduction of tocopheroxyl by various intracellular reductants. *In vitro* studies have demonstrated that this can occur in liposomes by ascorbic acid, in microsomal suspensions by NAD(P)H, and in mitochondrial suspensions by NADH and succinate, with the latter two systems showing synergism with reduced glutathione (GSH) or ubiquinones. Indeed, a membrane-bound tocopheroxyl reductase activity has been suggested. To constitute a physiologically significant pathway *in vivo*, such a multicomponent system may be expected to link the major reactants, which are compartmentalized within the cell (e.g., ascorbic acid in the cytosol, and tocopheroxyl in the membrane). Thus, it is possible that the recycling of tocopherol may be coupled to the shuttle of electrons between one or more donors in the soluble phase of the cell and the radical intermediate in the membrane, resulting in the reduction of the latter.

According to this model, tocopherols and tocotrienols would be retained through recycling until the reducing systems in both aqueous and membrane domains become rate limiting, whereupon lipid peroxidation and protein oxidation would increase. It is important to note that whereas the monovalent oxidation of tocopherol

to the tocopheroxyl radical is a reversible reaction (at least *in vitro*), further oxidation of the radical intermediate is unidirectional. Because tocopherylquinone lacks vitamin E activity, its production represents loss of the vitamin from the system. It can be reduced to α-tocopherylhydroquinone, which can be conjugated with glucuronic acid and secreted in the bile, thus making excretion with the feces the major route of elimination of the vitamin. Under conditions of intakes of nutritional levels of vitamin E, less than 1% of the absorbed vitamin is excreted with the urine.

Oxidation of the Phytyl Side Chain

Vitamin E is catabolized to water-soluble metabolites by a **cytochrome *P*-450**-mediated process initiated by hydroxylation of a terminal methyl group of the phytyl side chain.[40] This ω-hydroxylation step, which is catalyzed by a cytochrome *P*-450 isoform CYP4F2 that is also involved in leukotriene ω-hydroxylation, is followed by dehydrogenation to the 13′-chromanol and subsequent truncation of the phytyl side chain through the removal of two- and three-carbon fragments (Fig. 7.7). The resulting metabolites, 5′-**carboxymethylbutylhydroxychroman** (**CMBHC**) and, ultimately, **carboxyethylhydroxychroman** (**CEHC**),[41] are excreted in the urine, often as glucuronyl conjugates. This pathway catabolizes non-α-tocopherols much more extensively than it does α-tocopherol, resulting in more efficient catabolism and much faster turnover of those vitamers. The pathway, however, appears to be upregulated by high doses of

---

40. Sontag, T. J. and Parker, R. S. (2007). *J. Lipid Res.* 48, 1090.

41. These may also include **tocopheronic acid** and **tocopheronolactone**, referred to as **Simon's metabolites** after E. J. Simon, who described them in the urine of rabbits and humans. It is doubtful that these actually result from metabolism of the vitamin *in vivo* (e.g., by β oxidation in the kidney); they appear to be artifacts resulting from oxidation during their isolation.

TABLE 7.7 Plasma Tocopherols in Smokers and Non-Smokers

Metabolite	Non-Smokers (*n* = 19)	Smokers (*n* = 15)
α-tocopherol, μmol/l	16.0 ± 4.0	15.9 ± 5.0
γ-tocopherol, μmol/l	1.76 ± 0.98	1.70 ± 0.69
5-nitro-γ-tocopherol, nmol/l	4.03[*] ± 3.10	8.02 ± 8.33

Source: Leonard, S. W., Traber, M. G., and Zhao, Y. (2003). *Free Radical Biol. Med.* 12, 1560.
[*]*P* <0.05.

α-tocopherol, suggesting that it is also important in clearing that vitamer via conversion to the readily excreted α-CEHC. For this reason, increased maternal intakes during pregnancy have been found to increase α-CEHC levels in the fetal circulation.

Other Metabolism

The detection of small amounts of **α-tocopheryl phosphate** in tissues of vitamin E-fed animals suggests that the vitamin can be phosphorylated. The metabolic significance of this metabolite is unclear, and a kinase has not been identified. While the metabolic role of α-tocopheryl phosphate is not clear, evidence suggests that it may serve as an active lipid mediator of signal transduction and gene expression.[42]

That vitamin E can also be nitrated *in vivo* is indicated by the occurrence of **5-nitro-γ-tocopherol** in the plasma of cigarette smokers (Table 7.7), presumably due to the high amounts of reactive nitrogen species and the stimulatory effects of cigarette smoke on inflammatory responses. This reaction may be the basis for the enhanced turnover of tocopherols and reduced production of carboxyethylchromanyl metabolites in smokers.

6. METABOLIC FUNCTIONS OF VITAMIN E

Vitamin E as a Biological Antioxidant

The primary nutritional role of vitamin E is as a biological antioxidant. An **antioxidant** is an agent that inhibits oxidation, and thus prevents such oxidation reactions as the conversion of polyunsaturated fatty acids to fatty hydroperoxides, the conversion of free or protein-bound sulfhydryls to disulfides, etc. In this regard, vitamin E has functional importance in the maintenance of membrane integrity in virtually all cells of the body. Its antioxidant function involves the reduction of free radicals, thus protecting against the potentially deleterious reactions of such highly reactive oxidizing species.

Production of Free Radicals and Reactive Oxygen Species (ROS)

Free radicals (X·) are produced in cells under normal conditions either by homolytic cleavage of a covalent bond, as in the formation of a C-centered free radical of a polyunsaturated fatty acid, or by a univalent electron transfer reaction. It has been estimated that as much as 5% of inhaled molecular oxygen (O_2) is metabolized to yield the so-called reactive oxygen species, i.e., the one- and two-electron reduction products superoxide radical $O_2{\cdot}^-$ and H_2O_2, respectively. There appear to be three sources of ROS:

- *Normal oxidative metabolism.* The mitochondrial electron transport chain, which involves a flow of electrons from NADH and succinate through a series of electron carriers to cytochrome oxidase, which reduces oxygen to water, has been found to leak a small fraction of its electrons, which reduce oxygen to $O_2{\cdot}^-$.
- *Microsomal cytochrome P-450 activity.* Several xenobiotic agents are metabolized by the microsomal electron transport chain to radical species (e.g., the herbicide paraquat is converted to an N-centered radical anion) that can react with oxygen to produce $O_2{\cdot}^-$.
- *Respiratory burst of stimulated phagocytes.* Macrophages produce $O_2{\cdot}^-$ and H_2O_2 during phagocytosis.

The half-life of $O_2{\cdot}^-$ appears to be only about 1 second, and neither $O_2{\cdot}^-$ nor H_2O_2 is highly reactive. However, when exposed to transition metal ions (particularly Fe^{2+} or Cu^+), these two species can react to yield a very highly reactive free-radical species, **hydroxyl radical HO·**:

$$O_2^- + H_2O_2 + Fe^{2+} \rightarrow O_2 + HO{\cdot} + HO^- + Fe^{3+}$$

These divalent metals can also catalyze the decomposition of H_2O_2 or fatty acyl **hydroperoxide** (ROOH) produced by lipid peroxidation to yield the oxygen-centered radical RO· or HO·, respectively:

$$ROOH\ (HOOH) + Fe^{2+} \rightarrow RO{\cdot}\ (HO{\cdot}) + HO^- + Fe^{3+}$$

42. Zingg, J. M., Meydani, M., and Azzi, A. (2010) α-Tocopheryl phosphate – an active lipid mediator? *Mol. Nutr. Food Res.* 54, 679.

43. Base damage products such as 8-hydroxydeoxyguanosine (8OHdG), presumably resulting from DNA repair processes, are excreted in the urine. Smokers typically show elevated 8OHdG excretion.

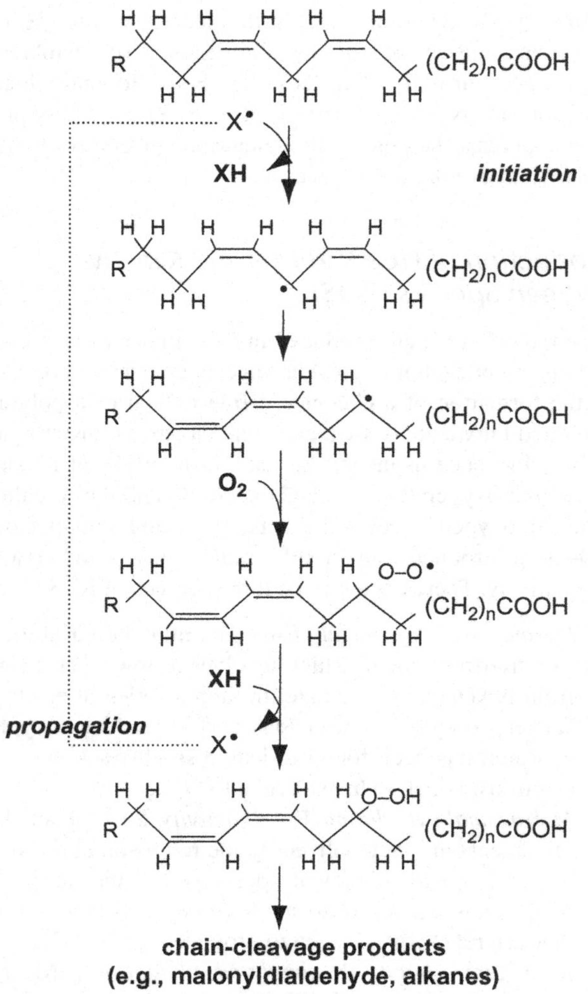

FIGURE 7.9　The self-propagating nature of lipid peroxidation.

Targets of HO· include:

- *DNA* – causing oxidative base damage[43]
- *Proteins* – causing the production of carbonyls and other amino acid oxidation products[44]
- *Lipids* – causing oxidation of PUFAs of membrane phospholipids, and the formation of lipid peroxidation products (e.g., malonyldialdehyde, isoprostanes, pentane, ethane), and resulting in membrane dysfunction.

Mechanism of Lipid Peroxidation

The PUFAs of biological membranes are particularly susceptible to attack by free radicals by virtue of their 1,4-pentadiene systems, which allow for the abstraction of a complete hydrogen atom (i.e., with its electron) from one of the –CH$_2$– groups in the carbon chain, and the consequent generation of a *C-centered free radical* (–C·–) (Fig. 7.9). This initiation of lipid peroxidation can

be accomplished by HO· and possibly HOO· (but *not* by H$_2$O$_2$ or O$_2$·$^-$). The C-centered radical, being unstable, undergoes molecular rearrangement to form a conjugated diene, which is susceptible to attack by molecular oxygen (O$_2$) to yield a peroxyl radical (ROO·). Peroxyl radicals are capable of abstracting a hydrogen atom from other PUFAs, and thus propagating a chain reaction that can continue until the membrane PUFAs are completely oxidized to hydroperoxides (ROOH).

Fatty acyl hydroperoxides so formed are degraded in the presence of transition metals (Cu^{2+}, Fe^{2+}) and heme and heme proteins (cytochromes, hemoglobin, myoglobin) to release radicals that can continue the chain reaction of lipid peroxidation,[45] as well as other chain-cleavage products, malonyldialdehyde,[46] pentane, and ethane.[47] This oxidative degradation of membrane phospholipid PUFAs is believed to result in physicochemical changes resulting in membrane dysfunction within the cell.[48] Cellular oxidant injury can also occur without significant lipid peroxidation, by oxidative damage to critical macromolecules (DNA, proteins) and decompartmentalization of Ca^{2+}.[49]

---

44. For example, methionine sulfoxide, 2-ketohistidine, hydroxylation of tyrosine to DOPA, formylkynurenine, *o*-tyrosine, protein peroxides.

45. Therefore, a single radical can initiate a chain reaction that may propagate over and over again.

46. Although malonyldialdehyde (MDA) is a relatively minor product of lipid peroxidation, it has received a great deal of attention, in part because of the ease of measuring it colorimetrically using 3-thiobarbituric acid (TBA). In fact, the TBA test has been widely used to assess MDA concentrations as measures of lipid peroxidation. It is important to note that the TBA test is subject to several limitations. Specifically, much of the MDA it detects may not have been present in the original sample, as lipid peroxides are known to decompose to MDA during the heating stage of the test, that reaction being affected by the concentration of iron salts.

47. The volatile alkanes, pentane and ethane, are excreted across the lungs and can be detected in the breath of vitamin E-deficient subjects. Pentane is produced from the oxidative breakdown of ω-6 fatty acids (linoleic acid family); whereas ethane is produced from the ω-3 fatty acids (linolenic acid family).

48. Whether lipid peroxidation does, indeed, occur *in vivo* has been surprisingly difficult to determine. Tissues of animals contain little or no evidence of lipid peroxides or their decomposition products. Although expired breath contains volatile alkanes that might have originated from the decomposition of fatty acyl hydroperoxides, it is difficult to exclude the possibility of their production by bacteria of the gut or skin. Perhaps the strongest evidence for *in vivo* lipid peroxidation is that biological systems have evolved redundant mechanisms of antioxidant protection, specifically involving the powerful chain-breaking antioxidant α-tocopherol in the membrane and several other antioxidants (glutathione, cysteine, ascorbate, uric acid, glutathione peroxidase, ceruloplasmin) in the soluble phase. Indeed, it is clear that deficiencies of this antioxidant defense system can greatly impair physiological function.

49. For example, pulmonary injury by the bipyridilium herbicide paraquat involves lipid peroxidation only as a late-stage event, rather than as a cause of it.

Vitamin E as a Scavenger of Free Radicals

Because of the reactivity of the phenolic hydrogen on its C-6 hydroxyl group and the ability of the chromanol ring system to stabilize an unpaired electron, vitamin E has antioxidant activity capable of terminating chain reactions among PUFAs in the membranes wherein it resides. This action, termed free-radical scavenging, involves the donation of the phenolic hydrogen to a fatty acyl free radical (or $O_2^{\cdot-}$) to prevent the attack of that species on other PUFAs. The antioxidant activities of the vitamers E thus relate to the leaving ability of their phenolic hydrogen. The tocopherols have greatest reactivities toward peroxyl and phenoxyl radicals, but can also quench such mutagenic electrophiles as reactive nitrogen oxide species (NO_x). When assessed *in vitro* in chemically defined systems, activities are greatest for α-tocopherol,[50] with the β- and γ-vitamers roughly comparable and greater than the δ-vitamer.[51] An exception to this rank order appears to be the case for NO_x, which is trapped more effectively by γ-tocopherol than by the α-vitamer. In view of this finding, it has been suggested that γ-tocopherol, the predominant form of the vitamin in diets, may be important *in vivo* as a trap for electrophilic nitrogen oxides and other electrophilic mutagens. The tocotrienols are also active in scavenging the chain-propagating peroxyl radical.

In serving its antioxidant function, tocopherols and tocotrienols are converted from their respective alcohol forms to semi-stable radical intermediates, the *tocopheroxyl* (or *chromanoxyl*) *radical* (Figs 7.7, 7.8). Unlike the free radicals formed from PUFAs, the tocopheroxyl radical is relatively unreactive, thus stopping the destructive propagative cycle of lipid peroxidation. In fact, tocopheroxyl is sufficiently stable to react with a second peroxyl radical to form inactive, non-radical products including *tocopherylquinone*.[52] Because α-tocopherol can compete for peroxyl radicals much faster than PUFAs, small amounts of the vitamin are able to effect the antioxidant protection of relatively large amounts of the latter (Fig. 7.10).

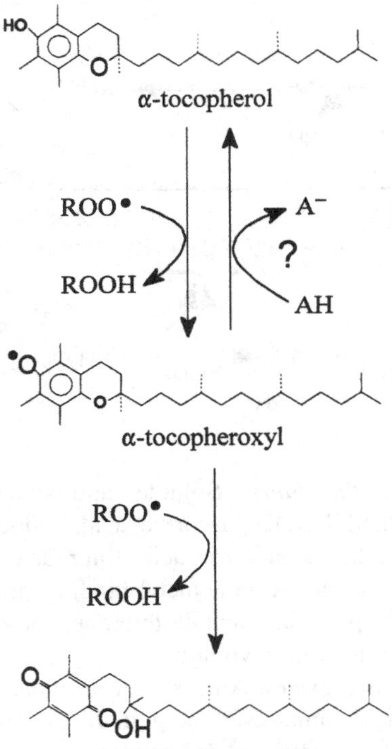

FIGURE 7.10 Oxidation of tocopherols by reaction with peroxyl radicals.

Roles of Tocotrienols

Tocotrienols are thought to have more potent antioxidant properties than tocopherols, as their unsaturated side chain facilitate more efficient penetration into tissues containing saturated fatty layers – e.g., brain, liver.

Interrelationships of Vitamin E and Other Factors in Antioxidant Defense

Factors that increase the production of ROS (e.g., xenobiotic metabolism, ionizing radiation, exposure to such pro-oxidants as ozone and NO_2) can be expected to increase the metabolic demand for antioxidant protection, including the need for vitamin E and the other nutrients involved in this system. Because it is hydrophobic, and therefore distributed in membranes, vitamin E serves as a lipid-soluble biological antioxidant with high specificity for loci of potential lipid peroxidation. However, in its antioxidant function vitamin E is but one of several factors in an antioxidant defense system that protects the cell from the damaging effects of oxidative stress. This system (Fig. 7.11) includes:

- *Membrane antioxidants.* The most important membrane antioxidants are the tocopherols, but the ubiquinones and carotenoids also participate in this function.

---

50. The antioxidant activity of α-tocopherol is about 200-fold that of the commonly used food antioxidant butylated hydroxytoluene (BHT).

51. The biological activities of the vitamers E are affected by both their intrinsic chemical antioxidant activities as well as their efficiencies of absorption and retention. Thus, γ-tocopherol has only 6–16% of the biological activity of the α-vitamer. An exception to this relationship has been identified in the case of sesame seed, lignans which act to potentiate the biopotency of γ-tocopherol such that the vitamin in sesame oil, which consists exclusively of the γ-vitamer, has a biopotency equivalent to that of α-tocopherol. The identity of the potentiating antioxidant factor is believed to be a lignan phenol, sesamolin, to which antiaging properties have been attributed.

52. Evidence indicates that tocopherylquinones can induce apoptosis in cancer cells.

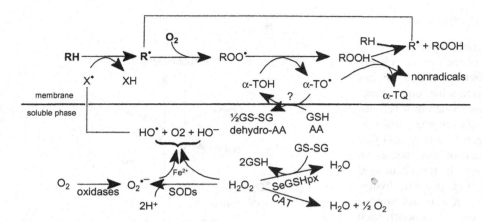

FIGURE 7.11 The cellular antioxidant defense system. CAT, catalase; GSH, reduced glutathione; GSSG, oxidized glutathione; α-TQ, α-tocopherylquinone; α-TO˙, α-tocopheryl radical; α-TOH, α-tocopherol; SeGSHpx, selenium-dependent glutathione peroxidase; SOD, superoxide dismutase.

- *Soluble antioxidants.* Soluble antioxidants include NADPH and NADH, ascorbic acid, reduced glutathione and other thiols, uric acid, thioredoxin, bilirubin, polyphenols, and several metal-binding proteins (copper: ceruloplasmin, metallothionein, albumin; iron: transferrin, ferritin, myoglobin).
- *Antioxidant enzymes.* Antioxidant enzymes include the superoxide dismutases,[53] the glutathione peroxidases,[54] thioredoxin reductase,[55] and catalase.[56]

In this multicomponent system, vitamin E scavenges radicals within the membrane, where it blocks the initiation and interrupts the propagation of lipid peroxidation. The group of metalloenzymes collectively blocks the initiation of peroxidation from within the soluble phase of the cell: the **superoxide dismutases** convert $O_2{}^-$ to H_2O_2; **catalase** and the **glutathione peroxidases** each further reduce H_2O_2. The aggregate effect of this enzymatic system is to clear $O_2{}^-$ by reducing it fully to H_2O, thus preventing the generation of other, more highly reactive oxygen species (e.g., HO· and singlet oxygen [$^1O_2$]). Glutathione

peroxidases can also reduce fatty acyl hydroperoxides to the corresponding fatty alcohols, thus serving to interrupt the propagation of lipid peroxidation.

Some components of this system are endogenous (e.g., NADPH, NADH, and, for most species, ascorbic acid), whereas other components must be obtained, at least in part, from the external chemical environment. The diversity of this system implies the ability to benefit from various antioxidants and other key factors obtained from dietary sources in variable amounts.

That the various components of the defense system function cooperatively is evidenced by the nutritional sparing observed particularly for vitamin E and selenium in the etiologies of several deficiency diseases (e.g., **exudative diathesis** in chicks, **liver necrosis** in rats, **white muscle disease** in lambs and calves). In those species, nutritional deprivation of either vitamin E or selenium alone is usually asymptomatic, whereas deficiencies of both nutrients are required to produce disease.

The activities of various components of this system, as well as the cellular redox state, have been found to change markedly during cellular differentiation, corresponding to increased oxidant production under those conditions. That gradients of reactive oxygen species and/or the redox state may be important in influencing genetic expression is indicated by several lines of evidence; oxidizing conditions have been found to affect cellular ion distribution, chromatin-controlling proteins, the cytoskeleton and nuclear matrix, which in turn affect chromatic configuration and pre-mRNA processing. Thus, it appears that the antioxidant defense system actually serves to control **peroxide tone**, such that the beneficial effects of pro-oxidizing conditions are realized and their deleterious effects are minimized.

Physiological Roles of ROS

Inflammation and Immunity

Studies in both animal models and humans have demonstrated anti-inflammatory effects of tocopherols.

---

53. The superoxide dismutases (SODs) are metalloenzymes. The mitochondrial SOD contains manganese at its active center, whereas the cytosolic SOD contains both copper and zinc as essential cofactors. Although not found in animals, an iron-centered SOD has been identified in blue-green algae.

54. The glutathione peroxidases contain selenium at their active centers and are dependent on adequate supplies of that element for their synthesis. There are at least four isoforms; each uses reducing equivalents from reduced glutathione (GSH) to reduce H_2O_2 to water or fatty acyl hydroperoxides to the corresponding fatty alcohols. One isoform is found in membranes and has specificity for esterified hydroperoxides; the others are soluble and have specificities for non-esterified hydroperoxide substrates including H_2O_2. The activities of these enzymes depend on a functioning glutathione cycle (i.e., the flavoenzyme **glutathione reductase**) to regenerate GSH from its oxidized form (GSSG).

55. Thioredoxin reductase contains selenium at its active center and depends on adequate selenium nutriture for its synthesis.

56. Catalase has an iron redox center. Because its distribution is almost exclusively limited to the peroxisomes/lysosomes, it is probably not important in antioxidant protection in the cytosol.

TABLE 7.8 Enhancement of Immune Responses by Vitamin E Supplementation of Healthy Adults

Treatment	Days of Treatment	Vit E in PMNs[a] (nmol)	DTH[b] Index (mm)	PMN Proliferation[c] ($\times 10^3$ cpm)	IL-2 Production[d] (kU/l)
Placebo	0	0.14 ± 0.04	16.5 ± 2.2	24.48 ± 2.73	31.8 ± 8.3
	30	0.19 ± 0.03	16.9 ± 2.1	21.95 ± 2.90	37.5 ± 12.5
Vitamin E[e]	0	0.12 ± 0.02	14.2 ± 2.9	20.55 ± 1.93	35.6 ± 9.1
	30	0.39 ± 0.05^f	18.9 ± 3.5^f	23.77 ± 2.99^f	49.6 ± 12.6^f

[a]Polymorphonucleocyte α-tocopherol content.
[b]Delayed-type hypersensitivity skin test.
[c]Conconavalin A-induced proliferation of polymorphonucleocytes.
[d]Conconavalin A-induced production of interleukin 2 by polymorphonucleocytes.
[e]A total of 800 IU all-rac-α-tocopheryl acetate per day.
[f]$p < 0.05$.
From Meydani, S. N., Barklund, M. P., and Liu, S. (1990). *Am. J. Clin. Nutr.* 52, 557.

α-Tocopherols have been found to modulate T cell function, reduce production of the pro-inflammatory mediator PGE_2 by macrophages and other pro-inflammatory cytokines by activated macrophages and monocytes, and reduce plasma levels of the inflammation marker C-reactive protein (CRP).[57] Similar responses have been shown for γ-tocopherol, although different potencies have been observed in different systems. For example, α-tocopherol appears to be a better inhibitor of PKC, while γ-tocopherol is a stronger inhibitor of cyclooxygenase.

While some anti-inflammatory actions of vitamin E appear to be unrelated to antioxidant function, the vitamin clearly serves as a biological antioxidant protecting immune cells from ROS produced during the inflammatory process by phagocytic cells attracted to the site of tissue injury. On encountering or ingesting a bacterium or other foreign particle, activated[58] neutrophils and macrophages produce large amounts of $O_2\cdot$ and H_2O_2 in a process referred to as the respiratory burst. This process involves myeloperoxidase, which catalyzes the H_2O_2-dependent oxidation of halide ions yielding such powerful oxidizing agents as hypochlorous acid, and xanthine oxidase, which catalyzes the reaction of xanthine or hypoxanthine with molecular oxygen to generate uric acid. These reactions are important in killing pathogens, but they can also be deleterious to immune cells themselves. If not controlled, they can contribute to the pathogenesis of disease. Therefore, it appears that adequate antioxidant status is required to maintain appropriate peroxide tone.

Vitamin E deficiency has been shown, in experimental animals and children, to compromise both humoral and cell-mediated immunity. Deficient individuals have polymorphonucleocytes with impaired phagocytic abilities, suppressed oxidative burst and bactericidal activities, and decreased chemotactic responses. They can also show generally suppressed lymphocyte production, impaired T cell functions, and decreased antibody production. Vitamin E deprivation has been found in animal models to increase susceptibility to viral infections, and to enhance the virulence of cardiophilic viruses passed through antioxidant-deficient hosts[59] These effects are thought to involve loss of peroxide tone, and appear to involve impaired cellular membrane fluidity and enhanced PGE_2 production by the host, creating an environment in which viral mutation rates increase. A recent study found low circulating α-tocopherol levels in older humans to be associated with chronic inflammation and lower reported quality of life.[60]

Supranutritional intakes of vitamin E have been found to stimulate many immune functions, including antibody production. Studies with experimental animals have shown these effects to result in increased resistance to infection. A randomized controlled trial found a high level of vitamin E (800 IU/day) to increase T cell-mediated responses (delayed-type hypersensitivity, mitogenesis, and IL-2 production), and to decrease phytohemagglutinin-stimulated prostaglandin production. These responses typically decline with aging, and another study found older subjects to show improvements in delayed-type hypersensitivity in response to vitamin E more frequently than younger subjects (Table 7.8). The optimal vitamin E intake for immune enhancement is not clear. Animal studies have found optimization of certain immune parameters to call for intakes

57. CRP is an acute-phase protein synthesized by the liver in response to cytokine adipocytes. Its metabolic role is to bind to phosphocholine on the surface of dying cells and some types of bacteria in activating the complement system. Its elevation in plasma indicates ongoing inflammation.
58. Various cytokines, such as tumor necrosis factor or interferon γ, can activate phagocytic cells to increase their $O_2\cdot$ generation.

59. This involved an increase in viral mutation rate (Levander and Beck [1997] *Biol. Trace Elem. Res.* 56, 5) and was also prevented by selenium.
60. Capuron, L., Moranis, A., and Combe, N. (2009). *Br. J. Nutr.* 102, 1390.

of at least an order of magnitude greater than the levels of the vitamin required to prevent clinical signs of deficiency.

Pro-Oxidant Potential of Vitamin E

α-Tocopherol can promote lipid peroxidation in LDLs in the absence of other antioxidants (e.g., ascorbic acid, coenzyme Q_{10}, urate). Under such conditions, the single-electron oxidation of tocopherol converts it to the tocopheroxyl radical, which moves into the particle's core where it can abstract hydrogen from a cholesteryl-PUFA ester to yield a peroxyl radical. Under such circumstances vitamin E serves as a chain-transfer agent to propagate lipid peroxidation in the lipid core, rather than as a chain-breaking antioxidant to block that process. Therefore, the presence of secondary antioxidants is needed to prevent LDL oxidation. Accordingly, one intervention trial found that a very high dose (1,050 mg/day) of α-tocopherol increased susceptibility to peroxidation.[61]

Non-Antioxidant Functions of Vitamin E

The recognition in the early 1990s that vitamin E could inhibit cell proliferation and protein kinase C (PKC) activity suggested that vitamin E may function *in vivo* in ways that are unrelated to its function as a biological antioxidant.

Enzyme Regulation

α-Tocopherol has been shown to be involved in the regulation of a number of enzymes. For several, the vitamin appears to participate in complex membrane-based recruitment processes affecting enzymatic function: inhibition of PKC, NADPH oxidase, phospholipase A_2, PKB/Akt, 5-lipoxygenase, cyclooxogenase A_2, and (at low doses) HMG-CoA reductase; activation of protein phosphatase 2A, diacylglycerol kinase, and (at high doses) HMG-CoA reductase.

Vitamin E also appears to participate in transcriptional regulation (Table 7.9). The underlying mechanisms of these diverse effects are not clear. It is likely that some may involve redox modulation due to free radical scavenging, some may involve modulation of translocation in membranes, and others may be secondary to effects on such factors as PKC and PPARγ. The transcriptional effects that have been demonstrated for α-tocopherol imply the existence of nuclear receptors for tocopherol and corresponding DNA responsive elements. One group of nuclear receptors, the pregname X receptors (PXR), has been found to bind vitamin E; however, the

61. Brown, K. M., Morrice, P. C., and Duthie, G. G. (1977). *Am. J. Clin. Nutr.* 65, 496.

TABLE 7.9 Genes Regulated by Vitamin E

Function	Gene Product
Tocopherol uptake	α-TTP
Metabolism	CYP3A, CYP4F2, HMG-CoA reductase, γ-glutamyl cysteine synthase, CRABP-II
Lipid uptake	SR-BI, CD36, SR-AI/II
Extracellular proteins	α-tropomyosin, collagen-α1, MMP-1, MMP-19, glycoprotein IIb
Cell adhesion	E-selectin, L-selectin, ICAM-1, VCAM-1, integrins
Cell growth	Connective tissue growth factor
Extracellular matrix formation/degradation	Collagen α1(1)
Inflammation	Il-2, Il-4, IL-1-ß, TGF-β
Transcriptional control	PPARγ
Cell cycle regulation	Cyclins D1 and E, Bcl12-L1, p27, CD95
Apoptosis	CD95L, Bcl2-L1
Lipoprotein receptors	CD36, SR-BI, SR-AI/II, LDL receptor
Other functions	Leptin, tropomyosin, a β-secretase

metabolic significance of such binding is not presently clear. Rats deprived of vitamin E show altered patterns of gene expression; studies have revealed a greater number of genes being downregulated, although a number of transport-related genes that were upregulated in liver were downregulated in the cerebral cortex. The emerging picture is one of α-tocopherol being involved in the assembly of membrane and vesicular transport with cells.

It is clear that supplementation with α-tocopherol can induce its own transport (via α-TTP) and catabolism (via CYP3A). The latter effect appears to be a general response also involving the upregulation of other cytochrome *P*-450 isoforms involved in the metabolism of foreign compounds; supplementation with α-tocopherol may affect the metabolism and contribute to the detoxification of potentially toxic xenobiotics. This includes CYP4F2, which is important for the regulation and activation of vitamin K.

7. VITAMIN E IN HEALTH AND DISEASE

The efficacy of vitamin E as a biological antioxidant appears to be dependent on the amount of the vitamin present at critical cellular loci, and the ability of the organism to maintain its supply and/or recycle it. Therefore, in the presence of other components of the cellular antioxidant defense system (deficiencies of which might result in tocopheroxyl assuming peroxidation chain-transfer activity), antioxidant protection would be expected to increase with increasing intake of vitamin E. High levels of vitamin E may therefore be appropriate in situations in which oxidative stress is increased.

ROS of Metabolic Origin

Exercise

Because most ROS are produced endogenously by mitochondria, which process 99% of the oxygen utilized by the cell, factors such as exercise that increase normal oxidative metabolism also increase the need for vitamin E.[62] Thus, it is thought that exercise-induced injuries to muscle membranes may be due to the enhancement of oxidative reactions. Tissue vitamin E levels drop as a result of exercise. Studies with humans have found that vitamin E supplementation can reduce the oxidative stress and lipid peroxidative damage induced by exhaustive exercise.

Metabolically produced ROS also appear to have essential metabolic functions as signaling molecules for the adaptation of skeletal muscle to accommodate the stresses presented by exercise training or periods of disuse. This signaling involves redox-sensitive kinases, phosphatases, and NFκB, which affect the rate of mitochondrial biogenesis, as well as the induction of genes related to insulin sensitivity (PGC1α/β, PPARγ) and ROS defense (SOD1/2, GPX1). This system of adaptive responses to oxidative stress facilitates the ultimate development of long-term resistance to that stress (Fig. 7.12). That this system, which has been called mitochondrial hormesis,[63] can be impaired by high-level antioxidant treatment was demonstrated by the finding that supplements of vitamins E and C (400 IU α-tocopheryl acetate plus two doses of 500 mg ascorbic acid per day) blocked the upregulation of muscle glucose uptake that is otherwise induced by exercise.[64] This finding raises several questions: Is this effect due to vitamin E, vitamin C, or the combination? What antioxidant dose is required for such effects? What level of "peroxide tone" is beneficial?

Aging

ROS are also thought to have causative roles in aging, which involves the accumulation of a wide array of oxidative lesions and chronic, low-grade inflammation. This

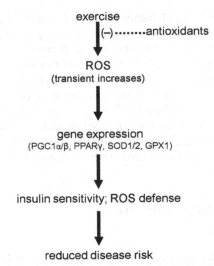

FIGURE 7.12 Mitochodrial hormesis: ROS signalling of insulin sensitivity. After *Ristow, M., Zarse, K., Oberbach, A.*, et al. *(2009)*. Proc. Nat. Acad. Sci., USA *106, 8665*.

includes oxidative damage to mitochondrial DNA and proteins, and lipid-soluble, brown to yellow, autofluorescent pigments[65] collectively called lipofuscin in several tissues (e.g., retinal pigment epithelium, heart muscle, brain). Studies in rats have shown that the age-related decline in the major glucose transporter in neurons (Glut3) is exacerbated by deprivation of vitamin E. Vitamin E deprivation has been shown to increase the accumulation of these so-called age pigments in experimental animals; but there is no evidence that supranutritional levels of vitamin E are any more effective than nutritionally adequate levels in reducing their build-up. That these changes are related causally to aging is suggested by interspecies observations that mammalian life-span potentials tend to correlate inversely with metabolic rate and directly with tissue concentrations of tocopherols and other antioxidants (carotenoids, urate, ascorbic acid, superoxide dismutase activity).[66] According to the free-radical theory of aging, it is proposed that cumulative damage by ROS is accompanied by gradual decreases in repair capacity likely due to changes in gene expression, diminished immune function, and enhanced programmed cell death[67] induced by increases in "peroxide tone" (i.e., the net amount of ROS within the cell).

---

62. Oxygen utilization increases 10- to 15-fold during exercise.
63. Hormesis is the term for a generally favorable biological response to low exposures to stressors/toxins.
64. Ristow, M., Zarse, K., Oberbach, A., *et al.* (2009.) Antioxidants prevent health-promoting effects of physical exercise in humans. *Proc. Nat. Acad. Sci. USA* 106, 8665.

65. Lipofuscins are generally thought to be condensation products of proteins and lipid. There is some evidence that the pigments isolated from the retinal pigment epithelium contain, at least in part, derivatives of vitamin A (e.g., *N*-retinylidene-*N*-retinylethanolamine).
66. It is likely that caloric restriction, which has been shown to increase longevity in animals, may have that effect by reducing the metabolic rate and, hence, the endogenous production of ROS.
67. That is, *apoptosis*.

Vitamin E supplements have been found to promote immune responsiveness, which is of particular benefit in older people. Supranutritional doses (up to 800 mg α-tocopherol per day) have been found to restore response to DTH, and increase induction of IL-2 and reduce synthesis of the pro-inflammatory lipid mediator prostaglandin E_2 (PGE_2) in lymphocytes.[68] Doses as low as 50 mg/day have been associated with reduced incidence of the common cold.[69] Trials evaluating effects of vitamin E on acute respiratory infection in older subjects have yielded inconsistent results.

ROS of Exogenous Origin

Radiation

Radiation damages cells by direct ionization of DNA and other cellular targets, and by indirect effects of ROS that are also produced. Vitamin E has been shown to decrease radiation-induced chromosome damage in animal models. This effect may be the basis of the links detected epidemiologically between consumption of antioxidant-rich foods and reduced cancer risk.

Air Pollution

Individuals living in smog-filled urban areas can be exposed to relatively high levels of ozone (O_3)[70] and nitrogen dioxide (NO_2), strong oxidants that provoke inflammatory responses of the airway. These include the activation of neutrophils and a respiratory burst resulting in overproduction of ROS leading to peroxidative damage. Vitamin E deprivation has been found to increase the susceptibility of experimental animals to the pathological effects of O_3 and NO_2.[71] Is has been suggested that supplements of the vitamin may protect humans against chronic exposure to smog.

Altitude

Individuals living at high altitudes can also be exposed to relatively high levels of ozone (O_3). One study found that a daily supplement of 400 IU vitamin E prevented decreases in anaerobic thresholds of high-altitude mountain climbers (Table 7.10). Collectively, these findings support the

TABLE 7.10 Antioxidant Protection by Vitamin E in High-Altitude Climbers

Treatment Group	% Change in Pentane Exhalation[a]		
	Median	Lower Quartile	Upper Quartile
Placebo	104	26	122
Vitamin E, 400 IU/day	−3	−7	3

[a]After 4 weeks at high altitude.
Source: Simon-Schnass, I. and Pabst, H. (1988). *J. Vit. Nutr. Res.* 58, 49.

hypothesis that exercise at altitude increases the need for vitamin E.

Smoking

Smoking constitutes an oxidative burden on the lungs and other tissues owing to the sustained exposure to free radicals from the tar and gas phases of tobacco smoke.[72] This is characterized by increased levels of peroxidation products in the circulation (e.g., malonyldialdehyde) and breath (e.g., ethane, pentane), with decreased levels of ascorbic acid in plasma and leukocytes, and of vitamin E in plasma and erythrocytes. The high amounts of reactive nitrogen species in cigarette smoke are thought to support the nitration of vitamin E, as 5-nitro-γ-tocopherol occurs at higher levels in the plasma of smokers than in non-smokers (Table 7.9). This may be the basis for the enhanced turnover of tocopherols and reduced production of CECH by smokers. One intervention trial found supranutritional doses of vitamin E (up to 560 mg of α-tocopherol per day) to reduce the peroxidation potential of erythrocyte lipids from smokers, although a very high level (1,050 mg/day) increased susceptibility to peroxidation for non-smokers (Table 7.11).

Disorders Involving Oxidative Stress

Oxidative stress plays a role in the pathogenesis of complications of several conditions for which it has therefore been reasonable to suggest that, in such conditions, vitamin E status may be a determinant of risk. Several such conditions have been found to be associated with relatively low vitamin E intake or status; however, in general, vitamin E intervention trials have failed to demonstrate benefits.

68. Meydani, S. N., Barklund, M. P. Kiu, S., *et al.* (1990). *Am. J. Clin. Nutr.* 52, 557; Meydani, S. N., Meydani, M., Blumberg, J. D., *et al.* (1997). *J. Am. Med. Assoc.* 277, 1380.
69. Hemila, H., Kaprio, J., Albanes, D., *et al.* (2002). *Epidemiology* 13, 32.
70. Ambient O_3 levels on the top of Mt Everest or in Los Angeles can be 70–80 ppb, while in Grand Forks, ND, they are <50 ppb.
71. Ozone (O_3) is produced in photochemical smog from nitrogen dioxide (NO_2), oxygen, and uncombusted gasoline vapors; NO_2 is produced in internal combustion engines. Both can generate unstable free radicals that damage lungs through oxidative attack on polyunsaturated membrane phospholipids.

72. Cigarette smoke contains a number of compounds that produce free radicals. It also increases the number of free radical-producing inflammatory cells in the lungs.

TABLE 7.11 Comparison of Effect of Vitamin E on Erythrocyte Lipid Peroxidation in Smokers and Non-Smokers

Group	Weeks of Vitamin E Administration[b]	Vitamin E in Erythrocytes (μmol/g Hb)	*In Vitro* Lipid Peroxidation (nmol MDA[a]/g Hb)
Non-smokers	0	20.0 ± 4.5	141 ± 54
	20	36.1 ± 8.2	86 ± 51
Smokers	0	18.0 ± 4.2	291 ± 102
	20	32.8 ± 8.2	108 ± 53

[a]MDA, Malonyldialdehyde.
[b]A total of 70 IU/day.
Source: Brown, K. M., Morrice, P. C., and Duthie, G. G. (1977). *Am. J. Clin. Nutr.* 65, 496.

TABLE 7.12 Diabetes and Tocopherol Status

Patient Group	Plasma Vitamin E, mg/l	LDL Vitamin E, mg/g
Non-diabetic	17.4 ± 3.7	6.4 ± 1.3
Diabetic	$12.9^* \pm 2.9$	5.5 ± 3.8

*$P < 0.05$.
Source: Quilliot, D., Walters, E., Bonte, J. P., et al. (2002) *Am. J. Clin. Nutr.* 81:1117.

Neurodegenerative Disorders

Neural tissues conserve vitamin E, apparently by maintaining a relatively larger portion in the less labile cellular pool. In fact, vitamin E appears to be redistributed to neural tissues under conditions of nutritional deficiency.[73] Several facts make it reasonable to expect neuronal tissue to be susceptible to oxidative stress: neurons contain large amounts of both PUFAs and iron, but do not have extensive antioxidant defense systems; neurons are terminally differentiated and do not replicate when damaged; redox cycling drugs (that generate ROS) can cause Parkinson-like neural damage in animal models; dopamine metabolism[74] in dopaminergic neurons generates ROS; exposure to hyperbaric oxygen can cause seizures; defects in α-TTP are manifest as ataxia (e.g., AVED); vitamin E deficiency is manifest by neurological signs. There is no question that vitamin E is essential for neurologic function.

Epidemiological studies have found estimated vitamin E intake to be inversely associated with risks of Alzheimer's disease[75] and Parkinson's disease,[76] both of which are thought to involve oxidative stress etiologically. Nevertheless, randomized clinical trials of vitamin E supplementation have yielded inconsistent results with respect to both diseases as well as Amyotrophic Lateral Sclerosis (ALS),[77] which is thought to involve mitochondrial stress.

Diabetes

In comparison with controls, diabetic erythrocytes have significantly more lipid peroxidation[78] which, by altering membrane fluidity, is thought to render erythrocytes hypercoagulable and more ready to adhere to endothelial cells. Membrane lipid peroxidation correlates with erythrocyte contents of glycated hemoglobin (HbA1c), and supplementation of non-insulin-dependent (type 2) diabetics with high levels of vitamin E has been found to reduce hemoglobin damage. Oxidative stress is also known to alter cellular serine/threonine kinase activities, resulting in the phosporylation of insulin receptor substrate-1, which reduces insulin signaling. However, insulin sensitivity also appears to be affected through translational regulatory functions of vitamin E. Supranutritional levels of either α- and γ-tocopherols have been shown to upregulate an endogenous ligand involved in activating PPARγ, which plays a role in insulin sensitivity by upregulating adiponectin, an adipokin that increases insulin sensitivity.[79]

Studies have found diabetics to have low plasma tocopherols (Table 7.12), and high-level vitamin E supplements (e.g., 900 mg α-tocopherol per day) to improve insulin responsiveness in both normal and diabetic individuals. One randomized clinical trial reported protection by vitamin E against the development of type 2 diabetes among subjects with impaired glucose tolerance.[80]

Vitamin E has been considered as a factor in protecting against diabetic complications (e.g., retinopathy, cardiac dysfunction), the etiologies of which are thought to involve oxidative stress. However, clinical trials have not found vitamin E supplementation to be beneficial in this regard.

73. Muller, D.P.R., Coss-Sampson, M. A., Burton, G. W., *et al.* (1992). *Free Radical Res. Commun.* 16, 10.
74. By monoamine oxidase B.
75. Alzheimer's disease is the world's most prevalent neurodegenerative disease, affecting an estimated 20–30 million, including almost half of people over the age of 85 years. It is characterized by memory dysfunction, loss of lexical access, temporal and spatial disorientation, and impaired judgment.
76. Parkinson's disease is characterized by progressive loss of postural stability, with slowness of movement and tremor.

77. ALS ("Lou Gherig's disease") is characterized by profound muscular weakness due to the selective death of upper and lower motor neurons; the disease is ultimately fatal, mostly due to respiratory failure.
78. That is, thiobarbituric acid-reactive substances.
79. Landrier, J. F., Gouranton, E., El Yazidi, C., *et al.* (2009). *Endocrinology* 150, 5318.
80. Mayer-Davis, E. J., Costacou, T., King, I., *et al.* (2002). *Diabetes Care* 25, 2172.

Cataracts

Cataracts result from the accumulation in the lens of damaged proteins that aggregate and precipitate, resulting in opacification of the lens.[81] Much of this damage involves oxidations characterized by the loss of sulfhydryls, the formation of disulfide and non-disulfide covalent linkages, and the oxidation of tryptophan residues. Several epidemiological studies have found circulating α-tocopherol level or vitamin E intake to be inversely associated with cataract risk. Vitamin E has been shown in animal models to reduce or delay cataracts induced by galactose or aminotriazol treatment, and to reduce the photoperoxidation of lens lipids. These effects are thought to involve its direct action as an antioxidant or its indirect antioxidant effect in maintaining lens glutathione in the reduced state. Nevertheless, large scale, randomized controlled trials have not found α-tocopherol supplementation at supranutritional levels (50–500 mg/day) to reduce risk of cataracts.

Ischemia–Reperfusion Injury

Vitamin E and other free-radical scavengers have been found to affect the functions of mitochondria and sarcoplasmic reticula in animal models of myocardial injury induced by **ischemia–reperfusion**. That injury, which occurs in tissues reperfused after a period of ischemia, appears to be due to the oxidative stress of reoxygenation, involving the production of ROS. The phenomenon has been demonstrated for several tissues (heart, brain, skin, intestine, and pancreas), and has relevance for the preservation of organs for transplantation. ROS are thought to contribute to milder forms of tissue injury at the time of reperfusion (e.g., myocardial stunning, reperfusion arrhythmias); however, it is not clear the extent to which free radicals are responsible for the acute tissue damage seen under those circumstances. Intervention trials with antioxidants have yielded conflicting results, but it is possible that preoperative treatment with vitamin E may be useful in reducing at least some of this type of injury. A disorder that appears to involve natural ischemia–reperfusion injury is intraventricular hemorrhage[82] of the premature infant. The results of a randomized clinical trial clearly demonstrate that vitamin E supplements (given by intramuscular injection) can be very effective in protecting premature infants against this disorder.

Lung Health

The lungs are continuously exposed to relatively high concentrations of O_2 as well as environmental oxidants and irritants. The first line of defense of the respiratory epithelium organ is the respiratory tract lining fluid, which contains a variety of antioxidants, including vitamin E, as well as relatively high concentrations of vitamin C, urate, reduced glutathione, extracellular superoxide dismutase, catalase, and glutathione peroxidase. Nevertheless, a meta-analysis of observational studies found no relationship of estimated dietary intake of vitamins E, C or β-carotene on risk of asthma.

Pre-Eclampsia

A key role of vitamin E in pregnancy is suggested by the fact that circulating α-tocopherol levels correlate positively with fetal growth rate, particularly during the last trimester when oxygen utilization is increased. That the resulting oxidative stress may increase risk of pre-eclampsia is indicated by the fact that increases in lipid peroxides of placental origin in the maternal circulation correlate with the severity of pre-eclampsia.[83] Nevertheless, a systematic review of clinical intervention trials conducted with vitamin E concluded that there was no evidence that supplemental vitamin E reduced pre-eclampsia risk.[84]

Skin Health

The skin is subject to the oxidizing effects resulting from exposure to ultraviolet light, which is known to generate ROS from the photolysis of intracellular water. Studies with animal models have shown that the tocopherol content of dermal tissues decreases with UV irradiation, presumably as a result of that oxidative stress. Vitamin E in skin is found in greatest concentrations in the lower levels of the strateum corneum, where it is released by sebum. Topical treatment with vitamin E may increase the hydration of the strateum corneum, and confer protection against UV-induced skin damage, as measured by reduced erythemal responses and delayed onset of tumorigenesis. One study reported that regular topical application of vitamin E reduced wrinkle amplitude and skin roughness in about half of cases. For these reasons, α-tocopherol and α-tocopheryl acetate are widely used in skin creams and cosmetics.

---

81. Cataracts constitute a significant public health problem in the US, where a million cataract extractions are performed annually at a cost of some $5 billion. The prevalence of cataracts among Americans increases from about 5% at age 65 years to about 40% at age 75 years. These rates are considerably (up to five-fold) greater in less developed countries.

82. Hemorrhage in and around the lateral ventricles of the brain occurs in about 40% of infants born before 33 weeks of gestation.

83. i.e., pregnancy-induced hypertension associated with proteinuria. Pre-eclampsia is thought to involve endothelial dysfunction of maternal blood vessels induced by factors released from the placenta. It can develop in the last trimester of pregnancy, and has the highest rates of morbidity and mortality of all complications of pregnancy.

84. Polyzos, N. P., Mauri, D., Tsappi, M., *et al.* (2007). *Obst. Gynecol. Surv.* 62, 202.

TABLE 7.13 High Vitamin E Intakes Associated with Reduced Coronary Heart Disease Risks in Two Cohorts of Americans

Parameter	Quintile Group for Vitamin E Intake[a]					P Value for Trend
	1	2	3	4	5	
Women[b]						
Total Vitamin E Intake (IU/day)						
Median	2.8	4.2	5.9	17	208	
Relative risk[c]	1.0	1.00	1.15	0.74	0.66	<0.001
Men[d]						
Total Vitamin E Intake (IU/day)						
Median	6.4	8.5	11.2	25.2	419	
Relative risk[e]	1.0	0.88	0.77	0.74	0.59	0.001
Dietary Vitamin E Intake (IU/day)						
Range	1.6–6.9	7.0–9.8	8.2–9.3	9.4–11.0	11.1	
Relative risk[e]	1.0	1.10	1.17	0.97	0.79	0.11
Supplemental Vitamin E Intake (IU/day)						
Range	0	<25	25–99	100–249	≥250	
Relative risk[e]	1.0	0.85	0.78	0.54	0.70	0.22

[a]Includes vitamin E from both foods and supplements.
[b]A total of 87,245 nurses (679,485 per-yrs follow-up); from Stampfer, M., Hennekens, C. H., Manson, J. E., et al. (1993). N. Engl. J. Med. 328, 1444.
[c]Adjusted for age and smoking.
[d]A total of 39,910 health professionals (139,883 per years follow-up); from Rimm, E. B., Stampfer, M. J., Ascherio, A., et al. (1993). N. Engl. J. Med. 328, 1450.
[e]Ratio of events in each quintile to those in the lowest quintile.

Diseases Involving Inflammation

Despite the clear anti-inflammatory effects that have been demonstrated for these vitamers, clinical trials have generally not found α-tocopherol supplements effective in reducing risks of inflammatory diseases (e.g., rheumatoid arthritis, asthma, and hepatitis) or diseases in which chronic inflammation contributes (e.g., cardiovascular disease, cancer, neurodegenerative diseases). The apparently different anti-inflammatory activities of vitamers E has caused some to suggest that mixed tocopherols may have potential value in this regard.

Arthritis

Rheumatoid arthritis is thought to be caused by antigenic triggering, in the articular joints, of an inappropriate immune response that leads to chronic inflammation. Indirect evidence suggests that the inflammatory production of ROS leads to the oxidation of lipids in the synovial fluid, which increases the viscosity of that fluid. Studies with animal models have found that vitamin E supplementation can reduce joint swelling, and randomized controlled trials have shown high-level supplementation with the vitamin

(100–600 IU/day) to relieve pain and be anti-inflammatory; however, α-tocopherol supplements have not been found effective in reducing rheumatoid arthritis risk.[85]

Cardiovascular Health

Observational epidemiologic studies have consistently demonstrated benefits of vitamin E on cardiovascular disease risk.[86] Seven of the nine major cohort studies conducted to date (and involving nearly a quarter of a million subjects) found inverse associations of vitamin E intake and cardiovascular disease incidence or associated mortality. Two large cohort studies found beneficial effects of vitamin E only for high vitamin E intakes achieved through the use of dietary supplements for at least 2 years' duration (Table 7.13).[87] The results of case–control and cohort studies have, however, been mixed. This is not

85. Karlson, E. W., Shadick, N. A., Cook, N. R., et al. (2008). Arthr. Rheum. 59, 1589.
86. See review: Cordero, Z., Drogen, D., Weikert, C., et al. (2010). Crit. Rev. Food Sci. Nutr. 50, 420.
87. Stampfer, M., Hennekens, C. H., Manson, J. E., et al. (1993). N. Engl. J. Med. 328,1444; Rimm, E. B., Stampfer, M. J., Ascherio, A., et al. (1993). N. Engl. J. Med. 328, 1450.

TABLE 7.14 Lipid and Antioxidant Contents of Human Low-Density Lipoproteins

Component	Moles per Mole LDL
Total phospholipids	700 ± 122
Fatty acids	
Free	26
Total	2,700
Triglycerides	170 ± 78
Cholesterol	
Free	600 ± 44
Esters	1,600 ± 119
Total	2,200
Antioxidants	
α-Tocopherol	6.52
γ-Tocopherol	1.43
Ubiquinonol-10	0.33
β-Carotene	0.27
Lycopene	0.21
Cryptoxanthin	0.13
α-Carotene	0.11

Adapted from Keaney, J. F. and Frei, B. (1994). Natural Antioxidants in Human Health and Disease (Frei, B., ed.). Academic Press, San Diego, CA, pp. 306–307.

TABLE 7.15 Reduced Low-Density Lipoprotein Susceptibility to Lipid Peroxidation by Oral Vitamin E in Humans

Treatment	Time of Sampling	LDL Oxidation[a]		
		LDL α-Tocopherol (μmol/g Protein)	Lag Phase (h)	Rate (μmol/g Protein/h)
Placebo	Baseline	14.3 ± 5.0	2.1 ± 0.9	396 ± 116
	8 weeks	15.8 ± 6.1	2.0 ± 0.8	423 ± 93
Vitamin E[b]	Baseline	13.8 ± 4.1	1.9 ± 0.6	373 ± 96
	8 weeks	32.6 ± 11.5[c]	2.9 ± 0.8[c]	367 ± 105

[a]Lipid peroxide formation.
[b]A total of 1,200 IU/day as RRR-α-tocopherol.
[c]Significantly different (P <0.05) from corresponding placebo value.
From Fuller, C. J., Chandalia, M., Garg, A., et al. (1996). Am. J. Clin. Nutr. 63, 753.

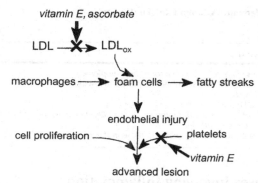

FIGURE 7.13   Model for prevention of atherogenesis by vitamin E.

surprising, given the many sources of variation in such studies: inherent errors in estimating vitamin E intake, variability in cardiovascular risk factors, variability in vitamin E utilization and baseline status, oxidative degradation of tocopherols during sample handling and storage, etc.

That vitamin E may protect against cardiovascular disease would appear likely, because it can function as a protective antioxidant in LDLs, as oxidative damage to LDLs appears to be a factor in the etiology of atherosclerosis.[88] Being rich in both cholesterol and PUFAs (Table 7.14), LDLs are susceptible to peroxidation by ROS. Research has shown that **oxidized LDLs** stimulate the recruitment, in the subendothelial space of the vessel wall, of monocyte-macrophages that can take up the oxidized particles

via **scavenger receptors**[89] to form the lipid-containing **foam cells** found in the early stages of atherogenesis. Studies *in vitro* have shown that the peroxidation of LDL PUFAs occurs only after the lag phase caused by the loss of LDL antioxidants. Enrichment of LDLs with vitamin E, the predominant antioxidant occurring naturally in those particles, increases the lag phase, thus indicating increased resistance to oxidation. This has been demonstrated for oral supplements of the vitamin (Table 7.15).

Protection of LDLs by vitamin E appears to depend on the presence of multiple antioxidants, as high tocopherol concentrations can act as pro-oxidants in the absence of

88. Atherosclerosis is the focal accumulation of acellular, lipid-containing material as plaques in the intima of the arteries. The subsequent infiltration of the intima by fatty substances (arteriosclerosis) and calcific plaques, and the consequent reduction in the lumenal cross-sectional area of the vessels, result in a reduction in blood flow to the organs served by the affected vessels, causing such symptoms as angina, cerebrovascular insufficiency, and intermittent claudication.

89. Monocyte-macrophages have very few LDL receptors, and the few they have are downregulated. Therefore, when incubated with non-oxidized LDLs they do not form foam cells, as the accumulation of cholesterol further reduces LDL receptor activity. On the other hand, these cells have specific receptors for modified LDLs; these are called *scavenger receptors*. It is thought that LDL lipid peroxidation products may react with amino acid side chains of apo B to form epitopes that have affinities for the scavenger receptor.

TABLE 7.16 Relationship of Haptoglobin Genotype and Cardioprotection by Vitamin E

Treatment	Hp Genotype	n	Events	Hazard Ratio (95% C.I.)	P Value
None	Hp2-1	1,248	25	1.0	
	Hp1-1	285	6	1.0 (0.4–2.5)	0.92
Placebo	Hp2-2	708	33	2.3 (1.4–3.9)	0.001
Vitamin E	Hp2-2	726	16	1.1 (0.6–2.00)	0.81

Source: Milman, U., Blum, S., Shapira, C., et al. (2008). Arterioscler. Thromb. Vasc. Biol. 28, 341.

water-soluble antioxidants (e.g., ascorbate, urate). Under such conditions the tocopheroxyl radical formed on the LDL surface moves into the particle's core, where it can abstract hydrogen from a cholesteryl-PUFA ester to yield a peroxyl radical, thus serving propagating lipid peroxidation in the lipid core.

Evidence indicates that vitamin E can also affect the adherence and aggregation of platelets, reductions of which would prevent the progression of a fatty streak and cell proliferation to advanced lesions (Fig. 7.13). Studies have shown that vitamin E supplementation at levels of about 200 IU/day reduced the adhesion of human platelets to a variety of adhesive proteins.

In light of the above view of vitamin E functions, it has been surprising that the majority of randomized clinical trials have not found vitamin E supplementation to reduce cardiovascular risk. Meta-analyses of those trials have detected no beneficial effects. As the intervention agent in most studies has been α-tocopheryl acetate, some have asked whether non-α vitamers, particularly γ-tocopherol and the tocotrienols, may be effective. It is possible that beneficial effects of vitamin E may have been missed because of failure to consider disease subtypes,[90] or sensitive sub-groups. Preventive effects of vitamin E may occur for individuals with polymorphisms with specific genes involved in cellular antioxidant protection. A recent randomized controlled trial found that adults with type 2 diabetes showed reduced cardiovascular disease risk in subjects with the *Hp2-2* genotype (Table 7.16).[91,92] It is likely that such genetic polymorphisms also determine responsiveness to vitamin E. Candidates include proteins involved in vitamin E transport/retention (α-TTP, TAPs, apolipoproteins E and A, SR-BI and CD36 scavenger receptors, LDL receptor, phospholipid transfer protein, microsomal triglyceride transfer protein, lipoprotein lipase, ATP binding cassettel transporter,

vitamin E-mediated gene expression (pregnane X receptor), vitamin E metabolism (CYP3A, CYP4F2), and antioxidant metabolism (dehydroascrobate reductase, sodium-coupled ascorbic acid transporters). Polymorphisms of each have been found to affect circulating α-tocopherol levels.

Anticarcinogenesis

Chemical carcinogenesis is thought to involve the electrophilic attack of free radicals with DNA. ROS are mutagenic to mammalian cells *in vitro*, and their generation has been found to correlate with the initiation, promotion, and progression of tumors in experimental animal models. A product of lipid peroxidation, malonyldialdehyde, has also been found to increase tumor production in animals. Therefore, it would seem reasonable to expect that, as an antioxidant, vitamin E may have a role in cancer prevention. Available evidence points to this possibility, at least for some types of cancer.

Studies with experimental animals, which have consisted mostly of two-stage, UV- or chemically-induced mammary, colon, oral or skin tumor models, have generally shown α-tocopherol to inhibit promotion. These include studies showing topically applied vitamin E to reduce UV-induced skin cancers by as much as 58%.

Evidence indicates that vitamin E can selectively stimulate apoptosis in neoplastic cells. These effects have been found greatest for α-, γ-, and δ-tocotrienols,[93] which have

90. For example, Shürks, M., Glynn, R. J., Rist, P. M., et al. ([2010] Br. J. Med. 341, 1) found supplemental WHAT not to affect stroke risk, but to reduce risk of ischemic stroke by 10% and increase risk of hemorrhagic stroke by 22%.

91. Milman, U., Blum, S., Shapira, C., et al. (2008). Arterioscler. Thromb. Vasc. Biol. 28, 341.

92. Haptoglobin (Hp) complexes with hemoglobin (Hb) to shield its iron center, thus reducing its pro-oxidative effect. The complex is recognized by the CD163 scavenger receptor and is thus cleared from the plasma endocytotically into the reticuloendothelial system. There are two common alleles at the Hp locus, 1 and 2, and functional differences between the Hp1 and Hp2 proteins. Individuals homozygous for Hp2 have Hp that forms cyclic polymers instead of the linear ones formed by the Hp1-1 and Hp1-2 genotypes. Hp2-2 polymers form complex Hb with 10-fold greater higher affinity than the linear Hp polymers, and bind less avidly to CD163, and are therefore less efficiently cleared from the plasma. Individuals with the Hp2-2 phenotype would be expected to have greater needs for the antioxidant effects of vitamin E.

93. McIntyre, B. S., Briski, K. P., Tirmenstein, M. S., et al. (2000). Lipids 35, 171.

been shown to cause receptor-induced caspase-8 and -3 activation (a typical response to oxidative stress) leading to apoptosis in some cancer cells, and apoptosis caspase-9 activation by mitochodrial stress in others.[94]

Several epidemiological studies have shown the consumption of vitamin E-rich foods to be inversely associated with cancer risk, and longitudinal studies that have compared circulating tocopherol levels and cancer risk have found those levels to be slightly lower (about 3%) in cancer patients than in healthy controls. A difference in this magnitude was observed between cancer patients and controls in a large trial in Finland,[95] which also found individuals with relatively low serum α-tocopherol levels to have a 1.5-fold greater risk of cancer than those with higher serum vitamin E levels. The results of most studies to date are inconsistent with respect to relationships of vitamin E and incidence of site-specific cancers. Available data suggest that vitamin E intake may be related to reduced risk of cancers of the lung, pancreas, head and neck, cervix, endometrium, and melanoma. Evidence is mixed for a relationship of α-tocopherol intake and prostate cancer, with risk reductions apparent in the early stage disease and perhaps in smokers.

Few clinical intervention trials have been conducted to evaluate the cancer-preventive potential of supplemental vitamin E. One study found the combination of vitamin E, selenium, and β-carotene to reduce lung cancer risk by an apparent 45%.[96] Some small trials have suggested that vitamin E may reduce the risk of breast cancer among women with mammary dysplasia; however, those results have not been confirmed in larger trials of benign breast disease. A study with more than 29,000 male smokers found vitamin E treatment (50 mg α-tocopherol per day) to be associated with a 34% reduction in prostate cancer incidence.[97] Two subsequent trials found supplemental vitamin E and/or vitamin C to be without effect on prostate cancer incidence.[98]

Other Conditions

Vitamin E has frequently proven effective as a therapeutic measure in several disorders in the etiologies of which it may not have direct involvement. These include hemolytic anemia of prematurity, intermittent claudication,[99] and chronic hemolysis in patients with glucose-6-phosphate dehydrogenase deficiency. In veterinary practice, vitamin E (most frequently administered with selenium) has had reported efficacy in the treatment of tying up[100] in horses, and postpartum placental retention in dairy cows. Formerly, vitamin E was thought to protect against retinopathy of prematurity;[101] however, a controlled clinical trial has shown that its use failed to reduce the prevalence of the syndrome.

Use in Food Animals

High-level vitamin E supplementation (e.g., α-tocopheryl acetate fed at levels 10- to 50-fold standard practice) of the diets of poultry, swine, and beef have been found effective in increasing the α-tocopherol contents of many tissues. High levels of the vitamin in edible tissues serve to inhibit *post-mortem* oxidative production of off-flavors (oxidative rancidity of lipids) and color (hemoglobin oxidation). This increases the effective shelf-life of retail cuts of meat.

8. VITAMIN E DEFICIENCY

Vitamin E deficiency can result from insufficient dietary intake or impaired absorption of the vitamin. Several other dietary factors affect the need for vitamin E. Two are most important in this regard: selenium and PUFAs. Selenium spares the need for vitamin E; accordingly, animals fed low-selenium diets generally require more vitamin E than animals fed the same diets supplemented with an available source of selenium. In contrast, the dietary intake of PUFAs directly affects the need for vitamin E; animals fed high-PUFA diets require more vitamin E than those fed low-PUFA diets.[102] Other factors that can be expected to increase vitamin E needs are deficiencies of sulfur-containing amino acids;[103] deficiencies of copper,

---

94. See review: Sylvester, P. W. (2007). Vitamin E and apoptosis. *Vitamins Hormones* 76, 329.

95. That is, the Finnish Mobile Clinic Health Survey, which involved more than 36,000 subjects; Kneckt, P., Aromaa, A., Maatela, J., *et al.* (1991). *J. Clin. Nutr.* 53, 283S.

96. Blot, W. J., Li, J. Y., Taylor, P. R., *et al.* (1994). *J. Natl Cancer Inst.* 85, 1483.

97. The ATBC Trial (Alpha-Tocopherol, Beta Carotene Cancer Prevention Study Group [1994] *N. Eng. J. Med.* 330, 1029).

98. The HOPE-TOO Trial (Lonn, E., Bosch, J., Yusuf, S., *et al.* [2005] *J. Am. Med. Assoc.* 293, 1338); The Physicians's Health Study II (Gaziano, J. M., Glynn, R. J., and Christen, W. G. [2009] *J. Am. Med. Assoc.* 301, 52.)

99. Nocturnal leg cramps.

100. That is, rhabdomyolosis involves the breakdown of striated muscle after exercise, characterized by soreness in the gluteal muscles and painful/stiff gait. It most frequently results from having had limited exercise or having been put in a stressful environment.

101. This disorder was formerly called *retrolental fibroplasia*. Its pathogenesis involves exposure to a hyperoxic environment during neonatal oxygen therapy. Retrolental fibroplasia can affect as many as 11% of infants with birth weights below 1500 g, resulting in blindness in about one-quarter of them.

102. Various researchers have suggested values for the incremental effects of dietary PUFA level on the nutritional requirement for vitamin E in the range of 0.18–0.60 mg of α-tocopherol per gram of PUFA. Although the upper end of that range is frequently cited as a guideline for estimating vitamin E needs, it is fair to state that there is no consensus among experts in the field as to the quantitation of this obviously important relationship.

103. Cysteine, which can be synthesized via transsulfuration from methionine, is needed for the synthesis of glutathione, i.e., the substrate for the selenium-dependent glutathione peroxidase.

zinc, and/or manganese;[104] and deficiency of riboflavin.[105] Alternatively, vitamin E can be replaced by several lipid-soluble synthetic antioxidants[106] (e.g., BHT,[107] BHA,[108] DPPD[109]) and, possibly, by vitamin C.[110]

Conditions involving the malabsorption of lipids can also lead to vitamin E deficiency (Table 7.17). Such conditions include those resulting in loss of pancreatic exocrine function (e.g., pancreatitis, pancreatic tumor, nutritional pancreatic atrophy in severe selenium deficiency), those involving a lumenal deficiency of bile (e.g., biliary stasis due to mycotoxicosis, biliary atresia), and those due to defects in lipoprotein metabolism (e.g., **abetalipoproteinemia**[111]). Premature infants, who are typically impaired in their ability to utilize dietary fats, are also at risk of vitamin E deficiency.

The clinical manifestations of vitamin E deficiency vary considerably between species. In general, however, the targets are the *neuromuscular*,[112] *vascular*, and *reproductive systems*. The various signs of vitamin E deficiency are believed to be manifestations of membrane dysfunction resulting from the oxidative degradation of polyunsaturated membrane phospholipids and/or the disruption of other critical cellular processes.[113] Many deficiency syndromes (e.g., **encephalomalacia** in the chick

---

104. These are essential cofactors of the superoxide dismutases.

105. Riboflavin is required for the synthesis of FAD, the coenzyme for glutathione reductase, which is required for regeneration of reduced glutathione via the so-called *glutathione cycle*.

106. The fact that vitamin E can be replaced by a variety of antioxidants was a point of debate concerning the status of the nutrient as a vitamin. It should be noted that although such replacement can occur, the effective levels of other antioxidants are considerably greater (e.g., two orders of magnitude) than those of α-tocopherol. Therefore, it is now generally agreed that although the metabolic role of vitamin E is that of an antioxidant that can be fulfilled by other reductants, tocopherol performs this function with high biological specificity, and is therefore, appropriately, considered a vitamin.

107. Butylated hydroxytoluene.

108. Butylated hydroxyanisole.

109. *N,N'*-Diphenyl-*p*-phenylenediamine.

110. The sparing effect of vitamin C may involve its function in the reductive recycling of tocopherol.

111. Humans with this rare hereditary disorder are unable to produce apoprotein B, an essential component of chylomicra, VLDLs, and LDLs. The absence of these particles from the serum prevents the absorption of vitamin E owing to the inability to transport it into the lymphatics. These patients show generalized lipid malabsorption with **steatorrhea** (i.e., excess fat in feces), and have undetectable serum vitamin E levels.

112. The skeletal myopathies of vitamin E-deficient animals entail lesions predominantly involving type I fibers.

113. It is interesting to note a situation in which vitamin E deficiency would appear advantageous: the efficacy of the antimalarial drug derived from Chinese traditional medicine, *qinghaosu* (artemisinin), is enhanced by deprivation of vitamin E. The drug, an endoperoxide, is thought to act against the plasmodial parasite by generating free radicals *in vivo*. Thus, depriving the patient of vitamin E appears to limit the parasite's access to the protective antioxidant.

TABLE 7.17 Signs of Vitamin E Deficiency

Organ System	Sign	Responds to: Vitamin E	Responds to: Selenium	Responds to: Antioxidants
General	Loss of appetite	+	+	+
	Reduced growth	+	+	+
Dermatologic	None			
Muscular	Myopathies			
	Striated muscles[a]	+	+	
	Cardiac muscle[b]	+		
	Smooth muscle[c]	+	+	
Skeletal	None			
Vital organs	Liver necrosis[d]	+	+	
	Renal degeneration[d]	+		+
Nervous system	Encephalo-malacia[e]	+		+
	Areflexia, ataxia[f]	+		
Reproduction	Fetal death[g]	+	+	+
	Testicular degeneration[h]	+	+	
Ocular	Cataract[i]	+		
	Retinopathy[j]	+ ?		
Vascular	Anemia[j,k]	+		
	RBC hemolysis[l]	+		
	Exudative diathesis[e]	+	+	
	Intraventricular hemorrhage[j]	+		

[a]*Nutritional muscular dystrophies (white muscle diseases) of chicks, rats, guinea pigs, rabbits, dogs, monkeys, minks, sheep, goats, and calves.*
[b]**Mulberry heart disease** *(congested heart failure) of pigs.*
[c]*Gizzard* **myopathy** *of turkeys and ducks.*
[d]*In rats, mice, and pigs.*
[e]*In chicks.*
[f]*In humans with abetalipoproteinemia.*
[g]*In rats, cattle, and sheep.*
[h]*In chickens, rats, rabbits, hamsters, dogs, pigs, and monkeys.*
[i]*Reported only in rats.*
[j]*Low vitamin E status is suspected in this condition in premature human infants.*
[k]*In monkeys, pigs, and humans.*
[l]*In chicks, rats, rabbits, and humans.*

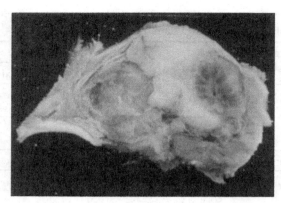

FIGURE 7.14 Encephalomalacia in a vitamin E-deficient chick.

FIGURE 7.15 Nutritional muscular dystrophy in a vitamin E-deficient chick; Zenker's degeneration of *m. pectorales* gives a striated appearance.

(Fig. 7.14), intraventricular hemorrhage in the premature human infant, and at least some myopathies (Fig. 7.15)) appear to involve local cellular anoxia resulting from primary lesions of the vascular system. Others appear to involve the lack of protection from oxidative stress. It has also been proposed that some effects (e.g., impaired immune cell functions) may involve loss of control of the oxidative metabolism of arachidonic acid in its conversion to leukotrienes; vitamin E is known to inhibit the 5'-lipoxygenase in that pathway.

9. VITAMIN E TOXICITY

Vitamin E has been viewed as one of the *least toxic* of the vitamins. Both animals and humans appear to be able to tolerate rather high levels. For animals, doses at least two orders of magnitude above nutritional requirements (e.g., to 1,000–2000 IU/kg) are without untoward effects. For humans, daily doses as high as 400 IU have been considered to be harmless, and large oral doses, as great as 3,200 IU, have not been found to have consistent ill effects. These views were challenged a few years ago by a meta-analysis (of 19 trials) suggesting that vitamin E supplements (≥400 IU/day) may increase all-cause mortality. A more recent meta-analysis, which included a larger number (57) of published trial results, concluded that supplemental vitamin E does does *not* affect all-cause mortality at doses up to 5,500 IU/day.[114]

It is known that at very high doses vitamin E can antagonize the functions of other fat-soluble vitamins. Hypervitaminotic E animals have been found to show impaired bone mineralization, reduced hepatic storage of vitamin A, and coagulopathies. In each case, these signs could be corrected with increased dietary supplements of the appropriate vitamin (i.e., vitamins D, A, and K, respectively), and the antagonism seemed to be based on the level of absorption. Isolated reports of negative effects in human subjects consuming up to 1,000 IU of vitamin E per day included headache, fatigue, nausea, double vision, muscular weakness, mild creatinuria, and gastrointestinal distress.

Potentially deleterious metabolic effects of high-level vitamin E status include inhibitions of retinyl ester hydrolase and vitamin K-dependent carboxylations. The former effect has been demonstrated in animals, where it results in impaired ability to mobilize vitamin A from hepatic stores. Supranutritional vitamin E treatment of rats has been shown to increase bleeding tendency; however, there is little evidence for comparable effects in humans. Patients given high doses of α-tocopherol (1,200 IU/day) have shown increased blood clotting times due to hypoprothrombinemia, but that effect may be of concern only for patients on anticoagulant therapy. Although the metabolic basis of the effect is not clear, it is likely to involve inhibition of vitamin K metabolism by virtue of their similar phytyl side chains (competing for the CYP4F2-dependent ω-hydroxylation of the vitamin K_1 side chain to form menadione) and/or of the structural similarities of the tocopherylquinone and vitamin K head groups (inhibiting the conversion of vitamin K_1 to MK-4; inhibiting the vitamin K-dependent carboxylase) (see Chapter 8, page 218 "Metabolism of Vitamin K").

10. CASE STUDIES

Instructions

Review the following case reports, paying special attention to the diagnostic indicators on which the treatments were based. Then answer the questions that follow.

114. Abner, E. L. Schmitt, F. A., and Mendiondo, M. S. (2011). *Curr. Aging Sci.* 4, 1.

Case 1

At birth, a male infant with *acidosis*[115] and **hemolytic anemia**[116] was diagnosed as having *glutathione (GSH) synthetase*[117] deficiency associated with *5-keto-prolinuria*;[118] he was treated symptomatically. During his second year, he experienced six episodes of bacterial *otitis media*.[119] His white cell counts fell to 3,000–4,000 cells/μl during two of these infections, with notable losses of polymorphonuclear leukocytes[120] (PMNs). Between infections the child had normal white and differential cell counts, and PMNs were obtained for study.

Functional studies of PMNs showed the following results:

Parameter	Finding
GSH synthetase activity	10% of normal
Phagocytosis of *Staphylococcus aureus*	Less than normal
Iodination of phagocytized zymosan particles	Much less than normal
H_2O_2 production during phagocytosis	Well above normal

The child was then treated daily with 30 IU of all-*rac*-α-tocopheryl acetate per kilogram body weight (about 400 IU/day). His plasma vitamin E concentration rose from 0.34 mg/dl (normal for infants) to 1.03 mg/dl. After 3 months of treatment, the same studies of his PMNs were performed. Although there were no changes in the activity of GSH peroxidase[121] or the concentration of GSH (which remained near 25% of normal during this study) in his plasma, the production of H_2O_2 by his PMNs had declined to normal levels, the iodination of proteins during phagocytosis had increased, and the bactericidal activity toward *S. aureus* had increased to the control level. Before his

vitamin E therapy, electron microscopy of his neutrophils had revealed defective cytoskeletal structure, with more than the usual number of *microtubules*[122] seen at rest, and a disappearance of microtubules seen during phagocytosis. This ultrastructural defect was corrected after vitamin E treatment.

Case 2

A 23-year-old woman with a 10-year history of neurologic disease was admitted complaining of severe ataxia,[123] titubation of the head,[124] and loss of proprioceptive sense in her extremities.[125] Her past history revealed that she had experienced difficulty in walking and was unsteady at age 10 years; there was no family history of ataxia, malabsorption or neurologic disease. At 18 years of age she had been hospitalized for her neurologic complaints; at that time, she had been below the 5th percentile for both height and weight. Her examination had revealed normal higher intellectual function, speech, and cranial nerve function, but her limbs had been found to be hypotonic[126] with preservation of strength and moderately severe ataxia. Her deep tendon reflexes were absent, plantar responses were abnormal, vibrational sense was absent below the wrists and iliac crests, and joint position sense was defective at the fingers and toes. Laboratory findings at that time had been negative – i.e., she showed no indications of hepatic or renal dysfunction. No etiologic diagnosis was made. Two years later, when she was 20 years old, the patient was re-evaluated. By that time, her gait had deteriorated and her proprioceptive loss had become more severe.

Over the next 3 years her symptoms worsened, and by age 23 years she had trouble walking unassisted. Still, she showed no sensory, visual, bladder, respiratory or cardiac signs, and ate a normal and nutritious diet. Her only gastrointestinal complaint was of constipation, with bowel movements only once per week. Nerve conduction tests revealed that the action potentials of both her sensory and motor nerves, recorded from the median and ulnar nerves, were normal. Electromyography of the biceps, vastas medialis, and tibialis anterior muscles was normal. However, her cervical and cortical somatosensory-evoked responses to median nerve stimulation were abnormal: there was no peripheral delay, and the nature of the response was abnormal. Further, no consistent cortical responses could be recorded after stimulation of the tibial nerve at the ankle. These findings were interpreted as

---

115. The condition of reduced alkali reserve.
116. Reduced number of erythrocytes per unit blood volume, resulting from their destruction.
117. This is the rate-limiting enzyme in the pathway of the biosynthesis of **glutathione (GSH)**, a tripeptide of glycine, cysteine, and glutamic acid, and the most abundant cellular thiol compound. **Oxidized glutathione (GSSG)** is a dimer joined by a disulfide bridge between the cysteinyl residues.
118. This is the condition of abnormally high urinary concentrations of 5-ketoproline, the intermediate in the pathway of GSH biosynthesis (the γ-glutamyl cycle).
119. Inflammation of the middle ear.
120. The PMN is a type of white blood cell, important in disease resistance, which functions by phagocytizing bacteria and other foreign particles.
121. An enzyme that catalyzes the reduction of hydroperoxides (including H_2O_2) with the concomitant oxidation of glutathione (two GSH converted to GSSG).

---

122. A subcellular organelle.
123. Loss of muscular coordination.
124. Unsteadiness.
125. Senses of position, etc., originating from the arms and legs.
126. Having abnormally low tension.

indicating spinocerebellar disease characterized by delayed sensory conduction in the posterior columns.

Routine screening tests failed to detect α-tocopherol in her plasma, although she showed elevated circulating levels of cholesterol (448 mg/dl versus normal: 150–240 mg/dl) and triglycerides (184 mg/dl versus normal: 50–150 mg/dl). Her plasma concentrations of 25-hydroxyvitamin D_3 [25-(OH)-D_3], retinol, and vitamin K-dependent clotting factors were within normal ranges. Tests of lipid malabsorption showed no abnormality. Her glucose tolerance and pancreatic function (assessed after injections of cholecystokinin and secretin) were also normal.

The patient was given 2 g of α-tocopheryl acetate with an ordinary meal; her plasma α-tocopherol level, which had been non-detectable before the dose, was in the subnormal range 2 hours later and she showed a relatively flat absorption[127] curve. She was given the same large dose of the vitamin daily for 2 weeks, at which time her plasma α-tocopherol concentration was 24 μg/ml. When her daily dose was reduced to 800 mg of α-tocopheryl acetate per day for 10 weeks, her plasma level was 1.2 mg/dl (i.e., in the normal range). During this time, she showed marked clinical improvement.

Case Questions and Exercises

1. What inborn metabolic error(s) was (were) apparent in the first patient?
2. What sign/symptom indicated a vitamin E-related disorder in each case?
3. Why are PMNs useful for studying protection from oxidative stress, as in the first case?
4. What inborn metabolic error might you suspect led to vitamin E deficiency in the second patient?

Study Questions and Exercises

1. Construct a concept map illustrating the nutritional interrelationships of vitamin E and other nutrients.
2. Construct a decision tree for the diagnosis of vitamin E deficiency in a human or animal.
3. What features of the chemical structure of vitamin E relate to its nutritional activity?
4. How might vitamin E utilization be affected by a diet high in polyunsaturated fat? A fat-free diet? A selenium-deficient diet?
5. What kinds of prooxidants might you expect people or animals to encounter daily?
6. How can nutritional deficiencies of vitamin E and selenium be distinguished?

127. That is, plasma α-tocopherol concentration versus time.

RECOMMENDED READING

Atkinson, J., Harroun, T., Wassall, S.R., et al. 2010. The location and behavior of α-tocopherol in membranes. Mol. Nutr. Food Res. 54, 641–651.
Azzi, A., 2004. The role of α-tocopherol in preventing disease. Eur. J. Nutr. 43, 18–25.
Azzi, A., 2007. Molecular mechanism of α-tocopherol action. Free Rad. Biol. Med. 46, 16–21.
Azzi, A., Gysin, R., Kempná, P., et al. 2003. The role of a-tocopherol in preventing disease: From epidemiology to molecular events. Mol. Aspects Med. 24, 325–336.
Azzi, A., Gysin, R., Kempná, P., et al. 2004. Regulation of gene expression by alpha-tocopherol. J. Biol. Chem. 385, 585–591.
Banks, R., Speakman, J.R., Selman, C., 2010. Vitamin E supplementation and mammalian lifespan. Mol. Nutr. Food Res. 54, 719–725.
Birringer, M., 2010. Analysis of vitamin E metabolites in biological specimen. Mol. Nutr. Food Res. 54, 588–598.
Brigelius-Flohé, R., 2009. Vitamin E: The shrew waiting to be tamed. Free Rad. Biol. Med. 46, 543–554.
Constantinou, C., Papas, A., Constantinou, A.I., 2008. Vitamin E and cancer: An insight into the anticancer activities of vitamin E isomers and analogs. Intl J. Cancer 123, 739–752.
Cutler, R.G., 2005. Oxidative stress profiling: Part I. Its potential importance in the optimization of human health. Ann. NY Acad. Sci. 1055, 93–135.
Cutler, R.G., Plummer, J., Chowdhury, K., Heward, C., 2005. Oxidative stress profiling. Part II. Theory, technology and practice. Ann. NY Acad. Sci. 1055, 136–158.
Di Donato, I., Bianchi, S., Federico, A., 2010. Ataxia with vitamin E deficiency: Update of molecular diagnosis. Neurol. Sci. 31, 511–515.
Eitenmiller, R., Lee, J., 2004. Vitamin E: Food Chemistry, Composition, and Analysis. Marcel Dekker, New York, NY., 530 pp.
Frei, B., 2004. Efficacy of dietary antioxidants to prevent oxidative damage and inhibit chronic disease. J. Nutr. 134, 3196S–3198S.
Fukuzawa, K., 2008. Dynamics of lipid peroxidation and antioxidation of α-tocopherol in membranes. J. Nutr. Sci. Vitaminol. 54, 273–285.
Gille, L., Staniek, K., Rosenau, T., Kozlov, A.V., 2010. Tocopheryl quinones and mitochondria. Mol. Nutr. Food Res. 54, 601–615.
Gohil, K., Vasu, V.T., Cross, C.E., 2010. Dietary α-tocopherol and neuromuscular health: Search for optimal dose and molecular mechanisms continues! Mol. Nutr. Food Res. 54, 693–709.
Gray, B., Swick, J., Ronnenberg, A.G., 2011. Vitamin E and adiponectin: Proposed mechanism for vitamin E-induced improvement in insulin sensitivity. Nutr. Rev. 69, 155–161.
Huebbe, P., Lodge, J.K., Rimbach, G., 2010. Implications of apolipoprotein E genotype on inflammation and vitamin E status. Mol. Nutr. Food Res. 54, 623–630.
Ju, J., Picinich, S.C., Yang, Z., et al. 2010. Cancer-preventive activities of tocopherols and tocotrienols. Carcinogenesis 31, 533–542.
Kelly, F., Meydani, M., Packer, L. (Eds.), 2004. Vitamin E and Health. New York Academy of Sciences, New York, NY, 463 pp.
Kline, K., Lawson, K.A., Yu, W., Sanders, B.G., 2007. Vitamin E and cancer. Vitam. Horm. 76, 435–461.
Lemaire-Ewing, S., Desrumaux, C., Néel, D., Lagrost, L., 2010. Vitamin E transport, membrane incorporation and cell metabolism: Is α-tocopherol in lipid rafts an oar in the lifeboat? Mol. Nutr. Food Res. 54, 631–640.

Manor, D., Morley, S., 2007. The α-tocopherol transfer protein. Vitam. Horm. 76, 45–65.

Mardones, P., Attilo, R., 2004. Cellular mechanisms of vitamin E uptake: Relevance in a-tocopherol metabolism and potential implications for disease. J. Nutr. Biochem. 15, 252–260.

Mène-Saffrané, L., DellaPenna, D., 2010. Biosynthesis, regulation and functions of tocochromanols in plants. Plant Physiol. Biochem. 48, 301–309.

Min, K.C., 2007. Structure and function of α-tocopherol transfer protein: Implications for vitamin E metabolism and AVED. Vitam. Horm. 76, 23–44.

Muller, D.P.R., 2010. Vitamin E and neurological function. Mol. Nutr. Food Res. 54, 710–718.

Munteanu, A., Zingg, J.M., 2007. Cellular, molecular and clinical aspects of vitamin E and atherosclerosis prevention. Mol. Aspects Med. 28, 538–590.

Ohnmacht, S., Nava, P., West, R., et al. 2008. Inhibition of oxidative metabolism of tocopherols with ω-N-heterocyclic derivatives of vitamin E. Bioorg. Med. Chem. 16, 7631–7638.

Pazdro, R., Burgess, J.R., 2010. The role of vitamin E and oxidative stress in diabetes complications. Mech. Ageing Dev. 131, 276–286.

Powers, S.K., Duarte, J., Kavazis, A.N., Talbert, E.E., 2009. Reactive oxygen species are signalling molecules for skeletal muscle adaptation. Exp. Physiol. 95, 1–9.

Quinn, P.J., 2007. Molecular associations of vitamin E. Vitam. Horm. 76, 67–98.

Reiter, E., Jiang, Q., Christen, S., 2007. Anti-inflammatory properties of α- and γ-tocopherol. Mol. Aspects Med. 28, 668–691.

Ricciarelli, R., Argellati, F., Pronzator, M.A., Domenicotti, C., 2007. Vitamin E and neurodegenerative diseases. Mol. Aspects Med. 28, 591–606.

Rigotti, A., 2007. Absorption, transport and tissue delivery of vitamin E. Mol. Aspects Med. 28, 423–436.

Rimbach, G., Minihane, A.M., Majewicz, J., et al. 2002. Regulation of cell signalling by vitamin E. Proc. Nutr. Soc. 61, 415–425.

Ristow, M., Zarse, K., 2010. How increased oxidative stress promotes longevity and metabolic health: The concept of mitochondrial hormesis (mitohormesis). Exp. Gerontol. 45, 410–418.

Schaffer, S., Müller, W.E., Eckert, G.P., 2005. Tocotrienols: Constitutional effects in aging and disease. J. Nutr. 135, 151–154.

Sen, C.K., Khanna, S., Roy, S., 2007. Tocotrienols in health and disease: The other half of the natural vitamin E family. Mol. Aspects Med. 28, 692–728.

Singh, U., Devaraj, S., 2007. Vitamin E: Inflammation and atherosclerosis. Vitam. Horm. 76, 519–549.

Sohal, R.S., Mockett, R.J., Orr, W.C., 2002. Mechanisms of aging: An appraisal of the oxidative stress hypothesis. Free Rad. Biol. Med. 5, 575–586.

Sylvester, P.W., 2007. Vitamin E and apoptosis. Vitam. Horm. 76, 329–356.

Takada, T., Suzuki, H., 2010. Molecular mechanisms of membrane transport of vitamin E. Mol. Nutr. Food Res. 54, 616–622.

Theile, J.J., Ekanayake-Mudiyanselage, S., 2007. Vitamin E in human skin: Organ-specific physiology and considerations for its use in dermatology. Mol. Aspects Med. 28, 646–667.

Traber, M.G., 2007. Vitamin E regulatory mechanisms. Ann. Rev. Nutr. 27, 347–362.

Traber, M.G., 2007. Vitamin E. In: Zempleni, J., Rucker, R.B., McCormick, D.B., Suttie, J.W. (Eds.), Handbook of Vitamins, fourth ed. CRC Press, New York, NY, pp. 154–174.

Traber, M.G., 2010. Regulation of xenobiotic metabolism, the only signaling function of α-tocopherol? Mol. Nutr. Food Res. 54, 661–668.

Traber, M.G., Atkinson, J., 2007. Vitamin E, antioxidant and nothing more. Free Rad. Biol. Med. 46, 4–15.

Traber, M.G., Hiroyuki, A., 1999. Molecular mechanisms of vitamin E transport. Ann. Rev. Nutr. 19, 343–355.

Villacorta, L., Azzi, A., Zingg, J.M., 2007. Regulatory role of vitamins E and C on extracellular matrix components of the vascular system. Mol. Aspects Med. 28, 507–537.

Wu, J.H., Croft, K.D., 2007. Vitamin E metabolism. Mol. Aspects Med. 28, 437–452.

Zingg, J.M., 2007. Modulation of signal transduction by vitamin E. Mol. Aspects Med. 28, 481–506.

Zingg, J.M., Azzi, A., Meydani, M., 2008. Genetic polymorphisms as determinants for disease-preventive effects of vitamin E. Nutr. Rev. 66, 406–414.

Zingg, J.M., Meydani, M., Azzi, A., 2010. α-Tocopheryl phosphate – an active lipid mediator? Mol. Nutr. Food Res. 54, 679–692.

Vitamin K

Anchoring Concepts

1. *Vitamin K* is the generic descriptor for 2-methyl-1, 4-naphthoquinone and all its derivatives exhibiting qualitatively the antihemorrhagic activity of phylloquinone.
2. The vitamers K are side chain homologs; each is hydrophobic, and thus insoluble in such aqueous environments as plasma, interstitial fluids, and cytoplasm.
3. The 1,4-naphthoquinone ring system of vitamin K renders it susceptible to metabolic reduction.
4. Deficiencies of vitamin K have a narrow clinical spectrum: hemorrhagic disorders.

... Then Almquist showed A substance, phthiocol, from dread T.B., Would cure the chicks ... And so the microbes of tuberculosis, That killed the poet Keats by hemorrhage, Has yielded forth the clue to save the lives Of infants bleeding shortly after birth.

T. H. Jukes

Learning Objectives

1. To understand the nature of the various sources of vitamin K.
2. To understand the means of absorption and transport of the vitamers K.
3. To understand the metabolic functions of vitamin K.
4. To understand the physiologic implications of impaired vitamin K function.

VOCABULARY

Atherocalcin
γ-Carboxyglutamate Chloro-K
Coagulopathies
Collagen
Coprophagy
Coumarins
Dicumarol
Dysprothrombinemia
Extrinsic clotting system
Factor II
Factor VII
Factor IX
Factor X
Fibrin
Fibrinogen
Gas6
Gla
Hemorrhage
Hemorrhagic disease of the newborn
Heparin
Hydroxyvitamin K
Hypoprothrombinemia
Intrinsic clotting system
Matrix Gla protein (MGP)
Menadione
Menadione pyridinol bisulfite (MPB)
Menadione sodium bisulfite complex
Menaquinones (MKs)
Naphthoquinone
Osteocalcin
Phylloquinones
PIVKA (protein induced by vitamin K absence)

The Vitamins. DOI: 10.1016/B978-0-12-381980-2.00008-6

Protein C
Protein M
Protein S
Protein Z
Serine protease
Stuart factor
Superoxide
Thrombin (factor IIa)
Thromboplastin
Vitamin K deficiency bleeding (VKDB)
Vitamin K-dependent carboxylase
Vitamin K epoxide
Vitamin K hydroquinone
Vitamin K oxide
Warfarin
Zymogens

1. THE SIGNIFICANCE OF VITAMIN K

Vitamin K is synthesized by plants and bacteria, which use it for electron transport and energy production. Animals, however, cannot synthesize the vitamin; still, they require it for blood clotting, bone formation, and other functions. These needs are critical to good health; yet the prevalent microbial synthesis of the menaquinones, including that occurring in the hindgut of humans and other animals (many of whom have coprophagous eating habits), results in frank deficiencies of this vitamin being rare. Nevertheless, vitamin K deficiency can occur in poultry and other monogastric animals when they are raised on wire or slatted floors and treated with certain antibiotics that reduce their hindgut microbial synthesis of the vitamin. Human neonates, particularly premature infants, can also be at risk of hemorrhagic disease by virtue of limited transplacental transfer of the vitamin.

The function of vitamin K in blood clotting is widely exploited to reduce risks to post-surgical thrombosis and cardiac patients. **Coumarin**-based drugs (e.g., **warfarin**, **dicumarol**) and other inhibitors of the vitamin K oxidation/carboxylation/reduction cycle are valuable in this purpose. In addition, vitamin K has clear roles in the metabolism of both calcified and non-calcified tissues. It may well prove that vitamin K functions with other vitamins in the regulation of intracellular Ca^{2+} metabolism, in signal transduction, and in cell proliferation, which functions have profound effects on health status.

2. SOURCES OF VITAMIN K

Vitamers K

There are two natural and one synthetic source of vitamin K:

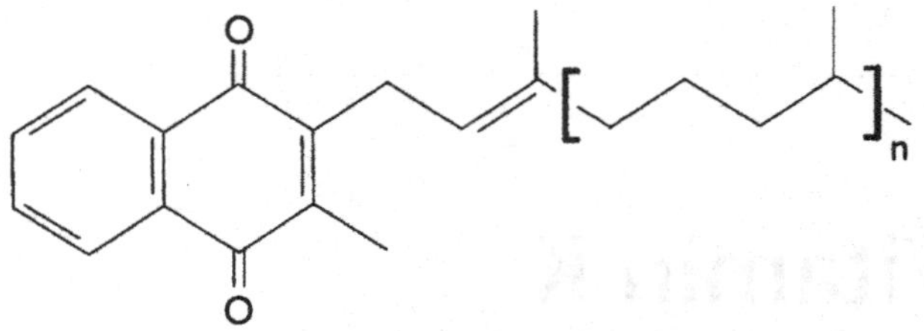

- *Phylloquinones* (formerly, K_1s) (Fig. 8.1). Green plants synthesize the phylloquinones (2-methyl-3-

Phylloquinones

FIGURE 8.1 Phylloquinones.

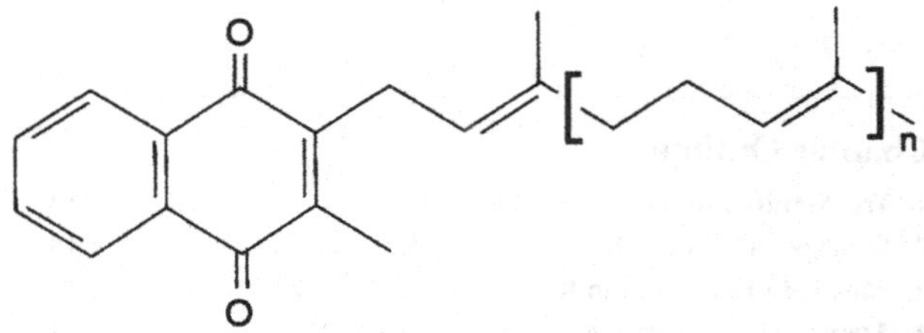

Menaquinones

FIGURE 8.2 Menaquinones.

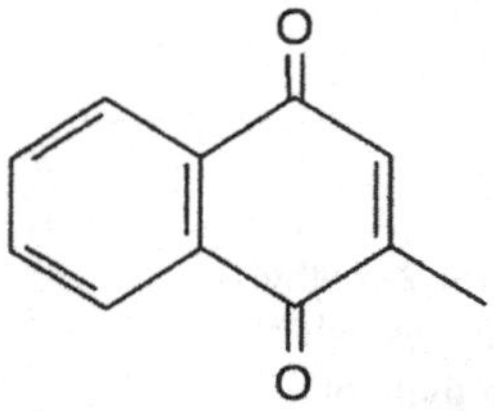

Menadione

FIGURE 8.3 Menadione.

phytyl-1,4-naphthoquinones) as a normal component of chloroplasts.
- *Menaquinones* (MKs, formerly K_2s) (Fig. 8.2). Bacteria (including those of the normal intestinal microflora) and some spore-forming *Actinomyces* spp. synthesize the menaquinones. The predominant vitamers of this series contain 6–10 isoprenoid units; however, forms with as many as 13 units have been identified.
- *Menadione* (Fig. 8.3). The formal parent compound of the menaquinone series does not occur naturally, but is a common synthetic form called menadione (2-methyl-1,4-naphthoquinone). This forms a water-soluble sodium bisulfite addition product, **menadione sodium bisulfite**, the practical utility of which is limited by its instability in complex matrices such as feeds. However, in the presence of excess sodium bisulfite it crystallizes as a complex with an additional mole of sodium bisulfite (i.e., **menadione sodium bisulfite complex**), which has greater stability and therefore is used as a supplement to poultry and swine feeds. A third water-soluble compound is **menadione pyridinol bisulfite** (**MPB**), a salt formed by the addition of dimethylpyridinol.

TABLE 8.1 Vitamin K Contents of Foods

Food	Vitamin K (µg/100 g)	Food	Vitamin K (µg/100 g)
Vegetables		**Fruits**	
Asparagus	39	Apples	4[a]
Beans (mung)	33	Bananas	0.5
Beans (snap)	28	Cranberries	1.4
Beets	5	Oranges	1.3
Broccoli	154	Peaches	3
Cabbage	149	Strawberries	14
Carrots	13	**Meats**	
Cauliflower	191	Beef	0.6
Chick peas	48	Chicken	0.01
Corn	4	**Liver**	
Cucumber	5	Beef	104
Kale	275	Chicken	80
Lettuce	113	Pork	88
Peas	28	**Dairy Products and Eggs**	
Potatoes	0.5	Milk, cow's	4
Spinach	266	Eggs	50
Sweet potatoes	4	Egg yolk	149
Tomatoes	48	**Grains**	
Oils		Oats	63
Canola 830		Rice	0.05
Corn	5	Wheat	20
Olive	58	Wheat bran	83
Peanut	2	Wheatgerm	39
Soybean	200		

[a] *Almost 90% in the skin (peeling).*
Source: USDA National Nutrient Database for Standard Reference, release 18 (http://www.nal.usda.gov/fnic/foodcomp/search/).

Dietary Sources

Green leafy vegetables tend to be rich in vitamin K, whereas fruits and grains are poor sources. The vitamin K activities of meats and dairy products tend to be moderate. Unfortunately, data for the vitamin K contents of foods (Table 8.1) are limited by the lack of good analytical methods. Nevertheless, it is clear that, because dietary needs for vitamin K are low, most foods contribute significantly to those needs.[1] This is not true for breast milk (Table 8.2), which has been found in most studies to be of very low

vitamin K content and insufficient to meet the vitamin K needs of infants.[2]

Bioavailability

Little is known about the bioavailability of vitamin K in most foods. It appears, however, that only about 10%

1. It is difficult to formulate an otherwise normal diet that does not provide about 100 µg of the vitamin per day.

2. In consideration of the low vitamin K contents of breast milk and to prevent hemorrhagic disease, the American Academy of Pediatrics recommended in the early 1960s the intramuscular administration of vitamin K (1 mg) at the time of birth. This practice is now required by law in the United States and Canada. All commercial infant formulas are supplemented with vitamin K at levels in the range of 50–125 ng/ml.

TABLE 8.2 Vitamin K Contents of Human Milk[a]

Sample	Vitamin K (nmol/l)
Colostrum (30–81 h)	7.52 ± 5.90
Mature Milk	
1 month	6.98 ± 6.36
3 months	5.14 ± 4.52
6 months	5.76 ± 4.48

From Canfield, L. M., Hopkinson, J. M., Lima, A. F., et al. (1991). Am. J. Clin. Nutr. 53, 730.

TABLE 8.3 Menaquinones Produced by Several Dominant Species of Enteric Bacteria

Species	Major Components	Minor Components
Bacteroides fragilis	MK-11, MK-10, MK-12	MK-9, MK-8, MK-7
Bacteroides vulgatus	MK-11, MK-10, MK-12	MK-9, MK-8, MK-7
Veillonella sp.	MK-7	MK-6
Eubacterium lentum	MK-6[a]	
Enterobacter sp.	MK-8	MK-8[a]
Enterococcus sp.	MK-9	MK-6,[a] MK-7,[a] MK8[a]

[a] Includes demethylated derivatives.
Source: Mathers, J. C., Fernandez, F., Hill, M. J. et al. (1990). Br. J. Nutr. 63, 639.

of the phylloquione in boiled spinach is absorbed by humans.[3] This may relate to its association with the thylakoid membrane in chloroplasts, as the free vitamer was well absorbed (80%). This suggests that vitamin K may be poorly bioavailable from the most quantitatively important sources of it in most diets: green leafy vegetables.

The relative biopotencies of the various vitamers K differ according to route of administration. Studies using restoration of normal clotting in the vitamin K-deficient chick have shown that, when administered orally, phylloquinone or MK homologs with three to five isoprenoid groups had greater activities than those with longer side chains owing to the relatively poor absorption of the long-chain vitamers. Studies with the vitamin K-deficient rat showed that the long-chain homologs (especially MK-9, i.e., menaquinone with nine isoprenoid units) had the greatest activities when administered intracardially. Of the three synthetic forms, some studies indicate MPB to be somewhat more effective in chick diets; however, each is generally regarded to be comparable in terms of biopotency to phylloquinone.

Intestinal Microbial Synthesis

Menaquinones can be synthesized from shikimic acid and α-ketoglutarate by many facultative and obligate anaerobic bacteria, including some in the human large intestine. These include the obligate anaerobes of the groups Bacteroides,[4] Eubacterium, Propionibacterium, and Arachnia, and the facultative anaerobe Escherichia coli, which produce mostly MK-10 and MK-11 (Table 8.3).

That enteric microbial MK synthesis has nutritional significance is indicated by observations that germ-free animals have greater dietary requirements for the vitamin

TABLE 8.4 Impaired Clotting in Germ-Free Rats

Treatment	Prothrombin Time (s)	Hepatic MK-4 (ng/g)
Germ-Free		
Vitamin K-deficient	∞	8.3 ± 2.3
+MK-4	5.8 ± 1.4[a]	66.5 ± 25.9
+K₃	11.1 ± 2.5 [a]	12.4 ± 2.0
Conventional		
Vitamin K-deficient	12.7 ± 1.6	103.5 ± 44.9
+MK-4	12.5 ± 2.0	207.6 ± 91.3
+K₃	12.6 ± 0.9	216.2 ± 86.5

[a] P <0.05.
Source: Komai, M., Shirakawa, H., and Kimura, S. (1987). Intl J. Vit. Nutr. Res. 58, 55.

than do animals with normal intestinal microfloras (Table 8.4), and that prevention of coprophagy is necessary to produce vitamin K deficiency in some species (e.g., chicks, rats). That the liver normally contains significant amounts of MKs is consistent with the absorption of MKs from the large intestine. In humans, the daily production of MKs by the gut microflora[5] has been thought to exceed the nutritionally required amount by a substantial margin,

3. Gijsbers, B. L., Jie, K. S., and Vermeer, C. (1996). Br. J. Nutr. 76, 223.
4. The most quantitatively important groups are Bacteroides and Bifidobacteria, which collectively comprise more than half of the intestinal microfloral mass. Bifidobacteria do not produce MKs.

5. A quantitative study found total gut MK content in humans to average 1.8 mg (ca. 2,100 nmoles) (Conley and Stein [1992] Am. J. Gastroenterol. 87, 311).

which would explain the difficulty in producing clear signs of vitamin K deficiency in normal subjects.

However, vitamin K-deficiency signs have been reported in individuals not taking antibiotics. This may be explained by interindividual variation in the preponderance of MK-producing species in the intestinal microflora. In fact, a recent study revealed three major human "entero-types" based on the species composition of the human gut microbiome.[6] This showed Bacteroides to be the most abundant and variable genus, being of particularly low abundance in one enterotype. Therefore, it is reasonable to suggest that the contributions of the gut microflora to the vitamin K nutriture of the host, and therefore the die-tary vitamin K needs of the host, may vary according to enterotype.

3. ABSORPTION OF VITAMIN K

Micellar Solubilization

The vitamers K are absorbed from the intestine into the lymphatic (in mammals) or portal (in birds, fishes, and rep-tiles) circulation by processes that first require that these hydrophobic substances be dispersed in the aqueous lumen of the gut via the formation of mixed micelles, in which they are dissolved. Vitamin K absorption therefore depends on normal pancreatic and biliary function. Conditions resulting in impaired lumenal micelle formation (e.g., dietary mineral oil, pancreatic exocrine dysfunction, bile stasis) therefore impair the enteric absorption of vitamin K. It can be expected that any diet will contain a mixture of MKs and phylloquinones. In general, such mixtures appear to be absorbed with efficiencies in the range of 40–70%; however, these vitamers are absorbed via different mechanisms.

Active Transport of Phylloquinone

Studies of the uptake of vitamin K by everted gut sacs of the rat show that phylloquinone is absorbed by an energy-dependent process from the proximal small intestine. The process is not affected by MKs or menadione; it is inhib-ited by the addition of short- and medium-chain fatty acids to the micellar medium.

Other Vitamers Absorbed by Diffusion

In contrast, the MKs and menadione are absorbed strictly via non-carrier-mediated passive diffusion, the rates of which are affected by the micellar contents of lipids and bile salts. This kind of passive absorption has been found

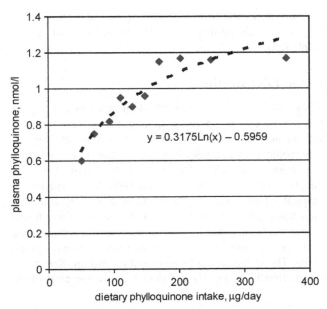

FIGURE 8.4 Relationship of plasma phylloquinone concentration to estimated dietary phylloquinone intake in healthy adults. After *McKeown, N. M., Jacques, P. F., Gundberg, C. M., et al. (2002). J. Nutr. 132, 1329.*

to occur in the distal part of the small intestine as well as in the colon. Thus, non-coprophagous[7] animals appear to profit from the bacterial synthesis of vitamin K in their lower guts by being able to absorb the vitamin from that location.[8]

4. TRANSPORT OF VITAMIN K

Absorbed Vitamin K Transferred to Lipoproteins

On absorption, vitamin K is transported in the lymph in association with chylomicra, whereby it is transported to the liver. Vitamin K is rapidly taken up by the liver via an apolipoprotein E receptor; it has a relatively short half-life there (about 17 hours) before it is transferred to very HDLs and LDLs, which carry it in the plasma. No specific carri-ers have been identified for any of the vitamers K. Plasma levels of phylloquinone reflect dietary intake (Fig. 8.4) and

---

6. Arumugam, M., Raes, J., Pelletier, E., *et al.* (2011). *Nature* 473, 174–180.

---

7. The term **coprophagy** describes the ingestion of excrement. This behavior is common in many species, and exposes them to nutrients such as vitamin K produced by the microbial flora of their lower guts. Coprophagy can be easily prevented in some species (e.g., chicks) by housing with raised wire floors; it is very difficult to prevent in others (e.g., rats) without the use of such devices as tail cups.

8. Humans appear to be able to utilize menaquinones produced by their lower gut microflora. Hypoprothrombinemia is rare *except* among patients given antibiotics, even when vitamin K-free purified diets have been used.

are correlated with those of triglycerides and α-tocopherol; those of healthy humans are in the range of 0.1–0.7 ng/ml (Table 8.5).

Tissue Distribution

When administered as either phylloquinone or MKs, vitamin K is rapidly taken up by the liver (Table 8.6), which is the site of synthesis of the vitamin K-dependent coagulation proteins. In contrast, little menadione is taken up by that organ; instead, it is distributed widely to other tissues. Hepatic storage of vitamin K has little long-term physiological significance, as the vitamin is rapidly removed from that organ and rapidly excreted. The vitamin is found at low levels in many organs; several tend to concentrate it: adrenal glands, lungs, bone marrow, kidneys, lymph nodes. The transplacental movement of vitamin K is poor; the vitamin is frequently not detectable in the cord blood from mothers with normal plasma levels. For this reason, newborn infants are susceptible to hemorrhage.[9]

Most of the vitamin K in liver is as MK-10 and MK-11, with a small portion (10%) as phylloquinones. Extrahepatic tissues of most animals ingesting plant materials contain phylloquinones as well as MKs with 6–13 isoprenoid units in their side chains. Tissues show such mixtures of vitamers K, even when the sole dietary form is vitamin MK-4, indicating that much of the vitamin in tissues is of enteric bacterial origin. Animals fed phylloquinone show MK-4 widely distributed in their tissues, which evidences interconversion of the phytyl side chain to a geranylgeranyl side chain. This metabolic capability has been demonstrated in rat tissues; thus, it appears that the accumulation of MK-4 in extrahepatic tissues of phylloquinone-fed animals may be due to local interconversion. In humans, adipose tissue has been found to store vitamin K at relatively high levels.[10] In each organ, vitamin K is found localized primarily in cellular membranes (endoplasmic reticulum, mitochondria). Under conditions of low vitamin K intake, the vitamin appears to be depleted from membranes more slowly than from cytosol.

TABLE 8.5 Vitamin K Transport in Humans

Fraction	Phylloquinones (% Serum Total)
Triglyceride-rich fraction	51.4 ± 17.0
LDLs	25.2 ± 7.6
HDLs	23.3 ± 10.9

Source: Kohlmeier, M., Solomon, A., Saupe, J., and Shearer, M. J. (1996). J. Nutr. 126, 1192S.

TABLE 8.6 Vitamers K Occurring in Livers of Several Animal Species

Vitamer	Human	Cow	Horse	Dog	Pig
K₁	+++	+	+++		+
MK-4					+
MK-6				+	
MK-7	+++			+	
MK-8	+			+++	+++
MK-9	+			+++	+++
MK-10	+	+++		+++	+++
MK-11	+	+++		+	
MK-12		+++		+	
MK-13				+	

5. METABOLISM OF VITAMIN K

Side Chain Modification

Dealkylation

Tissues contain MKs when the dietary source of vitamin K is phylloquinone. This was once taken as evidence for the metabolic dealkylation of the phylloquinone side chain (converting it to menadione), followed by its re-alkylation to the MKs. That the conversion appears to be greatest when phyolloquinone is taken orally[11] suggested that this dealkylation is performed during intestinal absorption and/or by gut microbes. However, recent findings of MK-4 in tissues of gnotobiotic animals given phylloquinone intraperitoneally (Fig. 8.5) make it clear that the conversion can be performed in several tissues.

Alkylation

The alkylation of menadione (either from a practical feed supplement or produced from microbial degradation of

9. Further, because human milk contains less vitamin K than cow's milk, infants who receive only their mother's milk are more susceptible to hemorrhage than are those who drink cow's milk.
10. Adults undergoing bariatric surgery were found to have phylloquinone levels of 148 ± 72 nmol/kg and 175 ± 112 nmol/kg in subcutaneous and visceral adipose tissue, respectively (Shea, M. K., Booth, S. L., Gundberg, C. M., et al. [2010] J. Nutr. 140, 1029).
11. Thijssen, H. H. W., Vervoort, L. M. T., Schurgers, L. J., and Shearer, M. J. H. (2006). Br. J. Nutr. 95, 260.

phylloquinone) does occur *in vivo*. This step has been demonstrated in chick liver homogenates, where it was found to use geranyl pyrophosphate, farnesyl pyrophosphate or geranylgeranyl pyrophosphate as the alkyl donor in a reaction inhibitable by O_2 or **warfarin**. The main product of the alkylation of menadione is MK-4.

Redox Cycling

Vitamin K is subject to a cycle of oxidation and reduction that is coupled to the carboxylation of peptidyl glutamyl residues to produce various functional γ-carboxylated proteins. The redox cycling of vitamin K (Fig. 8.6) occurs in three steps:

1. *Oxidation of dihydroxyvitamin K to vitamin K 2,3-epoxide.* The production of the 2,3-epoxide, also

called **vitamin K epoxide** or **vitamin K oxide**, is catalyzed by vitamin K γ-glutamyl carboxylase, a 94 kDa protein located in the endoplasmic reticulum and Golgi apparatus. Vitamin K 2,3-epoxide comprises about 10% of the total vitamin K in the normal liver, and can be the predominant form in the livers of rats treated with warfarin or other coumarin anticoagulants. Studies of human, bovine, and rat carboxylase show high (88–94%) sequence homology. The carboxylase itself appears to be a protein containing three γ-carboxyglutamate (Gla) residues per mole of enzyme.[12]

2. *Reduction of the 2,3-epoxide to vitamin K quinone.* This step is catalyzed by **vitamin K epoxide reductase**, a dithiol-dependent, microsomal enzyme inhibited by the coumarin-type anticoagulants. Genetic variability in the vitamin K epoxide reductase has been shown to account for the variability observed in patient responses to warfarin therapy, to affect circulating levels of phylloquinones and undercarboxylated osteocalcin[13] (see "Vitamin K-Dependent Gla Proteins" page 221), and to be associated with cardiovascular risk.[14] Studies have also demonstrated much greater hepatic concentrations

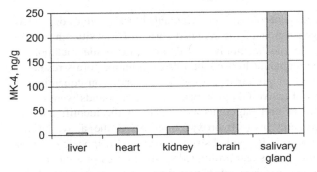

FIGURE 8.5 Evidence of conversion of phylloquinone to menaquinone-4 in gnotobiotic animals: MK-4 levels in tissues of vitamin K-deficient rats treated with phylloquinone by intraperitoneal injection. From *Davidson, R. T., Foley, A. L., Engelke, J. A., and Suttie, J. W. (1998).* J. Nutr. *128, 220.*

12. Berkner, K. L. and Pudota, B. N. (1998). *Proc. Natl. Acad. Sci. USA* 95, 466.
13. Individuals with the minor allele of rs8050894 (G) had significantly higher plasma phylloquinone levels than those with the C allele; GG homozygotes had slightly lower levels of undercarboxylated osteocalcin than other genotypes (Crossier, M. D., Peter, I., Booth, S. L., *et al.* [2009] *J. Nutr. Sci. Vitaminol.* 55, 112).
14. Individuals with the rs8050894 TT genotype had 60% fewer atherothrombotic events than those with other genotypes (Suh *et al.* [2009] *Am. Heart Assoc. J.* 157, 908).

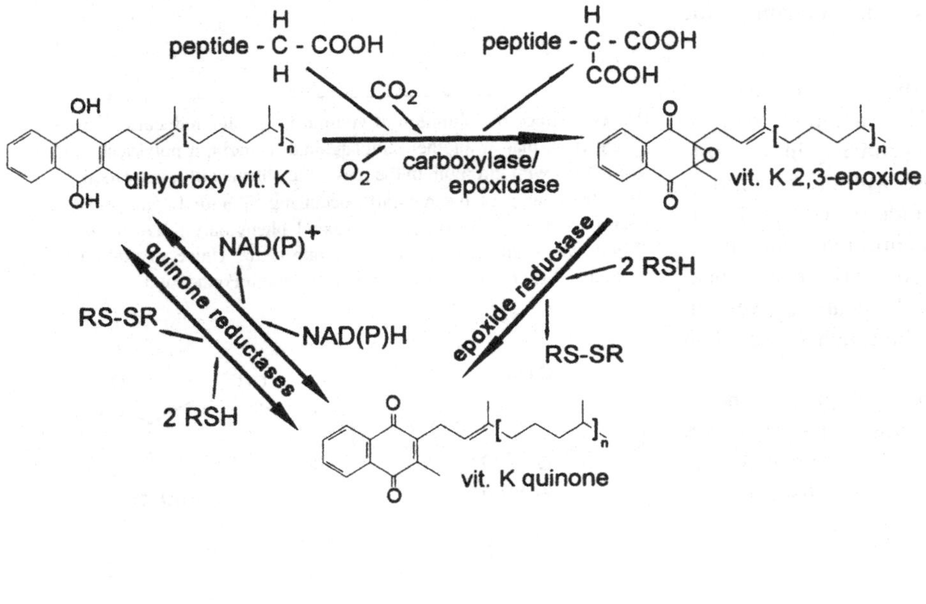

FIGURE 8.6 The vitamin K cycle.

TABLE 8.7 Species Differences in Vitamin K Metabolism[A]

Enzyme	Substrate	V_{max}[a] Chick	V_{max}[a] Rat
Carboxylase	Phylloquinone	14 ± 2	26 ± 1
	MK-4	41 ± 3	40 ± 2
Epoxide reductase	Phylloquinone	26 ± 2	280 ± 2
	MK-4	55 ± 7	430 ± 10

[a]V_{max} is expressed as $\mu mol/min/mg$ protein.
Source: Will, B. H., Usui, U., and Suttie, J. W. (1992). *J. Nutr.* 122, 2354.

of phylloquinone-2,3-epoxide in chicks compared with rats, corresponding (inversely) to a 10-fold difference in their hepatic vitamin K epoxide reductase activities (Table 8.7). These findings suggest that the relatively high dietary requirement of the chick is due to that species' inability to recycle the vitamin effectively.

3. *Reduction of the quinone to the active dihydroxyvitamin K.* This final reductive step can be catalyzed in two ways:

    a. by **vitamin K quinine reductase**, a dithiol-dependent microsomal enzyme that is inhibited by the coumarin-type anticoagulants; or

    b. by **DT-diaphorase**, a microsomal flavoprotein that uses NAD(P)H as a source of reducing equivalents. Unlike the dithiol-dependent enzyme, it is relatively insensitive to coumarin inhibition, such that reduction of vitamin K quinone persists in anticoagulant-treated individuals.

Catabolism

There is no evidence of catabolism of the napthoquinone ring; excretion of vitamers K involves metabolism of the isophytyl side chain.

- *Menadione.* Menadione is rapidly metabolized and excreted, leaving only a relatively minor portion to be converted to MK-4. It is excreted primarily in the urine (e.g., 70% of a physiological dose within 24 hours) as the phosphate, sulfate or glucuronide of menadiol. It is also excreted in the bile as the glucuronide conjugate.
- *Menaquinone.* Little is known about MK metabolism, but it is likely that extensive side chain conversion occurs. Its catabolism appears to be much slower than that of menadione.
- *Phylloquinone.* The total body pool of phylloquinone, *ca.* 100mg in an adult, appears to turn over in about 1.5 days. The most abundant phylloquinone metabolite is its 2,3-epoxide, formed by the vitamin K-dependent

γ-carboxylation of proteins. This and other phylloqinones undergo oxidative shortening of the side chain to 5- or 7-carbon carboxylic acids and a variety of other, more extensively degraded metabolites. A fifth of phylloquinone is ultimately excreted in the urine; however, the primary route of excretion of these metabolites is the feces, which contain glucuronic acid conjugates excreted via the bile.

Both vitamin K epoxide and dihydroxyvitamin K are thought to be degraded metabolically before excretion. That these may have different routes of degradation is suggested by the finding that warfarin treatment greatly increases the excretion of phylloquinone metabolites in the urine while decreasing the amounts of metabolites in the feces. Another ring-altered metabolite has been identified: 3-hydroxy-2,3-dihydrophylloquinone, also called hydroxyvitamin K (Fig. 8.7).

Vitamin K Antagonists

The coumarin-type anticoagulants block the thiol-dependent regeneration of the reduced forms of the vitamin, resulting in the accumulation of the 2,3-epoxide metabolites. The attendant loss of the dihydroquinone forms results in a loss of protein γ-carboxylation, and, consequently, the loss of active Gla-proteins. These compounds were developed as anticoagulants as a result of the identification of 3,3′methylbis-(4-hydroxycoumarin) as the active principle present in spoiled sweet clover, and responsible for the hemorrhages and prolonged clotting times of animals consuming that as feed. Compounds in this family block the thiol-dependent, redox recycling of the vitamin by inhibiting the dithiol-dependent reductases. This results in diminished synthesis of the Gla proteins involved in the clotting pathway (see "Blood Clotting," page 222).[15]

Several substituted 4-hydroxycoumarins have been widely used in anticoagulant therapy in clinical medicine, as well as in rodenticides (effective by causing fatal

---

15. This effect, by inhibition of vitamin K epoxide reductase, is very different from that of another anticoagulant, heparin, a polysaccharide that complexes with thrombin in the plasma to enhance its inactivation.
16. This analog of the naturally occurring vitamin K antagonist, dicumarol, warfarin (4-hydroxy-3-[3-oxo-1-phenylbutyl]-2*H*-1-benzopyran-2-one) was synthesized by Link's group at the University of Wisconsin and named for the Wisconsin Alumni Research Foundation.

Warfarin   dicumarol

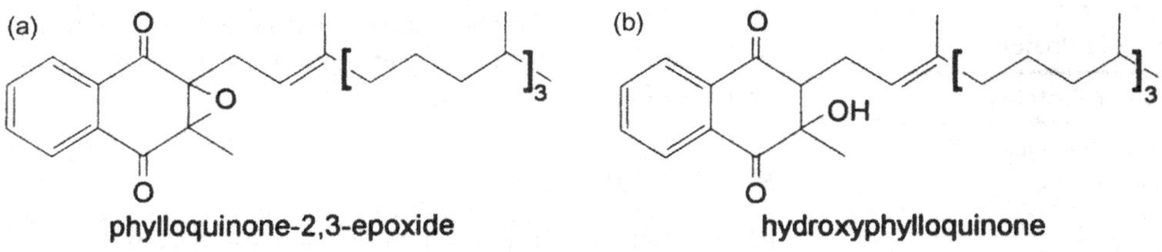

FIGURE 8.7 (a) Phyllo-quinone-2,3-epoxide; (b) Hydroxyphylloquinone.

phylloquinone-2,3-epoxide

hydroxyphylloquinone

hemorrhaging). The most widely used for each purpose has been warfarin (3-[a-acetonylbenzyl]-4-hydroxycoumarin),[16] an analog of the naturally occurring hemorrhagic factor dicumarol, and its sodium salt.[17] Warfarin therapy is prescribed for a million patients each year in the United States. Resistance to warfarin has been observed in both rats and humans, and has become a significant problem in Europe and North America. It involves an isoform of the vitamin K epoxide reductase that is not sensitive to the coumarin. While coumarin therapy has been important in preventing strokes, it is estimated that 12% of cases experience major bleeding episodes which are fatal in some 2%.[18] For this reason, there has been great interest in developing safer anticoagulants. The vitamin K_1 analog 2-chloro-3-phytyl-1,4-naphthoquinone (chloro-K) has proven to be very effective in warfarin-resistant rats, as it functions as a competitive inhibitor of the vitamin at the active site(s) in the vitamin K cycle. Other coumarins,[19] 2,3,5,6-tetrachloro-pyridinol, and several 2-phylloquinone derivatives[20] also act as competitive inhibitors of the vitamin K-dependent carboxylase.

6. METABOLIC FUNCTIONS OF VITAMIN K

Vitamin K-Dependent γ-Carboxylations

Vitamin K is the cofactor of a specific microsomal carboxylase that uses the oxygenation of **vitamin K hydroquinone** to drive the γ-carboxylation of peptide-bound glutamic acid residues (Fig. 8.6). This **vitamin K-dependent carboxylase** is found predominantly in liver, but also in several other organs.[21] In the reduced form, vitamin K provides reducing equivalents for the reaction, thus undergoing oxidation to the 2,3-epoxide form. This is coupled to the cleavage of a C–H bond and formation of a carbanion at the γ-position of a peptide-bound glutamyl residue, followed by carboxylation.

The vitamin K-dependent γ-carboxylation of specific glutamyl residues on the zymogen precursors of each blood clotting factor occurs post-translationally at the N-terminus of the nascent polypeptide. In the case of prothrombin, all 10 glutamyl residues in positions 7–33 (but none of the remaining 33 glutamyl residues) are γ-carboxylated. Carboxylation confers Ca^{2+}-binding capacity, facilitating the formation of Ca^{2+} bridges between the clotting factors and phospholipids on membrane surfaces of blood platelets and endothelial and vascular cells, as well as between Gla residues (i.e., glutamyl residues that have been carboxylated) to form internal Gla–Gla linkages.

The carboxylase requires reduced vitamin K (vitamin K hydroquinone), CO_2 as the carboxyl precursor, and molecular oxygen (O_2).[22] It is frequently referred to as vitamin K carboxylase/epoxidase to indicate the coupling of the γ-carboxylation step with the conversion of vitamin K to the 2,3-epoxide. Normally this coupling is tight; however, under conditions of low CO_2 levels or in the absence of peptidyl-Glu, the epoxidation of the vitamin proceeds without concomitant carboxylation.

Polymorphisms of the vitamin K-dependent γ-glutamyl carboxylase have been identified. These have been associated with variations in circulating levels of undercarboxylated osteocalcin (see "Vitamin K-Dependent Gla Proteins," below).

Vitamin K-Dependent Gla Proteins

Vitamin K functions in the post-translational modification of at least 20 proteins via γ-carboxylation of specific

17. Others include ethyl biscovmacetate (3,3′-carboxymethylenebis-[4-hyroxycoumarin] ethyl ester; and phenprocoumon (3-[1-phenylproyl]-4-hydroxycoumarin).

18. Landefeld, C.S., Anderson, P. A., and Goodnough, L. T. (1989). *Am. J. Med.* 87, 144.

19. For example, difenacoum (3-[3-p-diphenyl-1,2,3,4-tetrahydronaphth-1-yl]-4-hydroxycoumarin); bromodifenacoum (3-[3-{4′-bormodiphenyl-4-yl}-1,2,3,4-tetrahydronaphth-1-yl]-4-hydroxycoumarin).

20. e.g., desmethylphylloquinone, 2-ethylphylloquinone, 2-fluromethyl-phylloquinone, 2-hydroxymethylphylloquinone, 2-methoxymethylphyl-loquinone, 2-trifluromethylphylloquinone.

21. For example, lung, spleen, kidney, testes, bone, placenta, blood vessel wall, and skin.

22. The *in vitro* activity of the carboxylase is stimulated almost four-fold by pyridoxal phosphate when the substrate is a pentapeptide. It is doubtful whether that cofactor is important *in vivo*, as no stimulation was observed in the carboxylation of endogenous microsomal proteins.

TABLE 8.8 Vitamin K-Dependent Gla Proteins

Group	Member Proteins
Clotting regulatory proteins	Prothrombin (factor II)
	Factor VII
	Factor IX
	Factor X
	Protein C
	Protein M
	Protein S
	Protein Z
Bone proteins	Osteocalcin
	Matrix Gla protein (MGP)
	Protein S
Transmembrane proteins	Proline-rich Gla protein 1 (PRGP1)
	Proline-rich Gla protein 2 (PRGP2)
	Transmembrane Gla-protein 3
	Transmembrane Gla-protein 4
Other Gla proteins	Vitamin K-dependent carboxylase
	Gas6
	Transthyretin
	Atherocalcin

glutamate residues to produce **Gla** residues (Table 8.8).[23,24] The Gla proteins are thought to function by binding negatively charged phospholipids via Ca^{2+} held by their Gla residues. Each mammalian Gla protein contains a short, carboxylase-recognition sequence (cleaved after carboxylation) that binds covalently to glutamate-containing propeptides to enhance catalysis. The carboxylase binds these propetides with different affinities: compared to the most tightly bound Factor X, most other propeptides are bound 2–10 times less tightly, while protein C and prothrombin are bound 100 times less tightly.[25] This suggests

that competition for the carboxylase underlies variations in the amounts of Gla proteins produced that have been observed in warfarin-treated cells.

Blood Clotting

Blood clotting is produced by a complex system of proteins that functions to prevent hemorrhage and leads to thrombus formation. It contains coagulation at the site of injury and curtails the process upon formation of the clot. It is initiated by injury to tissues through the release of collagen fibers and tissue factor, a cell surface protein, whereupon they interact with vitamin K-dependent Gla proteins in the plasma. These signals are amplified via the clotting pathway, ultimately, to form the clot.

The eight vitamin K-dependent plasma proteins comprising this system (Table 8.9) have homologous amino acid sequences in their first 40 positions each containing 9–13 Gla residues in the amino-terminal domain. All require Ca^{2+} for activity. Most circulate as a zymogen, i.e., an inactive precursor of the respective functional form, which is a serine protease. Each participates in a cascade of proteolytic activation of a series of factors leading to the conversion of a soluble protein, fibrinogen, to insoluble fibrin, which cross-links with platelets to form the blood clot (Figs 8.8, 8.9). The activation of proteases in this cascade involves the Ca^{2+}-mediated association of the active protein, its substrate, and another protein factor with a phospholipid surface.

The key step in this system is the activation of factor X (also called Stuart factor), by the proteolytic removal of a short polypeptide from the zymogen. This can occur in two ways:

- by the actions of factor IX (also called Christmas factor or plasma thromboplastin component) which is activated by **plasma thromboplastin** as the result of a contact with a foreign surface in what is referred to as the **intrinsic clotting system**, or
- by the action of factor VII,[26] which is activated by **tissue thromboplastin** released as the result of injury in what is called the **extrinsic clotting system**.

Once activated, factor X,[27] binding Ca^{2+} and phospholipid, catalyzes the activation of several coagulation factors:

- prothrombin (factor II) to its active form, **thrombin (factor IIa)**, which catalyzes the proteolytic change in

---

23. This rare amino acid was discovered in studies of the molecular basis of abnormal clotting in vitamin K deficiency. While it had been known that vitamin K deficiency and 4-hydroxycoumarin anticoagulant treatment each caused hypoprothrombinemia, studies in the early 1970s revealed the presence in each condition of a protein that was antigenically similar to prothrombin but that did not bind Ca^{2+} and, therefore, was not functional. Studies of the prothrombin Ca^{2+}-binding sites revealed them to have Gla residues, which were replaced by glutamate residues in the abnormal prothrombin. Subsequently, Gla residues were found in each of the other vitamin K-dependent clotting factors, as well as in several other Ca^{2+}-binding proteins in other tissues.

24. Gla-rich proteins have also been found in the venomous cone snail, some poisonous snakes, and some urochordates. These Gla-proteins serve as paralyzing neurotoxins used to subdue prey.

25. Huber, P., Schmitz, T., and Griffen, J. (1990). *J. Biol. Chem.* 265, 12467.

---

26. Also called proconvertin, factor VII is also activated by a high-fat meal and has been associated with increased risk of ischemic heart disease in some studies (e.g., Junker, R., Heinrich, J., Schulte, H., *et al.* [1997] *Arterioscler. Thromb. Vasc. Biol.* 17, 1539–1544).

27. Polymorphisms of factor X have been identified; however, these do not appear to affect circulating factor X levels.

TABLE 8.9 Characteristics of the Vitamin K-Dependent Plasma Proteins

Parameter	II[a]	VII[b]	IX[c]	X[d]	C	M	S	Z
Level (µg/ml)	100	1	3	20	10	<1	1	<1
Molecular mass (kDa)	72	46	55	55	57	50	69	55
% Carbohydrate	8	13	26	13	8	+	+	+
Number of chains	1	1	1	2	2	1	1	1
Number of Gla residues	10	10	12	12	11	+	10	13

[a] Prothrombin.
[b] Proconvertin.
[c] Plasma thromboplastin component, also: Christmas factor, antihemophilic factor B, platelet cofactor II, antoprothrombin II.
[d] Stuart factor.

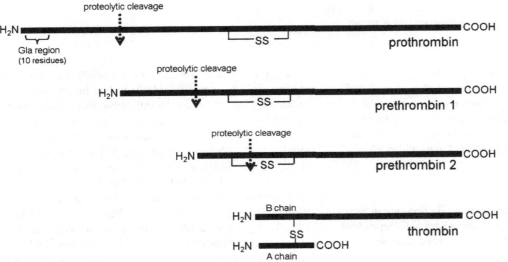

FIGURE 8.8 Thrombin is formed by the successive proteolytic removal of sequences from prothrombin.

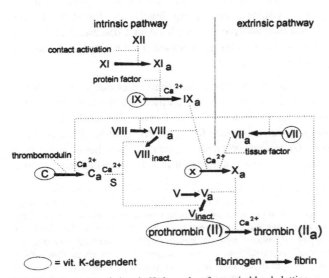

FIGURE 8.9 Roles of vitamin K-dependent factors in blood clotting.

fibrinogen that renders it insoluble (as fibrin) for clot formation.

- factor V to its active form, factor Va;
- factor VIII to its active form factor VIIIa.

Control of clotting is accomplished by the downregulation of thrombin production *via* thrombin binding to thrombomodulin, which complex activates protein C,[28] which in turn inactivates factors Va and VIIIa.

Two components of this system (proteins S and Z) are not serine proteases. Protein S is found in the plasma both in free form and as a bimolecular complex with a

28. Protein C also has anti-inflammatory and anti-apoptotic activities involving activation of a protease-activated receptor (PAR1) and inhibition of the NFκB pathway. Individuals with inherited protein C-deficiency have elevated risk of thrombosis.

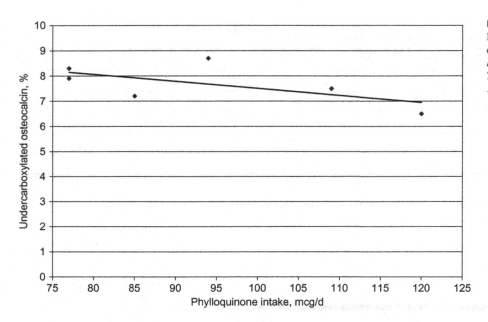

FIGURE 8.10 Relationship of vitamin K intake (for 3 weeks) and plasma under-carboxylated osteocalcin in humans. From *Binkley, N. C., Krueger, D. C., Kawahar, T. N., et al. (2002). Am. J. Clin. Nutr. 76, 1055.*

regulatory component (C4b-binding protein) of the complement system. Individuals with inherited protein S deficiency have been reported to have recurrent thromboses. Protein Z is a cofactor for inhibition of activated factor X. Protein Z deficiency has been associated with a bleeding tendency in patients with factor V Lieden mutation.[29]

Bone Mineralization

The presence of vitamin K-dependent Gla proteins in bone suggests a role in bone health. Three vitamin K-dependent Gla proteins have been identified in calcified tissues:

1. *Osteocalcin.* The best characterized vitamin K-dependent protein of calcified tissues is osteocalcin, which has also been referred to as "bone GLA protein." Osteocalcin shows no homology with the vitamin K-dependent plasma proteins. It has been strongly conserved between various species, consisting of a 5.7 kDa protein with three Gla residues in a 49- or 50-amino acid segment. The Gla protein binds Ca^{2+} weakly and hydroxyapatite strongly; that binding serves both to maintain its secondary structure and to allow it to bind mineralized bone matrix. Synthesized by osteoblasts, it is the second most abundant protein[30] in the bone matrix, comprising about 2% of total

bone protein and 10–20% of non-collagen protein. Osteocalcin is expressed relatively late in development with the onset of bone mineralization. Its carboxylation is inhibited by warfarin treatment and is stimulated by MK supplementation. Although osteocalcin carboxylation is stimulated *in vitro* by $1,25\text{-}(OH)_2$-vitamin D_3, oral supplementation of humans may not reduce circulating undercarboxylated osteocalcin levels (Fig. 8.10). An estimated 20% of osteocalcin is not bound to bone, and is free to enter the plasma. Because it is synthesized only by osteoblasts, plasma osteocalcin can be used as a marker of bone formation. Plasma levels of undercarboxylated osteocalcin are subject to genetic variability associated with polymorphisms in both the vitamin K-dependent carboxylase and epoxide reductase. Greatest circulating levels are found in young children and patients with Paget's disease,[31] and other disorders involving increased bone resorption/mineralization (Table 8.10).

It has been suggested that osteocalcin may function in the regulation of calcification, perhaps by acting as an attractant for osteoclast progenitor cells. However, supporting evidence for this hypothesis is inconsistent. Undercarboxylated osteocalcin has been associated with risk of hip fracture (Table 8.11). Studies have shown

29. This condition involves a point mutation in the gene encoding coagulation factor V and results in the expression of a form of the factor that is resistant to activated protein C and a relatively hypercoagulable state. The mutation occurs in 4–6% of the US population, and is associated with increased risk of venous thromboembolism.
30. Collagen is the most abundant protein in bone.

31. Paget's disease, also called osteitis, affects 3% of people over 40 years of age. It involves dysfunctional bone remodeling, with bone continually breaking down and rebuilding at rates faster than normal. This results in bone being replaced with soft, porous, highly vascularized bone that can be weak and easily bent, leading to shortening of the affected part of the body, or with excess bone that can be painful and easily fractured. The disease most commonly affects the spine, pelvis, skull, femur, and tibiae.

TABLE 8.10 Plasma Osteocalcin Concentrations in Humans

Group	Osteocalcin (ng/ml)
Children	10–40[a]
Adults	4–8
Women:	
60–69 years of age	7
80–89 years of age	8
Patients with Paget's disease	39
Patients with secondary hyperparathyroidism	47
Patients with osteopenia	9

[a] Highest levels observed in patients 10–15 years of age.

TABLE 8.11 Efficacy of Vitamin K Supplementation in Reducing Fracture Risk and Bone Mineral Loss in Older Adults

Study	Treatment	n	Fractures, %	Bone Mineral Density Change, %
Shiraki et al. (2000)[a]	MK[d]	86	14.3	−0.5
	Control	94	30.3	−3.3
Iwamoto et al. (2001)[b]	MK[d]	22	8.7	−0.1
	Control	20	25.0	−1.7
Ishida et al. (2004)[c]	MK[d]	63	14.3	−1.9
	Control	60	28.3	−3.3

[a] Shiraki, M., Shiraki, Y., Aoki, C. et al. (2000). J. Bone Min. Res. 15, 515.
[b] Iwamoto, J. Takeda, T., and Ichimura, S. (2001). J. Orthop. Sci. 6, 487.
[c] Ishida, Y. and Kawai, S. (2004). Am. J. Med. 117, 549.
[d] 45 mg/day.

increased bone mineralization in rats deficient in osteocalcin due to warfarin treatment, or made osteocalcin-null transgenically, suggesting that osetocalcin functions as a negative regulator of bone formation.

2. *Matrix Gla Protein (MGP)*. Sometimes referred to as periostin, this is a small (9.6 kDa), insoluble polypeptide structurally related to osteocalcin. It is expressed in many soft tissues, but accumulates only in calcified tissues, where it has clear affinities for demineralized bone matrix, and in non-mineralized cartilage, where it is thought to function as an inhibitor in the regulation of calcification. MGP-null mice show early onset periodontal disease.

3. *Protein*. That protein S is synthesized by osteoblasts suggests that it may have an activity in bone in addition to its role in clotting. A bone function was suggested by the finding of severe osteopenia, low bone mineral density, and vertebral compression fractures in two pediatric cases with very low protein S levels. The protein contains a thrombin-sensitive region, an epidermal growth factor-like domain, and a steroid hormone-binding domain.

Vascular System

Two Gla proteins appear to play anti-atherogenic roles by impairing calcification:

- *MGP*. This is expressed in vascular smooth muscle, and appears to play a dominant role in maintaining the rate of arterial calcification as low as possible. Its impaired carboxylation by warfarin treatment has been shown to lead to arterial calcification in animal models.
- *Atherocalcin*. This Gla protein was discovered in calcified atherosclerotic tissue. It has been suggested that atherocalcin may inhibit vitamin K-dependent γ-glutamyl carboxylase, which is found in the walls of arteries but not veins, and may be involved in the development of atherosclerosis.

Nervous System

Widely distributed in nervous tissue is a protein called **Gas6**,[32] which has 44% sequence homology with protein S and, like the latter, contains an epidermal growth factor-like domain. However, unlike protein S, Gas6 lacks thrombin-sensitive motifs. Gas6 functions as a ligand for the reception tyrosine kinases Ax1 and Sky/Rse, and protects cells from apoptosis by activating Ark phosphorylation and inducing MAP kinase[33] activity. It has also been found to facilitate growth of smooth muscle cells, and it has been suggested that Gas6 may function as a ligand for Tyro 3 and thus serve as a neurotrophic factor. That brain microsomes lack γ-carboxylase activity implies that the post-translational glutamation of Gas6 must occur in other tissues.

Other Functions

The finding of other Gla proteins makes it is clear that vitamin K functions widely in physiology. Several have

---

32. Named for its gene, Growth Arrest Specific gene 6.
33. Mitogen-activated protein kinase.

been identified, although their functions are presently unknown:

- *Proline-rich Gla proteins.* These small (**PRG1**, 23 kDa; **PRG2**, 17 kDa), single-pass, transmembrane proteins are expressed in a variety of extrahepatic proteins.
- *Transthyretin.* This retinol- and throxine-binding plasma protein has been reported to be a Gla-protein.
- *Renal Gla protein.* This protein is inhibited by PTH and vitamin D_3. It is thought to have some role in the transport/excretion of Ca^{2+} in the kidney.
- *Other Gla proteins.* Other Gla proteins have been reported in sperm, urine, hepatic mitochondria, shark skeletal cartilage, and snake venom.

7. VITAMIN K IN HEALTH AND DISEASE

The best understood metabolic roles of vitamin K are those involved in its anti-hemorrhagic function, which after all led to the vitamin's discovery. It has since become clear that vitamin K plays important metabolic roles beyond clotting.

Coagulation

Coagulopathies now associated with vitamin K deficiency were reported in the nineteenth century. These include those that were called "hemorrhagic disease of the newborn,"[34] which differed from hemophilia by its earlier presentation (within a couple of days after birth) and absence of family history. Routine prophylaxis of newborns with vitamin K has made this condition rare in countries with that practice.

Hereditary combined vitamin K-dependent clotting factor deficiency is a rare autosomal recessive disorder involving mutations in the genes for both vitamin K-dependent epoxide reductase and carboxylase. This involves suboptimal levels of coagulation factors II, VII, IX, and X, as well as proteins C, S, and Z. It is manifest as a range of spontaneous bleeding symptoms. High-level vitamin K supplementation is effective in managing the disorder.

Low vitamin K status appears to contribute to unstable anticoagulation control in warfarin treatment, which condition affects as many as half of patients. Interventions with vitamin K have been found to improve anticoagulation control.

Bone Health

Several studies have shown individuals with low circulating vitamin K levels or vitamin K intakes to be at elevated risk of osteoporosis or fracture.[35] The Nurses' Health

Study, a 10-year prospective study of more than 72,000 women, found the age-adjusted risk of hip fracture to be 30% less in women with vitamin K intakes >109 μg/d than in those consuming lower amounts.[36] A similar relationship was observed in the Framingham Heart Study: subjects in the highest quartile of vitamin K intake (median intake 254 μg/day) have significant reductions in hip fracture risk compared to those in the lowest quartile (median intake 56 μg/day).[37]

At least a dozen randomized clinical intervention studies have been conducted to determine the efficacy of vitamin K supplementation in reducing bone mineral loss and/or fracture risk. Supplementation with MK has been found to reduce fracture risk[38] and improve bone mineral density (Table 8.11).[39] Intervention studies with phylloquinone have yielded inconsistent results with respect to bone mineral density, although reductions in circulating levels of undercarboxylated osteocalcin have been observed. Undercarboxylated osteocalin has been found to be a predictor of low bone mineral density. Relatively high intakes of vitamin K appear to be required to support the maximal carboxylation of osteocalcin (Table 8.12).

Cardiovascular Health

Vitamin K-dependent Gla proteins may play roles in atherogenesis, which involves thrombus-induced coagulation as well as intimal calcification. Gene deletion studies in mice have suggested that MGP may be a regulator of this process. Its undercarboxylation (due to warfarin treatment) results in arterial calcification in the rat. Osteocalcin, normally expressed only in bone, is upregulated in arterial calcification. It has been suggested that Gas6 and protein S may inhibit calcification through enhancing apoptosis, which is extensive in atherosclerotic lesions. It is possible that increased dietary intakes of the vitamin may be useful in reducing atherosclerosis risk.

Anticarcinogenesis

That vitamin K status can play an anticarcinogenic role was suggested some six decades ago when MK treatment was found to increase the survival of inoperable bronchial carcinoma patients. Since then it has been observed that patients with hepatocellular carcinoma typically have

34. This condition is now called vitamin K deficiency bleeding (VKDB).
35. See review by Weber (2001) *Nutr.* 17, 880.

36. Feskanich, D., Weber, P., Willett, W. C., *et al.* (1999). *Am. J. Clin. Nutr.* 69, 74.
37. Booth, S. L., Tucker, K. L., Chen, H., *et al.* (2000) *Am. J. Clin. Nutr.* 71, 1201.
38. A meta-analysis found 7 studies to show an average risk reduction of 60% (Cockayne, S., Adamson, J, Lanham-New, S., *et al.* [2006] *Arch. Intern. Med.* 166, 1256).
39. Iwamoto, J., Takeda, T., and Sato, Y. (2006). *Nutr. Rev.* 64, 509.

TABLE 8.12 Total and Undercarboxylated Osteocalcin in Fracture and Non-Fracture Patients

Parameter	Non-Fracture	Fracture
n	153	30
Age, kg	82.5 ± 5.9	85.8 ± 6.5[a]
Body weight, kg	56.6 ± 11.5	49.4 ± 10.7[a]
Plasma osteocalcin, ng/ml:		
Total	6.18 ± 3.34	7.90 ± 4.34[a]
Undercarboxylated	0.89 ± 0.89 (14)[b]	1.47 ± 1.65[a] (19)[b]
Carboxylyated	5.29 ± 2.69 (86)[b]	6.43 ± 2.94[a] (81)[b]
25-OH-D$_3$, ng/ml	17.4 ± 14.1	15.9 ± 10.8
PTH, pg/ml	47.0 ± 23.8	60.0 ± 40.9
Alkaline phosphatase, IU/L	78 ± 37	92 ± 40[a]

From Szulc, P., Arlot M, Chapuy M.-C., et al. (1996). Bone 18, 487.
[a] P <0.05.
[b] % total.

abnormally high circulating levels of under-γ-carboxylated prothrombin (see "Signs of Vitamin K Deficiency," page 228). Recently, a large (24,340 subjects) prospective study showed that the intake of MKs, but not phylloquinone, was associated with reduced cancer incidence and mortality.[40] In an 8-year randomized clinical trial, MK (45 mg/day) reduced risk of hepatocellular carcinoma in 43 women with viral cirrhosis of the liver by 87% compared to controls.[41]

Studies with animal tumor models have shown all vitamers K capable of inhibiting cancer cell growth through several mechanisms:

- *Oxidative stress in malignant cells.* The actions of menadione have been attributed to the reactive oxygen species generated by the single-electron redox-cycling of the quinone, and to the depletion of cellular glutathione due to its arylation by the vitamer.
- *Modulation of transcription factors.* In cell culture, phylloquinone and MKs have been shown to induce proto-oncogenes, increasing the levels of c-myc, c-jun, and c-fos, delaying the cell cycle, and enhancing apoptosis. Menadione also appears to induce protein

tyrosine kinase activation and to inhibit, through direct interaction, extracellular signal-regulated kinase (ERK) protein tyrosine phosphatases. These effects are associated with reduced proliferation.

- *Cell cycle arrest.* Menadione has been shown to inhibit cyclin-dependent kinases (CDKs) by binding to sulfhydryls at the active site of those enzymes. This effect is associated with inhibition of malignant cell proliferation at the G1/S and S/G2 phases of the cell cycle. MKs have been found to affect cyclin function, also manifest as cell cycle inhibition.

Obesity

Obesity appears to impair the utilization of vitamin K. Adipose tissue stores vitamin K at relatively high levels,[42] and plasma phylloquinone levels have been found to vary inversely with percentage body fat in women who also showed increased circulating levels of undercarboxylated prothrombin. A 3-year intervention study found that phylloquinone supplementation reduced insulin resistance in men.[43] However, no significant benefits were found in women in that or another trial.[44]

Nervous Function

The vitamin is abundant in the rat brain, almost exclusively as MK-4. It is found in greatest concentrations in myelinated regions such as the pons medulla and midbrain. Still, the role of vitamin K in neural function is unclear. A function of the vitamin in sphingolipid metabolism has been shown in bacteria; such a role could have functional significance in mammalian neurons, as sphingolipids are important components of membranes and also serve as second messengers for intracellular signal transduction pathways. However, such a role has not been reported in animals, although studies have shown warfarin treatment to reduce brain sulfatides, and vitamin K depletion treatment to reduce the activity of brain glutathione S-transferase. Both Gas6 and protein S are widely expressed in the central nervous system of the rat. Gas6 has been shown to have neurotrophic activity toward hippocampal neurons, and to promote growth and survival of several types of neural cells.

40. The European Prospective Investigation into Cancer and Nutrition (EPIC) (Nimptsch, S., Rohrmann, S., and Linseise, J. [2008] *Am. J. Clin. Nutr.* 87, 985).
41. Habu, D., Shiomi, S., Tamori, A., *et al.* (2004). *J. Am. Med. Assoc.* 292, 358.

42. Adults undergoing bariatric surgery were found to have phylloquinone levels of 148 ± 72 nmol/kg and 175 ± 112 nmol/kg in subcutaneous and visceral adipose tissue, respectively (Shea, K. M., Booth, S. L., Gundberg, C. M., *et al.* [2010] J. Nutr. 140, 1029).
43. Yoshida, M., Jacques, P. F., Meigs, J. B., *et al.* (2008). *Diabetes Care* 31, 2092.
44. Kumar, R., Binkley, N., and Vella, A. (2010). *Am. J. Clin. Nutr.* 92, 1528.

8. VITAMIN K DEFICIENCY

Signs of Vitamin K Deficiency

Coagulopathy

The predominant clinical sign of vitamin K deficiency is hemorrhage (Table 8.13), which can lead to a fatal anemia. The blood shows prolonged clotting time and **hypoprothrombinemia**. Because a 50% loss of plasma prothrombin level is required to affect prothrombin time, prolongation of the latter is a useful biomarker for advanced subclinical vitamin D deficiency. Several congenital disorders of vitamin K-dependent proteins have been identified in human patients: at least a dozen forms of congenital **dysprothrombinemia**, at least three variants of factor VII, and a congenital deficiency of protein C. Patients with these disorders show **coagulopathies**; none responds to high doses of vitamin K.

Undercarboxylated Proteins

A more sensitive indicator of low vitamin K status is the presence in plasma of under-γ-carboxylated vitamin K-dependent proteins. The first was originally thought to be a distinct protein produced only in vitamin K deficiency; it was referred to as the **protein induced by vitamin K absence** (**PIVKA**). Subsequently, it became clear that PIVKA was actually inactive prothrombin lacking Gla residues required for its Ca^{2+} binding. It is thus useful as a marker for subclinical vitamin K deficiency. Another marker is **under-γ-carboxylated osteocalcin**, which is released from bone into the circulation. Studies have shown that increases in the serum concentrations of these factors are more sensitive to minidose warfarin therapy than are decreases in prothrombin or osteocalcin. Except for patients on anticoagulant therapy, undercarboxylation of blood-clotting factors is rare. The undercarboxylation of

osteocalcin, however, is frequent among postmenopausal women.[45]

Importance of Hindgut Microbial Biosynthesis

Vitamin K deficiency is rare among humans and most other animal species, the important exception being the rapidly growing chick raised in a wire-floored cage. This is due to the wide occurrence of vitamin K in plant and animal foods, and to the significant microbial synthesis of the vitamin that occurs in the intestines of most animals. In fact, for many species, including humans, the intestinal synthesis of vitamin K appears to meet normal needs. Species with short gastrointestinal tracts and very short intestinal transit times (e.g., about 8 hours in the chick), less than the generation times of many bacteria, do not have well-colonized guts. Being thus unable to harbor vitamin K-producing bacteria, they depend on their diet as the source of their vitamin K.

Risk Factors for Vitamin K Deficiency

The most frequent causes of vitamin K deficiency, thus, are factors that interfere with the microfloral production or absorption of the vitamin:

- *Lipid malabsorption.* Diseases of the gastrointestinal tract, biliary stasis, liver disease, cystic fibrosis, celiac disease, and *Ascaris* infection can interfere with the enteric absorption of vitamin K.
- *Anticoagulant therapy.* Certain types of drugs can impair vitamin K function. These include warfarin and other 4-hydroxycoumarin anticoagulants, and large doses of salicylates, which inhibit the redox cycling of the vitamin. In each case, high doses of vitamin K are generally effective in normalizing clotting mechanisms. In medical management of thrombotic disorders, over-anticoagulation with warfarin is common; this is reversed by warfarin dose reduction coupled with treatment with phylloquinone.[46]
- *Antibiotic therapy.* Sulfonamides and broad-spectrum antibiotic drugs can virtually sterilize the lumen of the intestine, thus removing an important source of vitamin K for most animals. Therefore, it has been thought that patients on antibiotic therapy can be at risk of vitamin K deficiency.[47] Indeed, cases of

TABLE 8.13 Signs of Vitamin K Deficiency

Organ system	Sign
General – growth	Decrease
Dermatologic	Hemorrhage
Muscular	Hemorrhage
Gastrointestinal	Hemorrhage
Vascular	
Erythrocytes	Anemia
Platelets	Decreased clotting

45. Clinical trials have found that members of this group respond to supplemental vitamin K with increases in bone formation and decreases in bone resorption.
46. Baker, P., Gleghorn, A., and Tripp, T. (2006). *Br. J. Haematol.* 133, 331–336.
47. This risk can be further increased by inanition. Because vitamin K is rapidly depleted from tissues, periods of reduced food intake, such as may occur post-surgically, can produce vitamin K deficiency in patients with reduced intestinal microbial synthesis of the vitamin.

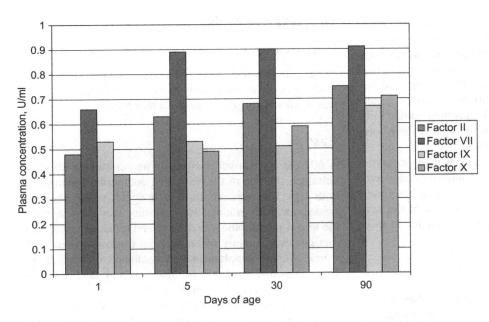

FIGURE 8.11 Vitamin K-dependent coagulation factors in infants. From *Zipursky, A. (1999).* Br. J. Haematol. *104, 430.*

hypoprothrominemia have been identified in association with the use of penicillin, semisynthetic penicillins, and cephalosporins. In fact, the prevalence of such cases appeared to increase in the 1980s with the introduction of the β-lactam antibiotics.[48] Although these drugs are administered intravenously, it is possible that they may affect enteric bacterial metabolism via biliary release. Studies have shown that not all patients treated with β-lactam antibiotics show altered fecal menaquinones, although they show significant increases in circulating vitamin K-2,3-epoxide levels when treated with vitamin K. This observation led to the demonstration that the cephalosporin-type antibiotics can inhibit the vitamin K-dependent carboxylase to produce coumarin-like depressions of the activities of the vitamin K-dependent clotting factors. Unlike the coumarins, however, the β-lactam antibiotics are very weak anticoagulants, the effects of which are observed only in patients of low vitamin K status.

Neonates

Neonates are at special risk of vitamin K deficiency for several reasons:

- *Placental transport of the vitamin is poor.* Infants have very limited reserves of vitamin K; their serum levels are typically about half those of their mothers.
- *The neonatal intestine is sterile for the first few days of life.* The neonatal intestine thus does not provide an enteric microbial source of the vitamin.

- *Hepatic biosynthesis of the clotting factors is inadequate in the young infant.* The plasma prothrombin concentrations of fetuses and infants are typically one-quarter those of their mothers (Fig. 8.11).
- *Human milk is an inadequate source of vitamin K.* The frequency of vitamin K-responsive hemorrhagic disease in 1-month-old infants is 1/4,000 overall, but 1/1,700 among breastfed infants.

For these reasons, some infants[49] will develop hemorrhage if continuing intake of vitamin K is not provided. This condition of **vitamin K deficiency bleeding (VKDB)**, also called **hemorrhagic disease of the newborn**, can present in different ways, depending on the age of the infant:

- newborns (first 24 hours) – cephalohematoma; intracranial, intrathoracic or intra-abdominal bleeding
- newborns (first week) – generalized ecchymoses[50] of the skin; bleeding from the gastrointestinal tract, umbilical cord stump or circumcision site
- infants (1–12 weeks) – intracranial, skin or gastrointestinal bleeding.

The major risk factors for VKDB are exclusive breastfeeding, failure to give vitamin K prophylaxis, and certain maternal drug therapies. Exclusively breastfed infants who have not received vitamin K or who have gastrointestinal

---

48. For example, cephalosporin, cefamandole, and the related oxa-β-lactam moxalactam.

49. Without vitamin K prophylaxis, the risk of hemorrhage for healthy, non-traumatized infants in the first two weeks of life has been estimated to be 1–2 per 1,000, and for older infants a third of that level (Hey, E., www.archdischild.com, July 13, 2006).

50. Sheet hemorrhages of the skin, *ecchymoses*, differ from the smaller *petechiae* only in size.

disorders involving lipid malabsorption (cystic fibrosis, biliary atresia, α_1-antitrypsin deficiency) can show signs within several weeks, as intracranial hemorrhage with liver disease, central nervous system damage, and high mortality due to bilirubinemia. That infants fed formula diets are at lower risk probably relates to the greater amounts of vitamin K in infant formulas than human milk. Hemorrhagic disease has also been reported for newborns of mothers on anticonvulsant therapy.

It has become a common practice in many countries to treat all infants at birth with parenterally administered vitamin K (1 mg phylloquinone). This practice has greatly reduced the incidence of hemorrhagic disease of the newborn, although much lower doses have been found to be effective.[51]

9. VITAMIN K TOXICITY

Phylloquinone exhibits no adverse effects when administered to animals in massive doses by any route. The menaquinones are similarly thought to have negligible toxicity. Menadione, however, can be toxic. At high doses, it can produce hemolytic anemia, hyperbilirubinemia, and severe jaundice. Accordingly, phylloquinone has replaced menadione for the vitamin K prophylaxis of neonates. The intoxicating doses of menadione appear to be at least three orders of magnitude above those levels required for normal physiological function. At such high levels, menadione appears to produce oxidative stress. This occurs as a result of the vitamer undergoing monovalent reduction to the semiquinone radical, which, in the presence of O_2, is reoxidized to the quinone, resulting in the formation of the **superoxide** radical anion. In addition, high levels of menadione are known to react with free sulfhydryl groups, thus depleting reduced glutathione (GSH) levels.

A review of the US Food and Drug Adminstration database revealed 2,236 adverse reactions reported for 1,019 patients receiving intravenous vitamin K in 1968–1997.[52] Of those cases, 192 were anaphylactoid reactions and 24 were fatalities; those numbers were only 21 and 4, respectively, for patients given vitamin K doses <5 mg. Persistent, localized eczematous plaque has been reported at the site of injection for some patients given vitamin K_1 intramuscularly or subcutaneously.[53]

51. Sutherland, J. M., Glueck, H. I., Gleser, G., *et al.* (1967). *Am. J. Dis. Child.* 113, 524–533.

52. Fiore, L. D., Scola, M. A., Cantillon, C. E., *et al.* (2001). *J. Thromb. Thromboysis* 11, 175–183.

53. Wilkins, K., DeKoven, J., and Assaad, D. (2000). *J. Cutan. Med. Surg.* 4, 164–168.

10. CASE STUDIES

Instructions

Review the following case report, paying special attention to the diagnostic indicators on which the treatments were based. Then answer the questions that follow.

Case 1

A 60-year-old woman involved in an automobile accident sustained injuries to the head, and compound fractures of both legs. She was admitted to hospital, where she was treated for acute trauma. Her recovery was slow, and for the next 4 months she was drowsy and reluctant to eat. Her diet consisted mainly of orange and glucose drinks, with a multivitamin supplement that contained no vitamin K. Her compound fractures became infected, and she was treated with a combination of antibiotics (penicillin, gentamicin, tetracycline, and cotrimoxazole). She then developed intermittent diarrhea, which was treated with codeine phosphate. After a month all antibiotics were stopped, and 46 days later (6 months after the injury) she experienced bleeding from her urethra. At that time other signs were also noted: bruising of the limbs, bleeding gums, and generalized *purpura*.[54] The clinical diagnosis was scurvy until it was learned that the patient was taking 25 mg of ascorbic acid per day via her daily vitamin supplement.

Laboratory Results:

Parameter	Patient	Normal Range
Hb	9.0 g/dl	12–16 g/dl
Mean RBC volume	79 fl	80–100 fl
White cells	$6.3 \times 10^9/l$	$5–10 \times 10^9/l$
Platelets	$320 \times 10^9/l$	$150–300 \times 10^9/l$
Plasma iron	22 µg/dl	72–180 µg/dl
Total iron-binding capacity	123 µg/dl	246–375 µg/dl
Calcium	7.6 mg/dl	8.4–10.4 mg/dl
Inorganic phosphate	2.4 mg/dl	2.4–4.3 mg/dl
Folate	1.6 ng/ml	3–20 ng/ml
Vitamin B_{12}	110 ng/l	150–1,000 ng/l
Prothrombin time	273 s	13 s (control)
Thrombin time	10 s	10 s (control)

When her abnormal prothrombin time was noted, specific coagulation assays were performed. These showed that the activity of each of the vitamin K-dependent factors (factors II, VII, IX, and X) was <1% of the normal level, and

54. Subcutaneous hemorrhages.

that the activity of factor V was 76% of normal. A xylose tolerance test (to measure small bowel absorption), performed with a single oral dose of 5 g of xylose, showed tolerance within normal limits. A stool culture showed normal fecal flora. The patient was then given phylloquinone (10 mg daily, administered intravenously, for 3 days), and showed a complete recovery of all coagulation factor activities to normal. She was given a high-protein/high-energy diet supplemented with $FeSO_4$ and, for a week, daily oral doses of 10 mg of phylloquinone. Her diarrhea subsided, her wounds healed, and she returned to normal health.

Case 2

A 55-year-old man with arteriosclerotic heart disease and type IV hyperlipoproteinemia was admitted to the hospital with a hemorrhagic syndrome. Six months earlier he had suffered a myocardial infarction[55] complicated by pulmonary embolism[56] for which he was treated with heparin[57] followed by warfarin. Two months earlier he had been admitted for a cardiac arrhythmia, at which time his physical examination was normal and chest radiograph showed no abnormalities, but his electrocardiogram showed first-degree atrioventricular block[58] with frequent premature ventricular contractions. At that time, he was taking 5 mg of warfarin per day.

Laboratory Findings 2 Months Before Third Admission:[a]

Parameter	Patient	Normal
Prothrombin time	16.6 s	12.7 s
Plasma triglycerides	801 mg/dl	20–150 mg/dl
Serum cholesterol	324 mg/dl	150–250 mg/dl

[a] Blood count, blood urea nitrogen, blood bilirubin, and urinalysis were all normal.

He was treated with warfarin (5 mg/day), digoxin,[59] diphenylhydantoin,[60] furosemide,[61] potassium chloride,[62] and clofibrate.[63] Within a month, quinidine gluconate[64] was substituted for diphenylhydantoin because the patient showed persistent premature ventricular beats, but that drug was discontinued because of diarrhea

55. Dysfunction due to necrotic changes resulting from an obstruction of a coronary artery.
56. Obstruction or occlusion of a blood vessel by a transported clot.
57. A highly sulfated mucopolysaccharide with specific anticoagulant properties.
58. Impairment of normal conduction between the atria and ventricles.
59. A cardiotonic.
60. A cardiac depressant (and anticonvulsant).
61. A diuretic.
62. That is, to correct for the loss of K^+ induced by the diuretic.
63. An antihyperlipoproteinemic.
64. A cardiac depressant (antiarrhythmic).

and procainamide was used instead. At that time his prothrombin time was 31.5 s, and his warfarin dose was reduced first to half the original dose and then to one-quarter of that level.

At the time of the third admission, the patient appeared well nourished, but had ecchymoses on his arms, abdomen, and pubic area. He had been constipated with hematuria[65] for the preceding 2 days. His physical examination was unremarkable except for occasional premature beats, and his laboratory findings were similar to those observed on his previous admission, with the exception that his prothrombin time had increased to 36.6 s. In questioning the patient, it was learned that he had been taking orally as much as 1,200 mg of all-rac-α-tocopheryl acetate each day for the preceding 2 months.

Both his warfarin and vitamin E treatments were discontinued, and 2 days later his prothrombin time had dropped to 24.9 s and his ecchymoses began to clear. The patient consented to participate in a clinical trial of vitamin E (800 mg of all-rac-α-tocopheryl acetate per day) in addition to the standard regimen of warfarin and clofibrate. The results were as follows.

Effect of Vitamin E on the Activities of the Patient's Coagulation Factors:

	Initial Value	(6 Weeks)	(1 Week)	Normal Range
Factor II (prothrombin)[a]	11	7	21	60–150
Factor VII[a]	27	16	20	50–150
Factor IX[a]	30	14	23	50–150
Factor X[a]	15	10	—	50–150
Prothrombin time (s)	20.7	29.2	22.3	11.0–12.5

[a] Values represent a percentage of normal means.

Case Questions and Exercises

1. What signs indicated vitamin K-related problems in each case?
2. What factors probably contributed to the vitamin K deficiency of the patient in Case 1? Why was phylloquinone, rather than menadione, chosen for treatment of that patient?
3. What factors may have contributed to the coagulopathy of the patient in Case 2? What might be the basis of the effect of high levels of vitamin E seen in that case?

65. The presence of blood in the urine.

1. Construct a concept map to illustrate the ways in which vitamin K affects blood coagulation.
2. Construct a decision tree for the diagnosis of vitamin K deficiency in a human or animal.
3. What features of the chemical structure of vitamin K relate to its metabolic function?
4. What relevance to their vitamin K nutrition would you expect of the rearing of experimental animals in a germ-free environment or fed a fat-free diet?
5. How does the concept of a coenzyme relate to vitamin K?

RECOMMENDED READING

Berkner, K.L., 2005. The vitamin K-dependent carboxylase. Ann. Rev. Nutr. 25, 127–149.

Berkner, K.L., 2008. Vitamin K-dependent carboxylation. Vit. Hormones 78, 131–156.

Booth, S.L., 2009. Roles for vitamin K beyond coagulation. Ann. Rev. Nutr. 29, 89–110.

Bügel, S., 2008. Vitamin K and bone health in adult humans. Vit. Hormones 78, 393–416.

Cashman, K.D., O'Connor, E., 2008. Does high vitamin K1 intake protect against bone loss in later life? Nutr. Rev. 66, 532–538.

Dahlbäck, B., Villoutreix, B.O., 2005. The anticoagulation protein C pathway. FEBS Letts 579, 3310–3316.

Danziger, J., 2008. Vitamin K-dependent proteins, warfarin, and vascular calcification. Clin. J. Am. Soc. Nephrol. 3, 1504–1510.

Denisova, N.A., Booth, S.A., 2005. Vitamin K and sphingolipid metabolism: Evidence to date. Nutr. Rev. 63, 111–121.

Ferland, G., 2006. Vitamin K. In: Bowman, B.A., Russell, R.M. (Eds.), Present Knowledge in Nutrition, ninth ed., Vol. I. ILSI Press, Washington, DC, pp. 220–232.

Greer, F.R., 2010. Vitamin K the basics – what's new? Early Human Dev. 86, S43–S47.

Iwamoto, J., Takeda, T., Sato, Y., 2006. Menatetrenone (vitamin K2) and bone quality in the treatment of postmenopausal osteoporosis. Nutr. Rev. 64, 509–517.

Kaneki, M., Hosoi, T., Ouchi, Y., Orimo, H., 2006. Pleiotropic actions of vitamin K: Protector of bone health and beyond? Nutr. 22, 845–852.

Lamson, D.W., Plaza, S.M., 2003. The anticancer effects of vitamin K. Alt. Med. Rev. 8, 303–318.

Mizuta, T., Ozaki, I., 2008. Hepatocellular carcinoma and vitamin K. Vit. Hormones 78, 435–442.

Napolitano., M., Mariani, G., Lapecorella, M., 2010. Hereditary combined deficiency of the vitamin K-dependent clotting factors. J. Rare Dis. 5, 21–29.

Nelsestuen, G.L., Shah, A.M., Harvey, S.B., 2000. Vitamin K-dependent proteins. Vit. Hormones 58, 355–389.

Oldenburg, J., Morinova, M., M ller-Reible, C., Watzka, M., 2008. The Vitamin K cycle. Vit. Hormones 78, 35–62.

Schurgers, L.J., Cranenburg, E.C.M., Vermeer, C., 2008. Matrix Gla-protein: The calcification inhibitor in need of vitamin K. Thromb. Haemost. 100, 593–603.

Shearer, M.J., 2008. Vitamin K deficiency bleeding (VKDB) in early infancy. Blood Rev. 23, 49–59.

Shearer, M.J., Newman, P., 2008. Metabolism and cell biology of vitamin K. Thromb. Haemost. 100, 530–547.

Siguret, V., Pautas, E., Gouin-Thibault, I., 2008. Warfarin therapy: Influence of pharmacogenetic and environmental factors on the anticoagulant response to warfarin. Vit. Hormones 78, 247–264.

Stafford, D.W., 2005. The vitamin K cycle. J. Thromb. Haemost. 3, 1873–1878.

Suttie, J.W., 2006. Vitamin K. In: Shils, M.E., Shike, M., Ross, A.C. (Eds.), Modern Nutrition in Health and Disease, tenth ed. Lippincott, New York, NY, pp. 412–425.

Suttie, J.W., 2007. Vitamin K. In: Zemplini, J., Rucker, R.B., McCormick, D.B., Suttie, J.W. (Eds.), Handbook of Vitamins, fourth ed. CRC Press, New York, NY, pp. 111–152.

Suttie, J.W., 2009. Vitamin K in Health and Disease. CRC Press, New York, NY, pp. 224.

Uprichard, J., Perry, D.J., 2002. Factor X deficiency. Blood Rev. 16, 97–110.

Vermeer, C., Shearer, M.J, Zitterman, A., et al. 2004. Beyond deficiency: Potential benefits of increased intakes of vitamin K for bone and vascular health. Eur. J. Nutr. 43, 325–335.

Vitamin C

Anchoring Concepts

1. Vitamin C is the generic descriptor for all compounds exhibiting qualitatively the biological activity of ascorbic acid.
2. Vitamin C-active compounds are hydrophilic, and have an oxidizable/reducible 2,3-enediol grouping.
3. Deficiencies of vitamin C are manifest as connective tissue lesions (e.g., capillary fragility, hemorrhage, muscular weakness).

I still had a gram or so of hexuronic acid. I gave it to [Svirbely] to test for vitaminic activity. I told him that I expected he would find it identical with vitamin C. I always had a strong hunch that this was so but never had tested it. I was not acquainted with animal tests in this field and the whole problem was, for me, too glamorous, and vitamins were, to my mind, theoretically uninteresting. "Vitamin" means that one has to eat it. What one has to eat is the first concern of the chef, not the scientist. Anyway, Svirbely tested hexuronic acid ... after one month the result was evident: hexuronic acid was vitamin C.

A. Szent-Györgyi

Learning Objectives

1. To understand the nature of the various sources of vitamin C.
2. To understand the means of vitamin C synthesis by most species.
3. To understand the means of enteric absorption and transport of vitamin C.
4. To understand the functions of vitamin C in connective tissue metabolism, in drug and steroid metabolism, and in mineral utilization.
5. To understand the physiologic implications of low and high intakes of vitamin C.

VOCABULARY

Antioxidant
Ascorbic acid
Ascorbigen
L-Ascorbic acid 2-sulfate
Ascorbyl free radical
Carnitine
Cholesterol 7α-hydroxylase
Collagens
Dehydroascorbic acid
Dehydroascorbic acid reductase
DNA oxidation
Dopamine β-monooxygenase
Elastin
Glucose transporters (GLUTs)
Glucuronic acid pathway
Guinea pig
L-Gulonolactone oxidase
Histamine
Homogentisate 1,2-dioxygenase
Hydroxylysine
4-Hydroxyphenylpyruvate
Hydroxyproline
Hypoascorbemia
Indian fruit bat
Insulin

The Vitamins. DOI: 10.1016/B978-0-12-381980-2.00009-8

Iron
Lipid peroxidation
Lysyl hydroxylase
Moeller–Barlow disease
Monodehydroascorbate
Monodehydroascorbate reductase
Nitric oxide
Oxalic acid
Peptidylglycine α-amidating monooxygenase
Petechiae
Prolyl hydroxylase
Pro-oxidant
Protein oxidation
Rebound scurvy
Red-vented bulbul
L-Saccharoascorbic acid
Scurvy
Sodium-dependent vitamin C transporters (SVCTs)
Systemic conditioning
Tropoelastin
Tyrosine
Vitamin C

1. THE SIGNIFICANCE OF VITAMIN C

Vitamin C is required by only a few species, which, by virtue of a single enzyme deficiency, cannot synthesize it. For most species, *ascorbic acid* is a normal metabolite of glucose, but it is not an essential dietary constituent. Whether it is synthesized or not, ascorbic acid is important for several physiological functions. Many, if not all, of these functions involve redox characteristics that allow ascorbic acid to play an important role, along with α-tocopherol, reduced glutathione, and other factors, in the antioxidant protection of cells. Thus, ascorbic acid represents the major water-soluble antioxidant in plasma and tissues. As such, it is thought to support the redox recycling of α-tocopherol, the bioavailability of non-heme iron, and the maintenance of enzyme-bound metals in oxidation states appropriate for several enzymatic functions. It is fairly well established that compromises of these effects underlie the pathophysiology of vitamin C deficiency. Other beneficial health effects of ascorbic acid have been reported: reductions in hypertension, atherogenesis, diabetic complications, colds and other infections, and carcinogenesis. Although some of these claims have become widely accepted, the empirical evidence remains incomplete for many.

2. SOURCES OF VITAMIN C

Distribution in Foods

Vitamin C is widely distributed in both plants and animals, occurring mostly (80–90%) as **ascorbic acid** but also as **dehydroascorbic acid**. The proportions of both species tend to vary with food storage time, due to the

time-dependent oxidation of ascorbic acid. Fruits, vegetables,[1] and organ meats (e.g., liver and kidney) are generally the best sources; only small amounts are found in muscle meats (Table 9.1). Plants synthesize L-ascorbic acid from carbohydrates; most seeds do not contain ascorbic acid, but start to synthesize it on sprouting. Some plants accumulate high levels of the vitamin (e.g., fresh tea leaves, some berries, guava, rose hips). Ascorbic acid-containing tissues of cruciferous vegetables of the family *Brassicaceae* typically contain β-thioglucopyranosides called glucosinolates, which degrade, when food is cut and cooked, to products[2] that react spontaneously with ascorbic acid to form a stable adduct without vitamin C activity, **ascorbigen**.

For practical reasons, citrus and other fruits are good daily sources of vitamin C, as they are generally eaten raw and are therefore not subjected to cooking procedures that can destroy vitamin C. Processed foods, such as cured meats and some beverages, can also contain the analog, erythorbic acid,[3] which is used as a preservative. While that analog has no vitamin C activity *in vivo*, it can yield false positives in some analyses for plasma ascorbic acid.[4]

Stability in Foods

The vitamin C contents of most foods decrease dramatically during storage owing to the aggregate effects of several processes by which the vitamin can be destroyed (Table 9.2). Ascorbic acid is susceptible to oxidation to dehydroascorbic acid, which itself can be irreversibly degraded by hydrolytic opening of the lactone ring to yield 2,3-diketogulonic acid, which is not biologically active. These reactions occur in the presence of O_2, even traces of metal ions, and are enhanced by heat and conditions of neutral to alkaline pH. The vitamin is also reduced by exposure to oxidases in plant tissues. Therefore, substantial losses of vitamin C can occur during storage and are enhanced greatly during cooking. For example, stored potatoes lose 50% of their vitamin C within 5 months, and 65% within 8 months, of harvest. Apples and cabbage stored for winter can lose 50% and 40%, respectively, of their original vitamin C content. Losses in cooking are usually greater with such methods as boiling, as the stability of ascorbic acid is much less in aqueous solution. For example, potatoes can lose 40% of their vitamin C content by boiling. Alternatively, quick heating methods can protect food vitamin C by inactivating oxidases.

---

1. Historically, the potato was the best source of vitamin C in North America and Europe.
2. Indole-3 carbinol or 3-indoacetylnitrile.
3. Also referred to as D-isoascorbic acid or D-araboascorbic acid.
4. This is not a problem for blood samples taken after an overnight fast, as erythorbic acid is cleared from the blood within 12 hours.

TABLE 9.1 Vitamin C Contents of Some Uncooked Foods

Food	Vitamin C, mg/100 g
Fruits	
Apples	10–30
Bananas	10
Cherries	10
Grapefruits	40
Guavas	300
Haw berries	160–800
Melon	13–33
Oranges, lemons	50
Peaches	7–14
Raspberries	18–25
Rose hips	1,000
Strawberries	40–90
Tangerines	30
Rhubarb	10
Vegetables	
Asparagus	15–30
Beans	10–30
Broccoli	90–150
Cabbage	30–60
Carrots	5–10
Cauliflower	60–80
Celery	10
Collard greens	100–150
Corn	12
Kale	120–180
Leeks	15–30
Onions	10–30
Peas	10–30
Parsley	170
Peppers	125–200
Potatoes	10–30
Cereals	
Wheat	0
Oats	0
Rice	0
Animal Products	
Meats	0–2
Liver, kidney	10–40
Milk, cow	1–2
Milk, human	3–6

TABLE 9.2 Two-Day Storage Losses of Vitamin C

Food	% Lost At:	
	4°C	20°C
Beans	33	53
Cauliflower	8	26
Lettuce	36	42
Parsley	13	70
Peas	10	36
Spinach	32	80
Spinach (winter)	7	22

TABLE 9.3 Vitamin C-Active Derivatives of Ascorbic Acid

Strongly Biopotent[a]	Weakly Biopotent[b]
Ascorbic acid 2-*O*-α-glucoside	L-ascorbyl palmitate
6-Bromo-6-deoxy-L-ascorbic acid	L-ascorbyl-2-sulfate
L-ascorbate 2-phosphate	L-ascorbate-*O*-methyl ether
L-ascorbate 2-triphosphate	

[a] >50% antiscorbutic activity of ascorbic acid.
[b] <50% antiscorbutic activity of ascorbic acid.

Vitamin C Bioavailability

Vitamin C in most foods appears to have biological activities comparable to that of purified L-ascorbic acid at doses in the nutritional range (15–200 mg). At higher doses bioavailability declines; doses great than 1,000 mg appear to be utilized with roughly 50% efficiency. Because dehydroascorbic acid can be reduced metabolically to yield ascorbic acid (after enteric absorption and subsequent cellular uptake), both forms present in foods have vitamin activity. Several synthetic ascorbic acid derivatives also have vitamin C activity and offer advantages of superior chemical stability. Forms such as ascorbate 2-sulfate, ascorbate 2-monophosphate, ascorbate 2-diphosphate, and ascorbate 2-triphosphate (mixtures of the latter three are referred to as **ascorbate polyphosphate**) are useful as vitamin C supplements for fish diets where the intrinsic instability of ascorbic acid in aqueous environments is a problem. The more highly biopotent of these vitamers appear to be effectively hydrolyzed in the digestive tract and tissues to yield ascorbic acid (Table 9.3).

Biosynthesis of Ascorbic Acid

In addition to probably all green plants, most higher animal species can synthesize vitamin C from glucose *via* the

TABLE 9.4 Estimated Rates of Ascorbic Acid Biosynthesis in Several Species

Species	Synthetic Rate (mg/kg BW)	$T_{1/2}{}^a$ (Days)	Turnover %/Day
Mouse	125	1.4	50
Golden hamster	20	2.7	26
Rat	25	2.6	26
Rabbit	5	3.9	18
Guinea pig	0	3.8	18
Human	0	10–20	3

^{a}Half-life in the body.

① multiple steps involving phosphorylation and reaction with UTP
② catalyzed by L-gulonolactone oxidase

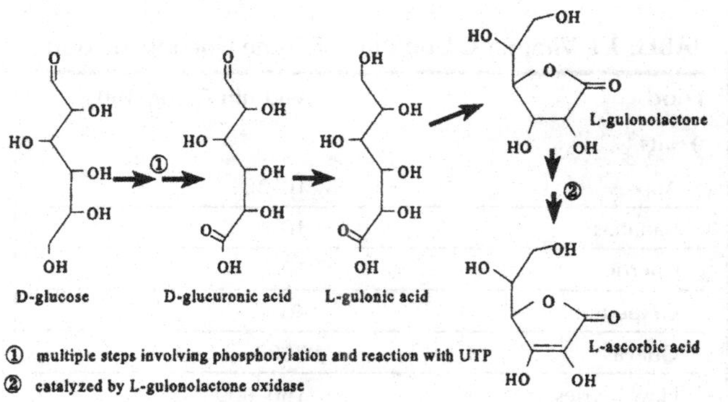

FIGURE 9.1 Biosynthesis of ascorbic acid.

glucuronic acid pathway (Table 9.4). The enzymes of this pathway are localized in the kidneys of amphibians, reptiles, and the more primitive orders of birds, but they occur in the livers of passerine birds and mammals. Egg-laying mammals synthesize ascorbic acid only in their kidneys, and many marsupials use both their liver and kidneys for this purpose. The transfer of ascorbic acid synthesis from the kidney to the larger liver has been interpreted as an evolutionary adaptation that provided increased synthetic capacity of the larger organ to meet the increased needs associated with homeothermy.

The biosynthesis of ascorbic acid may not occur early in fetal development, and can be inhibited by deficiencies of vitamins A and E, and of biotin. It can be stimulated under conditions of glycogen breakdown, and by certain drugs (e.g., barbiturates, aminopyrine, antipyrine, chlorobutanol) and carcinogens (e.g., 3-methylcholanthrene, benzo-α-pyrene). The stimulation of ascorbic acid biosynthesis that occurs owing to exposure to xenobiotic compounds appears to be due to a general induction of the enzymes of the glucuronic acid pathway, which produces glucuronic acid for conjugating foreign compounds as a means of their detoxification (Fig. 9.1).[5]

Dietary Need Due to Enzyme Deficiency

Some animal groups appear to have lost in evolution the capacity for ascorbic acid biosynthesis.[6] This involves the lost expression of the last enzyme in the biosynthetic

pathway, L-**gulonolactone oxidase**,[7,8] the microsomal flavoenzyme that catalyzes the oxidation of L-gulonolactone to L-2-ketogulonolactone, which yields L-ascorbic acid by spontaneous isomerization. While all species studied appear to have the gene, in some it is so highly mutated[9] that it yields no gene product. The loss of this single enzyme renders ascorbic acid, an otherwise normal metabolite, a vitamin. Such is the case for invertebrates, most fishes,[10] and a few species of birds (e.g., **red-vented bulbul**[11]) and mammals (humans, other primates, **guinea pigs**, **Indian fruit bat**, and mutant strains of rats).[12]

Therefore, **scurvy** can correctly be considered a congenital metabolic disease, **hypoascorbemia**. It is apparent that even the species capable of producing this key enzyme may not express it early in development. For example, the fetal rat has been found incapable of ascorbic acid biosynthesis until day 16 of gestation. This developmental lag may account for the perinatal declines in tissue ascorbic acid concentrations that have been observed in the species experimentally.

---

7. This has been established for primates, guinea pigs, and fruit bats; whether the loss of this enzyme activity is the basis of the inabilities of other species to synthesize ascorbic acid is still speculative.

8. That vitamin C deficiency results from the loss of this enzyme is indicated by the fact that enzyme replacement by injection with the substrate, L-gulonolactone, prevents scurvy in guinea pigs.

9. This may be due to the presence of retrovirus-like sequences, which have been identified in the human gene, that may have caused the activation of the gene. It has been suggested that mutations in this gene may have been driven by disadvantageous effects of H_2O_2 generated during the oxidation of gulono-1,4-lactone.

10. Although some fish appear to be able to synthesize ascorbic acid, only the carp and Australian lungfish appear to be able to do so at rates sufficient to meet their physiologic needs.

11. The bulbuls (Pycnonotidae) comprise 13 genera and 109 species distributed in Africa, Madagascar, and southern Asia. While the red-vented bulbul is often cited as being unable to biosynthesize ascorbic acid, it is not known how widely distributed in this family, as well as in the class Aves, is the dietary need for vitamin C.

12. A gulonolactone null mouse has been developed; and the osteogenic disorder Shionog rat (ODS-od/od) derived from the Wistar strain has a dysfunctional form of that enzyme.

---

5. Because ascorbic acid synthesis and excretion are increased by exposure to xenobiotic inducers of hepatic, cytochrome P-450-dependent, mixed-function oxidases (MFOs), it has been suggested that the urinary ascorbic acid concentration may be useful as a non-invasive screening parameter of MFO status.

6. That insects and other invertebrates are incapable of synthesizing L-gulono-1,4-lactone suggests that they do not synthesize ascorbic acid.

3. ABSORPTION OF VITAMIN C

Species without Dietary Needs: Passive Uptake

Species that can synthesize ascorbic acid do not have active transport mechanisms for its enteric absorption. They absorb ascorbic acid strictly by passive diffusion.

Species with Dietary Needs: Active Uptake

Species that do not synthesize ascorbic acid (e.g., humans, guinea pigs) absorb the vitamin by both passive and active means. Passive diffusion is important at high doses. At low doses, the most important means of absorbing the vitamin involves saturable, carrier-mediated active transport mechanisms. Thus, the efficiency of absorption of physiological doses (e.g., ≤ 180 mg/day for a human adult) of vitamin C is high, 80–90%, and declines markedly at vitamin C doses greater than about 1 g.[13] The reduced and oxidized forms of the vitamin are absorbed by different mechanisms of active transport which occur throughout the small intestine:

- *Sodium-dependent Vitamin C Transporters (SVCTs)*.[14] These carriers are glycoproteins that move L-ascorbic acid by an electrogenic, Na^+-dependent process with a stoichiometric ratio of two Na^+ ions per ascorbic acid molecule. Of two isoforms, SVCT1 is the dominant form expressed in the intestinal mucosa. It is inhibited by aspirin.[15] Because knockout of SVCT1 expression does not block ascorbic acid uptake,[16] it is clear that other means of absorbing the vitamin are available.
- *Glucose Transporters (GLUTs)*. The uptake of dehydroascorbic acid is 10- to 20-fold faster than that of ascorbic acid,[17] and involves isoforms of the glucose transporters GLUT1,[18] GLUT3, and, perhaps, GLUT4. Upon entry into the cell, dehydroascorbic acid is quickly reduced to ascorbic acid, probably by glutaredoxine reductase and/or reduced glutathione (GSH).

---

13. The efficiency of vitamin C absorption declines from about 75% of a 1 g dose, to about 40% of a 3 g dose and about 24% of a 5 g dose; net absorption plateaus at 1–1.2 g at doses of at least 3 g.

14. These are members of the SLC23 human gene family.

15. For example, in humans a 900 mg dose of aspirin blocks the expected rises in plasma, leukocyte, and urinary levels of ascorbic acid owing to a simultaneous dose of 500 mg of vitamin C.

16. Corpe, C.P., Tu, H., Eck, P. *et al.* (2010). *J. Clin. Invest.* 120,1069.

17. Studies with cultured cells have shown that D-isoascorbic acid has only 20–30% of the activity of L-ascorbic acid in stimulating collagen production. The basis of this difference involves the much slower cellular uptake of the D-form, as, once inside the cell, both vitamers behaved almost identically.

18. Congenital deficiency of GLUT1, a rare condition, is manifest in infancy as seizures and delayed development, presumably due to insufficient supply of glucose to the brain.

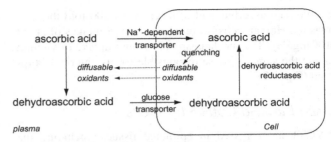

FIGURE 9.2 Redox cycling of ascorbic acid.

Several mechanisms have been proposed for the efflux of ascorbic acid from cells, one or more of which is likely to facilitate its movement across the basolateral side of the mucosal epithelial cell. These mechanisms include volume-sensitive or Ca^{2+}-sensitive anion channels that form pores in the plasma membrane, glutamate–ascorbate exchange, exocytosis of ascorbate-containing vesicles, and gap-junction hemichannels.[19]

4. TRANSPORT OF VITAMIN C

Transport Predominantly in Reduced Form

Vitamin C is transported in the plasma predominantly (80–90%) in the reduced form, ascorbic acid. Also present are small amounts of dehydroascorbic acid, which are thought to be formed by oxidation of ascorbic acid by diffusible oxidants of cellular origin (Fig. 9.2). Plasma ascorbic acid shows a sigmoid relationship with the level of vitamin C intake, saturation in humans being achieved at daily doses of 1,000 mg or more.[20] Plasma ascorbic acid levels in healthy humans are typically 30–70 μmol/l, and appear to be affected by body fat distribution[21] by mechanisms that may relate to its being a marker for the intake of vitamin C-rich fruits and vegetables.

Cellular Uptake by the Same Mechanisms

Simple diffusion of ascorbic acid and dehydroascorbic acid into cells is negligible due to the charge of the former, which is ionized under physiological conditions, and the oil:water partitioning characteristics of the latter, which excludes it from lipid bilayers. Nevertheless, cells

---

19. See review: Corti, A., Casini, A. F., and Pompella, A. (2010). *Arch. Biochem. Biophys.* 500, 107.

20. Levine, M., Conry-Cantilena, C., Wang, Y., *et al.* ([1996] *Proc. Natl Acad. Sci.* USA 93, 3704) found that 200 mg/day doses produced only 80% saturation and that RDA-level doses supported plasma ascorbic acid concentrations on the lower third of the response curve.

21. Plasma ascorbic acid levels were inversely related to waist-to-hip ratio, and to waist and hip circumferences, but not to body mass index, in a large European cohort (Canoy, D., Wareham, N., Welch, A., *et al.* [2005]. *Am. J. Clin. Nutr.* 82, 1203).

accumulate ascorbic acid to levels 5- to 100-fold those of plasma. Human cells become saturated at intakes of about 100 mg/day. The mechanisms of cellular uptake of vitamin C are the same as those responsible for its enteric absorption (Fig. 9.2).

Ascorbic Acid Uptake by SVCTs

SVCT1 is expressed in epithelial tissues, including the intestine, liver, and kidney; SVCT2 is expressed in brain, lung, heart, eye, and placenta, in neuroendocrine and exocrine tissues, and in endothelial tissues. Both isoforms contain multiple potential N-glycosylation and protein kinase C (PKC) phosphorylation sites, suggesting regulation via glycosylation and/or PKC pathways. In the absence of ascorbic acid, the SVCTs can facilitate the unitransport of Na^+, allowing that ion to leak from cells. The SVCTs are non-competitively inhibited by flavonoids, and can be affected by cytokines and steroids. Expression of SVCT1 appears to be reduced by exposure to high levels of ascorbic acid.[22] SVCT1 appears to be principally involved in the maintenance of whole-body vitamin C homeostasis by affecting enteric absorption and renal reabsorption;[23] while VCT2 appears to be of principal importance in the protection of metabolically active tissues from oxidative stress.

Dehydroascorbic Acid Uptake by GLUTs

By interacting at the level of these transporters, insulin can promote the cellular uptake of dehydroascorbic acid. By competing for uptake by the transporter, physiological levels of glucose can inhibit dehydroascorbic acid uptake by several cell types (adipocytes, erythrocytes, granulose cells, neutrophils, osteoblasts, and smooth muscle cells). Thus, diabetic patients can have abnormally high plasma levels of dehydroascorbic acid.[24]

Tissue Distribution

Nearly all tissues accumulate vitamin C, including some that lack ascorbic acid-dependent enzymes (Table 9.5). Certain cell types (e.g., peripheral mononuclear leukocytes) can accumulate concentrations as great as several millimolar. Tissue levels are decreased by virtually all forms of stress, which also stimulates the biosynthesis

TABLE 9.5 Ascorbic Acid Concentrations of Human Tissues

Tissue	Ascorbic Acid, mg/100 g
Adrenals	30–40
Pituitary	40–50
Liver	10–16
Thymus	10–15
Lungs	7
Kidneys	5–15
Heart	5–15
Muscle	3–4
Brain	3–15
Pancreas	10–15
Lens	25–31
Plasma	0.4–1

of the vitamin in those animals able to do so.[25] The concentration of ascorbic acid in the adrenals is very high (72–168 mg/100 g in the cow); approximately one-third of the vitamin is concentrated in the reduced form at the site of catecholamine formation in those glands, from which it is released with newly synthesized corticosteroids in response to stress.[26] The ascorbic acid concentration of brain tissue also tends to be high (5–28 mg/100 g); the greatest concentrations are found in regions that are also rich in catecholamines. Brain ascorbic acid levels are among the last to be affected by dietary deprivation of vitamin C in those animals unable to synthesize the vitamin. A relatively large amount of ascorbic acid is also found in the eye, where it is thought to protect critical sulfhydryl groups of proteins from oxidation.[27] The levels of ascorbic acid in plasma reach plateaus at vitamin C doses greater than 2 g/day (Fig. 9.3). White blood cells show similar thresholds, with lymphocytes, platelets, monocytes, and neutrophils showing decreasing plateau levels in that order.[28] Blood cells contain a substantial fraction of the ascorbic acid in the blood; of these, leukocytes have particular diagnostic value, as their ascorbic acid concentrations

22. MacDonald, L. (2002). *Br. J. Nutr.* 87, 97.

23. Knockout of SVCT1 resulted in massive losses of ascorbic acid from the plasma into the urine (Corpe, C. P., Tu, H., Eck, P., *et al.* [2010]. *J. Clin. Invest.* 120:1069).

24. In fact, it is thought that the impaired cellular uptake of vitamin C, owing to competition with glucose, may be one of the causes of pathology in diabetes.

25. The ascorbic acid content of brown adipose tissue of rats has been found to increase by about 60% during periods of cold stress.

26. That is, in response to the release of adrenocorticotropic hormone (ACTH).

27. Lenses of cataract patients have lower lens ascorbic acid concentrations (e.g., 0–5.5 mg/100 g) than those of healthy patients (e.g., 30 mg/100 g).

28. Levine, M., Wang, Y., Padayatty, S. J., *et al.* (2001). *Proc. Natl Acad. Sci. USA* 98, 9842.

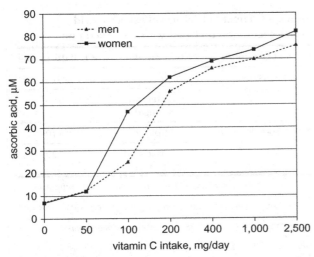

FIGURE 9.3 Relationship of plasma ascorbic acid (steady state) level and vitamin C intake. From *Levine, M., Wang, Y., Padayatty, S. J. et al. (2001)*. Proc. Natl Acad. Sci. USA *98, 9842.*

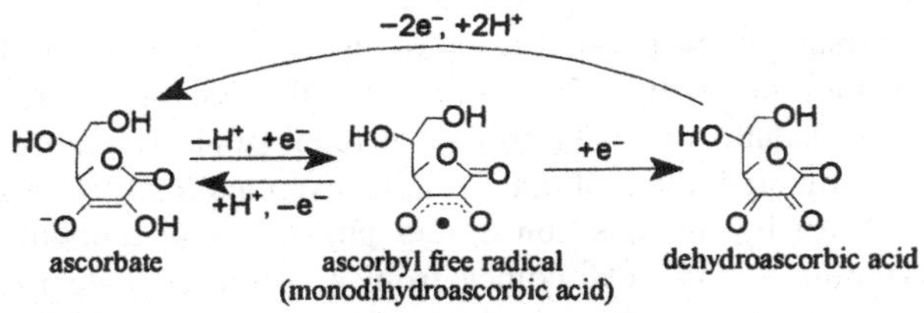

FIGURE 9.4 Oxidation–reduction reactions of vitamin C.

are regarded as indicative of tissue levels of the vitamin.[29] There is no stable reserve of vitamin C; excesses are quickly excreted. At saturation, the total body pool of the human has been estimated to be 1.5–5 g,[30] the major fractions being found in the liver and muscles, by virtue of their relatively large masses.

5. METABOLISM OF VITAMIN C

Oxidation

Ascorbic acid is oxidized *in vivo* by two successive losses of single electrons (Fig. 9.4). The first monovalent oxidation results in the formation of the **ascorbyl free radical**;[31] the ascorbyl radical forms a reversible electrochemical couple with ascorbic acid, but can be further oxidized irreversibly to dehydro-L-ascorbic acid.

---

29. Leukocyte ascorbic acid concentrations are usually greater in women than men, and normally decrease with age and in some diseases.
30. The first signs of scurvy are not seen until this reserve is depleted to 300–400 mg.
31. The ascorbyl radical is also called monodehydroascorbic acid; it is relatively stable, with a rate constant for its decay of about $10^5 M^{-1} s^{-1}$.

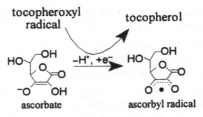

FIGURE 9.5 Coupling of ascorbate oxidation to reduction of tocopheroxyl radical.

Subsequent irreversible hydrolysis of dehydroascorbic acid yields 2,3-diketo-L-gulonic acid, which undergoes either decarboxylation to CO_2 and five-carbon fragments (xylose, xylonic acid, lyxonic acid), or oxidation to **oxalic acid** and 4-C fragments (e.g., threonic acid). In addition, the formation of L-**ascorbic acid 2-sulfate** from ascorbic acid occurs in humans, fishes, and perhaps rats, and the oxidation of the 6-position carbon (C-6) of ascorbic acid to form L-**saccharoascorbic acid** has been demonstrated in monkeys.

Ascorbic acid may also undergo oxidation by reaction with tocopheroxyl or urate radicals (Fig. 9.5). The former interaction, which is discussed in detail in Chapter 7, remains poorly supported by *in vivo* evidence.

Ascorbic Acid Regeneration

Ascorbic acid can also be regenerated in three ways:

- *Redox recycling.* Because dehyroascorbic acid can be reduced only by intracellular dehydroascorbic acid reductases, the presence of an uptake system specific for the oxidized form effectively establishes a redox cycle whereby plasma intracellular levels of the reduced vitamin are maintained for such functional purposes as the quenching of oxidants (Fig. 9.2). This system may be important in the ascorbate-stimulation of osteoid-forming activity of osteoblasts through the generation by osteoclasts of reactive oxygen species that oxidize ascorbic acid extracellularly. The existence of multiple dehydroascorbic acid reductase activities would appear to promote a favorable ascorbate redox potential, indirectly preserving other antioxidants (e.g., tocopherol). That the NADPH-dependent enzyme is found in the intestinal mucosa suggests that this function may be important in protecting against the many radical species generated there. Although dehydroascorbate appears to have no metabolic function *per se*, its regeneration to ascorbic renders it biologically active. Impairments in this recycling can occur in uncontrolled diabetes due to excessive plasma glucose, which competes with dehydroascorbic acid for cellular uptake by glucose transporters. Reduced cellular uptake of dehydroascorbic acid can lead to diminished

intracellular ascorbic acid levels, weakening antioxidant defenses in diabetes.

- *Reduction of dehydroascorbic acid.* Dehydroascorbate
  can react directly with the reduced forms of glutathione
  (GSH) or lipoic acid (dihydrolipoic acid).
- *Reduction of ascorbyl radical.* This is catalyzed by
  the widely distributed enzyme, semidehydroascorbyl
  reductase, which uses NADPH as a source of reducing
  equivalents.

Persistent, enhanced ascorbic acid catabolism after prolonged intake of large amounts of vitamin C (so-called **systemic conditioning**) has been proposed, on the basis of
uncontrolled observations of a few individuals and what is
now widely regarded as erroneous interpretations of experimental results.[32] Controlled studies have not consistently
demonstrated such effects. Kinetic studies in humans indicate
that high doses of ascorbic acid are mostly degraded to CO_2
by microbes, with major portions also excreted intact in the
urine. While such results would appear to refute the prospect
of induced catabolism, a well-controlled study with guinea
pigs found plasma ascorbic acid levels to drop transiently in
some animals removed from high-level vitamin C treatments,
which the authors attributed to upregulated catabolic enzymes.

Excretion

Ascorbate is thought to pass unchanged through the
glomeruli, and to be actively reabsorbed in the tubules by
a saturable, carrier-mediated process. Little, if any, ascorbic acid is excreted in the urine of humans consuming
less than 100 mg/day, and only one-fourth of the dose is
excreted at twice that intake. At doses greater than about
500 mg/day (i.e., when blood ascorbic acid concentrations
exceed 1.2–1.8 mg/dl), virtually all ascorbic acid above
that level is excreted unchanged in the urine, thus producing no further increases in body ascorbate stores. The
fractional excretion of a parenteral dose of ascorbic acid
approaches 100% at doses >2 g.

The epithelial cells of the renal tubules reabsorb dehydroascorbic acid after it has been filtered from the plasma.
Animal species vary in their routes of disposition of dehydroascorbic acid. Guinea pigs and rats degrade it almost
quantitatively to CO_2,[33] which is lost across the lungs.
Humans, however, normally degrade only a very small
amount via that route,[34] excreting the vitamin primarily as various urinary metabolites (mostly ascorbic acid,

TABLE 9.6 Effect of High-Level Ascorbic Acid
Supplementation on Urinary Oxalate Excretion

Subject Group (n)	Treatment[a]	Oxalate, μmole
Responders (19)	Control	513 ± 97
	Ascorbic acid, 1,000 mg/d	707 ± 165[b]
Non-responders (29)	Control	560 ± 110
	Ascorbic acid, 1,000 mg/d	551 ± 129[c]

[a]Each subject experienced alternating 6-day control and ascorbic acid treatments.
[b]Significantly different (P <0.05) from control treatment within responder group.
[c]Significantly different (P <0.05) from other responder group on same treatment.
From Massey, L. K., Liebman, M., Kynast-Gales, S. A., et al. (2005). J. Nutr. 135, 1673.

dehydroascorbic acid, and diketogulonic acid, with small
amounts of oxalate and ascorbate 2-sulfate).

Humans convert only 1.5% of ingested ascorbic acid
to oxalic acid within 24 hours. Nevertheless, the excretion of oxalate is relevant to risk of renal stone formation.
It is estimated that, of the oxalate excreted daily (e.g.,
30–40 mg) by humans consuming physiological amounts
of vitamin C, 35–50% comes from ascorbic acid degradation (the balance coming from glycine and glyoxylate).
Not all humans show increased urinary oxalate excretion
in response to ascorbic acid supplementation; about 40%
of adults consuming very high doses (1,000 mg/day) of
vitamin C increased their urinary oxalate levels by more
than 16% (Table 9.6). However, as oxalate excretion is not
a useful indicator of risk of renal calculi (e.g., a comparison of 75 patients with renal calculi and 50 healthy controls showed the same average oxalate excretion of about
28 mg/day), and as the contributions of high doses of vitamin C to oxalate formation are rather small, the physiological implications of these effects are unclear.

Ascorbic acid is also excreted in the gastric juice,
which typically has levels three times those of plasma.
Notable exceptions are in patients with atrophic gastritis or
Heliobacter pylori infection, which show low gastric juice
ascorbic acid.[35]

6. METABOLIC FUNCTIONS OF VITAMIN C

In its various known metabolic functions, vitamin C as
ascorbic acid serves as a classic enzyme cofactor (e.g., at
the active site of hydroxylating enzymes), as a protective

32. For example, enhanced $^{14}CO_2$ excretion from guinea pigs with larger body pools of ascorbic acid was taken as evidence of greater catabolism.
33. The C-1 carbon of ascorbic acid is the main source of CO_2 derived from the vitamin, whereas C-1 and C-2 are the precursors of oxalic acid.
34. Degradation by this path is increased greatly in some diseases, and can then account for nearly half of ascorbic acid loss.

35. Sobala, G. M., Schorah, C. J., Shires, S., *et al.* (1993). *Gut* 34, 1038.

agent (e.g., of hydroxylases in collagen biosynthesis), and as ascorbyl radical in reactions with transition metal ions. Each of these functions of the vitamin involves its redox properties.

Electron Transport

Ascorbic acid loses electrons easily, and, because of its reversible monovalent oxidation to the ascorbyl radical, it can serve as a biochemical redox system.[36] As such, it is involved in many electron transport reactions; these include reactions involved in the synthesis of collagen, the degradation of **4-hydroxyphenylpyruvate**,[37] the synthesis of norepinephrine,[38] and the desaturation of fatty acids. In many of these functions ascorbate is not required *per se* – i.e., it can be replaced by other reductants[39] – however, ascorbate is the most effective *in vitro*. In each, ascorbic acid is regenerated, as the electron acceptor ascorbyl radical is reduced by either of two microsomal enzymes, monodehydroascorbate reductase or ascorbate–cytochrome-b_5 reductase, the latter being part of the fatty acid desaturation system. Evidence has also been presented for a function of ascorbate in the transplasma membrane electron transport, which results in the net reduction of extracellular oxidants.[40]

Antioxidant Functions

Ascorbic acid can act as an **antioxidant** owing to its ability to react with free radicals, undergoing a single-electron oxidation to yield a relatively poorly reactive intermediate, the ascorbyl radical, which disproportionates to ascorbate and dehydroascorbate. Thus, ascorbic acid can reduce toxic, reactive oxygen species superoxide anion (O_2^-) and hydroxyl radical (OH·), as well as organic (RO_2·) and nitrogen (NO_2·) oxy radicals. Those reactions are likely to be of fundamental importance in all aerobic cells, which must defend against the toxicity of the very element depended on as the terminal electron acceptor for energy production via the respiratory chain enzymes. It is this type of reaction that appears to be the basis of most, if not all, of the essential biological functions of ascorbic acid. One of these is important in extending the antioxidant protection to the hydrophobic regions of cells: ascorbic acid

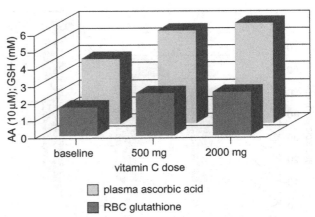

FIGURE 9.6 Enhancement of reduced glutathione (GSH) by vitamin C in men. After *Johnston, S. C., Meyer, C. G., and Srilakshmi, J. C. (1993). Am. J. Clin. Nutr. 58, 103.*

appears to be able to reduce the semistable chromanoxyl radical, thus regenerating the metabolically active form of the lipid antioxidant vitamin E.[41] Such quenching of oxidants protects glutathione in its reduced form (Fig. 9.6).

The antioxidant efficiency of ascorbic acid is significant at physiological concentrations of the vitamin (20–90 μmol/l). Under those conditions, the predominant reaction is a radical chain-terminating one of ascorbate (AH$^-$) with a peroxyl radical to yield a hydroperoxide and the **ascorbyl radical** (A·$^-$), which proceeds to reduce a second peroxyl radical and yield the vitamin in its oxidized form, dehydroascorbic acid (A). At low concentrations of the vitamin, 2 mol of peroxyl radical is reduced for every mole of ascorbate consumed:

$$AH^- + RO· \rightarrow A·^- + ROO· (+H^+ \rightarrow ROOH)$$

$$A·^- + R'OO· \rightarrow A + R'OO^- (+H^+ \rightarrow R'OOH)$$

Dehydroascorbic acid is inherently unstable, with a half-life of only minutes in physiological conditions and undergoing ring-opening to yield diketoglulonic acid. Therefore, exposure to free radicals can lead to the consumption of vitamin C. *In vitro* studies have confirmed that ascorbic acid is oxidized by free radicals in cigarette smoke.[42] This type of direct effect appears to be moderated by the presence of other antioxidants (e.g., reduced glutathione), but is exacerbated by inflammatory oxidants such

36. The redox potential of the dehydroascorbic acid–ascorbic acid couple is in the range of 0.06–0.1 V, but that of the ascorbyl radical–ascorbic acid couple is −0.17 V. These redox potentials result in the reduction of many oxidizing compounds.

37. The first product of tyrosine metabolism.

38. Noradrenaline.

39. For example, reduced glutathione, cysteine, tetrahydrofolate, dithiothreitol, 2-mercaptoethanol.

40. Lane, D. J. R. and Lawren, A. (2009). *Free Rad. Biol. Med.* 47, 485.

41. Evidence for such an effect comes from demonstrations *in vitro* of the reduction by ascorbic acid of the tocopheroxyl radical to tocopherol, as well as from findings in animals that supplemental vitamin C can increase tissue tocopherol concentrations and spare dietary vitamin E.

42. See Eiserich, J. P. *et al.* (1997). In *Vitamin C in Health and Disease* (Packer, L. and Fuchs, J., eds). Marcel Dekker, New York, NY, pp. 399–412.

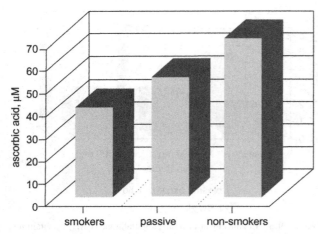

FIGURE 9.7 Effect of smoking on plasma ascorbic acid level. After *Tribble, D. L., Giuliano, L. J., and Fortmann, S. P. (1993). Am. J. Clin. Nutr. 58, 886.*

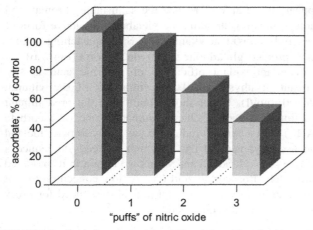

FIGURE 9.8 Oxidation of ascorbic acid by nitric oxide. After *Eiserich, J. P., Cross, C. E., and Van Der Vliet, A. (1997). In* Vitamin C in Health and Disease *(Packer, L. and Fuchs, J., eds). Marcel Dekkar, New York, NY, pp. 399–412.*

as $O_2\cdot$, H_2O_2, and hypochlorous acid (HOCl) produced by phagocytes recruited and activated in the lungs by cigarette smoke. Accordingly, smokers, who expose themselves to a variety of highly reactive free radicals in tobacco smoke[43,44] (Figs 9.7, 9.8), have been found to show a 40% greater turnover of ascorbic acid and higher plasma levels of lipid peroxidation products than do non-smokers with similar vitamin C intakes.[45] Even non-smokers exposed

passively to tobacco smoke have been found to have lower circulating ascorbic acid levels than non-exposed persons.

At relatively high vitamin C concentrations, a slower radical chain-propagating reaction of ascorbyl radical and molecular oxygen appears to become significant. It yields dehydroascorbic acid and superoxide radical, which in turn can oxidize ascorbate to return ascorbyl radical:

$$A^{\cdot-} + O_2 \rightarrow A + O_2^{\cdot-}$$

$$O_2^{\cdot-} + AH^- \rightarrow HOO^- + A^{\cdot-}$$

It is thought that, at high vitamin C concentrations, this two-reaction sequence can develop into a radical chain autoxidation process that consumes ascorbate, thus wasting the vitamin. Hence, in aerobic systems, the efficiency of radical quenching of ascorbate is inversely related to the concentration of the vitamin. At physiological concentrations, ascorbic acid thus serves as one of the strongest reductants and radical scavengers; it reduces oxy, nitro, and thyl radicals.

Pro-Oxidant Potential

In the presence of oxidized metal ions (e.g., Fe^{3+}, Cu^{2+}), high concentrations of ascorbic acid can have pro-oxidant functions at least *in vitro*. It does so by donating a single electron to reduce such ions to forms that, in turn, can react with O_2 to form oxygen radicals (the metal ions being reoxidized in the process):

$$AH^- + Fe^{3+} \rightarrow A^{\cdot-} + Fe^{2+}$$

$$Fe^{2+} + H_2O_2 \rightarrow Fe^{3+} + OH^{\cdot} + OH^-$$

Thus, ascorbate can react with copper or iron salts *in vitro* and lead to the formation of H_2O_2, $O_2^{\cdot-}$, and $OH\cdot$, which can damage nucleic acids, proteins, and polyunsaturated fatty acids. Accordingly, iron–ascorbate mixtures are often used to stimulate lipid peroxidation *in vitro*, and such pro-oxidative reactions with transition metals are likely to be the basis of the cytotoxic and mutagenic effects of ascorbic acid observed in isolated cells *in vitro*. It has been suggested that high serum ascorbic acid may reduce Fe^{3+} in ferritin[46] to the catalytically active form Fe^{2+}.

The physiological relevance of these pro-oxidative reactions is unclear. Under physiological conditions, tissue concentrations of ascorbic acid greatly exceed those of dehydroascorbic acid, which greatly exceed those of

---

43. For example, nitric oxide (NO·); nitrogen dioxide (·NO₂); and alkyl, alkoxyl, and peroxyl radicals.

44. Free radical-mediated processes are thought to be involved in the pathobiology of chronic and degenerative diseases associated with cigarette smoking, e.g., chronic bronchitis, emphysema, cancer, and cardiovascular disease.

45. Smith, J. L. and Hodges, R. E. (1987). *Ann. NY Acad. Sci.* 498, 144.

---

46. This would require that ascorbate enter the pores of the ferritin protein shell to react with iron on the inner surface.

ascorbyl radical. Thus, the redox potentials of the ascorbyl radical–ascorbic acid and dehydroascorbic acid–ascorbyl radical couples are sufficient for the reduction of most oxidizing compounds.

Cellular Antioxidant Functions

As the most effective aqueous antioxidant in plasma, interstitial fluids, and soluble phases of cells, ascorbic acid appears to be the first line of defense against reactive oxygen species arising in those compartments. Those species include superoxide and hydrogen peroxide arising from activated polymorphonuclear leukocytes or other cells, and from gas-phase cigarette smoke,[47] which can promote the oxidation of critical cellular components.

Lipid Peroxidation

The LDL-protective action of vitamin E (see Chapter 8) appears to be dependent on the presence of ascorbic acid which reduces the tocopheroxyl radical, thus preventing the latter from acting pro-oxidatively by abstracting hydrogen from a cholesteryl-polyunsaturated fatty ester to yield a peroxyl radical.

Protein Oxidation

It is clear that, at least *in vitro*, reactive oxygen species can oxidize proteins to produce carbonyl derivatives and other oxidative changes associated with loss-of-function. Whether ascorbic acid provides such protection *in vivo* is, however, not clear, although it has been suggested that the reductive repair of protein radicals may contribute significantly to the depletion of ascorbate under conditions of oxidative stress, and to the effects reported for vitamin C in reducing risks to cataracts and other illnesses.

DNA Oxidation

As a physiological antioxidant, ascorbic acid plays a role in the prevention of oxidative damage to DNA, which is elevated in cells at sites of chronic inflammation and in many pre-neoplastic lesions. In fact, the continuing attack of DNA by unquenched reactive oxygen species is believed to contribute to cancer. While most DNA damage is repaired metabolically, the frequency of elevated steady-state levels of oxidized DNA bases is estimated to be sufficient to cause mutational events.[48] The levels of one base

damage product, 8-hydroxy-2′-deoxyguanosine, have been found to be elevated in individuals with severe vitamin C deficiency[49] and to be reduced by supplementation with vitamins C and E.[50]

NO Oxidation

Ascorbic acid protects nitric oxide (NO) from oxidation, supporting the favorable effects of the latter on vascular epithelial function. This appears to be the basis for the effects of high-level vitamin C intake in lowering blood pressure.

Metabolic Switching of Neuronal Energy Metabolism

The brain and spinal cord are among the richest tissues in ascorbic acid content, with concentrations of 100–500 µmol/l. Astrocytes have been shown to release ascorbic acid, which can be taken up by neurons (via SVCTs) where it scavenges ROS generated by neurotransmission. Intracellular ascorbic acid also promotes a shift in the primary neuronal energy substrate by inhibiting glycolysis, which reduces pyruvate production and increases the NAD^+/NADH ratio, and leads to the oxidation of lactate.[51]

Metal Ion Metabolism

Promotion of Iron Bioavailability

Ascorbic acid increases the bioavailability of iron in foods. This effect is associated with increased enteric absorption[52] (which is normally low) of the mineral; it affects both non-heme iron as well as heme iron. The effect on non-heme iron involves reduction of the ferric form of the element (Fe^{3+}), predominant in the acidic environment of the stomach, to the ferrous form (Fe^{2+}), then forming a soluble stable chelate that stays in solution in the alkaline environment of the small intestine and is thus rather well absorbed. Studies with iron-deficient rats, which have upregulated enteric iron absorption, have shown ascorbic acid to affect the mucosal uptake of iron but not its mucosal transfer. This effect depends on the presence of both ascorbic acid and iron in the gut at the same time – e.g., the consumption of a vitamin C-containing food with the meal. Thus, the low bioavailability of non-heme iron

---

47. Indeed, genetically scorbutic rats have been found to have elevated levels of LDL lipid peroxidation products (i.e., thiobarbituric-reactive substances, TBARS), which respond to vitamin C supplementation.
48. About 1 per 10^5 bases (Halliwell, B. [2000] *Am. J. Clin. Nutr.* 72, 1082).

49. Rehman, A., Collis, C. S., and Yang, M. (1998). *Biochim. Biophys. Res. Commun.* 246, 293.
50. Moller, P., Viscovich, M., Lykkesfeldt, J., *et al.* (2004). *Eur. J. Nutr.* 43, 267.
51. Castro, M. A., Beltrán, F. A., Brauchi, S., *et al.* (2009). *J. Neurochem.* 110, 423.
52. This effect can be 200–600%.

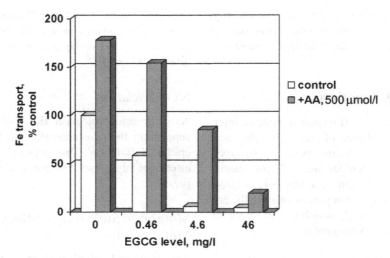

FIGURE 9.9   Effect of vitamin C on iron transport in the CaCo-2 cell model. Abbreviations: AA, ascorbic acid; EGCG, epigallocatchecin-3-gallate. After: *Kim, E. Y., Ham, S. K., Bradke, D.,* et al. *(2011). J. Nutr. 141, 828.*

and the iron-antagonistic effects of polyphenol- or phytate-containing foods, or of calcium phosphate, can be overcome by the simultaneous consumption of ascorbic acid (Fig. 9.9).[53] Similarly, ascorbic acid administered parenterally has been found useful as an adjuvant therapy to erythropoietin in hemodialysis patients.

Ascorbic acid also promotes the utilization of heme iron, which appears to involve enhanced incorporation of iron into its intracellular storage form, ferritin.[54] This effect involves facilitation of ferritin synthesis: ascorbate enhances the iron-stimulated translation of ferritin mRNA by maintaining the iron-responsive element-binding protein[55] in its enzymatically active form. Studies with cultured cells have shown that ascorbic acid also enhances the stability of ferritin by blocking its degradation through reduced lysosomal autophagy of the protein. Thus, the decline in ferritin and accumulation of **hemosiderin**[56] in scorbutic animals is reversed by ascorbic acid treatment.[57]

Interactions with Other Mineral Elements

Ascorbic acid can also interact with several other metallic elements of nutritional significance, reducing the toxicities of elements whose reduced forms are poorly absorbed or more rapidly excreted (e.g., selenium, nickel, lead, vanadium, cadmium). High dietary levels of the vitamin have been shown to increase tissue levels of manganese, but to reduce the efficiency of enteric absorption of copper.[58] The latter effect has been shown to stimulate mechanisms to preserve tissue copper stores, which are required for several enzymes, including two that require ascorbic acid as a co-substrate (dopamine β-monooxygenase, peptidylglycine α-amidating monooxygenase). Ascorbic acid can enhance the utilization of physiologic doses of selenium, increasing the apparent biologic availability of a variety of inorganic and organic forms of that essential nutrient.[59]

Enzyme Co-Substrate Functions

Ascorbate functions as a co-substrate for at least eight enzymes[60] that are either monooxygenases, which

53. Anemia, much of it due to iron deficiency, is an enormous global problem, affecting more than 40% of all women. Yet iron is the fourth most abundant element in the Earth's crust, and few diets do not contain the element at least in non-heme form. The problem of iron-deficiency anemia can be viewed as one of inadequate iron bioavailability; with that in mind, it has been suggested that the problem may be better thought of as one of vitamin C inadequacy.

54. A soluble, iron–protein complex found mainly in the liver, spleen, bone marrow, and reticuloendothelial cells. Containing 23% iron, it is the main storage form of iron in the body. When its storage capacity is exceeded, iron accumulates as the insoluble hemosiderin.

55. This is a dual-function protein that also has aconitase activity.

56. A dark yellow, insoluble, granular, iron-storage complex found mainly in the liver, spleen, and bone marrow.

57. The reverse relationship is apparently not significant, i.e., iron loading has been found to have no effect on ascorbic acid catabolism in guinea pigs.

58. This effect does not appear to be significant at moderate intakes of the vitamin. A study with healthy young men found that intakes up to 605 mg/person/day for 3 weeks had no effects on copper absorption or retention, or on serum copper or ceruloplasmin concentrations (although the oxidase activity of the protein was decreased by 23%).

59. This effect was a surprising finding, as it is known that ascorbic acid can reduce copper compounds to insoluble forms of little or no biological value. The biochemical mechanism of the stimulation of copper bioavailability by ascorbic acid appears to involve enhanced post-absorptive utilization of copper for the synthesis of selenoproteins, perhaps by creating a redox balance of glutathione in favor of its reduced form (GSH).

60. Three additional ascorbic acid-dependent enzymes have been identified in fungi.

TABLE 9.7 Enzymes That Require Ascorbic Acid as a Co-Substrate

Metabolic Role	Enzyme
Collagen synthesis	Prolyl 4-hydroxylase
	Prolyl 3-hydroxylase
	Lysine hydroxylase
Catecholamine synthesis	Dopamine β-monooxygenase
Peptide hormone synthesis	Eptidylglycine α-amidating monooxygenase
Carnitine synthesis	γ-Butyrobetaine 2-oxoglutarate 4-dioxygenase
	Trimethyllysine 2-oxoglutarate dioxygenase
Drug and steroid metabolism	Cholesterol-7α-hydroxylase
Tyrosine metabolism	4-Hydroxyphenylpyruvate dioxygenase

incorporate a single atom of oxygen into a substrate, or dioxygenases, which incorporate both atoms of molecular oxygen each in a different way (Table 9.7).

Collagen Synthesis

The best characterized metabolic role of ascorbic acid is in the synthesis of collagen proteins,[61] in which it is involved in the hydroxylation of specific prolyl and lysyl residues of the unfolded (non-helical) procollagen chain. These reactions are catalyzed by the enzymes prolyl 4-hydroxylase, prolyl 3-hydroxylase, and lysyl hydroxylase. Each is a dioxygenase[62] that requires O_2, Fe^{2+}, and ascorbate, and is stoichiometrically linked to the oxidative decarboxylation of α-ketoglutarate. It is thought that the role of ascorbate in each reaction is to maintain iron in the reduced state (Fe^{2+}), which dissociates from a critical region (an SH group) of the active site to reactivate the enzyme after catalysis. The post-translational hydroxylation of these procollagen amino acid residues is necessary for folding into the triple helical structure that can be secreted by fibroblasts. Hydroxyproline residues contribute to the stiffness of the collagen triple helix, and hydroxylysine

residues bind (via their hydroxyl groups) carbohydrates and form intramolecular cross-links that give structural integrity to the collagen mass. The underhydroxylation of procollagen, which then accumulates[63] and is degraded, appears to be the basis of the pathophysiology of scurvy, and vitamin C-deficient subjects usually show reduced urinary excretion of hydroxyproline.

Studies have indicated some modest effects of vitamin C deficiency on the hydroxylation of proline in the conversion of the soluble **tropoelastin** to the soluble **elastin**.[64] A component of complement, C1q, resembles collagen in containing **hydroxyproline** and **hydroxylysine**. Curiously, vitamin C deprivation reduces overall complement activity, but does not affect the synthesis of C1q. A role has been suggested for ascorbate in the synthesis of proteoglycans.

Catecholamine Biosynthesis

Ascorbic acid has important roles in several hydroxylases involved in the metabolism of neurotransmitters, steroids, drugs, and lipids. It serves as an electron donor for **dopamine β-monooxygenase**,[65] a copper enzyme located in the chromaffin vesicles[66] of the adrenal medulla and in adrenergic synapses, which hydroxylates dopamine to form the neurotransmitter norepinephrine. In this reaction ascorbate is oxidized to ascorbyl radical (**monodehydroascorbate**), which is returned to the reduced state by **monodehydroascorbate reductase** (Fig. 9.10).

Peptide Hormone Biosynthesis

A copper-dependent enzyme, **peptidylglycine α-amidating monooxygenase**, has been identified in the hypophyses of rats and cattle. It catalyzes the α-amidation of peptides, and requires ascorbate and O_2. It is thought to be involved in the amidation of the C terminals of physiologically active peptides.[67] That catalase inhibits its activity *in vitro* suggests that H_2O_2 is an intermediate in the reaction.

---

61. Collagens, secreted by fibroblasts and chondrocytes, are the major components of skin, tendons, ligaments, cartilage, the organic substances of bones and teeth, the cornea, and the ground substance between cells. Some 19 types of collagen have been characterized; collectively, they comprise the most abundant type of animal protein, accounting for 25–30% of total body protein.
62. One half of the O_2 molecule is incorporated into the peptidylprolyl (or peptidyllysyl) residue, and the other half is incorporated into succinate.

63. Accumulated procollagen also inhibits its own synthesis and mRNA translation.
64. About 1% of the prolyl residues in elastin are hydroxylated. This amount can apparently be increased by vitamin C, suggesting that normal elastin may by underhydroxylated.
65. The specific activity of this enzyme has been found to be abnormally low in schizophrenics with anatomical changes in the brain, suggesting impaired norepinephrine and dopamine neurotransmission in those patients.
66. These vesicles accumulate and store catecholamines in the adrenal medulla; they also contain very high concentrations of ascorbic acid, e.g., 20 mmol/l.
67. For example, bombesin (human gastrin-releasing peptide), calcitonin, cholecystokinin, corticotropin-releasing factor, gastrin, growth hormone-releasing factor, melanotropins, metorphamide, neuropeptide Y, oxytocin, vasoactive intestinal peptide, vasopressin.

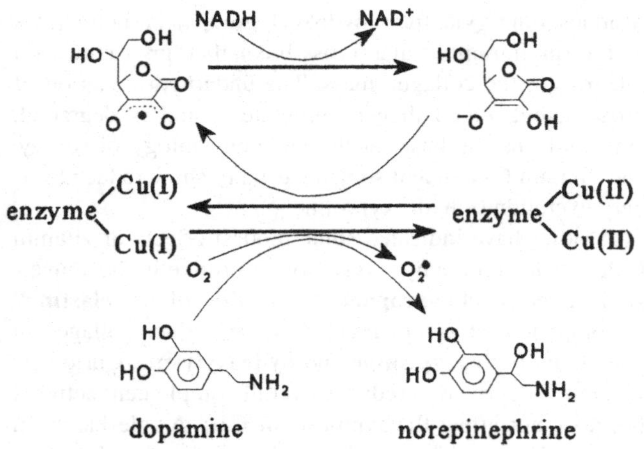

FIGURE 9.10 Role of vitamin C in the conversion of dopamine to norepinephrine by dopamine-ß-monooxygenase.

Carnitine Synthesis

Ascorbic acid is a cofactor of two Fe^{2+}-containing hydroxylases[68] involved in the synthesis of **carnitine**, which is required for the transport of fatty acids into mitochondria for oxidation to provide energy for the cell. Scorbutic guinea pigs show abnormally low carnitine levels in muscle and heart, and it is thought that the fatigue, lassitude, and hypertriglyceridemia observed in scurvy may be due to impaired formation of carnitine.

Drug and Steroid Metabolism

Ascorbic acid is thought to be involved in microsomal hydroxylation reactions of drug and steroid metabolism, i.e., those coupled to the microsomal electron transport chain. For example, the activity of **cholesterol 7α-hydroxylase**, the hepatic microsomal enzyme involved in the biosynthesis of bile acids, is diminished in the chronically vitamin C-deficient guinea pig and is stimulated by feeding that animal high levels of vitamin C.[69] Epidemiologic studies have detected significant positive correlations of ascorbic acid and HDL cholesterol in the plasma/serum of free-living humans.[70]

The metabolism of drugs and other xenobiotic compounds is similarly affected by ascorbic acid. Vitamin C-deficient guinea pigs showed significant increases in the half-lives of phenobarbital, acetanilide, aniline, and

antipyrine. Studies in animal models have clearly demonstrated positive correlations between ascorbic acid status, hepatic activity of cytochrome *P*-450, and drug metabolism (hydroxylations, demethylations). The activities of mitochondrial and microsomal steroid hydroxylases of the adrenal have been found to be impaired in scorbutic animals, in which they respond to vitamin C therapy. In these cases, ascorbic acid appears to function as a protective antioxidant. It is thought that reduced synthesis of corticosteroids accounts for the diminished plasma glucocorticoid responses to stress of vitamin C-deficient animals.

Tyrosine Metabolism

Vitamin C is involved in the oxidative degradation (which normally proceeds completely to CO_2 and water) of tyrosine via two mixed-function oxidases that are dependent on the presence of ascorbic acid. The first is 4-hydroxyphenylpyruvate dioxygenase, which catalyzes the oxidation and decarboxylation of the intermediate of tyrosine degradation, 4-hydroxylphenylpyruvic acid, to homogentisic acid. The second ascorbate-requiring enzyme catalyzes the next step in tyrosine degradation, **homogentisate 1,2-dioxygenase**. By impairing both reactions, vitamin C deficiency can result in tyrosinemia[71] and the excretion of tyrosine metabolites in the urine; both conditions respond to vitamin C supplements.

7. VITAMIN C IN HEALTH AND DISEASE

Tissue saturation of vitamin C occurs in humans at intakes of about 100 mg/day; higher intakes can result in elevated concentrations of the vitamin in extracellular fluids (plasma, connective tissue fluid, humors of the eye) where pharmacologic action of this antioxidant vitamin may be possible. Accordingly, vitamin C intakes greater than those required to prevent scurvy have been considered for a variety of health purposes. Many of the clinical studies in which supranutritional doses of vitamin C have been found to be of some benefit have compared treated subjects with controls who did not have tissue saturation with respect to the vitamin. Such studies therefore cannot indicate whether the effects of the vitamin C treatments were due simply to achieving tissue saturation or to pharmacologic actions of the vitamin in extracellular fluids. Thus, while there appear to be benefits associated with increasing vitamin C intakes to levels that effect tissue saturation, the evidence in support of benefits of vitamin C doses above that level is not clear.

68. ε-*N*-trimethyllysine hydroxylase and γ-butyrobetaine hydroxylase.
69. Guinea pigs fed a diet of 500 mg of ascorbic acid per kilogram show substantial reductions in plasma (about 40%) and liver (about 15%) cholesterol concentrations. Human studies have been inconsistent in showing similar effects.
70. A study of a healthy, elderly Japanese population found serum ascorbic acid to account for about 5 and 11% of the variation in serum HDL cholesterol concentrations in men and women, respectively.

71. Transient tyrosinemia (serum levels >4 mg/dl) occurs frequently in premature infants and involves reduced 4-hydroxyphenylpyruvate dioxygenase activity. Low doses of ascorbic acid usually normalize the condition.

TABLE 9.8 Relationship of Vitamin C Status and Biomarkers of Inflammation and Endothelial Function

Biomarker	Quartile[a] of Plasma Ascorbic Acid, µmol/l				P value
	<14.44	14.44–27.11	27.11–40.25	>40.25	
CRP[b], mg/l	1.88 (1.73–2.03)[c]	1.73 (1.60-1.80)	1.52 (1.40-1.63)	1.34 (1.23-1.44)	<0.001
Fibrinogen, g/l	3.30 (3.26–3.36)	3.29 (3.24–3.34)	3.18 (3.13–3.23)	3.12 (3.07–3.17)	<0.001
t-PA[d], ng/ml	10.92 (10.63–11.21)	10.66 (10.38–10.93)	10.70 (10.42–10.99)	10.31 (10.03–10.60)	0.01
Blood viscosity, mPa	3.41 (3.39–3.44)	3.41 (3.39–3.43)	3.40 (3.38–3.43)	3.35 (3.33–3.37)	<0.001

[a]3,019 subjects.
[b]C-reactive protein.
[c]Mean (95% confidence interval).
[d]Tissue plasminogen activator.
From Wannamethee, S. G., Lowe, G. D. O., Rumley, A., et al. (2006). Am. J. Clin. Nutr. 83, 567.

Immunity and Inflammation

Ascorbic acid has been found to affect immune function in several ways. It has been shown to modulate T cell gene expression, specifically affecting genes involved with signaling, carbohydrate metabolism, apoptosis, transcription, and immune function.[72] It can stimulate the production of interferons, the proteins that protect cells against viral attack. It can stimulate the positive chemotactic and proliferative responses of neutrophils. It can protect against free radical-mediated protein inactivation associated with the oxidative burst[73] of neutrophils. It can stimulate the synthesis of humoral thymus factor and antibodies of the IgG and IgM classes. Some studies have found massive oral doses of the vitamin (10 g/day) to enhance delayed-type hypersensitivity responses in humans, although somewhat lower doses (2 g/day) have shown no such effects.

Phagocytic cells of the immune system produce oxidants during infections that may play some role in the appearance of signs and symptoms. Therefore, it can be expected that ascorbic acid, which is present in high concentrations in phagocytes and lymphocytes, will provide some antioxidant protection. Indeed, studies have found that vitamin C increases the proliferative responses of lymphocytes, is associated with enhanced natural killer cell activity, increases the production of interferon, and decreases viral replication in cell culture systems.

A large, cross-sectional study indicated that plasma ascorbic acid was negatively associated with biomarkers of inflammation (C-reactive protein) and endothelial dysfunction (tissue plasminogen activator) (Table 9.8), suggesting that the vitamin has anti-inflammatory effects associated with reduced levels of endothelial dysfunction. Nevertheless, four of five studies of vitamin C in patients reported no anti-inflammatory effects. Therefore, this putative effect of the vitamin remains the subject of investigation.

Ascorbic acid is involved in histamine metabolism, acting with Cu^{2+} to inhibit its release and enhance its degradation. It does so by undergoing oxidation to dehydroascorbic acid with the concomitant rupture of the histamine imidazole ring. In tissue culture systems, this effect results in reductions of endogenous histamine levels as well as histidine decarboxylase activities, a measure of histamine synthetic capacity. It is also thought that ascorbic acid may enhance the synthesis of the prostaglandin E series (over the F series), members of which mediate histamine sensitivity. Circulating histamine concentration is known to be reduced by high doses of vitamin C, a fact that has been the basis of the therapeutic use of the vitamin to protect against histamine-induced anaphylactic shock. Further, blood histamine concentrations are elevated in several complications of pregnancy that are associated with marginal ascorbic acid status: pre-eclampsia,[74] abruption,[75] and prematurity. Because blood histamine and ascorbic acid concentrations were negatively correlated in women in preterm labor, it has been suggested that the combined effects of marginal vitamin C status and reduced plasma histaminase may result in the marked elevations of blood histamine levels seen in those conditions.

Studies have shown protective effects of vitamin C on several infectious diseases.

---

72. Grant, M. M., Mistry, N., Lunec, J., et al. (2007). Br. J. Nutr. 97, 19.
73. Neutrophils, when stimulated, take up molecular oxygen (O_2) and generate reactive free radicals and singlet oxygen, which, along with other reactive molecules, can kill bacterial pathogens. This process, called the oxidative burst because it can be observed in vitro as a rapid consumption of O_2, also involves the enzymatic generation of bactericidal halogenated molecules via myeloperoxidase. These killing processes are usually localized in intracellular vacuoles containing the phagocytized bacteria.

74. The non-convulsive stage of an acute hypertensive disease of pregnant and puerperal (after childbirth) women.
75. Premature detachment of the placenta.

TABLE 9.9 Large-Scale, Placebo-Controlled Clinical Trials do not Show Vitamin C Protection from Colds

Study	Vitamin C (g/day)	Duration (Months)	Vitamin C Group		Placebo Group		RR (95% CI)
			n	Colds/Year	n	Colds/Year	
1	1	3	407	5.5	411	5.9	0.93 (0.83–1.04)
2	1	3	339	6.7	349	7.2	0.93 (0.84–1.04)
3	3	9	101	1.7	89	1.8	0.93 (0.73–1.20)
4	2	2	331	11.8	343	11.8	1.00 (0.90–1.12)
5	1	3–6	265	1.2	263	1.2	1.03 (0.80–1.32)
6	1	3	304	8.6	311	8.0	1.08 (0.97–1.21)

Source: Hemilä, H. (1997). *Br. J. Nutr. 77*, 59.

Common Cold

The most widely publicized use of so-called megadoses of vitamin C are in prophylaxis and treatment of the common cold. Large doses (≥ 1 g) of vitamin C have been advocated for prophylaxis and treatment of the common cold, a use that was first proposed some 25 years ago by Irwin Stone and the Nobel laureate Dr Linus Pauling.[76] Since that time, many controlled clinical studies have been conducted to test that hypothesis (Table 9.9). Whereas many of these have yielded positive results, until recently few have been appropriately designed, with respect to blinding, controls, treatment randomization, and statistical power, to make such conclusions unequivocal. In general, most results have indicated only small positive effects in reducing the incidence, shortening the duration, and ameliorating the symptoms of the common cold.[77] A meta-analysis[78] of six large clinical trials (including more than 5,000 episodes) showed no detectable effects of gram doses of vitamin C on cold incidence, but some evidence of small protective effects in some subgroups of subjects. A more recent meta-analysis[79] of 29 randomized controlled trials noted a consistent benefit of vitamin C supplementation (≥ 200 mg/day) in reductions of cold duration by 8% in adults and 13.5% in children. That analysis also showed that six trials including a total of 642 athletes and soldiers showed a 50% reduction in risk of developing a cold.

Heliobacter pylori

Randomized trials have shown that vitamin C supplementation can reduce seropositivity for *H. pylori*[80] and protect against the progression of gastric atrophy in seropositive patients.[81] This appears to be related to reduced risk of gastric cancer, for which *H. pylori* is a risk factor.

Herpes

Topical application of ascorbic acid reduced the duration of lesions as well as viral shedding in patients with *Herpes simplex* virus infections.[82]

Other Infections

Ten of 14 randomized controlled trials found apparent reductions in incidence,[83] and 8 of 10 found apparent reductions in severity,[84] of infections other than colds (Table 9.10).

Nevertheless, the results of studies of vitamin C and infections have been inconsistent. Some studies with scorbutic guinea pigs, fishes, and rhesus monkeys have shown vitamin C deficiency to decrease resistance to infections,[85] but several studies have yielded negative results. Studies

76. Pauling received two Nobel Prizes: in Chemistry in 1954, and for Peace in 1962.

77. Chalmers, T. C. (1975). *Am. J. Med. 58*, 532.

78. Hemilä, H. (1997). *Br. J. Nutr. 77*, 59.

79. Douglas, R. M., Hemila, H., Chalker, E., *et al.* (2004). *Cochrane Database Syst. Rev.* 4:CD000980; Douglas, R.M., Hemila, H. Chalker, E. *et al.* (2007) *Cochrane Database Syst. Rev.* 18:CD000980.

80. Simon, J. A., Hudes, E. S., Perez-Perez, G. I. *et al.* (2003). *J. Am. Coll. Nutr. 22*, 283.

81. Sasazuki, S., Sasaki, S., Tsubono, Y., *et al.* (2003). *Cancer Sci. 94*, 378.

82. Hamuy, R. and Berman, B. (1998). *Eur. J. Dermatol. 8*, 310.

83. These involved posttransfusion hepatitis, pneumonia, tuberculosis, pharyngitis, laryngitis, tonsillitis, secondary bacterial infections after a common cold episode, and rheumatic fever.

84. These involved herpes labialis, bronchitis, tonsillitis, rubella, and tuberculosis.

85. For example, *Mycobacterium tuberculosis*, *Rickettsiae* spp., *Endamoeba histolytica*, and other bacteria, as well as *Candida albicans*.

TABLE 9.10 Results of Placebo-Controlled, Double-Blinded Studies of Vitamin C and Infections Other than Colds

Studies of Infection Incidence

Study	Infection	Vitamin C (g/day)	Cases/Total Subjects		OR (95% CI)
			Vitamin C	Placebo	
1[a]	Hepatitis	3.2	6/90	8/85	0.69 (0.26–1.80)
2[b]	Pneumonia	2	1/331	7/343	0.15 (0.01–0.74)
3[c]	Bronchitis	1	8/139	13/140	0.60 (0.27–1.30)
4[c]	Pharyngitis, laryngitis, tonsillitis	1	7/139	14/140	0.48 (0.21–1.10)

Studies of Infection Severity

Study	Infection	Vitamin C (g/day)	Outcome Value (n)		
			Outcome	Vitamin C	Placebo
5[d]	Herpes labialis	0.6	Days healing	4.21 ± 7^f (19)	9.72 ± 8 (10)
		1.0	Days healing	9.72 ± 8 (10)	
6[e]	Bronchitis	0.2	Decreased score	3.4 ± 1.8 (28)	$2.3 \pm .5$ (29)
7[a]	Hepatitis	3.2	SGOT units	474 ± 386 (6)	759 ± 907 (8)

[a]Kodell, R. G. et al. (1981). Am. J. Clin. Nutr. 34, 2023.
[b]Pitt, H. A. and Costrini, A. M. (1996). J. Am. Med. Assoc. 241, 908–911.
[c]Ritzel, G. (1961). Helv. Med. Acta 28, 63–68.
[d]Terzhalmy, G., Bottomley, W., and Pelleu, G. (1978). Oral Surg. 45, 56–62.
[e]Hunt, C. et al. (1994). Intl J. Vit. Nutr. Res. 64, 212–219.
[f]Significantly different from control value, $P < 0.05$.
Abbreviation: SGOT, Serum glutamic-oxaloacetic transaminase; now referred to as aspartate aminotransferase (ALT).

of ascorbic acid supplementation of species that do not require the vitamin (rodents, birds) have generally shown improved resistance to infection, as indicated by increased survival of infected animals, depressed parasitemia, enhanced bacterial clearance, and reduced duration of infection.

Cardiovascular Health

The antioxidant characteristics of ascorbic acid allow it to have an anti-atherogenic function in reducing the oxidation of low-density lipoproteins (LDLs), a key early event leading to atherosclerosis.[86] Being rich in both cholesterol and polyunsaturated fatty acids (PUFAs), LDLs are susceptible to lipid peroxidation by the oxidative attack of reactive oxygen species. Research has shown that oxidized LDLs stimulate the recruitment, in the subendothelial space of the vessel wall, of monocyte-macrophages that can take up the oxidized particles via scavenger receptors[87] to form the lipid-containing foam cells found in the early stages of atherogenesis.

According to this view, atherogenesis can be reduced by protecting LDLs from free-radical attack. The full protection of LDLs appears to involve both ascorbic acid and vitamin E, the latter being important in quenching radicals produced within the hydrophobic interior environment of the LDL particle. That vitamin C deficiency promotes the formation of atherosclerotic lesions in guinea pigs would support the hypothesis that vitamin C can reduce atherogenic risk; however, evidence of such an effect in humans is weak at present. Subjects in the first National Health and Nutrition Examination Survey (NHANES I) with the highest vitamin C intakes showed less cardiovascular death (standardized mortality ratio, 0.66; 95% confidence limits, 0.53–0.83) than subjects with lower estimated vitamin C

---

86. *Atherosclerosis* is the focal accumulation of acellular, lipid-containing material as plaques in the intima of the arteries. The subsequent infiltration of the intima by fatty substances (*arteriosclerosis*) and calcific plaques, and the consequent reduction in the vessel's lumenal cross-sectional area, result in a reduction in blood flow to the organs served by the affected vessel, causing such symptoms as *angina, cerebrovascular insufficiency*, and *intermittent claudication*.

87. Monocyte-macrophages have very few LDL receptors, and the few they have are downregulated. Therefore, when incubated with non-oxidized LDLs they do not form foam cells, as the accumulation of cholesterol further reduces LDL receptor activity. On the other hand, these cells have specific receptors for modified LDLs; these are called *scavenger receptors*. It is thought that LDL lipid peroxidation products may react with amino acid side chains of apo B to form epitopes that have affinities for the scavenger receptor.

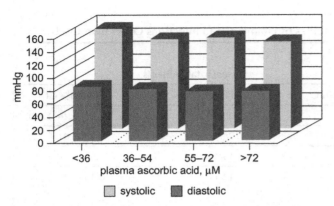

FIGURE 9.11 Relationship between blood pressure and plasma ascorbic acid. After *Choi, E. S. K., Jacques, P. F., Dallai, G. E., and Jacob, R. A. (1991).* Nutr. Res. *11, 1377.*

intakes.[88] Plasma ascorbic acid concentration was found to be highly negatively correlated with both the plasma malonyldialdehyde concentration and the values of several cardiovascular risk factors, including blood pressure, total serum cholesterol, and LDL cholesterol.[89] Other similar results have been reported; however, several other large observational studies[90] have not found such relationships.

Resting blood pressure in humans has been found to be inversely related to vitamin C intake or plasma ascorbic acid concentration (Fig. 9.11), and an intervention trial found vitamin C supplementation to reduce blood pressure and improve arterial stiffness in patients with non-insulin-dependent diabetes. While the metabolic basis of this relationship is still unclear, it has been suggested that ascorbic acid may serve to protect cell membrane pumps from oxidative damage in such ways as to promote ion flux and enhance the vasoactive characteristics of blood vessels. Ascorbic acid has also been shown to increase histamine receptor sensitivity.[91] The importance of hypertension as a risk factor for cerebrovascular and coronary heart diseases makes prospective blood pressure-lowering effects of vitamin C supplementation of considerable interest.

Risk of stroke has been found to be inversely related to vitamin C status. Serum vitamin C concentrations >45 μmol/l were associated with 30–50% reductions in the risk of cerebral infarction and hemorrhagic stroke in two large prospective studies in Japan and Finland.[92] These effects, however, were not associated with the use of vitamin

C supplements, suggesting a role for the consumption of vitamin C-containing fruits and vegetables.

Cohort studies have found high vitamin C intakes to be associated with reduced risk of non-fatal ischemic heart disease, and clinical trials have found vitamin C supplementation to enhance the protective effect of aspirin in reducing risk of ischemic stroke, and to retard atherosclerotic progression in hypercholesterolemic patients. A recent analysis of nine prospective trials concluded that high-level vitamin C supplements reduced the incidence of major coronary heart disease.[93]

Exercise Tolerance

Vigorous physical activity increases ventilation rates and produces oxidative stress,[94] which is thought to affect endothelial function. Studies have shown that antioxidant supplementation can alleviate muscle damage and protein oxidation induced by exercise. That vitamin C may be especially important in such protection is indicated by the finding that vitamin C prevented acute endothelial dysfunction induced by exercise in patients with intermittent claudication[95] (calf pain during walking). This effect is likely due to the protection of nitric oxide (NO), which mediates endothelium-dependent vasodilation.

Metabolically produced reactive oxygen species (ROS) also appear to have essential functions as signaling molecules for the adaption of skeletal muscle to accommodate the stresses presented by exercise training or periods of disuse. This signaling involves redox-sensitive kinases, phosphatases and NFκB, which affect the rate of mitochondrial biogenesis, as well as the induction of genes related to insulin sensitivity and ROS defense. This system of adaptive responses to oxidative stress, facilitating the ultimate development of long-term resistance to that stress, is called mitochondrial hormesis.[96]

Studies have shown that the responses of this system can be impaired by high-level antioxidant treatment. Administration of ascorbic acid has been found to block the oxidative effects of exercise in race horses, but to reduce performance in both dogs and human subjects undergoing training.[97] Combined supplements of vitamins C (500 mg/day ascorbic acid) and E (400 IU/day

88. Enstrom, J. E., Kanim, L. E., and Klein, M. A. (1992). *Epidemiol.* 5, 255.
89. Toohey, L., Harris, M. A., Allen, K. G., *et al.* (1996). *J. Nutr.* 126, 121.
90. For example, the National Heart Study (121,700 subjects), the Health Professionals Follow-Up Study (51,529 subjects), a prospective cohort study of 1,299 subjects, the Iowa Women's Health Study (34,486 subjects).
91. Dillon, P. F., Root-Bernstein, R.S., and Lieder, C. M. (2006). *Am. J. Cell Physiol.* 291, C977.
92. Yokoyama, T., Date, C., Kokuba, Y., *et al.* (2000). *Stroke* 31, 2287; Kurl, S., Tuomainen, T. P., Laukkanen, J. A., *et al.* (2002). *Stroke* 33, 1568.

93. Knekt, P., Pereira, M. A., O'Reilley, E. J., *et al.* (2004). *Am. J. Clin. Nutr.* 80, 1508.
94. Oxygen utilization increases 10- to 15-fold during exercise.
95. Silvestro, A., Scopacasa, F., Oliva, G., *et al.* (2002) *Atherosclerosis* 165:277.
96. Hormesis is the term for a generally favorable biological response to low exposures to stressors/toxins.
97. White, A., Estrada, M., Walker, K., *et al.* (2001). *Comp. Biochem. Physiol.* Pt A 128, 99; Marchall, R. J., Scott, K. C., Hill, R. C. *et al.* (2002). *J. Nutr.* 132, 1616S; Gomez-Cabrera, M. C., Domenech, E., Romangnoli, M. (2008). *J. Nutr.* 87, 142.

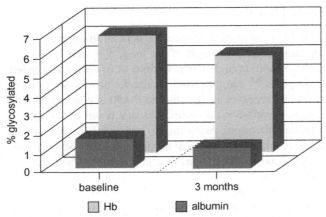

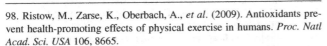

FIGURE 9.12 Protein glycosylation reduced by vitamin C (1 g/day). After *Davie, S. J., Gould, B. J., and Yudkin, J. S. (1992). Diabetes 41, 167.*

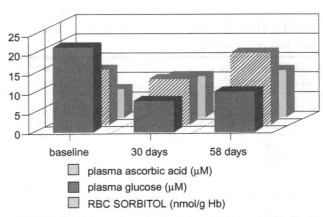

FIGURE 9.13 Effects of supplemental vitamin C on patients with diabetes. After *Cunningham, J. J., Mearkle, P. L., and Brown, R. G. (1994).* J. Am. Coll. Nutr. *13, 344.*

α-tocopheryl acetate) blocked the upregulation of muscle glucose uptake otherwise induced by exercise in non-trained humans.[98] This finding raises questions as to what level of "peroxide tone" may be beneficial.

Diabetes

Diabetic patients and subjects with metabolic syndrome typically show lower serum concentrations of ascorbic acid than non-diabetic controls. Accordingly, reduced serum antioxidant activity has been implicated in the pathogenesis of the disease. That vitamin C supplementation has been found to reduce the glycosylation of plasma proteins (Fig. 9.12) suggests that it may have a role in preventing diabetic complications. Controlled intervention trials have shown that vitamin C supplementation can be effective in reducing erythrocyte sorbitol accumulation[99] (Fig. 9.13) and urinary albumin excretion[100] in non-insulin-dependent diabetics, although one found no effect on microvascular reactivity.[101] Treatment with vitamin C has also been shown to prevent arterial hemodynamic changes induced by hyperglycemia.

Neurologic Function

Plasma ascorbic acid concentrations have been shown to be positively associated with memory performance

in patients with dementia, and cognitive performance in older subjects. Low vitamin C status is a frequent finding in patients with chronic schizophrenia. It has been suggested pathologic conditions may increase the oxidation of ascorbic acid, an estimated 2% of which turns over in the brain each hour. It is also possible that vitamin C supplementation may enhance protection from the effects of inflammatory mediators and free radicals involved in the pathogenesis of neurodegenerative disease. The results of two randomized clinical trials suggest that vitamin C supplementation can reduce symptoms of chronic schizophrenia.[102]

Pregnancy Outcomes

Low vitamin C status has been shown to be associated with increased risks of gestational diabetes and of premature delivery due to premature rupture of chorio-amniotic membranes. The latter responded to vitamin C supplementation.[103]

Skin Health

Ascorbic acid is critical for the health of the epidermis, by virtue of its essential role in collagen synthesis. It is well documented that vitamin C-deficient animals show prolonged wound healing times. This is thought to involve their diminished rates of collagen synthesis, as well as their increased susceptibility to infections. Rapid utilization of the vitamin occurs where relatively high levels of ascorbic

98. Ristow, M., Zarse, K., Oberbach, A., *et al.* (2009). Antioxidants prevent health-promoting effects of physical exercise in humans. *Proc. Natl Acad. Sci. USA* 106, 8665.
99. Because the sorbitol or its metabolites are thought to underlie the pathologic complications of diabetes, reduction of tissue sorbitol accumulation is a strategy for managing diabetes.
100. Gaede, P., Poulsen, H. E., Parving, H. H. *et al.* (2001). *Diabetic Med.* 18, 756.
101. Lu, Q., Bjorkhem, I., Wretlind, B., *et al.* (2005). *Clin. Sci.* 108, 507.

102. Milner, G. (1963). *Br. J. Psychiatry* 109, 294; Dakhale, G. N., Khanzode, S. D., Khanzode, S. S., *et al.* (2005). *Psychopharmacology* 182, 494.
103. Casanueva, E., Ripoll, C., Tolentino, M., *et al.* (2005). *Am. J. Clin. Nutr.* 81, 859; Borna, S., Borna, H., Daneshbodie, B., *et al.* (2005). *Intl J. Gynecol. Obstet.* 90, 16.

TABLE 9.11 Relationship of Dietary Vitamin C Intake and Risk of Cataracts

Quintile of Vitamin C Intake (mg/d)	Cataracts Risk, Odds Ratio[a,b]
≤102	1
>102–135	0.88 (0.56–1.40)[c]
>135–164	0.66 (0.41–1.07)
>164–212	0.60 (0.37–0.97)
>212	0.70 (0.44–1.13)

[a]Ratio of cataracts incidence in each quintile group to that of the lowest (reference) quintile group.
[b]P value for trend, 0.04.
[c]95% confidence interval.
From Valero, M. P., Fletcher, A. E., De Stavola, B. L. et al. (2002). J. Nutr. 132, 1299.

acid accumulate at wound sites.[104] Greater concentrations of ascorbic acid appear to be required for the maintenance of wound integrity than for collagen development. Topical application of ascorbic acid has been found useful in treating photo-damaged skin, as well as inflammatory conditions of the skin such as acne and eczema. That wound repair typically decreases with aging has been suggested as indicative of increasing needs for vitamin C by older individuals.

Cataracts

Cataracts, involving opacification of the ocular lens, are thought to result from the cumulative photo-oxidative effects of ultraviolet light from which the lens is protected by three antioxidants: ascorbic acid, tocopherol, and reduced glutathione. The lens typically contains relatively high concentrations of ascorbic acid (e.g., as much as 30-fold those of plasma), which are lower in aged and cataractous lens.[105] Scorbutic guinea pigs have been found to develop early cataracts, and ascorbate has been shown to protect against ultraviolet light-induced oxidation of lens proteins. Epidemiological studies have shown inverse associations of ascorbic acid status and cataract incidence (Table 9.11); however, an 8-year randomized trial with more than 11,500 subjects found intervention with both ascorbic acid (500 mg/day) and vitamin E (400 IU/day) to have no effect on cataract risk.[106]

Lung Health

Its redox properties give ascorbic acid an important role in the antioxidant protection of the lung,[107] which is consistently exposed to high concentrations of oxygen and inhaled toxic gases.[108] Lung parenchymal cells also generate ROS via such processes as cytochrome *P*-450-dependent mixed-function oxidase metabolism, and as a result of inflammatory cell invasion. Accordingly, patients with asthma or acute respiratory distress syndrome typically show lower than normal concentrations of ascorbic acid in both plasma and leukocytes.

The function of ascorbic acid in collagen synthesis makes the vitamin important in the synthesis of surfactant apoproteins, which have collagen-like domains that require ascorbic acid-dependent hydroxylation for proper folding and stability. Whether supplemental vitamin C can benefit pulmonary function is not clear. The results of five trials have suggested that vitamin C intake may be inversely related to susceptibility to pneumonia. Three controlled trials have found vitamin C supplements effective in preventing pneumonia; two found that treatment effective in reducing the symptoms of that condition.[109]

Half of the dozen clinical intervention trials of vitamin C to date have found improvements in parameters of respiratory function of asthma patients, but these studies have been very small (fewer than 160 patients in total). A meta-analysis of randomized, controlled trials revealed no evidence to recommend a role for vitamin C in the management of asthma.[110]

Anticarcinogenesis

Some epidemiological studies have suggested protective roles of vitamin C against cancers at several sites.[111] Interpreting such results, particularly those showing risk associations with estimated dietary vitamin C intake, is not straightforward as some vitamin C-containing foods also contain some potent carcinogens.[112] Clinical reports have indicated that high doses (10–60 g) can be useful in raising plasma ascorbic acid to levels (*ca.* 20 mmol/l) capable of

---

104. Studies with apparently vitamin C-adequate burn patients have shown their plasma ascorbic acid levels to drop to nearly zero after the trauma; this is presumed to reflect the movement of the vitamin to the sites of wound repair.
105. The ascorbic acid content of the oldest portion of the lens (the nucleus), where most senile cataracts originate, is typically only one-quarter the concentration in the lens cortex.
106. Christen, W. G., Glynn, R. J., Sesso, H. D., *et al.* (2010). *Arch. Ophthalmol.* 128, 1397.

107. See review by Brown, L. A. S. and Jones, D. P. (1997). In *Vitamin C in Health and Disease* (Packer, L. and Fuchs, J., eds). Marcel Dekker, New York, NY, pp. 265–278.
108. For example, ozone, nitric oxide, nitrogen dioxide, cigarette smoke.
109. Hemila, H. and Louhiala, P. (2007). *J. R. Soc. Med.* 100, 495.
110. Ram, F. S., Rowe, B. H., and Kaur, B. (2004), *Cochrane Database Syst. Rev.* 3:CD00993.
111. Statistically significant protective effects of dietary vitamin C have been detected in two-thirds of the epidemiologic studies in which a dietary vitamin C index was calculated. In several cases, high vitamin C intake was associated with half the cancer risk associated with low intake. Protective effects have also been detected in a similarly high proportion of studies in which the intake of fruit, but not vitamin C, was assessed.
112. For example, indole-3 carbinol (I3C), a glucosinolate hydrolysis product; ascorbigen, the adduct of ascorbic acid and I3C.

TABLE 9.12 Adrenal Depletion of Ascorbic Acid in Laying Hens under Simulated Adrenal Stress

Treatment	Renal Ascorbate (μg/g)	Adrenal Ascorbate (μg/g)	Adrenal Cholesterol (mg/g)	Adrenal Corticosterone (μg/g)	Serum Corticosterone (μg/l)
Control	1.41 ± 0.10	1.02 ± 0.05	6.93 ± 0.25	18 ± 2	
ACTH[a]	1.14 ± 0.14^c	0.77 ± 0.04^c		2.57 ± 0.36^c	32 ± 3^c
Dex[b]	1.30 ± 0.06^c	0.82 ± 0.08^c	8.02 ± 0.83^c	17 ± 2	2.4 ± 0.9^c

[a]*ACTH, Adrenocorticotropic hormone, 2.5 IU/day.*
[b]*Dex, Dexamethasone (a suppressor of adrenal corticosterone production), 50 μg/day.*
[c]*Significantly different from control, P < 0.05.*
From Rumsey, G. L. (1969). *Studies of the Effects of Simulated Stress and Ascorbic Acid upon Avian Adrenocortical Function and Egg Shell Metabolism. PhD Thesis, Cornell University, Ithaca, NY.*

killing cancer cells *in vitro*, and in improving outcomes of two small series of cancer patients.[113]

Ascorbic acid has been observed to reduce the binding of polycyclic aromatic carcinogens to DNA[114] and to reduce/delay tumor formation in several animal models. This effect is thought to involve quenching of radical intermediates of carcinogen metabolism. Ascorbic acid is also a potent inhibitor of nitrosamine-induced carcinogenesis, functioning as a nitrite scavenger. This action results from the reduction by ascorbate of nitrate (the actual nitrosylating agent of free amines) to NO, thus blocking the formation of nitrosamines.[115] There is evidence that ascorbic acid, normally secreted in relatively high concentrations in gastric juice,[116] is a limiting factor in nitrosation reactions in people. This appears to be particularly true for individuals with gastric pathologies that affect secretion. Dehydroascorbic acid can convert homocysteine thiolactone to mercaptopropionaldehyde, which kills cancer cells that make large amounts of that lactone.[117]

Bone Health

Impaired bone development is a diagnostic feature of juvenile scurvy. This sign is thought to reflect impaired collagen metabolism in the bone matrix. It is not clear, however, whether this function of ascorbic acid may be involved in the apparent value of vitamin C supplements in reducing bone mineral loss in non-deficient adults. This is indicated by the finding of low serum ascorbic acid being associated with markedly (three-fold) increased risk of hip fracture,

and of long-term use of vitamin C supplements being positively associated with bone mineral density in women.[118]

Environmental Stress

Vitamin C deficiency can have benefits under conditions of environmental stress in which the metabolic demands for ascorbic acid may exceed the rate of its endogenous biosynthesis. This is the case in commercial poultry production. Although the chicken does not require the vitamin in the classic sense, under practical conditions of poultry management, the species frequently benefits from ascorbic acid supplements[119] under stressful environmental conditions (e.g., extreme temperature, prevalent disease, crowding, inadequate ventilation) which stimulate depletion of ascorbic acid from adrenal glands (Table 9.12). Controlled experiments with laying hens have shown that supplemental ascorbic acid can improve egg production and eggshell characteristics in laying hens subjected to heat stress.

8. VITAMIN C DEFICIENCY

Determinants of Risk

Vitamin C deficiency can be caused by low dietary intakes, as well as by conditions in which the metabolic demands for ascorbic acid may exceed the rate of its endogenous biosynthesis, thus increasing the turnover of the vitamin in the body. Such conditions include smoking,[120]

113. Cameron, E. and Pauling, L. (1978). *Proc. Natl Acad. Sci.* USA 75, 4538; Drisko, J., Chapman, J., and Hunter, V. (2003). *J. Am. Coll. Nutr.* 22, 118; Padayatty, S., Riordan, H., Hewitt, S., *et al.* (2006). *Can. Med. Assoc. J.* 174, 937.
114. Reactions of this type are believed to compose the molecular basis of carcinogenesis. This vitamin C effect probably involves its function as a free-radical scavenger.
115. Vitamin E also has this effect.
116. The concentration of ascorbic acid in gastric juice has often been found to exceed that in the plasma.
117. Toohey, J. I. (2008). *Cancer Letts* 263,164.

118. Simon, J. A. and Hudes, E. S. (2001). *Am. J. Epidemiol.* 154, 427; Morton, D. J., Barrett-Conner, E. L., and Schneider, D. L. (2001). *J. Bone Miner. Res.* 16, 135.
119. For example, 150 mg/kg diet.
120. Ascorbic acid concentrations of serum and urine of smokers tend to be about 0.2 mg/dl less than those of non-smokers; these effects have been observed even after correcting for vitamin C intake, which was found to be about 53 mg/day less in smokers. Further, smokers have been found to have lower rates of vitamin C turnover (about 100 mg/day) than non-smokers (about 60 mg/day). It has been estimated that smokers require 52–68 mg of vitamin C per day *more* than non-smokers to attain comparable plasma ascorbic acid levels.

TABLE 9.13 Effect of Environmental Tobacco Smoke on Vitamin C Status of Children

Age (Years)	Plasma Ascorbic Acid, µmol/l[a]	
	Unexposed	Exposed[b]
2–4	53.0 (50.2–55.8)	47.9 (44.4–51.5)
5–8	53.6 (51.4–55.8)	51.0 (48.5–53.5)
9–12	49.7 (47.4–52.0)	47.7 (45.5–49.9)

[a]Mean (95% confidence interval).
[b]A multifactorial ANOVA, with plasma ascorbic acid level adjusted for dietary vitamin C intake, showed the effect of environmental tobacco smoke exposure to be significant across all age groups, $P = 0.002$. From Preston, A. M., Rodriguez, C., Rivera, C. E. et al. (2003). *Am. J. Clin. Nutr.* 77, 167.

TABLE 9.14 General Signs of Vitamin C Deficiency

Organ System	Signs
General:	
Appetite	Decrease
Growth	Decrease
Immunity	Decrease
Heat resistance	Decrease
Muscular	Skeletal muscle atrophy
Vascular	Increased capillary fragility, hemorrhage
Nervous	Tenderness

environmental/physical stress, chronic disease, and diabetes. Indeed, children exposed to environmental tobacco smoke show reduced plasma ascorbic acid concentrations (Table 9.13).

Assessment of vitamin C status may be biased by genetic factors. Glutathione-S-transferase (GST) genotype has been found to be a significant determinant of plasma ascorbic acid level, interacting with estimated dietary vitamin C intake: individuals with *GST* null genotypes had a higher risk of deficiency if their vitamin C intakes did not meet the Recommended Daily Allowance (RDA).[121]

General Signs of Deficiency

In individuals unable to synthesize the vitamin, acute dietary C deficiency is manifest as a variety of signs in the syndrome called scurvy (Table 9.14).

121. Chalill, L. E., Fontaine-Bisson, B., and El-Sohemy, A. (2009). *Am. J. Clin. Nutr.* 90, 1411.

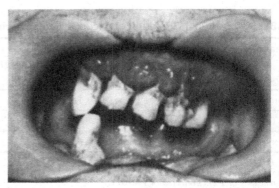

FIGURE 9.14 Scurvy in a middle-aged man. Note swollen, bleeding gums and tooth loss. *Courtesy J. Marks, University of Cambridge.*

Deficiency Signs in Humans

Classic scurvy is manifest in human adults after 45–80 days of stopping vitamin C consumption.[122] Signs of the disease occur primarily in mesenchymal tissues. Defects in collagen formation are manifested as impaired wound healing; edema; hemorrhage (due to deficient formation of intercellular substance) in the skin, mucous membranes, internal organs, and muscles; and weakening of collagenous structures in bone, cartilage, teeth, and connective tissues. Scorbutic adults may present with swollen, bleeding gums with tooth loss (Fig. 9.14), but that condition may signify accompanying periodontal disease. They also show lethargy, fatigue, rheumatic pains in the legs, muscular atrophy, skin lesions, massive sheet hematomas in the thighs, and ecchymoses[123] and hemorrhages in many organs, including the intestines, subperiosteal tissues, and eyes. These features are frequently accompanied by psychological changes: hysteria, hypochondria, and depression.

In children, the syndrome is called **Moeller-Barlow disease**; it is seen in non-breastfed infants usually at about 6 months of age (when maternally derived stores of vitamin C have been exhausted),[124] and is characterized by widening of bone–cartilage boundaries (particularly of the rib cage), stressed epiphyseal cartilage of the extremities, severe joint pain, and, frequently, anemia and fever. Scorbutic children may present with a limp or inability to walk, tenderness of the lower limbs, bleeding of the gums, and petechial hemorrhages. Response to vitamin C is

122. Clinical signs/symptoms of scurvy in humans become manifest when the total body pool of vitamin C is reduced to less than about 300 mg, from its normal level of about 1,500 mg. At that low level, patients show plasma vitamin C levels of 0.13–0.25 mg/dl (normal levels are 0.8–1.4 mg/dl).
123. Bluish patches caused by extravasation of blood into the skin (a "black-and-blue" spot); similar to petechiae except for their larger size.
124. In a retrospective study of the 28 cases diagnosed at the Queen Sirikit National Institute of Child Health from 1995 to 2002, 93% were 1–4 years of age (Ratanachu-Ek, S., Sukswai, P., Jeerathanyasakun, Y., et al. [2003] *J. Med. Assoc. Thai.* 86, S734).

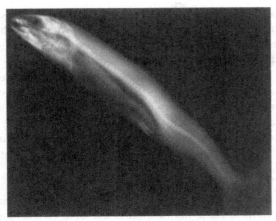

FIGURE 9.15 Radiograph of a vitamin C-deficient trout showing lordosis. *Courtesy G. L. Rumsey, Tunnison Laboratory of Fish Nutrition, USDI.*

dramatic; clinical improvements are seen within a week of vitamin C therapy.

Deficiency Signs in Animals

In guinea pigs, ascorbic acid deficiency is characterized by intermittent reductions in growth, hematomas (especially of the hind limbs), extremely brittle bones, and abnormalities of epiphyseal bone growth with calcification of bone–cartilage boundaries. Guinea pigs that are deprived of vitamin C also show reduced feed intake and growth, anemia, hemorrhages, altered dentin, and gingivitis. Continued deficiency results in disrupted protein folding and apoptosis in the liver. If not corrected, death usually occurs within 25–30 days. Ascorbic acid deficiency in at least some species of fishes (salmonids and carp) results in spinal curvature (scoliosis[125] and lordosis[126] [Fig. 9.15]), reduced survival, reduced growth rate, anemia, and hemorrhaging, especially in the fins, tail, muscles, and eyes. Similar signs have been reported in vitamin C-deficient shrimp and eels.

Subclinical Deficiency

Marginal vitamin status is characterized by oxidative effects in certain tissues, for example, loss of reduced glutathione in lymphocytes, and loss of α-tocopherol with accumulation of lipid peroxidation products in retinal tissues. In addition, elderly people typically show total body vitamin C pools of reduced size, perhaps owing to reduced enteric absorption and increased turnover. In humans, marginal vitamin C status characterized by plasma vitamin C level <0.75 mg/dl and a total body vitamin C pool <600 mg results in several non-specific pre-scorbutic

signs and symptoms: lassitude, fatigue, anorexia, muscular weakness, and increased susceptibility to infection. In the guinea pig, such conditions also result in hypertriglyceridemia, hypercholesterolemia, and decreased vitamin E concentrations in liver and lungs. Epidemiologic data indicate significant associations of low plasma ascorbic acid concentration with increased risk of ischemic heart disease or hypertension in humans.[127]

"Rebound Scurvy"

The possibility that marginal vitamin C intakes after chronically high intakes may lead to clinical signs was proposed as a logical consequence of "systemic conditioning" – i.e., enhanced ascorbic acid catabolism. That hypothesis is now widely regarded as being an erroneous interpretation of a particular set of clinical observations.[128]

9. VITAMIN C TOXICITY

Recent reviews have identified no significant adverse effects of ascorbic acid and its various salts and esters.[129] Nevertheless, the reported benefits of gram doses of vitamin C have sustained questions concerning the safety of the vitamin. Specifically, it has been proposed that megadoses of vitamin C may increase oxalate production, and thus increase the formation of renal stones, competitively inhibit the renal reabsorption of uric acid, enhance the destruction of vitamin B_{12} in the gut, enhance the enteric absorption of non-heme iron (thus leading to iron overload), produce mutagenic effects, and increase ascorbate catabolism that would persist after returning to lower intakes of the vitamin. Present knowledge indicates that most, if not all, of these concerns are not warranted.

Perhaps the greatest concern associated with high intakes of vitamin C has to do with increased oxalate production. In humans, unlike other animals, oxalate is a major metabolite of ascorbic acid, which accounts for 35–50% of the 35–40 mg of oxalate excreted in the urine each day.[130] The health concern, therefore, is that increases in vitamin C intake may lead to increased oxalate production, and thus to increased risk of formation of urinary calculi. Although metabolic studies indicate that the turnover of ascorbic

---

125. Lateral curvature of the spine.
126. Anteroposterior curvature of the spine.

127. Both effects may indicate relationships of antioxidant status to these diseases, as each is also associated with relatively low status with respect to vitamin E and/or copper. In the case of hypertension, a placebo-controlled, double-blind study showed that vitamin C supplements (1 g/day for 3 months) significantly reduced systolic and diastolic pressures in borderline hypertensive subjects with normal serum ascorbic acid levels.
128. Cochrance, W. A. (1965). *Can. Med. Assoc. J.* 93, 893.
129. Hathcock, J. N., Azzi, A., Blumberg, J., et al. (2005). *Am. J. Clin. Nutr.* 81, 736; Elmore, A. R. (2005). *Intl J. Toxicol.* 24, 51.
130. The balance of urinary oxalate comes mainly from the degradation of glycine (about 40% of the total); but some also can come from the diet (5–10%).

acid is limited and that high intakes of vitamin C therefore should not greatly affect oxalate production; clinical studies have revealed slight oxaluria in patients given daily multiple-gram doses of vitamin C. Whether this effect actually constitutes an increased risk of formation of renal calculi is not clear, however, as its magnitude is low and within normal variation.[131] Nevertheless, prudence dictates the avoidance of doses greater than 1,000 mg of vitamin C for individuals with a history of forming renal stones.

Concerns that uricosuria might be induced by megadoses of vitamin C are based on speculation that, because both ascorbic acid and uric acid are reabsorbed by the renal tubules by saturable processes, a common transport system may be involved, and thus high levels of ascorbic acid may competitively inhibit uric acid reabsorption. Experimental data, however, do not provide a basis for this concern. A recent randomized controlled trial with healthy subjects showed that vitamin C (500 mg/day) significantly reduced serum uric acid concentrations and increased the glomerular filtration rate.[132]

In the 1970s, it was claimed that high levels of ascorbic acid added to test meals that were held warm for 30 minutes resulted in the destruction (presumed to occur by chemical reduction) of food vitamin B_{12}, thus raising the concern that megadoses of vitamin C may antagonize the utilization of that important vitamin. More recent studies have not supported that claim; in fact, the only form of vitamin B_{12} that is sensitive to reduction and subsequent destruction by ascorbic acid is aquocobalamin, which is not a major form of the vitamin in foods. Further, the results of another study indicate that high doses of vitamin C can, in fact, partially protect rats from vitamin B_{12} deficiency.[133] The results of several clinical investigations have shown clearly that high doses of vitamin C do not affect vitamin B_{12} status.

That ascorbic acid can enhance the enteric absorption of dietary iron has led some to express concern that megadoses of vitamin C may lead to progressive iron accumulation in iron-replete individuals (iron storage disease). Such an effect is not to be expected, however, as optimal iron absorption is effected with rather low doses of the vitamin (25–50 mg of ascorbic acid per meal). Studies in mice have found ascorbic acid not to enhance the pro-oxidative effect induced by dietary iron; neither did high parenteral

doses[134] of ascorbic acid increase pro-oxidative biomarkers in human subjects. Nevertheless, patients with hemochromatosis or other forms of excess iron accumulation should avoid taking vitamin C supplements with their meals.

The suggestion that ascorbic acid might be mutagenic comes from studies with cultured cells that show a dependence on Cu^{2+} for such effects (the vitamin is not intrinsically mutagenic), which apparently involve the production of reactive oxygen radicals. No evidence of mutagenic effects *in vivo* has been produced; doses as great as 5,000 mg did not induce mutagenic lesions in mice.[135]

The only consistent deleterious effects of vitamin C megadoses in humans are gastrointestinal disturbances and diarrhea. Little information is available on vitamin C toxicity in animals, although acute LD_{50} (50% lethal dose) values for most species and routes of administration appear to be at least several grams per kilogram of body weight. A single study showed mink to be very sensitive to hypervitaminosis C; with daily intakes of 100–200 mg of ascorbic acid, pregnant females developed anemia and had reduced litter sizes.

10. CASE STUDIES

Instructions

Review each of the following case reports, paying special attention to the diagnostic indicators on which the treatments were based. Then answer the questions that follow.

Case 1

A 26-year-old man volunteered for a 258-day experiment of ascorbic acid metabolism. He was 184 cm tall and weighed 84.1 kg. His medical history, physical examination, vital signs, and past diet history revealed a healthy individual with no irregularities. During the experiment, his temperature, pulse, and respiration rates were recorded four times daily, and his blood pressure was measured twice daily. He was examined by an internist daily; periodically he was examined by an ophthalmologist, and had chest radiograms and electrocardiograms made. Twenty-four-hour collections of urine and feces were made daily in order to determine urinary and fecal nitrogen, and for the radioactive assay of ascorbic acid. Samples of expired air were collected for the measurement of radioactivity.

The subject was fed a control diet consisting of soy-based products. The diet provided 2.5 mg of ascorbic acid per day, which was supplemented by a daily capsule containing an

131. Forty percent of subjects given 2 g of ascorbic acid daily showed increases in urinary oxalate excretion by more than 10% (Chai, W., Liebman, M., Kynast-Gales, S., *et al.* [2004] *Am. J. Kidney Dis.* 44, 1060).
132. Huang, H. Y., Appel, L. J., Choi, M. J., *et al.* (2005). *Arthr. Rheumatism* 52, 1843. This finding suggests that vitamin C may be beneficial in the management of gout.
133. Rats given ascorbic acid (6 g/liter in the drinking water) showed greater hepatic vitamin B_{12} levels and lower urinary methylmalonic acid excretion, both indicators of enhanced vitamin B_{12} status, than controls.

134. Up to 7,500 mg (Mühlhöfer, A., Mrosek, S., Schlegel, B., *et al.* [2004] *Eur. J. Clin. Nutr.* 58, 1151).
135. Vojdani, A., Bazargan, M., Vojdani, E., and Wright, J. (2000). *Cancer Detect. Prev.* 24, 508.

additional 75 mg of ascorbic acid. The subject's body vitamin C pool was labeled with L-[1-^{14}C]ascorbate 1 week before initiating vitamin C depletion; it was calculated to be 1,500 mg. Beginning on day 14, the diet was changed to a liquid formula containing no vitamin C, as ascertained by actual analysis. This diet, based on vitamin-free casein, provided 3,300 kcal and supplied protein, fat, and carbohydrate as 15, 40, and 45% of total calories, respectively. It was fed from day 14 to day 104, during which time the subject developed signs of scurvy. Ascorbic acid was not detectable in his urine after 30 days of depletion. He showed **petechiae** on day 45, when his vitamin C pool was 150 mg and his plasma level was 0.19 mg/dl. Spontaneous ecchymoses occurred over days 36–103; these were followed by coiled hairs, gum changes, *hyperkeratosis*,[136] congested follicles and the Sjögren sicca syndrome,[137] dry mouth, and enlarged parotid salivary glands. The subject developed joint pains on day 68 and joint effusions[138] shortly thereafter, when his vitamin C pool was 100 mg and his plasma ascorbic acid level was less than 0.16 mg/dl. He also had the unusual complication of a bilateral femoral neuropathy, which began on day 71, when his vitamin C pool was 80 mg and his plasma ascorbic acid level was 0.15 mg/dl. This, accompanied by the joint effusions, was attributed to hemorrhage into the sheaths of both femoral nerves. On day 80 he experienced a rapid increase in weight, from 81 to 84 kg, in combination with dyspnea[139] on exertion and swelling of the legs. At this time, his vitamin C pool was 40 mg and his plasma level was 0.15 mg/dl.

Beginning on day 105, the subject was put on a vitamin C-repletion regimen involving daily doses of 4 mg of ascorbic acid. Immediately following this treatment, the edema worsened, urinary output dropped to 340 ml/day, and weight increased to 86.6 kg on day 109. There was no evidence of pulmonary congestion or cardiac failure. The ascorbic acid-repletion dose was increased to 6.5 mg/day on day 111. His edema persisted for 4 days, at which time he had a profound diuresis with complete disappearance of the edema by day 133, at which time his weight was 77.2 kg (he lost 9.4 kg of extracellular fluid). From days 101 to 133, his body ascorbic acid pool increased from 33 to 128 mg. The subject was given 6.5 mg of ascorbic acid per day from day 133 to day 227. During this time, all his scorbutic manifestations disappeared, and his plasma ascorbic acid fluctuated between 0.10 and 0.25 mg/dl. His body pool was restored slowly to an excess of 300 mg.

Beginning on day 228, he received 600 mg of ascorbic acid per day, which rapidly repleted his body pool. At the end of the study his weight was 81 kg, and he was discharged from the metabolic ward in excellent health.

Case 2

A 72-year-old man was admitted to the hospital with symptoms of increasing anorexia, epigastric discomfort unrelated to meals, and non-radiating *precordial*[140] pain. During the year before admission, he had become increasingly weak and easily fatigued, and had lost nearly 13 kg in weight. Six weeks before admission, he began to have sudden attacks of severe substernal pain followed by cough and dyspnea, and 1 month before admission he had a small *hematemesis*[141] and had noted bright red blood in his stools. He had been living alone, and his diet during the past year had consisted chiefly of bread and milk with various soups. For a considerable period, he had noted easy bruising of his skin. His past health had been good except for occasional seizures; these began 2 years before admission and involved loss of consciousness, spasmodic twitching of the limbs, and incontinence, preceded by abdominal discomfort.

Physical examination on admission revealed a thin, depressed, lethargic man with a rather gray complexion and numerous petechiae over the arms, legs, and trunk. His blood pressure was 140/80, his pulse was 68, his respiration was 19, and his temperature was 98.8°F. Examination of his head and neck showed an *edentulous*[142] mouth, foul breath, ulcerated palate, and retracted gums without hemorrhage. He had a large ecchymosis (15 cm in diameter) on his right thigh. Neurological examination was negative.

Laboratory Findings:

Parameter	Patient	Normal Range
Hb	13.2 g/dl	15–18 g/dl
WBC	8,000/μl	5,000–9,000/μl
Platelets	140,000/μl	150,000–300,000/μl
Clotting time	5.75 min	5–15 min
Blood urea	48 mg/dl	10–20 mg/dl
Serum protein	7 g/dl	6–8 g/dl
Serum albumin	3.9 g/dl	3.5–5.5 g/dl
Serum ascorbic acid	<0.1 mg/dl	0.4–1.0 mg/dl

---

136. A mouth disease with clinical characteristics usually of variously sized and shaped grayish-white, flat, adherent patches, having diffuse borders and a smooth surface with no papillary projections, fissures, erosions, or ulcerations.

137. Dry eyes due to reduction in tears, i.e., keratoconjunctivitis.

138. The escape of fluid from the blood vessels or lymphatics into the joint capsule.

139. Subjective difficulty or distress in breathing, frequently rapid breathing.

---

140. Relating to the diaphragm and anterior surface of the lower part of the thorax.

141. Vomiting of blood.

142. Toothless.

His heart was not enlarged and there were no heart murmurs; however, his electrocardiogram showed changes typical of an old myocardial infarction.[143] His chest radiograms showed emphysematous[144] and atheromatous[145] changes. His urine contained occasional pus cells with moderate growth of *Escherichia coli*; no abnormal bacilli were seen in the sputum. Sigmoidoscopy revealed no lesions in the distal 25 cm of the bowel. Because of his anorexia, epigastric discomfort, weight loss, and hematemesis, further investigation of the gastrointestinal tract was made using a barium bolus; this revealed a mass and ulcer crater in the prepyloric area of the stomach, suggesting a gastric neoplasm. A laparotomy[146] was planned. The tentative diagnoses were anterior myocardial infarction, suspected cancer of the stomach, epilepsy, and hemorrhagic diathesis[147] (probably scurvy). Accordingly, the patient was given a high-protein diet and ascorbic acid (1 g/day for 2 weeks, then 150 mg/day for a month).

The patient showed marked improvement following ascorbic acid treatment. He no longer showed an air of lassitude; he gained weight and began to relish his meals. His skin hemorrhages rapidly decreased, and no new ones appeared. Three weeks after admission, blood disappeared from his feces. At that time, his epilepsy was satisfactorily controlled using phenobarbital, and his liver function tests and blood chemistry were normal.

Laboratory Findings After Vitamin C Treatment:

Parameter	Patient
Blood urea	28 mg/dl
Serum protein	6.3 g/dl
Serum albumin	3.9 g/dl
Serum ascorbic acid	1.0 mg/dl

A second radiological examination, conducted 1 month after ascorbic acid treatment, indicated a normal pylorus; this was confirmed by gastroscopy. A biopsy of the previously involved area showed only a natural glandular pattern, with hemorrhage of the superficial layer of the gastric mucosa. The patient was discharged after 8 weeks of hospitalization, and was well when seen later in the outpatient clinic. The gastric lesion did not recur. It was concluded that what had appeared to be a prepyloric tumor and ulcer had actually been a bleeding site with a hematoma.[148]

Case Questions and Exercises

1. What thresholds are suggested by the results of the first case study for total body ascorbic acid pool size and plasma ascorbic acid concentration associated with freedom from signs of scurvy?
2. Compute the rate of reduction in ascorbic acid body pool size from the observations on the subject of the first case. Was it linear throughout the study?
3. What signs/symptoms did the patient in the second case show that indicated a problem related to vitamin C status?

Study Questions and Exercises

1. Construct a concept map illustrating the relationship of the chemical properties and physiological functions of vitamin C.
2. Construct a decision tree for the diagnosis of vitamin C deficiency in humans.
3. What health complications might you expect to be shown by scorbutic individuals?

RECOMMENDED READING

Aguirre., R., May, J.N., 2008. Inflammation in the vascular bed: Importance of vitamin C. Pharmacol. Ther. 119, 96–103.

Castro, M.A., Beltrán, F.A., Brauchi, S., Concha, I.I., 2009. A metabolic switch in brain: Glucose and lactate metabolism modulation by ascorbic acid. J. Neurochem. 110, 423–440.

Corti, A., Casini, A.F., Pompella, A., 2010. Cellular pathways for transport and efflux of ascorbate and dehydroascorbate. Arch. Biochem. Biophys. 500, 107–115.

Douglas, R., Hemila, H., D'Souza, R., et al. 2004. Vitamin C for preventing and treating the common cold. Cochrane Database Syst. Rev. (CD000980).

Duarte, T.L., Lunee, J., 2005. Review: When is an antioxidant not an antioxidant? A review of novel actions of vitamin C. Free Radic. Res. 39, 671–686.

Gebicki, J.M., Nauser, T., Domazou, A., et al. 2010. Reduction of protein radicals by GSH and ascorbate: Potential biological significance. Amino Acids 39, 1131–1137.

Harrison, F.E., May, J.M., 2009. Vitamin C function in the brain: Vital role of the ascorbate transporter SVCT2. Free Radic. Biol. Med. 46, 719–730.

Hathcock, J.N., Azzi, A., Blumberg, J., et al. 2005. Vitamins E and C are safe across a broad range of intakes. Am. J. Clin. Nutr. 81, 736–745.

---

143. Necrotic changes resulting from obstruction of an end artery.
144. Emphysema involves dilation of the pulmonary air vesicles, usually due to atrophy of the septa between the alveoli.
145. *Atheroma* refers to the focal deposit or degenerative accumulation of soft, pasty, acellular, lipid-containing material frequently found in intimal and subintimal plaques in arteriosclerosis (also called atherosclerosis).
146. A surgical procedure involving incision through the abdominal wall.
147. Any of several syndromes showing a tendency to spontaneous hemorrhage, resulting from weakness of the blood vessels and/or a clotting defect.

---

148. A localized mass of extravasated blood, usually clotted.

Johnston, C.S., Steinberg, F.M., Rucker, R.B., 2007. Ascorbic acid. In: Zempleni, J., Rucker, R.B., McCormick, D.B., Suttie, J.W. (Eds.), Handbook of Vitamins, fourth ed. CRC Press, New York, NY, pp. 489–520.

Kneckt, P., Ritz, J., Pereira, M.A., et al. 2004. Antioxidant vitamins and coronary heart disease risk: A pooled analysis of 9 cohorts. Am. J. Clin. Nutr. 80, 1508–1520.

Lane, D.J.R., Lawen, A., 2009. Ascorbate and plasma membrane electron transport – enzymes vs efflux. Free Radic. Biol. Med. 47, 485–495.

Lee, K.W., Lee, H.J., Surh, Y.J., Lee, C.Y., 2003. Vitamin C and cancer chemoprevention: Reappraisal. Am. J. Clin. Nutr. 78, 1074–1078.

Léger, D., 2008. Scurvy: Re-emergence of nutritional deficiencies. Can. Fam. Physician 54, 1403–1406.

Levine, M., 2003. Vitamin C as an antioxidant: Evaluation of its role in disease prevention. J. Am. Coll. Nutr. 22, 18–35.

Levine, M., Katz, A., Pakayatty, S.J., 2006. Vitamin C. In: Shils, M.E., Shike, M., Ross, A.C. (Eds.), Modern Nutrition in Health and Disease, tenth ed. Lippincott, New York, NY, pp. 507–524.

Linster, C.L., van Schaftingen, E., 2007. Vitamin C: Biosynthesis, recycling and degradation in mammals. FEBS J. 274, 1–22.

Lykkesfeldt, J., Poulsen, H.E., 2010. Is vitamin C supplementation beneficial? Lessons learned from randomized controlled trials. Br. J. Nutr. 103, 1251–1259.

Mandl, J., Szarka, A., Bánhegyi, G., 2009. Vitamin C: Update on physiology and pharmacology. Br. J. Pharmacol. 157, 1097–1110.

May, J.M., 2000. How does ascorbic acid prevent endothelial dysfunction? Free Radic. Biol. Med. 28, 1421–1429.

Pauling, L., 1970. Vitamin C and the Common Cold. W.H. Freeman and Company, San Francisco, CA.

Rice, M.E., 2000. Ascorbate regulation and its neuroprotective role in the brain. Trends Neurosci 23, 209–216.

Rivas, C.I., Zuniga, F.A., Salas-Burgos, A., et al. 2008. Vitamin C transporters. J. Physiol. Biochem. 64, 357–375.

Savini, I., Rossi, A., Pierro, C., et al. 2008. SVCT1 and SVCT2: Key proteins for vitamin C uptake. Amino Acids 34, 347–355.

Smirnoff, N., 2001. L-Ascorbic acid biosynthesis. Vit. Hormones 61, 241–266.

Toohey, J.I., 2008. Dehydroascrobid acid as an anti-cancer agent. Cancer Lett. 263, 164–169.

Verrax, J., Caderon, P.B., 2008. The controversial place of vitamin C in cancer treatment. Biochem. Pharmacol. 76, 1644–1652.

Wilson, J.X., 2002. The physiological role of dehyrdoascorbic acid. FEBS Letts. 527, 5–9.

Wilson, J.X., 2005. Regulation of vitamin C transport. Ann. Rev. Nutr. 25, 105–125.

Thiamin

Chapter Outline

Anchoring Concepts

1. Thiamin is the trivial designation of a specific compound, 3-(4-amino-2-methylpyrimidin-5-ylmethyl)-5-(2-hydroxyethyl)-4-methylthiazolium, which is sometimes also called **vitamin B$_1$**.
2. Thiamin is hydrophilic, and its protonated form has a quaternary nitrogen center in the thiazole ring.
3. Deficiencies of thiamin are manifest chiefly as neuromuscular disorders.

There is present in rice polishing a substance different from protein and salts, which is indispensable to health and the lack of which causes nutritional polyneuritis.

C. Eijkman and C. Grijns

Learning Objectives

1. To understand the chief natural sources of thiamin.
2. To understand the means of absorption and transport of thiamin.
3. To understand the biochemical function of thiamin as a coenzyme and the relationship of that function to the physiological activities of the vitamin.
4. To understand the physiologic implications of low thiamin status.

VOCABULARY

Acute pernicious beriberi
Alcohol
γ-Aminobutyric acid (GABA)
Anorexia
Ataxia
ATPase
Beriberi
Bradycardia
Cardiac beriberi
Cardiac hypertrophy
Chastak paralysis
Cocarboxylase
Confabulation
Dry beriberi
Dyspnea
Encephalopathy
Fescue toxicity
Hexose monophosphate shunt
Infantile (acute) beriberi
α-Ketoglutarate dehydrogenase
Maple syrup urine disease
Neuropathy
Nystagmus
Ophthalmoplegia
Opisthotonos
Oxythiamin
Pentose phosphate pathway
Perseveration
Phosphorylase
Polioencephalomalacia

The Vitamins. DOI: 10.1016/B978-0-12-381980-2.00010-4

Polyneuritis
Pyrimidine ring
Pyrithiamin
Pyruvate decarboxylase
Pyruvate dehydrogenase
Shoshin beriberi
Star-gazing
Sulfate
Sulfite
Tachycardia
Thiaminases
Thiamin-binding protein (TBP)
Thiamin diphosphate phosphotransferase
Thiamin disulfide
Thiamin monophosphatase
Thiamin monophosphate (TMP)
Thiamin pyrophosphatase
Thiamin pyrophosphate (TPP)
Thiamin pyrophosphokinase
Thiamin-responsive megaloblastic anemia (TRMA)
Thiamin transporters (ThTr1, ThTr2)
Thiamin triphosphate (TTP)
Thiazole ring
Thiochrome
TPP-ATP phosphoryltransferase
Transketolase
Vitamin B$_1$
Wernicke–Korsakoff syndrome
Wet (edematous) beriberi

1. THE SIGNIFICANCE OF THIAMIN

Thiamin is essential in carbohydrate metabolism and neural function. Severe thiamin deficiency results in the nerve and heart disease **beriberi**. Less severe deficiency results in non-specific signs: malaise, loss of weight, irritability, and confusion. Thiamin-deficient animals show inanition and poor general performance, and, in severe cases, *polyneuritis*, making thiamin status economically important in livestock production.

Historically, thiamin deficiency has been prevalent among peoples dependent on polished rice as the dominant staple food. Demographic trends indicate that, for many people, dependence on rice is likely to increase in the future. Rice and rice/wheat crop rotations are now the basis of the food systems currently supporting a fifth of the world's people – i.e., those in East, South, and Southeast Asia, where populations are expected to more than double within the next four decades. The irony is that whole-grain rice and other cereals are not particularly deficient in thiamin; however, the removal of their thiamin-containing aleurone cells renders the polished grains, which consist of little more than the carbohydrate-rich endosperm, nearly devoid of thiamin and other vitamins and essential elements. In fact, thiamin-containing rice polishings are often used to fuel the parboiling of the thiamin-deficient grain. Thus,

efforts are needed to reduce the need to polish rice[1] such that increased reliance on the new, high-yielding cultivars of rice will not lead to expansions of thiamin deficiency among the poor of South Asia.

2. SOURCES OF THIAMIN

Distribution in Foods

Thiamin is widely distributed in foods, but most contain only low concentrations of the vitamin. The richest sources are yeasts (e.g., dried brewer's and baker's yeasts) and liver (especially pork liver); however, cereal grains comprise the most important sources of the vitamin in most human diets. Thiamin in foods is considered to be readily available to healthy subjects except in cases of exposure to certain antagonists (see below).

Whole grains are typically rich in thiamin; however, the vitamin is distributed unevenly in grain tissues. The greatest concentrations of thiamin in grains are typically found in the scutellum (the thin layer between the germ and the endosperm) and the germ. The endosperm (the starchy interior) is quite low in the vitamin. Therefore, milling to degerminate grain, which, because it removes the highly unsaturated oils associated with the germ, yields a product that will not rancidify and, thus, has a longer storage life, but also has very low thiamin content. It is estimated that more than a third of thiamin in the US food supply is provided by grains and grain products, with meats providing about a quarter.

In foods derived from plants, thiamin occurs predominantly as free thiamin (Table 10.1). In contrast, thiamin occurs in animal tissues almost entirely (95–98%) in phosphorylated forms (*thiamin mono-*, *di-*, and *triphosphates*), the predominant form (80–85%) being the coenzyme thiamin diphosphate, also called **thiamin pyrophosphate (TPP)**.

Stability in Foods

Thiamin is susceptible to destruction by several factors including neutral and alkaline conditions, heat, oxidation, and ionizing radiation (Table 10.2). It is stable at low pH (pH <7), but decomposes when heated, particularly under non-acidic conditions.[2] Protein-bound thiamin, found in animal tissues, is more stable to such losses. Thiamin is stable during frozen storage; substantial losses occur during thawing, however, mainly due to removal via drip fluid.

---

1. Rice aleurone cells are rich in a highly polyunsaturated oil, which is prone to oxidative rancidity.
2. Therefore, the practice of adding sodium bicarbonate to peas or beans for retention of their color in cooking or canning results in large losses of thiamin.

TABLE 10.1 Thiamin Contents of Foods

Food	Thiamin, mg/100 g	Food	Thiamin, mg/100 g
Grains		Apricots	0.03
Cornmeal	0.20	Bananas	0.05
Oatmeal	0.55	Grapes	0.05
Rice		Oranges	0.10
Brown	0.29	Pears	0.02
White	0.07	Pineapples	0.08
White, cooked	0.02	**Meats**	
Rye		Beef	0.08
Whole grain	0.30	Duck	0.10
Degerminated	0.19	Pork	1.10
Wheat		Cured ham	0.74
Whole grain	0.55	Veal	0.18
White	0.06	Heart, veal	0.60
Vegetables		**Liver**	
Asparagus	0.18	Beef	0.30
Beans, green	0.07	Pork	0.43
Broccoli	0.10	**Fish**	
Cabbage	0.05	Trout	0.09
Carrots	0.06	Salmon	0.17
Cauliflower	0.11	**Dairy products and eggs**	
Kale	0.16	Cheese	0.02–0.06
Peas, green	0.32	Milk	0.04
Potatoes	0.11	Eggs	0.12
Tomatoes	0.06	**Other**	
Fruits		Brewer's yeast	15.6
Apples	0.04	Human milk	0.01

TABLE 10.2 Thiamin Losses in Food Processing

Procedure	Food	Loss (%)
Convection cooking	Meats	25–85
Baking	Bread	5–35
Heating with water	Vegetables	0–60
Pasteurization	Milk	9–20
Spray drying	Milk	10
Canning	Milk	40
Room temp. storage	Fruits, vegetables	0–20

Thiamin Antagonists

Thiamin in foods can be destroyed by several compounds that may occur naturally (Table 10.3). Cases of thiamin deficiency have been found to be related to the ingestion of food containing such antagonists (Table 10.4).[3] They include the following.

---

3. Perhaps the best known of these is the condition referred to as **Chastak paralysis**, a neurological disorder described in commercially raised foxes fed a diet containing raw carp. The syndrome, named for the fox producer, was found to be a manifestation of thiamin deficiency brought on by a microbial thiaminase present in fish gut tissue. Cooking the fish before feeding them to Mr Chastak's foxes did not produce the syndrome, apparently by heat denaturing the thiaminase.

TABLE 10.3 Types of Thiaminases

Type	Present in:	Mechanism
I	Fresh fish, shellfish, ferns, some bacteria	Displaces pyrimidine methylene group with a nitrogenous base or SH compound, to eliminate the thiazole ring
II	Certain bacteria	Hydrolytic cleavage of methylene–thiazole–nitrogen bond to yield the pyrimidine and thiazole moieties

TABLE 10.4 Thiaminase Activities in Seafoods

Seafood	Thiamin Destroyed, mg/100 g/hour
Marlin	0
Yellowfin tuna	265
Red snapper	265
Skipjack tuna	1,000
Mahi mahi	120
Ladyfish	35
Clam	2,640

From Hilker, D. M. and Peter, G. F (1966). J. Nutr. 89, 419.

- *Sulfites.* Sulfites react with thiamin to cleave its methylene bridge between the **pyrimidine ring** and **thiazole ring**; the reaction is slow at high pH, but is rapid in neutral and acidic conditions. Because rumen microbes can reduce sulfate to sulfite, high dietary levels of sulfate can have thiamin-antagonistic activities for ruminants.
- *Thiaminases.* Thiamin-destroying enzymes occur in a variety of natural products. Bacterial thiaminases are exoenzymes, i.e., they are bound to the cell surface; their activities depend on their release from the cell surface.

This can occur under acidotic conditions in ruminants. They are heat labile, but can be effective antagonists of the vitamin when consumed without heat treatment.

- *Thiamin antagonists in plants.* Several kinds of antithiamin compounds exist in foods:
  - o- and p-hydroxypolyphenols (e.g., caffeic acid, chlorogenic acid, tannic acid)[4] in ferns, tea, and betel nut react with thiamin to oxidize the thiazole ring yielding the non-absorbable form **thiamin disulfide**;
  - some plant flavonoids (quercetin, rutin) have been reported to antagonize thiamin;
  - hemin[5] in animal tissues is thought to bind the vitamin.
- *Thiamin analogs.* Several analogs are effective thiamin antagonists, each involving a substitution on either the pyrimidine or thiazole ring.[6] These include:
  - **oxythiamin**, which lacks the pyrimidine 4′-amino group essential for the release of aldehyde adducts from the C-2 of the thiazole ring. Oxythiamin does not cross the blood–brain barrier, and therefore does not affect thiamin-dependent enzymes in the central nervous system.
  - **2-methylthiamin**, which has a methyl group at the 2 position of the thiazole ring and forms an enzymatically inactive complex with TPP enzymes.
  - **pyrithiamin**, which has a pyridine ring structure in place of the thiazole ring. It is taken into cells by the thiamin transporter and competitively inhibits the conversion of thiamin to thiamin disphosphate, increasing the urinary excretion of thiamin.
  - **amprolium**, which also has a thiamin-like pyrimidine ring combined through a methylene bridge to a quarternary nitrogen of a pyridine ring. The absence of a hydroxyethyl side chain on the pyridine ring prevents the analog from forming diphosphate derivatives, making it a weak thiamin antagonist in animals. Still, it is an efficient inhibitor of thiamin uptake by bacteria, for which reason it is used as an anticoccidial drug for poultry.[7]

---

4. These and related compounds are found in blueberries, redcurrants, red beets, Brussels sprouts, red cabbage, betel nuts, coffee, and tea.

5. Ferriprotoporphyrin, the nonprotein, Fe^{3+}-containing portion of hemoglobin.

6.

thiamin   2-methylthiamin   oxythiamin   pyrithiamin   amprolium

7. These compounds are valuable in the protection of young poultry from coccidial infections by inhibiting the cellular uptake of thiamin by enteric coccidia. At low doses, this inhibition affects chiefly thiamin transport, but at higher doses it can affect the enteric absorption of the vitamin, producing clinical thiamin deficiency.

3. ABSORPTION OF THIAMIN

Thiamin released by the action of phosphatase and pyrophosphatase in the upper small intestine is absorbed in two ways.

- *Active transport.* At low luminal concentrations ($<2\,\mu mol/l$) the process is carrier-mediated. This saturable mechanism is located in the apical brush border of the mucosal epithelium, with greatest activity in the duodenum. Therefore, thiamin produced by the lower gut microflora is not utilized by non-coprophagous animals. Two **thiamin transporter** proteins have been identified, **ThTr1** and **ThTr2**.[8] Each is associated with microtubules, and is specific for free thiamin and some analogs. The mRNAs for the promoters for each are expressed throughout the gastrointestinal tract, as is that for ThTr1, which is also found (in highest concentrations) in the liver. In humans, ThTr2 appears to bind thiamin much more avidly than does ThTr1. Mutations in the *ThTr1* gene have been associated with **thiamin-responsive megaloblastic anemia** (**TRMA**, or Roger's disease).[9] Studies in the mouse have shown that thiamin deprivation results in the upregulation of ThTr2 but not ThTr1. Studies in the rat have shown thiamin transport to be reduced in the diabetic state and by exposure to ethanol; they also suggest that it may decline with age. Because adrenalectomized rats absorb thiamin poorly, it is thought that enteric absorption of the vitamin may also be subject to control by corticosteroid hormones.
- *Passive diffusion.* At higher concentrations (e.g., a 2.5 mg dose for a human), passive diffusion also occurs.

While most of the thiamin present in the intestinal mucosa is in the phosphorylated form, thiamin arriving on the serosal side of the intestine is largely in the free (non-phosphorylated) form. Therefore, the movement of thiamin through the mucosal cell appears to be coupled in some way to its phosphorylation/dephosphorylation. Evidence indicates that the serosal discharge of thiamin is dependent on an Na^+-dependent **ATPase** on that side of the enterocyte.

4. TRANSPORT OF THIAMIN

Thiamin Bound to Serum Proteins

Most of the thiamin in serum is bound non-specifically to protein, chiefly albumin. About 90% of the total thiamin in blood (typically $5-12\,\mu g/dl$) is contained in erythrocytes.[10] A specific binding protein, **thiamin-binding protein** (**TBP**), has been identified in rat serum.[11] With a molecular mass of 38 kDa, TBP binds free thiamin and forms a complex with the riboflavin-binding protein. Like the latter, TBP appears to be regulated by estrogens – i.e., it is inducible in male or ovariectomized rats by parenterally administered estrogen. It is believed that TBP is essential for the distribution of thiamin to critical tissues.

Cellular Uptake

Thiamin is taken up by cells of the blood and other tissues by active transport. Thiamin uptake and secretion appears to be mediated by the thiamin transporters ThTr1 and ThTr2, which depend on Na^+ and a transcellular proton gradient. The transporter ThTr1 has been cloned[12] and mapped to the human chromosome 1q24. This 497-amino acid protein has a high sequence homology with the reduced folate transporter, and is considered a member of that transporter family. Both ThTr1 and ThTr2 are expressed in a variety of tissues and appear to be responsible for the active transport of thiamin, which has been demonstrated in the erythrocyte, placenta, and renal tubular epithelium. Intracellular thiamin occurs predominantly (80%) in the phosphorylated form, most of which is bound to proteins, indicating that uptake of the vitamin is followed by its phosphorylation.

Tissue Distribution

The adult human stores only 25–30 mg thiamin, most of which is in skeletal muscle, heart, brain, liver, and kidneys. Plasma, milk, cerebrospinal fluid, and probably all extracellular fluids contain free (unesterified) thiamin and **thiamin monophosphate** (**TMP**), which, unlike the more highly phosphorylated forms (**thiamin diphosphate** [**TPP**], **thiamin triphosphate** [**TTP**]), appear capable of crossing cell membranes. Tissue levels of thiamin vary within and between species, with no appreciable storage in any tissue.[13] In infants, blood thiamin levels decline after

---

8. Also referred to as SLC19A2 and SLC19A3, respectively, these are members of the SLC family of soluble carriers, which also includes the folate transporter, SLC19A1.

9. Symptoms include megaloblastic anemia, sensory-neural deafness, and diabetes.

10. Several children who died of SIDS (sudden infant death syndrome) have been found to have very high plasma thiamin concentrations, e.g., five-fold those of infants who died of other diseases. The physiological basis of this effect is unknown, although thiamin deficiency is not thought to be a cause of death in SIDS.

11. TBP has also been identified in rat liver and hens' eggs (in both the yolk and albumin).

12. Ganapathy, V., Smith, S. B., and Prasad, P. D. (2004). *Pflügers Arch.* 447, 641.

13. Thiamin concentrations are generally greatest in the heart (0.28–0.79 mg/100 g), kidneys (0.24–0.58 mg/100 g), liver (0.20–0.76 mg/100 g), and brain (0.14–0.44 mg/100 g), and are retained longest in the brain.

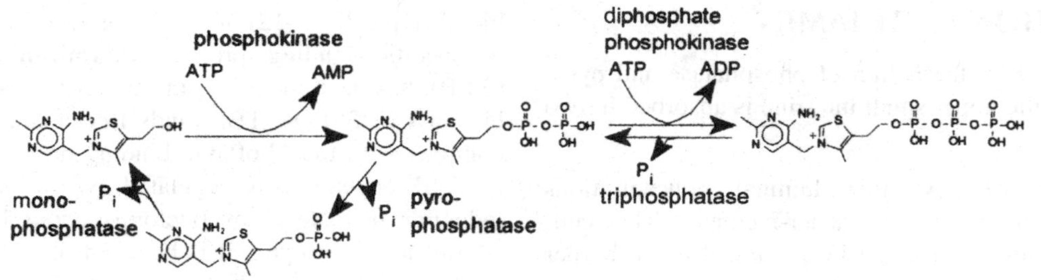

FIGURE 10.1 Metabolic activation of thiamin.

birth, owing initially to a decrease in free thiamin followed by a decrease in phosphorylated forms. In thiamin-deficient chickens fed the vitamin, heart tissues take up thiamin at much greater rates than liver or brain. In general, the thiamin contents of human tissue tend to be less than those of analogous tissues in other species, particularly the pig, which has relatively high tissue thiamin stores.

5. METABOLISM OF THIAMIN

Phosphorylation

Thiamin is phosphorylated in the tissues by two enzymes (Fig. 10.1):

- **thiamin diphosphokinase** catalyzes the formation of thiamin pyrophosphate (TPP) using ATP.
- **TTP-ATP phosphoryltransferase** catalyzes the formation of thiamin triphosphate (TTP) from TPP and ATP.

Several thiamin antagonists are thiazole ring analogs of the vitamin (see "Thiamin Antagonists," page 263). Those with hydroxyethyl groupings (oxythiamin, pyrithiamin, 2-methylthiamin) compete with the vitamin for phosphorylation.

Catabolism

The turnover of thiamin varies between tissues, but is generally high.[14] Thiamin in excess of that which binds in tissues is rapidly excreted. TTP and TPP are each catabolized by **thiamin pyrophosphatase**, yielding thiamin monophosphate (TMP). With an estimated half-life of 10–20 days in humans, thiamin deficiency states can deplete tissue stores within a couple of weeks. Studies with fasting and undernourished soldiers have shown that food restriction increases the rate of thiamin excretion.[15] Declines in tissue thiamin levels are thought to involve enhanced degradation

TABLE 10.5 Urinary Metabolites of Thiamin

Free thiamin
Thiamin disulfide
Thiamin monophosphate (TMP)
Thiamin diphosphate (TPP)
Thiochrome
Thiamin acetic acid
2-methyl-4-amino-5-pyrimidine carboxylic acid
4-methylthiazole-5-acetic acid
2-methyl-4-aminopyrimidine-5-carboxylic acid
2-methyl-4-amino-5-hydroxymethylpyrimidine
5-(2-hydroxyethyl)-4-methylthiazole
3-(2′-methyl-4-amino-5′-pyrimidinylmethyl)-4-methylthiazole-5-acetic acid
2-methyl-4-amino-5-formylaminomethylpyrimidine

of TPP-dependent enzymes in the absence of the vitamin. Numerous metabolites of thiamin have been identified (Table 10.5).

Excretion

Thiamin is excreted in the urine, chiefly as free thiamin and thiamin monophosphate, but also in smaller amounts as the diphosphate ester, the oxidation product **thiochrome**,[16] more than 20 other metabolites, and a 25 kDa thiamin-containing peptide. Metabolites retaining the pyrimidine–thiazole ring linkage account for increasing proportions of total thiamin excretion as thiamin status declines. Urinary losses of thiamin metabolites vary with

14. Thiamin turnover in rat brain was 0.16–0.55 μg/g per hour, depending on the region (Rindi, G., Patrini, C., Comincioli, V., *et al.* [1980] Brain Res. 181, 369).
15. Consolazio, C. F., Johnson, H. L, Krzywicki, J., *et al.* (1971). *Am. J. Clin. Nutr.* 24, 1060.

16. The strong fluorescence of thiochrome has been used in the chemical determination of thiamine, which can be oxidized to thiochrome using potassium ferricyanide or cyangogen bromide (Fujiwara, M. and Matsui, K. [1953] *Anal. Chem.* 25, 810).

plasma thiamin levels, but increase markedly when renal tubular reabsorption is saturated, which occurs in healthy adults at intakes of 0.3–0.4 mg thiamin per 1,000 kcal.[17] Above that threshold, excretion of the vitamin exceeds 100 μg/day, whereas urinary excretion in deficient individuals is <25 μg/day. Small amounts of the vitamin have also been reported to be lost in sweat.[18]

6. METABOLIC FUNCTIONS OF THIAMIN

Thiamin is an essential cofactor for five enzyme complexes. Three catalyze oxidative decarboxylation reactions, one catalyzes the transfer of a glycoaldehyde moiety between sugars, and one catalyzes the cleavage of 2-hydroxy straight-chain fatty acids and 3-methyl branched chain fatty acids after the α-oxidation of the latter. Each of these requires thiamin in the form of its diphosphate ester, TPP, also called cocarboxylase.

Cocarboxylase

TPP serves as a classic coenzyme, binding covalently to the respective holoenzyme by apoenzymic recognition of both the substituted pyrimidyl and thiazole moieties. Binding is facilitated by Mg^{2+} or some other divalent cation, which is therefore required for enzyme activity.

The general mechanism of TPP coenzyme action involves its deprotonation to form a carbanion at C-2 of the thiazole ring. That ion reacts with the polarized 2-carbonyl group of the substrate (an α-keto acid or α-keto sugar) to form a covalent bond, which results in the labilization of certain C–C bonds to release CO_2. The remaining adduct reacts by:

- protonation, to give an active aldehyde addition product (e.g., decarboxylases);
- direct oxidation with suitable electron acceptors, to yield a high-energy, 2-acyl product;
- reacting with oxidized lipoic acid, to yield an acyldihydrolipoate product (e.g., oxidases or dehydrogenases); or
- addition to an aldehyde carbonyl, to yield a new ketol (e.g., *transketolase*).

In higher animals, the decarboxylation is oxidative, producing a carboxylic acid. This involves transfer of the aldehyde from TPP to lipoic acid (forming a 6-S-acylated dihydrolipoic acid and free TPP) and then to coenzyme A.[19]

α-Keto Acid Dehydrogenases

In animals, TPP functions in the oxidative decarboxylation of α-keto acids by serving as an essential cofactor in multi-enzyme α-keto acid dehydrogenase complexes which share certain components. Each complex is composed of a decarboxylase that binds TPP, a core enzyme (that binds lipoic acid), and a flavoprotein dihydrolipoamide dehydrogenase (that regenerates lipoamide), as well as one or more regulatory components. There are three classes of this type of TPP-dependent enzyme:

- *Pyruvate dehydrogenase complex.* This enzyme complex converts pyruvate produced from gylcolysis to acetyl-CoA, a key intermediate in the synthesis of fatty acids and steroids and an acyl donor for numerous acetylation reactions. It consists of three components: TPP-dependent pyruvate dehyrdogenase; and two non-TPP-dependent enzymes, dihydrolipoylacetyltransferase and dihydrolipoyldehydrogenase.[20] The complex also includes a kinase and a phosphatase that regulate enzymatic activity by interconverting the dehydrogenase between active non-phosphorylated and inactive phosphorylated forms involving three specific serine residues that participate in TPP binding. Regulation also occurs via end-product inhibition (by acetyl CoA and NADH). Thiamin status appears to regulate pyruvate decarboxylase gene expression, as deprivation of thiamin has been shown to increase the mRNA for the β subunit of the enzyme.
- *α-Ketoglutaric dehydrogenase complex.* This multi-enzyme complex converts α-ketoglutarate to succinyl-CoA. Its TPP-dependent component, α-ketoglutaric dehydrogenase, appears to be regulated through stimulation by Ca^{2+} and inhibition by ATP, GTP, NADH, and succinyl CoA.
- *Branched-chain α-keto acid dehydrogenase complex.* This multi-enzyme complex converts branched-chain α-keto acids (produced by the transaminations of valine, leucine, and isoleucine) to the corresponding acyl-CoAs (isobutyryl-, isovaleryl-, and α-methylbutyryl-, respectively), which are subsequently oxidized to yield acetyl- and propionyl CoAs. The dehydrogenase is regulated by phosphorylation–dephosphorylation, involving a single serine residue. Deprivation of thiamin has been shown to increase the proportion of dephosphorylated (active) enzyme, thus serving to mitigate against the metabolic consequences of thiamin deficiency. A rare genetic defect in this enzyme complex results in

17. Interdepartmental Committee on Nutrition for National Defense (1963). *Manual for Nutrition Surveys*, 2nd edn. US Government Printing Office, Washington, DC.
18. Pearson, W. H. (1967). *Am. J. Clin. Nutr.* 20, 514; Sauberlich, H. E., Herman, Y. F., Stevens, C. O., *et al.* (1979). *Am. J. Clin. Nutr.* 32, 2237.
19. Coenzyme A is the metabolically active form of the vitamin pantothenic acid.
20. The complex in eukaryotes consist of a multi-unit structure containing 20–30 heterotetramers of pyruvate dehydrogenase, each with two α and two β subunits, associated with multiple units of the acyltransferase, a dihydrolipoyl dehydrogenase-binding protein, and homodimers of the dihydrolipoyl dehydrogenase.

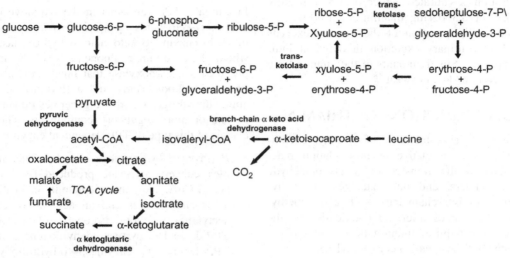

FIGURE 10.2 Role of TTP-dependent enzymes in metabolism.

the condition called **maple syrup urine disease**. Signs of this disorder are manifest in infancy as lethargy, seizures, and, ultimately, mental retardation. The condition can be detected by the maple syrup odor of the urine, which results from the presence of the ketoacid leucine. Some cases respond to high doses (10–200 mg/day) of thiamin.

Transketolase

Transketolase functions at two points in the **hexose monophosphate shunt**, which generates pentoses and NADPH from the oxidation of glucose (Fig. 10.2).[21] Both points involve the transfer of a 2-C "active glycoaldehyde" fragment from an α-keto sugar (xylulose-5-phosphate) to an aldose acceptor (either ribose-5-phosphate or erythrose-5-phosphate).

Transketolase is found in the cytosol of most tissues. It is present in remarkably high amounts in the cornea, where it has been reported to comprise some 10% of total soluble protein. It is present in high amounts in adipose tissue, mammary gland, adrenal cortex, and erythrocytes, all of which rely on carbohydrate metabolism. Transketolase activity depends on its binding to TPP; therefore, responses to thiamin may involve activation of the *apoenzyme*. That thiamin may also have a direct effect on the genetic expression of the enzyme is suggested by

the finding that thiamin deprivation increases the expression of transketolase mRNA.

In subjects adequately nourished with respect to thiamin, TPP-binding is at least 85% of saturation, whereas in thiamin deficiency the percentage of transketolase bound to TPP is much less. This phenomenon is exploited in the clinical assessment of thiamin status; the increase, on addition of exogenous TPP, in the activity of erythrocyte transketolase *in vitro* can be used to determine the percentage TPP saturation of the enzyme and, hence, thiamin status. The percentage stimulation in erythrocyte transketolase activity by the addition of TPP is called the *transketolase activity coefficient*. Subjects with activity coefficients <1.15 are considered to be at low risk of thiamin deficiency, whereas those with activity coefficients of 1.15–1.25 or >1.25 are considered to be at moderate and high risk, respectively. Transketolase isolated from patients with *Wernicke-Korsakoff syndrome* has been found to have an abnormally low binding affinity for TPP.

Peroxisomal Fatty Acid Oxidation

TPP has been found to be required by a peroxisomal enzyme complex involved in the catabolism of fatty acids. This complex catalyzes two types of reaction:

- *Cleavage of 2-hydroxy fatty acids* (e.g., 2-hydroxyoctadecanoic acid[22]). This involves the formation of a CoA derivative followed by the TPP-dependent cleavage to yield formate and a 1C-shortened aldehyde.

---

21. This pathway, also called the *pentose pathway*, is an important alternative to the glycolysis–Krebs cycle pathway, especially for the production of pentoses for RNA and DNA synthesis, and NADPH for the biosynthesis of fatty acids, etc.

---

22. This fatty acid is found in cerebrosides and sulfatides in brain.

- α-*Oxidation of 3-methyl branched chain fatty acids.* This process shortens these fatty acids (e.g., phytannic acid) to facilitate their subsequent β-oxidation. It involves four enzymatic steps, only one of which requires TPP: (1) activation by forming a CoA derivative; (2) hydroxylation of the 2-C; (3) TPP-dependent cleavage of a 1-C unit to release formyl CoA; and (4) dehydrogenation of the resulting long-chain aldehyde to yield an acyl group.

Thiamin Triphosphate

TTP can also serve as a phosphate donor for the phosphorylation of certain proteins; however, the physiological significance of this role remains unclear.

7. THIAMIN IN HEALTH AND DISEASE

Nervous Function

Thiamin has a vital role in nerve function, as the signs of thiamin deprivation are mainly neurologic ones. While the biochemical nature of this role remains unclear, several observations are informative in this regard. Thiamin has been identified in the mammalian brain, in synaptosomal membranes, and in cholinergic nerves. Nervous stimulation by either electrical or chemical means has been found to result in the release of thiamin (as free thiamin and TMP) associated with the dephosphorylation of its higher-phosphate esters. The antagonist pyrithiamin can displace thiamin from nervous tissue and change the electrical activity of the tissue. Irradiation with ultraviolet light at wavelengths absorbed by thiamin destroys the electrical potential of nerve fibers in a manner corrected by thiamin treatment. TTP, which appears to serve as a phosphate donor for phosphorylating synaptic proteins, has been shown to stimulate chloride transport.[23] Brain TTP concentrations tend to be resistant to changes with thiamin deprivation or parenteral thiamin administration, suggesting some degree of homeostatic control in that organ. Thiamin deprivation has been shown to cause oxidative stress, alter neurotransmitter metabolism, and cause dysfunction of the blood–brain barrier in experimental animals.

The neurologic function of thiamin is thought to involve its essential role in the metabolism of glucose, on which the brain depends for its energy source.[24] Clearly, TPP is required in sufficient amounts for the oxidative decarboxylations of pyruvate and α-ketoglutarate, which represent essential steps in energy production via the tricarboxylic

TABLE 10.6 Effects of Thiamin Deficiency on Brain Metabolism in the Rat

Parameter	Thiamin Fed	Thiamin Deficient
Body weight (g)	135 ± 4	96 ± 5
Erythrocyte transketolase (nmol/min/mg)		
Basal activity	6.4 ± 0.6	2.9 ± 0.4[a]
+ TPP	6.9 ± 1.4	4.7 ± 0.8
Activation coefficient	1.08 ± 0.08	1.63 ± 0.12[a]
Liver thiamin (nmol/g)	132.0 ± 8.2	6.0 ± 0.7[a]
Brain analytes		
Thiamin (nmol/g)	12.6 ± 0.6	6.2 ± 0.3[a]
ATP (μmol/g)	2.8 ± 0.1	2.9 ± 0.1
Glutamate (μmol/g)	13.8 ± 0.3	11.3 ± 0.2[a]
α-Ketoglutarate (μmol/g)	0.14 ± 0.01	0.09 ± 0.0[a]
GABA (μmol/g)	1.67 ± 0.05	1.58 ± 0.03[a]

[a]$P < 0.05$.
From Page, M. G., Ankoma-Say, V., Coulson, W. F., et al. (1989). Br. J. Nutr. 62, 245.

acid cycle. However, several experiments have indicated that the depressions of pyruvate and α-ketoglutarate dehydrogenase activities that occur in thiamin-deficient animals (Table 10.6) are not of sufficient magnitude to produce the neurological dysfunction associated with the deficiency. That brain ATP levels are unaffected by thiamin deprivation suggests that the metabolic flux though the alternative pathway, the **γ-aminobutyric acid (GABA)** shunt,[25] may be considerably increased in the brains of thiamin-deficient individuals. This suggests that, in addition to its role in the synthesis of that neurotransmitter, the GABA shunt may also yield energy under conditions of thiamin deprivation (Fig. 10.3). Such a phenomenon may explain the anorexia characteristic of thiamin deficiency, as increased GABA flux through the hypothalamus has been shown to inhibit feeding in animals.

It has been suggested that thiamin may involved the synthesis of myelin. However, the turnover of myelin is much slower (half-time 4–5 days) than the response of thiamin therapy (full recovery within 24 hours). Therefore, it has been proposed that thiamin/TPP has other functions related to nerve transmission. These include roles in

23. Bettendorf, L. (1993). J. Membr. Biol. 136, 281.
24. Unlike other tissues, the brain cannot use fatty acids as a source of energy.

25. GABA is synthesized by the decarboxylation of glutamate, which is produced by transamination of α-ketoglutarate. GABA, in turn, can be transaminated to form succinic semialdehyde, which is oxidized to succinate before entering the TCA cycle.

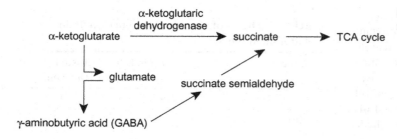

FIGURE 10.3 Flux through the GABA shunt is increased to maintain brain ATP levels in thiamin deficiency.

the regulation of Na^+ permeability and in maintaining the fixed negative charge on the inner surface of the membrane.

Several neurologic conditions have indicated key roles of thiamin in nervous function.

Wernicke–Korsakoff Syndrome

This syndrome involves Wernicke's encephalopathy and Korskoff psychosis. Signs range from mild confusion to coma. Those of the encephalopathy include ophthalmoplegia[26] with lateral or vertical nystagmus[27] and cerebellar ataxia. This psychosis includes severely impaired retentive memory and cognitive function, apathy and confabulation.[28] The pathology is limited to the central nervous system, with lesions limited to the submedial thalamic nucleus and parts of the cerebellum, particularly the superior cerebellar vermis. It is caused by thiamin insufficiency with excessive alcohol consumption.[29]

- Chronic alcohol consumption can lead to thiamin deficiency in two ways:
- **by reducing thiamin intake** due to the displacement of foods rich in thiamin (and other nutrients) by alcohol; and
- **by impairing thiamin absorption and utilization** due to inhibition of the ATPases involved in the enteric absorption and, probably, the cellular uptake of the vitamin.

The risk of Wenicke–Korsakoff syndrome is not limited to heavy alcohol users. The thiamin-responsive syndrome has been diagnosed in non-alcoholic patients with hyperemesis gravidarum[30] or undergoing dialysis.

Patients with Wernike–Korsakoff syndrome frequently have a transketolase isoform with abnormally low binding affinity for TPP. In most cases, it appears that this low affinity can be overcome by using high intramuscular doses of thiamin. While a quarter of patients can be cured by thiamin treatment, it has been suggested that those who fail to respond may have another aberrant transketolase (or other TPP-dependent enzyme) incapable of binding TPP.

Alzheimer's Disease

Comparisons of Alzheimer's disease patients and healthy controls have revealed modest (20%) reductions in the TPP contents of patients' brains, but dramatic differences in the brain activities of TPP-dependent enzymes: 55% less α-ketoglutaric dehydrogenase,[31] 70% less pyruvic dehydrogenase,[32] and markedly reduced transketolase.[33] It has been suggested that these associations reflect genetic variations in genes encoding portions of one or more of these enzymes. A limited number of small trials have been conducted to test the therapeutic value of thiamin for Alzheimer's disease; results have been inconclusive.

Parkinson's Disease

Limited studies have suggested that patients with Parkinson's disease may have lower cerebrospinal fluid levels of free thiamin (but not TPP or TMP), and reduced activities of α-ketoglutaric dehydrogenase.[34] In addition, α-ketoglutaric dehydrogenase was found to be inhibited by dopamine oxidation products,[35] which are known to be elevated in Parkinson's disease. A polymorphism in the gene of α-ketoglutaric dehydrogenase has been suggested as the basis of this association.

Vascular Function

Diabetic vascular complications appear to involve insufficiencies of thiamin/transketolase. TTP has an antidiabetic type role by virtue of its function in transketolase,

---

26. Paralysis of one or more of the motor nerves of the eye.

27. Rhythmical oscillation of the eyeballs, either horizontally, rotary or vertically.

28. Readiness to answer any question fluently with no regard whatever to facts.

29. This is also called Wernicke's encephalopathy. It has been seriously under-diagnosed, by as much as 80% (Harper, C. G. [1979] *J. Neurol. Neurosurg. Psychiatry* 46, 593).

30. Hyperemesis gravidarum is a severe and intractable form of nausea and vomiting in pregnancy.

31. Gibson, G. E. (1998). *Ann. Neurol.* 44, 676.

32. Gibson, G. E. (1981). *Am. J. Med.* 70, 1247.

33. Butterworth, R. F. and Besnard, A. M. (1990). *Metab. Brain Dis.* 5, 179.

34. Mizuno, Y. (1995). *Biochim. Biophys. Acta* 1271, 265.

35. Cohen, G., Farooqui, R., and Kesler, N. (1997). *Proc. Natl Acad. Sci.* 94, 4890.

which diverts cellular excesses of fructose-6-phosphate and glyceraldehyde-3-phosphate from glycolysis to the hexose monophosphate shunt. This serves to downregulate intracellular glucose levels, thus avoiding cellular damage. Diabetic subjects have been found to have lower circulating thiamin levels and lower erythrocyte transketolase activities than healthy controls. Both thiamin and the lipophilic thiamin-derivative benfotiamine[36] have been shown to reduce the accumulation of glycation products and prevent apoptosis in vascular cells cultured under hyperglycemic conditions. Supplementation of either of these compounds has been shown to prevent diabetic cardiomyopathy and neuropathy in the streptozotocin-induced diabetic rat model with moderate insulin treatment,[37] and both have been found effective in preventing vascular dysfunction, oxidative stress, and proteinuria in subjects with type 2 diabetes.[38] Therefore, it has been suggested that diabetes may appropriately be considered a thiamin-deficient state.

Thiamin-Responsive Megaloblastic Anemia

Thiamin-responsive megaloblastic anemia (TRMA) with diabetes and deafness is a rare, autosomal recessive disorder reported in fewer than three-dozen families. The disorder presents early in childhood with any of the above signs, plus optic atrophy, cardiomyopathy, and stroke-like episodes. The anemia responds to high doses of thiamin.[39] Defects in thiamin transport were reported in TRMA patients.[40] These have been found to be due to mutations of the SLC19A2 gene encoding the high-affinity thiamin transporter.

Thiamin Dependency

A few cases of apparent thiamin dependency have been reported. These involved cases of intermittent episodes of cerebral ataxia, pyruvate dehydrogenase deficiency, branched chain ketoaciduria, and abnormal transketolase – all of which responded to very high doses of thiamin.[41]

Other Conditions

In women with gestational diabetes, maternal thiamin deficiency correlates with macrosomia (abnormally high infant body weight).[42] Thiamin deficiency has been implicated in cases of sleep apnea and sudden infant death syndrome (SIDS). While this relationship is not elucidated, it would appear reasonable to expect thiamin to have a role in maintaining the brainstem function governing automatic respiration. In children admitted to a pediatric intensive care unit in Brazil, low blood thiamin level was associated with inflammation.[43]

8. THIAMIN DEFICIENCY

Groups at Risk

Low thiamin status has been reported in alcoholics, older adults, HIV$^+$/AIDS patients,[44] subjects with malaria, and pregnant women with prolonged hyperemesis gravidarum.

General Signs

Thiamin deficiency in humans and animals is characterized by a predictable range of signs/symptoms (see Table 10.7), including loss of appetite (**anorexia**), and cardiac and neurologic signs. Underlying these are a number of metabolic effects, including increased plasma concentrations of pyruvate, lactate, and to a lesser extent α-ketoglutarate (especially after a glucose meal), as well as decreased activities of erythrocyte transketolase. These effects result from diminished activities of TPP-dependent enzymes.

TABLE 10.7 General Signs of Thiamin Deficiency

Organ System	Signs
General:	
Appetite	Severe decrease
Growth	Decrease
Dermatologic	Edema
Muscular	Cardiomyopathy, bradycardia, heart failure, weakness
Gastrointestinal	Inflammation, ulcer
Vital organs	Hepatic steatosis
Nervous	Peripheral neuropathy, opisthotonos

36. S-benzoylthiamine O-monophoshate, a synthetic S-acyl derivative of thiamin that is dephosphorylated to yield the lipid-soluble S-benzoylthiamin.

37. Thornally, P. J., Jahan. I, and Ng, R. (2001). J. Biochem. 129, 543; Kohda, Y., Shirakawa, H., Yamane, K., et al (2008). J. Toxicol. Sci. 33, 459.

38. Stirban, A., Negrean, M, Mueller-Roesel, M., et al. (2006). Diabetes Care 29, 2064; Arora, S., Lidor, A., Abularrage, C. J., et al. (2006). Ann. Vasc. Surg. 20, 653; Riaz, S., Skinner, V., and Srai, S. K. (2011). J. Pharmaceut. Biomed. Anal. 54, 817.

39. Twenty to sixty times higher than RDA levels.

40. Poggi, V., Longo, G., DeVizia, B., et al. (1984). J. Inherit. Metab. Dis. 7, 153.

41. Lonsdale, D. (2006). eCAM 3:49.

42. Baker, H., Hockstein, S., DeAngelis, B., et al. (2000). Intl J. Vit. Nutr. Res. 70, 317.

43. De Lima, L. F. P., Leite, H. P., de A. C. Taddei, J. A. (2011). Am. J. Clin. Nutr. 93, 57.

44. Butterworth, R. F., Gaudreau, C., Vincelette, J., et al. (1991). Metab. Brain Dis. 6, 207.

Increased production of reactive species of oxygen and nitrogen has been reported in the brains of thiamin-deficient animals. This and the finding that antioxidants can attenuate the neurologic effects of thiamin deficiency[45] suggest that oxidative stress plays a role in the clinical manifestation of the nutritional deficiency.

The presentation of thiamin deficiency is variable, apparently affected by such factors as the subject age, caloric (especially carbohydrate) intake, and presence/absence of other micronutrient deficiencies.[46]

Polyneuritis in Animals

The most remarkable sign of thiamin deficiency in most species is anorexia, which is so severe and more specific than any associated with other nutrient deficiencies (apart from that of sodium) that it is a useful diagnostic indicator for thiamin deficiency. Other signs include the secondary effects of reduced total feed intake: weight loss, impaired efficiency of feed utilization, weakness, and hypothermia. The appearance of anorexia correlates with the loss of transketolase activity, and precedes changes in pyruvate or α-ketoglutarate dehydrogenase activities.

Animals also show neurologic dysfunction due to thiamin deficiency; birds, in particular, show a tetanic retraction of the head called **opisthotonos**, also **star-gazing** (Fig. 10.4).[47] Other species generally show **ataxia** and incoordination, which progresses to convulsions and death. These conditions are generally referred to as **polyneuritis**. Most species, but especially dogs and pigs, show cardiac **hypertrophy**[48] (Fig. 10.5), with slowing of the heart rate (**bradycardia**) and signs of congestive heart failure, including labored breathing and edema. Some species also show diarrhea and achlorhydria (rodents), gastrointestinal hemorrhage (pigs), infertility (chickens[49]), high neonatal mortality (pigs), and impaired learning (cats).

The rumen microflora are important sources of thiamin for ruminant species. The clinical manifestation of thiamin deficiency in young ruminants is the neurologic syndrome called **polioencephalomalacia**,[50] a potentially fatal condition involving inflammation of brain gray matter and presenting as opisthotonos (Fig. 10.6); it readily responds to thiamin treatment.[51]

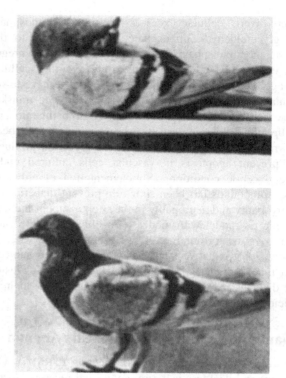

FIGURE 10.4 Opisthotonus in a thiamin-deficient pigeon before (*top*) and after (*bottom*) thiamin treatment. *Courtesy of Cambridge University Press.*

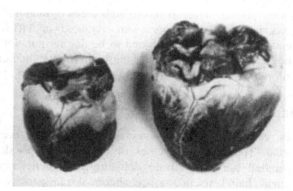

FIGURE 10.5 Hearts from normal (*left*) and thiamin-deficient (*right*) pigs. *Courtesy of T. Cunha, University of Florida.*

Thiamin deficiency can occur if microbial production of the vitamin is impaired. Such cases have occurred due to:

- *Depressed thiamin synthesis.* This has occurred as a result of a change in diet that disturbed rumen fermentation.

45. Pannunzio, P., Hazell, A. S., Pannunzio, M., *et al.* (2000) *J. Neurosci. Res.* 62, 286.
46. For example, deprivation of magnesium was shown to aggravate the signs of thiamin deficiency in the rat (Dyckner, T., Ek, B., Nyhlin, H., *et al.* [1985] *Acta Med. Scand.* 218, 129).
47. This sign occurs in young mammals, but it is not usual.
48. Enlargement of the heart.
49. Thiamin deficiency impairs the fertility of both roosters (via testicular degeneration) and hens (via impaired oviductal atrophy).
50. Cerebrocortical necrosis.

51. Polioencephalomalacia is thought to be a disease of thiamin deficiency that is induced by thiaminases such as those synthesized by rumen bacteria or present in certain plants. Affected animals are listless and have uncoordinated movements; they develop progressive blindness and convulsions. The disease is ultimately fatal, but responds dramatically to thiamin.

FIGURE 10.6 Opisthotonos in a thiamin-deficient sheep. *Courtesy of M. Hidiroglou, Agricultural Canada, Ottawa, Ontario, Canada.*

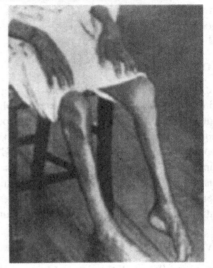

FIGURE 10.7 Neurologic signs of beriberi. *Courtesy of Cambridge University Press.*

- *Increased thiamin degradation.* This has occurred due to an alteration in the microbial population that increases total thiaminase activity. Most important in this regard are *Bacteroides thiaminolyticus, Clostridium sporogenes, Megasphaera elsdenii,* and *Streptococcus bovis,* other *Clostridium* spp., *Bacillus* spp., and gram-negative cocci, which have thiaminases bound to their cell surfaces as exoenzymes.[52] These can become significant sources of thiamin destruction when they are released into the rumen fluid, which can happen under conditions of sharply declining rumen pH. Accordingly, signs of thiamin deficiency have been observed in animals fed high-concentrate diets that tend to acidify the rumen.
- *Consuming thiamin antagonists.* This has occurred with the consumption of excess **sulfate** (which rumen microbes reduce to **sulfite**), excess amprolium, or factors contained in bracken fern or endophyte-infected fescue. Thiamin treatment has been found to reduce the signs of summer **fescue toxicity** (reduced performance, elevated body temperature, rough hair coat) in grazing beef cattle.

Beriberi in Humans

The classic syndrome resulting from thiamin deficiency in humans is beriberi. This disease is prevalent in Southeast Asia, where polished rice is the dietary staple. It appears to be associated with the consumption of diets high in highly digestible carbohydrates, but marginal or low in micronutrients. The general symptoms of beriberi are anorexia, cardiac enlargement, lassitude, muscular weakness (with resulting ataxia), paresthesia,[53] loss of knee and ankle jerk responses (with subsequent foot and wrist droop), and **dyspnea** on exertion (Fig. 10.7). Beriberi occurs in three clinical types.

- **Dry (or neuritic) beriberi** occurs primarily in adults; it is characterized by peripheral neuropathy consisting of symmetrical impairment of sensory and motor nerve conduction affecting the distal (more than proximal) parts of the arms and legs. It usually does not have cardiac involvement.
- **Wet** (edematous) **beriberi** (also called cardiac beriberi) involves as its prominent signs edema, tachycardia,[54] cardiomegaly, and congestive heart failure; in severe cases, heart failure is the outcome.[55] The onset of this form of beriberi can vary from chronic to acute, in which case it is called **shoshin beriberi** (also called **acute pernicious beriberi**), and is characterized by greatly elevated circulating lactic acid levels.

52. Of these, *B. thiaminolyticus* appears to be of greatest pathogenic importance, as it appears to occur routinely in the ruminal contents and feces of all cases of polioencephalomalacia.

53. An abnormal spontaneous sensation, such as burning, pricking, numbness, etc.
54. In contrast to thiamin-deficient animals, which show bradycardia (slow heart beat), beriberi patients show *tachycardia* (rapid heart rate, >100 beats/min).
55. Wet beriberi is also called cardiac beriberi.

- **Infantile (or acute) beriberi** occurs in breastfed infants of thiamin-deficient mothers, most frequently at 2–6 months of age. It has a rapid onset and may have both neurologic and cardiac signs, with death due to heart failure usually within a few hours. Affected infants are anorectic, and regurgitate ingested milk; they may experience vomiting, diarrhea, cyanosis, tachycardia, and convulsions. Their mothers typically show no signs of thiamin deficiency.

Other Conditions

Widespread thiamin depletion in Cuba was reported in 1992–1993 during an epidemic of optic and peripheral neuropathy that affected some 50,000 people in a population of 11 million.[56] A large portion (30–70%) of both the cases and the apparently unaffected population showed signs of low thiamin status. The incidence of new cases subsided with the institution of multivitamin supplementation. Still, it is not clear that thiamin deficiency, while widespread in that population, was the cause of the epidemic neuropathy.

9. THIAMIN TOXICITY

Thiamin is generally well tolerated. Most of the available information pertinent to its toxic potential is for thiamin hydrochloride. At very high doses (1,000-fold levels required to prevent deficiency signs), that form can be fatal by suppressing the respiratory center. Such doses of the vitamin to animals produce curare[57]-like signs, suggestive of blocked nerve transmission: restlessness, epileptiform convulsions, cyanosis, and dyspnea. In humans, parenteral doses of thiamin at 100-fold the recommended intake have been found to produce headache, convulsions, weakness, paralysis, cardiac arrhythmia, and allergic reactions. Lower levels (e.g., up to 300 mg/day) are used therapeutically (for example, to treat frank beriberi, Wernicke–Korsakoff syndrome, etc.) in humans without adverse reactions.

10. CASE STUDIES

Instructions

Review each of the following case reports, paying special attention to the diagnostic indicators on which the treatments were based. Then answer the questions that follow.

Case 1

A 35-year-old man with a history of high alcohol intake for 18 years was admitted to the hospital complaining of massive swelling and shortness of breath on exertion. For several months he had subsisted almost entirely on beer and whiskey, taking no solid food. He was grossly edematous, slightly jaundiced, and showed transient cyanosis[58] of the lips and nail beds. His heart showed gallop rhythm.[59] His left pleural cavity contained fluid. His liver was enlarged with notable ascites.[60] He had a coarse tremor of the hands, and reduced tendon reflexes. His electrocordiogram showed sinus tachycardia.[61] His radiogram showed pulmonary edema and cardiac enlargement. He was evaluated by cardiac catheterization.

Results:

Parameter	Patient	Normal Value
Systemic arterial pressure (mmHg)	100/55	120/80
Systemic venous pressure (cmH$_2$O)	300	<140
Pulmonary artery pressure (mmHg)	64/36	<30/<13
Right ventricular pressure (mmHg)	65/17	<30/<5
O$_2$ consumption (ml/min)	259	200–250
Peripheral blood O$_2$ (ml/liter)	148	170–210
Pulmonary arterial blood O$_2$ (ml/l)	126	100–160
Cardiac output (l/min)	11.8	5–7
Blood hemoglobin (g/dl)	11.0	14–19
Cyanide circulation time (s)	12	20
Femoral arterial pyruvate (mg/dl)	1.5	0.8
Femoral arterial lactate (mg/dl)	14.1	4.7
Femoral arterial glucose (mg/dl)	86	74

The patient was given thiamin intravenously (10 mg every 6 hours) for several days. Improvement was evident by 48 hours, and continued for 2 weeks. Thirty days later, he was free of edema, dyspnea, and cardiomegaly. Cardiac catheterization at that time showed that his blood, systemic venous, and all intracardiac pressures, as well as cardiac output, had returned to normal.

---

56. Macias-Matos, C., Rodriguez-Ojea, A., Chi, N., *et al.* (1996). *Am. J. Clin. Nutr.* 64, 347.

57. Curare is an extract of various plants (e.g., *Strychnos toxifera, S. castelraei, S. crevauxii, Chondodendron tomentosum*). Practically inert when administered orally, it is a powerful muscle relaxant when administered intravenously or intramuscularly, exerting its effect by blocking nerve impulses at the myoneural junction. Curare is used experimentally and clinically to produce muscular relaxation during surgery. It was used originally as an arrow poison by indigenous hunters of South America to kill prey by inducing paralysis of the respiratory muscles.

58. Dark bluish discoloration of the skin resulting from deficient oxygenation of the blood in the lungs or abnormally reduced flow of blood through the capillaries.

59. Triple cadence to the heart sounds at rates of ≥100 beats/min., indicative of serious myocardial disease.

60. Accumulation of serous fluid.

61. Rapid beating of the heart (≥100 beats/min.), originating in the sinus node.

Case 2

Fibroblasts were cultured from skin biopsies from four patients with Wernicke–Korsakoff syndrome and from four healthy control subjects. The properties of transketolase were studied.

Characteristics of Transketolase from Wernicke–Korsakoff Patients:

Parameter	Patients	Controls
V_{max} (nmol/min/mg protein)	27 ± 3	17 ± 1
K_m (μmol/l) TTP	195 ± 31	26 ± 2

The first patient was a 50-year-old woman with a history of chronic alcoholism. She had been admitted to the hospital with disorientation, nystagmus, sixth-nerve weakness,[62] ataxia, and malnutrition. Treatment with intravenous thiamin and large oral doses of multivitamins had improved her neurologic signs over a few months, but her mental state had deteriorated. She was readmitted with disorientation in both place and time, impaired short-term memory, nystagmus, ataxia, and signs of peripheral neuropathy. She was treated with parenteral thiamin and enteral B vitamins with thiamin; this had improved her general health but had not affected her mental status. The second patient, a 48-year-old man with a 20-year history of chronic alcoholism, was admitted in a severe confusional state. He was disoriented, had severe impairment of recent memory, confabulation, **perseveration**,[63] delusions, nystagmus, and ataxia. Treatment with thiamin and B vitamins had improved his behavior, without affecting his memory.

These results show that the affinity of transketolase for its coenzyme (TPP) in Wernicke–Korsakoff patients was less, by an order of magnitude, than that of controls. Further, this biochemical abnormality persisted in fibroblasts cultured for >20 generations in medium containing excess thiamin and no ethanol. The characteristics of pyruvate and α-ketoglutarate dehydrogenases were similar in fibroblasts from patients and controls.

Case Questions and Exercises

- What factors would appear to have contributed to the thiamin deficiencies of these patients?
- What defect in cardiac energy metabolism would appear to be the basis of the high-output cardiac failure observed in the first case?

- What evidence suggests that the transketolase abnormality of these patients was hereditary? Would you expect such patients to be more or less susceptible to thiamin deprivation? Explain.

Study Questions and Exercises

1. Construct a schematic map of intermediary metabolism showing the enzymatic steps in which TPP is known to function as a coenzyme.
2. Construct a decision tree for the diagnosis of thiamin deficiency in humans or animals.
3. How does the chemical structure of thiamin relate to its biochemical function?
4. What parameters might you measure to assess the thiamin status of a human or animal?
5. Construct a concept map illustrating the possible interrelationships of excessive alcohol intake and thiamin status.

RECOMMENDED READING

Alexander-Kaufman, K., Harper, C., 2009. Transketolase: Observations in alcohol-related brain damage research. Intl. J. Biochem. Cell Biol. 41, 717–720.

Balakumar, P., Rohilla, A., Krishan, P., Solairaj, P., 2010. The multifaceted therapeutic potential of benfotiamine. Pharmacol. Res. 61, 482–488.

Bates, C.J., 2007. Thiamine. In: Zemplini, J., Rucker, R.B., McCormick, D.B., Suttie, J.W. (Eds.), Handbook of Vitamins, fourth ed. Marcel Dekker, New York, NY, pp. 253–287.

Beltramo, E., Berrone, E., Tarallo, S., Porta, M., 2008. Effects of thiamine and benoftiamine on intracellular glucose metabolism and relevance in the prevention of diabetic complications. Acta Diabetol. 45, 131–141.

Bettendorff, L., Wins, P., 2009. Thiamin diphosphate in biological chemistry: New aspects of thiamin metabolism, especially triphosphate derivatives acting other than as cofactors. FEBS J. 276, 2917–2925.

Butterworth, R.F., 2003. Thiamin deficiency and brain disorders. Nutr. Res. Rev. 16, 277–283.

Butterworth, R.F., 2006. Thiamin. In: Shils, M.E., Shike, M., Ross, A.C., Caballero, B., Cousins, R. (Eds.), Modern Nutrition in Health and Disease, tenth ed. Lippincott, New York, NY, pp. 426–433.

del Arco, A., Satrústegui, J., 2005. New mitochondrial carriers: An overview. Cell Molec. Life Sci. 62, 2204–2227.

Frank, R.A.W., Leeper, F.J., Luisi, B.F., 2007. Structure, mechanism and catalytic duality of thiamine-dependent enzymes. Cell Molec. Life Sci. 64, 892–905.

Gibson, G.E., Zhang, H., 2002. Interactions of oxidative stress with thiamine homeostasis promote neurodegeneration. Neurochem. Intl. 40, 493–504.

Gregory III, J.F., 1997. Bioavailability of thiamin. Eur. J. Clin. Nutr. 51, S34–S37.

Kluger, R., Tittman, K., 2008. Thiamin diphosphate catalysis: Enzymic and nonenzymic covalent intermediates. Chem. Rev. 108, 1797–1833.

Lonsdale, D., 2006. A review of the biochemistry, metabolism and clinical benefits of thiamin(e) and its derivatives. eCAM 3, 49–59.

---

62. The sixth cranial nerve is the nervus abducens, the small motor nerve to the lateral rectus muscle of the eye.
63. The constant repetition of a meaningless word or phrase.

Malandrinos, G., Louloudi, M., Hadjiliadis, N., 2006. Thiamine models and perspectives on the mechanism of action of thiamine-dependent enzymes. Chem. Soc. Rev. 35, 684–692.

Martin, P.R., Singleton, C.K., Hiller-Sturmhöfel, S., 2003. The role of thiamine deficiency in alcoholic brain disease. Alcohol Res. Health 27, 134–142.

Pohl, M., Sprenger, G.A., Müller, M., 2004. A new persepective on thiamine catalysis. Curr. Opin. Biotechnol. 15, 335–342.

Rindi, G., Laforenza, U., 2000. Thiamine intestinal transport and related issues: Recent aspects. Proc. Soc. Exp. Biol. Med. 224, 246–255.

Rios, J.M., 2007. The THI-box riboswitch, or how RNA binds thiamin pyrophosphate. Cell Structure 15, 259–265.

Schenk, G., Duggleby, R.G., Nixon, P.F., 1998. Properties and functions of thiamin diphosphate dependent enzyme transketolase. Intl. J. Biochem. Cell Biol. 30, 1297–1318.

Shannon, B., Chipman, D.M., 2009. Reaction mechanisms of thiamin diphosphate enzymes: New insights into the role of a conserved glutamate residue. FEBS J. 276, 2447–2453.

Tittmann, K., 2009. Reaction mechanisms of thiamin diphosphate enzymes: Redox reactions. FEBS J. 276, 2454–2468.

Riboflavin

Anchoring Concepts

1. Riboflavin is the trivial designation of a specific compound, 7,8-dimethyl-10-(1′-D-ribityl)-isoalloxazine, sometimes also called **vitamin B$_2$**.
2. Riboflavin is a yellow, hydrophilic, tricyclic molecule that is usually phosphorylated (to FMN and FAD) in biological systems.
3. Deficiencies of riboflavin are manifested chiefly as dermal and neural disorders.

In retrospect – the discovery of riboflavin may be considered a scientific windfall. It opened the way to the unraveling of the truly complex vitamin B$_2$ complex. Perhaps even more significantly, it bridged the gap between an essential constituent and cell enzymes and cellular metabolism. Today, with the general acceptance of this idea, it is not considered surprising that water-soluble vitamins represent essential parts of enzyme systems.

P. György

Learning Objectives

1. To understand the chief natural sources of riboflavin.
2. To understand the means of enteric absorption and transport of riboflavin.
3. To understand the biochemical function of riboflavin as a component of key redox coenzymes, and the relationship of that function to the physiological activities of the vitamin.
4. To understand the physiologic implications of low riboflavin status.

VOCABULARY

Acyl-CoA dehydrogenase
Adrenodoxin reductase
Alkaline phosphatase
Amino acid oxidases
Cheilosis
Curled-toe paralysis
Dehydrogenase
Electron transfer flavoprotein (ETF)
Erythrocyte glutathione reductase
FAD
FAD-pyrophosphatase
FAD-synthetase
Flavin
Flavokinase
Flavoprotein
FMN
FMN-phosphatase
Geographical tongue
Glossitis
L-Gulonolactone oxidase
Hypoplastic anemia
Leukopenia
Lumichrome
Monoamine oxidase
NADH-cytochrome *P*-450 reductase
NADH dehydrogenase
Normocytic hypochromic anemia
Ovoflavin
Oxidase
Reticulocytopenia
Riboflavin-binding proteins (RfBPs)
Riboflavin-5′-phosphate

The Vitamins. DOI: 10.1016/B978-0-12-381980-2.00011-6

Riboflavinuria
Stomatitis
Subclinical riboflavin deficiency
Succinate dehydrogenase
Thrombocytopenia
Thyroxine
Ubiquinone reductase
Vitamin B$_2$

1. THE SIGNIFICANCE OF RIBOFLAVIN

Riboflavin is essential for the intermediary metabolism of carbohydrates, amino acids, and lipids, and also supports cellular antioxidant protection. The vitamin discharges these functions as coenzymes that undergo reduction through two sequential single-electron transfer steps. This allows the reactions catalyzed by **flavoproteins** (i.e., *flavoenzymes*) to involve single- as well as dual-electron transfers, and this versatility means that flavoproteins serve as switching sites between obligate two-electron donors such as the pyridine nucleotides and various obligate one-electron acceptors. Because of these fundamental roles of riboflavin in metabolism, a deficiency of the vitamin first manifests itself in tissues with rapid cellular turnover, such as skin and epithelium.

2. SOURCES OF RIBOFLAVIN

Distribution in Foods

Riboflavin is widely distributed in foods (Table 11.1), where it is present almost exclusively bound to proteins, mainly in the form of **flavin mononucleotide (FMN)** and **flavin adenine dinucleotide (FAD)**.[1,2] Rapidly growing, green, leafy vegetables are rich in the vitamin; however, meats and dairy products are the most important contributors of riboflavin to American diets, with milk products providing about one-half of the total intake of the vitamin.[3] Animal tissues have been found to contain small amounts of riboflavin-5′α-D-glucoside, which appears to be as well utilized as free riboflavin.

Stability

Riboflavin is stable to heat; therefore, most means of heat sterilization, canning, and cooking do not affect

---

1. Notable exceptions are milk and eggs, which contain appreciable amounts of free riboflavin.
2. It should be noted that, strictly speaking, FMN is not a nucleotide, neither is FAD a dinucleotide, because each is a D-ribityl derivative; nevertheless, these names have been accepted.
3. It is estimated that milk and milk products contribute about 50% of the riboflavin in the American diet, with meats, eggs, and legumes contributing a total of about 25%, and fruits and vegetables each contributing about 10%.

TABLE 11.1 Riboflavin Contents of Foods

Food	Riboflavin, mg/100 g
Dairy Products	
Milk	0.17
Yogurt	0.16
Cheese:	
American	0.43
Cheddar	0.46
Cottage	0.28
Ice cream	0.21
Meats	
Liver, beef	3.50
Beef	0.24
Chicken	0.19
Lamb	0.22
Pork	0.27
Ham, cured	0.19
Cereals	
Wheat, whole	0.11
Rye	0.08
Oat meal	0.02
Rice	0.01
Vegetables	
Asparagus	0.18
Broccoli	0.20
Cabbage	0.06
Carrots	0.06
Cauliflower	0.08
Corn	0.06
Lima beans	0.10
Potatoes	0.04
Spinach	0.14
Tomatoes	0.04
Fruits	
Apples	0.01
Bananas	0.04
Oranges	0.03
Peaches	0.04
Strawberries	0.07
Other	
Eggs	0.30

the riboflavin contents of foods. However, exposure to light (e.g., sun-drying, sunlight exposure of milk in glass bottles, cooking in an open pot) can result in substantial losses, as the vitamin is very sensitive to destruction by light. Thus, exposure of milk in glass bottles to sunlight can result in the destruction of more than half of its riboflavin within a day. Irradiation of food results in the production of reactive oxygen species (e.g., superoxide, hydroxyl radical) that react with riboflavin to destroy it. The short exposure of meat to sterilizing quantities of γ-irradiation destroys 10–15% of its riboflavin content. Riboflavin photodegradation can be exacerbated by sodium bicarbonate, which is used to preserve vegetable colors. Also, because riboflavin is water soluble, it leaches into water used in cooking and into the drippings of meats. As riboflavin in cereal grains is located primarily in the germ and bran, the milling of such materials,[4] which removes those tissues, results in considerable losses in their content of the vitamin. For example, about half of the riboflavin in whole-grain rice, and more than a third of riboflavin in whole wheat, is lost when these grains are milled. Parboiled ("converted") rice contains most of the riboflavin of the parent grain, as the steam processing of whole brown rice before milling this product drives vitamins originally present in the germ and aleurone layers into the endosperm, where they are retained.

Bioavailability

The non-covalently bound forms of riboflavin in foods, FMN, FAD, and free riboflavin, appear to be well absorbed. In contrast, covalently bound flavin complexes, such as are found in plant tissues, are stable to digestion, and thus unavailable. In general, riboflavin in animal products tends to have a greater bioavailability than that in plant products.

Role of Hindgut Microflora

Riboflavin is synthesized by the bacteria populating the hindgut. It has been suggested that microfloral riboflavin can be absorbed across the colon to contribute to the nutritional riboflavin status of the host.

3. ABSORPTION OF RIBOFLAVIN

Hydrolysis of Coenzyme Forms

Because riboflavin occurs in most foods as protein complexes of the coenzyme forms FMN and FAD, the utilization of the vitamin in foods depends on the hydrolytic conversion of these forms to free riboflavin. This occurs by the proteolytic activity of the intestinal lumen, which releases the riboflavin coenzymes from their protein complexes, and the subsequent hydrolytic activities of several brush border phosphatases that liberate riboflavin in free form. The latter enzymes include the relatively non-specific alkaline phosphatase,[5] as well as FAD-pyrophosphatase (which converts FAD to FMN) and FMN-phosphatase (which converts FMN to free riboflavin).

Active Transport of Free Riboflavin

Riboflavin is absorbed in the free form by an ATP-dependent, carrier-mediated process in the proximal small intestine and colon. This process has been found to be at least partially dependent on Na^+, and may also involve the Ca^{2+}/calmodulin-, protein kinase A and G pathways.[6] The upper limit of intestinal absorption has been estimated to be about 25 mg – an order of magnitude greater than the requirement. Riboflavin absorption is enhanced by riboflavin deficiency, bile salts,[7] and psyllium gum, and downregulated by high doses of the vitamin. The latter effect appears to involve decreased activity of the riboflavin carrier induced by increased intracellular concentrations of cyclic AMP.

The free form of riboflavin is transported into the intestinal mucosal cell; however, much of that form is quickly trapped within the enterocyte by phosphorylation to FMN. This is accomplished by an ATP-dependent **flavokinase**. Thus, riboflavin enters the portal circulation as both the free vitamin and FMN.

4. TRANSPORT OF RIBOFLAVIN

Protein Binding

Riboflavin is transported in the plasma as free riboflavin and FMN, both of which are bound in appreciable amounts (e.g., about half of the free riboflavin and 80% of FMN) to plasma proteins. This includes tight binding to immunoglobulins (IgA, IgG, and IgM) and fibrinogen, and weak binding to albumin. These proteins bind riboflavin and FMN by hydrogen bonding; the vitamin can be displaced readily by boric acid[8] or several drugs,[9] which thus inhibit its transport in peripheral tissues.

---

4. It is the practice in many countries to enrich refined wheat products with several vitamins, including riboflavin, which results in their actually containing *more* riboflavin than the parent grains (e.g., 0.20 mg/100 g vs. 0.11 mg/100 g). However, rice is usually *not* enriched with riboflavin, to avoid coloring the product yellow with this intensely colored vitamin.

5. Alkaline phosphatase appears to have the greatest hydrolytic capacity of the brush border phosphatases.
6. Huang, S. N. and Swaan, P. (2001). *J. Pharmacol. Exp. Ther.* 298, 264.
7. Children with biliary atresia (a congenital condition involving the absence or pathological closure of the bile duct) show reduced riboflavin absorption.
8. The feeding of boric acid to humans and rats has been shown to produce riboflavinuria and precipitate riboflavin deficiency. In addition, some effects of boric acid toxicity can be overcome by feeding riboflavin.
9. For example, ouabain, theophylline, penicillin.

Specific Binding Proteins

Riboflavin-binding proteins (**RfBPs**) have been identified in the plasma of the laying hen and pregnant cows, mice, rats, monkeys, and humans. The plasma RfBP of the hen has been well characterized; it is not found in the immature female, but is synthesized in the liver under the stimulus of estrogen with the onset of sexual maturity or with induction by estrogen treatment. The avian plasma RfBP is a 32 kDa phosphoglycoprotein with a single binding site for riboflavin. It appears to be one of three products of a single gene, which are variously modified post-translationally to yield the RfBPs in egg white and yolk. The hen plasma RfBP is antigenically similar to the RfBPs of pregnant mice and rats. In both species RfBP has vital functions in the transplacental/transovarian movement of riboflavin[10] and in the uptake of riboflavin by spermatozoa, as immunoneutralization of the protein terminates pregnancy in females and reduces sperm fertility in males.[11]

Cellular Uptake

Riboflavin uptake appears to occur by a manner similar to its enteric absorption, by a Na^+-dependent, carrier-mediated process probably also involving the Ca^{2+}/calmodulin-, protein kinase A and G pathways. Receptor-mediated endocytosis has also been implicated, particularly in the transport of riboflavin across the placental barrier.[12]

Tissue Distribution

Riboflavin is transported into cells in its free form. However, in the tissues, riboflavin is converted to the coenzyme form, predominantly as FMN (60–95% of total flavins) but also as FAD (5–22% of total flavins in most tissues but about 37% in kidney), both of which are found almost exclusively bound to specific flavoproteins. The greatest concentrations of the vitamin are found in the liver, kidney, and heart. In most tissues, free riboflavin comprises <2% of the total flavins. Significant amounts of free riboflavin are found only in retina, urine, and cow's milk,[13] where it is loosely bound to casein. Although the riboflavin content of the brain is not great, the turnover of the vitamin in that tissue is high and the concentration of the vitamin is relatively resistant to gross changes in riboflavin nutriture. These findings suggest a homeostatic mechanism for regulating the riboflavin content of the brain; such a mechanism has been proposed for the chorioid plexus,[14] in which riboflavin transport has been found to be inhibited by several of its catabolic products and analogs. It has been estimated that the total body reserve of riboflavin in the adult human is equivalent to the metabolic demands for 2–6 weeks. Riboflavin is found in much lower concentrations in maternal plasma than in cord plasma (in humans this ratio has been found to be 1:4.7), suggesting the presence of a transplacental transport mechanism.

Tissue RfBPs have been identified in the liver, egg albumen,[15] and egg yolk of the laying hen. Each is similar to the plasma RfBP[16] in that species, differing only in the nature of its carbohydrate[17] contents.[18] A hereditary abnormality in the chicken results in the production of defective RfBPs (in plasma as well as liver and egg). Affected hens show **riboflavinuria**, and produce eggs with about half the normal amount of riboflavin and embryos that fail to develop.[19]

5. METABOLISM OF RIBOFLAVIN

Conversion to Coenzyme Forms

After it is taken up by the cell, free riboflavin is converted to its coenzyme forms (Fig. 11.1) in two steps, both of which appear to be regulated by thyroid hormones:

1. *Conversion to FMN.* The ATP-dependent phosphorylation to yield **riboflavin-5′-phosphate** (flavin mononucleotide, FMN). This occurs in the cytoplasm of most cells, and is catalyzed by the enzyme **flavokinase**.[20]

---

10. Hens that do not express RfBP produce eggs that lack the normal faint yellow tinge of their otherwise clear albumen. This observation led to the discovery of RfBP as being essential to transferring riboflavin to the egg (Winter, W. P., Buss, E. G., Clagget, C. O., and Boucher, R. V. [1967] *Comp. Biochem. Physiol.* 22, 889).

11. Plasma RfBP or fragments have been suggested as having potential utility as a vaccine to regulate fertility in both sexes (Adiga, S. K. [1997] *Human Reprod. Update* 3, 325).

12. Foraker, A. M., Khantwal, C. M., and Swaan, P. W. (2002). *Adv. Drug Deliv. Rev.* 55, 1467.

13. It should be noted that cow's milk differs from human milk in both the amount and form of riboflavin. Cow's milk typically contains 1160–2020 μg of riboflavin per liter, which (like the milk of most other mammals studied) is present mostly as the free vitamin. In contrast, human milk typically contains 120–485 μg of riboflavin per liter (depending on the riboflavin intakes of the mother), which is present mainly as FAD and FMN.

14. The anatomical site of the blood–cerebrospinal fluid barrier.

15. This is the flavoprotein formerly called **ovoflavin**. Comprising nearly 1% of the total protein in egg white, it is the most abundant of any vitamin-binding protein. Unlike the plasma RfBP, which is normally saturated with its ligand, the egg white RfBP is normally less than half-saturated with riboflavin, even when hens are fed diets high in the vitamin. Still, its bound riboflavin is responsible for the faint yellow tinge of egg albumen.

16. It appears that the plasma RfBP, produced and secreted by the liver in response to estrogens, is the precursor to these other binding proteins found in tissues.

17. Primarily in their contents of sialic acid, which occurs in many polysaccharides.

18. It is interesting to note that egg white RfBP forms a 1:1 complex with the thiamin-binding protein (TBP) from the same source.

19. Embryos from hens that are homozygous recessive for the mutant *rd* allele die of riboflavin deficiency on day 13–14 of incubation. They can be rescued by injecting riboflavin or FMN into the eggs.

20. Therefore, hypothyroidism is associated with reduced flavokinase activity and, accordingly, reduced tissue levels of FMN and FAD. Hyperthyroidism, in contrast, results in increased flavokinase activity, although tissue levels of FMN and FAD, which appear to be regulated *via* degradation, do not rise.

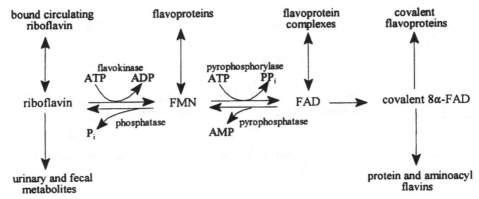

FIGURE 11.1 Riboflavin metabolism.

TABLE 11.2 Covalent Flavoproteins in Animals

Linkage	Enzyme
Histidinyl(N³)-8α-FAD	Succinate dehydrogenase
	Dimethylglycine dehyrogenase
	Sarcosine dehydrogenase
Histidinyl(N¹)-8α-FAD	L-Gulonolactone oxidase
Cysteinyl(s)-8α-FAD	Conoamine oxidase

Also called riboflavin kinase, this activity is regulated by **thyroxine**, which stimulates its synthesis. FMN produced by flavokinase can be complexed with specific apoproteins to form several functional flavoproteins.

2. *Conversion to FAD.* Most FMN is converted to the other coenzyme form, **flavin adenine dinucleotide (FAD)**, by a secondATP-dependent enzyme, **FAD-synthase.**[21] This step appears to be feedback-inhibited by FAD, which is complexed in tissues with a variety of dehydrogenases and oxidases mostly by noncovalent associations in discrete dinucleotide-binding domains apparently facilitated by a protein factor. This includes associations via hydrogen bonding with purines,[22] phenols, and indoles (e.g., to peptidyltryptophan in RfBPs). Less than 10% of FAD is covalently attached to certain apoenzymes (Table 11.2). Linkages of this type involve the riboflavin 8-methyl group, which can form a methylene bridge to the peptide histidyl imidazole function (e.g., in succinic dehydrogenase and sarcosine oxidase), or to the thioether function of a former cysteinyl residue (e.g., in **monoamine oxidase**).[23]

Glycosylation

The capacity to glycosylate riboflavin has been demonstrated in rat liver. Riboflavin 5α-α-D-glucoside appears to be a metabolically significant metabolite. It is found in the urine of riboflavin-fed rats, and has been shown to be comparable to riboflavin as a cellular source of the vitamin.

Catabolism

Flavins that are bound to proteins are resistant to degradation. However, unbound forms are subject to catabolism. Both FAD and FMN are catabolized by intracellular enzymes in ways directly analogous to the breakdown of these forms in foods during their absorption across the intestinal mucosal cell. Thus, FAD is converted to FMN by FAD-pyrophosphatase (releasing AMP), and FMN is degraded to free riboflavin by FMN-phosphatases. Both FAD and FMN are split to yield free riboflavin by alkaline phosphatase.

The degradation of riboflavin *per se* involves initially its hydroxylation at the 7α- and 8α-positions of the isoalloxazine ring by hepatic microsomal cytochrome *P*-450-dependent processes. It is thought that catabolism proceeds by the oxidation and then removal of the methyl groups. The liver, in at least some species, has the ability to form riboflavin α-glycosides. As a result of this metabolism, human blood plasma contains FAD and FMN as the major riboflavin metabolites, as well as small amounts of 7α-hydroxyriboflavin.[24] Side chain oxidation has been observed in bacterial systems, but not in higher animals.

Excretion

Riboflavin is rapidly excreted, primarily in the urine. Therefore, dietary needs for the vitamin are determined by its rate of excretion, not metabolism. In a riboflavin-adequate

---

21. This activity is also increased by thyroxine.
22. In FAD, the riboflavin and adenine moieties are predominantly (85%) hydrogen bonded in an intramolecular complex.
23. Another type of linkage involving the 8-methyl group, i.e., a thiohemiacetal linkage, is found in a microbial FAD-containing cytochrome.

---

24. This compound is also called 7-hydroxymethylriboflavin.

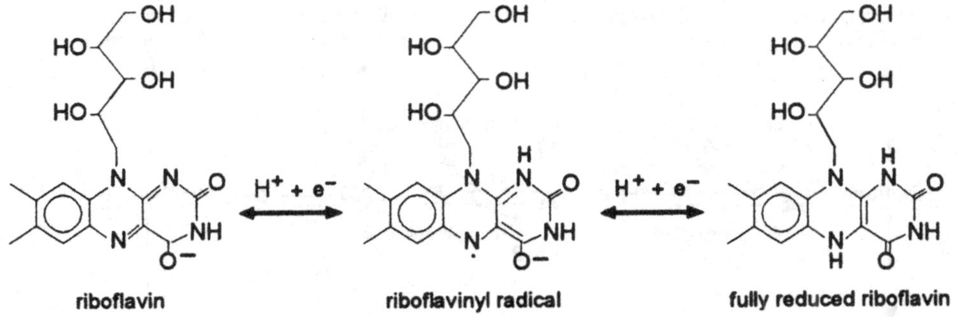

FIGURE 11.2   Two-step, single-electron, redox reactions of riboflavin.

human adult nearly all of a large oral dose of the vitamin will be excreted, with peak concentrations showing in the urine within a couple of hours. Studies in the rat have shown riboflavin to be turned over with a half-life of about 16 days in adequately nourished animals, and much longer in riboflavin-deficient animals. In normal human adults the urinary excretion of riboflavin is about $200\,\mu g/24$ hours, whereas riboflavin-deficient individuals may excrete only $40–70\,\mu g/24$ hours. Studies with a diabetic rat model[25] have shown riboflavin excretion to be significantly greater in diabetic individuals than in controls. Riboflavin excretion at $<27\,\mu g/mg$ creatinine is generally considered to indicate riboflavin deficiency in adults; however, this parameter tends to reflect current intake of the vitamin rather than total flavin stores.

The vitamin is excreted mainly (60–70%) as the free riboflavin, with smaller amounts of 7α- and 8α-hydroxyriboflavin,[26] 8α-sulfonylriboflavin, $5'$-riboflavinylpeptide, 10-hydroxyethylflavin, riboflavin $5'$-α-D-glucoside, **lumichrome**,[27] and 10-formylmethylflavin. Small amounts of riboflavin degradation products are found in the feces ($<5\%$ of an oral dose). As only about 1% of an oral dose of the vitamin is excreted in the bile by humans, most fecal metabolites are thought to be mainly of gut microbial origin. Little, if any, riboflavin is oxidized to CO_2.[28] Ingestion of boric acid, which binds to the riboflavin side chain, increases the urinary excretion of the vitamin.

Riboflavin is secreted into milk mostly as free riboflavin and FAD, and the antagonistic metabolite 10-(2'-hydroxyethyl)flavin, the amounts of which depend on the riboflavin intake of the mother. Milk also contains small amounts of other metabolites, including 7- and 8-hydroxymethyriboflavins, 10-formylmethylflavin, and lumichrome.

---

25. Streptozocin-induced diabetes.
26. This compound is also called 8-hydroxymethylriboflavin.
27. 7,8-Dimethylalloxazine, an irradiation product of riboflavin believed also to be produced by intestinal microbes.
28. Rats have been found to oxidize less than 1% of an oral dose of the vitamin.

6. METABOLIC FUNCTIONS OF RIBOFLAVIN

Coenzyme Functions

Riboflavin functions metabolically as the essential component of the coenzymes FMN and FAD, which act as intermediaries in transfers of electrons in biological oxidation–reduction reactions. More than 100 enzymes are known to bind FAD or FMN in animal and microbial systems. Most of these enzymes bind the flavinyl cofactors tightly but non-covalently; however, some[29] bind FAD covalently via histidinyl or cysteinyl linkages to the 8α-position of the isoalloxazine ring. These enzymes, called flavoproteins or flavoenzymes, include **oxidases**, which function aerobically, and **dehydrogenases**, which function anaerobically. Some involve one-electron transfers, whereas others involve two-electron transfers. This versatility allows flavoproteins to serve as switching sites between obligate two-electron donors (e.g., NADH, succinate) and obligate one-electron acceptors (e.g., iron–sulfur proteins, heme proteins). Flavoproteins serve this function by undergoing reduction through two single-electron transfer steps (Fig. 11.2) involving a riboflavinyl radical or semiquinone intermediate (with the unpaired electron localized at N-5). Because the radical intermediate can react with molecular oxygen, flavoproteins can also serve as cofactors in the two-electron reduction of O_2 to H_2O, and in the four-electron activation and cleavage of O_2 in monooxygenase reactions.

Collectively, the flavoproteins show great versatility in accepting and transferring one or two electrons with a range of potentials. This feature is attributed to the variation in the angle between the two planes of the isoalloxazine ring system (intersecting at N-5 and N-10), which is modified by specific protein binding. The flavin-containing dehydrogenases or reductases (their reduced forms) react slowly with molecular oxygen, in contrast

---

29. For example, succinate dehydrogenase, monoamine oxidase, monomethylglycine dehydrogenase.

TABLE 11.3 Important Flavoproteins of Animals

Flavoprotein	Flavin	Metabolic Function
One-Electron Transfers		
Mitochondrial electron transfer flavoprotein (ETF)	FAD	e^- acceptor for acyl-CoA, branched-chain acyl-CoA, glutaryl-CoA, and sarcosine and dimethylglycine dehydrogenases; links primary flavo-protein dehydrogenases with respiratory chain via ETF-ubiquinone reductase
Ubiquinone reductase	FAD	1-e^- transfer from ETF and coenzyme Q of respiratory chain
NADH-cytochrome *P*-450 reductase[a]	FMN	1-e^- transfer from FMN to cytochrome *P*-450 monooxygenase)
Pyridine-Linked Dehydrogenases		
NADP-cytochrome *P*-450 reductase	FAD	2-e^- transfer from NADP to FAD
Adrenodoxin reductase	FAD	2-e^- transfer from NADP to adrenodoxin[b] in steroid hydroxylation by adrenal cortex
NADP dehydrogenase	FMN	2-e^- transfer from NADP to FMN, then to ubiquinone[c]
NADP-dependent met-hemoglobin reductase	FAD	2-e^- transfer from NADP to FAD, then to met-hemoglobin
Non-Pyridine Nucleotide-Dependent Dehydrogenases		
Succinate dehydrogenase	FAD	Transfer reducing equivalents from succinate to ubiquinone yielding fumarate
Acyl-CoA dehydrogenases	FAD	2-e^- transfer from substrate to flavin, in oxidation of the N-methyl groups of choline and sarcosine
Pyridine Nucleotide Oxidoreductases		
Glutathione reductase	FAD	Reduces GSSG to GSH using NADPH
Lipoamide dehydrogenase[d]	FAD	Oxidizes dihydrolipoamide to lipoamide using NAD^+
Reactions of Reduced Flavoproteins with Oxygen		
D-Amino acid oxidase	FAD	Dehydrogenation of D-amino acid substrates to imino-acids, which are hydrolyzed to α-keto acids
L-Amino acid oxidase	FMN	Dehydrogenation of L-amino acid substrates to imino-acids, which are hydrolyzed to α-keto acids
Monoamine oxidase	FAD	Dehydration of biogenic amines[e] to corresponding imines with hydrogen transfer to O_2, forming H_2O_2
Xanthine oxidase	FAD	Oxidation of hypoxanthine and xanthine to uric acid with formation of H_2O_2
L-Gulonolactone oxidase	FAD	Oxidation of L-gulonolactone to ascorbic acid
Flavoprotein Monooxygenase		
Microsomal flavoprotein monooxygenase	FAD	Oxidation of N, S, Se, and I centers of various substrates in metabolism

[a]*A component of microsomal cytochrome P-450, it contains one molecule each of FAD and FMN.*
[b]*An iron–sulfur protein.*
[c]*Also has NADH-ubiquinone reductase activity, reductively releasing iron from ferritin.*
[d]*A component of the pyruvate dehydrogenase and α-ketoglutarate dehydrogenase complexes.*
[e]*For example, serotonin, noradrenaline, benzylamine.*

to the fast reactions of the flavin-containing oxidases and monooxygenases. In the former reactions, hydroperoxide derivatives of the flavoprotein are cleaved to yield super-oxide anion (O_2^-), but in the latter a heterolytic cleavage of the hydroperoxide group occurs to yield the peroxide ion (OOH^-). Many flavoproteins contain a metal (e.g., iron, molybdenum, zinc), and the combination of flavin and metal ion is often involved in the adjustments of these enzymes in transfers between single- and double-electron donors. In some flavoproteins, the means for multiple-electron transfers is provided by the presence of multiple flavins as well as metals.

Metabolic Roles

The flavoproteins, which are a large group of enzymes involved in biological oxidations and reductions, are essential for the metabolism of carbohydrates, amino acids, and lipids (Table 11.3). Some are also essential for the activation of the vitamins pyridoxine and folate to their respective coenzyme forms. Others participate in antioxidant protection by maintaining the glutathione redox cycle and providing reducing equivalents for neutralizing reactive oxygen species. Riboflavin has also been found to play a role in the regulation of gene expression in bacteria by forming mRNA structures called "riboswitches" that repress conformation to cause premature termination of transcription or inhibit initiation of translation. Analogous function in higher animals has not been reported.

7. RIBOFLAVIN IN HEALTH AND DISEASE

The fundamental metabolic roles played by flavoenzymes give riboflavin relevance to health in various ways. Accordingly, riboflavin status affects responses to pro-oxidants, homocysteine, carcinogenic factors, and malaria.

Oxidative Stress

Flavoenzymes participate in protection of erythrocytes and other cells against oxidative stress. This includes the support of intracellular levels of reduced glutathione via glutathione reductase, and the direct reduction of oxidized forms of hemoproteins by methemoglobin reductases.

Vascular Disease

There is evidence that riboflavin may play a role in reducing risk of vascular disease. Dietary riboflavin intake has been found to be inversely correlated with serum homocysteine levels.[30] The Framingham Offspring Study found elevated plasma homocysteine levels in subjects with relatively low plasma riboflavin levels (Table 11.4). That riboflavin status may prevent homocysteinemia would appear to involve its role in homocysteine metabolism, i.e., as the essential cofactor (FAD) for methyltetrahydrofolate reductase, which catalyzes the conversion of N-5,10-methylenetetrahydrofolate to N-5-methyltetrahydrofolate (see Chapter 16, Folate). Homocysteinemia has been associated with increased risks of occlusive vascular disease, total and cardiovascular disease-related mortality, stroke, dementia, Alzheimer's disease, fracture, and chronic heart failure.[31]

TABLE 11.4 Relationship of Plasma Riboflavin and Homocysteine Levels among Subjects in the Framingham Offspring Cohort Study

	Plasma Riboflavin Tertile, nmol/l		
	<6.89	6.89–10.99	≥11.0
Subjects	147	151	152
Plasma homocysteine – mean (95% CI)	10.3 (9.8–10.8)*	9.5 (9.1–10.0)*	9.5 (9.1–10.0)*

From Jacques, P. F., Bostom, A. G., Williams, R. R., et al. (2002). *J. Nutr.* 132, 283.
*P <0.05

Riboflavin may be most important in individuals with a polymorphism of methyltetrahydrofolate reductase (i.e., the heat-sensitive form of the enzyme) that leads to increased risk of vascular disease.[32] *In vitro* studies have shown that riboflavin can stabilize this isoform.[33]

Mineral Utilization

Riboflavin has the capacity to form complexes with divalent cations such as Fe^{2+} and Zn^{2+}. That correction of riboflavin deficiency has been shown to improve significantly the enteric absorption of iron and zinc in the mouse model[34] suggests that riboflavin can be a determinant of the bioavailability of such minerals.

Anticarcinogenesis

Riboflavin deprivation has been reported to enhance carcinogenesis.[35] This effect is thought to be due to diminished antioxidant protection, increasing the activation of carcinogens and oxidative damage to DNA, and/or diminished folate metabolism, reducing DNA synthesis, repair and methylation. Clinical data relative to this possibility are few: three observational studies found an inverse relationship of riboflavin status and cancer risk.[36]

32. That is, the C677T polymorphism (Hustad, S., Ueland, P. M., Vollset, S. E., *et al.* [2000] *Clin. Chem.* 46, 1065).
33. McNulty, H., McKinley, M. C., Wilson, B., *et al.* (2002). *Am. J. Clin. Nutr.* 76, 436.
34. Agte, V. V., Paknikar, K. M., Chiplonkar, S. A., *et al.* (1998). *Biol Trace Elem. Res.* 65, 109.
35. Webster, R. P., Gawde, M. D., and Bhattacharya, R. K. (1973). *Cancer Res.* 33, 1997.
36. Esophageal squamous cell cancer (He, Y., Shan, B., Song, G., *et al.* [2009] *Asian Pacific J. Cancer Prev.* 10, 619); colorectal cancer (Figueiredo, J. C., Levine, A. J., Grau, M. V., *et al.* [2008] *Cancer Epidemiol. Biomarkers Prev.* 17, 2137; de Vogel, S., Dindore, V., van Engeland, M., *et al.* [2008] *J. Nutr.* 138, 2372).

30. Ganji, G. and Kafai, M. R. (2004). *Am. J. Clin. Nutr.* 80, 1500.
31. Selhub, J. (2006) *J. Nutr.* 136, 1726S–1730S.

Malaria

Riboflavin deficiency appears to protect against malaria, decreasing parasitemia and signs of infection.[37] The metabolic basis of this protection is thought to involve increased vulnerability of erythrocytes to destructive lipid peroxidation caused by the oxidative stress resulting from the infection (*Plasmodium* sp.). Hence, they tend to autolyze before the plasmodia they contain can mature, reducing the parasitemia and decreasing the symptoms of infection. The infected erythrocyte has been found to have an increased need for riboflavin.[38] In addition, malarial parasites have been shown to be even more susceptible than erythrocytes to reactive oxygen species. Therefore, it has been suggested that marginal riboflavin deficiency may be preferentially deleterious for both the parasite and the infected cell. Flavin analogs[39] that antagonize riboflavin and inhibit glutathione reductase have been shown to have antimalarial activities.

Tryptophan Deficiency-Induced Cataract

Riboflavin deprivation has been shown to protect the rat from the cataractogenic effect of a low-tryptophan diet. The metabolic basis of this effect may involve the lack of formation of a riboflavinyl tryptophan adduct that accelerates the photo-oxidation of the amino acid to a pro-oxidative form.

Defects in Fat Metabolism

As essential coenzymes for acyl-CoA dehydrogenase and NADH dehydrogenase, riboflavin plays essential roles in lipid metabolism. Accordingly, riboflavin treatment has been found useful in treating cases of recurrent hypoglycemia and lipid storage myopathy in individuals with deficient expression of these flavoenzymes.

FAD-dependent pathways are involved in the oxidative folding of proteins due to the formation of disulfide bonds. This occurs for such secretory proteins in the endoplasmic reticulum as apolipoprotein B-100, the *in vitro* secretion of which has been shown to be impaired by riboflavin deprivation.

8. RIBOFLAVIN DEFICIENCY

Many tissues are affected by riboflavin deficiency (Table 11.5). Therefore, deprivation of the vitamin causes, in animals, such general signs as loss of appetite, impaired

TABLE 11.5 General Signs of Riboflavin Deficiency

Organ System	Signs
General	
Appetite	Decrease
Growth	Decrease
Dermatologic	Cheilosis, stomatitis
Muscular	Weakness
Gastrointestinal	Inflammation, ulcer
Skeletal	Deformities
Vital organs	Hepatic steatosis
Vascular	
Erythrocytes	Anemia
Nervous	Ataxia, paralysis
Reproductive	
Male	Sterility
Female	Decreased egg production
Fetal	Malformations, death
Ocular	
Retinal	Photophobia
Corneal	Decreased vascularization

growth, and reduced efficiency of feed utilization, all of which constitute significant costs in animal agriculture. In addition, both animals and humans experiencing riboflavin deficiency show specific epithelial lesions and nervous disorders. These manifestations are accompanied by abnormally low activities of a variety of flavoenzymes. The most rapid and dramatic loss of activity involves **erythrocyte glutathione reductase (EGR)**. Substantial losses also occur in the activities of flavokinase and FAD-synthetase; thus, the biosynthesis of flavoproteins is lost under conditions of riboflavin deprivation. In summary, then, riboflavin deficiency results in impairments in the metabolism of energy, amino acids, and lipids. These metabolic impairments are manifested morphologically as arrays of both general and specific signs/symptoms.

The rapid loss of erythrocyte glutathione reductase activity as a result of riboflavin deprivation makes this enzyme a useful marker of riboflavin status. Estimation of the degree of saturation of erythrocyte EGR by FAD has proven extremely useful in assessing riboflavin status, in a manner analogous to the use of erythrocyte transketolase saturation by thiamin pyrophosphate to assess thiamin status. Studies have shown that *in vitro* EGR activities of normal, riboflavin-adequate individuals is stimulated $\leq 20\%$

37. Das, B. S., *et al.* (1988). *Eur. J. Clin. Nutr.* 42, 227.
38. Dutta, P. (1991). *J. Protozool.* 38, 479.
39. For example, galactoflavin, 10-(4'-chlorophenyl)-3-methylflavin and some isoalloxazine derivatives.

by the addition of exogenous FAD. Individuals showing activity coefficients (native EGR activity÷EGR activity with added FAD) of 20–30% and >30% are considered to be at moderate and high risk, respectively, of riboflavin deficiency.

Riboflavin deficiency produces, in the small intestine, a hyperproliferative response of the mucosa, characterized by reductions in number of villi, increases in villus length, and increases in the transit rates of enterocytes along the villi. These morphological effects are associated with reduced enteric absorption of dietary iron, resulting in secondary impairments in nutritional iron status in riboflavin-deprived individuals.

Risk Factors for Riboflavin Deficiency

Several factors can contribute to riboflavin deficiency:

- *Inadequate diet.* Inadequate diet is the most important cause of riboflavin deficiency. Frequently, this involves the low consumption of milk,[40] which is the most important source of the vitamin available in most diets. In industrialized countries, riboflavin deficiency occurs most frequently among alcoholics, whose dietary practices are often faulty, leading to this and other deficiencies.
- *Enhanced catabolism.* Catabolic conditions associated with illness or vigorous physical exercise and involving nitrogen loss increase riboflavin losses.
- *Alcohol.* High intakes of alcohol appear to antagonize the utilization of FAD from foods.
- *Phototherapy.* Phototherapy of infants with hyperbilirubinemia often leads to riboflavin deficiency (by photodestruction of the vitamin)[41] if such therapy does not also include the administration of riboflavin.[42]
- *Exercise.* Physical exercise can produce abnormalities in a variety of biochemical markers of riboflavin status, such as increased erythrocyte glutathione reductase activity coefficient and reduced urinary riboflavin. Nevertheless, there is no evidence that such abnormalities lead to impairments in physiological performance.

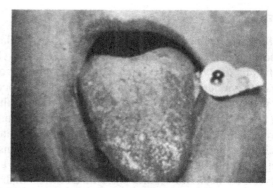

FIGURE 11.3 Geographical tongue in riboflavin deficiency. *Courtesy of Cambridge University Press.*

- *Other factors.* Although earlier studies purported to show reduced riboflavin status among some women using oral contraceptive agents, more recent critical studies have failed to detect any such interaction. Patients receiving diuretics or undergoing hemodialysis experience enhanced loss of riboflavin (as well as other water-soluble vitamins).

Although clinical signs of riboflavin deficiency are rarely seen in the industrialized world, **subclinical riboflavin deficiency** (that is, conditions wherein a subject's intake of the vitamin may be sufficient to prevent clinical signs but not to keep the flavoproteins saturated for optimal metabolism) is not uncommon. In fact, it has been estimated that as many as 27% of urban American teenagers of low socioeconomic status have subclinical riboflavin deficiency.

Deficiency Signs in Humans

Uncomplicated riboflavin deficiency becomes manifest in humans only after 3–4 months of deprivation of the vitamin. Signs include **cheilosis**,[43] angular stomatitis, **glossitis** (see Fig. 11.3),[44] hyperemia[45] and edema[46] of the oral mucosa, seborrheic dermatitis around the nose and mouth and scrotum/vulva, and a normocytic, normochromic anemia with **reticulocytopenia**,[47] **leukopenia**,[48] and **thrombocytopenia**.[49] Riboflavin-deficient humans also experience

---

40. Children consuming less than a cup of milk per week are likely to be deficient in riboflavin.
41. Phototherapy can be an effective treatment for infants with mild hyperbilirubinemia; however, the mechanism by which it leads to the degradation of bilirubin (to soluble substances that can be excreted) necessarily leads also to the destruction of riboflavin. It is the photoactivation of riboflavin in the patient's plasma that generates singlet oxygen, which reacts with bilirubin. Thus, plasma riboflavin levels of such patients have been found to drop as the result of phototherapy. Riboflavin supplementation prevents such a drop and has been shown to enhance bilirubin destruction.
42. For example, 0.5 mg of riboflavin sodium phosphate per kilogram body weight per day.

---

43. Lesions of the lips.
44. Inflammation of the tongue. This can involve disappearance of filiform papillae and enlargement of fungiform papillae, with the tongue color changing to a deep red. Subjects with this condition, called **geographical tongue**, have soreness of the tongue and loss of taste sensation.
45. Increased amount of blood present.
46. Accumulation of excessive fluid in the tissue.
47. Abnormally low number of immature red blood cells in the circulating blood.
48. Abnormally low number of white blood cells in the circulating blood (<5,000/ml).
49. Abnormally low number of platelets in the circulating blood.

neurological dysfunction involving peripheral neuropathy
of the extremities characterized by hyperesthesia,[50] cold-
ness, and pain, as well as decreased sensitivity to touch,
temperature, vibration, and position.

Deficiency Disorders in Animals

Riboflavin deficiency in animals is potentially fatal. In
addition to the general signs already mentioned, animals
show other signs that vary with the species. Riboflavin-
deficient rodents show dermatologic signs (alopecia,
seborrheic inflammation,[51] moderate epidermal hyperkeratosis[52] with atrophy of sebaceous glands), and a generally ragged appearance. Red, swollen lips and abnormal
papillae of the tongue are seen. Ocular signs may also
be observed (blepharitis,[53] conjunctivitis,[54] and corneal
opacity). Feeding a high-fat diet can increase the severity
of deficiency signs; high-fat-fed rats showed anestrus, multiple fetal skeletal abnormalities (shortening of the mandi-
ble, fusion of ribs, cleft palate, deformed digits and limbs),
paralysis of the hind limbs (degeneration of the myelin
sheaths of the sciatic nerves[55]), hydrocephalus,[56] ocular
lesions, cardiac malformations, and hydronephrosis.[57]

The riboflavin-deficient chick also experiences myelin
degeneration of nerves, affecting the sciatic nerve in particular. This results in an inability to extend the digits – a
syndrome called **curled-toe paralysis** (see Fig. 11.4). In
hens, the deficiency involves reductions in both egg production and embryonic survival (decreased hatchability of
fertile eggs). Riboflavin-deficient turkeys show severe dermatitis. The deficiency is rapidly fatal in ducks.

Riboflavin-deficient dogs are weak and ataxic. They
show dermatitis (chest, abdomen, inner thighs, axillae,
and scrotum) and **hypoplastic anemia**[58] with fatty infiltration of the bone marrow. They can have bradycardia and
sinus arrhythmia[59] with respiratory failure. Corneal opacity has been reported. The deficiency can be fatal, with
collapse and coma. Swine fed a riboflavin-deficient diet
grow slowly, and develop a scaly dermatitis with alopecia.

FIGURE 11.4   Curled-toe paralysis in a riboflavin-deficient chick.

They can show corneal opacity, cataracts, adrenal hemorrhages, fatty degeneration of the kidney, inflammation of
the mucous membranes of the gastrointestinal tract, and
nerve degeneration. In severe cases, deficient individuals
can collapse and die.

Riboflavin deficiency in the newborn calf[60] is manifested as redness of the buccal mucosa,[61] angular **stomatitis**,[62] alopecia, diarrhea, excessive tearing and salivation,
and inanition. Signs of riboflavin deficiency appear to
develop rather slowly in rhesus monkeys. The first signs
seen are weight loss (6–8 weeks), followed by dermatologic changes in the mouth, face, legs, and hands and a
normocytic hypochromic anemia[63] (2–6 months), and,
ultimately, collapse and death, with fatty degeneration of
the liver. Similar signs have been produced in baboons
made riboflavin deficient for experimental purposes.

9. RIBOFLAVIN TOXICITY

The toxicity of riboflavin is *very low*, and thus problems of
hypervitaminosis are not expected. Probably because it is
not well absorbed, high oral doses of riboflavin are essentially non-toxic. Oral riboflavin doses as great as 2–10 g/kg
body weight produce no adverse effects in dogs and rats.
The vitamin is somewhat more toxic when administered
parenterally. The LD_{50} (50% lethal dose) values for the rat
given riboflavin by the intraperitoneal, subcutaneous, and
oral routes have been estimated to be 0.6, 5, and >10 g/kg,
respectively.

50. Excessive sensibility to touch, pain, etc.
51. Involving excess oiliness due to excess activity of the sebaceous glands.
52. Hypertrophy of the horny layer of the epidermis.
53. Inflammation of the eyelids.
54. Inflammation of the mucous membrane covering the anterior surface of the eyeball.
55. The nerve situated in the thigh.
56. A condition involving the excessive accumulation of fluid in the cerebral ventricles, dilating these cavities and, in severe cases, thinning the brain and causing a separation of the cranial bones.
57. Dilation of one or both kidneys owing to obstructed urine flow.
58. Progressive non-regenerative anemia resulting from depressed, inadequate functioning of the bone marrow.
59. Irregular heartbeat, with the heart under control of its normal pacemaker, the sino-atrial (S-A) node.

60. Ruminants do not normally require a dietary source of riboflavin, as the bacteria in their rumens synthesize the vitamin in adequate amounts. However, newborn calves and lambs, whose rumen microflora is not yet established, require riboflavin in their diets. This is normally supplied by their mothers' milk or by supplements in their milk-replacer formula diets.
61. The mucosa of the cheek.
62. Lesions in the corners of the mouth.
63. Anemia involving erythrocytes of normal size but low hemoglobin content.

10. CASE STUDY

Instructions

Review the following summary of a research report, paying special attention to the diagnostic indicators on which the treatments were based. Then answer the questions that follow.

Case

An experiment was conducted to determine the basis of protection by riboflavin deficiency against malarial infection. An animal model, which previously showed such protection against *Plasmodium berghei*, was used. It involved depleting 3-week-old male rats of riboflavin by feeding them a sucrose-based purified diet containing <1 mg of riboflavin per kilogram. A control group was pair-fed[64] the same basal diet supplemented with 8.5 mg of riboflavin per kilogram.[65] At 6 weeks of age, several biochemical characteristics of erythrocytes (RBCs) were measured: reduced glutathione levels, activities of antioxidant enzymes, and stabilities of erythrocytes to hemolysis (measured by incubating 0.5% suspensions of RBCs with pro-oxidants [500 μmol/l H_2O_2 or 2.5 μmol/l ferriprotoporphyrin IX] or in a hypotonic medium [151 mOsm] for 1 hour at 37°C). Oxidative damage was assessed by measuring H_2O_2-induced production of malonyldialdehyde (MDA). Other studies with this and similar animal models have shown that the riboflavin-deficient group, when infected with the parasite, grows better and shows reduced parasitemia than pair-fed controls.

Case Questions and Exercises

1. What dependent variables did the investigators measure to confirm that riboflavin deficiency had been produced in their experimental animals?
2. Propose a hypothesis to explain the apparently discrepant results regarding the effects of riboflavin deficiency on erythrocyte stability.
3. Propose a hypothesis for the protective effect of riboflavin deficiency against malarial infection. What other nutrients might you expect to influence susceptibility to this erythrocyte-attacking parasite?

---

64. *Pair-feeding* is a method of controlling for the effects of reduced food intake that may be secondary to the independent experimental variable (e.g., a nutrient deficiency). It involves the matching of one animal from the experimental treatment group with one of similar body weight from the control group, and the feeding of the latter individual a measured amount of feed equivalent to the amount of feed consumed by the former individual on the previous day. In experiments of more than a few days' duration, this approach normalizes the feed intake of both the experimental and control groups.

65. This level is about three times the amount normally required by the rat.

Results of Biochemical Studies of Erythrocytes

Parameter	Riboflavin Deficient	Control	*P* value
Reticulocytes (% total RBCs)	1.50 ± 0.29	1.26 ± 0.37	NS[a]
Hemoglobin (g/dl blood)	14.7 ± 0.6	14.9 ± 0.3	NS
GSH (mmol/g Hb)	7.97 ± 2.89	6.19 ± 2.52	<0.001
Glutathione reductase (mU[b]/mg protein)	42 ± 6	124 ± 16	<0.001
Glutathione reductase activity coefficient	2.37 ± 0.19	1.20 ± 0.08	<0.01
Glutathione peroxidase (mU[b]/g Hb)	918 ± 70	944 ± 62	NS
In vitro hemolysis (%):			
H_2O_2-induced	32 ± 9	55 ± 9	<0.05
Hypotonicity	69 ± 4	53 ± 7	<0.01
Ferriprotoporphyrin IX	42 ± 3	29 ± 4	<0.001
MDA (nmol/g Hb):			
Before incubation	25.5 ± 3.8	25.9 ± 3.4	NS
Incubated with H_2O_2	34.8 ± 1.2	42.7 ± 1.8	<0.01

[a]*NS, not significant (P >0.05).*
[b]*1 mU = 1 nmol NADPH per min.*

Study Questions and Exercises

1. Diagram the general roles of FAD- and FMN-dependent enzymes in various areas of metabolism.
2. Construct a decision tree for the diagnosis of riboflavin deficiency in humans or an animal species.
3. What key feature of the chemistry of riboflavin relates to its biochemical functions in flavoproteins?
4. What diet and lifestyle factors would you expect to affect dietary riboflavin needs? Justify your answer.
5. What parameters might you measure to assess the riboflavin status of a human or animal?

RECOMMENDED READING

Bates, C.J., 1997. Bioavailability of riboflavin. Eur. J. Clin. Nutr. 51, S38–S42.

Bender, D.A., 2009. Nutritional Biochemistry of the Vitamins, second ed. Cambridge University Press, Cambridge. pp. 172–199.

Depeint, F., Bruce, W.R., Shangari, N., et al. 2006. Mitochondrial function and toxicity: Role of B vitamin family on mitochondrial energy metabolism. Chem.-Biol. Interactions 163, 94–112.

Foraker, A.B., Khantwal, C.M., Swaan, P.W., 2003. Current perspectives on the cellular uptake and trafficking of riboflavin. Adv. Drug Deliv. Rev. 55, 1467–1483.

Hefti, M.H., Vervoot, J., van Berkel, W.J.H., 2003. Deflavination and reconstitution of flavoproteins. Tackling fold and function. Eur. J. Biochim. 270, 4227–4242.

Leferink, N.G.H., Heuts, D.P.H.M., Fraaije, M.W., van Berkel, W.J.H., 2008. The growing VAO flavoprotein family. Arch. Biochem. Biophys. 474, 292–301.

Mansour, N.M., Sawhney, M., Tamang, D.G., et al. 2007. The bile/arenite/riboflavin transporter (BART) superfamily. FEBS J. 274, 612–629.

McCormick, D.B., 2006. Riboflavin. In: Shils, M.E., Shike, M., Ross, A.C., Caballero, B., Cousins, R.J. (Eds.), In: Modern Nutrition in Health and Disease, tenth ed. Lippincott, New York, NY, pp. 434–441.

Powers, H.J., 2003. Riboflavin (vitamin B-2) and health. Am. J. Clin. Nutr. 77, 1352–1360.

Rivlin, R.S., 2007. Riboflavin. In: Zemplini, J., Rucker, R.B., McCormick, B.B., Suttie, J.W. (Eds.), Handbook of Vitamins, fourth ed. CRC Press, New York, NY, pp. 233–251.

Said, H.M., 2004. Recent advances in carrier-mediated intestinal absorption of water-soluble vitamins. Ann. Rev. Physiol. 66, 419–446.

Stuehr, D.J., Tejero, J., Haque, M.M., 2009. Structural and mechanistic aspects of flavoproteins: Electron transfer through the nitric oxide synthase flavoprotein domain. FEBS J. 276, 3959–3974.

Niacin

Chapter Outline

Anchoring Concepts

1. Niacin is the generic descriptor for pyridine 3-carboxylic acid and derivatives exhibiting qualitatively the biological activity of nicotinamide.
2. The two major forms of niacin, nicotinic acid and nicotinamide, are active metabolically as the pyridine nucleotide coenzymes NAD(H) and NADP(H).
3. Deficiencies of niacin are manifest as dermatologic, gastrointestinal, and neurologic changes, and can be fatal.

So far as they have been studied, the foodstuffs that appear to be good sources of the blacktongue preventive also appear to be good sources of the pellagra preventive ... Considering the available evidence as a whole, it would seem highly probable, if not certain, that experimental black tongue and pellagra are essentially identical conditions and, thus, that the preventive of black tongue is identical with the pellagra preventive, or factor P-P. On the basis of the indications afforded by the test in the dog, liver, salmon and egg yolk are recommended for use in the treatment and prevention of pellagra in humans.

J. Goldberger

Learning Objectives

1. To understand the chief natural sources of niacin.
2. To understand the means of enteric absorption and transport of niacin.
3. To understand the biochemical function of niacin as a component of coenzymes of a variety of metabolically important redox reactions, and the relationship of that function to the physiological activities of the vitamin.
4. To understand the factors that can affect low niacin status, and the physiological implications of that condition.

VOCABULARY

ADP-ribotransferases (ARTs)
α-Amino-β-carboxymuconic-ε-semialdehyde (ACS)
α-Amino-β-carboxymuconic-ε-semialdehyde decarboxylase (ACSD)
Anthranilic acid
Black tongue disease
Casal's collar
Flushing
Formylase
N-Formylkynurenine
Four Ds of niacin deficiency
Hartnup disease
3-Hydroxyanthranilic acid
3-Hydroxyanthranilic acid oxygenase
3-Hydroxykynurenine
Kynurenine
Kynurenic acid
Kynurenine 3-hydroxylase
Leucine
1-Methylnicotinamide
1-Methylnicotinic acid
1-Methyl-6-pyridone 3-carboxamide
NAD(H)
NAD$^+$ kinase
NAD$^+$ synthetase

The Vitamins. DOI: 10.1016/B978-0-12-381980-2.00012-8

NADP(H)

NAD(P)$^+$ glycohydrolase

Niacin receptor

Nicotinamide

Nicotinamide methylase

Nicotinamide riboside (NR)

Nicotinate phosphoribosyltransferase

Nicotinic acid (NA)

Pellagra

Perosis

Phosphodiesterase

Picolinic acid

Poly(ADP-ribose) polymerase (PARP)

Pyridine nucleotide

Pyridoxal phosphate

Quinolinic acid

Schizophrenia

Transaminase

Transhydrogenase

Trigonelline

Tryptophan

Tryptophan pyrrolase

Xanthurenic acid

1. THE SIGNIFICANCE OF NIACIN

Niacin is required for the biosynthesis of the **pyridine nucleotides NAD(H)** and **NADP(H)**, through which the vitamin has key roles in virtually all aspects of metabolism. Historically, niacin deficiency was prevalent among people who relied on maize (corn) as their major food staple; before the availability of inexpensive supplements, the deficiency was also a frequent problem of livestock fed maize-based diets.

Great irony characterizes niacin deficiency. Unlike thiamin deficiency (which also involves a cereal-based diet), niacin deficiency more frequently results from poor bioavailability rather than scarcity *per se*. Hence, paradoxical questions have been asked:

- Why does niacin deficiency occur among individuals who can biosynthesize the vitamin?
- Why did pellagra occur among people eating maize (corn), whereas the disease was unknown in the Americas where maize was a historically important part of the diet?
- Why do maize-based diets produce pellagra, although maize contains an appreciable amount of niacin?
- Why does milk, which contains little niacin, prevent pellagra?
- Why does rice, which contains less niacin than maize, not produce pellagra?

Niacin is also of interest for its health value as a pharmacologic, i.e., multi-gram, dose levels. Understanding the bases of both the physiologic and pharmacologic activities

of niacin calls for an appreciation of its complexities, which are manifest differently in various species.

2. SOURCES OF NIACIN

Distribution in Foods

Niacin occurs in greatest quantities in brewers' yeasts and meats, but significant amounts are also found in many other foods (Table 12.1). The vitamin is distributed unevenly in grains, being present mostly in the bran fractions. Niacin occurs predominantly in bound forms, e.g., in plants mostly as protein-bound **nicotinic acid** (**NA**), and in animal tissues mostly as **nicotinamide** (**NAm**) **in nicotinamide-adenine dinucleotide** (**NAD**) and **nicotinamide-adenine dinucleotide phosphate** (**NADP**). Niacin is added by law to wheat flour and other grain products in the United States.[1]

Stability

Niacin in foods is very stable to storage and to normal means of food preparation and cooking (e.g., moist heat).

Bioavailability

Niacin is found in many types of foods in forms from which it is not released on digestion, thus rendering it unavailable to the eater. In grains, niacin is present in covalently bound complexes with small peptides and carbohydrates, collectively referred to as **niacytin**.[2] The esterified niacin in these complexes is not normally available; however, its bioavailability can be improved substantially by treatment with base to effect the alkaline hydrolysis of those esters. The tradition in Central American cuisine of soaking and cooking maize in lime[3]-water effectively renders available the niacin in that grain.[4] This practice appears to be responsible for effective protection against pellagra in that part of the world. In other foods, niacin is present as a methylated derivative (**1-methylnicotinic acid**, also called **trigonelline**) that functions as a plant

---

1. Fortification is mandated for niacin, thiamin, riboflavin, folate, and iron.
2. Polysaccharide extracted from wheat bran has been found to contain more than 1% nicotinic acid bound via an ester linkage to glucose in a complex also containing arabinose, galactose, and xylose. Although NAD$^+$ and NADP$^+$, both of which are biologically available to humans and animals, are present in early-stage corn, those levels decline as the grain matures and are replaced by nicotinamide and nicotinic acid as well as forms of very low bioavailability, such as bound niacin and trigonelline.
3. Calcium hydroxide.
4. This process, called nixtamalization, renders maize (corn) more easily ground, improves its flavor, and reduces mycotoxin content. It is used in making tortillas, hominy, and corn chips. The term itself derives from the Nahuatl (an Aztec dialect) words *nextli* (ashes) and *tamalii* (corn dough).

Chapter | 12 Niacin

TABLE 12.1 Niacin Contents of Foods

Food	Niacin, mg/100 g	Food	Niacin, mg/100 g
Dairy Products		Mushrooms	4.2
Milk	0.2	Yeast	50.1
Yogurt	0.1	**Vegetables**	
Cheeses	1.2	Asparagus	1.5
Meats		Beans	0.5–2.4
Beef	4.6	Broccoli	0.9
Chicken	4.7–14.7	Brussels sprouts	0.9
Lamb	4.5	Cabbage	0.3
Pork	0.8–5.6	Carrots	0.6
Turkey	8.0	Cauliflower	0.7
Calf heart	7.5	Celery	0.3
Calf kidney	6.4	Corn	1.7
Fish		Kale	2.1
Cod	2.2	Lentils	2.0
Flounder	2.5	Onions	0.2
Haddock	3.0	Peas	0.9–25.0
Herring	3.6	Peppers	1.7–4.4
Tuna	13.3	Potatoes	1.5
Cereals		Soy beans	1.4
Barley	3.1	Spinach	0.6
Buckwheat	4.4	Tomatoes	0.7
Cornmeal	1.4–2.9	**Fruits**	
Rice		Apples	0.6
Polished	1.6	Bananas	0.7
Unpolished	4.7	Grapefruits	0.2
Rye	0.9–1.6	Oranges	0.4
Wheat		Peaches	1.0
Whole grain	3.4–6.5	Strawberries	0.6
Wheat bran	8.6–33.4	**Nuts**	
Other		Most nuts	0.6–1.8
Eggs	0.1	Peanuts	17.2

hormone but is also not biologically available to animals. This form, however, is heat labile, and can be converted to NA by heating.[5]

5. Thus, the roasting of coffee beans effectively removes the methyl group from trigonelline, increasing the nicotinic acid content of that food from 20 to 500 mg/kg. This practice, too, appears to have contributed to the rarity of pellagra in the maize-eating cultures of South and Central America.

Importance of Dietary Tryptophan

A substantial amount of niacin can be synthesized from the indispensable amino acid **tryptophan** (see "Niacin Biosynthesis," page 295). Therefore, the niacin adequacy of diets involves both the level of the pre-formed vitamin and that of its potential precursor tryptophan (Table 12.2).

TABLE 12.2 The Niacin-Equivalent Contents of Several Foods

Food	Preformed Niacin (mg/1,000 kcal)	Tryptophan (mg/1,000 kcal)	Niacin Equivalents[a]
Cow's milk	1.21	673	12.4
Human milk	2.46	443	9.84
Beef	2.47	1280	23.80
Eggs (whole)	0.60	1150	19.80
Pork	1.15	61	2.17
Wheat flour	2.48	297	7.43
Corn meal	4.97	106	6.74
Corn grits	1.83	70	3.00
Rice	4.52	290	9.35

[a]Based on a conversion efficiency of 60:1 for humans.

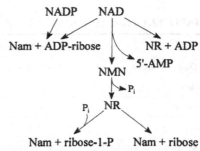

FIGURE 12.1 Metabolic disposition of absorbed niacin.

3. ABSORPTION OF NIACIN

Digestion of NAD/NADP

The predominant forms of niacin in most animal-derived foods are the coenzymes NAD(H) and NADP(H). These are digested to release NAm, in which form the vitamin is absorbed (Fig. 12.1). Both coenzyme forms can be degraded by the intestinal mucosal enzyme **NAD(P)⁺ glycohydrolase**, which cleaves the pyridine nucleotides into NAm and ADP-ribose. NAm can also be cleaved at the pyrophosphate bond to yield **nicotinamide mononucleotide (NMN)** and 5'-AMP, or by a **phosphodiesterase** to yield **nicotinamide riboside (NR)** and ADP. The dephosphorylation of NMN also yields NR, which can be converted to NAm either by hydrolysis (yielding ribose) or by phosphorylation (yielding ribose 1-phosphate). The cleavage of NAm to free NA appears to be accomplished by intestinal microorganisms, and is believed to be of quantitative importance in niacin absorption.

Facilitated Diffusion

Niacin is absorbed in the stomach and small intestine. Studies using everted intestinal sacs prepared from rats have demonstrated that both NA and NAm are absorbed at low concentrations via Na^+-dependent, carrier-mediated facilitated diffusion. Both forms of the vitamin are absorbed across the human buccal mucosa by the same mechanism. The rate of diffusion of NA is about half that of NAm. At high concentrations, however, each is absorbed *via* passive diffusion. Therefore,

at pharmacologic concentrations the vitamin is absorbed nearly completely.[6] The presence or absence of food in the gut appears to have no effect on niacin absorption. Because NR is not found in plasma, it appears not to be absorbed *per se*, but first converted to NAm.

4. TRANSPORT OF NIACIN

Free in Plasma

Niacin is transported in the plasma as both NA and NAm in unbound forms. Because the NA is converted to NAD(H) and subsequently to NAm in the intestine and liver, circulating levels of NAm tend to exceed those of NA.

Cellular Uptake

Both NA and NAm are taken up by most peripheral tissues by passive diffusion. However, some tissues have transport systems that facilitate niacin uptake. Erythrocytes take up NA by the anion transport system. Renal tubules do so by a Na^+-dependent, saturable transport system. The brain takes up the vitamin by energy-dependent transport systems; the site of the blood–cerebrospinal fluid barrier, the choroid plexus, appears to have separate systems for the accumulation/release of NA and NAm. Brain cells also have a high-affinity transport system for NAm. These two levels of control effect the homeostasis of niacin in the brain, with NAm, but not NA, entering readily.

Niacin Receptor

A high-affinity, G protein-coupled receptor for NA has been identified in adipose tissue.[7] Referred to as the **"niacin receptor,"** this is an orphan receptor. It binds NA only at unphysiologically high levels;[8] its natural ligand

6. In humans at steady state, consuming 3 g of nicotinic acid per day, 85% of the vitamin is excreted in the urine.
7. Lorensen, A. (2001). *Mol. Pharmacol.* 59, 349.
8. Wise, A., Foord, N. J., and Fraser, N. J. (2003). *J. Biol. Chem.* 278, 9869.

TABLE 12.3 Pyridine Nucleotide Contents (mg/kg) of Various Organs of Rats

Organ	NAD$^+$	NADH	NADP$^+$	NADPH
Liver	370	204	6	205
Heart	299	184	4	33
Kidney	223	212	3	54
Brain	133	88	<2	8
Thymus	116	35	<2	12
Lung	108	52	9	18
Pancreas	80	78	<2	12
Testes	80	71	<2	6
Blood	55	36	5	3

Source: Offermanns, K. et al. (1984). Kirk-Othmer Encycl. Chem. Technol. 24, 59.

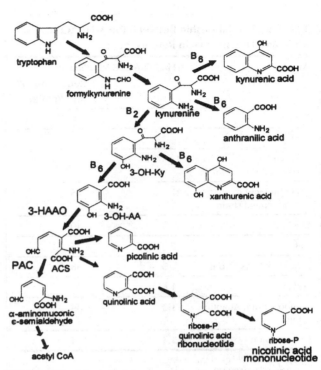

FIGURE 12.2 Metabolic interconversion of tryptophan to niacin.

may actually be β-hydroxybutyrate.[9] It is also expressed in spleen and immune cells; in the latter, it is regulated by various cytokines. The receptor appears to play roles in responses (flushing, antihyperlipidemic) to high doses of NA. It is referred to as HM74A in humans, GPR109A in the rat, and PUMA-G in the mouse.

Tissue Storage

Niacin is retained in tissues that take it up as NA and/or NAm by being trapped by conversion to the pyridine nucleotides NAD(H) and NADP(H) (see Table 12.3). By far the greater amount is found as NAD(H), most of which, in contrast to NADP(H), is found in the oxidized form (NAD$^+$).

5. METABOLISM OF NIACIN

Niacin Biosynthesis

Tryptophan–Niacin Conversion

All animal species (including humans) appear to be capable, to varying degrees, of the *de novo* synthesis of the metabolically active forms of niacin, NAD(H) and NADP(H). Biosynthesis occurs from the tryptophan metabolite quinolinic acid (Fig. 12.2). This conversion involves several steps:

1. Oxidative cleavage of the tryptophan pyrrole ring by **tryptophan pyrrolase** to yield **N-formylkynurenine**.

2. Removal of the formyl group by **formylase** to yield **kynurenine**.
3. Ring-hydroxylation of kyneurenine by the FAD-dependent **kynurenine 3-hydroxylase** to yield **3-hydroxykynurenine (3-OH-Ky)**.
4. Deamination of 3-OH-Ky by a Zn-activated, pyridoxal phosphate-dependent transaminase, to yield **xanthurenic acid**, which can be excreted in the urine or further metabolized.
5. Removal of an alanine residue from the xanthurenic acid side chain by the pyridoxal phosphate-dependent enzyme **kynureninase** to yield **3-hydroxyanthranilic acid (3-OH-AA)**.[10]
6. Oxidative ring-opening of 3-OH-AA by an Fe^{2+}-dependent dioxygenase, **3-hydroxyanthranilic acid oxygenase (3-HAAO)**, to yield the semi-stable **α-amino-β-carboxymuconic-ε-semialdehyde (ACS)**. ACS is a branch-point intermediate. It can:
   a. be converted to NAD$^+$ – as ACS accumulates, some can spontaneously cyclize with dehydration to form **quinolinic acid (QA)**, which is decarboxylated and phosphoribosylated by **quinolinate phosphoribosyltransferase** to yield NMN and, after phosphoadenylation by the ATP-dependent **NAD$^+$ synthetase**, NAD$^+$; or

9. β-hydroxybutyrate is one of three ketone bodies produced by the liver when glucose cannot be used as an energy substrate. Unlike the others (acetone, acetoacetate), β-hydroxybutyrate has lipolytic activity.

10. Kynureninase can also convert kynurenine to another urinary metabolite **anthranilic acid**.

TABLE 12.4 Relationship Between the 3-HAAO:ACSD Ratio and Dietary Niacin Requirement

Animal	3-HAAO:ACSD[a]	Niacin Requirement[b] (mg/kg diet)
Rat	273	0
Chick:		
Low-niacin requirement strain	48	5
High-niacin requirement strain	27	15
Duck	5.3	40
Cat	5	45
Brook trout, lake trout	2.5	88
Turkey	1.6	70
Rainbow trout, Atlantic salmon	1.3	88
Coho salmon	3.4	175

[a]i.e., piccolinic acid carboxylase, PAC.
[b]Animals fed tryptophan.
Source: Poston, H. A. and Combs, G. F. Jr (1980). *Proc. Soc. Exp. Biol. Med.* 163, 452.

TABLE 12.5 Variation in Hepatic α-Amino-β-Carboxymuconic-ε-Semialdehyde Decarboxylase (ACSD)[a] Activities in Animals

Animal	ACSD Activity, IU/g
Cat	50,000
Lizard	29,640
Duck	17,330
Frog	13,730
Turkey	9,230
Cow	8,300
Pig	7,120
Pigeon	6,950
Chicken:	
High-niacin requirement strain	5,380
Low-niacin requirement strain	3,200
Rabbit	4,270
Mouse	4,200
Guinea pig	3,940
Human	3,180
Hamster	3,140
Rat	1,570

[a]i.e., piccolinic acid carboxylase (PAC).
Source: DiLorenzo, R. N. (1972). PhD thesis. Cornell University, Ithaca, NY.

  b. **be catabolized to inactive metabolites** – ACS can spontaneously cyclize with decarboxylation to yield **picolinic acid**, which is converted by α-**amino-β-carboxymuconic-ε-semialdehyde decarboxylase (ACSD)**[11] to α-**aminomuconic-ε-semialdehyde**; that metabolite is subsequently reduced and further decarboxylated to yield **acetyl-CoA**.

Determinants of Tryptophan–Niacin Conversion

The conversion of tryptophan to NAD is a generally inefficient process. Humans appear normally to convert 60 mg of tryptophan to 1 mg of niacin;[12] this ratio is also wide for the chick (45:1) and the rat (50:1), and extremely wide for the duck (175:1). Conversion efficiency has been found in chicks to be depressed under conditions of nutritional iron deficiency, and in several species to be enhanced by niacin deprivation. Niacin-deficient humans are estimated to use nearly 3% of dietary tryptophan for niacin biosynthesis, and thus are able to satisfy two-thirds of their requirement for the vitamin from the metabolism of this indispensable amino acid.

Higher niacin-biosynthetic efficiencies (i.e., low tryptophan:niacin ratios) are associated with *high* activities of 3-HAAO (enhancing production of ACS, the branch-point intermediate in the pathway), and *low* activities of ACSD (which removes the first committed intermediate). Both hepatic activity of ACSD and the ratio of the hepatic activities of 3-HAAO and ACSD vary greatly between animal species and are inversely correlated with their dietary requirements for preformed niacin (see Tables 12.4, 12.5).

It would appear that protein turnover may pre-empt niacin synthesis under conditions of limiting tryptophan. In such circumstances the amount of tryptophan available for niacin synthesis would be expected to be low, rendering the calculation of niacin equivalents inaccurate.

Tryptophan–niacin conversion involves pyridoxal phosphate-dependent enzymes at four steps: two transaminases (which catalyze the conversions of kynurenine to kynurenic acid and of 3-hydroxykynurenine to xanthurenic acid) and kynureninase (which catalyzes the conversion of kynurenine to anthranilic acid as well as that of 3-hydroxykynurenine to 3-hydroxyanthranilic acid). While each

---

11. Also referred to as picolinic acid carboxylase (PAC).
12. Hence, food niacin value is defined in terms of niacin equivalents, one unit of which is defined as 1 mg niacin + 1/60 mg tryptophan.

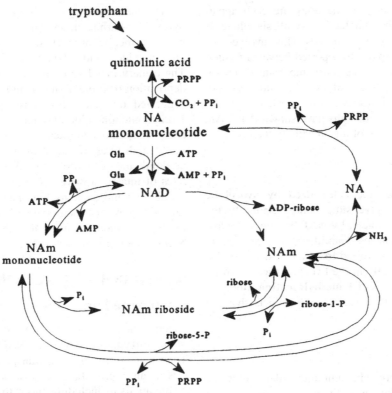

FIGURE 12.3 Niacin metabolism. PRPP, phosphoribosylpyrophosphate.

uses pyridoxal phosphate, only kynureninase is impaired by pyridoxine deprivation. Its affinity for pyridoxal phosphate (K_m $10^{-3}M$)[13] is five orders of magnitude less than those of the transaminases (K_m about $10^{-8}M$); this renders it stripped of its cofactor under conditions of pyridoxine deprivation that are not severe enough to reduce cofactor access of the transaminases. Thus, pyridoxine deficiency impairs the overall conversion of tryptophan to niacin by blocking the production of 3-hydroxyanthranilic acid.[14] It does not, however, block the excretion of the urinary metabolites kynurenic acid and xanthurenic acid. This phenomenon has been exploited for the assessment of pyridoxine status by monitoring the urinary excretion of xanthurenic acid after a tryptophan load.

The conversion of tryptophan to niacin is also reduced by high-fat diets or diets containing excess leucine.[15] These effects appear to be due to ketosis, which has been noted as a common feature of diets of individuals with

pellagra. NAD synthesis appears to be increased by such factors as caloric restriction and hypoxia, and to increase sirtuin[16] signaling, suggesting that NAD^+ levels may serve as indicators of physiological stress.

Three Sources of Pyridine Nucleotides

The metabolically active forms of niacin, the pyridine nucleotides NAD(H) and NADP(H), are produced from three precursors: NA, NAm, and tryptophan (Fig. 12.3). Whereas NA and NAm are formal intermediates in the biosynthesis of NAD^+ from tryptophan, that step (quinolinate phosphoribosyltransferase) actually leads directly to NAD^+ via NMN. Both NA and NAm are converted to NAD^+ by the same pathway after the latter is deamidated to yield NA. As the nicotinamide deamidase activities of animal tissues are low, this step is thought to be carried out by the intestinal microflora. The resulting NA is then phosphoribosylated (by nicotinate phosphoribosyltransferase), adenylated (by deamido-NAD^+ pyrophosphorylase), and amidated (by NAD synthetase) ultimately to yield NAD^+, which can be phosphorylated by an ATP-dependent NAD^+ kinase to yield $NADP^+$.

13. K_m: Michaelis constant; in this case, the concentration of pyridoxal phosphate necessary to support half-maximal enzyme activity.
14. It has also been suggested that the deficiency of zinc, an essential cofactor of pyridoxal kinase (see Chapter 13), may also impair tryptophan–niacin conversion by reducing the production of pyridoxal phosphate.
15. Shastri, N. V., Nayudu, S. G., and Nath, M. C. (1968). J. Vitaminol. 14, 198; Bender, D. A. (1983). Br. J. Nutr. 50, 25.
16. Sirtuins are protein deacetylases or ribosyltransferases that function in the regulation of transcription and apoptosis.

Although various tissues of the body are each apparently capable of pyridine nucleotide synthesis, there is clearly an exchange between the tissues. This involves primarily NAm, which is rapidly transported between tissues. In the rat, NA appears to be the most important precursor of these coenzymes in the liver, kidneys, brain, and erythrocytes; but in the testes and ovaries NAm appears to be a better precursor. Studies with chickens have shown that NAm can be a better dietary source of niacin activity than NA.[17]

Catabolism

The pyridine nucleotides are catabolized by hydrolytic cleavage of their two β-glycosidic bonds, primarily the one at the nicotinamide moiety, by NAD(P)$^+$ glycohydrolase. NAm so released can be deamidated to form NA, in which form it can be reconverted to NAD$^+$. Alternatively, it can be methylated (mainly in the liver) by **nicotinamide N-methyltransferase** to yield **1-methylnicotinamide**,[18,19] which can be oxidized to a variety of products that are excreted in the urine.

Excretion

Niacin is excreted in appreciable amounts under conditions of supranutritional intake, as both vitamers are actively reabsorbed by the renal glomerulus. Excretion involves a variety of water-soluble metabolites in the urine. At typical levels of intake of the vitamin, the major urinary metabolites are **1-methylnicotinamide**[20] and its oxidation product **1-methyl-6-pyridone-3-carboxamide**. Under such conditions, intact NA and NAm, as well as other oxidation products, are also excreted, but in much smaller amounts. Most mammals excrete several metabolites – nicotinamide 1-oxide, 1-methyl-4-pyridone-3-carboxamide, 1-methyl-6-pyridone-3-carboxamide, 6-hydroxynicotinamide, and 6-hydroxynicotinic acid; some species also excrete nicotinic acid/nicotinamide conjugates of ornithine (2,5-dinicotinyl ornithine by birds only) or glycine (nicotinuric acid by rabbits, guinea pigs, sheep, goats, and calves).

The major urinary metabolite in the rat is 1-methyl-4-pyridone-3-carboxamide. This metabolite is also found in human urine, but at levels substantially less than 1-methyl-nicotinamide and 1-methyl-2-pyridone-5-carboxamide. The urinary metabolite profile can be changed by dietary deprivation of protein and/or amino acids, and it has been suggested that the ratio of the pyridone metabolites to 1-methylnicotinamide may have utility as a biomarker for adequate amino acid intake.

At high rates of niacin intake, the vitamin is excreted predominantly (65–85% of total) in unchanged form. At all rates of intake, however, NAm tends to be excreted as its metabolites more extensively than is NA. Further, the biological turnover of each vitamer is determined primarily by its rate of excretion; thus, at high intakes, the half-life of NAm is shorter than that of NA.

6. METABOLIC FUNCTIONS OF NIACIN

Coenzyme Functions

Niacin functions metabolically as the essential component of the enzyme co-substrates NAD(H)[21] and NADP(H).[22] The most central electron transport carriers of cells, each acts as an intermediate in most of the hydrogen transfers in metabolism, including more than 200 reactions in the metabolism of carbohydrates, fatty acids, and amino acids according to the general reaction:

$$\text{substrate} + \text{NAD(P)}^+ \rightarrow \text{product} + \text{NAD(P)H} + \text{H}^+$$

The hydrogen transport by the pyridine nucleotides is accomplished by two-electron transfers in which the hydride ion (H$^-$) serves as a carrier for both electrons. The transfer is stereospecific, involving C-4 of the pyridine ring. The two hydrogen atoms at C-4 of NAD(H) and NADP(H) are not equivalent; each is stereospecifically transferred by the enzymes to the corresponding substrates.[23] In general, stereospecificity is independent of the nature of the substrate and the source of the enzyme, and few regularities are apparent except that dehydrogenases with phosphorylated and non-phosphorylated substrates tend to show opposite stereospecificities.[24]

---

17. Ohuho, M. and Baker, D. ([1993] *J. Nutr.* 123, 2201) showed NAm to be utilized some 24% better than NA by broiler chickens.
18. NAm methylase activity is very low in fetal rat liver, increasing only in mature animals or in animals in which hepatocyte proliferation has been stimulated (e.g., after partial hepatectomy or treatment with thioacetamide). Such increases in enzyme activity are accompanied by drops in tissue NAD$^+$ concentrations, as 1-methylnicotinamide reduces NAD$^+$ synthesis either by inhibiting NAD$^+$ synthetase and/or stimulating NAD(P)$^+$ glycohydrolase. Thus, it is thought that nicotinamide methylase and its product may be involved in the control of hepatocyte proliferation.
19. NA appears not to be methylated by animals. Trigonelline (1-methylnicotinic acid) does appear, however, in the urine of coffee drinkers, owing to its presence in that food.
20. Humans normally excrete daily up to 30 mg of total niacin metabolites, of which 7–10 mg is 1-methylnicotinamide.

21. Historically known as coenzyme I or diphosphopyridine nucleotide (DPN).
22. Historically known as coenzyme II or triphosphopyridine nucleotide (TPN).
23. Because of this phenomenon, the pyridine nucleotide-dependent enzymes are classified according to the side of the dihydropyridine ring to which each transfers hydrogen, i.e., class A and class B.
24. Dehydrogenases with phosphorylated substrates tend to be B-stereospecific (see footnote 21), whereas those with small (i.e., no more than three carbon atoms), nonphosphorylated substrates tend to be A-stereospecific.

The reactions catalyzed by the pyridine nucleotide-dependent dehydrogenases occur by the abstraction of the proton from the alcoholic hydroxyl group of the donor substrate, and the transfer of hydride ion from the same carbon atom to the C-4 of NAm. In many cases this reaction is coupled to a further reaction, such as phosphorylation or decarboxylation.

Metabolic Roles

Despite their similarities of mechanism and structure,[25] NAD(H) and NADP(H) have quite different metabolic roles, and most dehydrogenases have specificity for one or the other.[26]

NAD in Redox Reactions

The oxidized form NAD^+ serves as a hydrogen acceptor at the C-4 position of the pyridine ring, forming NAD(H), which, in turn, functions as a hydrogen donor to the mitochondrial respiratory chain (TCA cycle) for ATP production (Table 12.6). These reactions include:

- glycolytic reactions
- oxidative decarboxylations of pyruvate
- oxidation of acetate in the TCA cycle
- oxidation of ethanol
- β-oxidation of fatty acids
- other cellular oxidations.

NADP(H) in Reduction Reactions

The phosphorylation of NAD^+ facilitates the separation of oxidation and reduction pathways of niacin cofactors by allowing NADP(H) to serve as a co-dehydrogenase in the oxidation of physiological fuels.[27] Thus, NADP(H) is maintained in the reduced state, NADPH,[28] by the pentose phosphate pathway such that reduction reactions are favored. Many of these also involve flavoproteins.[29] These reactions involve reductive biosyntheses, such as those of fatty acids and steroids (Table 12.6). In addition, NADPH

TABLE 12.6 Some Important Pyridine Nucleotide-Dependent Enzymes of Animals

Role	Enzyme	
	NAD(H) Dependent	NADP(H) Dependent
Carbohydrate metabolism	3-Phosphoglyceraldehyde dehydrogenase	Glucose-6-phosphate dehydrogenase
	Lactate dehydrogenase	6-Phosphogluconate dehydrogenase
	Alcohol dehydrogenase	
Lipid metabolism	α-Glycerophosphate dehydrogenase	3-Ketoacyl ACP[a] reductase
	β-Hydroxyacyl-CoA dehydrogenase	Enoyl-ACP reductase
	3-Hydroxy-3-methylglutaryl-CoA reductase	
Amino acid metabolism	Glutamate dehydrogenase	Glutamate dehydrogenase
Other	NADH dehydrogenase/NADH-ubiquinone dihydrofolate reductase	Glutathione reductase
	Poly(ADP-ribose) polymerase	Thioredoxin-NADP reductase 4-Hydroxybenzoate hydroxylase NADPH-cytochrome *P*-450 reductase
		Mono-ADP-ribotransferases

[a]ACP, acyl carrier protein.

also serves as a co-dehydrogenase for the oxidation of glucose 6-phosphate in the pentose phosphate pathway.

ADP-Ribosylation

Mono(ADP-Ribosyl)ation

NAD^+ serves as the donor of an ADP-ribose moiety to an amino acid residue on an acceptor protein. Originally recognized as properties of bacterial toxins,[30] mono-**ADP-ribotransferases** (**ARTs**) have been identified in mammalian cells. Two groups have been found:

1. **Ecto-ARTs**, secreted or expressed on the outer surfaces of cells in skeletal and cardiac muscles, lung, testes, and lymphatic tissues. These show sequence homology with the bacterial ARTs. They have been

---

25. For example, each contains adenosine, which appears to serve as a hydrophobic *anchor*.
26. A small number of dehydrogenases can use either NAD(H) or NADP(H).
27. For example, glyceraldehyde-3-phosphate, lactate, alcohol, 3-hydroxybutyrate, pyruvate, and α-ketoglutarate dehydrogenases.
28. The $NADP^+$/NADPH couple is largely reduced in animal cells, owing to the **transhydrogenase** activity that catalyzes the energy-dependent exchange of hydride between the pyridine nucleotides coupled to proton transport across the mitochondrial membrane in which it resides (the so-called *redox-driven proton pump*).
29. The first step in most biological redox reactions is the reduction of a flavoprotein by NADPH.

30. Cholera, diphtheria, pertussis, and pseudomonas toxins use NAD^+ to catalyze the ADP-ribosylation of host G-proteins and disrupt host cell function.

found to ADP-ribophoshorylate integrins[31] in the control of myogenesis, defensins[32] in signaling macrophages, and an ATP-gated ion channel[33] to induce apoptosis. The puzzling aspect of the presence of ecto-ARTs is that extracellular concentrations of NAD are normally very low. Hence, it has been suggested their activity depends on NAD being released by damaged cells, which would make NAD a signaling molecule for the death of nearby cells.

2. **Endo-ARTs**, present in the cytosol or inner membranes of cells. These include one that inactivates, by ADP-ribosylation, excess G-protein β-subunits which serve in cell signaling, and one that similarly inactivates the protein-folding chaperone, the 78 kDa glucose-regulated protein (GRP78), which apparently reduces protein secretion under conditions of cellular stress. In both cases, the inactivation appears to be reversible through the activity of hydrolases which recycle the acceptor protein by removing the ADP-ribosyl moiety.

Poly(ADP-Ribosyl)ation

NAD(H) functions in the formation of ADP-ribose polymers by **poly(ADP-ribose) polymerases (PARPs)**,[34] which are activated by DNA single-strand breaks and, thus, serve as a DNA damage sensor. Activated PARPs add an ADP-ribose moiety to an initial glutamatyl or aspartatyl residue of PARP itself, or, to a much lesser extent, to some 30 other acceptor proteins. PARPs then catalyze the formation of chains of ADP-ribose sequences on that protein-bound monomer, creating branch points at 40- to 50-unit intervals, with new sites for subsequent elongation and taking on an increasingly negative charge.

Poly(ADP-ribose) polymers bind to high-affinity poly(ADP-ribose) binding sites on histones and many other nuclear proteins. This and/or polyanionic interactions have been shown to draw histones away from the DNA, and thus facilitate interactions of exposed DNA with other DNA-binding proteins (polymerases, ligases, helicases, topoisomerases) involved in replication and repair. The extent to which poly(ADP-ribose) may be involved directly in DNA excision repair is not clear; however, evidence indicates that it is involved in the prevention of non-homologous recombination between two sites of damage.

---

31. Integrins are extracellular receptors that mediate cell–cell attachment and control cell signaling.
32. Defensins are small, cysteine-rich peptides in neutrophils and most epithelial cells that participate in killing phagocytized bacteria by binding to the bacterial cell and forming a pore-like structure that allows loss of essential ions.
33. P_2X_7.
34. At least 18 PARP genes have been identified.

TABLE 12.7 Summary of Plasma Lipid Responses to Nicotinic Acid Treatment (>1.5 g/day) of Dyslipidemias

Parameter	% Reduction	% Increase
Triglycerides	21–44	—
VLDL	25–40	—
LDL cholesterol	2–22	—
HDL cholesterol	—	18–35
Total cholesterol	4–16	—
Lipoprotein Lp(a)	16–36	—

From Gille, A., Bodor, E. T., Ahmed, K., et al. (2008). Ann. Rev. Pharmacol. Toxicol. 48, 79.

Poly(ADP-ribosyl)ation causes extensive turnover of NAD^+ with concomitant production of NAm. Niacin deprivation appears to impair the activities of all PARPs, resulting in genomic instability.[35]

Glucose Tolerance Factor

Niacin has been identified as part of the chromium-containing **glucose tolerance factor** of yeast, which enhances the response to insulin. Its role, if any, in that factor is not clear, as free niacin is without effect. It is possible that this activity involves a metal-chelating capacity of NA such as has been reported for zinc and iron.[36]

7. NIACIN IN HEALTH AND DISEASE

Niacin has been associated with a number of health effects unrelated to the signs of niacin deficiency.

Cardiovascular Health

High doses of NA have been used in treating hyperlipidemia, reducing all major lipids and apolipoprotein B-containing lipoproteins (VLDL, LDL), and increasing apolipoprotein A1-containing lipoproteins (HDL) (Table 12.7). These effects appear to be unrelated to NAD(P) or NAm. They involve three metabolic phenomena (Fig. 12.4):

- *Reducing hepatic triglyceride synthesis.* Studies have shown that NA can non-competitively inhibit hepatic microsomal diacylglyceride transferase-2, which catalyzes the final reaction in triglyceride synthesis, thus limiting the amount of triglycerides available for the

---

35. Oei, S. L., Kel, C., and Ziegler, M. (2005). *Biochem. Cell Biol.* 83, 263.
36. Agte, W., Paknikar, K. M., and Chiplonkar, S. A. (1997). *Biometals* 10, 271.

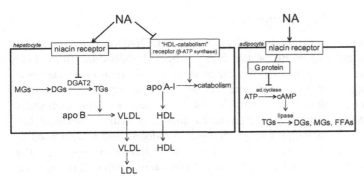

FIGURE 12.4 Schematic representation of apparent metabolic bases for the antihyperlipidemic effects of nicotinic acid. Dashed arrows indicate steps reduced by NA. *Abbreviations*: NA, nicotinic acid; MGs, monoglycerides; DG, diglycerides; TGs, triglycerides; apo B, aoplipoprotein B; apo A-I, apolipoprotein A-I; DGAT2, diacylglycerol acyltransferase 2; ATP, adenosine triphosphate; cAMP, cyclic adenosine monophosphate; ad. cyclase, adenylate cyclase; lipase, hormone-sensitive lipase.

assembly of VLDL.[37] This results in increased degradation of apolipoprotein B, and the consequent reduction in both VLDL and its catabolic product, LDL.

- *Reducing the removal of HDL apolipoprotein A1.* Studies in cultured cells have shown NA to inhibit the catabolism of HDL-apo A1 without affecting apo A1 synthesis. This results in increases in HDL and HDL cholesterol. It has been suggested that the response may involve the putative "HDL catabolism receptor," which may be a β-chain ATP synthase. These increases are thought to reflect reduced exchange of triglycerides and cholesterol esters,[38] and the retarded degradation of apolipoprotein A1.

- *Reducing adipocyte lipolysis.* NA can also bind to the niacin receptor, which is linked to a G-protein that inhibits adenylate cyclase. That inhibition leads to a decline in cAMP levels, which inhibits the hormone-sensitive lipase and consequently reduces the mobilization of fatty acids from triglycerides in adipose tissue. Reduced release of fatty acids is responsible for at least part of the reduction of hepatic synthesis and secretion of VLDLs, and the subsequent decline in circulating LDL levels. Decreased circulating levels of VLDLs are associated with decreased levels of triglycerides and cholesterol. Also contributing to reduced cholesterol levels is a decrease in cholesterol biosynthesis due to NA-inhibition of 3-hydroxy-3-methylglutaryl CoA reductase.[39]

High doses of NA have proven to be among the most useful treatments for hypercholesterolemia (Tables 12.7, 12.8); extended-release formulations of NA are effective and offer advantages of less frequently causing flushing (see page 305, "Niacin Toxicity"). These antihyperlipidemic effects provide the basis of current interest in nicotinic acid in the prophylaxis of coronary artery disease. A retrospective evaluation of results from the US Coronary Drug Project showed NA

TABLE 12.8 Results of Clinical Trials of Nicotinic Acid (≥1 g/day) in Patients on Statins

Study	Subjects (Duration)	Parameter	Placebo	Nicotinic Acid
1[a]	8,341 (5 years)	Myocardial infarction	12.2%	8.9%*
		Mortality	20.9%	21.2%
1[b]	8,441 (15-year follow-up)	Mortality	58.2%	52.0%*
2[c]	555 (5 years)	Mortality	29.7%	21.8%*
3[d]	146 (2.5 years)	Cardiovascular events	19.2%	4.2%*
6[e]	167 (1 year)	Carotid intima—media thickness	0.044 mm	0.014 mm*

*$P < 0.05$.
[a]*Coronary Drug Project Research Group (1975).* J. Am. Med. Assoc. *231, 360.*
[b]*Canner, P. L., Berge, K. G., and Wenger, N. K. (1986).* J. Am. Coll. Cardiol. *8, 1245.*
[c]*Carlson, L. A. and Rosenhamer, G. (1988),* Acta Med. Scan. *223, 405.*
[d]*Brown, G., Albers, J. J., and Fisher, L. D. (1990).* N. Engl. J. Med. *323, 1289.*
[e]*Taylor, A. J., Sullenberger, L. E., Lee, H. J., et al. (2004).* Circulation *110, 3512.*

treatment to have reduced lethal coronary events, resulting in highly significant reduction of mortality from all causes by 11% (vs. a placebo). A meta-analysis of 11 clinical trials with more than 6,600 patients found positive effects of NA, when given alone or in combination with statins, on cardiovascular events.[40] In 2011, a large trial[41] of combined NA–statin treatment was stopped after 32 months on the basis of its not showing reductions in cardiovascular events, despite showing the expected antihyperlidemic effects.

37. See review: Kammana, V. S. and Kashyap, M. L. (2008). *Am. J. Cardiol.* 101, 20B.
38. This process is mediated by the cholesterol ester transfer protein.
39. DiPalma, J. R. and Thayer, W. S. (2001). *Ann. Rev. Nutr.* 11, 169.

40. Bruckert, E., Labreuche, J., and Amarenco, P. (2010). *Atherosclerosis* 210, 353.
41. The Atherothrombis Intervention in Metabolic syndrome with low HDL/high triglycerides: Impact on Global Health outcomes (AIM-HIGH) trial (The AIM-HIGH Investigators [2011] *Am. Heart. J.* 161, 471).

Skin Health

Niacin supplementation has been shown to protect skin from DNA-damaging agents in animal models. This effect appears to be due to the inhibition of PARPs and sirtuins, which depend on NAD^+ to protect against DNA damage. Although niacin is included in several skin creams, most forms of the vitamin are water-soluble and therefore cannot cross the skin.

NA has vasodilatory activity, increasing microvascular blood flow due to changes in prostaglandin production. This effect is mediated by the niacin receptor.

Lung Health

Niacin treatment has been shown to reduce lung injury and fibrosis in animal models, including treatment with DNA-damaging agents (lipopolysaccharide, cyclophosphamide, bleomycin) and exposure to hyperoxic conditions which promote oxidative stress. Hyperoxia has been shown to induce poly(ADP-ribose) synthesis in the lung and to increase lung NAD levels in niacin-deprived animals.

Anticarcinogenesis

Epidemiological studies have associated marginal niacin intakes and/or the reliance on maize-based diets with increase risks of cancers of the esophagus.[42] One study found NA supplementation to reduce yields of esophageal tumors in the *N*-nitrosomethylbenzlamine-treated rat model.[43] Supranutritional doses of niacin have been shown to reduce dramatically the yield of skin tumors in UV-treated mice in a dose-dependent manner that correlated with skin NAD levels.[44]

Psychological Disorders

NAm has been shown to enhance the effect of tryptophan in supporting brain serotonin levels. It does so by reducing the urinary excretion of tryptophan metabolites and reducing the conversion of tryptophan to niacin. This increases the availability of tryptophan for the synthesis of serotonin, the general effect of which is antidepressive. High doses of NA have been found to benefit patients with certain psychological disorders:

Schizophrenia

Schizophrenia is associated with NAD-deficiency in critical areas of the brain. Affected individuals have been found to oxidize NAm more readily than unaffected people: they excrete greater amounts of 1-methyl-6-pyridone-3-carboxamide. As the excretion of this methylated product is increased by treatment with methylated hallucinogens (e.g., methylated indoles) and is decreased by treatment with tranquilizers, it has been suggested that schizophrenics suffer a depletion of NAm (via its methylation and excretion) which limits NAD^+ synthesis. Patients with first-episode schizophrenia can show diminished flushing responses to niacin. This response is mediated by vasodilators derived from arachidonic acid, which levels are typically low in such patients. High doses of NA (e.g., 1 g/day) given with ascorbic acid have been found to eliminate psychotic symptoms and prevent relapses of acute schizophrenics.[45]

Hartnup Disease[46]

This is a rare familial disorder involving malabsorption of tryptophan (and other amino acids). It is characterized by hyperaminoaciduria,[47] a pellagra-like skin rash (precipitated by psychological stress, sunlight or fever), and neurological changes including attacks of ataxia and psychiatric disorders ranging from emotional instability to delirium. Patients appear to have abnormally low capacities to convert tryptophan to niacin, apparently resulting from reduced enteric absorption and renal reabsorption of tryptophan and other monoamino-monocarboxylic acids. Non-reabsorbed tryptophan appears to be degraded by microbial tryptophanase to pyruvate and indole, the latter of which is reabsorbed from the intestine and is neurotoxic. Patients respond to treatment with NA.

Depression

Patients with depressive symptoms can show diminished flushing responses to niacin. Studies have shown that some 5% of depressive subjects do not show the niacin-induced flushing; recent studies suggest that the non-responders are likely to be severely ill, with depressed mood, anxiety, feelings of guilt, and physical symptoms.[48]

42. Van Rensburg, S. J., Bradshaw, E. S., Bradshaw, D., *et al.* (1985). *Br. J. Cancer* 51, 399; Wharendorf, J., Chang-Claude, Q. S., Lian, Y. G., *et al.* (1989). *Lancet* 2, 1239; Franceschi, S., Bidoli, E., Baron, A. E., *et al.* (1990). *J. Natl Cancer Inst.* 82, 1407.
43. Van Rensburg, S. J., Hall, J. M., and Gathercole, P. S. (1986). *Nutr. Cancer* 8, 163.
44. Gensler, H. L., Williams, T., Huang, A. C., *et al.* (1999). *Nutr. Cancer* 34, 36.
45. Osmond, H. and Hoffer, A. (1962). *Lancet* 1, 316.
46. The disease was named for the first case, described in 1951, involving a boy thought to have pellagra. Since that time, some 50 proved cases involving 28 families have been described.
47. The presence of abnormally high concentrations of amino acids in the urine.
48. Smensy, S., Baur, K., Rudolph, N., *et al.* (2010). *J. Affect. Disord.* 124, 335.

Diabetes

Diabetes is associated with a reduced state of the pyridine nucleotides in the cytosol and mitochondria due to the increased levels of glucose, FFA, lactate, and branched-chain amino acids. This reduced state affects gene expression via the NADH activated transcriptional co-repressor, C-terminal binding protein (CtBP), and the NADH enzyme, glyceraldehyde-3-phosphate dehydrogenase.

NAm has been found to delay or prevent the development of diabetic signs in the non-obese diabetic mouse model,[49] to decrease the severity of diabetic signs associated with β-cell proliferation induced by partial pancreatectomy,[50] and to protect against diabetes induced by agents[51] that cause DNA strand breakage in β-cells. In clinical trials, NAm has been found to protect high-risk children from developing clinically apparent insulin-dependent diabetes,[52] and to improve small artery vasodilatory function in statin-treated type 2 diabetic patients.[53] These actions of NAm are thought to involve support of pancreatic β-cell function through the maintenance of both NAD$^+$ and DNA-protective poly(ADP-ribose) polymerase activity. NA has also been shown to induce insulin resistance in short-term studies (see page 305, "Niacin Toxicity").

Fetal Alcohol Syndrome

Studies in animal models have demonstrated benefits of NAm treatment in reducing anxiety and preventing neural damage in progeny of alcohol-treated dams.[54]

Other Responses

Acute doses of NA have been shown to increase circulating levels of adiponectin, but without improving insulin sensivity or endothelial function.[55] High-dose NA has been shown to reduce blood pressure in hypertensive subjects, probably due to vasodilation; however it is not clear whether chronic treatment may have similar effects, as the results of clinical trials have been inconsistent.[56]

49. Reddy, S., Bibby, N. J., and Elliott, R. B. (1990). *Diabetes Res.* 15, 95.
50. Yonemura, Y., Takashima, T., Miwa, K., *et al.* (1984). *Diabetes* 33, 401.
51. Alloxan, streptozotocin.
52. Elliott, R. B. and Chase, H. P. (1991). *Diabetologia* 34, 362–365; Manna, R., Milgore, A., Martin, L. S., *et al.* (1992). *Br. J. Clin. Pract.* 46, 177.
53. Hamilton, S. J., Chew, G. T., Davis, M. E., *et al.* (2010). *Diabetes Vasc. Dis. Res.* 7, 296.
54. Feng, Y., Paul, I. A., and LeBlanc, M. H. (2006). *Brain Res. Bull.* 69, 117; Ieraci, A. and Herrara, D. G. (2006). *PLoS Med.* 3, e101.
55. Westphal, S., Borucki, K., Taneva, E., *et al.* (2007). *Atherosclerosis* 193, 361.
56. Bays, H. E. and Rader, D. J. (2009). *Intl J. Clin. Pract.* 63, 1.

TABLE 12.9 Signs of Niacin Deficiency

Organ system	Signs
General:	
Appetite	Decrease
Growth	Decrease
Dermatologic	Dermatitis, photosensitization
Gastrointestinal	Inflammation, diarrhea, glossitis
Skeletal	Perosis
Vascular:	
Erythrocytes	Anemia
Nervous	Ataxia, dementia

8. NIACIN DEFICIENCY

General Signs

Niacin deficiency in animals is characterized by a variety of species-specific signs that are usually accompanied by loss of appetite and poor growth (Table 12.9). The general progression of signs is captured as the **"Four Ds of niacin deficiency:"**

- Dermatitis
- Diarrhea
- Delirium
- Death.

Deficiency Signs in Humans

Niacin deficiency in humans results in changes in the skin, gastrointestinal tract, and nervous system. The dermatologic changes, which are usually most prominent (being called **pellagra**), are most pronounced in the parts of the skin that are exposed to sunlight (face, neck,[57] backs of the hands, and forearms) (Figs 12.5, 12.6). In some patients, lesions resemble early sunburn; in chronic cases the symmetric lesions feature cracking, desquamation,[58] hyperkeratosis, and hyperpigmentation.

Lesions of the gastrointestinal tract include angular stomatitis, cheilosis, and glossitis, as well as alterations of the buccal mucosa, tongue, esophagus, stomach (resulting in achlorhydria[59]), and intestine (resulting in diarrhea).[60] Pellagra almost always involves anemia.[61]

57. This is referred to as **Casal's collar**.
58. The shedding of the epidermis in scales.
59. The absence of hydrochloric acid from the gastric juice, usually due to gastric parietal cell dysfunction.
60. Many of these gastrointestinal changes also occur in schizophrenia.
61. The anemia associated with pellagra is of the macro- or normocytic, hypochromic types.

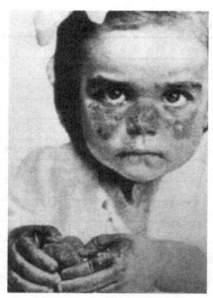

FIGURE 12.5   Pellagra: Affected child with facial *"butterfly wing."*
Courtesy of Cambridge University Press.

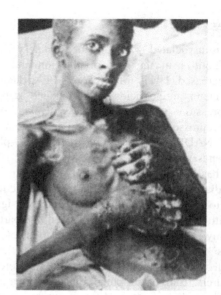

FIGURE 12.6   Pellagra: Affected woman with *"pellagra glove."*
Courtesy of Cambridge University Press.

Early neurological symptoms associated with pellagra include anxiety, depression, and fatigue;[62] later symptoms include depression, apathy, headache, dizziness, irritability, and tremors.

Deficiency Signs in Animals

Most niacin-deficient animals show poor growth and reduced efficiency of feed utilization. Pigs and ducks are particularly sensitive to niacin deficiency. Pigs show diarrhea, anemia, and degenerative changes in the intestinal mucosa and nervous tissue;[63] ducks show severely bowed and weakened legs, and diarrhea. Niacin-deficient dogs show necrotic degeneration of the tongue with changes of the buccal mucosa and severe diarrhea.[64] Rodents show alopecia and nerve cell histopathology. Chickens show inflammation of the upper gastrointestinal tract, dermatitis of the legs, reduced feather growth, and **perosis** (Fig. 12.7).[65]

It has been thought that ruminants are not susceptible to niacin deficiency, owing to the synthesis of the vitamin by their rumen microflora. Although that appears to be true for most ruminant species, evidence indicates that fattening beef cattle and some high-producing dairy cows

FIGURE 12.7   Perosis (left leg) in niacin-deficient chick.

can benefit from niacin supplements under some circumstances. Studies have shown niacin treatment of lactating cows to depress circulating levels of ketones, apparently by reducing lipolysis in adipocytes by a process involving increased cyclic 3′,5′-adenosine monophosphate (cAMP) and, consequently, the concentrations of non-esterified fatty acids in the plasma. That ruminal synthesis of the vitamin may not meet the nutritional needs of the host would appear most likely in circumstances wherein rumen fermentation is altered to enhance energy utilization, with associated reductions in rumen microbial growth.

Determinants of Niacin Status

Because a substantial amount of niacin can be synthesized from tryptophan, nutritional status with respect to niacin involves not only the level of intake of the preformed vitamin, but also that of its potential amino acid precursor.

---

62. Many of these symptoms also occur in schizophrenia.
63. The syndrome is called "pig pellagra."
64. **Black tongue disease**.
65. Inflammation and misalignment of the tibiotarsal joint (*hock*), in severe cases involving slippage of the Achilles tendon from its condyles, which causes crippling due to an inability to extend the lower leg.

Accordingly, the clinical manifestation of niacin deficiency includes evidence of an unbalanced diet with respect to both of these essential nutrients and, frequently, pyridoxine. Thus, the occurrence of pellagra, as well as niacin-deficiency diseases in animals, is properly viewed as the result of a multi-factorial dietary deficiency rather than of insufficient intake of niacin *per se*.

In addition to tryptophan and pyridoxine[66] supplies being important determinants of niacin status, it has been suggested that excess intakes of the branched-chain amino acid **leucine** may antagonize niacin synthesis and/or utilization, and thus also may be a precipitating factor in the etiology of pellagra. Excess leucine has been shown to inhibit the production of quinolinic acid from tryptophan by isolated rat hepatocytes; however, the magnitude of this effect is small in comparison with the K_m of quinolinate phosphoribosyltransferase for quinolinate, indicating that excess leucine (and/or its metabolites) is unlikely to affect the rate of NAD^+ biosynthesis by the liver. Some studies with intact animals (rats) have produced results supporting the view that excess leucine can impair the synthesis of NAD^+ from tryptophan (by inhibiting either the enzymatic conversion itself, or the cellular uptake of the amino acid); however, others have yielded negative results in this regard. Therefore, the role that high leucine intakes may have in the etiology of pellagra is not clear at present.

It has been suggested that zinc plays some role in the pyridoxine-dependent metabolic interconversion of tryptophan to niacin. Pellagra patients have been found to have low plasma Zn levels, and Zn supplementation increases their urinary excretion of 1-methylnicotinamide and 1-methyl-2-pyridone-5-carboxamide. Studies with rats have shown that treatment of niacin-deficient animals with the metabolic intermediate picolinic acid increases circulating Zn levels.

9. NIACIN TOXICITY

In general, the toxicity of niacin is low. Non-ruminant animals can tolerate oral exposures of at least 10- to 20-fold their normal requirements for the vitamin. The toxic potential of NAm appears to be greater than that of NA, probably by a factor of four. Side effects appear to result from metabolic disturbances due to the depletion of methyl groups as the result of the metabolism of the vitamin in high doses.

Nicotinic Acid

Short-term effects: The most common side effect of high-dose NA is **skin flushing** caused by cutaneous vasodilation. This response is transient (30–90 minutes), and accompanied by erythema, tingling, itching, and elevated skin temperature. It is seen at the beginning of NA therapy, and tends to subside over time with the development of tolerance. Still, for some it can be disagreeable to the point of discontinuing NA treatment for hyperlipidemia.[67] The flushing response can be evoked by either oral or topical exposure to NA. It appears to be mediated by prostanoids (prostacyclins, prostaglandins) and involves the niacin receptor, which is expressed by macrophages and bone marrow-derived cells of the skin. The response can be minimized by using a slow-release formulation of NA, or by using a cyclooxygenase inhibitor (e.g., aspirin, indomethacin) prior to taking NA.

High doses of NA have been reported to cause itching urticaria (hives), and gastrointestinal discomfort (heartburn, nausea, vomiting, rarely diarrhea) in humans. Animal studies have shown that high levels of NAm can raise circulating homocysteine levels, particularly on a high-methionine diet.

Chronic effects: The longer-term effects of high NA doses include cases of insulin resistance, which may involve a rebound in lipolysis that results in increased free fatty acid levels. A few cases of transient elevations in the plasma activities of liver enzymes without associated hepatic dysfunction have been reported, and chronic doses of NA have been reported to cause hepatic damage.[68]

Nicotinamide

Short-term effects: While acute adverse effects of NAm have not been reported for doses used to treat insulin-dependent diabetes (*ca.* 3 g/day), larger doses (10 g/day) have been found to cause hepatic damage.

Chronic effects: It is possible that chronic, high intakes of NAm may deplete methyl groups due to the increased demand for methylation to excrete the vitamin. Such effects would be exacerbated by low intakes of methyl donors, methionine, and choline, and

---

66. Zinc, which is required by the enzyme *pyridoxal phosphokinase*, is also related to the function of pyridoxine in this system. Alcoholics, who are typically of low zinc status, have been shown to excrete high levels of the niacin metabolites 1-methyl-6-pyridone-3-carboxamide and 1-methyl-nicotinamide. The excretion of these metabolites was increased by Zn supplementation, presumably owing to increased pyridoxal phosphokinase activities and the consequent activation of pyridoxine to the form (pyridoxal phosphate) that facilitates tryptophan–niacin conversion. It has also been suggested that zinc deficiency may reduce the availability of tryptophan for niacin biosynthesis by enhancing its oxidation, as has been shown for several other amino acids.

67. Drop-out for this reason has been seen in 5–20% of patients.
68. Rader, J. I., Calvert, R. J., and Hathcock, J. N. (1992). *Am. J. Med.* 92, 77.

suboptimal status with respect to folate and/or vitamin B_{12}. NAm can inhibit uricase, depressing intestinal microbial uricolysis, which could lead to uricemia.

10. CASE STUDY

Instructions

Review the following report, paying special attention to the responses to the experimental treatments. Then, answer the questions that follow.

Case

Fourteen patients with alcoholic pellagra and 7 healthy controls, all ranging in age from 21 to 45 years, were studied in the metabolic unit of a hospital. None had severe hepatic dysfunction on the basis of medical history, clinical examination, and routine laboratory tests. The nutritional status of each subject was evaluated at the beginning of the study by clinical examination, anthropometric measurements (body mass index [BMI; weight divided by the square of the height], triceps skinfold thickness, arm and muscle circumference), biochemical tests (24-hour urinary creatinine, serum albumin, total iron-binding capacity [TIBC]), and 24-hour recalls of food consumption. Results indicated that, before admission, the patients with alcoholic pellagra consumed a daily average of 270g of ethanol. Each showed signs of protein-calorie malnutrition (reduced BMI, skinfold thickness, arm and muscle circumference, serum albumin, and TIBC). In addition, their plasma zinc concentrations were significantly lower than those of controls, although their urinary zinc concentrations were not different from the control group.

The pellagra patients were assigned to one of two experimental treatment groups and the healthy controls to another (three treatments, each with n = 7). During the 7-day study, each group received enteral diets prepared from 10% crystalline amino acids (adequate amounts of each, except for tryptophan) and 85% sucrose, which supplied daily amounts of 0.8g of protein per kilogram of body weight and 200kcal/g N. In addition, each patient was given weekly, by vein, 500ml of an essential fatty acid emulsion as well as a vitamin–mineral supplement. The diets were administered by intubation directly to the midportion of the duodenum. The control diet was supplemented with tryptophan, and the vitamin–mineral supplement contained both niacin and zinc. The diets provided to each group of pellagra patients contained no tryptophan; neither did their vitamin supplement contain niacin. One group of pellagra patients received supplemental zinc (220mg of $ZnSO_4$) whereas the other did not. Several

biochemical measurements were made at the beginning of the experiment, and again after 4 days. Each of the biochemical measurements was repeated after 7 days of treatment. In most cases, the results showed the same effects but of greater magnitudes.

Results

Subject Group	Parameter	Initial Value	Day 4 Value
Healthy controls	Plasma Zn (mmol/l)	14.2 ± 1.5	16.0 ± 2.2
	Plasma tryptophan (mmol/l)	50.8 ± 12.5	74.3 ± 18.5
	Urine Zn (mmol/day)	7.34 ± 1.38	9.18 ± 2.91
	Urine 6-pyridone[a] (mmol/day)	70 ± 22	640 ± 235
	Urine CH_3-NAm[b] (mmol/day)	78 ± 32	143 ± 48
Pellagra patients	Plasma Zn (mmol/l)	9.9 ± 1.1	9.6 ± 2.0
	Plasma tryptophan (mmol/l)	33.3 ± 15.3	29.5 ± 6.1
	Urine Zn (mmol/day)	9.79 ± 3.06	11.93 ± 10.55
	Urine 6-pyridone[a] (mmol/day)	16 ± 10	19 ± 12
	Urine CH_3-NAm[b] (mmol/day)	6 ± 3	9 ± 6
Pellagra patients fed Zn	Plasma Zn (mmol/l)	9.8 ± 1.0	15.8 ± 3.2
	Plasma tryptophan (mmol/l)	37.3 ± 17.8	23.7 ± 7.6
	Urine Zn (mmol/day)	9.80 ± 3.10	24.02 ± 8.11
	Urine 6-pyridone[a] (mmol/day)	16 ± 11	55 ± 18
	Urine CH_3-NAm[b] (mmol/day)	6 ± 3	33 ± 20

[a] 1-Methyl-6-pyridone-3-carboxamide.
[b] 1-Methylnicotinamide.

Case Questions and Exercises

1. What signs support the diagnosis of protein-calorie malnutrition in these alcoholic patients with pellagra?
2. Propose a hypothesis for the mechanism of action of zinc in producing the responses that were observed in these patients with alcoholic pellagra. Outline an experiment (using either pellagra patients or a suitable animal model) to test that hypothesis.
3. List the probable contributing factors to the pellagra observed in these patients.

1. Diagram the several general areas of metabolism in which NAD(H)- and NADP(H)-dependent enzymes are involved.
2. In general, how do the pyridine nucleotides interact with the flavoproteins in metabolism? What is the fundamental metabolic significance of this interrelationship?
3. Construct a decision tree for the diagnosis of niacin deficiency in humans or an animal species.
4. What key feature of the chemistry of nicotinamide relates to its biochemical functions as an enzyme co-substrate?
5. What parameters might you use to assess niacin status of a human or animal?

RECOMMENDED READING

Al-Mohaissen, M.A., Pun, S.C., Frohlish, J.J., 2010. Niacin: From mechanisms of action to therapeutic uses. Mini Rev. Med. Chem. 10, 204–217.

Alsheikh-Ali, A.A., Karas, R.H., 2008. The safety of niacin in the US Food and Drug Administration adverse effect reporting database. Am. J. Cardiol. 101 (Suppl.), 9B–13B.

Bodor, E.T., Offermans, S., 2008. Nicotinic acid: An old drug with a promising future. Br. J. Pharmacol. 153, S68–S75.

Bogan, K.L., Brenner, C., 2008. Nicotinic acid, nicotinamide, and nicotinamide riboside: A molecular evaluation of NAD$^+$ precursor vitamins in human nutrition. Ann. Rev. Nutr. 28, 115–130.

Bourgeois, C., Cervantes-Laurean, D., Moss, J., 2006. Niacin. In: Shils, M.E., Shike, M., Ross, A.C. (Eds.), Present Knowledge in Nutrition, tenth ed. Lippincott, New York, NY, pp. 442–451.

Brooks, E.L., Kuvin, J.T., Karas, R.H., 2010. Niacin's role in the statin era. Expert Opin. Pharmacother. 11, 2291–2300.

Bruckert, E., Labreuche, J., Amarenco, P., 2010. Meta-analysis of the effect of nicotinic acid alone or in combination on cardiovascular events and atherosclerosis. Atherosclerosis 210, 353–361.

Capuzzi, D.M., Morgan, J.M., Brusco Jr, O.A., Intenzo, C.M., 2000. Niacin dosing: Relationship to benefits and adverse effects. Curr. Atheroscler. Rep. 2, 64–71.

Depeint, F., Bruce, W.R., Shangari, N., et al. 2006. Mitochondrial function and toxicity: Role of B vitamin family on mitochondrial energy metabolism. Chem. Biol. Interact. 163, 94–112.

DiPalma, J.R., Thayer, W.S., 1991. Use of niacin as a drug. Annu. Rev. Nutr. 11, 169–187.

Farmer, J.S., 2009. Nicotinic acid: A new look at an old drug. Curr. Atheroscler. Rep. 11, 87–92.

Gille, A., Bodor, E.T., Ahmed, K., Offermans, S., 2008. Nicotinic acid: Pharmacological effects and mechanisms of action. Annu. Rev. Pharmacol. Toxicol. 48, 79–106.

Guyton, J.R., Bays, H.E., 2007. Safety considerations with niacin therapy. Am. J. Cardiol. 99 (Suppl.), 22C–31C.

Harmeyer, J., Kollenkirchen, U., 1989. Thiamin and niacin in ruminant nutrition. Nutr. Res. Rev. 2, 201–225.

Hegyi, J., Schwarts, R.A., Hegyi, V., 2004. Pellagra: Dermatitis, dementia and diarrhea. Intl. J. Dermatol. 43, 1–5.

Ido, Y., 2007. Pyridine nucleotide redox abnormalities in diabetes. Antioxid. Redox Signal. 9, 931–942.

Jacob, R.A., 2006. Niacin. In: Bowman, B.A., Russell, R.M. (Eds.), Present Knowledge in Nutrition, ninth ed. ILSI Press, Washington, DC, pp. 260–268.

Kamanna, V.S., Kashyap, M.L., 2008. Mechanism of action of niacin. Am. J. Cardiol. 101 (Suppl.), 20B–26B.

Kamanna, V.S., Ganji, S.H., Kashyap, M.L., 2009. The mechanism and mitigation of niacin-induced flushing. Intl. J. Clin. Pract. 63, 1369–1377.

Kirkland, J.B., 2007. Niacin. In: Zemplini, J., Rucker, R.B., McCormick, D.B., Suttie, J.W. (Eds.), Handbook of Vitamins, fourth ed. CRC Press, New York, NY, pp. 191–232.

Maiese, K., Chong, Z.Z., Hou, J., Shang, Y.C., 2009. The vitamin nicotinamide: Translating nutrition into clinical care. Molecules 14, 3446–3485.

Messamore, E., Hoffman, W.F., Yao, J.K., 2010. Niacin sensitivity and the arachidonic acid pathway in schizophrenia. Schizophrenia Res. 122, 248–256.

Niehoff, I.D., Hüther, L., Lebzien, P., 2009. Niacin for dairy cattle: A review. Br. J. Nutr. 101, 5–19.

Sood, A., Arora, R., 2009. Mechanisms of flushing due to niacin, and abolition of these effects. J. Clin. Hyperten. 11, 685–689.

Suave, A.A., 2008. NAD$^+$ and vitamin B$_3$: From metabolism to therapies. J. Pharmacol. Exp. Therapeut. 324, 883–893.

Vosper, H., 2009. Niacin: A re-emerging pharmaceutical for the treatment of dyslipidemia. Br. J. Pharmacol. 158, 429–441.

Vitamin B$_6$

Anchoring Concepts

1. Vitamin B$_6$ is the generic descriptor for all 3-hydroxy-2-methylpyridine derivatives exhibiting qualitatively the biological activity of Pn [3-hydroxy-4,5-*bis*(hydroxymethyl)-2-methylpyridine].
2. The metabolically active form of vitamin B$_6$ is pyridoxal phosphate, which functions as a coenzyme for reactions involving amino acids.
3. Deficiencies of vitamin B$_6$ are manifested as dermatologic, circulatory, and neurologic changes.

Had we been able to afford Monel metal or stainless steel cages, we would have missed xanthurenic acid.

S. Lepkovsky

Learning Objectives

1. To understand the chief natural sources of vitamin B$_6$.
2. To understand the means of absorption and transport of vitamin B$_6$.
3. To understand the biochemical function of vitamin B$_6$ as a coenzyme of a variety of reactions in the metabolism of amino acids, and the relationship of that function to the physiological activities of the vitamin.
4. To understand the physiological implications of low vitamin B$_6$ status.

VOCABULARY

Acrodynia
Aldehyde dehydrogenase (NAD$^+$)
Aldehyde oxidase
Alkaline phosphatase
γ-Aminobutyric acid (GABA)
Cystathionuria
Decarboxylases
Epinephrine
Erythrocyte aspartate aminotransferase (EAAT)
Glycogen phosphorylase
Hemoglobin
Histamine
Homocysteinuria
Isonicotinic acid hydrazide (INH)
Kynureninase
Methionine load
Norepinephrine
Phosphorylases
Premenstrual syndrome
Pyridoxal (Pal)
Pyridoxal dehydrogenase
Pyridoxal kinase
Pyridoxal oxidase
Pyridoxal phosphate (PalP)
Pyridoxamine (Pm)
Pyridoxamine phosphate (PmP)
4-Pyridoxic acid
Pn (Pn)
Pyridoxol (Pol)
Racemases
Schiff base
Schizophrenia

The Vitamins. DOI: 10.1016/B978-0-12-381980-2.00013-X

Selenocysteine ß-lyase
Serotonin
Sideroblastic anemia
Steroid hormone–receptor complex
Transaminases
Tryptophan load
Xanthurenic acid

1. THE SIGNIFICANCE OF VITAMIN B_6

The biological functions of the three naturally occurring forms of vitamin B_6, Pyridoxal (Pal), Pn (Pn), and Pyridoxamine (Pm), depend on the metabolism of each to a common coenzyme form, pyridoxal phosphate (PalP). That coenzyme plays critical roles in several aspects of metabolism, giving the vitamin importance in such diverse areas as growth, cognitive development, depression, immune function, fatigue, and steroid hormone activity. Vitamin B_6 is fairly widespread in foods of both plant and animal origin; therefore, problems of primary deficiency are not prevalent. Still, vitamin B_6 status can be antagonized by alcohol and other factors that displace the coenzyme from its various enzymes to increase the rate of its metabolic degradation.

2. SOURCES OF VITAMIN B_6

Distribution in Foods

Vitamin B_6 is widely distributed in foods, occurring in greatest concentrations in meats, whole-grain products (especially wheat), vegetables, and nuts (Table 13.1). In the cereal grains, vitamin B_6 is concentrated primarily in the germ and aleuronic layer. Thus, the refining of grains in the production of flours, which removes much of these fractions, results in substantial reductions in vitamin B_6 content. White bread, therefore, is a poor source of vitamin B_6 unless it is fortified.

The chemical forms of vitamin B_6 tend to vary among foods of plant and animal origin; plant tissues contain mostly Pn (the free alcohol form, **pyrixodol [Pol]**), whereas animal tissues contain mostly Pal and Pm. A large portion of the vitamin B_6 in many foods is phosphorylated or bound to proteins via the ε-amino groups of lysyl residues or the sulfhydryl groups of cysteinyl residues. The vitamin is also found in glycosylated forms such as 5′-O-(β-D-glucopyranosyl) Pn. Vitamin B_6 glycosides are found in varying amounts in different foods but little, if at all, in animal products.

Stability

Vitamin B_6 in foods is stable under acidic conditions, but unstable under neutral and alkaline conditions, particularly when exposed to heat or light. Of the several vitamers, Pn is far more stable than either Pal or Pm. Therefore, the cooking and thermal processing losses of vitamin B_6 tend to be highly variable (0–70%), with plant-derived foods (which contain mostly Pn) losing little, if any, of the

vitamin, and animal products (which contain mostly Pal and Pm) losing substantial amounts. Milk, for example, can lose 30–70% of its inherent vitamin B_6 on drying. The storage losses of naturally occurring vitamin B_6 from many foods and feedstuffs, although they occur at slower rates, can also be substantial (25–50% within a year). Because it is particularly stable, **pyridoxine hydrochloride** is used for food fortification and in multivitamin supplements.

Bioavailability

The bioavailability of vitamin B_6 in most commonly consumed foods appears to be in the range of 70–80%. However, appreciable amounts of the vitamin in some foods are not biologically available. The determinants of the bioavailability of vitamin B_6 in a food include:

- *Pn glycoside content.* The pyridoxalal-5-β-D-glycosides are poorly digested, being taken up intact and converted to Pn by a hydrolase in the cytosol. Compared to free Pn, the bioavailabilities of the glycosides have been estimated to be 20–30% in the rat and about 60% in humans. In addition, the presence of Pn glycosides has been found to reduce the utilization of co-ingested free Pn.

- *Peptide adducts.* Vitamin B_6 can condense with peptide lysyl and/or cysteinyl residues during food processing, cooking or digestion; such products are less well utilized than the free vitamin. The reductive binding of Pal and Pal 5′-phosphate to ε-amino groups of lysyl residues in proteins or peptides produces adducts that are not only biologically unavailable, but also have vitamin B_6-antagonist activity.[1] For example, wheat bran contains vitamin B_6 in largely unavailable form(s), the presence of which reduces the bioavailability of the vitamin from other foods consumed at the same time.[2] Because plants generally contain complexed forms of Pn, bioavailability of the vitamin of plant foods tends to be greater than that of foods derived from animals.

Hindgut Microbial Synthesis

Although the microflora of the colon synthesize vitamin B_6, it is not absorbed there, and non-coprophagous animals derive no benefit from this microbial source of the vitamin. In contrast, ruminants benefit from their rumen microflora, which produces vitamin B_6 in adequate amounts to meet their needs proximal to where it is absorbed.

---

1. Gregory, J. F. (1980). *J. Nutr.* 110, 995.
2. Owing to the poor availability of vitamin B_6 from the bran fraction of the grain, the bioavailability of the vitamin from whole-wheat bread is less than that of Pn-fortified white bread.

TABLE 13.1 Vitamin B$_6$ Contents of Foods

Food	Vitamin B$_6$, mg/100 g	% Glycosylated	% Pn	% Pal	% Pm	Food	Vitamin B$_6$, mg/100 g	% Glycosylated	% Pn	% Pal	% Pm
Dairy Products						Beans	0.08–0.18	15–57	62	20	18
Milk	0.04		3	76	21	Broccoli	0.17	66			
Yogurt	0.05					Brussels sprouts	0.18				
Cheeses	0.04–0.08		4	8	88	Cabbage	0.16	46	61	31	8
Meats						Carrots	0.15	51–86	75	19	6
Beef	0.33		16	53	31	Cauliflower	0.21	66	16	79	5
Chicken	0.33–0.68	7	74	19		Celery	0.06				
Lamb	0.28					Corn	0.20		6	68	26
Pork	0.35		8	8	84	Onions	0.13				
Calf liver	0.84					Peas	0.16	15	47	47	6
Fish						Potatoes	0.25	32	68	18	14
Flounder	0.17		7	71	22	Spinach	0.28	50	36	49	15
Haddock	0.18					**Fruits**					
Herring	0.37					Apples	0.03		61	31	8
Oysters	0.05					Grapefruit	0.03				
Salmon	0.30		2	9	89	Oranges	0.06	47	59	26	15
Shrimp	0.10					Peaches	0.02	22	61	30	9
Tuna	0.43		19	69	12	Strawberries	0.06				
Cereals						Tomatoes	0.10	46	38	29	15
Barley, pearled	0.22		52	42	6	**Nuts**					
Corn meal	0.20		11	51	38	Almonds	0.10				
Oatmeal	0.14		12	49	39	Peanuts	0.40		74	9	17
Rice, polished	0.17	20	64	19	17	Pecans	0.18		71	12	17
Rice, brown	0.55	23	78	12	10	Walnuts	0.73	7	31	65	4
Wheat, whole	0.34	28	71	16	13	**Other**					
Wheat, white flour	0.06		55	24	21	Eggs	0.19		0	85	15
Vegetables						Human colostrum	0.001–0.002				
Asparagus	0.15					Human milk	0.010–0.025				

Sources: USDA data; Leklem, J. E. (1996). In *Present Understanding in Nutrition*, 7th edn (Ziegler, E. E. and Filer, L. J. Jr, eds), ILSI Press, Washington, DC, p. 75; Orr, M. L. (1969). In *Foods: Home Economics* Res. Rep. 36, USDA, Washington, DC, 52 pp.

3. ABSORPTION OF VITAMIN B$_6$

Digestion of Food Forms

The enteric absorption of protein-bound PalP and PmP involves their obligate dephosphorylation catalyzed by a membrane-bound alkaline phosphatase.

Diffusion Linked to Phosphorylation

Pn, Pal, Pm, and pyridoxine glycosides can be absorbed by passive diffusion throughout the gut. For Pn and Pal, the process is driven by the trapping of the intracellular vitamin as in the form of 5'-phosphates by the action of an ATP-dependent **Pal kinase** in cytosol. Pn glycosides taken up by diffusion are later converted to Pn by a cytosolic hydrolase and then oxidized to PalP.

4. TRANSPORT OF VITAMIN B$_6$

Plasma Vitamin B$_6$

Only a small portion (<0.1%) of total body vitamin B$_6$ is present in blood plasma. Most (>90%) occurs as PalP bound to albumin and derived from the hepatic turnover of flavoenzymes. Plasma PalP concentrations are typically <1 mmol; smaller amounts of other vitamers are also found. The circulating vitamin is tightly bound to albumin (and other plasma) via Schiff base linkages.[3]

The vitamin is present in erythrocytes at over six times more vitamin than plasma. In erythrocytes it is found as a Schiff base with hemoglobin by binding to the amino group of the N-terminal valine residue of the hemoglobin α-chain. This binding is twice as strong as that to albumin, which drives uptake of the vitamin by erythrocytes. Erythrocyte vitamin B$_6$ levels are particularly high in infants, but decline to adult levels by about 5 years of age. PalP content of erythrocytes is often used as a parameter of vitamin B$_6$ status.

In humans and other animals, plasma PalP concentrations decline during pregnancy, apparently as a result of a shift in the distribution of the vitamer in the blood to favor erythrocytes over the plasma, as the enteric absorption, excretion or hepatic uptake of the vitamin is unaffected. Renal failure has been found to reduce the plasma PalP level,[4] whereas submaximal exercise has been shown to increase it.

Cellular Uptake

Pal crosses cell membranes more readily than PalP. Thus, it appears to be the form taken up by the tissues, suggesting roles of phosphatases in the cellular retention and perhaps also the uptake of the vitamin. After being taken into the cell, the vitamin is again phosphorylated by Pal kinase to yield the predominant tissue form, PalP. Small quantities of vitamin B$_6$ are stored in the body, mainly as PalP but also as **Pm phosphate**.

Tissue Distribution

The greatest levels of vitamin B$_6$ are found in the liver, brain, kidney, spleen, and muscle, where it is bound to various proteins (Table 13.2). Muscle contains most (70–80%) of the body's vitamin B$_6$, in the form of Pal 5'-phosphate bound to **glycogen phosphorylase**. Glycogen phosphorylase binding accounts for much less of the Pal 5'-phosphate in other tissues (e.g., only 10% of the vitamin in liver) in which it also binds to various enzymes with which it has coenzyme functions. Protein binding is thought to protect PalP from hydrolysis while providing storage of the vitamin.

The total body pool of vitamin B$_6$ in the human adult is estimated to be 40–150 mg, constituting a supply sufficient to satisfy needs for 2,075 days. This amount is composed of two pools: one with a rapid turnover rate (0.5 day), and a second with a longer turnover rate (25–33 days).[5]

Moderate exercise has been found to increase plasma PalP concentrations substantially, e.g., by >20% within 20 minutes. This appears to be related to the increased need for gluconeogenesis which results in the release of PalP from glycogen phosphorylase. That plasma PalP levels increase so quickly under such conditions indicates either that the vitamer rapidly undergoes hydrolysis, discharge from the muscle, and then re-phosphorylation in the liver, or that it is released intact through interstial fluid.[6]

5. METABOLISM OF VITAMIN B$_6$

Interconversion of Vitamers

The vitamers B$_6$ are readily interconverted metabolically by phosphorylation/dephosphorylation, oxidation/reduction, and amination/deamination (Fig. 13.1). Because the non-phosphorylated vitamers cross membranes more readily than their phosphorylated analogs, phosphorylation appears to be an important means of retaining the vitamin intracellularly. Several enzymes are involved in this metabolism:

- **Pyridoxal kinase**. This hepatic enzyme catalyzes the phosphorylation of Pn, Pal, and Pm, yielding the corresponding phosphates. It requires a Zn–ATP complex, the formation of which is facilitated by Zn-metallothioneine (MT), and is stimulated by K$^+$. The role of MT in Pal kinase activity suggests that Zn status may be important in the regulation of vitamin B$_6$ metabolism. Erythrocyte Pal kinase activity in African-Americans has been reported to be about half that

---

3. Schiff bases are condensation products of aldehydes and ketones with primary amines; they are stable if there is at least one aryl group on either the N or the C that is linked. Vitamin B$_6$ forms Schiff base linkages with proteins by the bonding of the keto-C of PalP to a peptidyl amino (-NH$_2$) group. The vitamin also forms a Schiff base with the amino acid substrates of the enzymes for which it functions as a coenzyme; this occurs by the bonding of the amino nitrogen of PalP and the α-C of the substrate.
4. One study showed this depression to be >40% in rats.

5. Shane, B. (1978). *Human Vitamin B$_6$ Requirements*, National Academy Press, Washington, DC, pp. 111–128.
6. Crozier, P. G., Coredain, L., and Sampson, D. A. (1994). *Am. J. Clin. Nutr.* 40, 552.

TABLE 13.2 Concentrations (nmol/l) of Vitamers B$_6$ in the Plasma of Several Species

Species	Pal	PalP	Pol	Pm	PmP	Pyridoxic Acid
Pig	29	139	167	—	—	139
Human	62	13	33	6	<3	40
Calf	308	96	50		9	91
Sheep	626	57	43	—	466	318
Dog	417	268	66	—	65	109
Cat	2443	139	93	44	271	17

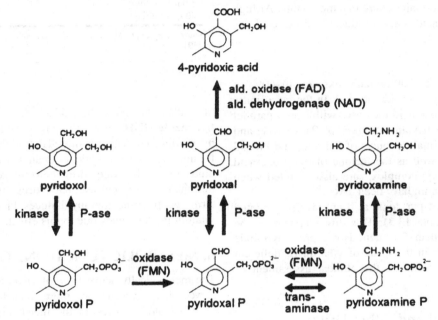

FIGURE 13.1 Metabolic interconversions of vitamin B$_6$.

of White Americans, although lymphocytes, granulocytes, and fibroblasts show no such differences. This may indicate reduced erythrocyte retention of vitamin B$_6$, which depends on the phosphorylation. Pyridoxal kinase also binds the anti-anxiety drug benzodiazepine,[7] suggesting that the mode of drug action may involve enhancement of neuronal γ-aminobutyrate levels (see page 314, "Metabolic Functions of Vitamin B$_6$").

- **Alkaline phosphatases**. Phosphorylated forms of the vitamin can be dephosphorylated by membrane-bound alkaline phosphatases in many tissues (e.g., liver, brain, and intestine).
- **Pyridoxamine phosphate oxidase**. This enzyme catalyzes the limiting step in vitamin B$_6$ metabolism.

It requires flavin mononucleotide (FMN); therefore, deprivation of riboflavin may reduce the conversion of Pn and Pm to the active coenzyme PalP.

- **Pyridoxal-5′-phosphate synthase**.[8] This enzyme catalyzes the oxidation of PnP and PalP to PalP.

The liver is the central organ for vitamin B$_6$ metabolism, containing all of the enzymes involved in its interconversions. The major forms of the vitamin in that organ are PalP and PmP, which are maintained at fairly constant intracellular concentrations in endogenous pools that are not readily accessible to newly formed molecules of those species. The latter instead comprise a second pool that is readily mobilized for metabolic conversion (mostly to PalP, Pal, and pyridoxic acid) and release to the blood.

7. Hanna, M. C., Turner, A. J., and Kirkness, E. F. (1997). *J. Biol. Chem.* 272, 10756.

8. That is, pyridoxal/pyridoxamine phosphate oxidase.

Enzyme Binding

Vitamin B_6 serves as an essential cofactor for several enzymes, to each of which it remains tightly bound as PalP or Pm phosphate. Binding occurs through the formation of a Schiff base between the keto-carbon of the coenzyme and the ϵ-amino group of a specific lysyl residue of the apoenzyme.

Catabolism

PalP is dephosphorylated and oxidized primarily in the liver by the FAD-dependent **aldehyde oxidase**, as well as the NAD-dependent **aldehyde dehydrogenase**, to yield **4-pyridoxic acid**, the major excretory metabolite. At high intakes, 5-pyridoxic acid is also produced and excreted.

Excretion

It has been estimated that humans oxidize 40–60% of ingested vitamin B_6 to 4-pyridoxic acid. In the rat, urinary excretion of 4-pyridoxic acid increases with age in parallel with increases in the hepatic activities of Pal oxidase and Pal dehydrogenase. Small amounts of Pal, Pm, and Pn and their phosphates, as well as the lactone of pyridoxic acid and a ureido–pyridoxyl complex,[9] are also excreted when high doses of the vitamin have been given.[10]

Urinary levels of 4-pyridoxic acid are inversely related to protein intake (Table 13.3). This effect appears to be greater for women than for men. However, 4-pyridoxic acid is not detectable in the urine of vitamin B_6-deficient subjects, making it useful in the clinical assessment of vitamin B_6 status.[11,12]

Effects of Alcohol and Other Drugs

Several drugs can antagonize vitamin B_6. Among these is alcohol; its degradation product, acetaldehyde, displaces PalP from proteins, resulting in enhanced catabolism of the coenzyme. Acetaldehyde also stimulates the activity of alkaline phosphatase, enhancing the dephosphorylation

of PalP. The anti-tuberculosis drug **isonicotinic acid hydrazide (INH)** also antagonizes vitamin B_6; it does so by binding the vitamin directly. For this reason, vitamin B_6 must be given to patients treated with INH. Pal kinase binds the anti-anxiety drug benzodiazepine, and can be inhibited by the anti-asthmatic drug theophylline. Short-term theophylline therapy induces biochemical signs of vitamin B_6 deficiency due to this effect.

6. METABOLIC FUNCTIONS OF VITAMIN B_6

The metabolically active form of vitamin B_6 is PalP, which serves as a coenzyme of more than 140 enzymes, most of which are involved in the metabolism of amino acids (Fig. 13.2), functioning in the following general mechanisms:

- decarboxylations
- transamination, racemization, elimination, and replacement reactions
- β-group interconversions.

Mechanisms of Action

Vitamin B_6-dependent enzymes have structural similarities in their coenzyme-binding regions at which PalP or Pm phosphate is bound through the formation of a Schiff base. Accordingly, the mechanisms of the reactions catalyzed by the vitamin B_6-dependent enzymes are also similar. Each involves the binding of an α-carbon of an α-amino acid substrate to the pyridine nitrogen of PalP. The delocalization of the electrons from the α-carbon by the action of the protonated pyridine nitrogen as an electron sink results in the conversion of the former to a carbanion (C^-) at the α-carbon and the labilization of its bonds. This results in

TABLE 13.3 Effect of Protein Intake on Vitamin B_6 Status

Dietary Treatment	Intakes		
Protein intake (g/kg)	0.5	1.0	2.0
Vitamin B_6 intake (mg/g protein)	0.04	0.02	0.01
Parameter (Adequate Value)	**% Subjects with Low Values**		
Urinary 4-pyridoxic acid (>3 mmol/day)	11	22	78
Urinary total vitamin B_6 (>0.5 mmol/day)	56	56	67
Plasma PalP (>30 nmol/l)	33	67	78
Urinary xanthurenic acid (<65 mmol/day)	11	11	44

Source: Hansen, C. M., Leklem, J. E., and Miller, L. T. (1996). *J. Nutr. 126*, 1891.

---

9. This is formed by the reaction of an amino group of urea with a hydroxyl group of the hemiacetal form of the aldehyde at position 4 of Pal.

10. For example, humans given 100 mg of Pal excrete about 60 mg 4-pyridoxic acid and 2 mg Pal over the next 24 hours.

11. In humans, excretion of less than 0.5 mg/day (men) or 0.4 mg/day (women) is considered indicative of inadequate intake of the vitamin. Typical excretion of total vitamin B_6 by adequately nourished humans is 1.2–2.4 mg/day. Of that amount, 0.5–1.2 mg (men) or 0.4–1.1 mg (women) is in the form of 4-pyridoxic acid.

12. Although no explanation has been offered for the correlation, it is of interest that excretion of relatively low amounts (<0.81 mg/24 hours) of 4-pyridoxic acid is associated with increased risk of relapse after mastectomy.

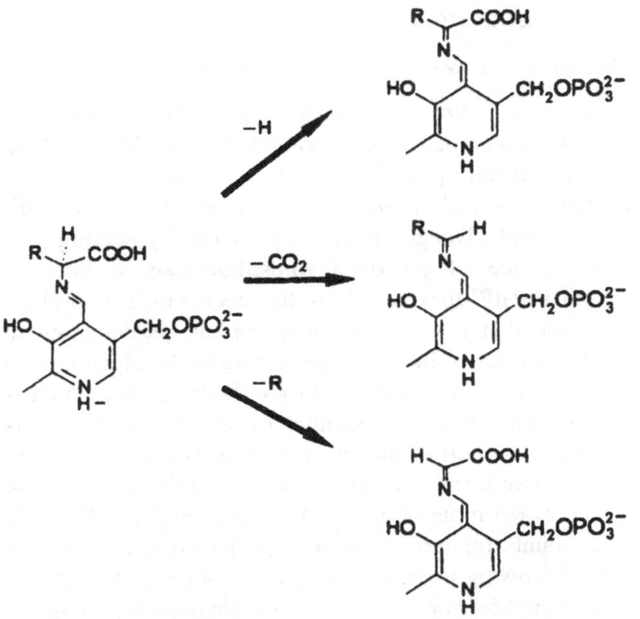

FIGURE 13.2 General reactions of PalP-dependent enzymes in amino acid metabolism.

the heterolytic cleavage of one of the three bonds to the α-carbon (Table 13.4, Fig. 13.2). The particular bond to be cleaved is determined by the particular PalP-dependent enzyme; each involves the loss of the cationic ligand of an amino acid.

Metabolic Roles

Amino Acid Metabolism

PalP is involved in practically all reactions involved in amino acid metabolism, being involved in their biosynthesis as well as their catabolism.

- *Transaminations.* PalP-dependent transaminases catabolize most amino acids.[13] The response of **erythrocyte aspartate aminotransferase (EAAT)** to *in vitro* additions of PalP has been used as a biochemical maker of vitamin B$_6$ status.[14]
- *Transsulfuration.* PalP-dependent enzymes cystathionine synthase and cystathionase catalyze the transsulfuration of methionine to cysteine. Vitamin B$_6$ deprivation therefore reduces the activities of these enzymes; affected individuals show **homocysteinuria** (due to the impaired conversion to cystathionine) and

---

13. The only amino acids that are not substrates for PalP-dependent transaminases are threonine, lysine, proline, and hydroxyproline.
14. However, EAAT activity coefficients can be affected by factors unrelated to vitamin B$_6$ status (e.g., intake of protein and alcohol, differences in body protein turnover, certain drugs, genetic polymorphism of the enzyme), which can compromise its use without careful controls.

TABLE 13.4 Important PalP-Dependent Enzymes of Animals

Type of Reaction	Enzyme
Decarboxylation	Aspartate 1-decarboxylase
	Glutamate decarboxylase
	Ornithine decarboxylase
	Aromatic amino acid decarboxylase
	Histidine decarboxylase
R-group interconversion	Serine hydroxymethyltransferase
	δ-Aminolevulinic acid synthase
Transamination	Aspartate aminotransferase
	Alanine aminotransferase
	γ-Aminobutyrate aminotransferase
	Cysteine aminotransferase
	Tyrosine aminotransferase
	Leucine aminotransferase
	Ornithine aminotransferase
	Glutamine aminotransferase
	Branched-chain amino acid aminotransferase
	Serine-pyruvate aminotransferase
	Aromatic amino acid transferase
	Histidine aminotransferase
Racemization	Cystathionine β-synthase
α,β-Elimination	Serine dehydratase
γ-Elimination	Cystathionine γ-lyase
	Kynureninase

---

cystathionuria (due to the impaired cleavage of cystathionine to cysteine and α-ketobutyrate). These conditions can be exacerbated for diagnostic purposes by the use of an oral methionine load. Plasma homocysteine concentrations, however, usually do not change in vitamin B$_6$ deficiency, and are therefore not suitable for assessment of vitamin B$_6$ status.

- *Selenoaminoacid metabolism.* Vitamin B$_6$ is essential for the utilization of selenium (Se) from the major dietary form, selenomethionine; after that, Se is transferred to selenohomocysteine. PalP is a cofactor for two enzymes, **selenocysteine β-lyase** and **selenocysteine γ-lyase**, which catalyze the elimination of the Se from selenohomocysteine to yield hydrogen selenide (H_2Se). Selenide is the obligate precursor for the incorporation of Se into selenoproteins in the form of selenocysteinyl residues produced during translation.[15]

Tryptophan–Niacin Conversion

Vitamin B_6 is required cofactor for two key enzymes in this pathway (see Chapter 12, Fig. 12.2):

- **Kyneureninase.** This enzyme catalyzes the removal of an alanyl residue from 3-hydroxykynurenine in the metabolism of tryptophan to the branch-point intermediate α-amino-β-carboxymuconic-ε-semialdehyde in the tryptophan–niacin conversion pathway. Kynureninase also catalyzes the analogous reaction (removal of alanine) using non-hydroxylated kynurenine as substrate and yielding the non-hydroxylated analog of 3-hydroxykynurenine, anthranilic acid.
- **Transaminases.** Vitamin B_6-dependent transaminases metabolize kynurenine and 3-hydroxykynurenine, yielding kynurenic and xanthurenic acids, respectively. The transaminases have much greater binding affinities for PalP than kynureninase,[16] and are therefore affected preferentially by vitamin B_6 deprivation. This results in blockage in the tryptophan–niacin pathway, with an accumulation of 3-hydroxykynurenine that gets diverted by transamination to yield xanthurenic acid, which appears in the urine.[17] This phenomenon is exploited in the assessment of vitamin B_6 status: deficiency is indicated by urinary excretion of xanthurenic acid after a tryptophan load.

---

15. These include: the Se-dependent glutathione peroxidases and thioredoxin reductases which have antioxidant functions; the iodothyronine 5′-deiodinases which are involved in thyroid hormone metabolism; selenophosphate synthase which is involved in selenoprotein synthesis; selenoproteins P and W which are major selenium-containing proteins in plasma and muscle, respectively; and at least a dozen other proteins.
16. The Michaelis constants (K_ms) for the transaminases are in the order of 10^{-8} mol/l, whereas the K_m for kynureninase is in the order of 10^{-3} mol/l.
17. Xanthurenic acid was discovered quite unexpectedly by Lepkovsky (University of California), who, during the Great Depression, sought to elucidate the nature of rat adermin. He wrote of his surprise in finding that the urine voided by his vitamin B_6-deficient rats was green, whereas that of his controls was the normal yellow color. In pursuing this observation, he found that urine from deficient animals was normally colored when voided, but turned green only on exposure to the rusty dropping pans their limited budget had forced them to use. Thus, he recognized that vitamin B_6-deficient rats excreted a metabolite that reacted with Fe^{3+} to form a green derivative. This small event, which might have been missed by someone "too busy" to observe the experimental animals, resulted in Lepkovsky's identifying the metabolite as xanthurenic acid and discovering the role of vitamin B_6 in the tryptophan–niacin conversion pathway. His message: *"The investigator has to do more than sit at his desk, outline experiments and examine data."*

Gluconeogenesis

Vitamin B_6 has two roles in gluconeogenesis:

- *Transaminations.* Amino acid catabolism depends on PalP is a cofactor for transaminases (see "Amino Acid Metabolism," page 315).
- *Glycogen utilization.* PalP is required for the utilization of glycogen to release glucose by serving as a coenzyme of **glycogen phosphorylase**. Unlike the other PalP enzymes, it is the coenzyme's phosphate group that is catalytically important, participating in the transfer of inorganic phosphate to the glucose units of glycogen to produce glucose-1-phosphate, which is released. That it is essential for enzymatic activity is clear; the shift of the enzyme from its inactive form to its active form involves an increase in the binding (2–4 moles per mole of enzyme) of the coenzyme. This role accounts for more than half of the vitamin B_6 in the body, owing to the abundance of both muscle and glycogen phosphorylase (5% of soluble muscle protein).

Neurotransmitter Synthesis

PalP-dependent enzymes function in the biosynthesis of the neurotransmitters **serotonin** as a cofactor for tryptophan decarboxylase, of **epinephrine** and **norepinephrine** as a cofactor for tyrosine carboxylase, as a source of energy for the brain, **γ-aminobutyric acid** (**GABA**), and as a cofactor for glutamate decarboxylase.[18]

Histamine Synthesis

PalP functions in the metabolism of the vasodilator and gastric secretagog histamine as a cofactor for histidine decarboxylase.

Hemoglobin Synthesis and Function

PalP functions in the synthesis of heme from porphyrin precursors as a cofactor for δ-aminolevulinic acid synthase. The vitamin also binds to hemoglobin at two sites on the β chains: the N-terminal valine and Lys-82 residues. Pal also binds at the N-terminal valine residues of the α chains. The binding of Pal and PalP enhances the O_2-binding capacity of that protein, and inhibits the physical deformation of sickle-cell hemoglobin.

Lipid Metabolism

Vitamin B_6 is required for the biosynthesis of sphingolipids via the PalP-dependent serine palmitoyltransferase and

---

18. It has been shown *in vitro* that PalP can inhibit the binding of GABA to brain synaptic membranes.

other enzymes in phospholipid synthesis. Diminution in the activities of these enzymes is thought to account for the changes observed in phospholipid contents of linoleic and arachidonic acids in vitamin B$_6$-deficient animals.

Other Enzymes

The catalytic properties of other enzymes are also affected (some stimulated,[19] others inhibited[20]) by interactions with PalP (usually via valine, histidine or lysine residues), suggesting that vitamin B$_6$ status may have effects beyond those involving its classic coenzyme function.

Gene Expression

PalP has been shown to modulate gene expression. Elevated intracellular levels of the vitamin are associated with decreased transcription in responses to glucocorticoid hormones (progesterone, androgens, estrogens). Such diminished responses include hydrocortisone-induction of rat liver cytosolic aspartate aminotransferase. Inhibition is caused by the formation of Schiff base linkages of the vitamin to the DNA-binding site of the receptor–steroid complex. This inhibits the ligand binding to the glucocorticoid-responsive element in the regulatory region of the gene. Vitamin B$_6$ deficiency increases the expression of albumin mRNA seven-fold. The effect appears due to the action of PalP inactivating tissue-specific transcription factors by directly interacting with DNA ligand-binding sites.[21] PalP appears to modulate glycoprotein IIb gene expression by interacting directly with tissue-specific transcription factors.[22] This results in inhibition of platelet aggregation due to impaired binding of fibrinogen or other adhesion proteins to glycoprotein complexes. PalP has also been shown to suppress mRNA levels for glycogen phosphorylase, apolipoprotein A-1, phenylalanine hydroxylase, glyceraldehyde-3-phosphate dehydrogenase, and β-actin, but to decrease mRNA levels for RNA polymerases I and II in the rat model.[23] The effect on glycogen phosphorylase appears to be tissue-specific, as deprivation of the vitamin was found to reduce phosphorylase mRNA levels in muscle, but to increase them in liver.[24]

7. VITAMIN B$_6$ IN HEALTH AND DISEASE

Vascular Disease

Low vitamin B$_6$ status has been associated with increased risk of coronary artery disease.[25] This relationship has been linked to altered platelet aggregation due to reduced Ca^{2+} influx caused by impaired adenosine-5'-diphosphate receptors, and to increased chronic inflammation marked by elevated plasma levels of C-reactive protein.[26]

Low vitamin B$_6$ status can cause homocyteinemia as a result of diminished conversion of that metabolite to cystathionine due to impaired activities of the PalP-dependent enzyme cystathionine β-synthase. Homocysteinemia has been associated with increased risks of occlusive vascular disease, total and cardiovascular disease-related mortality, stroke, and chronic heart failure.[27] Low plasma PalP levels have also been associated with increased risk of vascular disease independent of plasma homocysteine level.[28] A recent study suggested that treatment with high doses of vitamin B$_6$ and folic acid were effective in reducing both plasma homocysteine and the incidence of abnormal exercise electrocardiography tests, suggesting reductions in risk of atherosclerotic disease.[29] However, a meta-analysis of 12 randomized trials of vitamin (folate, vitamin B$_6$, vitamin B$_{12}$) supplements to lower homocysteine levels in free-living people found no evidence for vitamin B$_6$ supplements being effective.[30]

Deprivation of vitamin B$_6$ has been shown to produce moderate hypertension in the rat. These effects were associated with elevations in plasma levels of epinephrine and norepinephrine, and reduced levels of serotonin in the brain and 5-hydroxytryptophan in nerves.

Neurologic Function

Vitamin B$_6$ has a key role in the synthesis of the neurotransmitters dopamine, norepinephrine, serotonin, and GABA, as well as sphingolipids and polyamines. However, the apoenzymes involved in these various steps have widely different affinities for PalP. This means that vitamin B$_6$ deprivation affects preferentially those PalP-dependent decarboxylases with low affinities for the coenzyme. Accordingly, moderate deficiency of vitamin B$_6$ reduces brain serotonin levels without affecting other neurotransmitters.

19. For example, thymidylate synthase, vitamin K-dependent carboxylase/epoxidase.

20. For example, glucose-6-phosphate dehydrogenase, glycerol-3-phosphate dehydrogenase, ribulose bisphosphate carboxylase, ribosomal peptidyltransferase.

21. See review: Oka, T. (2001). *Nutr. Res. Rev.* 14, 257.

22. Chang, S. J., Chuang, H. J., and Chen, H. H. (1999). *J. Nutr. Sci. Vitaminol.* 45, 471.

23. Oka, T., Komori, N., Kuwahata, M., *et al.* (1993). *FEBS Letts* 331, 162.

24. Oka, T., Komori, N., Kuwahata, M., *et al.* (1994). *Experientia* 50, 127.

25. Robinson, K., Arheart, D., and Refsum, H. (1998). *Circulation* 97, 437.

26. Morris, M. S., Sakakeeny, L., and Jacques, P. F. (2010). *J. Nutr.* 140, 103.

27. Selhub, J. (2006). *J. Nutr.* 136, 1726S.

28. Robinson, K., Arheart, K., Refsum, H., *et al.* (1998). *Circulation* 97, 437.

29. Vermuelen, E. G. J., Stehouwer, C. D., Twisk, J. W., *et al.* (2000). *Lancet* 355, 517.

30. Folic acid produced an average 25% reduction, and vitamin B$_{12}$ produced an average 7% reduction (Clarke, R. and Armitage, J. [2000] *Semin. Thromb. Hemost.* 26, 341).

Animal studies of long-term potentiation, a synaptic model of learning and memory, have revealed that maternal deprivation of the vitamin during gestation and lactation specifically reduces the development of the N-methyl-D-aspartate receptor subtype in the young. Although the metabolic basis is not understood, these effects appear to be related to the loss of dendritic arborization in vitamin B_6 deficiency. These lesions are thought to underlie reported effects of impaired learning on the part of the progeny of vitamin B_6-deficient animals and humans.

Pyridoxine-dependent seizures have been described. These appear to involve aberrant forms of glutamic acid decarboxylase with impaired conversion of the excitatory neurotransmitter glutamate to the inhibitory neurotransmitter GABA. Vitamin B_6 has also been recommended (0.1–1 g/day alone or in combination with tryptophan or Mg) for reducing seizures in alcoholics, and for the treatment of schizophrenia.

There is no evidence for vitamin B_6 affecting mood, depression or cognitive functions.

Immune Function

Vitamin B_6 has a role in the support of immune competence that has not been elucidated. Animal and human studies have demonstrated effects of vitamin B_6 deprivation on both humoral (diminished antibody production) and cell-mediated immune responses (increased lymphocyte proliferation, reduced delayed-type hypersensitivity responses, reduced T cell-mediated cytotoxicity, reduced cytokine production), and suboptimal status of the vitamin has been linked to declining immunologic changes among the elderly (Table 13.5), persons with human immunodeficiency virus (HIV), and patients with uremia or rheumatoid arthritis. These effects appear to be due to reduced activities of such PalP enzymes as serine transhydroxymethylase and thymidylate synthase, which would impair single-carbon metabolism and reduce DNA synthesis. The vitamin has also been shown to bind to a glycoprotein surface receptor (CD4) on T-helper cells by which it affects photoimmunosuppression.[31] It has also been shown to noncompetitively inhibit HIV-1 reverse transcriptase.[32]

Anticarcinogenesis

Epidemiological studies have demonstrated inverse associations of projected vitamin B_6 intake and colon cancer risk.[33] Recent studies in animal models have found that supranutritional doses of the vitamin can reduce tumorigenesis through effects on cell proliferation, production of reactive oxygen and nitrogen species, and angiogenesis.[34]

TABLE 13.5 Effects of Vitamin B_6 Status on Mitogenic Responses and Interleukin 2 Production by Peripheral Blood Mononucleocytes of Elderly Humans

Parameter	Baseline	B_6-Deprived	B_6-Supplemented
Mitogenic response to:			
Conconavalin A	120	70	190
Phytohem-agglutinin	100	70	100
Staphylococcus aureas	115	60	200
IL-2 production (kU/l)	105	40	145

Source: Meydani, S. N., Ribaya-Mercado, J. D., Russwel, R. M., *et al.* (1991). *Am. J. Clin. Nutr.* 53, 1275.

Effects Associated with High Vitamin B_6 Doses

Vitamin B_6 at relatively high doses has been reported to produce positive effects in a number of conditions affecting individuals who were not apparently deficient in the vitamin.

- *Sideroblastic anemia.* Dosages as great as 200 mg/day (usually as Pn HCl) have been found to stimulate δ-aminolevulinic acid synthase activity and thus enhance hematopoiesis in patients.

- *Sickle cell anemia.* A small study found patients to have lower plasma PalP levels than controls, which responded to oral supplementation (100 mg/day) of Pn within 2 months. Both Pn and PalP have been found to protect sickle cells *in vitro*,[35] but it is not clear whether supplementation with the vitamin may benefit sickle cell anemia patients.

- *Iron storage disease.* Complexes of Pal, which chelate iron (e.g., the isonicotinyl and benzoyl hydrazones), have been found effective in stimulating the excretion of iron in patients with iron-storage disease.

- *Suppression of lactation.* A few studies have reported vitamin B_6 as effective in suppressing lactation, probably through the stimulation of dopaminergic activity in the hypothalamus and, thus, the suppression of prolactin.

- *Asthma.* Low circulating PalP levels have been reported in patients with asthma, and one small study found vitamin B_6 treatment (100 mg/day) to reduce the severity and frequency of attacks.[36] These effects may be secondary to those of theophylline, which inhibits Pal kinase.

31. Salhany, J. M. and Schopper, L. M. (1993). *J. Biol. Chem.* 268, 7643.
32. Mitchell, L. L. W. and Cooperman, B. S. (1992). *Biochem.* 31, 7707.
33. Ishihara, J., Otani, T., Inoue, M., *et al.* (2007). *J. Nutr.* 137, 1808.
34. Komatsu, S., Yanaka, N., Matsubara, K., *et al.* (2003). *Biochem. Biophys. Acta* 1647, 127.

35. Kark, J. A., Tarassoff, P. G., and Bongiovanni, R. (1983). *J. Clin. Invest.* 71, 1224.
36. Simon, R. A. and Reynolds, R. D. (1988). In: *Clinical and Physiological Applications of Vitamin B₆* (Leklem, J. E. and Reynolds, R. D., eds). Alan R. Liss, New York, NY, pp. 307–315.

- *Adverse drug effects.* Vitamin B$_6$ is used at doses of 3–5 mg/kg body weight to counteract adverse effects of several types of drugs. The anti-tuberculin drug isonicotinic acid hydrazide (isoniazid) produces a peripheral neuropathy similar to that of vitamin B$_6$ deficiency by inhibiting the activities of PalP-dependent glutamate-decarboxylase and γ-aminobutyrate aminotransferase, which produce and degrade GABA, respectively, in nerve tissue. Two antibiotics antagonize vitamin B$_6$ by reacting with PalP to form inactive products. Cycloserine reacts with the coenzyme to produce an oxime. Penicillamine produces thiazolidine. L-3,4-dihydroxyphenylalanine antagonizes vitamin B$_6$ by reacting with PalP to form tetrahydroquinolines. Ethanol increases PalP catabolism. Synthetic estrogens can alter tryptophan-niacin conversion by increasing the synthesis of PalP-dependent enzymes in that pathway, thus increasing the need for vitamin B$_6$.
- *Herpes.* Pn (400–4,000 mg/day) has been used to treat *herpes gestationis.*
- *Carpal tunnel syndrome.* This disorder, involving pain and paresthesia of the hand, is caused by irritation and compression of the medial nerve by the transverse ligaments of the wrist in ways that are exacerbated by redundant motions. The condition has been associated with low circulating levels of PalP and low erythrocyte glutamic-oxaloacetic transaminase activities. It has been suggested that such deficiencies lead to edematous changes to and proliferation of the synovia, causing compression of the nerve in the carpal tunnel. Some investigators have reported high doses (50–300 mg/day for 12 weeks) of Pn to be effective as treatment;[37] however, there is no evidence from randomized clinical trials supporting such use of the vitamin.
- *Diabetes.* Several studies have found vitamin B$_6$ supplementation to improve glucose tolerance. It has been suggested that this may involve the reactivation of kynureninase, which leads to inactivation of insulin through complexation with xanthurenic acid. Studies have also shown vitamin B$_6$ supplements (100 mg/day) useful in preventing complications of diabetes mellitus caused by the non-enzymatic glycation of critical proteins.[38] Pm has been shown to be a potent inhibitor of the formation of advance glycation products from glycated proteins by scavenging reactive carbonyls; this would appear to be the basis of its protection against the development of renal disease in the diabetic rat model.[39]

- *Chinese restaurant syndrome.* This syndrome, which involves headache, sensation of heat, altered heartbeat, nausea, and tightness of the neck induced by oral intake of monosodium glutamate, has been reported to respond to vitamin B$_6$ (50 mg/day).
- *Premenstrual syndrome.* This syndrome affects some 40% of women 2–3 days before their menstrual flow. It involves tension of the breasts, pain in the lumbar region, thirst, headache, nervous irritability, pelvic congestion, peripheral edema, and, usually, nausea and vomiting. Premenstrual syndrome has been reported to respond to vitamin B$_6$, presumably by affecting levels of the neurotransmitters, serotonin and γ-aminobutyric acid, that control depression, pain perception, and anxiety. Women experiencing premenstrual symptoms appear to have circulating PalP levels comparable to unaffected women; nevertheless, high doses of the vitamin have been found to alleviate at least some symptoms in many cases. A review of randomized clinical trials concluded that Pn doses of up to 100 mg/day are likely to be of benefit in treating these symptoms.[40]
- *Morning sickness.* A randomized clinical trial showed that the use of Pn (25 mg every 8 hours for 3 days) significantly reduced vomiting and nausea in pregnant women.[41] However, there is no evidence that women who experience nausea and vomiting in pregnancy are of abnormal vitamin B$_6$ status.

Congenital Disorders of Vitamin B$_6$ Metabolism

Several rare familial disorders have been identified, each thought to be caused by the expression of deficient amounts or dysfunctional forms of PalP-dependent enzyme (Table 13.6).

- *Homocysteinuria.* Hereditary deficiency of the cystathionine β-synthase occurs at a rate of 3/1,000,000. The resulting impairment in homocysteine catabolism is manifest as elevations in plasma levels of homocysteine, methionine, and cysteine with dislocation of the optic lens,[42] osteoporosis, and abnormalities of long bone growth, mental retardation, and thromboembolism. The condition is treated with a low-methionine diet. Half of cases respond to high doses (250–500 mg/day) of Pn.[43] Of more than a hundred alleles that have been studied, mutations of the cystathionine β-synthase

37. Aufiero, E., Stitik, T. P., Foye, P. M., *et al.* (2004). *Nutr. Rev.* 62, 96; Goodyear-Smith, F. and Arroll, B. (2004). *Ann. Fam. Med.* 2, 267; Ellis, J. M. and Pamplin, J. (1999). *Vitamin B6 Therapy.* Avery Publishing Group, New York, NY, pp. 47–56.
38. Solomon, L. R. and Cohen, K. (1989). *Diabetes* 38, 881.
39. Metz, T. O., Alderson, N. L., and Thorpe, S. R. (2003). *Arch. Biochem. Biophys.* 419, 41.

40. Wyatt, K.M., Dimmock, P.W., Jones, P.W., *et al.* (1999) *Br. Med. J.* 318, 1375.
41. Sahakian, V., Rouse, D., Sipes, S., *et al.* (1991). *Obstet. Gynecol.* 78, 33.
42. *Ectopia lentis.*
43. Berber, G. and Spaeth, G. (1969). *J. Pediatr.* 75, 463.

TABLE 13.6 Congenital Disorders of Vitamin B_6-Dependent Metabolism

Disorder	Enzyme Deficiency	Clinical Manifestations
Homocysteinuria[a]	Cystathionine β-synthase	Dislocation of lenses, thromboses, malformation of skeletal and connective tissue, mental retardation
Cystathionuria	Cystathionine γ-lyase	Mental retardation
GABA deficiency	Glutamate decarboxylase	Seizures
Sideroblastic anemia	δ-Aminolevulinic acid synthase	Anemia, cystathionuria, xanthurenic aciduria

[a]Another form is caused by impaired vitamin B_{12}-dependent methionine synthesis.

associated with disease phenotypes have been found in almost a third.[44] Some mutations, including some of the most frequent ones in the human populations studied to date,[45] have been shown to correlate with Pn-responsiveness; these would appear to involve the expression of a mutant enzyme with low affinity for PalP.

- *Vitamin B_6-responsive seizures.* This is a rare,[46] autosomal recessive disorder involving impaired synthesis of the inhibitory neurotransmitter, GABA. The defect was originally thought to involve the inability of an abnormal form of glutamic acid decarboxylase, GAD-65, in the nerve terminal[47] to bind PalP. Recent studies have shown the defect to involve, instead, mutations in Pn/Pm oxidase.[48] The disorder is manifest as intractable seizures appearing within hours after birth. Patients show normal circulating levels of PalP, but their seizures stop immediately upon administration of high levels (100–500 mg) of Pn intravenously, and are controlled using daily oral doses of Pn (0.2–3 mg/kg body weight).[49] If untreated, progressive cerebral atrophy ensues.

- *Hyperoxaluria.* Type 1 primary hyperoxaluria has been found to be due to a mutant hepatic alanine glyoxylate transferase with abnormally low PalP-binding capacity. In such cases high doses of vitamin B_6 (e.g., 400 mg/day) reduce hyperoxaluria and, thus, the risk of formation of oxalate stones and renal injury.

8. VITAMIN B_6 DEFICIENCY

Severe deficiency of vitamin B_6 results in dermatologic and neurologic changes in most species. Less obvious are the metabolic lesions associated with insufficient activities of the coenzyme PalP. The most prominent of these is impaired tryptophan–niacin conversion, which can be detected on the basis of urinary excretion of xanthurenic acid after an oral **tryptophan load**. Vitamin B_6 deficiency also results in impaired transsulfuration of methionine to cysteine, which can be detected as homocysteinuria and cystathionuria after an oral **methionine load**. The PalP-dependent transaminases and glycogen phosphorylase give the vitamin a role in gluconeogenesis. Deprivation of vitamin B_6 therefore impairs glucose tolerance, although it may not affect fasting glucose levels.

Deficiency Syndromes in Animals

Vitamin B_6 deficiency in animals is generally manifest as symmetrical scaling dermatitis. In rodents the condition is called **acrodynia**, and is characterized by hyperkeratotic[50] and acanthotic[51] lesions on the tail, paws, face, and upper thorax, as well as by muscular weakness, hyper-irritability, anemia, hepatic steatosis,[52] increased urinary oxalate excretion, insulin insufficiency,[53] hypertension, and poor growth. Neurological signs include convulsive seizures (epileptiform type) that can be fatal.[54] Reproductive disorders

44. Kraus, J. P., Janosik, M., Kozich, V., *et al.* (1999). *Human Mutation* 13, 362.

45. Such studies have been conducted only in Europe; no information is available for other populations.

46. Only a hundred cases had been documented; it is likely that the disorder is under-reported.

47. Another, non-PalP-dependent glutamic acid decarboxylase, GAD-67, is present in neuronal soma and dendrites.

48. Gospe, S. M. Jr (2006). *Curr. Opin. Neurol.* 19, 148.

49. Gupta, V., Mishra, D., Mathur, I., *et al.* (2001). *J. Pediatr. Child Health* 37, 592; Gospe, S. M. Jr (2002). *Ped. Neurol.* 26, 181.

50. Involving hypertrophy of the horny layer of the epidermis.

51. Involving an increase in the prickle cell layer of the epidermis.

52. This can be precipitated by feeding a vitamin B_6-deficient diet rich in protein.

53. This is believed to be due to reduced pancreatic synthesis of the hormone.

54. Nervous dysfunction is believed to be due to nerve tissue deficiencies of GABA due to decreased activities of the PalP-dependent glutamate decarboxylase. The seizures can be controlled by administering either the vitamin or GABA.

include infertility, fetal malformations,[55] and reduced fetal survival. Some reports indicate effects on blood cholesterol levels and immunity. That tissue carnitine levels are depressed in vitamin B$_6$-deficient animals has been cited as evidence of a role of the vitamin in carnitine synthesis.

Similar changes are observed in vitamin B$_6$-deficient individuals of other species. Chickens and turkeys show reduced appetite and poor growth, dermatitis, marked anemia, convulsions, reduced egg production, and low fertility. Pigs show paralysis of the hind limbs, dermatitis, reduced feed intake, and poor growth. Monkeys show an increased incidence of dental caries and altered cholesterol metabolism with arteriosclerotic lesions. Vitamin B$_6$ deficiency has been reported to cause hyper-irritability, hyperactivity, abnormal behavior, and performance deficits in several species. These signs accompany an underlying neuropathology that reduces axonal diameter and dendritic arborization, and thus impairs nerve conduction velocity.

Ruminants are rarely affected by vitamin B$_6$ deficiency, as their rumen microflora appear to satisfy their need for the vitamin. Exceptions are lambs and calves, which, before their rumen microfloras are established, are susceptible to dietary deprivation of vitamin B$_6$, showing many dermatologic and neurologic changes observed in non-ruminant species.

Deficiency in Humans

Vitamin B$_6$-deficiency occurs in free-living populations. Prevalences in the range of 0.8–68% have been reported for signs of vitamin B$_6$ deficiency in developed countries.

Vitamin B$_6$-deficient humans exhibit symptoms that can be quickly corrected by administration of the vitamin: weakness, sleeplessness, nervous disorders (peripheral neuropathies), **cheilosis**,[56] **glossitis**,[57] **stomatitis**, and impaired cell-mediated immunity (Table 13.7). Behavioral differences have been associated with low vitamin B$_6$ status: a study in Egypt found that mothers of marginal (subclinical) vitamin B$_6$ status were less responsive to their infants' vocalizations, showed less effective response to infant distress, and were more likely to use older siblings as caregivers than were mothers of better vitamin B$_6$ status. In addition, studies with volunteers fed a vitamin B$_6$-free diet or a vitamin B$_6$ antagonist[58] have shown elevated urinary

TABLE 13.7 Signs of Vitamin B$_6$ Deficiency

Organ System	Signs
General	
Appetite	Decrease
Growth	Decrease
Dermatologic	Acrodynia, cheilosis, stomatitis, glossitis
Muscular	Weakness
Skeletal	Dental caries
Vital organs	Hepatic steatosis
Vascular	
Vessels	Arteriosclerosis
Erythrocytes	Anemia
Nervous	Paralysis, convulsions, peripheral neuropathy
Reproductive	
Female	Decreased egg production
Fetal	Malformations, death

xanthurenic acid concentrations[59] and increased susceptibility to infection. Because plasma concentrations of PalP decrease with age, it is expected that elderly people may be at greater risk of vitamin B$_6$ deficiency than younger people.

Infants consuming less than 100 mg of vitamin B$_6$ per day are at risk of developing seizures, which are thought to result from insufficient activities of the PalP-dependent enzyme glutamate carboxylase required for the synthesis of GABA.

9. VITAMIN B$_6$ TOXICITY

The toxicity of vitamin B$_6$ appears to be relatively low, although high doses of the vitamin (several grams per day) have been shown to induce sensory neuropathy marked by changes in gait and peripheral sensation. The primary target thus appears to be the peripheral nervous system; although massive doses of the vitamin have produced convulsions in rats, central nervous abnormalities have not been reported frequently in humans. The potential for toxicity resulting from therapeutic or pharmacologic doses of the vitamin for human disorders (which rarely exceed 50 mg/day) must be considered small. Reports of individuals taking massive doses of the vitamin (>2 g/day) indicate that the earliest detectable signs were ataxia and loss of small motor control. Many of the signs of vitamin B$_6$ toxicity resemble those of vitamin B$_6$ deficiency; it has

55. For example, omphalocele (protrusion of the omentum or intestine through the umbilicus), exencephaly (defective skull formation with the brain partially outside of the cranial cavity), cleft palate, micrognathia (impaired growth of the jaw), splenic hypoplasia.
56. The lesion is morphologically indistinguishable from that produced by riboflavin deficiency.
57. The lesion is morphologically indistinguishable from that produced by niacin deficiency.
58. For example, 4'-deoxyPn.

59. After tryptophan loading, vitamin B$_6$-deficient subjects also had elevated urinary concentrations of kynurenine, 3-hydroxykynurenine, kynurenic acid, acetylkynurenine, and quinolinic acid.

been proposed that the metabolic basis of each condition involves the tissue-level depletion of PalP. Doses up to 750 mg/day for extended periods of time (years) have been found safe.[60]

In doses of 10–25 mg, vitamin B_6 increases the conversion of L-dopa to dopamine[61] which, unlike its precursor, is unable to cross the blood–brain barrier. The vitamin can thus interfere with L-dopa in the management of Parkinson's disease; it should not be administered to individuals taking L-dopa without the concomitant administration of a decarboxylase inhibitor.

10. CASE STUDIES

Instructions

Review the following case reports, paying special attention to the diagnostic indicators on which the treatments were based. Then, answer the questions that follow.

Case 1

A 16-year-old boy was admitted with dislocated lenses and mental retardation. Four years earlier, an ophthalmologist had found dislocation of the lenses. On the present occasion, he was thin and blond-headed, with ectopia lentis,[62] an anterior thoracic deformity (pectus excavatum[63]), and normal vital signs. His palate was narrow, with crowding of his teeth. He had mild scoliosis[64] and genu valgum,[65] which caused him to walk with a toe-in, Chaplin-like gait. His neurological examination was within normal limits. On radiography, his spine appeared osteoporotic. His performance on the Stanford-Binet Intelligence Scale gave him a development quotient of 60. His hematology, blood glucose, and blood urea nitrogen values were all within normal limits. His plasma homocystine level (undetectable in normal patients) was 4.5 mg/dl, and his blood methionine level was 10-fold normal; the levels of all other amino acids in his blood were within normal limits. Both homocystine and methionine were increased in his urine, which also contained traces of S-adenosylhomocystine.

The patient was given oral Pn HCl in an ascending dose regimen. Doses up to 150 mg/day were without effect, but after the dose had been increased to 325 mg/day for 200 days his plasma and urinary homocystine and methionine levels decreased to normal. These changes were accompanied by a striking change in his hair pigmentation: dark hair grew out from the scalp (the cystine content of the dark hair was nearly double that of the blond hair, 1.5 versus 0.8 mEq/mg). On maintenance doses of Pn he attained relatively normal function, although the connective tissue changes were irreversible.

Case 2

A 27-year-old woman had experienced increasing difficulty in walking. Some 2 years earlier, she had been told that vitamin B_6 prevented premenstrual edema and she began taking 500 mg/day of Pn HCl. After a year, she had increased her intake of the vitamin to 5 g/day. During the period of this increased vitamin B_6 intake, she noticed that flexing her neck produced a tingling sensation down her neck, and to her legs and the soles of her feet.[66] During the 4 months immediately before this examination, she had become progressively unsteady when walking, particularly in the dark. Finally, she had become unable to walk without the assistance of a cane. She had also noticed difficulty in handling small objects, and changes in the feeling of her lips and tongue, although she reported no other positive sensory symptoms and was not aware of any weaknesses. Her gait was broad-based and stamping, and she was not able to walk at all with her eyes closed. Her muscle strength was normal, but all of her limb reflexes were absent. Her sensations of touch, temperature, pin-prick, vibration, and joint position were severely impaired in both the upper and lower limbs. She showed a mild subjective alteration of touch-pressure and pin-prick sensation over her cheeks and lips, but not over her forehead. Laboratory findings showed the spinal fluid and other clinical tests to be normal. Electrophysiologic studies revealed that no sensory nerve action potentials could be elicited in her arms and legs, but that motor nerve conduction was normal.

The patient was suspected of having vitamin B_6 intoxication, and was asked to stop taking that vitamin. Two months after withdrawal, she reported some improvement and a gain in sensation. By 7 months, she could walk steadily without a cane and could stand with her eyes closed. Neurologic examination at that time revealed that, although her strength was normal, her tendon reflexes were absent. Her feet still had severe loss of vibration sensation, despite definite improvements in the senses of joint position, touch, temperature, and pin-prick. Electrophysiologic examination revealed that her sensory nerve responses were still absent.

60. Mpofus, C., Alani, S. M., Whitehouse, C., *et al.* (1991). *Arch. Dis. Child.* 66, 1081.

61. It has been claimed that, via its effect on dopamine, vitamin B_6 can inhibit the release of prolactin, thus inhibiting lactation in nursing mothers. Although this proposal is still highly disputed, there is no evidence that daily doses of less than about 10 mg of the vitamin (in multivitamin preparations) has any such effect on lactation.

62. Dislocated lenses.

63. Funnel chest.

64. Lateral curvature of the spine.

65. Knock-knee.

66. Lhermitte's sign.

Case Questions and Exercises

1. Propose a hypothesis consistent with the findings in Case 1 for the congenital metabolic lesion experienced by that patient.
2. Would you expect supplements of methionine and/or cystine to have been effective in treating the patient in Case 1? Defend your answer.
3. If the toxicity of Pn involves its competition, at high levels, with PalP for enzyme-binding sites, which enzymes would you propose as potentially being affected in the condition described in Case 2? Provide a rationale for each of the candidate enzymes on your list.

Study Questions and Exercises

1. Diagram schematically the several steps in amino acid metabolism in which PalP-dependent enzymes are involved.
2. Construct a decision tree for the diagnosis of vitamin B$_6$ deficiency in humans or an animal species.
3. What key feature of the chemistry of vitamin B$_6$ relates to its biochemical functions as a coenzyme?
4. What parameters might you measure to assess vitamin B$_6$ status of a human or animal?
5. What factors might be expected to affect the dietary need for vitamin B$_6$?

RECOMMENDED READING

Allen, G.F.G., Land, J.M., Heales, S.J.R., 2009. A new perspective on the treatment of aromatic l-amino acid decarboxylase deficiency. Molec. Genet. Metab. 97, 6–14.

Aufiero, E., Stitk, T.P., Foye, P.M., Chen, B., 2004. Pn hydrochloride treatment of carpal tunnel syndrome: A review. Nutr. Rev. 62, 96–104.

Baxter, P., 2003. Pn-dependent seizures: A clinical and biochemical conundrum. Biochim. Biophys. Acta 1647, 36–41.

Bender, D.A., 1999. Non-nutritional uses of vitamin B$_6$. Br. J. Nutr. 81, 7–20.

Dakshinamurti, S., Dakshimanurti, K., 2007. Vitamin B$_6$. In: Zempleni, J., Rucker, R.B., McCormick, D.B., Suttie, J.W. (Eds.), Handbook of Vitamins, fourth ed. CRC Press, New York, NY, pp. 315–359.

Depeint, F., Bruce, W.R., Shangari, N., et al. 2006. Mitochondrial function and toxicity: Role of B vitamins in the one-carbon transfer pathways. Chem. Biol. Interact. 163, 113–132.

Gregory, J.F., 1997. Bioavailability of vitamin B-6. Eur. J. Clin. Nutr. 51, S43–S48.

Gregory, J.F., 1998. Nutritional properties and significance of vitamin glycosides. Ann. Rev. Nutr. 18, 277–296.

Lheureux, P., Penaloza, A., Gris, M., 2005. Pn in clinical toxicology: A review. Eur. J. Emerg. Med. 12, 78–85.

Mackey, A.D., Davis, S.R., Gregory, J.F., 2006. Vitamin B$_6$. In: Shils, M.E., Shike, M., Ross, A.C. (Eds.), Modern Nutrition in Health and Disease, tenth ed. Lippincott, New York, NY, pp. 452–461.

McCormick, D.B., 2006. Vitamin B-6. In: Bowman, B.A., Russell, R.M. (Eds.), Present Knowledge in Nutrition, ninth ed. ILSI, Washington, DC, pp. 269–277.

Oka, T., 2001. Modulation of gene expression by vitamin B$_6$. Nutr. Res. Rev. 14, 257–265.

Sánchez-Moreno, C., Jiménez-Excrig, A., Martín, A., 2009. Stroke: Roles of B vitamins, homocysteine and antioxidants. Nutr. Res. Rev. 22, 49–67.

Schneider, G., Kack, H., Lindquist, Y., 2000. The manifold of vitamin B6 dependent enzymes. Structure 8, R1–R6.

Rall, L.C., Meydani, S.N., 1993. Vitamin B$_6$ and immune competence. Nutr. Rev. 51, 217–225.

Biotin

Anchoring Concepts

1. Biotin is the trivial designation of the compound hexahydro-2-oxo-1H-thieno[3,4-d]imidazole-4-pentanoic acid.
2. Biotin functions metabolically as a coenzyme for carboxylases, to which it is bound by the carbon at position 2 (C-2) of its thiophene ring via an amide bond to the ϵ-amino group of a peptidyllysine residue.
3. Deficiencies of biotin are manifested predominantly as dermatologic lesions.

We started with a bushel of corn, and at the end of the purification process, when the solution was evaporated in a small beaker, nothing could be seen, yet this solution of nothing greatly stimulated growth (of propionic acid bacteria). We now know that the factor was biotin, which is one of the most effective of all vitamins.

H. G. Wood

Learning Objectives

1. To understand the chief natural sources of biotin.
2. To understand the means of absorption and transport of biotin.
3. To understand the biochemical function of biotin as a component of coenzymes of metabolically important carboxylation reactions.
4. To understand the metabolic bases of biotin-responsive disorders, including those related to dietary deprivation of the vitamin and those involving inherited metabolic lesions.

VOCABULARY

Acetyl-CoA carboxylase 1
Acetyl-CoA carboxylase 2
Achromotrichia
Alopecia
Apocarboxylase
Avidin
Biocytin
Biotin-binding proteins
Biotin sulfoxide
Biotinidase
Biotinyl 5′-adenylate
Bisnorbiotin
Egg white injury
Fatty liver and kidney syndrome (FLKS)
Foot pad dermatitis
Glucokinase
Holocarboxylase synthetase (HCS)
3-Hydroxyisovalerate
3-Hydroxyisovaleryl carnitine
Kangaroo gait
β-Methylcrotonoyl-CoA carboxylase
Monocarboxylate transporter (MCT1)
Multiple carboxylase deficiencies
Ornithine transcarbamylase

The Vitamins. DOI: 10.1016/B978-0-12-381980-2.00014-1

Phosphoenolpyruvate carboxykinase (PEPCK)
Propionyl-CoA carboxylase
Pyruvate carboxylase
Sodium-dependent vitamin transporter (SMVT).
Spectacle eye
Streptavidin
Sudden infant death syndrome (SIDS)
Transcarboxylase

1. THE SIGNIFICANCE OF BIOTIN

Biotin was discovered in the search for the nutritional factor that prevents **egg white injury** in experimental animals, and the use of the biotin antagonist in egg white, the biotin-binding protein **avidin**, remains useful in producing biotin deficiency in animal models. Practical cases of biotin deficiency were encountered with the advent of total parenteral nutrition before the vitamin was routinely added to tube-feeding solutions, and biotin-responsive cases of foot pad dermatitis remain a problem in commercial poultry production. Biotin deficiency manifests itself differently in different species, but most often involves dermatologic lesions to some extent. In addition to primary deficiencies of the vitamin, genetic disorders in biotin metabolism have been identified; some of these respond to large doses of biotin. A key feature of biotin metabolism is that it is recycled by proteolytic cleavage from the four biotinyl carboxylases by the enzyme **biotinidase**. This recycling, and the prevalent hindgut microbial synthesis of the vitamin, allows quantitative dietary requirements for biotin to be relatively small. Nevertheless, inborn errors of metabolism have been identified which are associated with impaired absorption of, and increased need for, the vitamin.

2. SOURCES OF BIOTIN

Distribution in Foods

Biotin is widely distributed in foods and feedstuffs, but mostly in very low concentrations (Table 14.1). Only a couple of foods (royal jelly[1] and brewers' yeast) contain biotin in large amounts. Milk, liver, egg (egg yolk), and a few vegetables are the most important natural sources of the vitamin in human nutrition; the oilseed meals, alfalfa meal, and dried yeasts are the most important natural sources of the vitamin for the feeding of non-ruminant animals. The biotin contents of foods and feedstuffs can be highly variable;[2] for the cereal grains at least it is influenced by such factors as plant variety, season, and yield (endosperm-to-pericarp ratio). Biotin is found in human milk at concentrations in the range of 30–70 nmol/l; it occurs almost exclusively as free biotin in the skim fraction.[3] Most foods contain the vitamin as free biotin and as **biocytin** – i.e., biotin covalently bound to protein lysyl residues.

Stability

Biotin is unstable to oxidizing conditions and therefore is destroyed by heat, especially under conditions that support simultaneous lipid peroxidation.[4] For this reason, such processing techniques as canning, heat curing, and solvent extraction can result in substantial losses of biotin. These losses can be reduced by the use of a food-grade antioxidant (e.g., vitamin C, vitamin E, butylated hydroxytoluene [BHT], butylated hydroxyanisole [BHA]).

Bioavailability

Studies of the bioavailability of food biotin to humans have not been conducted. However, biotin bioavailability has been determined experimentally using two types of bioassay: healing of skin lesions in avidin-fed rats; and support of growth and maintenance of pyruvate carboxylase activity in chicks. Such assays have shown that the nutritional availability of biotin can be low and highly variable among different foods and feedstuffs (Table 14.2). In general, less than one-half of the biotin present in feedstuffs is biologically available. Although all of the biotin in corn is available, only 20–30% of that in most other grains, and none in wheat, is available. Biotin in meat products also tends to be very low.

Differences in biotin bioavailability appear to be due to differential susceptibilities to digestion of the various biotin–protein linkages in which the vitamin occurs in foods and feedstuffs. Those linkages involve the formation of covalent bonds between the carboxyl group of the biotin side chain with free amino groups of proteins. Such amide linkages constitute the means by which biotin binds to the enzymes for which it serves as an essential prosthetic group, in which cases they involve the ε-amino group of a peptidyl

---

1. Royal jelly is a substance produced by the labial glands of worker honeybees, and has been found to contain more than 400 μg of biotin per 100 g. Female honeybee larvae that are fed royal jelly develop reproductive ability as queens, whereas those fed a mixture of honey and pollen fail to develop reproductive ability and become workers. Although the active factor in royal jelly has not been identified, it appears to be associated with the lipid fraction of that material (one known component, 10-hydroxy-Δ^2-decenoic acid, has been suggested). It is interesting to speculate about the apparent survival value to the honeybee colony of biotin as a component of this unique food that is necessary for the sexual development of the female.

2. In one study, the biotin contents of multiple samples of corn and meat meal were found to be 56–115 μg/kg ($n = 59$) and 17–323 μg/kg ($n = 62$), respectively.

3. These levels are 20- to 50-fold greater than those found in maternal plasma. Human milk also contains biotinidase, which is presumed to be important in facilitating infant biotin utilization.

4. About 96% of the pure vitamin added to a feed was destroyed within 24 hours after the addition of partially peroxidized linolenic acid.

TABLE 14.1 Biotin Contents of Foods

Food	Biotin, μg/100 g	Food	Biotin, μg/100 g
Dairy Products		Kale	0.5
Milk	2	Lentils	13
Cheeses	3–5	Onions	4
Meats		Peas	9
Beef	3	Potatoes	0.1
Chicken	11	Soybeans	60
Pork	5	Spinach	7
Calf kidney	100	Tomatoes	4
Cereals		**Fruits**	
Barley	14	Apples	1
Cornmeal	7.9	Bananas	4
Oats	24.6	Grapefruit	3
Rye	8.5	Grapes	2
Sorghum	28.8	Oranges	1
Wheat	10.1	Peaches	2
Wheat bran	36	Pears	0.1
Oilseed Meals		Strawberries	1.1
Rapeseed meal	98.4	Peanuts	34
Soybean meal	27	Walnuts	37
Vegetables		Watermelons	4
Asparagus	2	**Other**	
Brussels sprouts	0.4	Eggs	20
Cabbage	2	Brewers' yeast	80
Carrots	3	Alfalfa meal	54
Cauliflower	17	Molasses	108
Corn	6		

From USDA National Nutrient Database for Standard Reference, Release 18.

lysine residue. That form, biotinyl lysine, is referred to as biocytin. The utilization of biotin bound in such forms thus depends on the hydrolytic digestion of the proteins and/ or the hydrolysis of those amide bonds. Biotin from purified preparations, such as are used in dietary supplements, should be well utilized.

Synthesis by Intestinal Microflora

In both rats and humans it has been found that total fecal excretion of biotin exceeds the amount consumed in the diet. This is due to the biosynthesis of significant amounts of biotin by the microflora of the colon. However, the extent to which hindgut microbial biotin can be a determinant of the biotin status of the host is not clear. That biotin can be absorbed across the colon is indicated by the demonstration of transport capacity at levels in the rat as great as 25% that of the jejunum,[5] and the identification of a Na^+-dependent biotin carrier in human-derived colonic epithelial cells. In fact, the dietary requirement of the rat for biotin has been determined *only* under gnotobiotic (germ-free) conditions *or* with avidin feeding. Nevertheless, that intracecal treatment of pigs with antibiotics to inhibit microbial growth or with lactulose to stimulate microbial growth failed to affect

5. Bowman, B. B. and Rosenberg, I. H. (1987). *J. Nutr.* 117, 2121.

TABLE 14.2 Biotin Availability in Several Feedstuffs

Feedstuff	Total Biotin[a] (μg/100 g)	Available Biotin[b] (μg/100 g)	Bioavailability (%)
Barley	10.9	1.2	11
Corn	5.0	6.5	133
Wheat	8.4	0.4	5
Rapeseed meal	93.0	57.4	62
Sunflower seed meal	119.0	41.5	35
Soybean meal	25.8	27.8	108

[a]Determined by microbiological assay.
[b]Determined by chick growth assay.
Source: Whitehead, C. C., Armstrong, J. A., and Waddington, D. (1982). *Br. J. Nutr.* 48, 81.

TABLE 14.3 Inhibition of Enteric Biotin Transport by Ethanol

Ethanol (%, v/v)	Biotin Transport (pmol/g Tissue/15 min)
0	16.89 ± 0.80
0.5	15.16 ± 1.02
1	12.56 ± 1.03[a]
2	11.59 ± 1.16
5	6.61 ± 0.42[a]

[a]$P > 0.05$.
Source: Said, H. M., Sharifian, A., Bagerzadeh, A., and Mock, D. (1990). *Am. J. Clin. Nutr.* 52, 1083.

plasma biotin[6] would suggest that hindgut sources may not always make significant contributions to biotin nutriture.

3. ABSORPTION OF BIOTIN

Liberation from Bound Forms

In the digestion of food proteins, protein-bound biotin is released by the hydrolytic action of the intestinal proteases to yield the e-N[1]-biotinyllysine adduct, biocytin, from which free biotin is liberated by the action of an intestinal biotin amide aminohydrolase, **biotinidase**.

Two Types of Transport

Free biotin is absorbed in the proximal small intestine by two mechanisms, depending on its lumenal concentration:

- *Facilitated transport.* At low concentrations, it is absorbed by a saturable, facilitated mechanism dependent on Na^+ located on the epithelial brush border, particularly villus cells. This appears to be a function of a **sodium-dependent multivitamin transporter** (**SMVT**). Biotin uptake is inhibited by pantothenic acid, lipoic acid, certain anticonvulsant drugs,[7] and ethanol. The inhibitory effect of ethanol or its major metabolite acetaldehyde, has been demonstrated with solutions as dilute as 1% (v/v) (Table 14.3). Similar inhibition has been demonstrated for ethanol against biotin transport in human placental basolateral membrane vesicles, which also occurs by a Na^+-dependent, carrier-mediated process. Biotin uptake is downregulated by biotin deprivation through a transcriptionally mediated mechanism involving the biotinylation of histone 4 (H4) at an SMVT promoter to silence SMVT expression. It can also be inhibited by activation of protein kinase C, which apparently phosphorylates the SMVT. Suboptimal SMVT expression is thought to underly the low biotin absorption observed in alcoholics, pregnant women, and patients with inflammatory bowel disease, seborrheic dermatitis, or on anti-convulsants or long-term parenteral nutrition. Biotin uptake is adaptively downregulated by biotin deprivation, which limits H4 biotinylation, thus de-repressing SMVT expression.

- *Passive diffusion.* At high lumenal concentrations, free biotin is also absorbed by non-saturable, simple diffusion.

4. TRANSPORT OF BIOTIN

Unbound Biotin

Biotin is transported across the basolateral side of the enterocyte by a Na^+-dependent process. Less than half of the total biotin present in plasma appears to be free biotin, the balance being composed of **bisnorbiotin**, **biotin sulfoxide**, and other biotin metabolites yet to be identified. Most of this total is not protein bound; only 12% of the total biotin in human plasma is covalently bound.

Protein-Bound Biotin

Some 7% of plasma biotin is reversibly and non-specifically bound to plasma proteins including albumin and α- and β-globulins. Two biotin-binding proteins have been identified in human plasma: biotinidase, which has two high-affinity binding sites for the vitamin; and an immunoglobulin, which has been found to be elevated in patients with Graves disease.[8] Biotinidase also occurs in human milk (and at particularly high levels in colostrum), where it is

---

6. Kopinski, J. S., Leibholz, J., and Love, R. J. (1989). *Br. J. Nutr.* 62, 781.

7. Carbamazepine, primidone.

8. Graves disease is an autoimmune disorder involving thyroid over-activity.

thought to function in the transport of biotin by the mammary gland to the nursing infant.

Cellular Uptake

Biotin is taken up by cells by multiple mechanisms, two of which have been characterized:

- *Na-dependent multivitamin transporter.* The SMVT mediates biotin uptake and transport through the enterocyte, the apical membrane being targeted by a C-terminal tail of the polypeptide. The intracellular trafficking of SMVT involves distinct trafficking vesicles, the microtubular network, and the microtubule motor protein dynein.[9] The process is not specific for the vitamin, as SMVT also functions in the cellular uptake of pantothenic acid and lipoic acid, which it binds with similar affinities.
- *Monocarboxylate transporter (MCT1).* That this member of the monocarboxylate tansporter family can facilitate the cellular uptake of biotin into peripheral blood mononuclear cells explained the fact that biotin is taken up by those cells by a process with a K_m three orders of magnitude less than that for SMVT-mediated transport, and is not competitively inhibited by either pantothenic or lipoic acids. It remains to be seen whether MCT1 is expressed in other tissues, and whether other members of this family of transporters may be involved in biotin uptake.

The routing of biotin within cells would appear to be related to the activity of transporters, as well as binding to proteins and incorporation into biotin-dependent carboxylases. In fact, the intracellular distribution of biotin closely parallels that of its carboxylases: mostly in the cytoplasm (the primary location of acetylCoA carboxylase) and mitochondria (in which MCT1 has been detected). A small amount (<1%) is found in the nucleus;[10] because this fraction has been found to increase (to *ca.* 1% of total cellular biotin) in response to proliferation, it would appear to reflect binding of the vitamin to histones.

Tissue Distribution

Appreciable storage of the vitamin appears to occur in the liver, where concentrations of 800–3,000 ng/g have been found in various species.[11] Most of this appears to be in mitochondrial acetyl CoA carboxylase. Hepatic stores, however, appear to be poorly mobilized during biotin deprivation, and thus do not show the reductions measurable in plasma under such conditions.

Biotin is transported to the fetus by specific carriers, including SMVT. Concentrations of the vitamin in fetal plasma are 3- to 17-fold greater than maternal levels. That milk biotin levels exceed those of maternal plasma by 10- to 100-fold evidences a mammary transport system. **Biotin-binding proteins** have been identified in the egg yolks of many species of birds, where they are believed to function in transporting the vitamin into the oocyte as their binding is weak enough to be reversible.[12] The yolk biotin-binding protein also occurs in the plasma of the laying hen.

5. METABOLISM OF BIOTIN

Linkage to Apoenzymes

Free biotin is attached to its apoenzymes via the formation of an amide linkage to the ε-amino group of a specific lysyl residue. In each of the biotin-dependent enzymes, this binding occurs in a region containing the same amino acid sequence: -Ala-Met-biotinyl-Lys-Met-. It is catalyzed by **biotin holocarboxylase synthetase**.[13]

Recycling the Vitamin

The normal turnover of the biotin-containing holocarboxylases involves their degradation to yield **biocytin**. The biotinyl–lysine bond is not hydrolyzed by cellular proteases; instead, it is cleaved by biotinidase to yield free biotin. Biotinidase is the major biotin-binding protein in plasma; it is also present in breast milk, in which its activity is particularly high in colostrum. The proteolytic liberation of biotin from its bound forms is essential for the reutilization of the vitamin, which is accomplished by its reincorporation into another holoenzyme. It is thought that biotinidase also catalyzes the de-biotinylation of histones.

A storage system for biotinyl enzymes has been suggested. This would involve the mitochondrial biotinyl acetyl-CoA carboxylase serving as a reservoir to maintain hepatic acetyl-CoA at appropriate levels in the cytosol. This would also provide biotin indirectly to support other biotinyl mitochondrial enzymes under low-biotin conditions.

9. Subramanian, V. S., Marchant, J. S., and Said, H. M. (2007). *Gastroenterology* 132, A583.
10. Stanley, J. S., Griffen, J. B., and Zempleni, J. (2001). *Eur. J. Biochem.* 268, 5424.
11. These levels contrast with those of plasma/serum which, in humans and rats, are typically *ca.* 300 ng/liter.
12. The yolk biotin-binding protein is a glycoprotein with a molecular mass of 74.3 kDa and a homologous tetrameric structure, each subunit of which binds a biotin molecule. This protein is not to be confused with avidin, the biotin-binding protein of egg white, which irreversibly binds biotin with an affinity three orders of magnitude greater.
13. This activity is sometime referred to as "biotin holoenzyme synthetase" and, in microorganisms, as "biotin protein ligase."

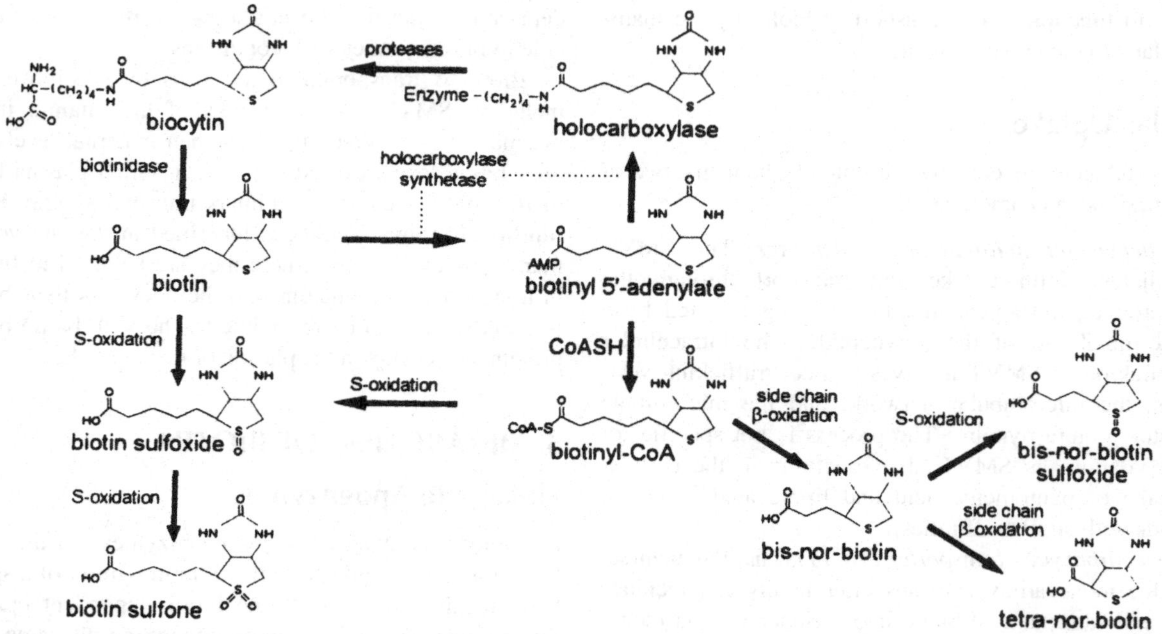

FIGURE 14.1   Biotin metabolism and recycling.

Catabolism

Little catabolism of biotin seems to occur in mammals. A small fraction is oxidized to biotin D- and L-sulfoxides, but the ureido ring system is not otherwise degraded. The side chain of a larger portion is metabolized via mitochondrial β-oxidation to yield bisnorbiotin and its degradation products. Biotin catabolism appears to be greater in smokers than non-smokers.[14]

Excretion

Biotin is rapidly excreted in the urine (Fig. 14.1). Studies have shown the rat to excrete about 95% of a single oral dose (5 mg/kg) of the vitamin within 24 hours. Half of urinary biotin occurs as free biotin, the balance being composed of bisnorbiotin, bisnorbiotin methyl ketone, biotin sulfone, tetranorbiotin-*L*-sulfoxide, and various side chain products.[15] Although unabsorbed biotin appears in the feces, much fecal biotin is of gut microbial origin and benefits the host by way of hindgut absorption. Thus, at low dietary levels of the vitamin, urinary excretion of biotin can exceed intake. Only a small amount (<2% of an intravenous dose) of biotin is excreted in the bile.

---

14. Sealey, W. M., Teague, A. Q. M., Stratton, S. L., *et al.* (2004). *Am. J. Clin. Nutr.* 80, 932.

15. The urinary excretion of patients with achlorhydria is very low. It is thought that this reflects impaired release of bound biotin for absorption.

TABLE 14.4 The Biotin-Dependent Carboxylases of Animals

Enzyme	Location	Metabolic Function
Pyruvate carboxylase	Mitochondria	Formation of oxaloacetate from pyruvate; requires acetyl-CoA
Acetyl-CoA carboxylase 1	Cytosol	Formation of malonyl-CoA from acetyl-CoA for carboxylase fatty acid synthesis; requires citrate
Acetyl-CoA carboxylase 2	Mitochondrial	Formation of malonyl-CoA from acetyl-CoA for carboxylase fatty acid synthesis; requires citrate
Propionyl-CoA carboxylase	Mitochondria	Formation of methylmalonyl-CoA from propionyl-CoA produced by catabolism of some amino acids (e.g., isoleucine) and odd-chain fatty acids
β-methylcrotonyl-CoA carboxylase	Mitochondria	Part of the leucine degradation pathway

6. METABOLIC FUNCTIONS OF BIOTIN

Biotin functions in carboxylations with key roles in the metabolism of lipids, glucose, some amino acids, and energy (Table 14.4). It is also involved in the regulation of gene expression.

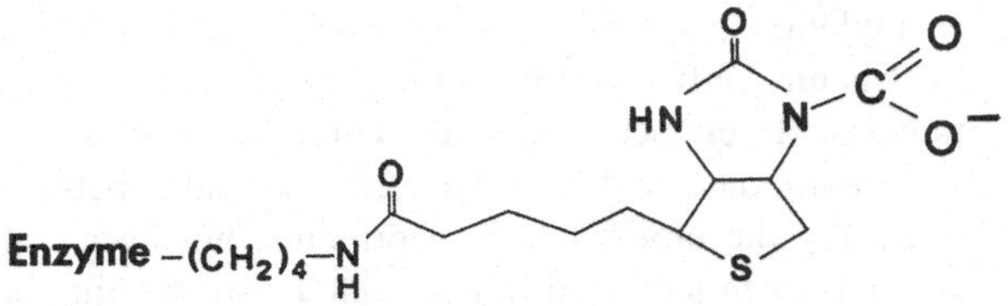

FIGURE 14.2   Biotin is bound covalently to its carboxylases.

Carboxylations

Biotin functions in fatty acid metabolism, gluconeogenesis, and amino acid metabolism in key steps involving the transfer of covalently bound, single-carbon units in the most oxidized form, CO_2. This function is implemented by five biotin-dependent carboxylases and transcarboxylases[16,17] to which the biotin prosthetic group is attached by **biotin holocarboxylase synthetase**. All but one of these carboxylases is mitochondrial; acetyl CoA carboxylase 2 is cytosolic. Biotin has additional metabolic functions as a regulator of gene expression, as a substrate for the modification of proteins by biotinylation, and as a cell cycle regulator.

Biotin Holocarboxylase Synthetase (HCS)

This enzyme catalyzes the formation of the linkage of the biotin prosthetic group to each apoenzyme covalently to the ε-amino group of a lysyl residue (Fig. 14.2). The process is driven thermodynamically by the hydrolysis of ATP. It occurs in two steps:

1. Activation of the vitamin as **biotinyl 5′-adenylate**.
2. Covalent attachment to the **apocarboxylase** by an amide bond with the lysyl residue, with the release of AMP.

A high percentage of cellular HCS is present in the nucleus, where it is thought to function in the biotinylation of histones (see below). Its expression is dependent on adequate biotin status.

The Biotin-Dependent Carboxylases

The catalytic action of each biotin-dependent carboxylase proceeds by a non-classic, two-site, ping-pong mechanism, with partial reactions being performed by dissimilar subunits:

1. The first reaction (carboxylation) occurs at the carboxylase subsite; it involves the addition of the carboxyl moiety to the biotin ureido-N opposite the side chain, using the bicarbonate/ATP system as the carboxyl donor.

2. The subsequent step (carboxyl transfer) occurs at the carboxyl transferase subsite; it involves transfer of the carboxyl group from carboxybiotin to the acceptor substrate.

These two subsites appear to be spatially separated in each enzyme. The physicochemical mechanisms proposed for these reactions entail the transfer of biotinyl CO_2 between subsites via movement of the prosthetic group back and forth by virtue of rotation of at least one of the 10 single bonds (probably C-2–C-6) on the valeryl lysyl side chain.

- **Pyruvate carboxylase** catalyzes the incorporation of bicarbonate into pyruvate to form oxaloacetate, a key step in gluconeogenesis. It also serves to replenish the mitochondrial supply of oxaloacetate to support the tricarboxylic acid (TCA) cycle and the formation of citrate for transport to the cytosol for lipogenesis. Dysfunction caused by biotin deprivation can lead to fasting hypoglycemia, lactic acidosis, and ketosis.
- **Acetyl-CoA carboxylase 1** catalyzes the incorporation of bicarbonate into acetyl CoA to yield malonyl CoA in the cytosol. This is the first committed step in the synthesis and elongation of fatty acids; disruption due to biotin deprivation impairs lipid synthesis.
- **Acetyl-CoA carboxylase 2** controls mitochondrial fatty acid oxidation by way of the inhibitory effects of malonyl CoA, which it produces.
- **Propionyl-CoA carboxylase** catalyzes the incorporation of bicarbonate into propionyl CoA to form methylmalonyl CoA, which isomerizes to succinyl CoA and enters the TCA cycle for energy and glucose production.
- **3-Methylcrotonoyl-CoA carboxylase** catalyzes the degradation of the ketogenic branched-chain amino acid, leucine. Disruption by biotin deprivation results in the shunting of the leucine degradation product 3-methylcrotonyl CoA through an alternate catabolic pathway to **3-hydroxyisovaleric acid**, which is excreted in the urine.[18]

Biotin Sensing

Biotin appears to sense and regulate its own levels within cells. This involves regulation of its cellular uptake through a novel function of HCS, which moves into the nucleus when biotin is available and silences the biotin transporter SMVT through biotinylation of H4 at its promoter. Gene silencing results from changes in chromatin structure that

---

16. **Transcarboxylase** has been called the "Mickey Mouse enzyme," as the electron micrograph image of the bacterial enzyme, with its large single subunit and two smaller flanking subunits, resembles the head of the famous rodent.
17. Biotin also functions in a number of decarboxylases in microorganisms.

---

18. Urinary 3-hydroxyisovalerate has, therefore, been proposed as a screening parameter for the detection of biotin deficiency.

affect specific loci.[19] HCS translocation is thought to be regulated by tyrosine kinases and zinc-finger proteins that direct the protein to specific chromatin regions.

Gene Expression

Biotin participates in the regulation of gene expression. This appears to be mediated in several ways:

- *Translocation of NFκB*. Biotin deprivation has been shown to increase the nuclear translocation, binding, and transcriptional activity of NFκB, and increase the nuclear contents of p50 and p65 and activities of IκB kinases and activating genes involved in suppression of apoptosis.[20]
- *cGMP signaling*. Biotin has been shown to increase intracellular production of the key second messenger nitrous oxide (NO), which activates a soluble guanylate cyclase to increase cellular levels of cGMP. Those increases stimulate protein kinase G, and lead to phosphorylation and activation of proteins that increase the transcription of genes encoding HCS and some carboxylases.[21] It also suppresses endo/sarcoplasmic reticular ATPase to increase cytosolic concentrations of Ca^{2+}, which stimulates protein unfolding.[22]
- *Sp1 and Sp3*. Biotin appears necessary for the expression of these transcription factors associated with the expression of a cytochrome *P*-450 gene and decreased transcription of the sarco/endoplasmic reticulum ATPase 3 gene.[23]
- *jun/fos signaling*. Biotin has been shown to activate jun/fos signaling, which in turn activates AP1-dependent pathways.[24]
- *Receptor tyrosine kinases*. Biotin deficiency has been shown to activate signaling by tyrosine kinases, which may contribute to increasing SMVT-mediated biotin uptake.
- *Histone biotinylation*. The biotinylation of histones by HCS has been shown to suppress the expression of several genes by epigenetic means. Three histones (H2A, H3, and H4) can be covalently modified by biotinylation at specific lysyl residues to affect chromatin structure, influencing stability of repeat regions and transposable elements, and regulation of expression.[25] Both biotinidase and HCS have been found capable of catalyzing the biotinylation of proteins; however, only HCS has been localized in the nucleus, suggesting that it is the physiologically relevant factor in histone biotinylation. While de-biotinylation of histones occurs, the mechanism is not clear. It is possible that this process is catalyzed by an isoform of biotinidase.

Only a small fraction (<0.1%) of histones appear to be biotinylated in humans.[26] Nevertheless, the epigenetic impact on the mammalian genome is significant, as a third of H4 molecules have been found to be biotinylated at lysyl residue 12 in telomeric repeats,[27] and enrichments in H3 (at lysyl residues 9 or 18) and H4 (at lysyl residue 8) have also been found.

Biotin plays roles in the expression of genes encoding enzymes involved in glucose metabolism (glucokinase, phosphoenoylpyruvated carboxykinase, ornithine transcarbamylase), cytokines (interleukin-2, interleukin-2 receptor γ; interleukin-1β, interferon-γ, interleukin-4[28]), amino acid metabolism (β-methylcrotonyl CoA carboxylase), and regulation of biotin status (SMVT,[23] holocarboxylase synthetase).

Cell Cycle

Biotin has been shown to be necessary for the normal progression of cells through the cell cycle, with biotin-deficient cells arresting in the G1 phase. Proliferating lymphocytes increase their uptake of biotin, as well as their activities of β-methylcrotonyl-CoA carboxylase and propionyl-CoA carboxylase.

7. BIOTIN IN HEALTH AND DISEASE

Birth Defects

It has been suggested that marginal biotin status may be teratogenic. Fetal malformations have been produced in mice (Table 14.5) and poultry by feeding maternal diets containing marginal biotin levels – i.e., amounts that did not

19. Gralla, M., Campreale, G., and Zempleni, J. (2008). *J. Nutr. Biochem.* 19, 400; Bao. B., Pestinger, V., Hassan, Y. I., *et al.* (2011). *J. Nutr. Biochem.* 22, 470.

20. Rodriquez-Melendez, R., Schwab, L. D., and Zempleni, J. (2004). *Intl J. Vitaminol. Nutr. Res.* 74, 209.

21. Solorzano-Vargas, R. S., Pacheco-Alvarez, D., and Leon-Del-Rio, A. (2002). *Proc. Natl Acad. Sci. USA* 99, 5325.

22. Griffen, J. B., Rodriguez-Melendez, R., and Dode, L. (2006). *J. Nutr. Biochem.* 17, 272.

23. Griffen, J. B., Rodriguez-Melendez, R., and Dode, L. (2006). *J. Nutr. Biochem.* 17, 272.

24. Rodriguez-Melendez, R., Griffen, J. B., and Zempeni, J. (2006). *J. Nutr.* 135, 1659.

25. Reduction of histone biotinylation by biotin-deprivation reduced lifespan and heat tolerance in *Drosophila melanogaster* (Camporeale, G., Giordano, E., Rendina, R., *et al.* [2006] *J. Nutr.* 136, 2735).

26. Bailey, L. M., Ivanov, R. A., Wallace, J. C., *et al.* (2008). *Anal. Biochem.* 373, 71.

27. Wijeratne, S. S., Camporeale, G., and Zempleni, J. (2010). *J. Nutr. Biochem.* 21, 310.

28. Unlike the other proteins listed, which depend on biotin for expression, IL-4 and SMVT are downregulated by biotin.

TABLE 14.5 Effect of Egg White Injury on Fetal Malformations in Pregnant Mice

Malformation	Dietary Egg White, %						
	0	1	2	3	5	10	25
Cleft palate	0.10 ± 0.13	0.25 ± 0.25	2 ± 2	4 ± 2	10 ± 1	11 ± 1	12 ± 0.4
Micrognathia	0	0	0.2 ± 0.2	3 ± 2	9 ± 1	11 ± 1	12 ± 1
Microglossia	0	0	0	0.5 ± 0.5	2 ± 1	6 ± 2	9 ± 3
Hydrocephaly	0	0	0	0.3 ± 0.2	2 ± 1	3 ± 1	3 ± 2
Open eye	0	0	0	0	0.8 ± 0.5	4 ± 1	5 ± 2
Forelimb hypoplasia	0	0	0	7 ± 2	9 ± 2	11 ± 1	12 ± 0.4
Hindlimb hypoplasia	0	0	0	5 ± 2	9 ± 2	10 ± 0.8	12 ± 0.4
Pelvic girdle hypoplasia	0	0	0	5 ± 2	9 ± 2	10 ± 1	12 ± 0.5

From Mock, D. M., Mock, N. I., Stewart, C. W., et al. (2003). J. Nutr. 133, 2519.

produce clinical signs of deficiency in the dams. Because humans appear to have relatively poor transport of biotin across the placenta,[29] it has been suggested that human fetuses may be predisposed to biotin deficiency when maternal intakes of the vitamin are marginal. Support for this hypothesis comes from observations that the production of arachidonic acid and prostaglandins, which depend on acetyl-CoA carboxylase and propionyl-CoA carboxylase activities, is required for normal palatal plate growth, elevation, and fusion in mice, and skeletal development in chicks. That marginal biotin status may be prevalent was suggested by the finding that apparently healthy pregnant women had increased rates of biotin excretion, and abnormally low activities of propionyl-CoA carboxylase.[30]

Sudden Infant Death Syndrome (SIDS)

It has been suggested that marginal biotin status may play a role in the etiology of SIDS, which occurs in human infants at 2–4 months of age. In many ways SIDS resembles fatty liver and kidney syndrome (FLKS) in the chick, which is caused by biotin deprivation. Studies have shown that infants who died of SIDS had significantly lower hepatic concentrations of biotin than did those who died of unrelated causes.[31]

Disorders of Biotin Metabolism

Genetic defects in all of the known biotin enzymes have been identified in humans (Table 14.6). These are rare,

affecting infants and children, and usually have serious consequences. Some involve mutations in genes encoding the three proteins involved in biotin homeostasis: biotinidase, HCS, and SMVT. Defects in these cause **multiple carboxylase deficiencies**.

Biotinidase deficiency occurs at a frequency of 1 in 60,000 live births, with affected individuals having <30% of normal serum biotinidase activity, compromising their ability to release biotin from dietary proteins and recycle it from endogenous biotinylated proteins. They show the neurological and dermatological symptoms of biotin deficiency, with onset in the first year of life (sometimes within weeks). They also experience additional signs, including hearing loss and optic atrophy, which has been interpreted as suggestive of other, still-unidentified functions of biotinidase. This can lead to irreversible neurologic damage, but can be prevented by life-long high doses (5–20 mg/day) of biotin.

Defects involving the absence of an apocarboxylase do not respond to supplements of the vitamin. They are treated by restricting dietary protein to limit the production of metabolites upstream of the metabolic lesion.

Egg White Injury

When, in the mid-1930s, it was found that biotin supplements prevented the dermatitis and alopecia produced in experimental animals by feeding uncooked egg white, the damaging factor was isolated and named *avidin*. Avidin is a water-soluble, basic glycoprotein with a molecular mass of 67 kDa. It is a homologous tetramer, each 128-amino acid subunit of which binds a molecule of biotin, apparently by linking to two to four tryptophan residues and an adjacent lysine in the subunit binding site. The binding of biotin to avidin is the strongest known non-covalent bond in

29. Schenker, S., Hu, Z., Johnson, R. F., et al. (1993). Alcohol Clin. Exp. Res. 17, 566.
30. Mock, D. M. and Stadler, D. D. (1997). J. Am. Coll. Nutr. 16, 252.
31. Johnson, A. R., Hood, R. L., and Emergy, J. L. (1980). Nature 285, 159.

TABLE 14.6 Congenital Disorders of Biotin Enzymes

Defect	Metabolic Basis	Physiological Effect	Treatment
Propionyl CoA carboxylase deficiency	Autosomal recessive lack of enzyme[a]	Propionate accumulation: acidemia, ketoacidosis, hyperammonemia; high urine citrate, 3-OH-propionate, propionyl glycine *Symptoms*[b]: vomiting, lethargy, hypotonia, mental retardation, cramps	Restrict protein
Pyruvate carboxylase deficiency	Autosomal recessive lack of enzyme[c]	Changes in energy production, gluconeogenesis, and other pathways *Symptoms*: metabolic acidosis (lactate), hypotonia, mental retardation	None
3-Methylcrotonyl-CoA carboxylase deficiency	Defective enzyme (basis unknown[d])	High urine 3-CH_3-crotonylglycine and 3-OH-isovaleric acid *Symptoms*: cramps	Restrict protein
Acetyl-CoA carboxylase deficiency	Lack of enzyme (basis unknown[e])	Aciduria *Symptoms*: myopathy, neurologic changes	None
Multiple Carboxylase Deficiency			
Neonatal type	Autosomal recessive lack of holocarboxylase synthase[f]	Deficiencies of all biotin-containing holocarboxylases; acidosis and aciduria *Symptoms*: vomiting, lethargy, hypotonia	None[g]
Juvenile type	Autosomal recessive lack of biotinidase[h]	Deficiencies of all biotin-containing holocarboxylases; acidosis and aciduria *Symptoms*: skin rash, alopecia, conjunctivitis, ataxia, developmental anomalies, neurological signs	Massive doses of biotin

[a]*Incidence: 1 in 350,000.*
[b]*There is a wide variation in the clinical expression.*
[c]*Fewer than two dozen patients described.*
[d]*Three confirmed cases.*
[e]*One case described.*
[f]*Involves failure to link biotin to the apocarboxylases.*
[g]*Fatal early in life.*
[h]*Involves failure to release biotin from its bound forms in holocarboxylases; this reduces use of biotin in foods and blocks endogenous recycling of the vitamin.*

nature.[32] Avidin is secreted by the oviductal cells of birds, reptiles, and amphibians, and thus is found in the whites of their eggs, in which it is thought to function as a natural antibiotic as it is resistant to a broad range of bacterial proteases. It antagonizes biotin by forming with the vitamin a non-covalent complex[33] that is also resistant to pancreatic proteases, thus preventing the absorption of biotin.[34] The avidin–biotin complex is unstable to heat; heating to at least

100°C denatures the protein and releases biotin available for absorption. Therefore, although raw egg white is antagonistic to the utilization of biotin, the cooked product is without effect. The consumption of raw or undercooked whole eggs is probably of little consequence to biotin nutrition, as the biotin-binding capacity of avidin in the egg white is roughly comparable to the biotin content of the egg yolk. However, as a tool to produce experimental biotin deficiency, avidin in the form of dried egg white has been very useful.[35]

Deficiency Syndromes in Animals

Avidin-induced biotin deficiency causes the syndrome originally referred to as **egg white injury**. The major

32. $K_a = 10^{15}$ mol/l.
33. Two very similar biotin-binding proteins have been identified, both of which show considerable sequence homology with avidin at the biotin-binding site. One, from *Streptomyces avidinii*, is called **streptavidin**. The other is an epidermal growth factor homolog found in the purple sea urchin *Strongylocentrotus purpuratus*.
34. It is of interest to note that some cultured mammalian cells (e.g., fibroblasts and HeLa cells) are able to absorb the biotin–avidin complex, using it as a source of the vitamin.

35. Other structural analogs of biotin are also antagonistic to its function: α-dehydrobiotin, 5-(2-thienyl)valeric acid, acidomycin, α-methylbiotin, and α-methyldethiobiotin, several of which are antibiotics.

lesions appear to involve impairments in lipid metabolism and energy production. In rats and mice, this is characterized by seborrheic dermatitis and **alopecia**, and a hind-limb paralysis that results in **kangaroo gait**. In mice and hamsters, it involves congenital malformations (cleft palate, micrognathia,[36] micromelia[37]). Fur-bearing animals (mink and fox) show general dermatitis with hyperkeratosis, circumocular alopecia (**spectacle eye**), **achromotrichia** of the underfur, and unsteady gait. Pigs and kittens show weight loss, digestive dysfunction, dermatitis, alopecia, and brittle claws. Guinea pigs and rabbits show weight loss, alopecia, and achromotrichia. Monkeys show severe dermatitis of the face, hands, and feet, alopecia, and watery eyes with encrusted lids. The dermatologic lesions of biotin deficiency relate to impairments of lipid metabolism; affected animals show reductions in skin levels of several long-chain fatty acids (16:0,[38] 16:1, 18:0, 18:1, and 18:2) with concomitant increases in certain others (in particular, 24:1 and 26:1). All species show depressed activities of the biotin-dependent carboxylases, which respond rapidly to biotin therapy.

Biotin deficiency can be produced in chicks by dietary deprivation, and seems to occur from time to time in practical poultry production, particularly in northern Europe.[39] This results in impaired growth and reduced efficiency of feed utilization, and is characterized by dermatitis mainly at the corners of the beak, but also of the foot pad.[40] In some instances, death occurs suddenly without gross lesions; this condition usually involves hepatic and renal steatosis with hypoglycemia, lethargy, paralysis, and hepatomegaly, and is thus referred to as **fatty liver and kidney syndrome (FLKS)**. The etiology of FLKS appears to be complex, involving such other factors as choline, but seems to involve a marginal deficiency of biotin that impairs gluconeogenesis by limiting the activity of pyruvate carboxylase, especially under circumstances of glycogen depletion brought on by stress.

Deficiency Signs in Humans

Few cases of biotin deficiency have been reported in humans. Most have involved nursing infants whose mother's milk contained inadequate supplies of the

TABLE 14.7 Effect of Biotin Treatment on Abnormalities in Serum Fatty Acid Concentrations in a Biotin-Deficient Human

Fatty Acid	Normal Values	Biotin-Deficient Patient Values	
		Before Biotin	After Biotin
18:2ω6	21.56 ± 6.65	9.85[a]	5.36[a]
18:3ω6	0.21 ± 0.27	0.45	0.40
20:3ω6	3.67 ± 1.39	8.66[a]	10.62[a]
20:4ω6	12.49 ± 3.79	9.26	11.72
22:4ω6	1.87 ± 1.01	0.52[a]	0.71[a]
20:3ω9	1.30 ± 1.25	1.05	1.67
18:3ω3	0.21 ± 0.19	0.33	0.18
Total ω6 acids	41.08 ± 5.86	29.42[a]	29.61[a]
Total ω3 acids	5.23 ± 2.16	5.24	4.97
Total ω9 acids	13.14 ± 3.98	17.59[a]	16.4

[a]$P > 0.05$.
Source: Mock, D., Henrich-Shell, C. L., Carnell, N., et al. (1988). J. Nutr. 118, 342.

vitamin[41] or patients receiving incomplete parenteral nutrition. One case involved a child fed raw eggs for 6 years. The signs and symptoms included dermatitis, glossitis, anorexia, nausea, depression, hepatic steatosis, and hypercholesterolemia. The impairments of lipid metabolism respond to biotin therapy (Table 14.7).

8. BIOTIN DEFICIENCY

Because biotin is rather widespread among foods and feedstuffs, and is synthesized by the intestinal microflora,[42] simple deficiencies of biotin in animals or humans are rare (Table 14.8). However, biotin deficiency can be induced by certain antagonists.

Subclinical Deficiency

The frequency of marginal biotin status (deficiency without clinical manifestation) is not known, but the incidence

36. Underdevelopment of the jaw (usually the lower jaw).
37. Undergrowth of the limbs.
38. That is, a 16-carbon fatty acid with no double bonds.
39. In that part of the world, barley and wheat, each of which has little biologically available biotin, are frequently used as major ingredients in poultry diets.
40. **Foot pad dermatitis** caused by biotin deficiency is often confused with the dermatologic lesions of the foot caused by pantothenic acid deficiency. Unlike the latter, biotin deficiency lesions are limited to the foot pad, and do not involve the toes and superior aspect of the foot.

41. The biotin content of human milk, particularly early in lactation, is often insufficient to meet the demands of infants. Therefore, it is recommended that nursing mothers take a biotin supplement. When this is done, substantial increases in the biotin concentrations of breast milk are observed (e.g., a 3-mg/day supplement increases milk biotin concentrations from 1.2–1.5 μg/dl to >33 μg/dl).
42. Although it is clear that the gut microflora synthesize biotin, the quantitative importance of this source to the biotin nutrition of humans and other non-ruminant species is highly speculative.

TABLE 14.8 General Signs of Biotin Deficiency

Organ System	Change/Signs
General:	
Appetite	Decrease
Growth	Decrease
Dermatologic	Dermatitis
	Alopecia
	Achromotrichia
Skeletal	Perosis
Vital organs	Hepatic steatosis
	FLKS*

Fatty liver and kidney syndrome (poultry).

of low circulating biotin levels has been found to be substantially greater among alcoholics than the general population.[43] Relatively low levels of biotin (vs. healthy controls) have also been reported in the plasma or urine of patients with partial gastrectomy or other causes of achlorhydria, burn patients, epileptics,[44] elderly individuals, and athletes. Because animal products figure prominently as dietary sources of biotin, it has been suggested that vegetarians may be at risk for deficiency. Studies have failed to support that hypothesis; in fact, both plasma and urinary biotin levels of strict vegetarians (vegans) and lacto-ovovegetarians[45] have been found to exceed those of persons eating mixed diets, indicating that the biotin status of the former groups was not impaired relative to the latter group.

Because biotin is required in several aspects of intermediary metabolism, deficient individuals show abnormal metabolic profiles. This includes accumulation of propionyl-CoA, which can react with oxaloacetate to form methylcitrate, and the accumulation of β-methylcrotonyl-CoA, which can deplete mitochrondria of glycine. One of the first indicators of biotin deficiency in humans is an increase in circulating concentrations of **3-hydroxyisovaleryl carnitine**, which changes as a result of the alternative metabolism of β-methylcrotonyl-CoA by enoyl-CoA hydratase with declining biotin-dependent β-methylcrotonyl-CoA carboxylase activity.[46]

Studies with validated biomarkers of biotin status indicate that subclinical biotin deficiency may be common in pregnancy – perhaps as frequent as one-third of pregnancies. The increased urinary excretion of 3-hydroxyisovaleric acid, which can occur late in pregnancy, has been found to respond to biotin supplementation.[47] This finding, and the detection of increases in the urinary excretion of bisnorbiotin, biotin sulfoxide, and other biotin metabolites, suggests that pregnant women may experience marginal biotin deficiency due to increased catabolism of the vitamin.

9. BIOTIN TOXICITY

The toxicity of biotin appears to be very low. No cases have been reported of adverse reactions by humans to high levels (doses as high as 200 mg orally or 20 mg intravenously) of the vitamin, as are used in treating seborrheic dermatitis in infants, egg white injury or inborn errors of metabolism. Animal studies have revealed few, if any, indications of toxicity, and it is probable that animals, including humans, can tolerate the vitamin at doses at least an order of magnitude greater than their respective nutritional requirements. Biotin excess appears to provide effective therapy to reduce the diabetic state, lowering postprandial glucose and improving glucose tolerance.

10. CASE STUDY

Instructions

Review the following case report, paying special attention to the diagnostic indicators on which the treatments were based. Then, answer the questions that follow.

Case

A 12-month-old girl had experienced malrotation[48] and midgut volvulus,[49] resulting in extensive infarction[50] of the small and large bowel, at 4 months of age. Her bowel was resected, after which her clinical course was complicated by failure of the anastomosis[51] to heal, peritoneal infection, and intestinal obstruction. After several subsequent surgeries, she was left with only 30 cm of jejunum, 0.5 cm

43. About 15% of alcoholics have plasma biotin concentrations <140 pmol/l, whereas only 1% of randomly selected hospital patients have plasma biotin levels that low.

44. This may be due to anticonvulsant drug therapy, known side effects of which are dermatitis and ataxia. Some anticonvulsants (e.g., carbamazepine, primidone) have been shown to be competitive inhibitors of biotin transport across the intestinal brush border.

45. Individuals eating plant-based diets that include dairy products and eggs.

46. Stratton, S. L., Horvath, T.D., Bogusiewicz, A., *et al.* (2010). *Am. J. Clin. Nutr.* 92, 1399.

47. Mock, D. M., Stadler, D. D., Stratton, S. L., *et al.* (1997). *J. Nutr.* 127, 710.

48. Failure of normal rotation of the intestinal tract.

49. Twisting of the intestine, causing obstruction.

50. Necrotic changes resulting from obstruction of an end artery.

51. An operative union of two hollow or tubular structures, in this case the divided ends of the intestine.

of ileum, and approximately 50% of colon. By 5 months of age she had lost 1.5 kg in weight, and total parenteral nutrition[52] (TPN) was initiated (providing 125 kcal/kg/day). By the third month of TPN, she had gained 2.9 kg; thereafter, her energy intake was reduced to 60 kcal/kg per day, which sustained her growth within the normal range. Soybean oil emulsion[53] was administered parenterally at least twice weekly in amounts that provided 3.9% of total calories as linoleic acid. Repeated attempts at feeding her orally failed because of vomiting and rapid intestinal transit; therefore, her only source of nutrients was TPN. She had repeated episodes of sepsis and wound infection; broad-spectrum antibiotics were administered virtually continuously from 4 to 11 months of age. Multiple enteroenteric and enterocutaneous fistulas[54] were formed; over 8 months, they provided daily fluid losses >500 ml.

During the third month of TPN, an erythematous[55] rash was noted on the patient's lower eyelids adjacent to the outer canthi.[56] Over the next 3 months the rash spread, became more exfoliative, and exuded clear fluid. New lesions appeared in the angles of the mouth, around the nostrils, and in the perineal region.[57] This condition did not respond to topical application of various antibiotics, cortisone, and safflower oil.

During the fifth and sixth months of TPN, the patient lost all body hair, developed a waxy pallor, irritability, lethargy, and mild hypotonia.[58] That she was not deficient in essential fatty acids was indicated by the finding that her plasma fatty acid triene-to-tetraene ratio was normal (0.11). During the period from the third to the sixth month, the patient was given parenteral zinc supplements at 7, 30, and 250 times the normal requirement (0.2 mg/day). Her serum zinc concentration increased from 35 to 150 μg/dl (normal, 50–150 μg/dl) and, finally, to greater than 2,000 μg/dl without any beneficial effect. Intravenous zinc supplementation was then reduced to 0.4 mg/day. Biotin was determined by a bioassay using *Ochromonas danica*; urinary organic acids were determined by HPLC[59] and GC/MS.[60]

Laboratory Results:

Parameter	Patient	Normal Range
Plasma biotin	135 pg/ml	215–750 pg/ml
Urinary biotin excretion	<1 μg/24 hours	6–50 μg/24 hours
Urinary organic acid excretion:		
Methylcitrate	0.1 μmol/mg creatinine	<0.01 μmol/mg creatinine
3-Methylcrotonylglycine	0.7 μmol/mg creatinine	<0.2 μmol/mg creatinine
3-Hydroxyisovalerate	0.35 μmol/mg creatinine	<0.2 μmol/mg creatinine

Treatment with biotin (10 mg/day) was initiated and, after 1 week, the plasma biotin concentration increased to 11,500 pg/ml and organic acid excretion dropped to <0.01 μmol/mg creatinine. After 7 days of biotin supplementation, the rash had improved strikingly and the irritability had resolved. After 2 weeks of supplementation, new hair growth was noted, waxy pallor of the skin was less pronounced, and hypotonia improved. During the next 9 months of biotin therapy, no symptoms and signs of deficiency recurred. The patient's rapid transit time and vomiting did not improve.

Case Questions and Exercises

1. What signs were first to indicate a problem related to biotin utilization by the patient?
2. What is the relevance of aciduria to considerations of biotin status?
3. How were problems involving essential fatty acids and zinc ruled out in the diagnosis of this condition as biotin deficiency?

Study Questions and Exercises

1. Diagram the areas of metabolism in which biotin-dependent carboxylases are involved.
2. Construct a decision tree for the diagnosis of biotin deficiency in humans or an animal species.
3. What key feature of the chemistry of biotin relates to its biochemical function as a carrier of active CO_2?
4. What parameters might you measure to assess biotin status of a human or animal?

RECOMMENDED READING

Beckett, D., 2007. Biotin sensing: Universal influence of biotin status on transcription. Ann. Rev. Genet. 41, 443–464.

Beckett, D., 2009. Biotin sensing at the molecular level. J. Nutr. 139, 167–170.

Brownsey, R.W., Boone, A.W., Elliot, J.E., et al. 2006. Regulation of acetyl-CoA carboxylase. Biochem. Soc. Trans. 34, 223–227.

52. Feeding by means other than through the alimentary canal, referring particularly to the introduction of nutrients into veins.
53. For example, intralipid.
54. Passages created between one part of the intestine and another (an enteroenteric fistula) or between the intestine and the skin of the abdomen (an enterocutaneous fistula).
55. Marked by redness of the skin owing to inflammation.
56. Corners of the eye.
57. The area between the thighs extending from the coccyx to the pubis.
58. A condition of reduced tension of any muscle, leading to damage by overstretching.
59. High-performance liquid–liquid partition chromatography.
60. Gas–liquid partition chromatography with mass spectrometric detection.

Chapman-Smith, A., Cronan Jr, J.E., 1999. Molecular biology of biotin attachment to proteins. J. Nutr. 129, 477S–484S.

Cronan Jr, J.E., Waldrop, G.L., 2002. Multi-subunit acetyl-CoA carboxylases. Progr. Lipid Res. 41, 407–435.

Dakshinamurti, K., 2005. Biotin – a regulator of gene expression. J. Nutr. Biochem. 16, 419–423.

Fernandez-Mejia, C., 2005. Pharmacological effects of biotin. J. Nutr. Biochem. 16, 424–427.

Hassam., Y.I., Zemplini, Y., 2008. A novel, enigmatic histone modification: Biotinylation of histones by holocarboxylase synthetase. Nutr. Rev. 66, 721–725.

McMahan, R.J., 2002. Biotin in metabolism and molecular biology. Ann. Rev. Nutr. 22, 221–239.

Mock, D.M., 2007. Biotin. In: Zemplini, J., Rucker, R.B., McCormick, D.B., Suttie, J.W. (Eds.), Handbook of Vitamins, fourth ed. CRC Press, New York, NY, pp. 361–383.

Mock, D.M., 2009. Marginal biotin deficiency is common in normal human pregnancy and is highly teratogenic in mice. J. Nutr. 139, 154–157.

Rodrigues-Melendez, M., Zemplini, J., 2003. Regulation of gene expression by biotin. J. Nutr. Biochem. 14, 680–690.

Said, H.M., 2009. Cell and molecular aspects of human intestinal biotin absorption. J. Nutr. 139, 158–162.

van den Berg, H., 1997. Bioavailability of biotin. Eur. J. Clin. Nutr. 51, S60–S61.

Wolf, B., 2005. Biotinidase: Its role in biotinidase deficiency and biotin metabolism. J. Nutr. Biochem. 16, 441–445.

Wolf, B., 2010. Clinical issues and frequent questions about biotinidase deficiency. Mol. Gen. Metab. 100, 6–13.

Zempleni, J., 2005. Uptake, localization and noncarboxylase roles of biotin. Ann. Rev. Nutr. 25, 175–196.

Zempleni, J., 2009. Biotin. Biofactors 35, 36–46.

Zempleni, J., Chew, Y.C., Hassan, Y.I., Wijeratne, S.K., 2008. Epigenetic regulation of chromatin structure and gene function by biotin: Are biotin requirements being met? Nutr. Rev. 66, S46–S48.

Zempleni, J., Hassan., Y.I., Wijeratne, S.S.K., 2008. Biotin and biotinidase deficiency. Expert Rev. Endocrinol. Metab. 3, 715–724.

Pantothenic Acid

Anchoring Concepts

1. Pantothenic acid is the trivial designation for the compound dihydroxy-β,β-dimethylbutyryl-β-alanine.
2. Pantothenic acid is metabolically active as the prosthetic group of coenzyme A (CoA) and the acyl-carrier protein.
3. Deficiencies of pantothenic acid are manifested as dermal, hepatic, thymic, and neurologic changes.

A pellagrous-like syndrome in chicks has recently been obtained ... in an experiment which was originally designed to throw added light upon an unusual type of leg problem occurring in chicks fed semi-synthetic rations ... The data obtained in this experiment demonstrate the requirement in another species of the vitamin or vitamins present in autoclaved yeast, occasionally called vitamin B$_2$, vitamin G or the P-P factor, and indicate that the chick may be a more suitable animal than the white rat for delineating the quantities of this vitamin present in feedstuffs.

L. C. Norris and A. T. Ringrose

Learning Objectives

1. To understand the chief natural sources of pantothenic acid.
2. To understand the means of absorption and transport of pantothenic acid.
3. To understand the biochemical functions of pantothenic acid as components of coenzyme A and the acyl-carrier protein.
4. To understand the physiological implications of low pantothenic acid status.

VOCABULARY

Acetyl-CoA
Acyl carrier protein (ACP)
Acyl-CoA synthase
Burning feet syndrome
Coenzyme A (CoA)
Dephospho-CoA kinase
Fatty acid synthase
Malonyl-CoA
ω-Methylpantothenic acid
Pantetheine
Pantetheinase
Pantothenate kinase (PanK)
Pantothenic acid
Phosphopantetheine adenyltransferase
Phosphopantethenylcysteine decarboxylase
Phosphopantethenylcysteine synthase
4′-Phosphopantetheine
4′-Phosphopantothenic acid
4′-Phosphopantothenylcysteine
Propionyl-CoA
Succinyl-CoA

1. THE SIGNIFICANCE OF PANTOTHENIC ACID

Pantothenic acid is widely distributed in many foods. Therefore, problems of deficiency of the vitamin are rare. The vitamin has critical roles in metabolism, being an integral part of the acylation factors **coenzyme-A (CoA)** and **acyl-carrier protein (ACP)**. In these forms, pantothenic acid is required for the normal metabolism of fatty acids, amino

The Vitamins. DOI: 10.1016/B978-0-12-381980-2.00015-3

acids, and carbohydrates, and has important roles in the acylation of proteins. Because pantothenic acid is required for the synthesis of CoA, it has been a surprising observation that the rates of tissue CoA synthesis are not affected by deprivation of the vitamin. From such observations, it can be inferred that the vitamin is recycled metabolically; however, the regulation of pantothenic acid remains to be elucidated.

2. SOURCES OF PANTOTHENIC ACID

Distribution in Foods

As its name implies, pantothenic acid is widely distributed in nature (Table 15.1). It occurs mainly in bound forms (CoA, CoA esters, ACP). A glycoside has been identified in tomatoes. Therefore, it must be determined in foods and feedstuffs after enzymatic hydrolysis to liberate the vitamin from CoA. This is done in a two-step procedure using alkaline phosphatase followed by avian hepatic pantotheinase, yielding "total" pantothenic acid.

The most important food sources of pantothenic acid are meats (liver and heart are particularly rich). Mushrooms, avocados, broccoli, and some yeasts are also rich in the vitamin. Whole grains are also good sources; however, the vitamin is localized in the outer layers, thus it is largely (up to 50%) removed by milling. The most important sources of pantothenic acid for animal feeding are rice and wheat brans, alfalfa, peanut meal, molasses, yeasts, and condensed fish solubles. The richest sources of the vitamin in nature are cold-water fish ovaries,[1] which can contain more than 2.3 mg/g, and royal jelly,[2] which can contain more than 0.5 mg/g.

Stability

Pantothenic acid in foods and feedstuffs is fairly stable to ordinary means of cooking and storage. It can, however, be unstable to heat and either alkaline (pH >7) or acid (pH <5) conditions.[3] Reports indicate losses of 15.50% from cooking meat, and of 37–78% from heat-processing vegetables. The alcohol derivative, pantothenol, is more stable; for this reason it is used as a source of the vitamin in multivitamin supplements.

Bioavailability

The biologic availability of pantothenic acid from foods and feedstuffs is a function of the efficiency of the enteric hydrolysis of its food forms and the absorption of those

products. This area has not been well investigated. One study indicated "average" bioavailability of the vitamin in the American diet in the range of 40–60%;[4] similar results were obtained for maize meals in another study.[5]

3. ABSORPTION OF PANTOTHENIC ACID

Hydrolysis of Coenzyme Forms

Because pantothenic acid occurs in most foods and feedstuffs as CoA and ACP, the utilization of the vitamin in foods depends on the hydrolytic digestion of these protein complexes to release the free vitamin. Both CoA and ACP are degraded in the intestinal lumen by hydrolases (pyrophosphatase, phosphatase) to release the vitamin as **4-phosphopantetheine** (Fig. 15.1). That form is dephosphorylated to yield **pantetheine**, which is absorbed or converted to **pantothenic acid** by another intestinal hydrolase, **pantotheinase**.

Two Types of Transport

- *Facilitated transport.* At low concentrations, pantothenic acid is absorbed by a saturable, facilitated mechanism dependent on Na^+ located on the epithelial brush border, particularly villus cells. This appears to be a function of a **sodium-dependent multivitamin transporter** (**SMVT**), which also transports biotin and can be inhibited by biotin, lipoic acid, certain anticonvulsant drugs,[6] and ethanol.
- *Passive diffusion.* At high lumenal concentrations, pantothenic acid is also absorbed by non-saturable, simple diffusion throughout the small intestine.[7] The alcohol form, **panthenol**, which is oxidized to pantothenic acid *in vivo*, appears to be absorbed somewhat faster than the acid form.

4. TRANSPORT OF PANTOTHENIC ACID

Plasma and Erythrocytes

Pantothenic acid is transported in both the plasma and erythrocytes. Plasma contains the vitamin only in the free acid form, which erythrocytes take up by passive diffusion. While erythrocytes carry some of the vitamin unchanged, they convert most of the vitamin to 4′-phosphopantothenic

1. Tuna, cod.
2. Royal jelly, the food responsible for the diet-induced reproductive development of the queen honeybee, is also the richest natural source of biotin.
3. Pasteurization of milk, owing to its neutral pH, does not affect its content of pantothenic acid.

4. Tarr, J. B., Tamura, T., and Stokstad, E. L. (1981). *Am. J. Clin. Nutr.* 34, 1328.
5. Yu, B. H. and Kies, C. (1993). *Plant Food Hum. Nutr.* 43, 87.
6. Carbamazepine, primidone.
7. Earlier studies in which unphysiologically high concentrations of the vitamin were employed failed to detect the carrier-mediated mechanism of its enteric absorption. This led to the conclusion that the vitamin is absorbed only by simple diffusion.

TABLE 15.1 Pantothenic Acid Contents of Foods

Food	Pantothenic Acid, mg/100 g	Food	Pantothenic Acid, mg/100 g
Dairy Products		Carrots	0.27
Milk	0.2	Cauliflower	1.0
Cheeses	0.1–0.9	Lentils	1.4
Meats		Potatoes	0.3
Beef	0.3–2	Soybeans	1.7
Pork	0.4–3.1	Tomatoes	0.3
Calf heart	2.5	**Fruits**	
Calf kidney	3.9	Apples	0.1
Chicken liver	9.7	Bananas	0.2
Pork liver	7.0	Grapefruits	0.3
Cereals		Oranges	0.2
Cornmeal	0.9	Strawberries	0.3
Rice, unpolished	1.1	**Nuts**	
Oatmeal	0.9	Cashews	1.3
Wheat	1.0	Peanuts	2.8
Wheat bran	2.9	Walnuts	0.7
Barley	1.1	**Other**	
Vegetables		Eggs	2.9
Avocado	1.1	Mushrooms	2.1
Broccoli	1.2	Bakers' yeast	5.3–11
Cabbage	0.1–1.4		

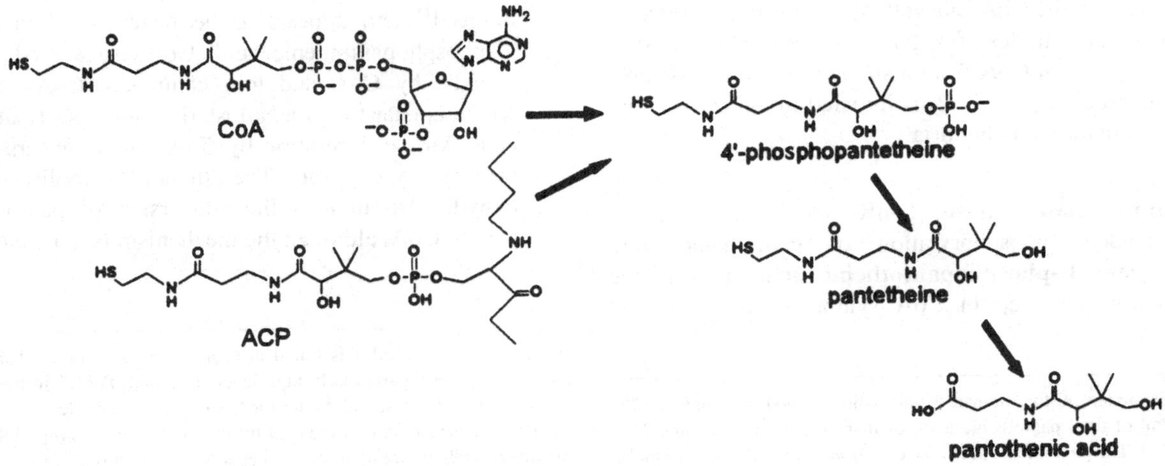

FIGURE 15.1 Liberation of pantothenic acid from coenzyme forms in foods.

acid and pantetheine. Erythrocytes carry most of the vitamin in the blood.[8]

Cellular Uptake

Pantothenic acid is taken into cells in its free acid form. This appears to be mediated by the same mechanism involved in enteric absorption, i.e., SMVT. Upon cellular uptake, most of the vitamin is converted to CoA, the predominant intracellular form.

Tissue Distribution

The greatest concentrations of CoA are found in the liver,[9] adrenals, kidneys, brain, heart, and testes. Much of this (70% in liver, 95% in heart) is located in the mitochondria. Tissue CoA concentrations are not affected by deprivation of the vitamin. This surprising finding has been interpreted as indicating a mechanism for conserving the vitamin by recycling it from the degradation of pantothenate-containing molecules.

Pantothenic acid is taken up in the choroid plexus by a specific transport process, which, at low concentrations of the vitamin, involves the partial phosphorylation of the vitamin. The cerebrospinal fluid, because it is constantly renewed in the central nervous system, requires a constant supply of pantothenic acid, which, as CoA, is involved in the synthesis of the neurotransmitter acetylcholine in brain tissue.

5. METABOLISM OF PANTOTHENIC ACID

CoA Synthesis

All tissues have the ability to synthesize CoA from pantothenic acid of dietary origin. At least in rat liver, all of the enzymes in the CoA biosynthetic pathway are found in the cytosol. Four moles of ATP are required for the biosynthesis of a mole of CoA from a single mole of pantothenic acid. The process (Fig. 15.2) is initiated in the cytosol and completed in the mitochondria.

Cytosol:

a. **Pantothenate kinase** (PanK) catalyzes the ATP-dependent phosphorylation of pantothenic acid to yield **4′-phosphopantothenic acid**. This is the rate-limiting step in CoA synthesis; under normal

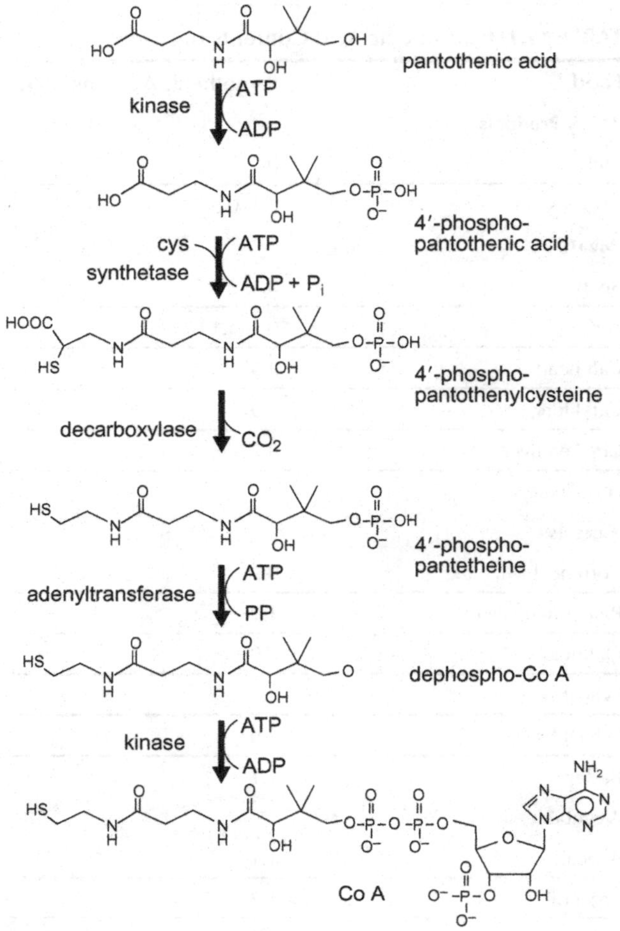

FIGURE 15.2   Biosynthesis of coenzyme A.

conditions it functions far below its capacity. Four isoforms of PanK have been identified; these show different tissue distributions.[10] PanK can be induced[11] and appears to be feedback-inhibited by 4′-phosphopantothenic acid, CoA esters, and, more weakly, by CoA and long-chain acyl-CoAs, all of which appear to act as allosteric effectors of pantothenate kinase. Inhibition by CoA esters appears to be reversed by carnitine. The ethanol metabolite acetaldehyde also inhibits the conversion of pantothenic acid to CoA, although the mechanism is not clear.[12]

---

8. For example, in the human adult, whole blood contains 1,120–1,960 ng/ml of total pantothenic acid; of that, the plasma contains 211–1,096 ng/ml. The pantothenic acid concentration of liver is about 15 mol/l; that of heart is about 150 μmol/l. Blood pantothenic acid levels are generally lower in elderly individuals, e.g., 500–700 ng/ml.

9. The human liver typically contains about 28 mg of total pantothenic acid.

10. In the mouse, PanK1 is found in heart, liver, and kidney; PanK2 in most tissues, with particularly high levels in retina; PanK3 in liver; and PanK4 in most tissues, with highest levels in skeletal muscle.

11. Pantothenate kinase is induced by the antilipidemic drug clofibrate. Treatment with clofibrate increases hepatic concentrations of CoA, apparently owing to increased synthesis.

12. Alcoholics have been reported to excrete in their urine large percentages of the pantothenic acid they ingest, a condition corrected on ethanol withdrawal.

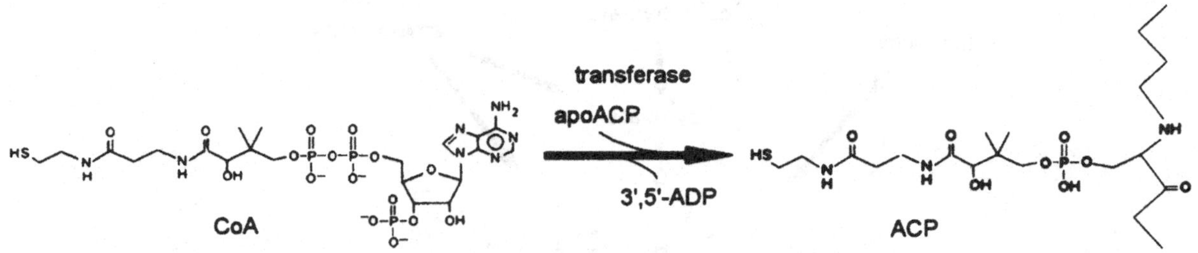

FIGURE 15.3 Coenzyme A provides 4-phosphopantetheine in the biosynthesis of the acyl-carrier protein.

b. **Phosphopantothenylcysteine synthetase** catalyzes the ATP-dependent condensation of 4′-phosphopantothenic acid with cysteine to yield **4′-phosphopantothenylcysteine**.

c. **Phosphopantothenylcysteine decarboxylase** catalyzes the decarboxylation of 4′-phosphopantothenylcysteine to yield **4′-phosphopantetheine** in the cytosol, which is transported into the mitochondria.

Mitochondrial inner membrane:

a. **Phosphopantetheine adenyltransferase** catalyzes the ATP-dependent adenylation of 4′-phosphopantetheine to yield **dephospho-CoA**. Because this reaction is reversible, at low ATP levels dephospho-CoA can be degraded to yield ATP.

b. **Dephospho-CoA kinase** catalyzes the ATP-dependent phosphorylation of dephospho-CoA to yield CoA.

The concentration of non-acylated CoA determines the rate of oxidation-dependent mitochondrial energy production. CoA has been shown to enter mitochondria by non-specific membrane-binding as well as by an energy-dependent membrane transporter.[13] That CoA synthesis occurs within the mitochondria is suggested by the finding that PanK2 is localized there.

ACP Synthesis

In higher animals ACP is associated with a large fatty acid synthetase complex composed of two 250 kDa subunits which contain several functional domains.[14] The ACP complex is modified post-translationally by the addition of the 4′-phosphopantetheine prosthetic group via a phosphoeseter linkage at a serinyl residue on the apo-ACP. The modification is catalyzed by 4′-phosphopantetheine-apoACP transferase using CoA as the donor (Fig. 15.3). It is therefore likely that ACP synthesis serves as a regulator of intracellular CoA levels.

---

13. Tahiliani, A. G. and Neely, J. R. (1987). *J. Mol. Cell Cardiol.* 19, 1161.

14. Fatty acid synthase complex has several catalytic sites: acetyl transferase, malonyl transferase, 2-oxoacyl synthase, oxoacyl reductase, 3-hydroxyacyl dehydratase, enoyl reductase, thioester hydrolase.

Catabolism of CoA and ACP

The pantothenic acid components of both CoA and ACP are released metabolically ultimately in the free acid form of the vitamin. An ACP hydrolase has been identified that releases 4′-phosphopantetheine from holo-ACP to yield apo-ACP. The catabolism of CoA appears to be initiated by a non-specific phosphate-sensitive lysosomal phosphatase, which yields dephospho-CoA. That metabolite appears to be degraded by a pyrophosphatase in the plasma membrane. 4′-Phosphopantetheine produced from either source is degraded to 4′-pantothenylcysteine and, finally, to pantothenic acid by microsomal and lysosomal phosphatases.

Excretion

Pantothenic acid is excreted mainly in the urine as free pantothenic acid and some 4′-phosphopantethenate; no catabolic products are known. The renal tubular secretion of pantothenic acid, probably by a mechanism common to weak organic acids, results in urinary excretion of the vitamin correlating with dietary intake. An appreciable amount (~15% of daily intake) is oxidized completely, and is excreted across the lungs as CO_2. Humans typically excrete, in the urine, 0.8–8.4 mg of pantothenic acid per day. There appear to be two renal mechanisms for regulating the excretion of pantothenic acid:

- *Active transport.* At physiological concentrations of the vitamin in the plasma, pantothenic acid is reabsorbed by active transport;
- *Tubular secretion.* At higher concentrations, tubular secretion of pantothenic acid occurs. Tubular reabsorption appears to be the only mechanism for conserving free pantothenic acid in the plasma.

6. METABOLIC FUNCTIONS OF PANTOTHENIC ACID

General Functions

Both CoA and 4′-phosphopantetheine in ACP function metabolically as carriers of acyl groups and activators of

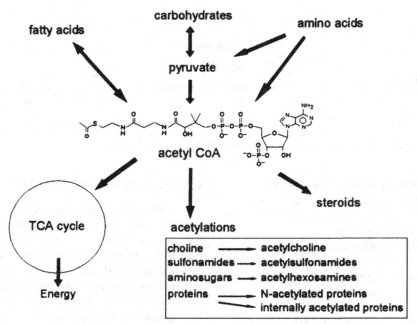

FIGURE 15.4 The central role of acetyl-CoA in metabolism.

carbonyl groups in a large number of vital metabolic trans-
formations, including the tricarboxylic acid (TCA) cycle[15]
and the metabolism of fatty acids. In each case, the linkage with the transported acyl group involves the reactive
sulfhydryl of the 4′-phosphopantetheinyl prosthetic group.
There is a clear distinction between the metabolic roles of
CoA and ACP:

- CoA is involved in a broad array of acyl transfer reactions related to oxidative energy metabolism and
catabolism;
- ACP is involved in synthetic reactions.

CoA

CoA serves as an essential cofactor for some 4% of known
enzymes, including at least 100 enzymes involved in intermediary metabolism. In these reactions CoA forms high-
energy thioester bonds with carboxylic acids, the most
important of which is acetic acid, which can come from
the metabolism of fatty acids, amino acids, or carbohydrates (Fig. 15.4). CoA functions widely in metabolism in
reactions involving either the carboxyl group (e.g., formation of acetylcholine, acetylated amino sugars, acetylated
sulfonamides[16]) or the methyl group (e.g., condensation with oxaloacetate to yield citrate) of an acyl-CoA.

Acetyl-CoA, the "*active acetate*," group has many metabolic uses:

- synthesis of fatty acids,[17] and isopranoids (e.g., cholesterol, steroid hormones)
- acetylations of alcohols, amines, and amino acids (e.g.,
choline, sulfonamides, *p*-aminobenzoate, proteins[18])
- oxidation of amino acids
- N-terminal acetylation in more than half of eukaryotic
proteins, including the processing of peptide hormones
from their polyprotein precursors (e.g., processing
ACTH to α-melanocyte-stimulating hormone, and
β-lipotropin to β-endorphin)
- internal acetylation of proteins including histones[19] and
α-tubulin[20]
- post-translational modification of a large number of
proteins (GTP-binding proteins, protein kinases, membrane receptors, cytoskeletal proteins, mitochondrial
proteins) by addition of long-chain fatty acids – most

---

15. Often called the citric acid cycle or Krebs cycle.
16. Coenzyme A was discovered as an essential factor for the acetylation
of sulfonamide by the liver and for the acetylation of choline in the brain;
hence, *coenzyme A* stands for *coenzyme for acetylations*.

17. Studies with liver slices *in vitro* have demonstrated a correlation
between hepatic CoA content and lipid biosynthetic capacity, suggesting
that CoA may be a limiting factor in lipogenesis.
18. Many CoA-dependent reactions modify protein structure and function via acetylations at N termini or at internal sites (particularly, at the
ε-amino groups of lysyl residues).
19. Acetylated histones are enriched in genes that are being actively
transcribed.
20. Acetylation occurs in the α-tubulin after it has been incorporated into the microtubule. It can be induced by such agents as taxol.
Acetylated microtubules are more stable to depolymerizing agents such
as colchicines.

frequently involving palmitic acid[21] added in a reversible ester bond and myristic acid[22] added in an irreversible amide linkage

- activation of fatty acids for incorporation into triglycerides, membrane phospholipids, and regulatory sphingolipids
- transacylation to carnitine to form energy-equivalent acylcarnitines capable of being transported into the mitochondria where β-oxidation occurs
- production of the "*ketone body*" acetoacetate derived from fat metabolism when glucose is limiting.

ACP

ACP is a component of the multienzyme complex **fatty acid synthase**.[23] In ACP, the cofactor functions in two domains, acetyl transferase and malonyl transferase, that transfer the respective acyl groups between 4'-phosphopantetheine at different active sites with successive cycles of condensations and reductions.[24] The nature of the fatty acid synthase complex varies considerably among different species. However, in each, 4'-phosphopantetheine is the prosthetic group for the binding and transfer of the acyl units during catalysis. The sulfhydryl group of the cofactor serves as the point of temporary covalent attachment of the growing fatty acid via a thiol linkage each time an acyl group is added by transfer to the cofactor. In this way, the cofactor appears to function as a swinging arm, allowing the growing fatty acid to reach the various catalytic sites of the enzyme.

7. PANTOTHENIC ACID IN HEALTH AND DISEASE

Benefits have been reported for the use of supplements of pantothenic acid and/or metabolites.

Reduced Serum Cholesterol Level

High doses (500–1,200 mg/day) of pantothine, the dimer of pantetheine, have been shown to reduce serum concentrations of total and LDL cholesterol and triglycerides,

with increases in HDL cholesterol.[25] While the underlying mechanism is unclear, it is thought to involve roles of pantetheine as a cofactor in shunting acetyl groups away from steroid synthesis to oxidative metabolism, and/or in reducing triglyceride synthesis through inhibition of hydroxymethylglutyryl-CoA reductase.

Rheumatoid Arthritis (RA)

Patients with RA have been found to show lower blood pantothenic acid levels than healthy controls. Nearly 50 years ago, an unblinded trial found relief of symptoms in 20 patients treated with pantothenic acid.[26] A subsequent randomized, controlled trial showed that high doses (up to 2 g/day) of calcium pantothenate reduced the duration of morning stiffness, the degree of disability, and the severity of pain for rheumatoid arthritis patients.[27]

Athletic Performance

While pantothenic acid deficiency is known to reduce exercise endurance in animal models, results of the few studies conducted in humans have been inconsistent. Some showed improved efficiency of oxygen utilization and reduced lactate acid accumulation in athletes;[28] others showed no benefits.[29]

Wound Healing

Studies in animal models have found pantothenic acid, given orally or topically as pantothenol, to promote the closure of wounds of the skin. Studies with humans given high, combined doses of pantothenic acid and ascorbic acid have shown no benefits,[30] although a derivative, dexapanthenol, has been found useful in reducing skin dehydration and irritation.[31]

Other Outcomes

It has been suggested that pantothenic acid may have value in treating the systemic autoimmune disease lupus erythromatosus, the argument being a theoretical one based on the observation that lupus can be caused by drugs that impair

21. *n*-hexadecanoic acid, C16:0.
22. *n*-tetradecanoic acid, C14:0.
23. Fatty acid synthase is the name used to identify the multienzyme complex on which the several reactions of fatty acid synthesis (condensations and reductions) occur. The best studied complex is that of *Escherichia coli*; it consists of seven separate enzymes plus the small (10 kDa) protein ACP.
24. The seven functional activities of the fatty acid synthase complex are acetyltransferase, malonyltransferase, 3-ketoacyl synthase, 3-ketoacyl reductase, 3-hydroxyacyl dehydratase, enoyl reductase, and thioester hydrolase.

25. Binaghi, P., Cellina, G., Lo Cicero, G. *et al.* (1990). *Minerva Med.* 81, 475.
26. Subjects were given calcium pantothenate i.m. (Barton-Wright, E. C. and Elliot, W. A. [1963] *Lancet* 2, 862).
27. US Practioner Research Group (1980). *Practitioner* 224, 208.
28. Litoff, D. (1985). *Med. Sci. Sports Exer.* 17(Suppl.), 287.
29. Nice, C., Reeves, A. G., Brinck-Johnson, T., *et al.* (1984). *J. Sports Med. Phys. Fitness* 24, 26.
30. Vaxman, F., Olender, S., Lambert, A., *et al.* (1995). *Eur. Surg. Res.* 27, 158.
31. Biro, K., Thaci, D., Oschendorf, F. R., *et al.* (2003). *Contact Dermatitis* 49, 80.

TABLE 15.2 General Signs of Pantothenic Acid Deficiency

Organ System	Signs
General:	
Appetite	Decrease
Growth	Decrease
Vital organs	Hepatic steatosis, thymic necrosis, adrenal hypertrophy
Dermatologic	Dermatitis, achromotrichia, alopecia
Muscular	Weakness
Gastrointestinal	Ulcers
Nervous	Ataxia, paralysis

pantothenic acid metabolism. No relevant clinical data have been reported. It has also been proposed that pantothenic acid may have value in the prevention of graying hair.[32] That, too, is without substantiating evidence.

Disorders of Pantothenic Acid Metabolism

A polymorphism of PanK2 has been identified as the metabolic basis of an autosomal recessive neurodegenerative disorder, Hallervorden-Spatz syndrome. Affected subjects show dystonia and optic atrophy or retinopathy, with the deposition of iron in basal ganglia.[33]

8. PANTOTHENIC ACID DEFICIENCY

Deficiencies Rare

Deprivation of pantothenic acid results in metabolic impairments, including reduced lipid synthesis and energy production. Signs and symptoms of pantothenic acid deficiency vary among different species; most frequently, they involve the skin, liver, adrenals, and nervous system. Owing to the wide distribution of the vitamin in nature, dietary deficiencies of pantothenic acid are rare; they are more common in circumstances of inadequate intake of basic foods and vitamins, and are often associated with (and mistakenly diagnosed as) deficiencies of other vitamins. Understanding of the presentation of pantothenic acid deficiency comes mostly from studies with experimental animals. These have shown a pattern of general deficiency signs (Table 15.2).

Antagonists

Pantothenic acid deficiency has been produced experimentally using purified diets free of the vitamin or by administering an antagonist. One antagonist is the analog **ω-methylpantothenic acid**, which has a methyl group in place of the hydroxymethyl group of the vitamin; this change prevents it from being phosphorylated and inhibits the action of pantothenic acid kinase. Other antagonists include desthio-CoA, in which the terminal sulfhydryl of the active metabolite is replaced with a hydroxyl group, and hopantenate, in which the three-carbon β-alanine moiety of the vitamin is replaced with the four-carbon γ-aminobutyric acid (GABA).

Deficiency Signs in Animals

Pantothenic acid deficiency in most species results in reduced growth and reduced efficiency of feed utilization. In rodents, the deficiency results in a scaly dermatitis, achromotrichia, alopecia, and adrenal necrosis. Congenital malformations of offspring of pantothenic acid-deficient dams have been reported. Excess amounts of porphyrins[34] are excreted in the tears of pantothenic acid-deficient rats, in a condition called **blood-caked whiskers**. Pantothenic acid-deficient chicks develop skin lesions at the corners of the mouth, swollen and encrusted eyelids, dermatitis of the entire foot (with hemorrhagic cracking),[35] poor feathering, fatty liver degeneration, thymic necrosis, and myelin degeneration of the spinal column with paralysis and lethargy. Chicks produced from deficient hens show high rates of embryonic and post-hatching mortality. Pantothenic acid-deficient dogs develop hepatic steatosis, irritability, cramps, ataxia, convulsions, alopecia, and death. Deficient pigs show similar nervous signs, develop hypertrophy and steatosis of the adrenals, liver, and heart, and show ovarian atrophy with impaired uterine development. Deficient fish show anorexia.

Marginal deficiency of pantothenic acid in the rat has been found to produce elevated serum levels of triglycerides and free fatty acids. The metabolic basis of this effect is not clear; however, it is possible that it involves a somewhat targeted reduction in cellular CoA concentrations, affecting the deposition of fatty acids in adipocytes (via impaired acyl-CoA synthase), but not the hepatic production of triglycerides.

---

32. That pantothenic acid deficiency can cause achromatrichia in rodents does not imply that graying of human hair is necessarily caused by insufficient pantothenic acid status, or that supplemental pantothenic acid could have any effect. Indeed, none has been demonstrated.

33. Gordon, N. (2002). *Eur. J. Paediatr. Neurol.* 6, 243.

34. For example, protoporphyrin IX.

35. These lesions are often confused with the foot pad dermatitis caused by biotin deficiency. Unlike the latter, in which lesions are limited to the foot pad (i.e., plantar surface), the lesions produced by pantothenic acid deficiency also involve the toes and superior aspect of the foot. Prevention of foot pad dermatitis is economically important in poultry production; the US–European market for chicken and duck feet has been estimated to exceed $250 million.

Deficiency Signs in Humans

Pantothenic acid deficiency in humans has been observed only in severely malnourished patients and in subjects treated with the antagonist ω-methylpantothenic acid. In cases of the former type, neurologic signs (paresthesia in the toes and soles of the feet) have been reported.[36] Subjects made deficient in pantothenic acid through the use of ω-methylpantothenic acid also developed burning sensations of the feet; in addition, they showed depression, fatigue, insomnia, vomiting, muscular weakness, and sleep and gastrointestinal disturbances. Changes in glucose tolerance, increased sensitivity to insulin, and decreased antibody production have also been reported.

Some evidence suggests that pantothenic acid intakes may not be adequate for some individuals. Urinary pantothenic acid excretion has been found to be low for pregnant women, adolescents, and the elderly, compared with the general population.

9. PANTOTHENIC ACID TOXICITY

The toxicity of pantothenic acid is negligible. No adverse reactions have been reported in any species following the ingestion of large doses of the vitamin. Massive doses (e.g., 10 g/day) administered to humans have not produced reactions more severe than mild intestinal distress and diarrhea. Similarly, no deleterious effects have been identified when the vitamin was administered parenterally or topically. It has been estimated that animals can tolerate, without side effects, doses of pantothenic acid as great as at least 100 times their respective nutritional requirements for the vitamin.

10. CASE STUDY

Review the following experiment, paying special attention to the independent and dependent variables in the design. Then, answer the questions that follow.

Experiment

To evaluate the possible role of pantothenic acid and ascorbic acid in wound healing, a study was conducted of the effects of these vitamins on the growth of fibroblasts. Human fibroblasts were obtained from neonatal foreskin; they were cultured in a standard medium supplemented with 10% fetal calf serum and antibiotics.[37] The medium contained no ascorbic acid, but contained 4 mg of pantothenic acid per liter. Cells were used between the third and ninth passages. Twenty-four hours before each experiment, the basal medium was replaced by medium supplemented with pantothenic acid (40 mg/l) or pantothenic acid (40 mg/l) plus ascorbic acid (60 mg/l). Cells (1.5×10^5) were plated in 3 ml of culture medium in 28 cm^2 plastic dishes. After incubation, they were collected by adding trypsin and then scraping; they were counted in a hemocytometer. The synthesis of DNA and protein was estimated by measuring the rates of incorporation of radiolabel from [^{3}H]thymidine and [^{14}C]proline, respectively. Total protein was measured in cells (lyzed by sonication and solubilized in 0.5 N NaOH) and in the culture medium.

Results After 5 Days of Culture:

Treatment	Cells ($\times 10^5$)	^{3}H (10^3 cpm)	^{14}C (10^3 cpm)	Cell Protein (mg/dish)	Protein in Medium (mg/ml)
Control	2.90 ± 0.16	11.6 ± 0.4	1.7 ± 1.0	10.0 ± 1.0	1.93 ± 0.01
+Pantothenic acid	3.83 ± 0.14[a]	18.7 ± 0.5[a]	2.9 ± 0.1[a]	14.5 ± 0.9[a]	1.93 ± 0.02
+Pantothenic acid and ascorbic acid	3.74 ± 0.19[a]	18.1 ± 0.8[a]	2.8 ± 0.1[a]	8.1 ± 0.9	2.11 ± 0.01[a]

[a] Significantly different from control value, P <0.05.

Case Questions and Exercises

1. Why were thymidine and proline selected as carriers of the radiolabels in this experiment?
2. Why were fibroblasts selected (rather than some other cell type) for use in this study?
3. Assuming that the protein released into the culture medium is largely soluble procollagen, what can be concluded about the effects of pantothenic acid and/or ascorbic acid on collagen synthesis in this system?
4. What implications do these results have regarding wound healing?

36. Burning feet syndrome was described during World War II in prisoners in Japan and the Philippines, who were generally malnourished. That large oral doses of calcium pantothenate provided some improvement suggested that the syndrome involved, at least in part, deficiency of pantothenic acid.

37. Gentamicin and amphotericin B (Fungizone).

Study Questions and Exercises

1. Diagram the areas of metabolism in which CoA and ACP (via fatty acid synthase) are involved.
2. Construct a decision tree for the diagnosis of pantothenic acid deficiency in humans or an animal species.
3. What key feature of the chemistry of pantothenic acid relates to its biochemical functions as a carrier of acyl groups?
4. What parameters might you measure to assess pantothenic acid status of a human or animal?

RECOMMENDED READING

Folmes, C.D.L., Lopaschuk, G.D., 2007. Role of malonyl-CoA in heart disease and the hypothalamic control of obesity. Cardiovas. Res. 73, 278–287.

Leonoardi, R., Zhang, Y.M., Rock, C.O., Jackowski, S., 2005. Coenzyme A: Back in action. Progr. Lipid Res. 44, 125–153.

Li, L.O., Klett, E.L., Coleman, R.A., 2010. Acyl-CoA synthesis, lipid metabolism and lipotoxicity. Biochim. Biophys. Acta 1801, 246–251.

Miller, J.W., Rogers, L.M., Rucker, R.B., 2006. Pantothenic acid. In: Bowman, B.A., Russell, R.M. (Eds.), Present Knowledge in Nutrition, ninth ed. ILSI Press, Washington, DC, pp. 327–339.

Rucker, R.B., Bauerly, K., 2007. Pantothenic acid. In: Zempleni, J., Rucker, R.B., McCormick, D.B., Suttie, J.W. (Eds.), Handbook of Vitamins, fourth ed. CRC Press, New York, NY, pp. 289–313.

Shindow, H., Hishikawa, D., Harayama, T., Yuki, K., Shoizu, T., 2009. Recent progress on acyl CoA:lysophospholipid aceltransferase research. J. Lipid Res. 50, S46–S51.

Spry, C., Kirk, K., Saliba, K.J., 2008. Coenzyme A biosynthesis: An antimicrobial drug target. FEMS Microbiol. Rev. 32, 56–106.

van den Berg, H., 1997. Bioavailability of pantothenic acid. Eur. J. Clin. Nutr. 51, S62–S63.

Watkins, P.A., 2008. Very-long-chain acyl-CoA synthetases. J. Biol. Chem. 283, 1773–1777.

Wolfgang, M.J., Lane, M.D., 2006. The role of hypothalamic malonyl-CoA in energy homeostasis. J. Biol. Chem. 281, 37265–37269.

Folate

Chapter Outline

Anchoring Concepts

1. Folate is the generic descriptor for folic acid (pteroylmonoglutamic acid) and related compounds exhibiting qualitatively the biological activity of folic acid. The term folates refers generally to the compounds in this group, including mono- and polyglutamates.
2. Folates are active as coenzymes in single-carbon metabolism.
3. Deficiencies of folate are manifested as anemia and dermatologic lesions.

Using Streptococcus lactis R *as a test organism, we have obtained in a highly concentrated and probably nearly pure form an acid nutrilite with interesting physiological properties. Four tons of spinach have been extracted and carried through the first stages of concentration … This acid, or one with similar chemical and physiological properties, occurs in a number of animal tissues of which liver and kidney are the best sources … It is especially abundant in green leaves of many kinds, including grass. Because of this fact, we suggest the name "folic acid" (Latin, folium − leaf). Many commercially canned greens are nearly lacking in the substance.*

H. K. Mitchell, E. S. Snell, and R. J. Williams

Learning Objectives

1. To understand the chief natural sources of folates.
2. To understand the means of absorption and transport of the folates.
3. To understand the biochemical functions of the folates as coenzymes in single-carbon metabolism, and the relationship of that function to the physiological activities of the vitamin.
4. To understand the metabolic interrelationship of folate and vitamin B_{12}, and its physiological implications.

VOCABULARY

p-Acetaminobenzoylglutamate
Betaine
Cervical paralysis
7,8-Dihydrofolate reductase
Dihydrofolic acid (FH_2)
Folate
Folate-binding proteins (FBPs)
Folate export pump
Folate receptor (FRs)
Folic acid
Folyl conjugase (folyl γ-glutamyl carboxypeptidase)
Folyl polyglutamates
Folyl polyglutamate synthetase
5-Formimino-FH_4
Formiminoglutamate (FIGLU)
5-Formyl-FH_4
10-Formyl-FH_4
γ-Glutamyl hydrolase
Homocysteine (Hcy)
Homocysteinemia
Leukopenia
Macrocytic anemia
Megaloblasts
5,10-Methenyl-FH_4
Methionine synthetase

The Vitamins. DOI: 10.1016/B978-0-12-381980-2.00016-5

Methotrexate
5-Methyl FH$_4$
5,10-Methylene FH$_4$
5,10-Methylene FH$_4$ dehydrogenase
5,10-Methylene FH$_4$ reductase (MTHFR)
Methyl-folate trap
Methylmalonic acid (MMA)
Organic anion transporter (OAT)
Pernicious anemia
Proton-coupled folate transporter (PCFT)
Pteridine
Pterin ring
Pteroylglutamic acid
Purines
Reduced folate carrier (RFC)
S-adenosylhomocysteine (SAH)
S-adenosylmethionine (SAM)
Serine hydroxymethyltransferase
Single-carbon metabolism
Sulfa drugs
Tetrahydrofolate reductase
Tetrahydrofolic acid (FH$_4$)
Thymidylate
Thymidylate synthase
Vitamin B$_{12}$

1. THE SIGNIFICANCE OF FOLATE

Folate is a vitamin that has only recently been appreciated for its importance beyond its essential role in normal metabolism, especially for its relevance to the etiologies of chronic diseases and birth defects. Widely distributed among foods, particularly those of plant foliar origin, this abundant vitamin is underconsumed by people whose food habits do not emphasize plant foods. Intimately related in function with vitamins B$_{12}$ and B$_6$, its status at the level of subclinical deficiency can be difficult to assess, and the full extent of its interrelationships with these vitamins and with amino acids remains incompletely elucidated. Folate deficiency is an important problem in many parts of the world, particularly where there is poverty and malnutrition. It is an important cause of anemia, second only to nutritional iron deficiency.

Evidence shows that marginal folate intakes can support apparently normal circulating folate levels while still limiting single-carbon metabolism. Thus, folate emerged as having an important role in the etiology of homocysteinemia, which was identified as a risk factor for occlusive vascular disease, cancer, and birth defects, particularly **neural tube defects** (**NTDs**). In 1998, the US Food and Drug Administration mandated that folic acid be added to all "enriched" cereal grain products (breads, pastas, wheat flours, breakfast cereals, rice) to reduce the prevalence of NTDs. The food system-wide measure increased the folate intakes of Americans, more than doubled circulating levels of the vitamin, and was expected to reduce both NTDs and coronary artery disease deaths, while also driving folate supplementation efforts in other countries.

More than a decade of population-based folate supplementation has seen the prevalence of NTDs decline, indicating that this strategy has been successful. Still, concerns remain about potential risks of treating individuals who are not in need. The first of these was the prospective masking of **macrocytic anemia** of vitamin B$_{12}$ deficiency, which will lead to neuropathy if not corrected. Additional concerns have been added in recent years, with growing doubt about the causal role of **homocysteinemia** in the etiology of cardiovascular disease, with reports of enhanced cognitive impairment and colorectal cancer risk as a consequence of folate supplementation. For these several reasons, it is important to understand the role of folate in nutrition and health.

2. SOURCES OF FOLATE

Distribution in Foods

Folates (**folyl polyglutamates**) occur in a wide variety of foods of both plant and animal origin (Table 16.1). Liver, mushrooms, and green, leafy vegetables are rich sources of folate in human diets, while oilseed meals (e.g., soybean meal) and animal by-products are important sources of folate in animal feeds. The folates in foods and feedstuffs are almost exclusively in reduced form as polyglutamyl derivatives of **tetrahydrofolic acid** (**FH$_4$**). Very little free folate (folyl monoglutamate) is found in foods or feedstuffs.

Analyses of foods have revealed a wide distribution of general types of polyglutamyl folate derivatives, the predominant forms being **5-methyl-FH$_4$** and **10-formyl-FH$_4$**. The folates found in organ meats (e.g., liver and kidney) are about 40% methyl derivatives, whereas that in milk (and erythrocytes) is exclusively the methyl form. Some plant materials also contain mainly 5-methyl-FH$_4$ (e.g., lettuce, cabbage, orange juice), but others (e.g., soybean) contain relatively little of that form (~15%), the rest occurring as the 5- and 10-formyl derivatives. Most of the folates in cabbage are hexa- and heptaglutamates, whereas half of those in soybean are monoglutamates. More than a third of the folates in orange juice are present as monoglutamates, and nearly half are present as pentaglutamates. Liver and kidney contain mainly pentaglutamates, and ~60% of the folates in milk are monoglutamates (with only 4–8% each of di- to heptaglutamates).

Stability

Most folates in foods and feedstuffs (that is, folates other than **folic acid**[1] and **5-formyl-FH$_4$**) are easily oxidized,

---

1. Throughout this text, the term *folic acid* is used as the specific trivial name for the compound **pteroylglutamic acid**.

TABLE 16.1 Folate Contents of Foods

Food	Folate, μg/100 g
Dairy Products	
Milk	5–12
Cheese	20
Meats and Fish	
Beef	5–18
Liver	
Beef	140–1070
Chicken	1,810
Tuna	15
Cereals	
Barley	15
Corn	35
Rice, polished	15
Rice, unpolished	25
Wheat, whole	30–55
Wheat bran	80
Vegetables	
Asparagus	70–175
Beans	70
Broccoli	180
Brussels sprouts	90–175
Cabbage	15–45
Cauliflower	55–120
Peas	90
Soybeans	360
Spinach	50–190
Tomatoes	5–30
Fruits	
Apples	5
Bananas	30
Oranges	25
Other	
Eggs	70
Brewer's yeast	1,500

the corresponding derivatives of **dihydrofolic acid (FH$_2$)** (partially oxidized) or folic acid (fully oxidized), some of which can react further to yield physiologically inactive compounds. For example, the two predominant folates in fresh foods, 5-methyl-FH$_4$ and 10-formyl-FH$_4$, are converted to 5-methyl-5,6-FH$_2$ and 10-formylfolic acid, respectively. For this reason, 5-methyl-5,6-FH$_2$ has been found to account for about half of the folate in most prepared foods. Although it can be reduced to the FH$_4$ form (e.g., by ascorbic acid), in the acidity of normal gastric juice it isomerizes to yield 5-methyl-5,8-FH$_2$, which is completely inactive. It is of interest to note that, owing to their gastric anacidosis, this isomerization does not occur in pernicious anemia patients, who are thus able to utilize the partially oxidized form by absorbing it and subsequently activating it to 5-methyl-FH$_4$. Because some folate derivatives of the latter type can support the growth responses of test microorganisms used to measure folates,[2] some information in the available literature may overestimate the biologically useful folate contents of foods and/or feedstuffs. Substantial losses in the folate contents of food can occur as the result of leaching in cooking water when boiling (losses of total folates of 22% for asparagus and 84% for cauliflower have been observed), as well as oxidation, as described above. Due to such losses, green leafy vegetables can lose their value as sources of folates despite their relatively high natural contents of the vitamin.

Bioavailability

The biological availability of folates in foods has been difficult to assess quantitatively. Estimates are variable among foods, but generally indicate bioavailabilities of about half that of folic acid; a recent study found a relatively high (80%) aggregate bioavailability of a mixed diet.[3] In general, folates appear to be less well utilized from plant-derived foods than from animal products (Table 16.2). Several factors affect the biologic availability of food folates:

- *Anti-folates in the diet.* Folates can bind to the food matrices; many foods contain inhibitors of the intestinal brush border folate conjugase and/or folate transport.
- Inherent characteristics of various folates. Folate vitamers vary in biopotency.

2. *Lactobacillus casei*, *Streptococcus fecium* (formerly, *S. lactis* R. and *S. fecalis*, respectively), and *Pediococcus cerevisiae* (formerly *Leuconostoc citrovorum*) have been used. Of these, *L. casei* responds to the widest spectrum of folates.

3. That diet contained folates from fruits, vegetables, and liver (Winkels, R. M., Brouwer, I. A., Sieblink E., *et al.* [2007] *Am. J. Clin. Nutr.* 85, 465).

and therefore are unstable to oxidation under aerobic conditions of storage and processing. Under such conditions (especially in the added presence of heat, light, and/or metal ions), FH$_4$ derivatives can readily be oxidized to

TABLE 16.2 Biologic Availability to Humans of Folates in Foods

Food/Feedstuff	Bioavailability, %[a] (Reported Range)
Bananas	0–148
Cabbage	0–127
Eggs	35–137
Lima beans	0–181
Liver (goat)	9–135
Orange juice	29–40
Spinach	26–99
Tomatoes	24–71
Wheat germ	0–64
Brewers' yeast	10–100
Soybean meal	0–83

[a]*Results expressed relative to folic acid.*
Sources: Baker, H., Jaslow, F. P., and Frank, O. (1978). *J. Am. Geriatr. Soc.* 26, 218; Baker, H. and Srikantia, S. G. (1976). *Am. J. Clin. Nutr.* 29, 376; Tamura, T. and Stokstad, E. R. L. (1973). *Br. J. Haematol.* 25, 513.

- *Nutritional status of the host.* Deficiencies of iron and vitamin C status are associated with impaired utilization of dietary folate.[4] Vitamin C has also been shown to enhance the utilization of 5-methyl-FH$_4$ by preventing its oxidative degradation to 5-methy-FH$_2$, which does not enter the folate metabolic pool.

Interactions of these factors complicate the task of predicting the bioavailability of dietary folates (Table 16.2). This problem is exacerbated by the methodological difficulties in evaluating folate utilization, which can be done through bioassays with animal models,[5] balance studies with humans, or isotopic methods to measure the appearance of folates in blood, excreta, and tissues.

---

4. Some anemic patients respond optimally to oral folate therapy only when they are also given iron. Patients with scurvy often have megaloblastic anemia, apparently owing to impaired utilization of folate. In some scorbutic patients vitamin C has an anti-anemic effect; others require folate to correct the anemia.

5. As with any application of information from studies with animal models, the validity of extrapolation is an issue important in assessing folate bioavailability. For example, the rat and many other species have little or no brush border conjugase activity, these species rely on the pancreatic conjugase for folate deconjugation. This contrasts with the pig and human, which deconjugate folates primarily by brush-border activity.

Synthesis by Intestinal Microflora

The microflora of the hindgut, particularly *Bacteriodes* spp., can synthesis folate in amounts that approximate daily dietary needs, although the extent to which that metabolism may benefit the monogastric host is not clear. Studies have shown that folate can be absorbed across the human colon.[6] However, the feeding of pigs with prebiotics known to be preferentially used by *Bacteroides* spp. markedly increased their colonic microbial biosynthesis of folate without affecting their circulating levels of the vitamin.[7]

3. ABSORPTION OF FOLATE

The absorption efficiency of dietary folates appears to be about 50%, but can be highly variable (10–90%). While the process is not completely elucidated, it involves several steps.

Deconjugation of Polyglutamyl Folates

Under fasting conditions, folic acid, 5-methyl-FH$_4$, and 5-formyl-FH$_4$ are virtually completely absorbed, and most polyglutamyl folates are absorbed at efficiencies in the general range of 60–80%. Because the majority of food folates occur as reduced polyglutamates, they must be cleaved to the mono- or diglutamate forms for absorption. This is accomplished by the action of folyl γ-glutamyl carboxypeptidases, more commonly called folyl conjugases.

Conjugase activities are widely distributed in the mucosa of the small intestine, both intracellularly and in association with the brush border.[8] These appear to be different enzymes:

- a 700 kDa brush border enzyme exocarboxypeptidase with an optimum of pH 6.5–7.0. Although present in lower amounts, it appears to be most important for the hydrolysis of dietary folyl polyglutamates. A genetic variant of this enzyme has been associated with low serum folate concentrations and homocysteinemia.
- a 75 kDa intracellular (lysosomal) enzyme with an optimum of pH 4.5–5.0.

Loss of conjugase activity results in impaired folate absorption. This can be produced by nutritional zinc deficiency or by exposure to naturally occurring inhibitors in foods (Table 16.3). Studies with several animal models have demonstrated that chronic ethanol feeding can

---

6. Aufreiter, S., Gregory, J. F., Pfeiffer, C. M., *et al.* (2009). *Am. J. Clin. Nutr.* 90, 116.

7. Aufreiter, S., Kim, J. H., and O'Connor, D. L. (2011). *J. Nutr.* 141, 366.

8. Conjugase activities have also been identified in bile, pancreatic juice, kidney, liver, placenta, bone marrow, leukocytes, and plasma, although the physiological importance of these activities is unclear. In the uterus, conjugase activity is induced by estrogen.

TABLE 16.3 Inhibition of Jejunal Folyl Conjugase Activities *in vitro* by Components of Selected Foods

Food	Pig Conjugase (% Inhibition)	Human Conjugase (% Inhibition)
Red kidney beans	35.5	15.9
Pinto beans	35.1	33.2
Lima beans	35.6	35.2
Black-eyed peas	25.9	19.3
Yellow cornmeal	35.3	28.3
Wheat bran	−2.0	0
Tomato	8.1	14.2
Banana	45.9	46.0
Cauliflower	25.2	15.3
Spinach	21.1	13.9
Orange juice	80.0	73.4
Egg	11.5	5.3
Milk	13.7	–
Cabbage	12.1	–
Whole wheat flour	0.3	–
Medium rye flour	2.2	–

From Bhandari, S. D. and Gregory, J. F. (1990). *Am. J. Clin. Nutr.* 51, 87.

decrease intestinal hydrolysis of folyl polyglutamates and can impair the absorption, transport, cellular release, and metabolism of folates.[9] Folyl conjugase inhibitors have been identified in certain foods: cabbage, oranges, yeast, beans (red kidney, pinto, lima, navy, soy), lentils, and black-eyed peas.[10] This appears to be the basis for the low bioavailability of folate in orange juice. Folate absorption can also be reduced by certain drugs, including cholestyramine (which binds folates), salicylazosulfapyridine,[11] diphenylhydantoin,[12] aspirin, and other salicylates, as well as several non-steroidal anti-inflammatory drugs.

Active Uptake by the Enterocyte

Dietary folates are absorbed in deconjugated form, i.e., as folic acid, 5-methyl-FH_4, and 5-formyl-FH_4.[13] These vitamers are actively transported across the brush border by a Na^+-coupled, carrier-mediated process facilitated by three protein transporters (Fig. 16.1):

- the **reduced folate carrier (RFC)**, which is stimulated by glucose and shows a pH optimum at about pH 5; it is a member of the solute carrier 19 (*SLC19*) family of transporters.[14] It is expressed in many tissues in which it binds reduced folates preferentially to non-reduced forms; however, as it is expressed in the intestinal mucosa, the RFC binds reduced and oxidized forms of the vitamin with comparable affinities. The transporter function appears to be driven by a transmembrane pH gradient. The expression of RFC appears to be upregulated by folate deficiency and downregulated by folate excess.
- a **proton-coupled folate transporter (PCFT)**, which is active under more acidic conditions. This proton-coupled, Na^+-dependent, high-affinity transporter was originally described as a low-affinity heme carrier, which is no longer regarded as its primary metabolic function.[15] Hereditary folate malabsorption, apparently involving failure of PCFT expression, has been reported.[16]
- **folate receptors (FRs)**, which form a family of high-affinity FRs (also called **folate-binding proteins, FBPs**) that bind folic acid stoichiometrically, and also bind 5-methyl-FH_4 and some antifolate drugs. Cellular internalization of folate is thought to involve the endocytosis of membrane-bound FR and/or the clustering of FRs in membrane-associated vesicles which release the ligand into the cytosol. FR and the folate transporters are localized on opposite aspects of polarized cells; this appears to facilitate the movement of folate across the apical membrane into the cell.

Passive Diffusion into the Enterocyte

Folic acid can also be absorbed passively, apparently by diffusion. This non-saturable process is linearly related to lumenal folate concentration, and can account for 20–30%

---

9. Other factors may contribute to this phenomenon: enterocytes are known to be sensitive to ethanol toxicity; many chronic alcoholics can have inadequate folate intakes.

10. The conjugase inhibitors in beans and peas reside in the seed coats and are heat labile.

11. This drug, also called azulfidine and sulfasalazine, is used to treat inflammatory bowel disorder.

12. This drug, also called dilantin, is an anticonvulsant.

13. The dog appears to absorb folyl polyglutamates.

14. This family also includes two thiamin transporters, SLC19A2 and SLC19A3. RFC (SLC19A1) also transports the folate antagonist and antineoplastic agent, methotrexate.

15. For this reason, it is referred to as PCFT/HCP1 or SLC46A1.

16. Hereditary folate malabsorption has been reported in some 30 patients, presenting at 2–6 months of age as megaloblastic anemia, mucositis, diarrhea, failure to thrive, recurrent infections, and seizures.

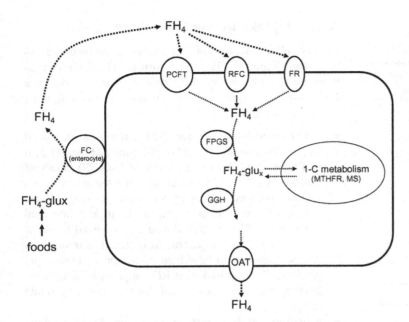

FIGURE 16.1 Protein-facilitated uptake and utilization of folate. *Abbreviations:* FC, folyl conjugase; PCFT, proton-coupled folate transporter; RFC, reduced folate carrier; FR, folate receptor; FPGS, folylpolyglutamte synthetase; GGH, γ-glutamyl hydrolase; OAT, organic anion transporter; MTHFR, methylenetetrahydrofolate reductase; MS, methionine synthase. *After Devos, L., Chanson, A., Liu, Z., et al. (2008). Am. J. Clin. Nutr. 88, 1149.*

of folate absorption at high folate intakes. It has been proposed that elevated folate absorption seen in individuals with pancreatic exocrine insufficiency may be due to their reduced excretion of bicarbonate, which loss of buffering capacity renders the lumenal milieu slightly more acidic and reduces the charge on the folate molecule. This could facilitate diffusion of the vitamin across the brush border. Under more basic conditions (i.e., pH >6.0), folate absorption is known to fall off rapidly.

Re-Conjugation

Absorbed monoglutamyl folates are converted to higher-glutamyl forms by the action of **folylpolyglutamate synthetase**. In an immediate sense, this prevents the vitamin being pumped out of the cell; it is thus part of the regulation of folate uptake.

Exportation

Absorbed FH_4 can be exported without further metabolism to the portal circulation or after being first alkylated (e.g., by methylation to 5-methyl-FH_4). Both forms are exported across the basolateral cell membrane apparently by a **folate export pump**, which is likely to be a member of the family of multispecific organic anion transporters (OATs), one of which has been shown to transport the folate antagonist methotrexate (Fig. 16.1). Because folate pumps do not transport polyglutamates, those forms are retained within the cell until being deconjugated by a **γ-glutamyl hydrolase**.

4. TRANSPORT OF FOLATE

Free in Plasma

In most species, folate is transported to the tissues mainly as monoglutamyl derivatives in free solution in the plasma. The notable exception is the pig, in which FH_4 is the predominant circulating form of the vitamin.[17] The predominant form in portal plasma is the reduced form, tetrahydrofolic acid (FH_4). This is taken up by the liver, which releases it to the peripheral plasma after converting it primarily to 5-methyl-FH_4, but also to 10-formyl-FH_4. The concentration of 10-formyl-FH_4 is tightly regulated,[18] whereas that of 5-methyl-FH_4 is not. Thus, the latter varies in response to folate meals, etc. Thus, folates of dietary origin are absorbed and transported to the liver as FH_4, which is converted to the methylated form and transported to the peripheral tissues.

Plasma folate concentrations in humans are typically in the range of 10–30 nmol/l. Most circulates in free solution, but some is bound to low-affinity protein binders such as albumin, and some is bound to a soluble form of the high-affinity FBP. The latter is also found in high levels in milk. Erythrocytes contain greater concentrations of folate than plasma; typically 50–100 nmol/l. These stores are accumulated during erythropoiesis; the mature erythrocyte does not take up folate.

---

17. The metabolic basis for this anomaly is not clear.
18. In humans, the plasma level is held at about 80 ng/dl.

The folate levels of both plasma and erythrocytes are reduced by cigarette smoking; smokers show plasma folate levels that are more than 40% less than those of non-smokers.[19] While serum folates have been found to be normal among middle-class drinkers of moderate amounts of ethanol, more than 80% of impoverished chronic alcoholics show abnormally low serum levels and some 40% show low erythrocyte levels. This corresponds to a similar incidence (34–42%) of megaloblastosis of the bone marrow in alcoholic patients. These effects probably relate to the displacement of foods containing folates by alcoholic beverages, which are virtually devoid of the vitamin, as well as to direct metabolic effects: inhibition of intestinal folyl conjugase activity and decreased urinary recovery of the vitamin.

Folate-Binding Proteins (FBP)

Folate is also bound to albumin and two other proteins in the plasma. Of the latter, one is a high-affinity FBP present in low concentrations (binding <10ng folic acid per deciliter) and elevated in folate-deficient subjects and pregnant women. That protein is thought to be a solubilized form of folate-binding proteins in tissues. FBPs have also been identified in milk and other tissues.[20] Each binds folates non-covalently with high affinity in a stable complex.[21] The FBPs in milk[22] have been shown to stimulate the enteric absorption of folate. Liver contains two FBPs with dimethylglycine dehydrogenase and sarcosine dehydrogenase activity. It has been suggested that tissue FBPs, which bind polyglutamate forms of the vitamin, may play important roles in stabilizing folates within cells, thus reducing their rates of metabolic turnover and increasing their intracellular retention.

Cellular Uptake

Circulating monoglutamyl folates are taken up by the same process by which they were taken up by enterocytes. This involves the folate transporters and receptors (Fig. 16.1):

- **RFC**, which is ubiquitously expressed in tissues in which its functional characteristics vary. In tissues other than intestine, RFCs bind reduced folates preferentially to non-reduced forms. They facilitate the bi-directional movement of folates across cell membranes.

- **PCFT**, which is ubiquitously expressed in the liver and other tissues. It is also thought to play a role in folate receptor-mediated endocytosis.

- **FRs**, which facilitate the uni-directional movement of folates across cell membranes. Isoforms of FR cDNAs have been cloned: three from humans (FR-α, -β and -γ) and two from murine L1210 cells. The regulation of FR expression is not well understood, but it is clear that extracellular folate concentration plays an important role in that process, serving as an inverse stimulus to FR expression.

Tissue Distribution

In humans, the total body content of folate is 5–10mg, about half of which resides in the liver in the form of tetra-, penta-, hexa-, and heptaglutamates of 5-methyl-FH_4 and 10-formyl-FH_4.[23] The relative amounts of these single-carbon derivatives vary among tissues, depending on the rate of cell division. In tissues with rapid cell division (e.g., intestinal mucosa, regenerating liver, carcinoma) relatively low concentrations of 5-methyl-FH_4 are found, usually with concomitant elevations in 10-formyl-FH_4. In contrast, in tissues with low rates of cell division (e.g., normal liver), 5-methyl-FH_4 predominates. Brain folate (mostly 5-methyl-FH_4) levels tend to be very low, with a subcellular distribution (penta- and hexaglutamates mostly in the cytosol and polyglutamates mostly in the mitochondria) the opposite of that found in liver.

Folate-deficient animals show relatively low hepatic concentrations of shorter chain-length folyl polyglutamates compared with longer chain-length folates. This suggests that the longer chain-length metabolites are better retained within cells. In the rat, uterine concentrations of folates show cyclic variations according to the menstrual cycle, with maxima coincident with peak estrogenic activity just before ovulation.[24]

The concentration of folates in erythrocytes is widely used in the assessment of folate status, as it responds to changes in intake of the vitamin (Fig. 16.2). That response, however, tends to be greater for women that for men; half of this difference is explained on the basis of differences in body size.

---

19. These findings probably relate to the inactivation of cobalamins by factors (cyanides, hydrogen sulfide, nitrous oxide) in cigarette smoke (see Chapter 17).
20. Urinary FBP is presumed to be of plasma origin.
21. Other proteins (e.g., albumin) bind folates non-specifically, forming complexes that dissociate readily.
22. There are two FBPs in milk. Each is a glycoprotein; one may be a degradation product of the other.

23. The hepatic reserve of folate should be sufficient to support normal plasma concentrations of the vitamin (>400ng/dl) for at least 4 weeks. (Signs of megaloblastic anemia are usually not observed within 2–3 months of folate deprivation.) However, some evidence suggests that the release of folate from the liver is independent of nutritional folate status, resulting instead from the deaths of hepatocytes.
24. On the basis of this type of observation, it has been suggested that estrogen enhancement of folate turnover in hormone-dependent tissues may be the basis of the effects of pregnancy and oral contraceptive steroids in potentiating low folate status.

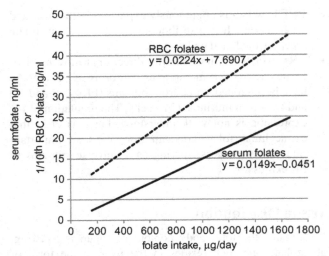

FIGURE 16.2 Relationships of folates in serum and erythrocytes (RBC) to dietary folate intake. *From population data from National Health and Nutrition Examination Surveys (NHANES III and annual NHANES from 1999–2004); Quinlivan, E. P. and Gregory, J. F. (2007). Am. J. Clin. Nutr. 86, 1773.*

5. METABOLISM OF FOLATE

There are three aspects of folate metabolism:

- *Reduction of the pteridine ring system.* Reduction of the pterin ring from the two non-reduced states, folic acid and dihydrofolic acid (FH_2), to the fully reduced form tetrahydrofolic acid (FH_4) that is capable of accepting a single-carbon unit is accomplished by the cytosolic enzyme **7,8-dihydrofolate reductase** (Fig. 16.3).[25] This activity is found in high amounts in liver and kidney, and in rapidly dividing cells (e.g., tumor). The reductase is inhibited by several important drugs, including the cancer chemotherapeutic drug **methotrexate**,[26,27] which appears to exert its antitumor action by inhibiting the reductase activity of tumor cells.

- *Reactions of the polyglutamyl side chain.* The folyl monoglutamates that are taken up by cells are trapped therein as polyglutamate derivatives that cannot cross cell membranes. Polyglutamate forms are also mobilized by side chain hydrolysis to the monoglutamate. These conversions are catalyzed by two enzymes:

- *Polyglutamation.* The conversion of 5-methyl-FH_4 to folyl polyglutamate is accomplished by the action of the ATP-dependent **folyl polyglutamate synthetase**,[28] which links glutamyl residues to the vitamin by peptide bonds involving the γ-carboxyl groups.[29] The enzyme requires prior reduction of folate to FH_4 or demethylation of the circulating 5-methyl-FH_4 (by vitamin B_{12}-dependent methionine synthetase). It is widely distributed at low concentrations in many tissues. That folyl polyglutamate synthetase is critical in converting the monoglutamyl transport forms of the vitamin to the metabolically active polyglutamyl forms was demonstrated by the discovery of a mutational loss of the synthetase activity, which produced lethal folate deficiency. In most tissues, the activity of the low-abundance enzyme is rate-limiting for folate accumulation and retention. Cells that lack the enzyme are unable to accumulate the vitamin.[30] Those lacking the mitochondrial enzyme cannot accumulate the vitamin in that subcellular compartment, and consequently are deficient in mitochondrial single-carbon metabolism.

- *Cellular conjugases.* Folyl polyglutamates are converted to derivatives of shorter chain length by lysosomal[31] γ-glutamyl carboxypeptidases, also referred to as cellular conjugases, some of which are zinc-metalloenzymes.

- *Acquisition of single-carbon moieties.* Folate is metabolically active as a variety of derivatives with single-carbon units at the oxidation levels of formate, formaldehyde, or methanol[32] substituted at the N-5 and/or N-10 positions of the pteridine ring system (Fig. 16.4). The main source of single-C fragments is **serine hydroxymethyltransferase** (Table 16.4), which uses the

---

25. Also called *tetrahydrofolate dehydrogenase*, this 65 kDa NADPH-dependent enzyme can reduce folic acid to FH_2 and, of greater importance, FH_2 to FH_4. The enzyme is potently inhibited by the drug methotrexate, a 4-aminofolic acid analog.

26. 4-Amino-10-methylfolic acid.

27. Other inhibitors include the antimalarial drug pyrimethamine and the antibacterial drug trimethoprim.

28. The mitochondrial and cytosolic forms of the enzyme are encoded by a single gene, the transcription of which has alternate start sites and the mRNA of which has alternative translation sites.

29. This enzyme also catalyzes the polyglutamation of the anticancer folate antagonist methotrexate, which enhances its cellular retention. Tumor cells, which have the greatest capacities to perform this side chain elongation reaction, are particularly sensitive to the cytotoxic effects of the antagonist.

30. Because polyglutamation is also necessary for the cellular accumulation and cytotoxic efficacy of anti-folates such as methotrexate, decreased folyl polyglutamate synthase activity is associated with clinical resistance to those drugs.

31. Lysosomes also contain a folate transporter which is thought to be active in bringing folypolyglutamates into that vesicle.

32. It should be noted that single carbons at the oxidation level of CO_2 cannot be transported by folates; such fully oxidized carbon is transported by biotin and thiamin pyrophosphate.

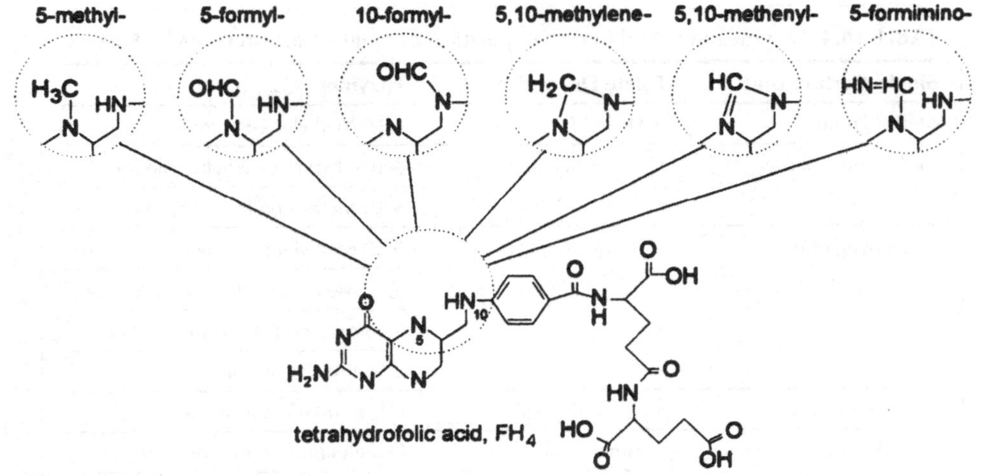

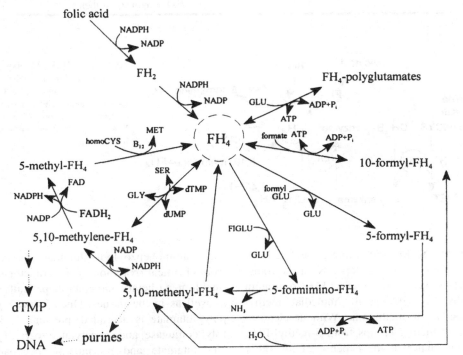

FIGURE 16.3 Single-carbon units carried by folate.

FIGURE 16.4 Role of folate in single-carbon metabolism.

dispensable amino acid serine[33] as the single-C donor. Each folyl derivative is a donor of its single-C unit in metabolism;[34] thus, by cycling through the acquisition/loss of single-C units, each derivative delivers these species to a variety of metabolic uses. Most single-C folate derivatives in cells are bound to enzymes or FBPs; the concentrations of the free pools of folate coenzymes are therefore low, in the nanomolar range.

Methyl-Folate Trap

The major cycle of single-C flux in mammalian tissues appears to be the serine hydroxymethyltransferase/5,10-methylene-FH$_4$ reductase/methionine synthase cycle, in which the latter reaction is rate limiting (Fig. 16.5). The committed step (5,10-methylene-FH$_4$ reductase) is

33. Serine is biosynthesized from glucose in non-limiting amounts in most cells.
34. Although the route of its biosynthesis is unknown, eukaryotic cells contain significant amounts of 5-formyl-FH$_4$. That folyl derivative, also called leucovorin, folinic acid, and citrovorum factor, is used widely to reverse the toxicity of methotrexate and, more recently, to potentiate the cytotoxic effects of 5-fluorouracil.

TABLE 16.4 Enzymes Involved in the Acquisition of Single-Carbon Units by Folates

Single-Carbon Unit	Folate Derivative	Enzymes
Methyl group	5-Methyl-FH$_4$	5,10-Methylene-FH$_4$ reductase
Methylene group	5,10-Methylene-FH$_4$	Serine hydroxymethyltransferase
		5,10-Methylene-FH$_4$ dehydrogenase
Methenyl group	5,10-Methenyl-FH$_4$	5,10-Methylene-FH$_4$ dehydrogenase
		5,10-Methenyl-FH$_4$ cyclohydrolase
		5-Formimino-FH$_4$ cyclohydrolase
		5-Formyl-FH$_4$ isomerase
Formimino group	5-Formimino-FH$_4$	FH$_4$ formiminotransferase
Formyl group	5-Formyl-FH$_4$	FH$_4$:glutamate transformylase
	10-Formyl-FH$_4$	5,10-Methenyl-FH$_4$ cyclohydrolase
		10-Formyl-FH$_4$ synthetase

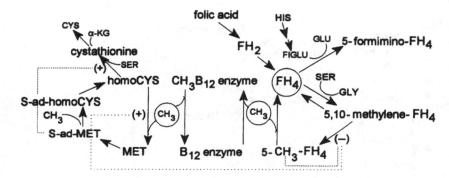

FIGURE 16.5 Metabolic interrelationships of folate and vitamin B$_{12}$; basis of the *"methyl-folate trap"*. *Abbreviations*: CYS, cysteine; α-KG, α-ketoglutarate; SER, serine; homoCYS, homocysteine; S-ad-homoCYS, S-adenosylhomocysteine; S-ad-MET, S-adenosylmethionine; MET, methionine.

feedback-inhibited by SAM and product-inhibited by 5-methyl-FH$_4$. Methionine synthase depends on the transfer of labile methyl groups from 5-methyl-FH$_4$ to vitamin B$_{12}$, which, as methyl-B$_{12}$, serves as the immediate methyl donor for converting Hcy to methionine. Without adequate vitamin B$_{12}$ to accept methyl groups from 5-methyl-FH$_4$, that metabolite accumulates at the expense of the other metabolically active folate pools. This is known as the "methyl-folate trap." The loss of FH$_4$ that results from this blockage in folate recycling blocks transfer of the histidine-formino group to folate (as 5-formimino-FH$_4$) during the catabolism of that amino acid. This results in the accumulation of the intermediate FIGLU. Thus, elevated urinary FIGLU levels after an oral histidine load are diagnostic of vitamin B$_{12}$ deficiency.

Catabolism

Tissue folates turn over by the cleavage of the polyglutamates at the C-9 and N-10 bonds to liberate the pteridine and *p*-aminobenzoylpolyglutamate moieties. This results from chemical oxidation of the cofactor both in the intestinal lumen (dietary and enterohepatically recycled folates) as well as in the tissues. Once formed, *p*-aminobenzoyl-polyglutamate is degraded, presumably by the action of folyl conjugase, and is acetylated to yield *p*-acetaminobenzoylglutamate and *p*-acetoaminobenzoate. The rate of folate catabolism appears to be related to the rate of intracellular folate utilization. Accordingly, urinary levels of *p*-acetoaminobenzoate are correlated with the total body folate pool. Folate breakdown is greatest under hyperplastic conditions (e.g., rapid growth, pregnancy[35]). Studies of the kinetics of folate turnover using isotopic labels have indicated at least two metabolic pools of the vitamin in non-deficient individuals: a larger, slow-turnover pool, and

35. This effect may be an important contributor to the folate-responsive megaloblastic anemia common in pregnancies in parts of Asia, Africa, and Central and South America (where rates as high as 24% have been reported), as well as in the industrialized world (where rates of 2.5–5% have been reported).

TABLE 16.5 Folate Catabolic Rates in Humans

Folate Intake (μg/Day)	Total Body Catabolic Rate	
	mg	% body pool/day
200	28.5	0.47
300	31.5	0.61
400	32.2	0.82

From Gregory, J. F., Williamson, J., Liao, J. F., *et al.* (1998). *J. Nutr.* 128, 1896.

TABLE 16.6 Polymorphisms of Proteins Related to Folate Absorption and Metabolism

Enzyme	Polymor-phism	Genotype: Frequency		
MTHFR	T6557C	CC: 41%	TT: 18%	CT: 41%
	C1289A	AA: 53%	CC: 9%	AC: 37%
Folyl conjugase	T1561C	CC: 92%	TT: 0[a]	CT: 8%
Reduced folate transporter	G80A	AA: 35%	GG: 18%	AG: 47%
Methionine synthase	G2756A	AA: 59%	GG: 3%	AG: 38%
Methionine synthase reductase	G66A	AA: 28%	GG: 23%	AG: 49%

[a] 1 in 625 reported.
From Molloy, A. M. (2002). *J. Vit. Nutr. Res.* 72, 46.

a smaller, rapid-turnover pool. At low to moderate intakes, 0.5–1% of total body folate appears to be catabolized and excreted per day (Table 16.5).

Excretion

Intact folates and the water-soluble side chain metabolites *p*-acetaminobenzoylglutamate and *p*-acetaminobenzoate are excreted in the urine and bile. The total urinary excretion of folates and metabolites is small (e.g., ≤1% of total body stores per day). Folate conservation is effected by the reabsorption of 5-methyl-FH$_4$ by the renal proximal tubule; renal reabsorption appears to be mainly a non-specific process, although an FBP-mediated process has also been demonstrated.

Fecal concentrations of folates are usually rather high – comparable to that excreted in the urine. However, fecal folates represent mainly those folates of intestinal microfloral origin, as enterohepatically circulated folates appear to be absorbed quantitatively.

Polymorphisms of Enzymes in Folate Metabolism

Genetic polymorphisms have been identified in most of the major enzymes involved in folate utilization (Table 16.6):[36]

- *Folyl Conjugase T1561TC.* Polymorphism involving a T→C substitution of base pair 1561 has been described.[37] Individuals with the T allele have been found to have higher circulating folate levels than those with the C allele.[38]

- *RFC G80A.* Polymorphism involving a G→A substitution of base pair 80 has been described. Individuals homozygous for the G allele have been found to have a tendency toward elevated circulating Hcy levels and reduced erythrocyte folate levels.[39]

- *MTHFR polymorphisms.* Three polymorphisms have been identified in the human MTHFR: a C→T substitution of base pair 677; an A→C substitution of base pair 1298.; and a G→A substitution of base pair 1793.

  - *C677T.* Some 89% of the American population have the C allele. Only 20% of Mexican Americans, 12% of non-Hispanic Whites but only 1% of non-Hispanic Blacks are homozygous for the T variant (TT);[40] they have a form of MTHFR with 70% lower enzyme activity, lower affinity for the flavin cofactor, and lower thermal stability[41] than the C/C form of the enzyme. They also show lower plasma folate concentrations, mild homocysteinemia,[42] and lower global DNA methylation than other genotypes (Table 16.7) – effects that appear to be exacerbated by low folate intakes. Individuals of the TT

36. Polymorphisms of other enzymes involved in folate and single-carbon metabolism have also been described, although none has been shown to affect folate utilization. These include: folylpolyglutamate synthetase C303T and A2006G, methionine synthase G2756A, and folyl γ-glutamyl hydrolase upstream T→C and C→T polymorphisms.
37. The enzyme is encoded by the glutamate carboxypeptidase II gene (*GCPII*).
38. Afman, L. A., Trijbels, F. J. M., and Blom, H. J. (2003). *J. Nutr.* 133, 75.

39. Chango, A., Emery-Fillon, N., de Courcy, G.P., *et al.* (2000). *Mol. Genet. Med.* 70, 310.
40. Yang, Q. H., Botto, L. D., Gallagher, M., *et al.* (2008). *Am. J. Clin. Nutr.* 88, 232.
41. For this reason the variant is frequently referred to as the "thermolabile enzyme."
42. Kauwell, G. P. A., Wilsky, C. E., Cerda, J. J., *et al.* (2000). *Metabolism* 49,1440.

TABLE 16.7 Effect of MTHFR C677T Genotype on Folate, Homocysteine, and Vitamin B_{12} Status in Humans

Metabolite	MTHFR C677T Genotype		
	CC	TC	TT
Plasma folate (nmol/l)	12.8 ± 6.5	12.8 ± 6.7	9.5 ± 3.1
Erythrocyte folate (nmol/l)	541 ± 188	517 ± 182	643 ± 186
Plasma Hcy (μmol/l)	13.4 ± 3.4	13.2 ± 3.1	17.1 ± 11.5
Plasma vitamin B_{12} (pmol/l)	246 ± 130	271 ± 121	233 ± 94

From van der Put, N. M., Gabreels, F., Stevens, E. M., *et al.* (1995). *Lancet* 346, 1070–1071.

genotype show the greatest Hcy-lowering responses to folate supplements.

- *A1298C.* More than 90% of Americans have the A allele.[43] Having the C allele appears to be without significant physiological consequence unless combined with the MTHFR C677T polymorphism; doubly heterozygous individuals have been found to have MTHFR-specific activities two-thirds those of doubly homozygous individuals, with lower circulating levels of folate and increased circulating levels of Hcy.[44]
- *G1793A.* This polymorphism is less frequent, about 4% of American women have the GG genotype. The functional significance of this polymorphism is unclear.
- *Dihydrofolate reductase deletion.* A polymorphism has been described involving a 19-bp deletion/insertion in the first intron, 60 bases from the splice donor site. Individuals homozygous for the deletion have been shown to have impaired folate metabolism at both low and high intakes of the vitamin, apparently mostly affecting tissue stores.[45]
- *Methionine synthase reductase A66A.* A polymorphism involving an A→G substitution of base pair 66 has been described. This appears to be without physiologic consequence except for individuals doubly homozygous with MSR 66GG and MTHFR 677TT, who have been found to have serum Hcy levels 26% less than those of other genotypes.

6. METABOLIC FUNCTIONS OF FOLATE

Folates function as enzyme co-substrates in many reactions of the metabolism of amino acids and nucleotides, as well as the formation of the primary donor for biological methylations, **S-adenosylmethionine** (**SAM**). In each of these functions, the fully reduced form (tetrahydrofolic acid, FH_4) of the vitamin serves as an acceptor or donor of a single-carbon unit (Table 16.8). Collectively, these reactions are referred to as **single-carbon metabolism**.

Single-Carbon Metabolism

Methionine and SAM Synthesis

The active metabolite 5-methyl-FH_4, which is freely produced from 5,10-methenyl-FH_4, provides labile methyl groups for methionine synthesis from **homocysteine** (**Hcy**).[46] Methionine is essential for the synthesis of proteins and polyamines, and is the precursor of SAM, which serves as a donor of "labile" methyl groups for more than 100 enzymatic reactions that have critical roles in metabolism.[47] SAM[48] also serves as a key regulator of the transsulfuration and remethylation pathways. This metabolism is catalyzed by two enzymes: **methylenetetrahydrofolate reductase** (**MTHFR**) and **methionine synthase**.[49]

Histidine Catabolism

Cytosolic forminonotransferase catalyzes the final reaction in the catabolism of histidine by transferring the formimino group from **formiminoglutamate** (**FIGLU**) to FH_4.

Nucleotide Metabolism

While not required for the *de novo* synthesis of pyrimidines, folates are required for the synthesis of thymidylate, and they are also required for purine synthesis. Both roles are necessary for the *de novo* synthesis of DNA, and thus for DNA replication and cell division. Accordingly, disruption of these functions impairs cell division and results in the macrocytic anemia of folate deficiency.

43. Chango, A., Emery-Fillon, N., de Courcy, G.P., *et al.* (2000). *Mol. Genet. Med.* 70, 310.

44. Chango, A., Boisson, F. Barbe, F., *et al.* (2000). *Br. J. Nutr.* 83, 593.

45. Kalmbach, R. D., Choumenkovitch, S. F., Troen, A. P., *et al.* (2008). *J. Nutr.* 138, 2323.

46. This is one of two pathways for the synthesis of methionine from Hcy, the other using betaine as a methyl donor.

47. By loss of the flux of methyl groups via 5-methyl-FH_4:Hcy methyltransferase, folate deficiency causes a secondary hepatic choline deficiency.

48. It is also an allosteric inhibitor of 5,10-methylene-FH_4 reductase and an activator of the pyridoxal phosphate-dependent enzyme cystathionine β-synthase, which catalyzes the condensation of Hcy and serine to form cystathionine.

49. Two genetic polymorphisms in methionine synthase have been identified. It is not clear whether either has functional significance.

TABLE 16.8 Metabolic Roles of Folate

Folate Coenzyme	Enzyme	Metabolic Role
5,10-Methylene-FH$_4$	Serine hydroxymethyltransferase	Receipt of a formaldehyde unit in serine catabolism (mitochondrial enzyme important in glycine synthesis)
	Thymidylate synthetase	Transfers formaldehyde to C-5 of dUMP to form dTMP in pyrimidines
10-Formyl-FH$_4$	10-Formyl-FH$_4$ synthetase	Accepts formate from tryptophan catabolism
	Glycinamide ribonucleotide transformylase	Donates formate in purine synthesis
	5-Amino-4-imidazolecarboxamide transformylase	Donates formate in purine synthesis
	10-Formyl-FH$_4$ dehydrogenase	Transfers formate for oxidation to CO_2 in histidine catabolism
5-Methyl-FH$_4$	Methionine synthetase	Provides methyl group to convert Hcy to methionine
	Glycine *N*-methyltransferase	Transfers methyl group from *S*-adenosylmethionine to glycine in the formation of Hcy
5-Formimino-FH$_4$	Formiminotransferase	Accepts formimino group from histidine catabolism

- *Thymidylate synthesis.* Folates transport formaldehyde (as 5,10-methylene-FH$_4$) for **thymidylate** synthesis. The enzyme **thymidylate synthase**[50] catalyzes the the transfer of the single-C unit from 5,10-methylene-FH$_4$ conversion of deoxyuridine monophosphate (dUMP), thus converting the latter to deoxythymidine monophosphate (dTMP). This step is rate-limiting to DNA replication, and thus to the normal progression of the cell cycle. Thymidylate synthase is expressed only in replicating tissues; highest expression occurs during the S phase of the cell cycle.

- *Purine synthesis.* Folates transport formate (as 10-formyl-FH$_4$, produced from 5,10-methylene-FH$_4$) in the synthesis of adenine and guanine. In this way formyl groups are used to provide the C-2 and C-8 positions of the purine ring, and FH$_4$ is regenerated. These reactions are catalyzed by aminoimidazolecarboxamide ribonucleotide tranformylase and glycinamide ribonucleotide tranformylase, respectively.

Regulation

The regulation of single-C metabolism is effected by the interconversion of oxidation states of the folate intermediates. In mammalian tissues, the β-carbon of serine is the major source of single-C units for these aspects of metabolism. The C-fragment is accepted by FH$_4$ to form 5,10-methylene-FH$_4$ (by **serine hydroxymethyltransferase**), which has a central role in single-C metabolism. It can be used directly for the synthesis of thymidylate

(by thymidylate synthetase);[51] it can be oxidized to **5,10-methenyl-FH$_4$** (by **5,10-methylene-FH$_4$ dehydrogenase**) for the *de novo* synthesis of purines or it can be reduced to 5-methyl-FH$_4$ (by MTHFR) for use in the synthesis of methionine. The result is the channeling of single-C units in several directions: to methionine, to thymidylate (for DNA synthesis) or to purine synthesis.

Because folyl polyglutamates have been found to inhibit a number of the enzymes of single-C metabolism, it has been suggested that variation in their polyglutamate chain lengths (observed under different physiological conditions) may play a regulatory role.

DNA Methylation

Epigenetic regulation of specific genes is emerging as a major means by which different phenotypes are generated, with differing risks of diseases, including neurological diseases and tumorigenesis. This occurs by the methylation of cytosine bases in DNA as well as of the histone proteins associated with the chromatin. Hypomethylation of these

---

50. The anti-folate 5-fluorodeoxyuridylate is active by forming a complex with the enzyme and its folate co-substrate.

51. This is the sole *de novo* path of thymidylate synthesis. It is also the only folate-dependent reaction in which the cofactor serves both as a single-carbon donor and as a reducing agent. Thymidylate synthase is the target of the anticancer drug 5-fluorouracil (5-FU). The enzyme converts the drug to 5-fluorodeoxyuridylate, which is incorporated into RNA and also is a suicide inhibitor of the synthetase. Inhibition of the synthetase results in the cellular accumulation of deoxyuridine triphosphate (dUTP), which is normally present at only very low concentrations, and the incorporation of deoxyuridine (dU) into DNA. DNA with this abnormal base is enzymatically cleaved at sites containing dU, leading to enhanced DNA breakage.

factors is believed to alter chromatin structure in ways that affect transcription and can increase the rate of C→T transition mutation.[52] DNA hypermethylation is associated with gene silencing.

Folate has a role in the methylation of DNA by virtue of its function in single-C metabolism, i.e., 5-methyl-FH_4 supporting the continued adequate supply of SAM, the primary methylating agent. Folate deprivation has been found to produce chromosomal breaks in megaloblastic bone marrow, reflecting DNA strand breaks and hypomethylation. Studies have found the MTHFR 677TT genotype, which limits the activity of that enzyme by some 70%, to have reduced genomic DNA methylation.[53] One study found folate deprivation to produce hypomethylation within the *p53* gene; although genome-wide hypomethylation was not observed, genome-wide DNA strand breakage was increased. Such results suggest that folate deficiency may affect carcinogenesis by creating fragile sites in the genome and potentially inducing proto-oncogene expression. Another study showed that low-folate subjects with the MTHFR 677TT genotype had hypomethylation of whole blood DNA, compared to those with the CC genotype; those differences were not apparent among folate-adequate subjects of both genotypes.[54] That folate status affects DNA integrity is supported by findings that the circulating folate level is directly related to the length of peripheral lymphocyte telomeres, i.e., the capping chromosomal segments characterized by tandem repeats of DNA and associated proteins, the dysfunction of which is associated with age-related disease.[55]

Erythropoiesis

Folate is required for the production of new erythrocytes through its functions in the synthesis of purines and thymidylate required for DNA synthesis, as well as through its function in regenerating methionine and SAM, the methyl donor for DNA methylation. Folate deficiency results in the arrest of erythropoiesis prior to the latter stages of differentiation, resulting in apoptotic reduction of cells surviving to post-mitotic, terminal stages in the condition called megaloblastic anemia. Anemia can have folate-responsive components in subjects with apparently normal plasma folate levels; therefore, addition of folate to iron supplements can improve the treatment of anemia in pregnancy as well as in undernourished individuals.

7. FOLATE IN HEALTH AND DISEASE

Pernicious Anemia

High doses of folate (e.g., 400 μg/day intramuscular; 5 mg/day oral) have been shown to correct the **megaloblastic anemia** of pernicious anemia patients, who are deficient in vitamin B_{12}. This phenomenon renders megaloblastic anemia not useful for diagnosing either vitamin deficiency without accompanying metabolic measurements: FIGLU (elevations indicate folate deficiency) and **methylmalonic acid** (**MMA**) (elevations indicate vitamin B_{12} deficiency).

Supplemental folate does not mask the irreversible progression of neurological dysfunction and cognitive decline of vitamin B_{12} deficiency; however, those signs develop over a longer period of time than the anemia produced by the same deficiency. In fact, folate supplementation has been shown to exacerbate the cognitive symptoms of vitamin B_{12} deficiency.[56] Because vitamin B_{12} deficiency is estimated to affect 10–15% of the American population over 60 years of age, the amount of folate for the fortification of wheat flour (140 μg/100 g flour) was chosen to provide an amount of added folate (100 μg/person per day) sufficient for only a small proportion of the general population to receive a level (>1 mg/day) capable of masking vitamin B_{12} deficiency.

Infant Health

Pregnancy increases the need for folate to meet the increasing demands of the placenta and fetus, which can put women at increased risk of folate deficiency. Accordingly, several studies in apparently well-nourished populations have demonstrated the value of periconceptional folate supplementation in increasing placental weight and fetal growth rate, and reducing the risk of low birth weight. Folate, an essential factor in the support of normal cell division, is a key factor affecting embryogenesis and normal development.

Neural Tube Defects (NTDs)

For nearly four decades, adequate folate status has been linked to reduced risks of abnormalities in early embryonic development and, specifically, to risk of malformations of the embryonic brain and/or spinal cord, collectively

52. Methylated CpG sites appear to be at particularly high risk for C→T changes, which occur in the *p53* tumor suppressor gene.
53. Casto, R., Rivera, I., Ravasco, P., *et al.* (2004). *J. Med. Genet.* 41, 454.
54. Frisco, S., Choi, S. W., Girelli, D., *et al.* (2002). *Proc. Natl Acad. Sci. USA* 99, 5606.
55. Paul, L., Catterneo, M., D'Angelo, A., *et al.* (2009). Telomere length in peripheral blood mononuclear cells is associated with folate status in men. *J. Nutr.* 139, 1273.

56. Morris, M. S., Jacques, P. F., and Rosenberg, I. H. (2007). *Am. J. Clin. Nutr.* 85, 193.

TABLE 16.9 Results of Placebo-Controlled, Clinical Intervention Trials of Folate Supplements in the Prevention of Neural Tube Defects

Trial	Folate Treatment	NTD Rates, Cases/Total Pregnancies		RR (95% CI)
		Placebo	Treatment	
1[a]	4 mg	4/51	2/60	0.42 (0.04–2.97)
2[b]	4 mg ± multivitamins	21/602	6/593	0.34 (0.10–0.74)
3[c]	0.8 mg + multivitamins	2/2104	0/2052	0.00 (0.00–0.85)

[a]Lawrence, K. M., James, N., Miller, M., and Campbell, H. (1980). Br. Med. J. 281, 1,542 (women with NTD histories).
[b]Milunsky, A., Jick, H., Jick, S. S. et al. (1989). J. Am. Med. Assoc. 262, 2847 (women with NTD histories).
[c]Czeizel, A. E. and Fritz, I. (1992). J. Am. Med. Assoc. 262, 1,634 (women without previous NTD births).

referred to as NTDs.[57] These linkages have involved observations of high incidences of low folate status among women with NTD birth compared to women with normal birth outcomes. Additional support came from the production of NTDs in animal models by the folate antagonist aminopterin.

Several clinical intervention trials have tested the hypothesis that periconceptional supplemental folate can reduce NTD risk. One of these, a large, well-designed, multi-centered trial conducted by the British Medical Research Council, found that a daily oral dose of 4 mg of folic acid reduced significantly the incidence of confirmed NTDs among the pregnancies of women at high risk for such disorders.[58] Several subsequent studies (Table 16.9) have shown that periconceptional supplementation of folate can reduce the risk of NTDs. These have included trials conducted in the US, which found folate supplements (400–4,000 µg) effective in preventing NTDs in women with prior NTD pregnancies.[59] While folate supplements do not appear to affect NTD case-fatality rates, reductions in NTD incidence are associated with reductions in neonatal deaths. A meta-analysis of eight observational studies indicated that folate supplementation was associated with a 46% reduction in NTD risk, which was associated with a 13% reduction in neonatal deaths.[60]

Folate intervention trials conducted in several countries have shown greatest benefits of folate supplementation in circumstances with relatively high NTD rates (Fig. 16.6). A trial conducted in China with some 250,000 subjects showed that a daily supplement containing 400 µg folic acid consumed with ≥80% compliance during the periconceptional period was associated with reductions in NTD risk of 85% in a high NTD prevalence area and of 40% in a low prevalence area.[61] A systematic review of 14 folate intervention trials pointed out that not all NTD cases can be prevented by folate; that analysis suggested that a rate of 8–10/10,000 live births or abortions would appear to involve factors not affected by increasing folate intakes.[62]

The protective effect of folate supplementation may be limited to a subset of subjects with defective folate metabolism due to MTHFR mutations. The MTHFR 677TT genotype is associated with NTD risk, but the effect appears to be dependent on folate status. In a recent analysis, TT homozygosity was associated with a five-fold increase in NTD risk for mothers not using multivitamin supplements, but was without effect for mothers using supplements.[63]

57. NTDs comprise the most common forms of congenital malformations, with an annual global incidence estimated at >300,000 new cases, >41,000 deaths, and the loss of 2.3 million disability-adjusted life years. These involve developmental failures of the neural structures (brain, spinal cord, cranial bones, vertebral arches, meninges and overlying skin) formed from the embryonic neural tube in humans 20–28 days after fertilization. The most prominent NTDs are anencephaly and spina bifida. More than 95% of NTD pregnancies occur in families with no history of such defects, but women with one affected pregnancy or with spina bifida themselves face a risk of 3–4% of an NTD in a subsequent pregnancy.
58. The double-blind, randomized clinical trial involved 1,817 women, each with a previous affected pregnancy, who were followed in 33 clinics in 7 countries. Each subject was randomly assigned to treatments consisting of a placebo or a multivitamin supplement (A, D, C, B6, thiamin, riboflavin, and nicotinamide) and/or a placebo or folic acid (4 mg/day) in a complete factorial design, and the outcomes of their pregnancies were confirmed. Of a total of 1,195 completed pregnancies, 27 had confirmed NTDs; these included 21 cases in both groups not receiving folate, but only 6 cases in both folate groups (relative risk, 0.28; 95% CI, 0.12–0.71). The multivitamin treatment did not significantly affect the incidence of NTDs (MRC Vitamin Research Group [1991] Lancet 338, 131).

59. Centers for Disease Control and Prevention (2000). Morbid. Mortal. Weekly Rep. 49, 1; Stevenson, R. E., Allen, R. E., Pai, G. S., et al. (2000). Pediatrics 106, 677.
60. Blencowe, H., Cousens, S., Modell, B., et al. (2010). Intl J. Epidemiol. 39, i110.
61. Berry, R.J., Li, Z., Erickson, J. D., et al. (1999). N. Engl J. Med. 341, 1485.
62. Heseker, H. B., Mason, J. B., Selub, J., et al. (2009). Br. J. Nutr. 102, 173.
63. Botto, L. D., Moore, C. A., Khoury, M. J., et al. (1999). N. Engl. J. Med. 341, 1509.

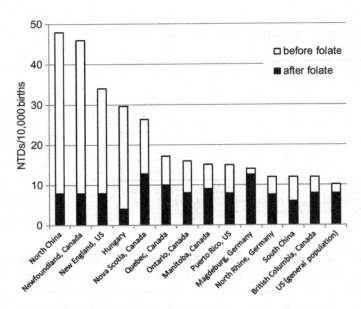

FIGURE 16.6 Reductions in NTD risk achieved in folate intervention trials and several countries. *After Heseker, H. B., Mason, J. B., Selub, J., et al. (2009). Br. J. Nutr. 102, 173.*

The mechanism of folate action in reducing the incidence of these birth defects remains unclear. Folate deficiency does not produce neural tube changes in animal models, although Hcy treatment of chick and mouse embryos has been found to increase the frequencies of a wide range of congenital malformations, including damage to cells of the neural crest.[64] That the protective effect of folate may involve its role in single-C metabolism is supported by the finding that exencephaly was produced by knocking out serine hydroxymethyltransferase, which transfers a single-C unit from 5,10-methylene FH_4 in the synthesis of thymidylate,[65] although knocking out MTHFR appears to be without neural developmental effect. It has also been suggested that folate may function in preventing the misexpression of micro-RNAs,[66] which play key roles in development by regulating the expression of certain target mRNAs. Micro-RNAs show distinct expression patterns in the developing brain and are highly regulated by genomic methylation, deficiencies of which have also been proposed to contribute to NTDs.

Food System-Based Folate Supplementation

In consideration of this evidence, in 1992 the US Public Health Service issued a recommendation that all women of childbearing age consume 0.4 mg folic acid daily to reduce their risks of an NTD pregnancy. In the following year, the US Food and Drug Administration ruled that all cereal grain products be fortified with 140 µg folic acid per 100 g, and that additions of folic acid be allowed for breakfast cereals, infant formulae, medical and special dietary foods, and meal replacement products. Other countries developed similar policies; those that increased the folate in their food systems experienced significant reductions in the incidence of NTDs: the US, by 19–31%; Costa Rica, by 63–87%; Canada, by 47–54%.[67]

The US folate-fortification program increased folate intakes and more than doubled circulating levels of the vitamin (see Table 16.12, page 367). Great increases in folate status occurred early in the program, i.e., between 1988–1994 and 1999–2000, but a slight drop occurred thereafter. That drop was driven by decreases in the highest folate intakes, and had minimal affects on NTD risk, which is affected primarily by low folate intakes (Fig. 16.7).

Other Birth Defects

Evidence is inconsistent for associations of low folate status and risks of other congenital defects, including orofacial clefts, and defective development of limbs and the heart. Homocysteinemia has been associated with increased risks of hypertension, pre-eclampsia, and placental abruption. Evidence suggests that MTHFR 677TT genotype elevates risk of Down syndrome in individuals also carrying a mutation in methionine synthase reductase.[68] Mothers and fetuses with the heterozygous 677CT genotype appear to have the best chances for viable

64. Van Hill, N. H., Oostervaan, A. M., and Steegers-Theunisswen, R.P. M. (2010). *Repro. Toxicol.* 30, 520.

65. Beaudin, A. E., Abarinov, E. V., Noden, D. M., *et al.* (2011). *Am. J. Clin. Nutr.* 93, 789.

66. miRNAs are small (*ca.* 22 nucleotides), non-coding transcripts that repress the expression of target mRNAs.

67. Yetley, E. A. and Rader, J. I. (2004) *Nutr. Rev.* 62, S50; Chen, L. T. and Rivera, M. A. (2004). *Nutr. Rev.* 62, S40; Mills, J. L. and Signore, C. (2004). *Birth Defects Res.* (Pt A) 70, 844.

68. Hobbs, C. A., Sherman, S. L., Yi, P., *et al.* (2000). *Am. J. Hum. Genet.* 67, 623.

69. Laanpere, M., Altmäe, S., Straveus-Evers, A., *et al.* (2009). *Nutr. Rev.* 68, 99.

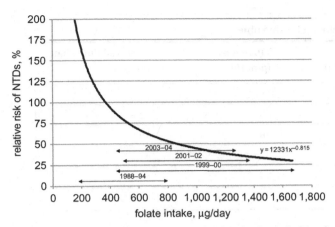

FIGURE 16.7 Relationships of estimated folate intake and risk of neural tube defects showing estimated ranges (10–90% of population) of folate intakes in National Health and Nutrition Examination Surveys (NHANES III 1988–1994 and annual NHANES 1999–2000, 2001–2002, and 2003–2004). *Based on combined population data from those surveys (Quinlivan, E. P. and Gregory, J. F. [2007] Am. J. Clin. Nutr. 86, 1773).*

TABLE 16.10 Plasma Homocysteine Levels in Myocardial Infarction Cases and Controls

Parameter	Cases	Controls
Plasma Hcy, nmol/l	11.1 ± 4.0	10.5 ± 2.8[a]
Cases ≥95th percentile of Hcy	31	13
Total number of cases	271	271

[a]$P = 0.026$.
From Stampfer, M. J., Malinow, M. R., Willett, W. C., *et al.* (1992). *J. Am. Med. Assoc.* 268, 877.

pregnancy and live birth.[69] Women with MTHFR 1298CC have a lower chance of producing a healthy embryo, i.e., capable of live birth after *in vitro* fertilization, compared to those with the 1298AA genotype.[70]

Cardiovascular Health

Homocysteine Hypothesis

In 1969, McCully pointed out a relationship between elevated plasma homocysteine levels and risk of occlusive vascular disease. Epidemiologic studies subsequently showed associations of moderately elevated plasma Hcy and risks of coronary, peripheral, and carotid arterial thrombosis and atherosclerosis; venous thrombosis; retinal vascular occlusion; carotid thickening; and hypertension (Table 16.10).[71] These observations are supported by studies in animal models that have shown folate deprivation to be thrombogenic. A meta-analysis of the results of 27 cross-sectional and case–control studies[72] attributed 10% of total coronary artery disease to **homocysteinemia**.[73] That analysis suggested that a 5 μmol/l increase in plasma Hcy level was associated with an increase in the risk of coronary artery disease comparable to a 0.5 mmol/l (20 mg/dl) increase in plasma total cholesterol.

Folate and Homocysteinemia

Homocysteinemia and the resulting homocysteinuria have been associated with increased risks of cardiovascular disease, recurrent early pregnancy loss, and hip fracture. Accumulation of Hcy can occur through its elevated production from methionine and, probably to a lesser extent, its impaired disposal through transsulfuration to cystathionine. Both result in homocysteinemia, can have congenital causes,[74] and can be related to nutritional status with respect to vitamin B_6, vitamin B_{12}, and folate.[75] Hcy can be converted to a thiolactone by methionyl tRNA synthetase, in an error-editing reaction that prevents its incorporation into the primary structure of proteins, but at high levels the thiolactone can react with protein lysyl residues. Protection against the damage to high-density lipoproteins that results from homocysteinylation is effected by a Ca^{2+}-dependent Hcy-thiolactonase associated with those particles.[76] Homocysteinemia also causes displacement of protein-bound cysteine, which changes redox thiol status, probably via thiol–disulfide exchange and redox reactions. Such changes have also been observed in patients with cardiovascular disease, renal failure or HIV infection.

Experimental folate deprivation has been shown to cause elevated plasma Hcy concentrations, and the use of folate-containing multivitamin supplements has been associated with low mean plasma Hcy levels. One prospective, community-based study found plasma Hcy to be strongly inversely associated with plasma folate level (which was positively associated with the consumption of folate-containing breakfast cereals, fruits, and vegetables) but not plasma levels of vitamin B_{12} and pyridoxal phosphate (Table 16.11).

70. Haggarty, P., McCallum, H. McBain, H., *et al.* (2006). *Lancet* 367, 1513.
71. The low prevalence of coronary heart disease among South African Blacks has been associated with their typically lower plasma Hcy levels and their demonstrably more effective Hcy clearance after methionine loading.
72. Boushey, C. J. Beresford, S. A., Omenn, G. S. *et al.* (1995). *J. Am. Med. Assoc.* 274, 1049.
73. Defined as a plasma Hcy concentration >14 μmol/l.

74. Inherited deficiencies of cysteine β-synthase and 5,10-methylene FH_4 reductase have been identified in humans.
75. Chronic alcoholics have been found to have mean serum Hcy levels about half those of non-alcoholics. Chronic ethanol intake appears to interfere with single-C metabolism, and alcoholics are at risk of folate deficiency.
76. Jakubowski, H. (2000). *J. Nutr.* 130, 377S.

TABLE 16.11 Plasma Levels of Homocysteine and Vitamin in Elderly Subjects

Subject Age (years)	Hcy		Folate (nmol/l)	Vitamin B$_{12}$ (pmol/l)	Pyridoxal Phosphate (nmol/l)
	(μmol/l)	(% elevated)			
Men:					
67–74	11.8	25.3	9.3	265	52.6
75–79	11.9	26.7	9.3	260	49.6
80+	14.1	48.3	10.0	255	47.6
Trend, P	<0.001	<0.001	NS[a]	NS	NS
Women:					
67–74	10.7	19.5	10.4	302	59.9
75–79	11.9	28.9	10.2	289	52.2
80+	13.2	41.1	9.7	290	52.1
Trend, P	<0.001	<0.001	NS	NS	NS

[a]NS, not significant.
From Selhub, J., Jacques, P. F., Wilson, P. W., et al. (1993). J. Am. Med. Assoc. 270, 2693.

Role of MTHFR Genotype

Individuals with the MTHFR 677TT genotype have slightly lower levels of folate and slightly higher levels of Hcy compared to other genotypes (Table 16.7). The TT genotype has been identified as a risk factor for carotid intima-media thickening, itself a risk factor to vascular disease. Low riboflavin status has been shown to be a significant determinant of another risk factor, high blood pressure, in the TT genotype; this presumably involves a lower affinity of the TT enzyme, which is, after all, a flavoprotein.[77] One meta-analysis of 53 studies showed the TT genotype to be associated with a 20% greater risk of venous thrombosis compared to the CC genotype,[78] but another meta-analysis found the available evidence inconclusive.

Dietary intakes of polyunsaturated fatty acids (PUFAs) have been found to be a significant covariate with the MTHFR C677T and A1298T genotype in affecting plasma Hcy level.[79] Individuals with 1298AA showed homocysteinemia only with consuming a high-PUFA diet, whereas individuals with the 677TT and the 1298C allele had low Hcy levels even on the high-PUFA diet. Individuals with the 1298CC genotype have been found to have a relatively higher risk of cardiovascular disease than those carrying the 1298A allele.

Effects of Folate Supplementation

Homocysteinemia can respond to supplementation of folic acid. Folate supplementation has been shown to reverse endothelial dysfunction independent of its effect in lowering plasma Hcy level, and to reduce arterial pressure and increase coronary dilation. It is likely that these effects involve the stimulation of nitric oxide production by 5-methylfolic acid and, perhaps, the inhibition of lipoprotein oxidation by folic acid. Reduction of serum Hcy in response to supplemental folate is linear up to daily folate intakes of about 0.4 mg, particularly for individuals with relatively high serum Hcy levels (Fig. 16.8). Greater efficacy may be realized when folate is given in combination with vitamin B$_{12}$. A meta-analysis of 25 randomized controlled trials showed that daily intakes of 0.8 mg folic acid are required to realize maximal reductions in plasma Hcy levels.

The food system-based measure that was implemented in the US and several other countries was successful in increasing folate intakes, more than doubling circulating levels of the vitamin, and reducing Hcy levels (Table 16.12). An analysis showed that countries using folate fortification in 1999–2002 experienced significantly greater reductions in stroke incidence than those countries without such programs.[80]

77. Yamada, K., Chen, Z., Rozen, R., et al. (2001). Proc. Natl Acad. Sci. USA 98, 14853.
78. Den Heijer, M., Lewwington, S., and Clarke, R. (2005). J. Thrombosis Haemostasis 3, 292.
79. Huang, T., Tucker, K. L., Lee, Y. C., et al. (2011). J. Nutr. 141, 654.

80. Yang, Q., Botto, L. D., Erickson, D., et al. (2006). Circulation 113, 1335.

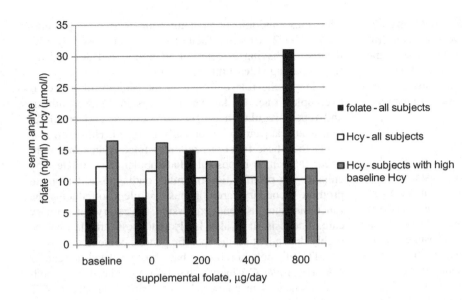

FIGURE 16.8 Effects of folate supplementation on serum levels of folate and homocysteine: results of a 26-week intervention. *After: Tighe, P., Ward, M., McNulty, H., et al. (2011). Am. J. Clin. Nutr. 93, 11.*

TABLE 16.12 Changes in Plasma Folate and Homocysteine Levels in the US Since Implementation of Folate Fortification of Cereal Foods

Survey Year	Plasma Folate		Plasma Hcy	
	ng/ml	% ≤6.8 nmol/l	µmol/l	% ≤13 µmol/l
1988–1994	12.1 ± 0.3[a]	18.4	8.7 ± 0.1	13.2
1999–2000	30.2 ± 0.7	0.8	7.0 ± 0.1	4.5
2001–2002	27.8 ± 0.5	0.2	7.3 ± 0.1	4.7

[a]*mean ± SE.*
From Ganji, V. and Kafi, M. R. (2006). J. Nutr. 136, 153.

Nevertheless, the lowering of circulating Hcy levels has not been found consistently to reduce risk of cardiovascular disease. A meta-analysis of 8 trials involving 37,485 subjects randomized to folate and/or other B vitamins were used as the intervention agents showed that a 25% reduction in circulating Hcy level for 5 years was not associated with any reductions in cardiovascular events (or death from any cause).[81] In one trial, cardiovascular disease patients who received a combined supplement of folate, vitamin B_6, and vitamin B_{12} showed increased risk for subsequent myocardial infarction.[82] Meta-analyses of trials that used folate as the single intervention agent yielded inconsistent results: one found the vitamin to improve flow-mediated dilatation,

suggestive of enhanced vascular function;[83] the other found folate supplementation to be of no benefit in reducing stroke risk.[84]

Homocysteine Hypothesis Turned Controversy

These findings have cast doubt on the role of Hcy in the etiology of cardiovascular disease. The controversy involves questions as to whether homocysteinemia may be a cause or a marker of disease, and whether serum Hcy may be an indicator of metabolic status. It has been suggested that the cardiovascular pathogenesis may actually involve another factor: the metabolic precursor of Hcy, *S*-adenosylhomocysteine (SAH). SAH is produced by the loss of the methyl group from the methyl donor SAM in some 50 methylation reactions. SAH is a potent inhibitor of methyltransferases; accordingly, the intracellular ratio of

81. Clarke, R., Halsey, J., and Bennett, D. (2011). *J. Inherit. Metab. Dis.* 34, 83.

82. In the Norwegian Vitamin Trial, 3,749 men and women with histories of heart attack were randomized to four combinations of folate and vitamins B_6 and B_{12} (Bønaa, K. H., Njølstadn I, Ueland, P. M., *et al.* [2006] *N. Engl. J. Med.* 354, 1578).

83. De Bree, A., van Mierol, L. A., and Draijer, R. (2007). *Am. J. Clin. Nutr.* 86, 610.

84. Lee, M. Hong, K. S., and Chang, S. C. (2010). *Stroke* 41, 1205.

SAH:SAM is taken as an indictor of methylation capacity – i.e., the "methylation index." SAH is reversibly converted with Hcy, the equilibrium favoring SAH. Therefore, homocysteinemia would be expected to lead to elevated levels of SAH, which appears to be present at only very low amounts (an estimated <0.2% of Hcy levels) in the circulation in normal circumstances.

Anticarcinogenesis

Folate deprivation, by enhancing genomic instability and dysregulated gene transcription, appears to enhance predisposition to neoplastic transformation. Accordingly, low folate status has been associated with increased risk of cancers of the colon, cervix, lung, pancreas, prostate, mouth and pharynx, head and neck, stomach, and brain. A case–control study of women infected with human papilloma virus showed a five-fold increase in risk of cervical dysplasia when they also had low serum folates.[85] Two large epidemiological studies have indicated that folate adequacy may reduce the effect of alcohol consumption in elevating breast cancer risk.[86] Meta-analyses of cohort studies have found food-folate intakes to be associated with reductions in colorectal cancer risk.[87] Studies in animal models have shown folate deprivation to promote colon carcinogenesis. However, a meta-analysis of five randomized clinical trials failed to find folate supplementation effective in reducing risk of recurrent colorectal cancers.[88]

These findings must be rationalized with the fact that folate is required for cell proliferation, for which reason its deprivation would be expected to impair the tumor growth. Thus, the question has been raised as to whether population-wide folate fortification may have cancer-promoting effects.[89] In fact, the results of trials support that concern. The Prostate, Lung, Colorectal and Ovarian Cancer Screening Trial, a prospective study involving 25,400 American women 55–74 years of age, found the incidence of breast cancer to be 20% greater for subjects reporting folate intakes ≥400 μg/day compared to those with lower intakes.[90] The Aspirin/Folate Polyp Prevention Trial, which involved 1,000 subjects with histories of colorectal adenomas randomized to 1 mg folic acid or a placebo, found after 2 years that folate treatment increased by 67% the risk of having a recurrent adenoma, and doubled the risk of having at least three adenomas.[91]

Clearly, the role of folate in the biology of cancer is a complex one. Its known activities in DNA synthesis and repair, regulation of gene expression, and proliferation would appear to be logically involved. However, it is likely that these functions relate to folate status in a non-linear fashion. It is possible that supplemental folate may enhance the proliferation of small, undiagnosed lesions to promote secondary carcinogenesis, while also reducing the functions, such as gene expression, necessary for primary carcinogenesis. It is also likely that these effects may be highly dependent on the specific genetic context.

MTHFR polymorphisms have been related to cancer risk. The 1298CC genotype has been associated with moderate reductions in colorectal cancer risk,[92] and the 677TT genotype does not appear to affect risk of colorectal adenoma unless folate status is low, in which case it increases it.[93] The 677TT genotype has also been associated with increased risk of esophageal cancer and of lymphocytic leukemia (which were greatest in 677TC/1298GC individuals).[94] Individuals of the 677CC or 1298AA genotype have been shown to be more susceptible to malignant lymphoma. In contrast, MTHFR polymorphisms do not appear to affect risk of cancers of the prostate or bladder. Limited studies indicate that low folate intakes may negate these genotype-associated risks.

Immune Function

That folate status may affect immune cell function is suggested by findings that folate deprivation of lymphocytes *in vitro* caused depletion of interleukin-2 and stimulated p53-independent apoptosis. Similar effects have not been observed *in vivo*. It has been noted that folate supplementation can stimulate natural killer (NK) cell cytotoxicity among subjects of low folate status. Curiously, the opposite effect was observed among subjects consuming a high-folate diet, in whom NK cytotoxicity was inversely related to the plasma concentration of the non-metabolized form of the vitamin, folic acid.[95] The MTHFR 677T allele has been associated with reduced risk of hepatitis B virus (HBV) infection in an HBV-endemic area.[96]

85. Butterworth, C. E. J., Haatsh, K. D., Macaluso, M., *et al.* (1992). *J. Am. Med. Assoc.* 267, 528; Liu, T., Soong, S. J., Wilson, N. P., *et al.* (1993). *Cancer Epidemiol. Biomarkers Prev.* 2, 525.
86. Zhang, S., Hunter, D. J., Hankinson, S. E., *et al.* (1999). *J. Am. Med. Assoc.* 281, 1632; Rohan, T. E., Jain, M. G., Howe, G. R., *et al.* (2000). *J. Natl Cancer Inst.* 92, 266.
87. Sanjoaquin, M. A., Allem, N., Couto, E., *et al.* (2005). *Intl J. Cancer* 113, 825; Kim, D. H., Smith-Warner, S. A., and Spiegelman, D. (2010). *Cancer Causes Control* 21, 1919.
88. Ibrahim, E. M. and Zekri, J. M. (2010). *Med. Oncol.* 27, 915.
89. Kim, Y. I. (2004). *Am. J. Clin. Nutr.* 80, 1123.
90. Stolzenberg-Solomon, R. Z., Chang, S. C., Leitzmann, M. F., *et al.* (2006). *Am. J. Clin. Nutr.* 83, 895.

91. Cole, B. F., Baron, J. A., Sandler, R. S., *et al.* (2007). *J. Am. Med. Assoc.* 297, 2351.
92. Kono, S. and Chen, K. (2005). *Cancer Sci.* 96, 535.
93. Kono, S. and Chen, K. (2005). *Cancer Sci.* 96, 535.
94. Skibola, C. F., Smith, M. T., Kane, E., *et al.* (1999). *Proc. Natl Acad. Sci. USA* 96, 12810.
95. Troen, A. M., Mitchell, B., Sorenson, B., *et al.* (2006). *J. Nutr.* 136, 189.
96. Bronowicki, J. P., Abdelmouttaleb, I., and Peyrin-Biroulet, L. (2008). *J. Hepatol.* 48, 532.

Neurological Conditions

Folate is required to maintain Hcy at low levels in the central nervous system. Hyperhomocysteinemia has been associated with increased risks of psychiatric and neurodegenerative disorders. Studies have shown several ways Hcy can be neurotoxic: through inhibition of methyltransferases involved in catecholamine methylation by the metabolite S-adenosylhomocysteine; through enhanced excitotoxicity leading to cell death by Hcy oxidation products that act as agonists of the N-methyl-D-aspartate receptor; and through oxidative stress resulting from the production of reactive oxygen species generated by the oxidation of Hcy. Folate may also be directly involved in the regulation of neurotransmitter metabolism, as neuropsychiatric subjects with low erythrocyte folate levels and homocysteinemia have been found to show low cerebral spinal fluid levels of the serotonin metabolite 5-hydroxyindole acetic acid, and reduced turnover of dopamine and noradrenaline. It has been suggested that folate may affect the metabolism of monamine neurotransmitters as a structural analog of tetrahydrobiopterin,[97] an essential cofactor in that metabolic pathway. MTHFR and dihydrofolate reductase are thought to function in tetrahydrobiopterin metabolism; the MTHFR 677 TT genotype appears to be associated with increased risk of neurological disorders. Such patients are also likely to be exposed to anticonvulsants that antagonize folate metabolism.[98]

Cognitive Function

Homocysteinemia is associated with age-related cognitive decline and risk of developing dementia; however, analyses of the NHANES data suggest that these outcomes are associated with low vitamin B_{12} status and *not* low folate status.[99] In fact, they suggest that high folate status may exacerbate the neuropsychiatric effects of low vitamin B_{12} status. A meta-analysis of randomized clinical trials concluded that folic acid yielded no beneficial effects on measures of cognition within 3 years of supplementation.[100]

Depression

Mood changes and other symptoms of depression have frequently been observed in folate deficiency. These symptoms are associated with homocysteinemia; it has been suggested that they reflect Hcy-induced cerebral vascular disease and neurotransmitter deficiency. That such symptoms can be caused by folate deficiency is suggested by the results of Victor Herbert, who experienced depressive mood, irritability, insomnia, fatigue, and forgetfulness after consuming a folate-deficient diet for several months.[101] When he took a folate supplement, those symptoms resolved within 48 hours. Studies have shown that individuals of the MTHFR 677TT genotype have increased risks of depression.[102] A systematic review of randomized clinical trials suggested that folic acid may benefit the treatment of depression.[103]

Schizophrenia

High doses of 5-methylfolate have been found to benefit patients with acute psychiatric disorders.[104] Both depressed and schizophrenic patients responded to daily doses of 15 mg of 5-methylfolate with improved clinical and social outcomes compared with placebo controls.[105] Studies have shown that individuals of the MTHFR 677TT genotype have increased risks of schizophrenia.

Bone Health

Homocysteinemia caused by a deficiency of cystathionine β-synthase, which converts Hcy to cysteine, can produce skeletal signs. These include knock-knees (genu valgum) and unusually high arches of the foot (pes cavus) in children, with subsequent development of Marfanoid features (long limbs) and osteoporosis. Homocysteinemia induced in animals by feeding large amounts of methionine or Hcy is accompanied with severe trabecular bone loss, with attendant changes in microarchitecture and strength. These changes appear to ensue from ROS-dependent activation of osteoclastic bone resorption.

Studies of the relationship of plasma Hcy level and bone health in humans have yielded inconsistent results, although some studies have found high plasma Hcy to be associated with reduced bone mineral density and increased fracture risk.[106] A randomized trial that used a combined supplement of folate, vitamin B_6, and vitamin B_{12} found no affect on biomarkers of bone turnover in

97. Both contain a pterin moiety.

98. For example, phenytoin, carbamazeprine, primidone, phenobarbital, valproic acid products, pamotrigine.

99. Selhub, J., Morris, M. S., Jacques, P. F., *et al.* (2009). *Am. J. Clin. Nutr.* 89(Suppl.), 702S.

100. Wald, D. S., Kasturiratne, A., and Simmonds, M. (2010). *Am. J. Med.* 123, 522.

101. Herbert, V. and Zalucky, R. (1962). *J. Clin. Invest.* 41, 1263.

102. Almedia, O. P., McCaul, K., and Hankey, G. J. (2008). *Arch. Gen. Psychiat.* 65, 1286.

103. Taylor, M. J., Carney, S. M., Goodwin, G. M., *et al.* (2004). *J. Psychopharmacol.* 18, 251.

104. Nearly two-thirds of patients with megaloblastic anemia due to vitamin B_{12}/folate deficiency show neuropsychiatric complications.

105. Godfrey, P. S., Toone, B. K., Carney, M. W., *et al.* (1990). *Lancet* 336, 392.

106. Levasseur, R. (2009). *Joint Bone Spine* 76, 234.

older adults, despite the efficacy of that treatment in reducing plasma Hcy levels.[107]

Some studies have indicated that the MTHFR genotype may be related to bone health. Individuals with the 677TT genotype have been found to have higher risks of fracture and lower bone mineral density, independent of any effects of Hcy.[108] However, not all studies have detected such relationships. That other factors are involved is suggested by a study that found 677TT women to show reduced bone mineral density only if they also had relatively low intakes of folate, vitamin B_{12}, and riboflavin.[109]

Malaria

The malarial parasite, *Plasmodium falciparum*, requires folate for its own metabolism, including the DNA synthesis required for growth and proliferation.[110] The parasite is capable of synthesizing folate from p-aminobenzoic acid and L-glutamate, but also uses exogenous supplies of folate, such as it finds within the host erythrocyte it invades in order to continue its life cycle. Accordingly, anti-folates[111] are the first-line drugs in the treatment of malaria.[112]

That the host and parasite are in competition for folate appears to contribute to the anemia observed in malarial patients, and malaria-induced hemolysis appears to increase the host's need for folate. While prenatal supplements of folate and iron have been found effective in reducing neonatal mortality in malaria-endemic regions, it is likely that those benefits may be limited to situations in which anemia is prevalent and the actual malaria prevalence is low. A robust study in a population with a high prevalence of malaria found supplementation with folate and iron to increase risk of severe illness and death.[113]

Arsenicosis

Studies in animal models have shown folate status to be a determinant of the metabolism and tissue distribution of arsenic, which must be methylated to be excreted. Accordingly, urinary concentrations of dimethylarsenate of arsenic-exposed subjects in Bangladesh were found to correlate positively with plasma folate level and negatively with plasma Hcy level, and serum arsenic levels were reduced by supplementation with folate.[114] This relationship is relevant, as many of the millions of people exposed to arsenic in drinking and irrigation waters in South Asia are inadequately nourished and can be expected to be of low folate status.

Effects of Drugs on Folate Utilization

Several drugs can impair the absorption and metabolism of folates. These have proven to have clinical applications ranging from the treatment of autoimmune diseases, to cancer and malaria – all conditions in which cell proliferation can be suppressed through the inhibition of a folate-dependent step in single-C metabolism.

Methotrexate

Also called amethopterin, this is an analog of folate, differing from the vitamin by the presence of an amino group replacing the 4-hydroxyl group on the pteridine ring, and a methyl group at the N-10 position. These differences give methotrexate a greater affinity than the natural substrate for dihydrofolate reductase, resulting in its inhibition. Accordingly, the drug produces an effective folate deficiency, with reductions in thymidine synthesis and purine levels. This anti-proliferative effect is the basis of the role of the drug in the treatment of cancer, rheumatoid arthritis, psoriasis, asthma, and inflammatory bowel disease. Because its side effects include those of folate deficiency, methotrexate is usually used with accompanying and carefully monitored folate supplementation to reduce the incidence of mucosal and gastrointestinal side effects. Individuals with the MTHFR 677TT genotype have a higher risk of discontinuing methotrexate treatment for rheumatoid arthritis due to adverse effects. In contrast, individuals with the MTHFR 1298CC genotype typically show improved efficacy of methotrexate therapy without increased side effects.

Other Drugs

Impairments in folate status/metabolism have been reported for anticonvulsants (diphenylhydantoin, phenobarbital), anti-inflammatory drugs (sulfasalazine), glycemic control drugs (metformin), and alcohol.

107. Green, T. J., McMahon, J. S., Skeaff, C. M., *et al.* (2007). *Am. J. Clin. Nutr.* 85, 460.

108. Villadsen, M. M., Bunger, M. H., Carstens, M., *et al.* (2005). *Osteoporos. Intl* 16, 411; Abrahamsen, B., Jorgensen, H. L., Nielsen, T. L., *et al.* (2006). *Bone* 38, 215; Hong, X., Hsu, Y. U., Terwedow, H., *et al.* (2007). *Bone* 40, 737.

109. Abrahamsen, B., Madsen, J. S., Tofteng, C. L., *et al.* (2005). *Bone* 36, 577.

110. Globally, malaria causes an estimated 200 million morbid episodes and 2–3 million deaths each year. During pregnancy, the disease also contributes to low birth weight and intrauterine growth retardation.

111. For example, pyrimethamine, sulfadoxine.

112. Unfortunately, strains of *P. falciparum* have developed resistance through mutations in their dihydrofolate reductase.

113. Sazawal, S., Black, R. E., Rarnsan, M., *et al.* (2006). *Lancet* 367, 133.

114. Gamble, M. V., Liu, X., Ashan, H., *et al.* (2005). *Environ. Health Perspect.* 113, 1683.

TABLE 16.13 Contributing Factors to Folate Deficiency

Poor dietary intake

Malabsorption:

    Intestinal diseases: tropical sprue, celiac disease

    Gastric diseases: atrophic gastritis

    Drugs: alcohol, methotrexate, sufasalazine, p-aminosalicylic acid, gastric acid-suppressants

    Genetic: hereditary folate malabsorption

Metabolic impairments:

    Acquired: dihydrofolate reductase inhibitors (e.g., methotrexate); alcohol; sulfasalazine

    Genetic: methylenetetrahydrofolate reductase deficiency; glutamate formiminotransferase deficiency; methyelenetetrahydrofolate reductase polymorphisms

Increased requirements: hemodialysis; prematurity; pregnancy; lactation

8. FOLATE DEFICIENCY

Determinants of Folate Deficiency

The most important factor contributing to folate deficiency is insufficient dietary intake. Other factors can contribute, particularly when combined with low dietary intakes of the vitamin. Those factors are shown in Table 16.13.

Methionine–Folate Linkage

Folate utilization is impaired by insufficient supplies of vitamin B_{12} and/or the indispensable amino acid methionine; therefore, dietary deficiencies of either of those nutrients can produce signs of folate deficiency. Thus, patients with pernicious anemia[115] generally have impaired folate utilization, and show signs of folate deficiency. The metabolic basis of this effect involves the methionine synthetase reaction, which is common to the functions of both folate and vitamin B_{12}. Methionine supplements cannot correct the low circulating folate levels caused by vitamin B_{12} deficiency. The amino acid appears to exert its action via S-adenosylmethionine, which inhibits 5,10-methylene-FH_4 reductase (thus reducing the *de novo* synthesis of 5-methyl-FH_4) and activates methionine synthetase.

Relationship with Zinc

Folate utilization can be impaired by depletion of zinc. Dietary Zn deficiency has been found in humans to reduce the absorption of folyl polyglutamates (*not* monoglutamates), and in animals to reduce liver folates. This is thought to indicate a need for Zn by the enzymes of folate metabolism. Studies have yielded inconsistent but mostly negative results concerning effects of folate deficiency on Zn metabolism.

Effects of drugs

Several drugs can impair the absorption and metabolism of folates. These include the anti-proliferative drug methotrexate, anti-convulsants (diphenylhydantoin, phenobarbital), anti-inflammatory drugs (sulfasalazine), glycemic control drugs (metformin), and alcohol (*see* "Effects of Drugs on Folate Utilization," page 370).

General Signs of Folate Deficiency

Deficiencies of folate result in impaired biosynthesis of DNA and RNA, and thus in reduced cell division, which is manifested clinically as anemia, dermatologic lesions, and poor growth in most species. The anemia of folate deficiency is characterized by the presence of large, nucleated erythrocyte-precursor cells called macrocytes, and of hypersegmented[116] polymorphonuclear neutrophils, reflecting decreased DNA synthesis and delayed maturation of bone marrow. While clinical deficiency is usually detected as anemia, megaloblastic changes occur in other cells. These involve increased cellular DNA content, with defects characterized by DNA strand breakage and growth arrest in the G2 phase of the cell cycle just prior to mitosis. These defects are thought to result from the mis-incorporation of uracil into DNA in place of thymidylate – a potentially mutagenic situation.

---

115. Pernicious anemia is vitamin B_{12} deficiency resulting from the lack of the intrinsic factor required for the enteric absorption of that vitamin (*see* Chapter 17).

116. Neutrophil hypersegmentation is defined as >5% five-lobed or any six-lobed cells per 100 granulocytes.

TABLE 16.14 General Signs of Folate Deficiency

Organ System	Signs
General:	
Appetite	Decrease
Growth	Decrease
Dermatologic	Alopecia, achromotrichia, dermatitis
Muscular	Weakness
Gastrointestinal	Inflammation
Vascular:	
Erythrocytes	Macrocytic anemia
Nervous	Depression, neuropathy, paralysis

Severely anemic individuals show weakness, fatigue, difficulty in concentrating, irritability, headache, palpitations, and shortness of breath. Folate deficiency also affects the intestinal epithelium, where impaired DNA synthesis causes megaloblastosis of enterocytes. This is manifested clinically as malabsorption and diarrhea, and is a contributor to the clinical picture of tropical sprue.

Signs and/or symptoms of folate deficiency are observed among individuals consuming inadequate dietary levels of the vitamin (Table 16.14). These effects are exacerbated by physiological conditions that increase folate needs (e.g., pregnancy, lactation, rapid growth), by drug treatments that reduce folate utilization, by aging, and by diseases of the intestinal mucosa.[117]

Deficiency Syndromes in Animals

Folate deficiency in animals is generally associated with poor growth, anemia, and dermatologic lesions involving skin and hair/feathers (Table 16.14). In chicks, severe anemia is one of the earliest signs of the deficiency. The anemia is of the macrocytic (megaloblastic) type, involving abnormally large erythrocyte size (the normal range in humans is $82–92\,\mu m^3$) due to the presence of **megaloblasts**, which are also seen among the hyperplastic erythroid cells in the bone marrow. Anemia in folate deficiency is followed by **leukopenia** (abnormally low numbers of

white blood cells), poor growth, very poor feathering, perosis, lethargy, and reduced feed intake.

Poultry with normally pigmented plumage[118] show achromotrichia due to the deficiency. Folate-deficient turkey poults show a spastic type of **cervical paralysis** in which the neck is held rigid.[119] Folate-deficient guinea pigs show leukopenia and depressed growth; pigs and monkeys show alopecia, dermatitis, leukopenia, anemia, and diarrhea; mink show ulcerative hemorrhagic gastritis, diarrhea, anorexia, and leukopenia. The deficiency is not easily produced in rodents unless a **sulfa drug**[120] or folate antagonist is fed, in which case leukopenia is the main sign.[121] Folate-responsive signs (reduced weight gain, macrocytic anemia) can be produced in catfish by feeding them succinylsulfathiazole. Folate deficiency in the rat has been shown to reduce exocrine function of the pancreas, in which single-carbon metabolism is important.[122]

Folate deficiency is not expected in ruminants with functioning microflora, which produces the vitamin in amounts that are apparently adequate to meet the needs of the host. In fact, nearly all supplemental folate to a dairy diet appears to be degraded; high doses of the vitamin (e.g., 0.5mg/kg of host body weight) are needed to increase serum and milk folate levels.

High-level folate supplementation of the diets of laying hens (e.g., 16mg/kg of diet) has been effective in producing eggs enriched in the vitamin for marketing purposes.

Deficiency Signs in Humans

Folate deficiency in humans is characterized by a sequence of signs, starting with nuclear hypersegmentation of circulating polymorphonuclear leukocytes[123] within about 2 months of deprivation of the vitamin. This is followed by megaloblastic anemia, then by general weakness, depression, and polyneuropathy. In pregnant women, the deficiency can lead to birth defects or spontaneous abortion. Elderly humans tend to have lower circulating levels

---

117. Examples include *tropical sprue* (inflammation of the mucous membranes of the alimentary tract) and other types of enteritis that involve malabsorption and, usually, diarrhea.

118. Such breeds include the barred Plymouth Rock, the Rhode Island Red, and the Black Leghorn.

119. Poults with cervical paralysis may not show anemia; the condition is fatal within a couple of days of onset, but responds dramatically (within 15 minutes) to parenteral administration of the vitamin.

120. For example, sulfanilamide.

121. Although leukopenia was manifested relatively soon after experimental folate depletion, rats kept alive with small doses of folate eventually also developed macrocytic anemia.

122. Experimental pancreatitis can be produced in that species by treatment with ethionine, an inhibitor of cellular methylation reactions, or by feeding a diet deficient in choline.

123. These cytological changes do not become manifest until well after circulating folate levels drop (by 6–8 weeks).

of folate, indicating that they may be at increased risk of folate deficiency. Although the basis of this finding is not fully elucidated, it appears to involve age-related factors such as food habits that affect intake of the vitamin, rather than its utilization.

That folate-responsive homocysteinemia can be demonstrated in apparently healthy free-living populations suggests the prevalence of undiagnosed suboptimal vitamin status. For example, twice-weekly treatments of elderly subjects with folate (1.1 mg) in combination with vitamin B_{12} (1 mg) and vitamin B_6 (5 mg) have been shown to reduce plasma concentrations of Hcy by as much as half, and also to reduce methylmalonic acid, 2-methylcitric acid, and cystathionine, despite the fact that pretreatment plasma levels of those vitamins were not low.

9. FOLATE TOXICITY

No adverse effects of high oral doses of folate have been reported in animals, although parenteral administration of pharmacologic amounts (e.g., 250 mg/kg, which is about 1,000 times the dietary requirement) has been shown to produce epileptic responses and renal hypertrophy in rats. Inconsistent results have been reported concerning the effects of high folate levels (1 to 10 mg doses) on human epileptics; some have indicated increases in the frequency or severity of seizures and reduced anticonvulsant effectiveness,[124] whereas others have shown no such effects. It has also been suggested that folate may form a non-absorbable complex with zinc, thus antagonizing the utilization of that essential trace element at high intakes of the vitamin. Because studies with animal models have shown that folate treatment can exacerbate teratogenic effects of nutritional Zn deficiency, which is thought to be prevalent but seldom detected in some areas and population groups, some have urged caution in recommending folate supplements to pregnant women.

10. CASE STUDY

Instructions

Review the following case report, paying special attention to the diagnostic indicators on which the treatments were based. Then, answer the questions that follow.

Case

A 15-year-old girl was admitted to the hospital because of progressive withdrawal, hallucinations, anorexia, and tremor. Her early growth and development were normal, and she had done average schoolwork until she was 11 years old, when her family moved to a new area. The next year, she experienced considerable difficulty in concentrating and was found to have an IQ of 60. She was placed in a special education program, where she began to fight with other children and have temper tantrums; when punished, she became withdrawn and stopped eating. A year earlier she had experienced an episode of severe abdominal pain for which no cause could be found, and she was referred to a mental health clinic. Her psychologic examination at that time had revealed inappropriate giggling, poor reality testing, and loss of contact with her surroundings. Her verbal and performance IQs were then 46 and 50, respectively. She was treated with thioridazine,[125] and within 2 weeks she was eating and sleeping better and was helpful around the house. However, over the succeeding months, while she continued taking thioridazine, her functioning fluctuated and the diagnosis of catatonic schizophrenia was confirmed. Three months before the present admission, she had become progressively withdrawn and drowsy, and needed to be fed, bathed, and dressed. She also experienced visual hallucinations, feelings of persecution, and night terrors. On having a seizure, she was taken to the hospital.

Her physical examination on admission revealed a tall, thin girl with fixed stare and catatonic posturing, but no neurologic abnormalities. She was mute and withdrawn, incontinent, and appeared to have visual and auditory hallucinations. Her muscle tone varied from normal to diffusely rigid.

On the assumption that her homocysteinuria was due to cystathionase deficiency, she was treated with pyridoxine HCl (300 mg/day, orally) for 10 days. Her homocysteinuria did not respond; however, her mental status improved, and within 4 days she was able to conduct some conversation and her hallucinations seemed to decrease. She developed new neurological signs: foot and wrist droop, and gradual loss of reflexes. She was then given folate (20 mg/day orally) for 14 days because of her low serum folate level. This resulted in a marked decrease in her urinary homocystine, and a progressive improvement in intellectual function over the next 3 months. She remained severely handicapped by her peripheral neuropathy, but she showed no psychotic symptoms. After 5 months of folate and pyridoxine treatment, she was tranquil and retarded, but showed no psychotic behavior; she left the hospital against medical advice and without medication.

---

124. High doses of folate appear to interfere with diphenylhydantoin absorption.

125. An antischizophrenic drug.

Laboratory Findings:

Parameter	Patient	Normal Range
Electroencephalogram	Diffusely slow	
Spinal fluid:		
Protein (mg/dl)	42	15–45
Cells	None	None
Urine:		
Homocystine	Elevated	
Methionine	Normal	
Serum:		
Homocystine	Elevated	
Methionine	Normal	
Folate (ng/ml)	3	5–21
Vitamin B_{12} (pg/ml)	800	150–900
Hematology:		
Hemoglobin (g/dl)	12.1	11.5–14.5
Hematocrit (%)	39.5	37–45
Reticulocytes (%)	1	~1
Bone marrow	No megaloblastosis	

Enzyme Activities:

Enzyme	Tissue or Cell Type	Enzyme Activity[a]	
		Patient	Normal
Methionine adenyltransferase	Liver	20.6	4.3–14.5
Cystathionine-β-synthetase	Fibroblasts	25.9	3.7–65.0
Betaine:Hcy methyltransferase	Liver	26.7	1.2–16.0
5-Methyl FH_4:Hcy methyltransferase	Fibroblasts	3.5	2.9–7.3
5,10-Methylene FH_4 reductase	Fibroblasts	0.5	1.0–4.6

[a]Enzyme units.

The girl was readmitted to the hospital 7 months later (a year after her first admission) with a 2-month history of progressive withdrawal, hallucinations, delusions, and refusal to eat. The general examination was the same as at her first admission, with the exceptions that she had developed hyperreflexia and her peripheral neuropathy had improved slightly. Her mental functioning was at the 2-year-old level.

She was incontinent, virtually mute, and had visual and auditory hallucinations. She was diagnosed as having simple schizophrenia of the childhood type. Folate and pyridoxine therapy was started again; it resulted in decreased Hcy excretion and gradual improvement in mental performance. After 2 months of therapy in the hospital, she was socializing, free of hallucinations, and able to feed herself and recognize her family. At that time, the activities of several enzymes involved in methionine metabolism were measured in her fibroblasts and liver tissue (obtained by biopsy). (See table opposite.)

Thereafter, she was maintained on oral folate (10 mg/day). She has been free of homocysteinuria and psychotic manifestations for several years.

Case Questions and Exercises

1. On admission of this patient to the hospital, which of her symptoms were consistent with an impairment in a folate-dependent aspect of metabolism?
2. What finding appeared to counterindicate an impairment in folate metabolism in this case?
3. Propose a hypothesis for the metabolic basis of the observed efficacy of oral folate treatment in this case.

Study Questions and Exercises

1. Diagram the metabolic conversions involving folates in single-carbon metabolism.
2. Construct a decision tree for the diagnosis of folate deficiency in humans or an animal species. In particular, outline a way to distinguish folate and vitamin B_{12} deficiencies in patients with macrocytic anemia.
3. What key feature of the chemistry of folate relates to its biochemical function as a carrier of single-carbon units?
4. What parameters might you measure to assess folate status of a human or animal?

RECOMMENDED READING

Bailey, L.B., 2007. Folic acid. In: Zempleni, J., Rucker, R.B., McCormick, D.B., Suttie, J.W. (Eds.), Handbook of Vitamins, fourth ed. CRC Press, New York, NY, pp. 385–412.

Bailey, L.B., Gregory, J.F., 2006. Folate. In: Bowman, B.A., Russell, R.M. (Eds.), Present Knowledge in Nutrition, ninth ed. ILSI Press, Washington, DC, pp. 278–301.

Blom, H.J., Smulders, Y., 2011. Overview of homocysteine and folate metabolism, with special references to cardiovascular disease and neural tube defects. J. Inherit. Metab. Dis. 34, 75–81.

Brustolin, S., Giugliani, R., Félix, T.M., 2010. Genetics of homocysteine metabolism and associated disorders. Br. J. Med. Biol. Res. 43, 1–7.

Carmel, R., 2006. Folic acid. In: Shils, M.E., Shike, M., Ross, A.C. (Eds.), Modern Nutrition in Health and Disease, tenth ed. Lippincott, New York, NY, pp. 470–481.

Caudill, M.A., 2010. Folate bioavailability: Implications for establishing dietary recommendations and optimizing status. Am. J. Clin. Nutr. 91 (Suppl.), 1455S–1460S.

Cuskelly, G.J., Mooney, K.M., Young, I.S., 2007. Folate and vitamin B12: Friendly or enemy nutrients for the elderly. Proc. Nutr. Soc. 66, 548–558.

Day, O., 2009. Nutritional interpretation of folic acid interventions. Nutr. Rev. 67, 235–244.

Devos, L., Chanson, A., Liu, Z., et al. 2008. Associations between single nucleotide polymorphisms in folate uptake and metabolizing genes with blood folate, homocysteine, and DNA uracil concentrations. Am. J. Clin. Nutr. 88, 1149–1158.

Di Minno, M.N.D., Tremoli, E., Coppola, A., et al. 2010. Homocysteine and arterial thrombosis: Challenge and opportunity. Thromb. Haemost. 103, 942–961.

Duthie, S.J., 2011. Folate and cancer: How DNA damage, repair and methylation impact colon carcinogenesis. J. Inherit. Metab. Dis. 34, 101–109.

Eicholzer, M., Tönz, O., Zimmermann, R., 2006. Folic acid: A public health challenge. Lancet 367, 1352–1361.

Fingas, P.M., Wright, A.J.A., Wolfe, C.A., et al. 2003. Is there more to folates than neural-tube defects? Proc. Nutr. Soc. 62, 591–598.

Folstein, M., Liu, T., Peter, I., et al. 2007. The homocysteine hypothesis of depression. Am. J. Psychiatry 164, 861–867.

Giovannucci, E., 2002. Epidemiological studies of folate and colorectal neoplasia: A review. J. Nutr. 132, 2350S–2355S.

Gregory, J.F., Quinlivan, E.P., 2002. *In vivo* kinetics of folate metabolism. Ann. Rev. Nutr. 22, 199–220.

Hamid, A., Wani, N.A., Kaur, J., 2009. New perspectives on folate transport in relation to alcoholism-induced folate malabsorption – association with epigenome stability and cancer development. FEBS J. 276, 2175–2191.

Heseker, H.B., Mason, J.B., Selub, J., et al. 2009. Not all cases of neural-tube defect can be prevented by increasing the intake of folic acid. Br. J. Nutr. 102, 173–180.

Kaman, B.A., Smith, A.K., 2004. A review of folate receptor alpha cycling and 5-methyltetrahydrofolate accumulation with an emphasis on cell models *in vitro*. Adv. Drug Deliv. Rev. 56, 1085–1097.

Kim, Y.I., 2003. Role of folate in colon cancer development and progression. J. Clin. Nutr. 133, 3731S–3739S.

Kono, S., Chen, K., 2005. Genetic polymorphisms of methylenetetrahydrofolate reductase and colorectal cancer and adenoma. Cancer Sci. 96, 535–542.

Koury, M.J., Ponka, P., 2004. New insights into erythropoiesis: The roles of folate, vitamin B12, and iron. Ann. Rev. Nutr. 24, 105–131.

Lawrence, M.A., Chai, W., Kara, R., et al. 2009. Examination of selected national policies towards mandatory folic acid fortification. Nutr. Rev. 67, S73–S78.

Manolescu, B.N., Oprea, E., Farcasanu, I.C., et al. 2010. Homocysteine and vitamin therapy in stroke prevention and treatment: A review. Acta Biochim. Pol. 57, 467–477.

McCully, K.S., 2007. Homocysteine, vitamins, and vascular disease prevention. Am. J. Clin. Nutr. 86 (Suppl.), 1563S.

McNulty, H., Pentieva, K., 2004. Folate bioavailability. Proc. Nutr. Soc. 63, 526–529.

Moat, S.J., Lang, D., McDowell, I.F.W., et al. 2004. Folate, homocysteine, endothelial function and cardiovascular disease. J. Nutr. Biochem. 15, 64–79.

Picciano, M.F., Yetley, E.A., Coates, P.M., McGuire, M.K., 2009. Update on folate and human health. Nutr. Today 44, 142–152.

Pitkin, R.M., 2007. Folate and neural tube defects. Am. J. Clin. Nutr. 85 (Suppl.), 285S–288S.

Quinlivan, E.P., Hanson, A.D., Gregory, J.F., 2006. The analysis of folate and its metabolic precursors in biological samples. Anal. Biochem. 348, 163–184.

Rush, D., 1994. Periconceptional folate and neural tube defect. Am. J. Clin. Nutr. 59, 511S–516S.

Smith, A.D., Kim, Y.I., Refsum, H., 2008. Is folic acid good for everyone? Am. J. Clin. Nutr. 87, 517–533.

Smulders, Y.M., Blom, H.J., 2011. The homocysteine controversy. J. Inherit. Metab. Dis. 34, 93–99.

Spence, J.D., 2007. Homocysteine-lowering therapy: A role in stroke prevention? Lancet Neurol. 6, 830–838.

Stover, P.J., 2009. One-carbon metabolism–genome interactions in folate-associated pathologies. J. Nutr. 139, 2402–2405.

Tamura, T., Picciano, M.F., 2006. Folate and human reproduction. Am. J. Clin. Nutr. 83, 993–1016.

Tibbetts, A., Appling, D.R., 2010. Compartmentalization of mammalian folate-mediated one-carbon metabolism. Ann. Rev. Nutr. 30, 57–81.

Ulrich, C.M., 2008. Folate and caner prevention – where to next: Counterpoint. Cancer Epidemiol. Biomarkers Prev. 17, 226–230.

Wilson, C.P., McNulty, H., Scott, J.M., et al. 2010. The MTHFR C677T polymorphism, B-vitamins and blood pressure. Proc. Nutr. Soc. 69, 156–165.

Xia, W., Low, P.S., 2010. Folate-targeted therapies for cancer. J. Med. Chem. 53, 6811–6824.

Vitamin B$_{12}$

Chapter Outline

Anchoring Concepts

1. *Vitamin B$_{12}$* is the generic descriptor for all *corrinoids* (compounds containing the cobalt-centered corrin nucleus) exhibiting qualitatively the biological activity of cyanocobalamin.
2. Deficiencies of vitamin B$_{12}$ are manifested as anemia and neurologic changes, and can be fatal.
3. The function of vitamin B$_{12}$ in single-carbon metabolism is interrelated with that of folate.

Patients with Addisonian pernicious anemia have ... a "conditioned" defect of nutrition. The nutritional defect in such patients is apparently caused by a failure of a reaction that occurs in the normal individual between a substance in the food (extrinsic factor) and a substance in the normal gastric secretion (intrinsic factor).

W. B. Castle and T. H. Hale

Learning Objectives

1. To understand the chief natural sources of vitamin B$_{12}$.
2. To understand the means of enteric absorption and transport of vitamin B$_{12}$.
3. To understand the biochemical functions of vitamin B$_{12}$ as a coenzyme in the metabolism of propionate and the biosynthesis of methionine.
4. To understand the metabolic interrelationship of vitamin B$_{12}$ and folate.
5. To understand the factors that can cause low vitamin B$_{12}$ status, and the physiological implications of that condition.

VOCABULARY

Adenosylcobalamin
S-Adenosylhomocysteine (SAH)
Anemia
Aquocobalamin
Cobalamins
Cobalt
Cubulin
Cyanocobalamin
Deoxyadenosylcobalamin
Gastric parietal cell
Haptocorrin
Homocysteine (Hcy)
Homocysteinemia
Hydroxycobalamin
Hypochlorhydria
IF (intrinsic factor)
IF receptor
IF–vitamin B$_{12}$ complex
Imerslund-Gräsbeck syndrome
Lipotrope
Megaloblastic anemia
Megaloblastic transformation
Methionine synthase
Methionine synthase reductase
Methylcobalamin
Methyl-FH$_4$ methyltransferase
Methylfolate trap
Methylmalonic acid
Methylmalonic acidemia
Methylmalonic aciduria
Methylmalonyl-CoA mutase
Ovolactovegetarian

The Vitamins. DOI: 10.1016/B978-0-12-381980-2.00017-7

Pepsin
Peripheral neuropathy
Pernicious anemia
Pseudovitamin B_{12}
R proteins
S-adenosylmethionine (SAM)
Schilling test
TC receptor
Transcobalamin (TC)
Vegan
Vitamin B_{12} coenzyme synthetase

1. THE SIGNIFICANCE OF VITAMIN B_{12}

Vitamin B_{12} is synthesized by bacteria. It is found in the tissues of animals, which require the vitamin for critical functions in cellular division and growth; in fact, some animal tissues can store the vitamin in very appreciable amounts that are sufficient to meet the needs of the organism for long periods (years) of deprivation. The vitamin is seldom found in foods derived from plants; therefore, animals and humans that consume strict vegetarian diets are very likely to have suboptimal intakes of vitamin B_{12} which, if prolonged and uncorrected, will lead to anemia and, ultimately, to peripheral neuropathy. However, most vegetarians are not strict **vegans** (who exclude all foods of animal origin), and many consume foods and/or supplements containing at least some vitamin B_{12}. For this reason, frank vitamin B_{12} deficiency is not common. Nevertheless, low vitamin B_{12} status occurs in individuals with deficiencies in proteins involved in vitamin B_{12} transport and/or metabolism, and in individuals with compromised gastric parietal cell function. Low vitamin B_{12} status impairs the metabolic utilization of folate and contributes to homocysteinemia, a risk factor for occlusive vascular disease.

2. SOURCES OF VITAMIN B_{12}

Distribution in Foods

Because the synthesis of vitamin B_{12} is limited almost exclusively to bacteria, the vitamin is found only in foods that have been bacterially fermented and those derived from the tissues of animals that have obtained it from their ruminal or intestinal microflora, ingesting it either with their diet or coprophagously. Animal tissues that accumulate vitamin B_{12} (e.g., liver[1]) are therefore excellent food sources of the vitamin (Table 17.1). The richest food sources of vitamin B_{12} are dairy products, meats, eggs, fish, and shellfish. The principle vitamers in foods are **methylcobalamin**, **deoxyadenosylcobalamin**, and **hydroxycobalamin**. The richest sources of vitamin B_{12} for

---

[1]. Vitamin B_{12} was discovered as the anti-pernicious anemia factor in liver.

TABLE 17.1 Sources of Vitamin B_{12} in Foods

Foods	Vitamin B_{12}, µg/100 g
Meats	
Beef	1.9–3.6
Beef brain	7.8
Beef kidney	38
Beef liver	69–122
Chicken	0.3
Chicken liver	24
Ham	0.8
Pork	0.6
Turkey	0.4
Dairy Products	
Milk	0.4
Cheeses	0.4–1.7
Yogurt	0.1–0.6
Fish and Seafood	
Herring	4.3
Salmon	3.2
Trout	7.8
Tuna	2.8
Clams	19.1
Oysters	21.2
Lobster	1.28
Shrimp	1.9
Vegetables	–
Grains	–
Fruits	–
Other	
Eggs, whole	1.26
Egg whites	0.09
Egg yolk	9.26
Edible algae, nori	32–78
Tempe	0.7–8

animal feedstuffs are animal by-products such as meat and bone meal, fish meal, and whey.

Compounds with vitamin B_{12}-like activity, known as the "alkali-resistant factor," have been found in bamboo, cabbage, spinach, celery, lily bulb, bamboo shoots, and taro. Trace amounts of the vitamin have been found in several vegetables, including broccoli, asparagus, and mung bean sprouts. This appears to reflect the ability of plants to take up vitamin B$_{12}$ from organic fertilizers. While soybeans contain little, if any, vitamin B$_{12}$, fermented soy products (e.g., tempe, natto) can contain significant amounts, likely due to bacterial contamination of the fermentation process.

Some species of edible green algae (*Enteromorpha* sp.) and purple laver (*Porphyra* sp., i.e., nori) contain large amounts of vitamin B$_{12}$; however, studies have indicated that the vitamin is not available from those foods to humans.[2] Edible cyanobacteria (*Spirulina*, *Apahnizomenon*, *Nostoc*) are often cited as containing vitamin B$_{12}$; however, they often contain large amounts of **pseudovitamin B$_{12}$** (7-adeninyl cyanocobamide),[3] which is biologically inactive and may antagonize the utilization of vitamin B$_{12}$.[4]

Human milk contains vitamin B$_{12}$ almost exclusively bound to an R protein. Initial levels, 260–300 pmol/l, decline by half after the first 12 weeks of lactation. Breast-milk vitamin B$_{12}$ levels of women consuming strict vegetarian diets are less than those of women consuming mixed diets, and tend to be inversely correlated with the length of time on the vegetarian diet. Infant urinary methylmalonic acid levels are increased when maternal milk vitamin B$_{12}$ concentrations are <360 pmol/l, indicating vitamin B$_{12}$ deficiency.

The microbial synthesis of vitamin B$_{12}$ by long-gutted animals depends on an adequate supply of cobalt, which must be ingested in the diet. If the supply of cobalt is sufficient, the rumen microbial synthesis of vitamin B$_{12}$ in ruminants is substantial. For that reason, those species not only have no need for preformed vitamin B$_{12}$ in the diet, but their tissues also tend to contain appreciably greater amounts of the vitamin than those of non-ruminant species.

Stability

Vitamin B$_{12}$ is very stable in crystalline form and aqueous solution. High levels of ascorbic acid have been shown to catalyze the oxidation of vitamin B$_{12}$ in the presence of iron to forms that are poorly utilized.

Bioavailability

Vitamin B$_{12}$ is bound to the two vitamin B$_{12}$-dependent enzymes and carrier proteins in foods. Therefore, the

utilization of ingested vitamin B$_{12}$ depends on the nature of the food/meal matrix and the host's ability to release the vitamin and bind it to proteins that facilitate its enteric absorption. In practice, the bioavailability of vitamin B$_{12}$ in foods is very difficult to determine. Bioassays in animal models fed vitamin B$_{12}$-deficient diets always leave questions about applicability to humans, and studies in non-deficient humans require the use of the vitamin labeled with an intrinsic tracer. Further, the microbiological method commonly used to measure vitamin B$_{12}$ in foods (i.e., *Lactobaccillus delbrueckii* growth) appears to yield overestimates by some 30% due to responses to non-vitamin active corrinoids.

With those caveats, the bioavailability of vitamin B$_{12}$ from most foods appears to be moderate. Studies have found that about half of the vitamin in most foods is absorbed by individuals with normal gastrointestinal function (Table 17.2). Bioavailability falls off rapidly at intakes (1.5–2 µg/day) that saturate the mechanism for actively transporting the vitamin across the gut, as greater amounts depend on absorption by passive diffusion, a process with only 1% efficiency. Accordingly, about 1% of the vitamin is absorbed from vitamin B$_{12}$ supplements.

3. ABSORPTION OF VITAMIN B$_{12}$

Digestion

The naturally occurring vitamin B$_{12}$ in foods is bound in coenzyme form to proteins. The vitamin is released from such complexes on heating, gastric acidification, and/or proteolysis (especially by the action of pepsin). Thus, impaired gastric parietal cell function, as in achlorhydria or with chronic use of proton pump inhibitors, impairs vitamin B$_{12}$ utilization.

Protein Binding

Free vitamin B$_{12}$ is bound to proteins secreted by the gastric mucosa.

TABLE 17.2 Bioavailability of Vitamin B$_{12}$ in Common Foods

Food	% Bioavailable
Eggs	4–9
Fish	42
Chicken	61–66
Lamb	56–89
Milk	55–65

From Watanabe, F. (2007). *Exp. Biol. Med.* 232, 1266.

2. Dagnelie, P. C., van Staveren, W. A., and van den Berg, H. (1991). *Am. J. Clin. Nutr.* 53, 695.
3. Pseudovitamin B$_{12}$ differs from the vitamin by having an adenine moiety replacing the dimethylbenimidazole.
4. Herbert, V. (1988). *Am. J. Clin. Nutr.* 48, 852.

R Proteins[5]

The binding of vitamin B_{12} to these glycoproteins[6] may be adventitious. They are found in human saliva, gastric juice, and intestinal contents, as well as in several other body fluids and certain cells,[7] and probably only in a few other species. Members of a family of proteins called haptocorrins, they show structural and immunologic similarities, with differing carbohydrate contents. The salivary R binder is the first to bind vitamin B_{12} released from the food matrix. Vitamin B_{12} binds preferentially to R proteins in the acidic conditions of the stomach, but the salivary R binder is normally digested proteolytically in the small intestine to release the vitamin to be bound by intrinsic factor (IF). Patients with pancreatic exocrine insufficiency, and consequent deficiencies of proteolytic activities in the intestinal lumen, can achieve high concentrations of R proteins that render the vitamin poorly absorbed.

Intrinsic Factor

IF is a glycoprotein synthesized and secreted in most animals (including humans) by the gastric parietal cells[8] in response to histamine, gastrin, pentagastrin, and the presence of food. Individuals with loss of gastric parietal cell function may be unable to use dietary vitamin B_{12}, as these cells produce both IF and acid, both of which are required for the enteric absorption of the vitamin.[9] For this reason, geriatric patients, many of whom are hypoacidic, may be at risk of low vitamin B_{12} status. A relatively small protein (human 44–63 kDa, pig 50–59 kDa, according to the carbohydrate moiety isolated with particular preparations), IF binds the four cobalamins (**methylcobalamin, adenosylcobalamin, cyanocobalamin,** and **aquocobalamin**) with comparable, high affinities under alkaline conditions. It does not bind other R protein-binding corrinoids (cobamamides, cobinamides), which remain bound to those proteins.

IF also binds a specific receptor in the ileal mucosal brush border, a site reached by the IF–vitamin B_{12} complex traveling the length of the small intestine. The binding of vitamin B_{12} by IF produces a complex with a smaller molecular radius than that of IF alone; this appears to protect the vitamin from side chain modification of the corrin ring by intestinal bacteria, while also protecting IF from hydrolytic attack by pepsin and chymotrypsin. Cobalamin-binding appears to have an allosteric effect on the ileal receptor-binding center of IF, causing the protein complex to dimerize and increasing its binding to the receptor.

Mutations in the IF gene have been identified that result either in failure of its expression or in expression of a defective protein incapable of binding vitamin B_{12}. Affected individuals show normal gastric mucosa and acid production. They show megaloblastic anemia within the first 3 years of life, which responds to large doses of vitamin B_{12} administered orally or by intramuscular injection.

Mechanisms of Absorption

Active Transport

The carrier-mediated absorption of vitamin B_{12} is highly efficient and quantitatively important at low doses (1–2 µg). Such doses of appear in the blood within 3–4 hours of consumption. The active transport of vitamin B_{12} depends on the interactions of the IF–vitamin B_{12} complex with a specific receptor in the microvilli of the ileum. That receptor[10] consists of two components: the multi-ligand apical membrane protein **cubulin**,[11] which binds IF–vitamin B_{12}; and a product of the receptor-associated protein (RAP),[12] which contributes structure necessary for membrane anchorage, trafficking to the plasma membrane, and signaling of endocytosis and receptor recycling.[13]

Absorption of vitamin B_{12} by the enterocyte involves the cellular uptake of the vitamin–IF complex, with IF being degraded within enterocyte lysosomes and the vitamin being transferred to a specific carrier protein, **transcobalamin (TC)** (see below), for secretion into the portal circulation. Patients lacking IF have very poor abilities to absorb vitamin B_{12}, excreting 80–100% of oral doses in the feces (vs. 30–60% fecal excretion rates in individuals with adequate IF).

Congenital deficiencies result in **Imerslund-Gräsbeck syndrome**, a common cause of vitamin B_{12}-associated megaloblastic anemia. This syndrome involves dysfunction

---

5. These vitamin B_{12}-binding glycoproteins were named for their high electrophoretic mobilities, *rapid*.

6. They contain sialic acid and fucose.

7. For example, plasma, saliva, tears, bile, cerebrospinal fluid, amniotic fluid, leukocytes, erythrocytes, and milk.

8. That is, the same cells that produce gastric acid.

9. Individuals lacking IF are unable to absorb vitamin B_{12} by active transport. Such individuals can be given the vitamin by intramuscular injection (1 µg/day) or in high oral doses (25–2,000 µg) to prevent deficiency. Randomized controlled trials have shown that an oral dose regimen of 1,000 µg daily for a week, followed by the same dose weekly, and then monthly, can be as effective as intramuscular administration of the vitamin for controlling short-term hematological and neurological responses in deficient patients (Butler, C. C. [2006] *Fam. Pract.* 23, 279).

---

10. Genetic defects in these proteins occur in Imerslund-Gräsbeck's disease, which is characterized by vitamin B_{12} malabsorption leading to megaloblastic anemia.

11. Cubulin is a large (460 kDa) membrane protein with no apparent transmembrane segment. It is expressed at high levels in the kidney, where it appears to function in the reabsorption of several specific nutrient carriers, including albumin, vitamin D-binding protein, transferrin, and apolipoprotein A.

12. This was discovered as a product of the amnionless gene *AMN*.

13. Fyfe, J. C., Madsen, M., Højrup, P., *et al.* (2004). *Blood* 103, 1573.

of the ileal IF receptor, caused by defects in either cubulin or RAP. Affected individuals malabsorb vitamin B$_{12}$.

Passive Diffusion

Diffusion of the vitamin occurs with low efficiency (~1%) throughout the small intestine, and becomes significant only at higher doses. Such doses appear in the blood within minutes of comsumption. This passive mechanism is utilized in therapy for pernicious anemia, in which patients are given high doses (>500 µg/day) of vitamin B$_{12}$ *per os*. For such therapy, the vitamin must be given an hour before or after a meal to avoid competitive binding of the vitamin food.

4. TRANSPORT OF VITAMIN B$_{12}$

Transport Proteins

On absorption from the intestine, vitamin B$_{12}$ is initially transported in the plasma as adenosylcobalamin and methylcobalamin bound to two proteins:

- *Plasma haptocorrin.* Most (70–80 %) of the vitamin B$_{12}$ in plasma is bound to a 60 kDa R protein, plasma **haptocorrin** (previously referred to as transcobalamin I). Vitamin B$_{12}$ bound to this carrier turns over very slowly (half-life 9–10 days), becoming available for cellular uptake only over fairly long timeframes. Plasma haptocorrin is typically 80–90% saturated with its ligand. A minor form of this protein (previously referred to as transcobalamin II), differing only in carbohydrate content, can also be found in plasma. Congenital defects in plasma haptocorrin are asymptomatic, suggesting that this form of the vitamin is not physiologically important. Affected individuals show normal absorption and distribution of vitamin B$_{12}$ to their tissues; however, they show low circulating levels of the vitamin, and can be wrongly diagnosed as vitamin B$_{12}$-deficient if other parameters (MMA, Hcy, FIGLU) are not considered. The prevalence of plasma haptocorrin defects may be relatively high; one study noted that 15% of apparently healthy subjects had low plasma vitamin B$_{12}$ levels.
- *Transcobalamin (TC).* Most of the remaining vitamin B$_{12}$ in plasma (10–20% of the total) is bound to TC, a smaller (38 kDa) protein (previously, transcobalamin II) synthesized in several tissues, including the intestinal mucosa, liver, seminal vesicles, fibroblasts, bone marrow, and macrophages. TC binds the vitamin stoichiometrically.[14] The rapid turnover (half-life 60–90

minutes) of the protein–ligand complex renders TC the primary functional source of vitamin B$_{12}$ for cellular uptake. TC is typically 10–20% saturated with its ligand. Congenital TC deficiencies have been described. Cases show normal circulating levels of vitamin B$_{12}$, most of which is bound to plasma haptocorrin, but develop severe megaloblastic anemia as infants within a few weeks of birth. They respond to vitamin B$_{12}$ administered in large doses by intramuscular injection – e.g., 1 mg three times per week. A single-nucleotide polymorphism has been identified in TC: C→G at base position 766. It is thought that the protein encoded by the 766G allele, with a prevalence of about 45% in White populations and somewhat more in Blacks and Asians, may have a lower affinity for vitamin B$_{12}$ than the protein encoded by the 766C allele; TC 766G is associated with lower circulating levels of both apo- and holo-TC, and higher levels of MMA.

The movement of vitamin B$_{12}$ from the intestinal mucosal cells into the plasma depends on the formation of the TC–vitamin B$_{12}$ complex, which turns over rapidly (half-life *ca.* 6 minutes). Within hours of absorption, however, much of the vitamin originally associated with TC becomes bound to plasma haptocorrin and, in humans, to other plasma R proteins. Whereas deficiency of plasma haptocorrin does not appear to impair cobalamin metabolism, TC is clearly necessary for normal cellular maturation of the hematopoietic system.[15] Because cobalamin is lost within days from TC, the amount bound to that protein can be a useful parameter of early stage vitamin B$_{12}$ deficiency.

In humans, most of the recently absorbed vitamin B$_{12}$ is transferred to the plasma haptocorrin,[16] which binds methylcobalamin preferentially. Therefore, the predominant circulating form of the vitamin in humans is methylcobalamin. Most other species lack R proteins; they transport the vitamin exclusively as the TC complex. Therefore, in species other than the human, the dominant circulating form is adenosylcobalamin.

Transcobalamin Receptor

Membrane-bound receptor proteins for holotranscobalamin[17] occur in all cells. The **TC receptor** is structurally similar to TC; it is a 50 kDa glycoprotein with a single binding site for the holo-TC. Binding is of high affinity, and requires Ca^{2+}. It is thought that the cellular uptake of vitamin B$_{12}$ involves TC receptors mediating the pinocytotic uptake of the holo–TC complex (Fig. 17.1).

14. About half of patients with acquired immunodeficiency syndrome (AIDS) have been found to have subnormal levels of holo-TC$_I$.

15. A rare autosomal recessive deficiency in TC$_{II}$ has been described.
16. Due to their affinity for R proteins, the TCs are grouped in a heterogeneous class of proteins called *R binders.*
17. That is, the transcobalamin–vitamin B$_{12}$ complex, holo-TC.

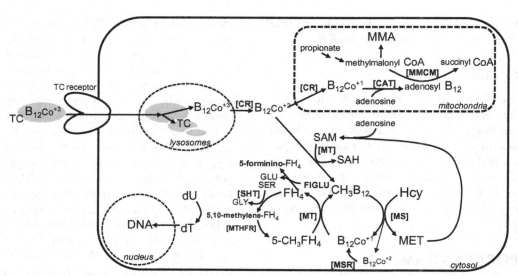

FIGURE 17.1 Uptake and metabolism of vitamin B_{12}, and its relationship with folate in single-carbon metabolism. *Abbreviations:* TC, transcobalamin; MMA, methylmalonic acid; SAM, *S*-adenosylmethionine; SAH, *S*-adenosylhomocysteine; Hcy, homocysteine; MET, methionine; FH4, tetrahydrofolic acid; CH3B12, methylcobalamin; 5-CH3-FH4, methyltetrahydrofolic acid; FIGLU, formiminoglutamic acid; GLU, glutamic acid; SER, serine; dU, deoxyuridylate; dT, deoxythimidylate; CR, cobalamin reductases; CAT, cobalamin adenosyl transferase; MMCM, methylmalonyl CoA mutase; MT, methyltransferases; MS, methinine synthase; MSR, methionine synthase reductase; SHT, serine hydroxymethyltransferase, MTHFR, methylenetetrahydrofolate reductase.

Intracellular Protein Binding

After its cellular uptake, the TC–receptor complex is degraded in the lysosome to yield the free vitamin, which can be converted to methylcobalamin in the cytosol. Virtually all of the vitamin in the cell is ultimately bound to two vitamin B_{12}-dependent enzymes:

- **Methionine synthetase** (also called **methyl-FH$_4$ methyltransferase**) in the cytosol
- **Methylmalonyl-CoA** mutase in mitochondria.

Distribution in Tissues

Vitamin B_{12} is the best stored of the vitamins. Under conditions of non-limiting intake, the vitamin accumulates to very appreciable amounts in the body, mainly in the liver (about 60% of the total body stores) and muscles (about 30% of the total). Body stores vary with the intake of the vitamin, but tend to be greater in older subjects. Hepatic concentrations approaching 2,000 ng/g have been reported in humans; however, a total hepatic reserve of about 1.5 mg is typical. Mean total body stores of vitamin B_{12} in humans are in the range of 2–5 mg. The greatest concentrations of vitamin B_{12} occur in the pituitary gland; the kidneys, heart, spleen, and brain also contain substantial amounts – in humans, these organs each contain 20–30 μg of vitamin B_{12}. The great storage and long biological half-life (350–400 days in humans) of the vitamin provide substantial protection against periods of deprivation. The low reserve of the human infant (~25 μg) is sufficient to meet physiological needs for about a year.

The predominant form in human plasma is methylcobalamin (60–80% of the total),[18] owing to the presence of TC$_I$ that selectively binds that vitamer (Table 17.3). However, the predominant vitamer in the plasma of other species, and in other tissues of all species, is adenosylcobalamin (in humans, this form accounts for 60–70% of the total vitamin in liver, and about 50% of that in other tissues). Whereas methylcobalamin is the main form bound by haptocorrin, both it and adenosylcobalamin are bound in similar amounts by TC. Normal plasma vitamin B_{12} concentrations vary widely among various mammalian species, from only hundreds (humans) to thousands (rabbits) of picomoles per liter.

The vitamin B_{12} concentration of human milk varies widely (330–320 pg/ml), and is particularly great (10-fold that of mature milk) in colostrum. Although those products

TABLE 17.3 Cobalamins in Normal Human Plasma

	Range, pmol/l
Total cobalamins	173–545
Methylcobalamin	135–427
Adenosylcobalamin	2–77
Cyanocobalamin	2–48
Aquocobalamin	5–67

18. Methylcobalamin is lost preferentially, in comparison with the other forms of the vitamin, in pernicious anemia patients.

contain TC, most of the vitamin (mainly methylcobalamin) is bound to R proteins, which they also contain in large amounts. In contrast, cow's milk, which does not contain R proteins, typically shows lower concentrations of the vitamin, present in that product mainly as adenosylcobalamin.

Enterohepatic Circulation

Secretion of vitamin B$_{12}$ in the bile is a major means by which it is recycled by reabsorption. Significant amounts (0.5–5 µg) of the vitamin enter the bile each day, and are thus made available for binding to luminal IF for reabsorption from the gut.[19]

5. METABOLISM OF VITAMIN B$_{12}$

Activation to Coenzyme Forms

Vitamin B$_{12}$ is delivered to cells in the oxidized form, hydroxycob(III)alamin, where it is reduced by thiol- and reduced flavin-dependent reduction of the **cobalt** center of the vitamin (to Co$^+$) to form cob(I)amin, also called vitamin B$_{12s}$ (Fig. 17.1). However, the vitamin is active in metabolism *only* as methyl or 5'-deoxyadenosyl derivatives that have either respective group attached covalently to the cobalt atom. The conversion to these coenzyme forms involves two different enzymatic steps:

- *Methylcobalamin.* Methylation of the vitamin is catalyzed by the cytosolic enzyme **5-methyl-FH$_4$:homocysteine methyltransferase**. This renders the vitamin, as methylcobalamin, a carrier for the single-carbon unit used in the regeneration of **methionine** from **homocysteine (Hcy)**. Methylcobalamin is also produced by re-charging the reduced vitamin (Co^{1+}) with a methyl group transferred from 5-methyl FH$_4$. This cycling risks the occasional oxidation of cobalamin-cobalt (to Co^{2+}), in which case it is reduced back to Co^{1+} by the enzyme **methionine synthase reductase**.

- *Adenosylcobalamin.* Adenosylation of the vitamin occurs in the mitochondria due to the action of **vitamin B$_{12}$ coenzyme synthetase**, which catalyzes the reaction of cob(II)amin with a deoxyadenosyl moiety derived from ATP. This step depends on the entry of hydroxycobalamin into the mitochondria and its subsequent reduction in sequential, one-electron steps involving NADH- and NADPH-linked aquacobalamin reductases[20] to yield cob(II)alamin.

Catabolism

Little, if any, metabolism of the corrinoid ring system is apparent in animals, and vitamin B$_{12}$ is excreted as the intact cobalamin. Apparently, only the free cobalamins (not the methylated or adenosylated forms) in the plasma are available for excretion.

Excretion

Vitamin B$_{12}$ is excreted via both renal and biliary routes at the daily rate of about 0.1–0.2% of total body reserves (in humans this is 2–5 µg/day, thus constituting the daily requirement for the vitamin). Although it is found in the urine, glomerular filtration of the vitamin is minimal (<0.25 µg/day in humans), and it is thought that urinary cobalamin is derived from the tubular epithelial cells and lymph. Urinary excretion of the vitamin after a small oral dose can be used to assess vitamin B$_{12}$ status; this is called the **Schilling test**. The biliary excretion of the vitamin is substantial, accounting in humans for the secretion into the intestine of 0.5–5 µg/day. Most (65–75%) of this amount is reabsorbed in the ileum by IF-mediated active transport. This enterohepatic circulation constitutes a highly efficient means of conservation, with biliary vitamin B$_{12}$ contributing only a small amount to the feces.

Congenital Disorders of Vitamin B$_{12}$ Metabolism

Several congenital deficiencies in proteins involved in vitamin B$_{12}$ metabolism, each an autosomal recessive trait, have been reported in humans (Table 17.4). Most of these disorders result in signs alleviated by high parenteral doses of the vitamin. Several congenital disorders of intracellular vitamin B$_{12}$ utilization have been identified in humans. These have been categorized into complementation groups that relate to eight different human genes (Table 17.5). Most of these disorders respond to high parenteral doses of the vitamin.

TABLE 17.4 Congenital Disorders of Vitamin B$_{12}$ Absorption and Transport

Condition	Missing/ Deficient Factor	Signs/Symptoms
Lack of intrinsic factor	Intrinsic factor	Megaloblastic anemia presenting at 1–3 years
Imerslund-Gräsbeck syndrome	IF receptor	Specific malabsorption of vitamin B$_{12}$
Lack of transcobalamin	Transcobalamin	Severe (fatal) megaloblastic anemia presenting early in life
Lack of R proteins	R proteins	None

19. Green, R., Jacobsen, D. W., van Tonder, S. V., *et al.* (1981). *Gastroenterology* 81, 773.
20. These activities are derived from a cytochrome b$_5$/cytochrome b$_5$ reductase complex, and from a cytochrome P-450 reductase complex and an associated flavoprotein.

TABLE 17.5 Categories of Congenital Disorders of Vitamin B_{12} Metabolism

Group	Defect	Missing/Deficient Factor (Gene)	Signs/Symptoms
Mitochondrial			
cblA	B_{12}-Co^{3+} reduction to B_{12}-Co^{2+}	Mitochondrial cobalamin reductase (*MMAA*)	Methylmalonic aciduria
cblB	Adenosyl-B_{12} production	Adenosyl transferase (*MMAB*)	Methylmalanic aciduria
cblC[a]	Production of ado-/methyl B_{12}	Cobalamin reductase (*MMACHC*)	Homocysteinuria, methylmalonic aciduria
cblD	B_{12} entry into mitochondria	B_{12} chaperone (*MMADHC*)	Homocysteinuria, methylmalonic aciduria
mut	Isomerization of methylmalonyl CoA	Methylmalonyl Co mutase (*MCM*)	Methylmalanic aciduria
Cytosolic			
cblG	Methionine synthase	Methyl transferase activity (*MTR*)	Homocysteinemia, hypomethioninemia, megaloblastic anemia, developmental delay
cblE	B_{12}-Co^{3+} reduction to B_{12}-Co^{2+}	Methionine synthase reductase (*MTRR*)	Homocysteinemia, hypomethioninemia, megaloblastic anemia, developmental delay
Lysosomal			
cblF	Lysosome to cytosol B_{12} export	Lysosomal membrane protein (*LMBRD1*)	Developmental delay, homocysteinuria, methylmalonic aciduria

[a]The most common disorder of vitamin B_{12} metabolism, with some 400 patients described to date.

6. METABOLIC FUNCTIONS OF VITAMIN B_{12}

Coenzyme Functions

Vitamin B_{12} functions in metabolism in two coenzyme forms: adenosylcobalamin and methylcobalamin (Fig. 17.1). While several vitamin B_{12}-dependent metabolic reactions have been identified in microorganisms,[21] only two have been discovered in animals. These play key roles in the metabolism of propionate, amino acids, and single carbon.

Methylmalonyl-CoA Mutase

The adenosylcobalamin-dependent enzyme methylmalonyl-CoA mutase catalyzes the conversion of methylmalonyl-CoA to succinyl-CoA in the degradation of propionate formed from odd-chain fatty acids (and an important energy source for ruminants, in which it is produced by rumen microflora). This reaction involves splitting a carbon–carbon bond of the coenzyme with the formation of a free radical on the coenzyme that can be transferred through an amino acid residue to the substrate. That the propionic acid pathway is also important in nerve tissue *per se* is suggested by the delayed onset of the neurological signs of vitamin B_{12} deficiency effected in animals by dietary supplements of direct (valine, isoleucine) or indirect (methionine) precursors of propionate.

The mutase is a mitochondrial matrix enzyme, the dimer of which binds two adenosylcobalamin molecules. In humans, it is the first vitamin B_{12}-dependent enzyme to be affected by deprivation of vitamin B_{12}. Owing to loss of this activity, vitamin B_{12}-deficient subjects show **methylmalonic aciduria**, especially after being fed odd-chain fatty acids. The accumulation of **methylmalonic acid (MMA)** can disrupt normal glucose and glutamic acid metabolism, apparently by inhibiting the tricarboxylic acid (TCA) cycle. Vitamin B_{12} deficiency can also cause a reversal of propionyl-CoA carboxylase activity, leading to the incorporation of the three-carbon propionyl-CoA in place of the two-carbon acetyl-CoA, and resulting in the production of small amounts of odd-chain fatty acids.

---

21. The following microbial enzymes require adenosylcobalamin: glutamate mutase, 2-methylene-glutarate mutase, L-β-lysine mutase, D-α-lysine mutase, D-α-ornithine mutase, leucine mutase, 1,2-diolde-hydratase, glyceroldehydratase, ethanolamine deaminase, and ribonucleotide reductase; methylcobalamin is also required for the bacterial formation of methane and acetate.

Increased levels of methylmalonyl-CoA can also lead to its incorporation in place of malonyl-CoA, resulting in the synthesis of small amounts of methyl branched-chain fatty acids. It has been suggested that the neurological signs of vitamin B$_{12}$ deficiency may result, at least in part, from the production of these abnormal fatty acids in neural tissues.

Several inborn metabolic errors result in decreases in methylmalonyl-CoA mutase activity, leading to methyl-malonic aciduria. These errors include mutations affecting the gene that encodes the enzyme, which results in the absence of the enzyme or the expression of a defective protein. Other mutations reduce the synthesis of its cofactor adenosylcobalamin; individuals with these defects respond to vitamin B$_{12}$ treatment.

Methionine Synthase

Methionine synthetase catalyzes the methylation of Hcy to regenerate methionine, serving as the methyl group carrier (via methylcobalamin) between the donor 5-methyl-tetrahydrofolate (5-methyl-FH$_4$) and the acceptor Hcy (Fig 17.1). This reaction is a simple transfer of the single-C moiety. Because of diminished methionine synthetase activity, vitamin B$_{12}$-deficient subjects show reduced availability of methionine. Methionine is essential for the synthesis of proteins and polyamines, and is the precursor of **S-adenosylmethionine (SAM)**, which serves as the primary donor of "labile" methyl groups for more than 100 enzymatic reactions that have critical roles in metabolism.[22] SAM also serves as a key regulator of the transsulfuration and re-methylation pathways, which involve the folate-dependent methylenetetrahydrofolate reductase (MFTHR).

Losses of SAM lead to impairments in the synthesis of creatine, phospholipids, and the neurotransmitter acetylcholine, all of which have broad impacts on physiological function. Low vitamin B$_{12}$ status thus results in the accumulation of both Hcy and 5-methyl-FH$_4$ (via the methyl-folate trap; see Chapter 16), the latter resulting in the loss of FH$_4$, the key functional form of folate. Methionine synthase can also catalyze the reduction of nitrous oxide to elemental nitrogen; in doing so it generates a free radical that inactivates the enzyme.

Methionine synthase expression is induced by vitamin B$_{12}$. This process appears to involve the vitamin binding to a transactivating protein to induce a conformational change that allows it to bind to an internal site on the methionine synthase mRNA to enhance ribosomal recruitment and promote translation.[23]

Genetic defects in methionine synthase and in the production of its cofactor methylcobalamin have been identified. Each results in homocysteinemia and, commonly, megaloblastic anemia. Individuals with these defects do not respond to vitamin B$_{12}$ treatment, but their anemia can respond to folate supplementation. An A→G polymorphism at base position 2756 of methionine synthase has been identified; women with the AG genotype have been found to have double the risk of having a child with NTDs, and the risk of having a child with Down syndrome is increased 3.5-fold.[24] Polymorphisms of methionine synthase reductase, also involving this functioning pathway (see Fig. 17.1), have been associated with similar effects.

Interrelationship with Folate

The major cycle of single-C flux in mammalian tissues appears to be the serine hydroxymethyltransferase/5,10-methylene-FH$_4$ reductase/methionine synthase cycle, in which the latter reaction is rate-limiting (Fig. 17.1). The committed step (5,10-methylene-FH$_4$ reductase) is feedback inhibited by SAM, and product-inhibited by 5-methyl-FH$_4$. Methionine synthase depends on the transfer of labile methyl groups from 5-methyl-FH$_4$ to vitamin B$_{12}$. Methyl-B$_{12}$ serves as the immediate methyl donor for converting Hcy to methionine. Without adequate vitamin B$_{12}$ to accept methyl groups from 5-methyl-FH$_4$, that metabolite accumulates at the expense of the other metabolically active folate pools. This is known as the "methyl-folate trap." This blockage results in the accumulation of the intermediate formiminoglutamic acid (FIGLU). Thus, an elevated urinary FIGLU level after an oral histidine load is diagnostic of vitamin B$_{12}$ deficiency.

DNA Methylation

Different phenotypes appear to be generated through processes including the epigenetic regulation of specific genes. This involves the methylation of DNA cytosine bases and histone proteins. Hypomethylation of these factors is believed to alter chromatin structure in ways that affect transcription and can increase the rate of C→T transition mutation.[25] DNA hypermethylation is associated with gene silencing. Vitamin B$_{12}$ and folate play key roles in the provision of single-C units for these processes. Thus, suboptimal status with respect to either nutrient, or to their metabolic functions, would be expected to affect gene expression and stability. Chromosomal aberrations have also been reported for some patients with pernicious anemia.[26] A cross-sectional study showed that the vitamin B$_{12}$ levels of buccal

22. By loss of the flux of methyl groups via 5-methyl-FH$_4$:Hcy methyl-transferase, folate deficiency causes a secondary hepatic choline deficiency.
23. Oltean, S. and Banerjee, R. (2005). *J. Biol. Chem.* 280, 32662.

24. Doolin, M. T., Barbaux, S., McDonnel, M., *et al.* (2002). *Am. J. Hum. Genet.* 71, 1222; Bosco, P., Guéant-Rodriguez, R. M., Anello, G., *et al.* (2003). *Am. J. Med. Genet.* 121A, 219.
25. Methylated CpG sites appear to be at particularly high risk for C→T changes, which occur in the *p53* tumor suppressor gene.
26. Jensen, M. K. (1977). *Mutat. Res.* 45, 249.

cells were significantly lower in smokers than non-smokers, and that elevated levels of the vitamin were associated with reduced frequency of micronucleus formation.[27]

7. VITAMIN B$_{12}$ IN HEALTH AND DISEASE
Neural Tube Defects (NTDs)

It is possible that vitamin B$_{12}$ may be involved in the residual incidence of NTDs not prevented by folate supplementation (*see* Chapter 16, Fig. 16.6). Studies have shown lower vitamin B$_{12}$ levels in amniotic fluid from NTD pregnancies compared to healthy ones, while mothers' serum vitamin B$_{12}$ levels are comparable.[28] This suggests a limitation in the maternal capacity to provide the fetus with an adequate supply of the vitamin. That women with NTD pregnancies are more likely to have the methionine synthase 66AG genotype, which presumably produces an aberrant enzyme, is consistent with this hypothesis.

Cardiovascular Health

Owing to lost methionine synthetase activity, subclinical vitamin B$_{12}$ deficiency can result in a moderate to intermediate elevation of plasma Hcy concentrations (*see* Fig. 17.1). This condition, called homocysteinemia, can also be produced by folate deficiency, which, by the methyl-folate trap, also limits methionine synthase activity. Vitamin B$_{12}$ deficiency may be the primary cause of homocysteinemia in many people; almost two-thirds of elderly subjects with homocysteinemia also show methylmalonic acidemia, indicative of vitamin B$_{12}$ deficiency (Table 17.6). Still, less than a third of individuals with low circulating vitamin B$_{12}$ levels also show homocysteinemia.

TABLE 17.6 Vitamin B$_{12}$ and Folate Status of Elderly Subjects Showing Homocysteinemia

Parameter	Serum Hcy		Serum MMA	
	>3 SD	≤3 SD	>3 SD	≤3 SD
Serum vitamin B$_{12}$ (pmol/l)	197 ± 77[a]	325 ± 145	217 ± 83[a]	332 ± 146
Serum folate (nmol/l)	12.7 ± 8.2[a]	22.9 ± 19.0	18.1 ± 12.5[a]	22.7 ± 19.5

[a]$P > 0.05$; SD, standard deviation.
From Lindenbaum, J., Rosenberg, I.H., Wilson, P. W., *et al.* (1994). Prevalence of cobalamin deficiency in the Framingham elderly population. *Am. J. Clin. Nutr.* 60, 2.

Homocysteine Hypothesis

Epidemiologic studies subsequently showed associations of moderately elevated plasma Hcy and risks of coronary, peripheral, and carotid arterial thrombosis and atherosclerosis; venous thrombosis; retinal vascular occlusion; carotid thickening; and hypertension.[29] A meta-analysis of the results of 27 cross-sectional and case–control studies[30] attributed 10% of total coronary artery disease to homocysteinemia.[31] However, the results of clinical trials have shown that Hcy-lowering by folate treatment does not reduce cardiovascular disease risk, casting doubt on the role of Hcy in the etiology of cardiovascular disease. It has been suggested that cardiovascular pathogenesis may actually involve the metabolic precursor of Hcy, *S*-adenosylhomocysteine (SAH). SAH is a potent inhibitor of methyltransferases; accordingly, the intracellular ratio of SAH:SAM is taken as an indictor of methylation capacity – i.e., the "methylation index." SAH is reversibly converted with Hcy, the equilibrium favoring SAH. Therefore, homocysteinemia would be expected to lead to elevated levels of SAH, which appears to be present at only very low amounts (an estimated <0.2% of Hcy levels) in the circulation in normal circumstances.

Neurological Function

Insufficient vitamin B$_{12}$ status is thought to lead to neurodegeneration (mostly of glial cells, myelin, and interstitium) as a result of abnormal incorporation of MMA into neuronal lipids including those in myelin sheaths, stimulation of the inflammatory cytokine tumor necrosis factor-α and/or reduced synthesis of choline, the precursor of the neurotransmitter acetylcholine.

Cognition

Serum Hcy levels have been found to be negatively correlated with neuropsychological test scores. Low serum vitamin B$_{12}$ levels have been associated with poor or declining cognition in older subjects, and have been observed more frequently in patients with senile dementia of the Alzheimer's type than in the general population. The hypothesis that low vitamin B$_{12}$ status may have a role in that disorder is supported by observations of inverse relationships of serum vitamin B$_{12}$ levels and platelet monoamine oxidase activities, and the finding that vitamin

27. Piyathilke, C. J., Macaluso, M., Hine, R. J., *et al.* (1995). *Cancer Epid. Biomarkers Prev.* 4, 751.
28. Ray, J. G. and Blom, H. J. (2003). *Q. J. Med.* 96, 289.

29. The low prevalence of coronary heart disease among South African Blacks has been associated with their typically lower plasma Hcy levels and their demonstrably more effective Hcy clearance after methionine loading.
30. Boushey, C. J. Beresford, S. A., Omenn, G. S., *et al.* (1995). *J. Am. Med. Assoc.* 274, 1049.
31. Defined as a plasma Hcy concentration >14 μmol/l.

B$_{12}$ treatment reduced the latter. Nevertheless, there is little evidence that vitamin B$_{12}$ treatment can improve cognitive function in most impaired patients, although a review of clinical experience in India suggested value of the vitamin in improving frontal-lobe and language function in patients.[32]

Depression

Low plasma levels of vitamin B$_{12}$ (and folate) have been reported in nearly a third of patients with depression, who also tend to show homocysteinemia. A recent study[33] suggested that patients with high vitamin B$_{12}$ status may have better treatment outcomes, but randomized clinical trials of vitamin B$_{12}$ treatment have not been reported.

Schizophrenia

That patients with schizophrenia generally have elevated circulating levels of methionine, with many also having homocysteinemia, suggests perturbations in single-carbon metabolism that naturally suggests a role of vitamin B$_{12}$. Randomized clinical trials of vitamin B$_{12}$ treatment have not been reported.

Multiple Sclerosis

It has been suggested that low vitamin B$_{12}$ status may exacerbate multiple sclerosis by enhancing the processes of inflammation and demyelination, and by impairing those of myelin repair. Pertinent to this hypothesis are the results of a study that found combination therapy with interferon-β and vitamin B$_{12}$ to produce dramatic improvements in an experimental model of the disease.

Anticarcinogenesis

That low, asymptomatic vitamin B$_{12}$ status produced aberrations in base substitution and methylation of colonic DNA in the rat model[34] suggests that subclinical deficiencies of the vitamin may enhance carcinogenesis. This hypothesis is supported by the results of a prospective study that found significantly increased breast cancer risk among women ranking in the lowest quintile of plasma vitamin B$_{12}$ concentration,[35] and the finding of significantly different risks for esophageal squamous cell carcinoma and gastric cardia adenocarcinoma associated with

polymorphisms of MTHFR[36] which, because it catalyzes the production of 5'-methyl-FH$_4$, can limit the activity of the vitamin B$_{12}$-dependent methionine synthase. The only studies to date of prostate cancer risk indicate positive associations of serum vitamin B$_{12}$ level and vitamin B$_{12}$ intake;[37] however, the causal directionality of such relationships is not clear.

Osteoporosis

That vitamin B$_{12}$ may affect bone health was suggested by the finding that osteoporosis was more prevalent among elderly Dutch women of marginal or deficient status with respect to the vitamin (Table 17.7). Subsequent studies have revealed positive associations of serum vitamin B$_{12}$ level and bone mineral density, markers of bone turnover, and risks of osteoporosis and hip fracture. It has been suggested that homocysteinemia may underlie these relationships. Intervention with both vitamin B$_{12}$ and folate, which reduced plasma Hcy, reduced hip fracture risk.[38]

Hearing Loss

Serum vitamin B$_{12}$ levels have been reported to be lower in subjects with hearing loss compared to normal-hearing controls, and vitamin B$_{12}$ supplementation has been reported to lessen tinnitus[39] in chronically affected subjects.[40]

TABLE 17.7 Relationship of Vitamin B$_{12}$ Status and Osteoporosis Risk Among Elderly Women

Plasma Vitamin B$_{12}$ (pmol/l)	n	Relative Risk
>320	34	1.0
210–320	43	4.8 (1.0–23.9)[a]
<210	35	9.5 (1.9–46.1)

[a]95% confidence interval.
From Dhonukshe-Rutten, R. A. M., Lips, M., de Jong, N., et al. (2003). J. Nutr. 133, 801.

32. Rita, M. (2004). Neurol. India 52, 310.
33. Levitt, A. J., Wesson, V. A., and Joffe, R. T. (1998). Psychiat. Res. 79, 123.
34. Choi, S. W., Friso, S., Ghandour, H., et al. (2004). J. Nutr. 134, 750.
35. Wu, K., Helzlsouer, K. J., Comstock, G. W., et al. (1999). Cancer Epidemiol. Biomarkers Prev. 8, 209.
36. Stolzenberg-Solomon, R. Z., Qiao, Y. L, Abnet, C. C., et al. (2003). Cancer Epidemiol. Biomarkers Prev. 12, 1222.
37. Collin, S. M., Metcalfe, C., Refsum, H., et al. (2010). Cancer Epidemiol. Biomarkers Prev. 19, 1632.
38. Sato, Y., Honda, Y., and Iwamoto, J. (2005). J. Am. Med. Assoc. 293, 1082.
39. That is, the perception of sound within the ear in the absence of corresponding external sound.
40. Shemesh, Z., Attias, J., Ornan, M., et al. (1993). Am. J. Otolaryngol. 2, 94.

Cyanide Metabolism

Cobalamins can bind cyanide, to produce the non-toxic cyanocobalamin. For that reason hydroxocobalamin is a well recognized cyanide antidote. Thus, it has been proposed that vitamin B_{12} may have a role in the inactivation of the low levels of cyanide consumed in many fruits, beans, and nuts.

8. VITAMIN B_{12} DEFICIENCY

Determinants of Vitamin B_{12} Deficiency

Low vitamin B_{12} status has been estimated to affect a large portion of the general population, including 30–40% of older adults. Deficiency of vitamin B_{12} can be produced by several factors.

Vegetarian Diets

Strict vegetarian diets, containing no meats, fish, animal products (e.g., milk, eggs) or vitamin B_{12} supplements (e.g., multivitamin supplements, nutritional yeasts), contain practically no vitamin B_{12} (Tables 17.8, 17.9). Therefore, individuals consuming such diets typically show very low circulating levels of the vitamin; one study found 56% of vegetarian American women to have low serum concentrations (<148 pmol/l) of vitamin B_{12}. Nevertheless, clinical signs among such individuals appear to be rare and may not become manifest for many years, although they are more common among breastfed infants (Table 17.10). The vitamin B_{12} content of breast milk has been found to vary inversely with the length of maternal vegetarian practice. Serum vitamin B_{12} concentrations have also been found to vary inversely with the length of time of vegetarian practice, showing progressive declines through about 7 years (Fig. 17.2). This time compares very favorably with the estimated drawdown of hepatic stores of the vitamin.

It should be remembered that not all vegetarians are strict vegans; many (called **ovolactovegetarians**) consume plant-based diets that also contain servings of dairy products, eggs or fish to varying extents. Studies have shown that the occasional consumption of animal products (e.g., once per month) will support serum vitamin B_{12} levels comparable to those of people eating traditional mixed diets (Table 17.11). In addition, such foods as *Nori* sp. and *Chlorella* sp. seaweeds, which may be eaten by vegetarians, appear to contain vitamin B_{12}.

Malabsorption

A study of elderly Americans found >40% to show elevations in urinary MMA levels; half also showed low serum vitamin B_{12} levels.[41] Low cobalamin levels have been observed in 10–15% of apparently healthy elderly Americans with apparently adequate vitamin B_{12} intakes,

TABLE 17.8 Vitamin B_{12} and Folate Status of Thai Vegetarians and Mixed Diet Eaters

Group	Vitamin B_{12} (pg/ml)	Folate (ng/ml)
Mixed diet		
Males	490	5.7
Females	500	6.8
Vegetarian		
Males	117[a]	12.0[a]
Females	153[a]	12.6[a]

[a]$P > 0.05$.
From Tungtrongchitr, V., Pongpaew, P., Prayurahong, B., *et al.* (1993). *Intl J. Vit. Nutr. Res.* 63, 201.

TABLE 17.9 Plasma Indicators of Vitamin B_{12} Status in Vegetarians and Non-Vegetarians

Plasma Analyte	Omnivorous Subjects	Lacto- and Ovolacto-Vegetarians		Vegans	
		Vitamin Users	Vitamin Non-Users	Vitamin Users	Vitamin Non-Users
Vitamin B_{12}, pmol/l	287 (190–471)[a]	303 (146–771)	179 (124–330)	192 (125–299)	126 (92–267)
Transcobalamin, pmol/l	54 (16–122)	26 (30–235)	23 (4–84)	14 (3–53)	4 (2–35)
Methylmalonic acid, nmol/l	161 (95–357)	230 (120–1,344)	368 (141–2,000)	708 (163–2,651)	779 (222–3,480)
Hcy, μmol/l	8.8 (5.5–16.1)	9.6 (5.5–19.4)	10.9 (6.4–27.7)	11.1 (5.3–25.9)	14.3 (6.5–52.1)
Folate, nmol/l	21.8 (14.5–51.5)	30 (14.8–119)	27.7 (16.0–76.9)	29.5 (18.8–71.8)	34.3 (20.7–72.7)

[a]*Mean, 95% confidence interval.*
From Herrmann, W., Schorr, H., and Obeid, R. (2003). *Am. J. Clin. Nutr.* 78, 131.

41. Norman, E. J. and Morrison, J. A. (1993). *Am. Med. J.* 94, 589.

TABLE 17.10 Ranges of Urinary Methylmalonic Acid Excretion by Breastfed Infants of Vegetarian and Omnivorous Mothers

Group	MMA (μmol/mmol Creatinine)
Vegetarian	2.6–790.9
Mixed-diet	1.7–21.4

From Specker, B. L., Miller, D., Norman, E. J., et al. (1988). *Am. J. Clin. Nutr.* 47, 89.

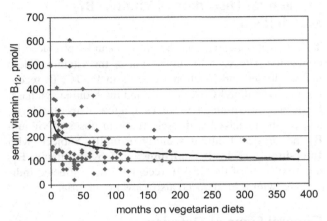

FIGURE 17.2 Inverse relationship of serum vitamin B$_{12}$ concentrations and time following vegetarian eating practices in people in the northeastern United States. (*From Miller, D. R., Specker, B. L., Ho, M. L., et al. (1991). Am. J. Clin. Nutr. 53, 524.*)

TABLE 17.11 Impact of Occasional Consumption of Animal Products on Vitamin B$_{12}$ Status in a Macrobiotic Community

Food	Consumed	Serum Vitamin B$_{12}$ (pmol/l)	Urine MMA (mmol/mol Creatinine)
Dairy	Never	122	5.3
	≤1/week	183[a]	2.8[a]
	>1/week	179[a]	2.1[a]
Eggs	Never	139	4.8
	≤1/week	167	3.1
	>1/week	157	2.2
Seafoods[b]	Never	111	4.4
	≤1/week	145	5.3
	>1/week	161	2.6

[a]$P > 0.05$.
[b]Includes various sea vegetables (e.g., wakame, kombu, hijiki, arame, nori, dulse).
From Miller, D. R., Specker, B. L., Ho, M. L., et al. (1991). *Am. J. Clin. Nutr.* 53, 524.

and in 60–70% of those with low vitamin B$_{12}$ intakes.[42] The prevalence of low plasma vitamin B$_{12}$ concentrations in all Central American age groups was found to be 35–90%.[43] It would appear that poor absorption of the vitamin accounts for a third of such cases. Vitamin B$_{12}$ malabsorption can be caused by inadequate production of IF by gastric parietal cells, and/or defective functioning of ileal IF-receptors.

Loss of Gastric Parietal Cell Function

Vitamin B$_{12}$ malabsorption occurs when IF production by gastric parietal cells is inadequate.[44] Such conditions can have causes of four general types:

- *Pernicious anemia.* Pernicious anemia affects at least 2–3% of Americans, with women showing a higher

prevalence than men. It is a disease of later life, 90% of cases being diagnosed in individuals >40 years of age. The anemia is the end result of autoimmune gastritis, also called type A chronic atrophic gastritis or gastric atrophy. It involves the destruction of the fundus and body of the stomach by antibodies to the membrane H$^+$/K$^+$-ATPase of the parietal cells. This causes progressive atrophy of parietal cells leading to hypochlorhydria and loss of IF production, and resulting in severe vitamin B$_{12}$ malabsorption. The condition presents as megaloblastic anemia within 2–7 years. The disorder is likely to be widely underdiagnosed, as affected subjects may have neurological rather than hematological disease.

- *Heliobacter pylori.* *Heliobacter pylori* infection produces damage mostly to the stomach, referred to as Type B chronic atrophic gastritis, affecting an estimated 9–30% of Americans. The condition involves hypochlorhydria, which facilitates the proliferation of bacteria in the intestine. Both conditions limit and compete for the enteric absorption of vitamin B$_{12}$; however, reports of the effect of the condition on vitamin B$_{12}$ status have been conflicting.

- *Other gastric diseases.* Vitamin B$_{12}$ utilization can be affected by disease involving damage to the gastric parietal cells, and thus reduced production of stomach acid and IF. Such damage can result in megaloblastic anemia or, frequently, hypochromic anemia due to

42. Carmel, R., Green, R., Jacobsen, D. W., *et al.* (1999). *Am. J. Clin. Nutr.* 70, 904; Carmel, R. (2000). *Annu Rev. Med.* 51, 357.
43. Allen, L. H. (2004). *Nutr. Rev.* 62, S29.
44. Chronic atrophic gastritis can be a precancerous lesion, involving progressive metaplasia of the gastric mucosa leading to carcinoma.

impaired iron absorption caused by the resulting hypo-acidic condition. Such damage occurs in patients with simple (non-autoimmune) atrophic gastritis as well as those undergoing gastrectomy. After bariatric surgery, 10–15% of patients develop vitamin B_{12} deficiency within a few years; all patients undergoing complete gastrectomy are placed in need of supplemental vitamin.

- *Chronic use of proton-pump inhibitors.* Chronic inhibition of parietal call acid production reduces the dissociation of vitamin B_{12} from the proteins to which it is bound in food matrices.

Pancreatic Insufficiency

The loss of pancreatic exocrine function can impair the utilization of vitamin B_{12}. For example, about one-half of all human patients with pancreatic insufficiency show abnormally low enteric absorption of the vitamin. This effect can be corrected by pancreatic enzyme replacement therapy, using oral pancreas powder or pancreatic proteases. Thus, the lesion appears to involve specifically the loss of proteolytic activity, resulting in the failure to digest intestinal R proteins, which thus retain vitamin B_{12} bound in the stomach instead of freeing it for binding by IF.

Intestinal Diseases

Tropical sprue[45] and ileitis involve damage to the ileal epithelium that can cause the loss of ileal IF receptors. Intestinal parasites such as the fish tapeworm *Diphyllobothrium latum* can effectively compete with the host for uptake of the vitamin. Explosively growing bacterial floras can do likewise. Protozoal infections such as *Giardia lamblia*, which cause chronic diarrhea, appear to cause vitamin B_{12} malabsorption in malnourished individuals.

Chemical Factors

Several other factors can impair the utilization of vitamin B_{12}:

- *Xenobiotics.* Biguanides,[46] alcohol, and smoking can damage the ileal epithelium and cause the loss of ileal IF receptors.
- *Nitrous oxide.* Animal models of vitamin B_{12} deficiency have been developed using exposure to nitrous oxide

as the precipitating agent.[47] Nitrous oxide oxidizes the reduced form of cobalamin, cob(I)alamin, to the inactive form cob(II)alamin, causing rapid inactivation of the methylcobalamin-dependent enzyme and the excretion of the vitamin. Thus, repeated exposure to nitrous oxide results in the depletion of body cobalamin stores.

- *Oral contraceptive agents.* The use of contraceptive steroids has been shown to cause a slight drop in plasma vitamin B_{12} concentration; however, no signs of impaired function have been reported.

Congenital Disorders of Vitamin B_{12} Metabolism

Tissue-level deficiency in vitamin B_{12} can be produced by congenital defects in the utilization of the vitamin. These include losses and functional defects in IF, TC, IF receptor, methylmalonyl-CoA mutase, and methionine synthase. Signs present within a few weeks to a few years after birth, and can be managed with high, frequent doses of vitamin B_{12} administered intramuscularly or orally. Subjects with **Imerslund-Gräsbeck syndrome**[48] malabsorb vitamin B_{12} due to defects of the ileal IF receptor; many affected individuals (with RAP defects) also show proteinuria.

General Signs of Deficiency

Vitamin B_{12} deficiency causes delay or failure of normal cell division, particularly in the bone marrow and intestinal mucosa. Because the biochemical lesion involves arrested synthesis of DNA precursors, the process depends on the availability of single carbon units. The vitamin B_{12} deficiency-induced decreases in methionine biosynthesis, folate coenzymes (due to methylfolate "trapping"), and thymidylate synthesis all lead to a failure of DNA replication. This appears to be accompanied by uracil-misincorporation into DNA (due to the use of deoxyuridine instead of thymidine pyrophosphate by DNA polymerase), apparently resulting in double-stranded DNA damage and apoptosis. The reduced mitotic rate results in the formation of abnormally large, cytoplasm-rich cells. This is called a **megaloblastic transformation**; it manifests itself as a characteristic type of anemia in which such enlarged cells are found (**megaloblastic anemia**).

Neurological abnormalities develop in most species. These appear to result from impaired methionine biosynthesis; however, some investigators have proposed that

---

45. Tropical sprue is endemic in south India, occurs epidemically in the Philippines and the Caribbean, and is frequently a source of vitamin B_{12} malabsorption experienced by tourists to those regions.
46. That is, guanylguanidine, amidinoguanidine, and diguanidine, the sulfates of which are used as reagents for the chemical determination of copper and nickel.

47. Much of the toxicity of N_2O may actually be due to impaired vitamin B_{12} function. Indeed, it is known that excessive dental use of laughing gas (which is N_2O) can lead to neurologic impairment.
48. This syndrome is a common cause of vitamin B_{12}-associated megaloblastic anemia.

neurological signs result instead from the loss of adenosylcobalamin. They are typically manifest with relatively late onset, due to the effective storage and conservation of the vitamin. Neurological lesions of vitamin B$_{12}$ deficiency involve diffuse and progressive nerve demyelination, manifested as progressive neuropathy (often beginning in the peripheral nerves) and progressing eventually to the posterior and lateral columns of the spinal cord (Table 17.12).

Marginal (Subclinical) Deficiencies

Marginal deficiencies of vitamin B$_{12}$ are estimated to be at least 10-fold more prevalent than frank deficiencies characterized by classical signs. Marginal status involves metabolic changes resulting from losses of vitamin B$_{12}$ coenzyme functions; these are marked by elevated circulating levels of FIGLU, MMA, and Hcy.

Deficiency Syndromes in Animals

Vitamin B$_{12}$ deficiency in animals is characterized most frequently by reductions in rates of growth and feed intake, and impairments in the efficiency of feed utilization. In a few species (e.g., swine), a mild anemia develops. In growing chicks and turkey poults, neurologic signs may appear. Vitamin B$_{12}$ deficiency has also been related to the etiology of perosis in poultry, but this effect seems to be secondary to those of methionine and choline, and related to the availability of labile methyl groups. Also related to limited methyl group availability (for the synthesis of phosphatidylcholine) in poultry is increased lipid deposition in the liver, heart, and kidneys. For this reason, vitamin B$_{12}$ is known as a **lipotrope** for poultry. Vitamin B$_{12}$ deficiency also causes embryonic death in the chicken, with embryos showing myopathies of the muscles of the leg, hemorrhage, myocardial hypertrophy, and perosis. Impaired utilization of dietary protein and testicular lesions (decreased numbers of seminiferous tubules showing spermatogenesis) have been observed in vitamin B$_{12}$-deficient rats.

Ruminants do not have dietary requirements for vitamin B$_{12}$, relying instead on their rumen microorganisms to produce the vitamin in amounts adequate for their needs. Synthesis of the vitamin in the rumen is dependent on a continuous supply of dietary cobalt, as microbes can synthesize the organic portion of the corrin nucleus in the presence of cobalt to attach at the center. Sheep fed a diet containing less than about 70 µg of cobalt per kilogram of diet show signs of deficiency: inappetence, wasting, diarrhea, and watery lacrimation. Cobalt deficiency reduces hepatic cobalamin levels and increases plasma methylmalonyl-CoA levels, but does not produce clinical signs; therefore, it is generally accepted that cattle are less susceptible than sheep to cobalt deficiency. Rumenal production of vitamin B$_{12}$ can also be affected by the composition of dietary roughage, the ratio of roughage to concentrate, and the level of dry matter intake. Most microbially produced vitamin B$_{12}$ appears to be bound to rumen microbes, and is released for absorption only in the small intestine.

Deficiency Signs in Humans

Vitamin B$_{12}$ deficiency in humans is characterized by megaloblastic anemia and abnormalities of lipid metabolism; after prolonged periods, neurological signs affect approximately one-quarter of affected individuals. The neurological signs of vitamin B$_{12}$ deficiency in humans include **peripheral neuropathy**, characterized by numbness of the hands and feet, and losses of proprioreception and vibration sense of the ankles and toes. Associated psychiatric signs can also be seen: memory loss, depression, irritability, psychosis, and dementia.

Distinguishing Deficiencies of Vitamin B$_{12}$ and Folate

Some clinical signs (e.g., macrocytic anemia) can result from deficiencies of either vitamin B$_{12}$ or folate. The only metabolic process that is common to the two vitamins is the methyl group transfer from 5′-methyl-FH$_4$ to methylcobalamin for the subsequent methylation of Hcy to yield methionine and the return of folate to its most important central metabolite, FH$_4$. Thus, deficiencies of either vitamin will reduce the FH$_4$ pool either directly by deprivation of folate or indirectly via the methyl-folate trap, resulting from deprivation of vitamin B$_{12}$. In either case, the availability of FH$_4$ is reduced, and, consequently, its conversion ultimately to 5,10-methylene-FH$_4$ is also reduced. This limits the production of thymidylate and thus of DNA, resulting in impaired mitosis, which manifests itself as macrocytosis

TABLE 17.12 General Signs of Vitamin B$_{12}$ Deficiency

Organ System	Signs
General:	
Growth	Decrease
Vital organs	Hepatic, cardiac, and renal steatosis
Fetus	Hemorrhage, myopathy, death
Circulatory:	
Erythrocytes	Anemia
Nervous	Peripheral neuropathy

and anemia. Similarly, the urinary excretion of the histidine metabolite formiminoglutamic acid (FIGLU) is elevated by deficiencies of either folate or vitamin B_{12}, as FH_4 is required to accept the formimino group, yielding $5'$-formimino-FH_4. Studies have revealed that about half of subjects with either vitamin B_{12} or folate deficiencies, and more than half of those with the combined deficiencies, show plasma methionine concentrations below normal.[49] Serum vitamin B_{12} levels are highly correlated with those of methionine in vitamin B_{12}-deficient subjects.

While supplemental folate can mask the anemia or FIGLU excretion (especially after histidine loading) associated with vitamin B_{12} deficiency by maintaining FH_4 in spite of the methyl-folate trap, supplemental vitamin B_{12} does not affect the anemia (or other signs) of folate deficiency. Although such signs as macrocytic anemia, urinary FIGLU, and subnormal circulating folate concentrations are therefore not diagnostic for either vitamin B_{12} or folate deficiency (these deficiencies cannot be distinguished on the basis of these signs), the urinary excretion of MMA can be used for that purpose. **Methylmalonic aciduria** (especially after a meal of odd-chain fatty acids or a load of propionate) occurs only in vitamin B_{12} deficiency (methylmalonyl-CoA mutase requires adenosylcobalamin). Therefore, patients with macrocytic anemia, increased urinary FIGLU, and low blood folate levels can be diagnosed as being vitamin B_{12}-deficient if their urinary MMA levels are elevated, but as being folate-deficient if they are not (Table 17.13).

9. VITAMIN B_{12} TOXICITY

Vitamin B_{12} has no appreciable toxicity. Results of studies with mice indicate that it is innocuous when administered parenterally in very high doses. Localized, injection-site sclerodermoid reaction[50] secondary to vitamin B_{12} injection has been reported. Dietary levels of at least several hundred times the nutritional requirements are safe. High plasma levels of the vitamin are indicative of disease,[51] rather than hypervitaminosis B_{12}.

TABLE 17.13 Distinguishing Vitamin B_{12} and Folate Deficiencies

Deficiency	Urinary FIGLU	Urinary MMA	Serum Hcy	Serum Folate
Vitamin B_{12}	Elevated	Elevated	Increased	Decreased
Folate	Elevated	Normal	Increased	Decreased

49. Humans typically show plasma methionine concentrations in the range of 37–136 μmol/l.

50. Such reactions are not common, but have been reported for various drugs and for vitamin K.

10. CASE STUDY

Instructions

Review the following case report, paying special attention to the diagnostic indicators on which the treatment was based. Next, answer the questions that follow.

Case

A 6-month-old boy was admitted in comatose condition. He had been born at term, weighing 3 kg, the first child of an apparently healthy 26-year-old *vegan*.[52] The mother had knowingly eaten no animal products for 8 years, and took no supplemental vitamins. The infant was exclusively breastfed. He smiled at 1–2 months of age, and appeared to be developing normally. At 4 months, his development began to regress; this was manifested by his loss of head control, decreased vocalization, lethargy, and increased irritability. Physical examination revealed a pale and flaccid infant who was completely unresponsive even to painful stimuli. His pulse was 136/min, respiration 22/min, and blood pressure 100 mmHg by palpation. His length was 65 cm (50th percentile for age) and his weight was 5.6 kg (<3rd percentile, and at the 50th percentile for 3 months of age). His head circumference was 41 cm (3rd percentile). His optic disks[53] were pale. There were scattered ecchymoses[54] over his legs and buttocks. He had increased pigmentation over the dorsa of his hands and the feet, most prominently over the knuckles. He had no head control and a poor grasp. He showed no deep tendon reflexes. His liver edge was palpable 2 cm below the right costal margin.

Laboratory Results:

Parameter	Patient	Normal Range
Hemoglobin, g/dl	5.4	10.0–15.0
Hematocrit, %	17	36
Erythrocytes, $\times 10^6$/μl	1.63	3.9–5.3
White blood cells, $\times 10^3$/μl	3.8	6–17.5
Reticulocytes, %	0.1	<1
Platelets, $\times 10^3$/μl	45	200–480

51. Elevated cobalamin levels are typical of myelogenous leukemia and promyelocytic leukemia, and are used as diagnostic criteria for polycythemia vera and hypereosinophilic syndrome. Several liver diseases (acute hepatitis, cirrhosis, hepatocellular carcinoma, and metastatic liver disease) can cause similar increases, which are due to increased levels of TC_I.

52. A strict vegetarian.

53. Circular area of thinning of the sclera (the fibrous membrane forming the outer envelope of the eye) through which the fibers of the optic nerve pass.

54. Purple patches caused by extravasation of blood into the skin, differing from petechiae only in size (the latter being very small).

A peripheral blood smear revealed mild macrocytosis,[55] and some hypersegmentation of the neutrophils.[56] Bone marrow aspiration showed frank megaloblastic changes in both the myeloid[57] and the erythroid[58] series. Megakaryocytes[59] were decreased in number. The sedimentation rate, urinalysis, spinal fluid analysis, blood glucose, electrolytes, and tests of renal and liver function gave normal results. An electroencephalogram was markedly abnormal, as manifested by minimal background Θ activity and epileptiform transients in both temporal regions. Analysis of the urine obtained on admission demonstrated a markedly elevated excretion of methylmalonic acid, glycine, methylcitric acid, and Hcy. Shortly after admission, respiratory distress developed, and 5 mg of folic acid was given, followed by transfusion of 10 ml of packed erythrocytes per kilogram body weight. Four days later, a repeat bone marrow examination showed partial reversal of the megaloblastic abnormalities.

Other Laboratory Results:

Parameter	Patient	Normal Range
Serum vitamin B$_{12}$ (pg/ml)	20	150–1,000
Serum folates (ng/ml)	10	3–15
Serum iron (µg/dl)	165	65–175
Serum iron-binding capacity (µg/dl)	177	250–410

Cyanocobalamin (1 mg/day) was administered for 4 days. The patient began to respond to stimuli after the transfusion; however, the response to vitamin B$_{12}$ was dramatic. Four days after the initial dose he was alert, smiling, responding to visual stimuli, and maintaining his body temperature. As he responded, rhythmical twitching activity in the right hand and arm developed that persisted despite anticonvulsant therapy, and despite a concomitant resolution of electroencephalographic abnormalities. The mother showed a completely normal hemogram. Her serum vitamin B$_{12}$ concentration was 160 pg/ml (normal, 150–1,000 pg/ml), but she showed moderate methylmalonic aciduria. Her breast milk contained 75 pg of vitamin B$_{12}$/ml (normal, 1–3 ng/ml).

With vitamin B$_{12}$ therapy, the infant's plasma vitamin B$_{12}$ rose to 600 pg/ml and he continued to improve

clinically. The abnormal urinary acids and homocysteine disappeared by day 10; cystathionine persisted until day 20. On day 14, the Hb was 14.4 g/dl, hematocrit was 41%, and the WBC was 5,700/ml. The platelet count had become normal 20 days after admission. The unusual pigment on the extremities had improved considerably 2 weeks after he received the parenteral vitamin B$_{12}$, and disappeared gradually over the next month. The liver was no longer palpable. The twitching of the hands disappeared within a month of therapy. Developmental assessment at 9 months of age revealed him to be functioning at the 5-month age level. A month later, he was sitting and taking steps with support. Head circumference had exhibited catch-up growth, and at 44 cm was in the normal range for the first time since admission. His length was 70 cm (10th percentile) and weight 8.4 kg (10th percentile). By this time, the mother's serum vitamin B$_{12}$ had dropped to only 100 pg/ml, and she began taking supplemental vitamin B$_{12}$.

Case Questions and Exercises

1. Which clinical findings suggested that two important coenzyme forms of vitamin B$_{12}$ were deficient or defective in this infant? How do the clinical findings relate specifically to each coenzyme?
2. What findings allow the distinction of vitamin B$_{12}$ deficiency from a possible folic acid-related disorder in this patient?
3. Offer a reasonable explanation for the fact that the mother, who had avoided vitamin B$_{12}$-containing foods for 8 years before her pregnancy, did not show overt signs of vitamin B$_{12}$ deficiency.

Study Questions and Exercises

1. Construct a decision tree for the diagnosis of vitamin B$_{12}$ deficiency in humans or an animal species and, in particular, the distinction of this deficiency from that of folate.
2. What key feature of the chemistry of vitamin B$_{12}$ relates to its coenzyme functions?
3. What parameters might you measure to assess vitamin B$_{12}$ status of a human or animal?
4. What is the relationship of normal function of the stomach and pancreas with the utilization of dietary vitamin B$_{12}$?

RECOMMENDED READING

Bailey, L.B., 2004. Folate and vitamin B12 recommended intakes and status in the United States. Nutr. Rev. 62, S14–S20.
Birn, H., 2006. The kidney in vitamin B12 and folate homeostasis: Characterization of receptors for tubular uptake of vitamins and carrier proteins. Am. J. Physiol. Renal Physiol. 291, F22–F36.

---

55. Occurrence of unusually large numbers of *macrocytes* (large erythrocytes) in the circulating blood; also called *megalocytosis*, *magalocythemia*, and *macrocythemia*.
56. A type of mature white blood cell in the granulocyte series.
57. Related to myocytes.
58. Related to erythrocytes.
59. An unusually large cell thought to be derived from the primitive mesenchymal tissue that differentiates from hematocytoblasts.

Bridden, A., 2003. Homocysteine in the context of cobalamin metabolism and deficiency states. Amino Acids 24, 1–12.

Brown, K.L., 2005. Chemistry and enzymology of vitamin B_{12}. Chem Rev. 105, 2075–2149.

Carmel, R., 2006. Cobalamin (vitamin B_{12}). In: Shils, M.E., Shike, M., Ross, A.C. (Eds.), Modern Nutrition in Health and Disease, tenth ed. Lippincott, New York, NY, pp. 482–497.

Elmadfa, I., Singer, I., 2009. Vitamin B-12 and homocysteine status among vegetarians: A global perspective. Am. J. Clin. Nutr. 89 (Suppl.), 1693S–1698S.

Ermens, A.A.M., Vlasveld, L.T., Lindemans, J., 2003. Significance of elevated cobalamin (vitamin B_{12}) levels in blood. Clin. Biochem. 36, 585–590.

Froese, D.S., Gravel, R.A., 2010. Genetic disorders of vitamin B_{12} metabolism: Eight complementation groups – eight genes. Exp. Rev. Molec. Med. 12, 1–20.

Green, R., Miller, J.W., 2007. Vitamin B_{12}. In: Zempleni, J., Rucker, R.B., McCormick, D.B., Suttie, J.W. (Eds.), Handbook of Vitamins, fourth ed. CRC Press, New York, NY, pp. 413–457.

Herrmann, W., Geisel, J., 2002. Vegetarian lifestyle and monitoring of vitamin B-12 status. Clin. Chim. Acta 326, 47–59.

Koury, M.J., Ponka, P., 2004. New insights into erythropoiesis: The roles of folate, vitamin B_{12} and iron. Ann. Rev. Nutr. 24, 105–131.

Kräutler, B., 2005. Vitamin B_{12}: Chemistry and biochemistry. Biochem. Soc. Trans. 33, 806–810.

Li, F., Watkins, D., Rosenblatt, D.S., 2009. Vitamin B12 and birth defects. Molec. Cell Gen. Metab. 98, 166–172.

Monsen, A.L.B., Ueland, P.M., 2003. Homocysteine and methylmalonic acid in diagnosis and risk assessment from infancy to adolescence. Am. J. Clin. Nutr. 78, 7–21.

Rita, M., Paola, T., Rodolfo, A., et al. 2004. Vitamin B_{12} and folate depletion in cognition: A review. Neurol. India 52, 310–318.

Scalabrino, G., 2009. The multi-faceted basis of vitamin B12 (cobalamin) neurotrophism in adult central nervous system: Lessons learned from its deficiency. Progr. Neurol. 88, 203–220.

Scott, J.M., 1997. Bioavailability of vitamin B_{12}. Eur. J. Clin. Nutr. 51, S49–S53.

Smith, A.D., Refsum, H., 2009. Vitamin B-12 and cognition in the elderly. Am. J. Clin. Nutr. 89 (Suppl.), 707S–711S.

Stabler, S.P., 2006. Vitamin B_{12}. In: Bowman, B.A., Russell, R.M. (Eds.), Present Knowledge in Nutrition, eighth ed. ILSI Press, Washington, DC, pp. 302–313.

Stabler, S.P., Allen, R.H., 2004. Vitamin B12 deficiency as a worldwide problem. Ann. Rev. Nutr. 24, 299–326.

Stover, P.J., 2004. Physiology of folate and vitamin B_{12} in health and disease. Nutr. Rev. 62, S3–S12.

Toh, B.H., van Driel, I.R., Gleeson, P.A., 2006. Pernicious anemia. N. Engl. J. Med. 337, 1441–1448.

Watanabe, F., 2007. Vitamin B12 sources and bioavailability. Exp. Biol. Med. 232, 1266–1274.

Zetterberg, H., 2004. Methylenetetrahydrofolate reductase and transcobalamin genetic polymorphisms in human spontaneous abortion: Biological and clinical implications. Reprod. Biol. Endocrinol. 2, 7–15.

Quasi-Vitamins

Anchoring Concepts

1. The designation "vitamin" is specific for animal species, stage of development or production, and/or particular conditions of the physical environment and diet.
2. Each of the presently recognized vitamins was initially called an accessory factor or an unidentified growth factor, and these terms continue to be used to describe biologically active substances, particularly for species of lower orders.

Have all the vitamins been discovered? From all indications in the extensive recent and current publications in the scientific literature dealing with the purification and effects of "unidentified factors," the answer appears to be "no." It is from such studies that new vitamins may be recognized and characterized.

A. F. Wagner and K. Folkers

Learning Objectives

1. To understand that the designation of a compound as a vitamin is biased in favor of dietary essentials for humans.
2. To understand that other substances have been proposed as vitamins.
3. To understand that choline and carnitine are vitamins for certain animal species.
4. To understand the metabolic functions of other conditionally essential nutrients: *myo*-inositol, pyrroloquinoline quinine, the ubiquinones, and orotic acid.
5. To understand why flavonoids, non-provitamin A carotenoids, *p*-aminobenzoic acid, and lipoic acid are not called vitamins.

VOCABULARY

Acetylcholine
Acylcarnitine esters
Acylcarnitine translocase
p-Aminobenzoic acid
Antioxidant response element (ARE)
Arachidonic acid
Betaine
Betaine aldehyde dehydrogenase
Betaine:homocysteine methyltransferase
γ-Butyrobetaine hydroxylase
Ca^{2+} channel
Calcisomes
Carnitine
Carnitine acyltransferases I and II
Carnitine translocase
Catechins
Chelates
Choline
Choline acetyltransferase
Choline dehydrogenase
Choline kinase

The Vitamins. DOI: 10.1016/B978-0-12-381980-2.00018-9

Choline oxidase
Choline phosphotransferase
Coenzyme Q_{10} (CoQ_{10})
Conditional nutrient
Cyanogenic glycoside
Cytidine diphosphorylcholine (CDP–choline)
Dimethylglycine
Eicosanoids
Ethanolamine
Flavanols
Flavonoids
Flavonols
Gerovital
Glycerylphosphorylcholine
Glycerylphosphorylcholine diesterase
Inositol 1,4,5-triphosphate (IP_3)
Intestinal lipodystrophy
Isoflavones
Labile methyl groups
Laetrile
Lecithin
Lipoamide
Lipofuscin
Lipoic acid
Lycopene
Lysolecithin
Lutein
Macula
Macular degeneration
Methionine
myo-Inositol
Orotic acid
Pangamic acid
Perosis
Phosphatidylcholine
Phosphatidylcholine glyceride choline transferase
Phosphatidylethanolamine
Phosphatidylethanolamine *N*-methyl transferase
Phosphatidylinositol (PI)
Phosphatidylinositol 4-phosphate (PIP)
Phosphatidylinositol 4,5-biphosphate (PIP_2)
Phospholipases A_1, A_2, B, C, and D
Phosphorylcholine
Phytic acid
Phytoestrogens
Proanthocyanidins
Pyrimidines
Pyrroloquinoline quinone (PQQ)
Quinoproteins
Second messenger
Sphingomyelin
Stearic acid
Tannic acid
Thiotic acid
Trimethylamine
ϵ-*N*-Trimethyllysine
Ubiquinones
Vitamin B_T
Vitamin B_X

Xanthophyll-binding protein (XBP)
Xanthophylls
Zeaxanthin

1. IS THE LIST OF VITAMINS COMPLETE?

Common Features in the Recognition of Vitamins

Reflection on the ways in which the traditional vitamins were recognized reveals a process of discovery involving both empirical and experimental phases (see Chapter 2). That is, initial associations between diet and health status were the sources of hypotheses that could be tested in controlled experiments. As is generally true in science, where hypotheses were clearly enunciated and adequate experimental approaches were available, insightful investigators were able to make remarkable progress in identifying these essential nutrients. Those endeavors, of course, also revealed some unidentified factors not to be new at all,[1] some to be identical or otherwise related to each other,[2] some to be biologically active but not essential in diets,[3] and some to be without basis in fact.[4] The apparently irregular and often confusing array of informal names of the vitamins[5] reveals this history of discovery.

Limitations of Traditional Designations of Vitamins

The development of the Vitamine Theory was instrumental in conditioning thought such that the discovery of the vitamins could occur. Indeed, it provided the basis for the evolution of the operating definition that has been used to designate vitamin status for biologically active substances. However, after several decades of learning more and more about the metabolism and biochemical actions of the substances called vitamins, it has become clear that the traditional criteria for that designation[6] are, in several cases (e.g., vitamins D and C, niacin, and choline), inappropriate unless they are used with specific reference to animal species, stage

---

1. An example is vitamin T (also called termitin, penicin, torutilin, insectine, hypomycin, myocoine or sesame seed factor). This extract from yeast, sesame seeds or insects appeared to stimulate the growth of guppies, hamsters, baby pigs, chicks, mice, and insects, promoted wound healing in mice, and improved certain human skin lesions. It was found to be a varied mixture containing folate, vitamin B_{12}, and amino acids.
2. For example, vitamins M, B_c, B_{10}, T, and B_x were found to be various forms of folate.
3. For example, the ubiquinones.
4. For example, pangamic acid (vitamin B_{15}), laetrile (vitamin B_{17}), orotic acid (vitamin B_{13}), and vitamins H_3 and U.
5. See Appendix A.
6. See Chapter 1.

of development, diet or nutritional status, and physical environment. However, because it is often more convenient to consider nutrients by general group without such referents, the designation of vitamin status has become a bit arbitrary as well as anthropocentric in that the traditional designation of the vitamins reflects, to a large extent, the nutritional needs of humans, and, to a lesser extent, those of domestic animals. For example, although it is now very clear that 1,25-dihydroxycholecalciferol [1,25-(OH)$_2$-cholecalciferol] is actually a hormone produced endogenously by all species exposed to sunlight, the parent compound cholecalciferol continues to be called vitamin D$_3$ in recognition of its importance to the health of many people whose minimal sunlight exposure renders them in need of it in their diets.

Other species may have obligate dietary needs for substances that are biosynthesized by humans and/or higher animals. Such substances can be considered to be vitamins in the most proper sense of the word. Available evidence indicates this to be true for at least three substances, and some reports have suggested this for others.

It is also the case that specific physiological states resulting from specific genetics, disease or dietary conditions may place an individual incapable of synthesizing a metabolite, resulting in a need for the pre-formed nutrient from exogenous sources such as the diet. Several such conditionally essential nutrients have characteristics of vitamins for affected individuals.

Finally, as more is learned, it is certainly possible that more vitamins may be discovered.

Quasi-Vitamins

It is useful to recognize the other factors that appear to satisfy the criteria of vitamin status for only a few species or only under certain conditions, and to do so without according them the full status of a vitamin. This is done using the term *quasi-vitamin*. The list of quasi-vitamins includes the following.

- Factors required for some species:
  - **choline**, required in the diets of young growing poultry for optimal growth and freedom from leg disorders
  - **carnitine**, required for growth of certain insects
  - *myo*-**inositol**, clearly required for optimal growth of fishes and to prevent intestinal lesions in gerbils.
- Factors for which evidence of nutritional essentiality is less compelling:
  - **pyrroloquinoline quinine**
  - **ubiquinones**
  - **flavonoids**
  - some **non-provitamin A carotenoids**
  - **orotic acid**
  - *p*-**aminobenzoic acid**
  - **lipoic acid**.

In addition, the biologically active properties of certain foods/feedstuffs are worthy of consideration, as most of the consensus vitamins were initially recognized as **unidentified growth factors** (UGFs) in such natural products. Also included in this chapter are discussions of ineffective factors often confused with vitamins.

2. CHOLINE

Metabolite Acting Like a Vitamin

The discovery of insulin by Banting and Best in the mid-1920s led to studies of the metabolism of de-pancreatized dogs that showed dietary lecithin (phosphatidylcholine) to be effective in mobilizing the excess lipids in the livers of insulin-deprived animals. Best and Huntsman showed that choline was the active component of lecithin in mediating this effect. Choline had been isolated by Strecker in 1862, and its structure had been determined by Bayer shortly after that. Yet it was these latter findings, revealing choline as a lipotropic factor, that stimulated interest in its nutritional role.

In 1940, Jukes showed that choline is required for normal growth and the prevention of the leg disorder called **perosis**[7] in turkeys; he found that the amount required to prevent perosis is greater than that required to support normal growth. Further studies by Jukes and by Norris's group at Cornell showed that **betaine**, the metabolic precursor to choline, was not always effective in preventing choline-responsive perosis in turkeys and chicks. These findings stimulated further interest in the metabolic roles and nutritional needs for choline, as it was clear that its function was more than simply that of a lipotrope.

Chemical Nature

Choline is the trivial designation for the compound 2-hydroxy-*N,N,N*-trimethylethanaminium (also, [β-hydroxyethyl]trimethylammonium) (Fig. 18.1). It is freely soluble in water and ethanol, but insoluble in organic solvents. It is a strong base, and decomposes in alkaline solution with the release of trimethylamine. The prominent feature of its chemical structure is its triplet of methyl groups, which enables it to serve as a methyl donor.

Distribution in Foods and Feedstuffs

All natural fats contain some choline; therefore, the vitamin is widely distributed in foods and feedstuffs (Table 18.1). The factor occurs naturally mostly in the form of

---

7. Perosis occurs in rapidly growing heavy-bodied poultry, and involves the misalignment of the tibiotarsus and consequent slippage of the Achilles tendon. This impairs ambulation and can reduce feeding, consequently impairing growth. Perosis can also be caused by dietary deficiencies of niacin or manganese.

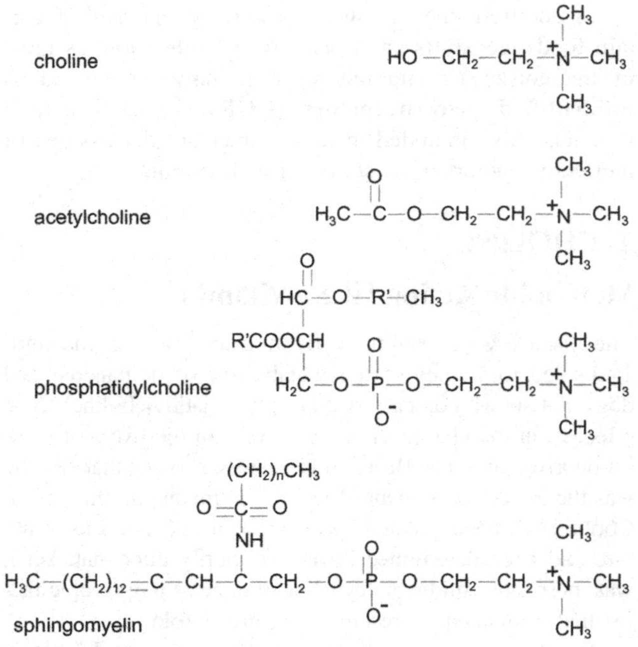

FIGURE 18.1 Choline and its functional metabolites.

phosphatidylcholine[8] (Fig. 18.1, also called **lecithin**), which, because it is a good emulsifying agent, is used as an ingredient or additive in many processed foods and food supplements. Some dietary choline (<10%) is present as the free base and **sphingomyelin** (Fig. 18.1).[9] The richest sources in human diets[10] are egg yolk, glandular meats (e.g., liver, kidney, brain), pork (meat, bacon), soybean products, wheatgerm, and peanuts. The choline metabolite betaine also occurs in foods; rich sources include wheat and shellfish. Choline is added (as choline chloride and choline bitartrate) to infant formulas as a means of fortification.

The best sources of choline for animal feeding are the germs of cereals, legumes, and oilseed meals (e.g., soybean meal). Corn is notably low in choline (half the levels found in barley, oats, and wheat). Because wheat is rich in the choline-sparing factor betaine, the choline needs of livestock fed diets based on wheat are much lower than those of animals fed diets based on corn.

The bioavailability of choline in foods and feedstuffs appears to be generally good. It is dependent on those factors that affect the utilization of dietary fats.

Naturally occurring choline, as well as the choline salts used as supplements, has good stability. The processing of foods/feedstuffs can enhance choline bioavailability, as mechanical disruption of plant cells by chopping and grinding, etc., can activate phospholipases to release choline in free form.

Absorption

Choline is released from its phosphatidylcholine by hydrolysis in the intestinal lumen. This is accomplished enzymatically through the action of phospholipases produced by the pancreas (**phospholipase A_2**, which cleaves the β-ester bond) and the intestinal mucosa (**phospholipases A_1 and B**, both of which cleave the α-ester bond to yield **glycerylphosphorylcholine**). The mucosal enzymes are much less efficient than the pancreatic enzyme. Therefore, most of the phosphatidylcholine that is ingested is absorbed as **lysolecithin** (deacylated only in the α position), which is reacylated to yield phosphatidylcholine. This reaction involves the dismutation of two molecules of lysolecithin to yield one molecule of glycerylphosphorylcholine and one molecule of phosphatidylcholine. Analogous reactions occur with sphingomyelin, which, unlike phosphatidylcholine, is not degraded in the intestinal lumen, but is taken up intact by the intestinal mucosa.

When free choline or one of its salts is consumed, a large amount (e.g., nearly two-thirds) is catabolized by intestinal microorganisms to the end product **trimethylamine**,[11] much of which is absorbed and excreted in the urine. The remaining portion is absorbed intact. Phosphatidylcholine is not subject to such extensive microbial metabolism, and therefore produces less urinary trimethylamine.

Choline is absorbed in the upper portion of the small intestine by a saturable, carrier-mediated process localized in the brush border, and efficient at low lumenal concentrations (<4 mmol/l). At high lumenal concentrations, it is also absorbed by passive diffusion. This process appears to involve a member of the intermediate-affinity choline transporter-like proteins (CTLs). The oxidized metabolite betaine is absorbed by way of a different carrier, the IMINO proline transporter.

Transport and Cellular Uptake

Recently absorbed choline is transported into the lymphatic circulation (or the portal circulation in birds, fishes, and reptiles) primarily in the form of phosphatidylcholine bound to chylomicra, which are subject to clearance to the lipoproteins that circulate to the peripheral tissues. Thus, choline is transported to the tissues predominantly as phospholipids associated with the plasma lipoproteins (Table 18.2).

---

8. Phosphatidylcholine comprises 95% of total choline in eggs and 55–70% of total choline in meats and soy products.
9. That is, phosphatidylcholine analogs containing, instead of a fatty acid, sphingosine [2-amino-4-octadecene-1,3-diol] at the glycerol α-carbon
10. Choline intakes of Americans have been estimated to be 443 ± 88 mg/day for women and 631 ± 157 mg/day for men (Fischer, L. M., Scearce, J. A., Mar, M. H., *et al.* [2005] *J. Nutr.* 135, 826).
11. The characteristic fishy odor of this product is identifiable after consumption of a choline supplement.

TABLE 18.1 Total Choline and Betaine Contents (mg/100 g) of Common Foods

Food	Choline	Total Choline Equivalents[a]	Betaine
Meats			
Beef	3.57	78.15	10.12
Beef liver	56.67	418.22	5.6
Chicken	5.27	65.83	4.95
Chicken liver	47.87	290.03	11.43
Pork	2.19	102.8	1.39
Bacon	12.06	124.9	3.14
Fish and Fish Products			
Shrimp	5.56	70.60	218.7
Cod	17.73	83.63	8.58
Salmon	8.62	65.45	1.87
Vegetables			
Beans	4.00	13.46	0.08
Broccoli	8.45	40.06	0.11
Cabbage	6.87	15.45	0.31
Carrots	6.82	8.79	0.34
Corn	8.93	21.95	0.15
Cucumber	3.99	5.95	0.07
Lettuce	4.80	6.70	0.08
Mushrooms	5.93	16.86	9.52
Onions	4.39	6.10	0.07
Peas	2.16	27.51	0.13
Potatoes	8.44	14.36	0.38
Spinach	1.69	22.08	599.8
Soybeans	47.27	115.9	1.85
Tomatoes	4.40	6.74	0.06
Fruits			
Apples	0.33	3.44	0.09
Avocados	8.64	14.18	0.58
Bananas	3.20	9.76	0.07
Blueberries	3.00	6.04	0.16
Cantaloupe	4.12	7.58	0.07
Grapefruit	3.56	7.53	0.13
Grapes	4.80	5.63	0.12
Oranges	4.68	8.38	0.11
Peaches	0.78	6.10	0.24
Strawberries	0.63	5.65	0.14
Cereals			
Oats	1.25	7.42	2.72
Oat bran	4.41	58.57	0.27
Rice, polished	0.72	2.08	0.27
Rice, unpolished	4.66	9.22	0.43
Wheat	17.98	26.53	201.4
Wheat bran	50.89	74.39	1339
Other			
Milk	3.67	14.29	0.54
Eggs	0.62	251.00	0.53
Peanuts	17.59	52.47	0.56

[a]Includes free choline, phosphocholine, glyerophosphocholine, phosphatidylcholine, and sphingomyelin.
From Zeisel, S. H., Mar, M. H., Howe, J. C., et al. (2003). *J. Nutr.* 133, 1302.

TABLE 18.2 Distribution of Phospholipids in Plasma Lipoproteins

Lipoprotein Class	Phospholipid Content (%)
High-density lipoproteins (HDLs)	~30
Low-density lipoproteins (LDLs)	~22
Very low-density lipoproteins (VLDLs)	10–25
Chylomicra	3–15

Free choline is a positively charged quaternary amine that does not cross biological membranes without a carrier. Three transport systems have been identified:

- *High-affinity, Na^+-dependent transporter (CHT1)*. This system is unique to the cholinergic neurons of the brain, brain stem, and spinal column, where it serves to provide choline for the synthesis of the neurotransmitter acetylcholine.
- *Intermediate-affinity, Na^+-dependent transporter*. These are CTLs; five gene products have been identified.
- *Low-affinity, Na^+-independent transporters*. Three organic cation transporters (OCTs), members of the solute carrier family, are expressed in many tissues.

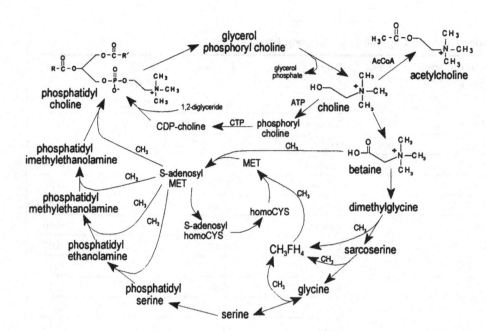

FIGURE 18.2 The biosynthesis and utilization of choline. *Abbreviations*: AcCoA, acetyl-CoA; CH_3FH_4, 5-methyl tetrahydrofolate; CTP, cytidine triphosphate; homoCys, homocysteine; MET, methionine.

Choline is present in all tissues as an essential component of phospholipids in membranes of all types. It is not stored, but occurs in relatively great concentrations in the essential organs (e.g., brain, liver, kidney), mostly as phosphatidylcholine and sphingomyelins. Placental tissues are unique in that they accumulate large amounts of acetylcholine, presumably to meet fetal needs, which is otherwise present only in the parasympathetic nervous system.

Choline Metabolism

The biosynthesis and metabolism of choline is outlined in Fig. 18.2.

Production of Choline

- *Release from phosphatidylcholine.* Choline is released in free form in the tissues by the actions of **phospholipases A$_2$, B, C,** and **D**. The primary means of phosphocholine degradation is by phospholipase A$_2$, which yields lysophosphatidylcholine (and arachidonic acid). The latter is converted to glycerophosphorylcholine (by lysophospholipases) and then to choline (by glycerophosphodiester phosphodiesterases). Phospholipase B yields lysophosphatidylcholine and arachidonic acid. Phospholipase C yields a diglyceride and phosphorylcholine, which is converted to choline (by alkaline phosphatase). Phospholipase D yields phosphatidic acid.
- De novo *synthesis.* Choline is synthesized through the sequential, SAM-dependent methylation of ethanolamine (by phosphatidylethanolamine *N*-methyltransferase) to produce posphatidylcholine.

- *Phospholipid base exchange.* Choline (and phosphatidylserine) can be produced from phosphatidylcholine and serine.

Oxidation

Most (60–90%) of the metabolism of free choline involves its oxidation, which frees its methyl groups to enter single-C metabolism. This metabolism is induced by dietary choline; it occurs in several tissues, but is notably absent from brain, muscle, and blood. It is accomplished in three steps, the first two of which comprise a dual enzyme system collectively referred to as **choline oxidase**:

- *Oxidation of choline.* Choline is oxidized in the mitochondria by **choline dehydrogenase** to yield betaine aldehyde.
- *Production of betaine.* Betaine aldehyde is converted to betaine by betaine aldehyde dehydrogenase, most of which is in the cytosol. Choline is oxidized to betaine at a rate an order of magnitude greater than that of its incorporation into phosphorylcholine.
- *Transfer of methyl groups to homocysteine (Hcy).* Betaine cannot be reduced to form choline, but it can donate its methyl groups to Hcy to produce **dimethylglycine** and **methionine** by the action of the enzyme **betaine:homocysteine methyltransferase**.[12] Therefore, while the choline oxidase pathway removes choline from the body, it also serves as a source of **labile methyl groups**.

---

12. This methyltransferase is one of the most abundant proteins in liver, comprising as much as 2% of total soluble protein.

Synthesis of Phosphatidylcholine

Humans and other animals synthesize phosphatidylcholine in two ways:

- *Methylation of ethanolamine.* Phosphatidylcholine is produced by the sequential methylation of phosphatidylethanolamine by phosphatidylethanolamine *N*-methyltransferase (Fig. 18.1). This activity is actually due to two enzymes: a cell inner membrane enzyme adds the first methyl group; a cell outer membrane enzyme adds the second and third methyl groups. Each enzyme uses *S*-adenosylmethionine (SAM) as the methyl donor. The first step is rate-limiting to choline synthesis. This biosynthetic capacity occurs in many tissues, but is greatest in liver, accounting for some 15–40% of phosphatidylcholine in that organ.

- *Cytidine diphosphate (CDP)–choline pathway.* Free choline is phosphorylated by the cytosolic enzyme choline phosphotransferase (also called choline kinase), using ATP as the phosphate donor. This is the rate-limiting step in the synthesis of phosphatidylcholine, which proceeds with the generation of cytidine diphosphatylcholine (CDP–choline; by the action of CTP:phosphocholine cytidyltransferase), which combines with diacylglycerol (by the action of choline phosphotransferase in the endoplasmic reticulum) to form phosphatidylcholine. The CDP–choline pathway can account for some 70% of phosphatidylcholine in liver. Phosphatidylcholine appears to feedback-inhibit the cytidyltransferase step, while diacylglycerol enhances it.

Synthesis of Acetylcholine

Only a small fraction of choline is acetylated, but that amount provides the important neurotransmitter **acetylcholine**. This step involves the reaction of choline with acetyl-CoA catalyzed by an enzyme **choline acetyltransferase** localized in cholinergic nerve terminals, as well as in certain other non-nervous tissues (e.g., placenta). Because brain choline acetyltransferase appears not to be saturated with either substrate, it is likely that the availability of choline (and acetyl-CoA) may determine the rate of synthesis of acetylcholine.

Metabolic Functions of Choline

Choline has seven essential physiological functions:

- *Phosphatidylcholine.* Phosphatidylcholine is a structural element of biological membranes. It is a precursor to ceramide,[13] the basic structure of sphingolipids, which play roles in transmembrane signal transduction.[14] It also promotes lipid transport (as a lipotrope).

- *Acetylcholine.* Acetylcholine is a neurotransmitter occurring primarily in the parasympathetic nervous system. It is also important for the development of the brain; choline deprivation of rat pups produces permanent memory impairment.

- *Platelet-activating factor.* Choline is a component of platelet-activating factor (1-*O*-alkyl-2-acetyl-*sn*-glycero-3-phosphocholine), which is important in clotting, inflammation,[15] uterine ovum implantation, fetal maturation, and induction of labor.

- *Plasmalogen.* Choline is a component of plasmalogen, which occurs at high levels in the sarcolemma, important in myocardial function.[16]

- *Phosphocholine.* Studies have shown phosphocholine to act as a signaling molecule in cell division, including in tumorigenesis.[17]

- *Source of labile methyl groups.* Choline is, after its oxidation to betaine, a source of labile methyl groups for transmethylation reactions in the formation of methionine from Hcy or of creatine from guanidoacetic acid. This function links choline to folate metabolism. When the 5-methyl-FH$_4$ cannot meet intracellular demands for SAM, choline is an important dietary source of labile methyl groups for Hcy transmethylation.

- *Organic osmolytes.* Choline, after its oxidation to betaine, and glycerophosphocholine produced from phosphatidylcholine, serve as organic regulators of intracellular osmolarity, particularly in the renal cortex and medulla, which are exposed to high extracellular osmolarity as a consequence of concentrating urine for excretion.

Conditions of Need

Developmental deficiencies of the inner membrane phosphatidylethanolamine *N*-methyltransferase create needs for dietary sources of choline. Choline deprivation in other animal species can cause signs under circumstances in which the biosynthesis of methionine by the methylation of Hcy by 5-methyl-FH$_4$ cannot meet all of the intracellular demands for SAM. Those signs include depressed growth, hepatic

---

13. Ceramide is formed by adding a fatty acid to the amino group of sphingosine. Among the biological activities of ceramide is the stimulation of *apoptosis*, i.e., programmed cell death.

14. Phospholipid-mediated signal transduction involves membrane phospholipases that trigger the generation of inositol-1,4,5-triphosphate, which acts to release Ca^{2+} from stores in the endoplasmic reticulum.

15. Overproduction of platelet-activating factor has been shown to produce a hyperresponsive condition, as occurs in asthma.

16. It is thought that the adverse effects of myocardial ischemia may involve the breakdown of plasmalogen.

17. Janardhan, S., Srivani, P., and Sastry, G. N. (2006). *Curr. Med. Chem.* 13, 1169.

steatosis, and hemorrhagic renal degeneration. The fatty infiltration of the liver that occurs in choline-deficient animals is presumed to be due to the need for phosphatidylcholine for hepatic lipoprotein (mainly, VLDL) synthesis, which in turn is necessary for the export of triglycerides.

Poultry

Due to developmental deficiencies of phosphatidylethanolamine *N*-methyltransferase, the chick has an absolute need for dietary choline until about 13 weeks of age. Choline deprivation can therefore produce signs in young poultry, and also in older poultry fed diets deficient in methyl groups (e.g., methionine-deficient). Deficiency signs include fatty liver and perosis, which is thought to be secondary to impaired lipid transport.

Fish

Clear needs for dietary choline have been demonstrated for fish.[18] The signs of deficiency include impaired weight gain and efficiency of feed utilization, and hepatic steatosis. It is generally assumed that most fishes cannot synthesize choline at levels sufficient to meet their physiological needs.

Rats

Rats require choline only if their capacity to methylate phosphatidylethanolamine is limited by the availability of methyl groups. In such cases, the feeding of methyl donors (methionine, betaine) spares the need for choline. Rats fed a choline-deficient diet show 30–40% reductions in hepatic and brain levels of folate, resulting in a shift toward longer folylpolyglutamate metabolites, and in the undermethylation of DNA. Pregnancy and lactation result in significant decreases in hepatic choline levels. Betaine can spare the need for choline to support growth and prevent fatty liver, apparently by providing single-C units.

Humans

Choline deficiency has not been reported in humans, but this may reflect the adequate intakes of other methyl donors in the subjects frequently studied. One study, in which healthy adults were fed a diet adequate (but not excessive) in methionine and folacin, showed that deprivation of choline produced signs of hepatic dysfunction (increased serum transaminase activities) that were corrected by feeding choline. Further, some evidence supports benefits of choline in the treatment of diseases involving hepatic steatosis and liver dysfunction associated with total parenteral feeding of low-choline fluids.

Single nucleotide polymorphisms (SNPs) in several enzymes have been found to affect the need for dietary choline. Women with a SNP in phosphatidylethanolamine-*N*-methyltransferase (PEMT) that yields that enzyme unresponsive to induction by estrogen have been found to need a dietary source of choline.[19] Individuals, particularly women, with an SNP in 5,10-methylenetetrahydrofolate dehdrogenase have increased sensitivity to choline deprivation compared to those with other MTHD genotypes.[20]

In 1998, the Food and Nutrition Board of the Institute of Medicine set recommended intakes for choline.[21]

Health Effects

Neurologic Function

The intake of choline can affect the concentrations of the neurotransmitter acetylcholine in the brain, suggesting that choline loading may be beneficial to patients with diseases involving deficiencies of cholinergic neurotransmission. Indeed, studies with animal models have shown choline supplementation during development to enhance cognitive performance, particularly on more difficult tasks; to increase electrophysiological responsiveness; and to provide some protection against alcohol and other neurotoxic agents. A large cohort study found plasma choline level to be inversely associated with the incidence of symptoms of anxiety (but not depression).[22]

In humans, large doses (multiple-gram quantities) of choline have been used to increase brain choline concentrations above normal levels, thereby stimulating the synthesis of acetylcholine in nerve terminals. Such supplementation has been found to help in the treatment of tardive dyskinesia, a movement disorder involving inadequate neurotransmission at striatal cholinergic interneurons.[23] Choline supplements have also been used with some success to improve free memory in subjects without

---

18. For example, red drum (*Sciaenops ocellutus*), striped bass (*Morone* spp.).

19. da Costa, K. A., Kozyreva, O. G., and Song, J. (2006). *FASEB J.* 20, 1336.

20. Kohlmeier, M. da Costa K. A. Fischer, L. M., *et al.* (2005). *Proc. Natl Acad. Sci. USA* 102, 16025.

21. Food and Nutrition Board (1998). Choline. In *Dietary Reference Intakes of Thiamin, Riboflavin, Niacin, Vitamin B₆, Folate, Vitamin B₁₂, Biotin and Choline.* National Academy Press, Washington, DC, pp. 390–422.

22. Bjelland, I., Tell, G. S., and Vollset, S. E. (2009). *Am. J. Clin. Nutr.* 90, 1056.

23. Tardive dyskinesia is prevalent among patients treated with neuroleptic drugs (drugs affecting the autonomic nervous system) and is characterized by choreoathetotic movements (involuntary movements resembling both *chorea* [irregular and spasmodic] and *athetosis* [slow and writhing]) of the face, the extremities, and, usually, the trunk.

dementia,[24] and to diminish short-term memory losses associated with Alzheimer's disease,[25] a disorder involving deficiency of hippocampal cholinergic neurons. It has been suggested that autocannibalism of membrane phosphatidylcholine may be an underlying defect in that disease; this is supported by the fact that patients treated with anticholinergic drugs develop short-term memory deficits resembling those associated with hippocampal lesions.

Phosphatidylcholine has been reported to reduce manic episodes in patients, suggesting that it can be centrally active; however, such treatment has been found to exacerbate depression among tardive dyskinesia patients.

Anticarcinogenesis

Choline deficiency in animal models has been found to increase the incidence of spontaneous hepatocarcinomas in the absence of any known carcinogen, and to enhance hepatocarcinogenesis induced chemically. These effects have been shown to involve both the initiation and promotion phases of carcinogenesis. They would appear to result from metabolic responses to choline deprivation, in particular, decreases in tissue levels of S-adenosylmethionine resulting in hypomethylation of DNA and the consequent changes in gene transcription, including modified expression of p53 protein. They may also be related to the progressive increase in hepatocyte proliferation that occurs after parenchymal cell death in the regenerating choline-deficient liver. Similar results have been found in mice (but not rats) treated with the choline precursor diethanolamine.

Safety

The toxicity of choline appears to be very low. However, deleterious effects have been reported for the salt choline chloride; these have included growth depression, impaired utilization of vitamin B_6, and increased mortality. The cause of these effects is not clear, however; the apparent toxicity of that form of the vitamin may actually have been due to the perturbation of acid–base balance caused by the high level of chloride administered with large doses of the salt. In humans, high doses (e.g., 20 g) have produced dizziness, nausea, and diarrhea.

3. CARNITINE

In the 1950s, a group of entomologists at the University of Illinois, Fraenkel and colleagues, found that the successful growth of the yellow mealworm *Tenebrio molitor* in culture required the feeding of a natural substance that they found to be present in milk, yeast, and many animal tissues. They purified the growth factor, which they named "**vitamin B$_T$**," from whey solids and identified it as **carnitine**,[26] a known metabolite isolated at the turn of the century from extracts of mammalian muscle. Although this finding established carnitine as a biologically active substance, the first indication of its metabolic role came a decade later when Fritz found it to stimulate the *in vitro* oxidation of long-chain fatty acids by subcellular fractions of heart muscle. More recently, research interest in carnitine has increased dramatically, stimulated by the finding of Broquist and colleagues that carnitine is biosynthesized by mammals from the amino acid lysine, which is limiting in the diets of many third-world populations, and by the description by Engel and colleagues of clinical syndromes (of apparently genetic origin) associated with carnitine deficiency.

Chemical Nature

Carnitine is the generic term for a number of compounds including L-carnitine (β[-]-β-hydroxy-γ-[N,N,N-trimethylaminobutyrate][27]) and its acetyl and propionyl esters (Fig. 18.3). Only the L-isomer is made by and is biologically active for eukaryotes. At physiological pH, carnitine exists as a zwitterion,[28] with a positively charged quaternary amine and a negatively charged carboxyl. It forms esters with fatty acids by virtue of its hydroxyl.

Dietary Sources

The available data concerning the carnitine contents of foods are scant and must be considered suspect, owing to the use of non-standard analytical methods. Nevertheless, it is apparent that materials of plant origin tend to be low in carnitine, whereas those derived from animals tend to be rich in the factor (Table 18.3). Red meats and dairy products are particularly rich sources. Typical mixed diets can be expected to provide 1–16 μg carnitine per day, the actual amount depending mainly on the intake of meats.

Absorption

Carnitine appears to be absorbed across the gut by an active process dependent on Na^+ co-transport, as well as by a passive, diffusional process that may be important for the absorption of large doses of the factor. The efficiency

24. Spiers, P., Myers, D., Hochanadel, G. S., *et al.* (1996). *Arch. Neurol.* 53, 441; Ladd, S. L., Sommer, S. A., LaBerge, S. *et al.* (1993). *Clin. Neuropharmacol.* 16, 540; Sitram, N., Weingartner, H., Caine, E.D. *et al. Life Sci.* (1978). 22, 1555.
25. Alvarez, X. A., Laredo, M., Corzo, D., *et al.* (1997). *Methods Find. Exp. Clin. Pharmacol.* 19, 201.

26. Carter, H. E., Bhattacharyya, P. K., Weidman, K. R., *et al.* (1952). *Arch. Biochem. Biophys.* 38, 4056.
27. Molecular weight, 161.5.
28. That is, a molecule with a positive and a negative electrical charge at different locations.

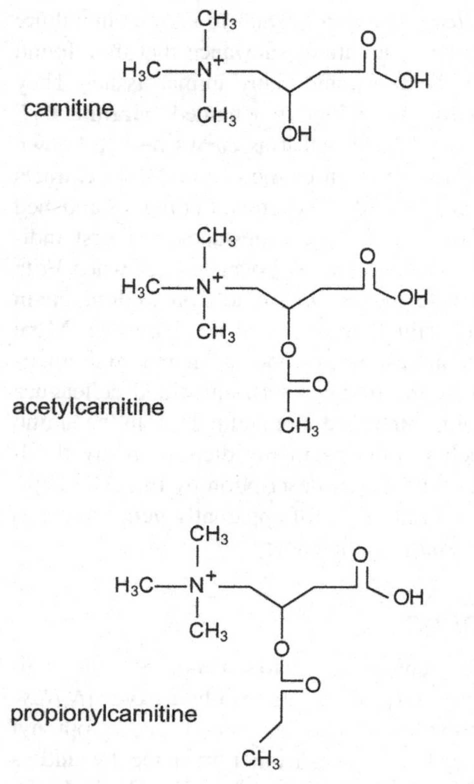

FIGURE 18.3 Carnitine and functional metabolites.

TABLE 18.3 Approximate Amounts of Total Carnitine in Selected Foods and Feedstuffs

Food	Carnitine[a] (µg/100 g)
Vegetables	
Alfalfa	2.00
Avocados	1.25
Cauliflower	0.13
Cereals	
Wheat	0.35–1.22
Bread	0.24
Meats	
Beef	59.8–67.4
Beef liver	2.6
Beef kidney	1.8
Beef heart	19.3
Chicken	4.6–9.1
Lamb, muscle	78.0
Other	
Cow's milk	0.53–3.91
Casein, acid washed	0.4
Peanuts	0.76
Torula yeast	1.60–3.29

Source: Mitchell, M. (1978). *Am. J. Clin. Nutr.* 31, 293.
[a]*None detected in: cabbage, spinach, orange juice, barley, corn, egg.*

of absorption appears to be high, *ca.* 55–95%. High doses are absorbed at lower efficiencies (≤25%), with <1% appearing in the urine and very little appearing in the feces. The uptake of carnitine from the intestinal lumen into the mucosa is rapid, and about half of that taken up is acetylated in that tissue.

Transport and Cellular Uptake

Carnitine is released slowly from tissues (e.g., erythrocytes) into the plasma in both the free and acetylated forms, which are found there in simple solution. Plasma total carnitine concentrations in healthy adults are 30–89 µmol/l, with men typically showing slightly greater (by ~15%) concentrations than women. Carnitine is taken up, against concentration gradients, by peripheral tissues, most of which can also synthesize it.

Cellular uptake of carnitine and its short-chain acyl esters is facilitated by Na^+-dependent organic cation transporters (OCTNs). A high-affinity transporter (OCTN2) is expressed in kidney, skeletal muscle (which contains some 95% of the body's carnitine), heart, pancreas, testis, and placenta, but is notably low in liver, brain, and lung. Subjects with inborn errors in OCTN2 have been identified; they readily develop signs of carnitine deficiency unless they are given supplemental carnitine. Another high-affinity, OTCN-related transporter (CT-2) is expressed only in testis. Carnitine concentrations of skeletal muscles are typically 70-fold that of plasma; somewhat smaller differences have been reported between other tissues and extracellular fluids.

Short-chain carnitine derivatives are better utilized than the parent molecule by some tissues. Acetylcarnitine, which is structurally similar to acetylcholine, crosses the blood–brain barrier more readily than carnitine, whereupon it is readily converted to carnitine. Propionylcarnitine, which is lipophilic, has high affinities for skeletal and cardiac muscles.

Metabolism

Biosynthesis

Carnitine is synthesized in mammals by the post-translational modification of protein lysyl residues,[29] which are thrice methylated to form **ε-*N*-trimethyllysine**, using

29. These include: actin, myosin, ATP synthase, calmodulin, and histones.

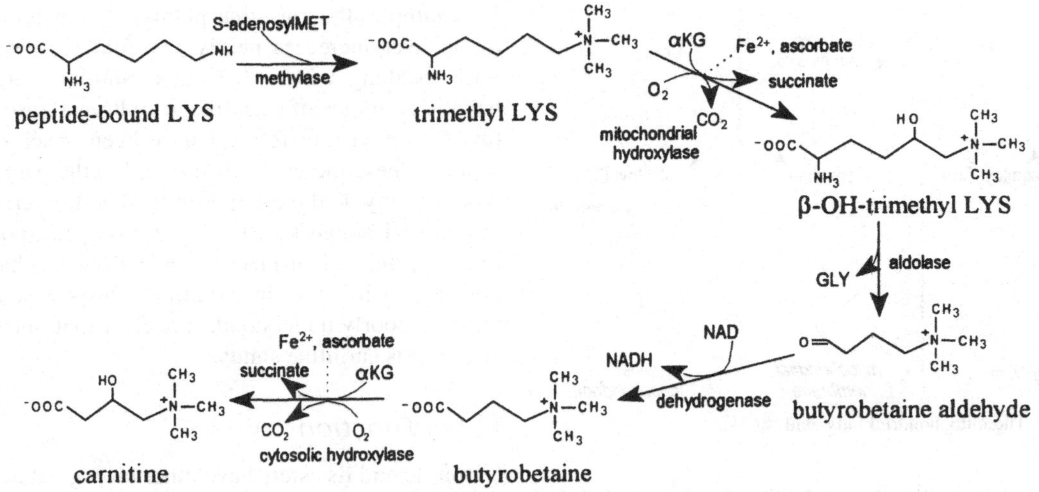

FIGURE 18.4 The biosynthesis of carnitine. *Abbreviations:* αKG, α-ketoglutarate; LYS, lysine; GLY, glycine.

SAM as the methyl donor. With protein turnover, ε-N-trimethyllysine is released to be converted to L-carnitine through a sequence of reactions catalyzed by two Fe^{2+}- and ascorbate-dependent hydroxylases, a pyridoxal phosphate-dependent aldose, and an NAD^+-dependent dehydrogenase (Fig. 18.4).

The enzymes that catalyze the conversion of ε-N-trimethyllysine to γ-butyrobetaine appear to occur in all tissues. In contrast, the last enzyme in the biosynthetic pathway, **γ-butyrobetaine hydroxylase**, is present only in liver, kidney, and brain. In human liver and kidney, it is present as multiple isoenzymes; the activity increases during development,[30] reaching maxima in the mid-teens. Tissue carnitine levels are depressed under conditions of vitamin B_6 deprivation, and stimulated by the catabolic state (e.g., fasting), thyroid hormone, and the peroxisome proliferator clofibrate.[31]

Turnover

The turnover of carnitine in muscle is relatively slow; however, it is increased substantially by exercise, which reduces muscle carnitine concentrations. Exercise appears to produce a preferential mobilization of free carnitine, thus resulting in an apparent shift toward fatty acid esters of carnitine in the muscle. The turnover times for carnitine in human tissues have been estimated to be ~8 days in skeletal muscle and heart, 11.6 hours in liver and kidney, and 68 minutes in extracellular fluid. Whole-body turnover time has been estimated to be 66 days, indicating

significant reutilization of carnitine among the various tissues of the body.

Excretion

Carnitine is not degraded in tissues, but is metabolized by the gastrointestinal microflora to γ-butyrobetaine or trimethylamine and malic semialdehyde, which appear in the feces. Trimethylamine so produced can be absorbed across the gut and metabolized in the liver to yield trimethylamine oxide, which appears in the urine.

Carnitine is highly conserved by the human kidney, which reabsorbs >90% of filtered carnitine, being the dominant means of regulating plasma carnitine concentration. Renal reabsorption is facilitated by a brush border protein, OTCN2, which recovers carnitine as well as its short-chain acyl esters. Renal tubular excretion of carnitine, its short-chain acyl esters and γ-butyrobetaine, adapt to circulating carnitine concentrations; some of this may come from the renal secretion of carnitine either in free form or as short-chain acylcarnitine esters. That urinary carnitine is typically comprised of a higher proportion of acylcarnitine esters than the general circulation suggests selective secretion of carnitine esters by renal tubular cells.

Metabolic Function

Mitchondrial Fatty Acid Shuttle

Carnitine functions in the transport of long-chain fatty acids (fatty acyl-CoA) from the cytosol into the mitochondrial matrix for oxidation as sources of energy (Fig. 18.5). The mitochondrial inner membrane is impermeable to long-chain fatty acids and their CoA derivatives, which are therefore dependent on activation as carnitine esters for entry into that organelle. This transport process, referred to as the

---

30. For example, the hepatic γ-butyrobetaine hydroxylase activities of three infants and a 2.5-year-old boy were 12% and 30%, respectively, of the mean adult activity.
31. This effect appears to involve the peroxisome proliferator-activated nuclear receptor α (PPARα).

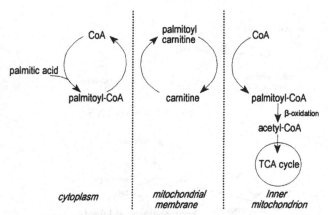

FIGURE 18.5 The mitochondrial fatty acid shuttle.

carnitine transport shuttle, is effected by two transesterifications involving fatty acyl-CoA esters and carnitine, and the action of three mitochondrial enzymes: **carnitine acyltransferases I** and **II**, and **carnitine translocase**.

Carnitine acyltransferase I resides on the outer side of the inner mitochondrial membrane, while carnitine acyltransferase II is located on the matrix side and acylcarnitine translocase spans the inner membrane. The acyltransferases catalyze the formation and hydrolysis of fatty acylcarnitine esters.[32] The translocase catalyzes the exchange of carnitine and acylcarnitines (produced by carnitine acyltransferase I) across the membrane. The result of the concerted action of these enzymes is that long-chain fatty acids are brought into the mitochondrion by being esterified to carnitine and transported as fatty acylcarnitine esters, after which carnitine is released and returned to the outer side of the membrane, thus rendering the free fatty acid available for β-oxidation within the mitochrondrion. It has been suggested that the carnitine transport shuttle may also function in the reverse direction by transporting acetyl groups back to the cytoplasm for fatty acid synthesis.[33] Under normal metabolic conditions, it appears that short-chain acyl-CoAs are generated at rates comparable to their use rates such that acylcarnitine does not accumulate. However, under conditions of propionic acidemia or methylmalonic acidemia and aciduria, which occur in vitamin B_{12} deficiency, the urinary excretion of acylcarnitine is enhanced owing to the increased formation of short-chain acylcarnitines.

The activity of the carnitine transport shuttle is typically low at birth but increases dramatically after birth –

for example, the carnitine palmitoyl transferase I activity in rat liver increases nearly five-fold within 24 hours of birth, peaking within 2–3 days. Similar increases in the hepatic activities of carnitine acetyltransferase I and carnitine palmitoyl transferase I have been observed in human infants. These increases correspond to the parallel development of fatty acid oxidation in the heart, liver, and adipose tissue, and suggest that the carnitine transport shuttle is rate-limiting to that process. While the mechanism of the postpartum increase in carnitine transport shuttle activity remains poorly understood, it is clear that one factor influencing it is carnitine status.

Other Functions

Carnitine and its esters have other biological actions:

- *Peroxisomal fatty acid shuttle.* Chain-shortened fatty acyl CoAs are transesterified to carnitine within peroxisomes from which they are transported into the cytoplasm for mitochondrial uptake.
- *Glucocorticoid-like actions.* Evidence suggests that carnitine can bind the glucocorticoid receptor and cause receptor-mediated release of cytokines.
- *Phospholipid remodeling.* Studies have shown carnitine-bound fatty acids to be incorporated into erythrocyte membrane phospholipids during repair after oxidative attack, and into dipalmitoylphosphatidylcholine, which serves as a lung surfactant.

Conditions of Need

Insects

Carnitine is an essential nutrient for some insect species, including beetles of the family *Tenebrionidei*,[34] the beetle *Oryzaephilus surinamensis*, and the fly *Drosophila melanogaster*. It is presumed that carnitine plays the same essential role in the metabolism of fatty acids in insects that it does in mammals, and that their special requirement for the factor as an essential nutrient is due to their inability to synthesize it from endogenous sources. For these species, carnitine clearly is a vitamin.[35]

Mammals

Neonatal rabbits fed a carnitine-free colostrum replacer or a carnitine-free weaning diet showed abnormally low tissue and urinary carnitine levels, decreased plasma total and

---

32. The carnitine acyltransferases are actually a family of related enzymes. Six carnitine acyltransferases with different but overlapping chain-length specificities have been isolated from mitochondria (three each from the inner and outer sides of the inner membrane).

33. Even if such a reverse shuttle were to function, its contribution to fatty acid synthesis would be insignificant in comparison with that of the citrate shuttle, which transports acetyl-CoA to the cytoplasm by the action of a citrate cleavage enzyme.

34. A family of mealworms.

35. Carnitine is an amino acid, but because it has no role in protein synthesis it meets the criteria of vitamin status (an organic compound that is distinct from fats, carbohydrates, and proteins; see Chapter 1).

VLDL cholesterol, and increased apolipoprotein levels. Carnitine needs have been demonstrated in rats under conditions in which carnitine biosynthesis is impaired by nutritional deprivation of the amino acids lysine or methionine. It is estimated that about 0.1% of the lysine required by the rat may be consigned to carnitine biosynthesis; rats fed lysine-deficient diets have been shown to develop mild depressions in tissue carnitine concentrations and to suffer growth depression and fatty liver, both of which are at least partially alleviated by feeding carnitine.[36]

Carnitine supplementation of the diets of pregnant and lactating sows has been shown to improve the growth and nursing of piglets,[37] indicating suboptimal endogenous carnitine biosynthesis in the gestating and neonatal periods of that species.

Humans

Healthy adults can synthesize carnitine at rates sufficient for their needs. This is indicated by findings that subjects in Southeast Asia whose cereal-based diets provide very little preformed carnitine show plasma carnitine concentrations[38] comparable to those of Western subjects whose diets typically provide abundant amounts of the factor. The same appears to be true for animals, as attempts to produce carnitine deficiency by restricting carnitine in the diet have proved unsuccessful.

It is likely that the carnitine biosynthetic capacities of humans and other animals may be insufficient to meet the needs of the developing fetus during gestation. This has been demonstrated in the pig: carnitine supplementation of sows' diets improves placental development, enhances fetal glucose oxidation, and results in higher birthweights.[39]

Neonates appear to have compromised endogenous carnitine synthesis (i.e., very low hepatic γ-butyrobetaine hydroxylase activities). Their carnitine status is dependent on that of the mother, on the placental transfer of carnitine

in utero, and on the availability of exogenous sources[40] after birth. Appreciable amounts of carnitine are found in the milk of several species.[41] Infants fed soy-based formulas, which contain little or no carnitine, are unable to maintain normal plasma carnitine levels, whereas intravenous administration of L-carnitine allows them to do so. Preterm infants can be at special risk in this regard; although their plasma carnitine levels tend to be nearly normal, they can be depleted rapidly during the course of intravenous feeding with solutions unsupplemented with carnitine. One study found the plasma carnitine concentrations of preterm infants (gestational age <36 weeks) to drop from 29 to 13 μmol/l during total parenteral feeding.

The consequences of suboptimal carnitine status would appear to be great for the infant, who at birth changes from a pattern of energy metabolism based on glucose as the major fuel to one based on the utilization of fats.[42] Thus, for the newborn, free fatty acids appear to be the preferred metabolic fuels, especially for the heart and skeletal muscle (tissues depending on the oxidation of fatty acids for more than half of their total energy metabolism), when glucose availability is limited. Accordingly, carnitine is an important cofactor for neonatal energy metabolism.

Carnitine deficiency in infants has been found to produce several subclinical biochemical changes. Infants fed soy-based formula diets for as long as 2 weeks after birth have shown reduced hepatic carnitine concentrations with associated reductions in hepatic fatty acid oxidation and ketogenesis. Hypertriglyceridemia has also been reported in infants fed soy-based diets not supplemented with carnitine. However, the long-term physiological consequences of these reductions are unknown, and no clinical symptoms of carnitine deficiency in infants have been described.

Abnormally low circulating carnitine levels have been found in humans with severe protein malnutrition (Table 18.4). Low plasma carnitine concentrations, which responded to nutritional therapy, have also been found

36. The feeding of a lysine-deficient diet to rats produced a severe depression in growth, but only marginal depressions in the carnitine concentrations of some tissues (e.g., the carnitine concentrations of skeletal muscle and heart were each about 70% of normal); in addition, the carnitine concentration in the liver rose. Despite that increase, hepatic steatosis occurred unless L-carnitine was fed. Experiments have shown that such mild carnitine deficiency produced in lysine-deficient rats can, indeed, reduce palmitic acid oxidation in homogenates of those tissues. Other experiments showed that rats fed low-protein diets limiting in methionine had reduced growth and developed fatty livers; supplementation of that diet with 0.2% L-carnitine overcame part of the growth depression and markedly reduced the hepatic signs.
37. Ramanau, A. (2005). *Br. J. Nutr.* 93, 717.
38. Mean ± SD: men, 59 ± 12 μmol/l ($n = 40$); women, 52 ± 12 μmol/l ($n = 45$).
39. Eber, K. (2009). *Br. J. Nutr.* 102, 645.

40. Examples are human milk, prepared infant formulas and milk replacers. It has been suggested that natural selection has resulted in mother's milk containing carnitine in proportion to the needs of the infant. In fact, the greatest concentrations of carnitine in human milk occur during the first 2–3 days of suckling. During the first 3 weeks of lactation, the carnitine content of human milk varies from 50 to 70 nmol/ml; after that time, it declines to about 35 nmol/ml by 6–8 weeks. Most milk-based infant formulas contain comparable or slightly greater amounts of carnitine; however, formulas based on soybean protein or casein and casein hydrolyzate contain little or no carnitine. Lipid emulsions also contain no appreciable carnitine.
41. For example, human milk, 28–95 nmol/ml; cow's milk, 190–270 nmol/ml.
42. At birth, plasma free fatty acids and β-hydroxybutyrate concentrations are rapidly elevated owing to the mobilization of fat from adipose tissue. These elevated levels are maintained by the utilization of high-fat diets such as human milk and many infant formulas, which typically contain more than 40% of total calories as lipid.

TABLE 18.4 Apparent Carnitine Deficiency in Protein-Malnourished Children

Group	Plasma Carnitine (μmol/dl)	Plasma Albumin (g/dl)
Healthy controls	9.0 ± 0.6 (8)[a]	3.5 ± 0.1 (8)
Undernourished patients	6.4 ± 0.9 (10)	2.7 ± 0.2 (5)
Marasmus patients	3.7 ± 0.5 (12)	2.7 ± 0.2 (8)
Kwashiorkor patients	2.6 ± 0.5 (13)	1.7 ± 0.1 (9)

Source: Khan, L. and Bamji, M. S. (1977). *Clin. Chim. Acta* 75, 163.
[a]Mean ± SD for (n) children.

renal tubular loss of total carnitine and of acylcarnitine esters in particular are elevated. It is thought that, in these abnormal metabolic conditions, carnitine may function to remove excess organic acids.

Health Effects

Hepatic Function

Hypocarnitinemia (plasma concentrations <55 μmol/l) and tissue carnitine depletion appear to be common in patients with advanced cirrhosis, who not only tend to have marginal intakes of carnitine and its precursors, but also have loss of hepatic function, including the capacity to synthesize carnitine. Carnitine supplementation has been found to protect against ammonia-induced encephalopathy in cirrhotics.[45]

in children with schistosomiasis and associated signs of anemia and protein malnutrition (low serum albumin) in the Middle East. Two steps in the biosynthesis of carnitine require Fe^{2+} (the mitochondrial ε-N-trimethyllysine hydroxylase and the cytosolic γ-butyrobetaine hydroxylase). Therefore, it is possible that both the iron deficiency (manifested as anemia) and the protein deficiency (manifested as a low serum albumin concentration) of these patients may have reduced their abilities to synthesize carnitine.

Carnitine deficiencies can also result from genetic disorders that affect the tissue utilization of carnitine. These are of two general types:

- *Muscle carnitine deficiency.* This is thought to involve defective transport of carnitine into skeletal muscle due to deficiency of carnitine palmitoyltransferase II or acylcarnitine translocase. The major clinical features include mild to severe muscular weakness, and, frequently, excessive lipid accumulation in skeletal muscle fibers.
- *Systemic carnitine deficiency.* This is thought to involve errors in OTCN2 and/or renal carnitine reabsorption. It has a much more heterogeneous clinical picture, including such features as multiple episodes of metabolic encephalopathy, cardiomyopathy, hypoglycemia, hypoprothrombinemia, hyperammonemia, and hepatic steatosis.

Carnitine-deficient individuals, diagnosed by low muscle and/or plasma carnitine levels, typically show lipid accumulation in muscle, with high risk of encephalopathy, progressive muscular weakness, and cardiomyopathy. Carnitine deficiency has also been recognized as a secondary feature of a variety of other genetic disorders, such as organic acidurias[43] and Fanconi syndrome,[44] in which the

Renal Function

Patients with renal disease managed with chronic hemodialysis can be depleted of carnitine[46] owing to the loss of carnitine in the dialysate, which greatly exceeds the amount normally lost in the urine.[47] Tissue depletion of carnitine has been related to the complications attendant on hemodialysis: hyperlipidemia, cardiomyopathy, skeletal muscle asthenia, and cramps. While randomized controlled trials have not been conducted, clinical experience and the results of small, open-label studies have suggested that carnitine administration to dialysis patients can increase hematocrit, allow a lower erythropoietin dose, and reduce intradialytic hypotension and fatigue.[48]

Diabetes

Reduction in the carnitine-dependent transport of fatty acids into the mitochondria, resulting in cytosolic triglyceride accumulation, has been implicated in the pathogenesis of insulin resistance; and findings from studies with animal models suggest that carnitine status affects the control of glycolysis and/or gluconeogenesis. Carnitine

43. Examples include isovaleric, glutaric, propionic, and methylmalonic acidemias, which result from long- and medium-chain acyl-CoA dehydrogenase deficiencies.
44. Fanconi syndrome is a renal disease characterized by the excessive renal excretion of a number of metabolites that are normally reabsorbed (e.g., amino acids).

45. Malaguarnera, M., Pistone, G., Elvira, R., *et al.* (2005). *World J. Gastroenterol.* 11, 7197.
46. In one study, the muscle carnitine concentrations of eight patients after hemodialysis were only 10% of those of healthy controls. It is of interest, however, that not all hemodialysis patients experience carnitine depletion. Some show chronic hypocarnitinemia, whereas others show a return of plasma carnitine concentrations to normal or higher than normal within about 6 hours after dialysis. The recovery of the latter group is hastened (to about 2 hours) if each patient is given 3g of D,L-carnitine orally at the end of the dialysis period.
47. This can be prevented by adding carnitine to the dialysate, e.g., 65nmol/ml.
48. Handleman, G. J. (2006). *Blood Purif.* 24, 140.

supplementation has been found to stimulate the insulin-mediated disposal of glucose, and confer protection of vascular function in insulin resistance.[49]

A clinical trial found carnitine to reduce fasting plasma glucose levels and increase fasting triglycerides in type 2 diabetics.[50] Intramuscular administration of acetylcarnitine has been shown to reduce pain in patients with diabetic neuropathy, as well as in type 2 diabetics with poorly controlled blood glucose.[51] These effects appear to be associated with enhanced nerve regeneration. Studies in the rat have shown experimentally induced diabetes to result in the depletion of lens carnitine and acetylcarnitine, and carnitine supplementation to reduce the development of cataracts.

Cardiovascular Health

Studies have shown that carnitine supplementation can benefit cardiac function.[52] Supplementation of rats with carnitine has been shown to produce effects on prostaglandins that are associated with cardioprotection (i.e., lowering of the ratios of 6-keto-prostaglandin $F_{1\alpha}$ to thromboxane B_2 and leukotriene B_4), and to reduce myocardial injury after ischemia and reperfusion. Randomized controlled trials with cardiac patients have shown carnitine treatment to reduce hypertension, enhance vascular function, reduce left ventrical dilatation, and prevent ventricular remodeling. Long-term studies with chronic heart-failure patients showed that carnitine supplementation improved exercise capability.

Studies with animal models have found propionyl-carnitine treatment to have a prostacyclin-like effect in countering vasoconstrictor activity, to promote endothelium-dependent arterial dilation in hypertensive individuals, and to have anti-hyperlipidemic and anti-atherosclerotic effects.

Administration of propionylcarnitine to animal models has been found to improve cardiac function, and to improve the functional recovery of the myocardium after ischemia,[53] which otherwise involves a decline in cardiac carnitine levels. A multi-center randomized trial found propionylcarnitine to enhance the exercise capacity of chronic heart-failure patients with relatively intact myocardial function.[54]

Neurologic Health

Because animal studies have shown that acetylcarnitine treatment can induce the release of acetylcholine in the striatum and hippocampus, a multi-center, randomized, clinical intervention trial with Alzheimer's disease patients found attenuated progression for several parameters of behavior, disability, and cognitive performance.[55] One study found carnitine supplementation to reduce attention problems and aggressive behavior in boys with attention-deficit hyperactivity disorder.[56]

Studies have found acetylcarnitine supplements to reduce memory loss in older animals, which effects are associated with improved function of brain mitochondria and reduced accumulation of lipid oxidation products. Trials with Alzheimer's disease patients have shown that acetylcarnitine treatment can reduce deterioration in reaction time, reduce depressive symptoms, and improve cognitive performance.[57] A meta-analysis of 21 studies with patients with mild congnitive impairment or mild Alzheimer's disease concluded that treatment with acylcarnitine (1.5–2 g/day) improved both clinical and psychomotor assessments.[58] A randomized clinical trial found acylcarnitine to be as effective as, but better tolerated than, the anti-depressive drug amisulpride in reducing depressive symptoms in dysthymia patients.[59] Studies have found acetylcarnitine effective in improving adaptive behavior and socialization skills in subjects with attention-deficient hyperactivity disorder and Fragile X syndrome.[60]

Male Reproductive Health

Epididymal tissue and spermatozoa typically contain high concentrations of carnitine. Studies in rodent models as well as humans indicate that carnitine levels are related to sperm count, motility, and maturation, and that carnitine supplementation can improve sperm quality.[61] Two randomized trials found treatment with carnitine and/or

49. Mingorance, C., del Pozo, M. G., Herra, M. D., et al. (2009). Br. J. Nutr. 102, 1145.
50. Rahbar, A.R., Shakerhosseini, R., Saadat, N. et al. (2005). Eur. J. Clin. Nutr. 59, 592.
51. DeGrandis, D. and Minardi, C. (2002). Drugs R. D. 3, 223; Sima, A. A., Calvani, M., Mehra, M., et al. (2005). Diabetes Care 28, 89.
52. Ferrari, R., Merli, E., Cicchitelli, G., et al. (2004). Ann. NY Acad. Sci. 1033, 79.
53. Lango, R., Smoleński, R. T., Rogowski, J., et al. (2005). Cardiovasc. Drug Ther. 19, 267.
54. That is, ejection fractions of 30–40%; Anonymous (1999). Eur. Heart J. 20, 70.

55. Spagnoli A., Lucca, U., Menasce, G., et al. (1991). Neurology 41, 1726.
56. Van Oudheusden, L. J. and Scholte, H. R. (2002). Prostagland. Leukotr. Essent. Fatty Acids 67, 33.
57. Rai, G., Wright, G., and Scott, L. (1990). Curr. Med. Res. Opin. 11, 638.
58. Montgomery, S. A., Thal, L. J., and Amrein, R. (2003). Intl Clin. Psychopharmacol. 18, 61.
59. Zanardi, R. and Smeraldi, E. (2006). Eur. Neuropsychopharmacol. 16, 281.
60. Torrioli, M. G., Vernacotola, S., Mariotti, P., et al. (1999). Am. J. Med. Genet. 87, 366; Torrioli, M. G., Vernacotola, S., Peruzzi, L., et al. (2008). Am. J. Med. Genet. 146, 803.
61. Ng, C. M., Blackman, M. R., Wang, C., et al. (2004). Ann. NY Acad. Sci. 1033, 177.

acylcarnitine to improve sperm motility in low-fertility males. Propionylcarnitine has been found to enhance the efficacy of the drug sildenafil in treating erectile dysfunction in diabetic patients and in post-prostatectomy patients.[62]

Thyroid Function

Carnitine appears to be a peripheral agonist of thyroid hormone. A randomized controlled trial found administration of carnitine effective in reversing symptoms of hyperthyroidism.[63]

Human Performance

It has been suggested that oral carnitine supplementation may improve performance (i.e., act as an ergogenic aid) and attenuate the deleterious effects of hypoxic training, and thus hasten recovery from strenuous exercise. Implicit in such proposals is the notion that carnitine supply may be rate-limiting to fatty acid oxidation in muscle, a phenomenon demonstrated only when muscle carnitine drops to less than half normal levels. A review of the published literature concluded that there is no evidence that carnitine supplements can improve athletic performance.[64]

Weight Management

Studies with animals have demonstrated that supplementation with carnitine or propionylcarnitine can enhance the loss of body fat under conditions of negative energy balance.[65] Controlled trials with humans, however, have not found such effects.

Safety

A systematic review of published literature concluded that evidence of safety is strong for carnitine intakes as great as 2,000 mg/day on a chronic basis, although studies using higher levels have not observed adverse effects.[66]

4. MYO-INOSITOL

Although **myo-inositol** had been discovered in extracts of animal tissues almost 100 years earlier, interest in its potential nutritional role was first stimulated in the 1940s, when

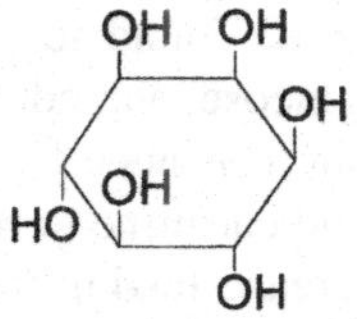

FIGURE 18.6   Chemical structure of *myo*-inositol.

Wolley reported it to be a new vitamin required for normal growth and for the maintenance of normal hair and skin of the mouse (the "mouse anti-alopecia factor"). Subsequently, that original report was questioned regarding the adequacy of the diet with respect to other known vitamins. Nevertheless, several groups found dietary supplements of *myo*-inositol to stimulate growth of chicks, turkeys, rats, and mice, in ways that appeared to depend on other factors such as biotin and folate. Whether the observed effects were actually responses to a missing nutrient was debated owing to the observation by Needham that the daily urinary excretion of *myo*-inositol by the rat exceeds the amount ingested, which led to the conclusion that the factor was synthesized by that species. However, *myo*-inositol was found to be essential for the growth of most cells in culture. More recently, it has been found that deprivation of *myo*-inositol can render the hepatic triglyceride accumulation by the rat susceptible to influence by the fatty acid composition of dietary fat, indicating a function resembling that of an essential nutrient. Further, Hegsted and colleagues found that the female Mongolian gerbil[67] develops **intestinal lipodystrophy** when depleted of the factor. In fact, that group demonstrated a dietary requirement for *myo*-inositol to prevent this disorder in the gerbil fed a diet containing adequate levels of all other known nutrients.

Chemical Nature

myo-Inositol is a water-soluble, hydroxylated, cyclic six-carbon compound (*cis*-1,2,3,5-*trans*-4,6-cyclohexanehexol) (Fig. 18.6). It is the only one of the nine possible stereoisomeric forms of cyclohexitol with biological activity.

Dietary Sources

myo-Inositol occurs in foods and feedstuffs in three forms: free *myo*-inositol, **phytic acid**,[68] and inositol-containing

62. Gentile, V., Vicini, P., Prigiotti, G., *et al.* (2004). *Curr. Med. Res. Opin.* 20, 1377; Cavalini, G., Modenini, F., Vitali, G., *et al.* (2005). *Urology* 66, 1080.
63. Benvenga, S., Amato, A., Calvani, M., and Trimarchi, F. (2004). *Ann. NY Acad. Sci.* 1033, 158.
64. Brass, E.P. (2004). *Ann. NY Acad. Sci.* 1033, 67.
65. Heo, K., Odle, J., and Han, I. K. (2000). *J. Nutr.* 130, 1809; Mingorance, C., del Pozo, M. G., Herra, M. D., *et al.* (2009). *Br. J. Nutr.* 102, 1145.
66. Hathcock, J. N. and Shao, A. (2006). *Reg. Toxicol. Pharmacol.* 46, 23.
67. *Meriones unguiculatus.*
68. Inositol hexaphosphate:

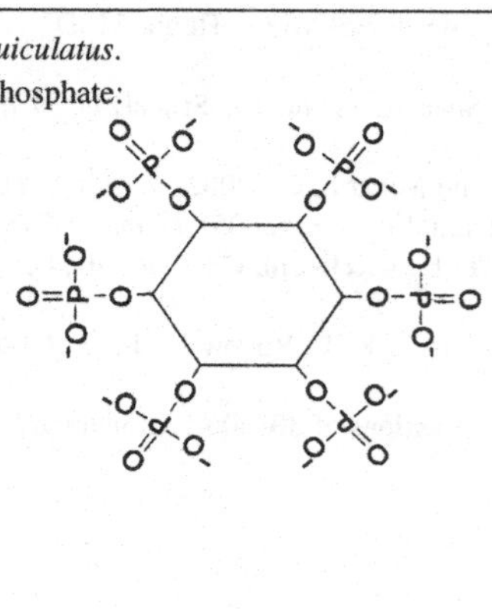

phospholipids. The richest sources of *myo*-inositol are the seeds of plants (e.g., beans, grains, and nuts) (Table 18.5). However, the predominant form occurring in plant materials is phytic acid (which can comprise most of the total phosphorus present in materials such as cereal grains[69]). Because most mammals have little or no intestinal phytase activity, phytic acid is poorly utilized as a source of either *myo*-inositol or phosphorus.[70,71] In animal products, *myo*-inositol occurs in free form as well as in inositol-containing phospholipids (primarily **phosphatidylinositol**); free *myo*-inositol predominates in brain and kidney, whereas phospholipid inositol predominates in skeletal muscle, heart, liver, and pancreas. The richest animal sources of inositol are organ meats. Human milk is relatively rich in *myo*-inositol (colostrum, 200–500 mg/l; mature milk, 100–200 mg/l) in comparison with cow's milk (30–80 mg/l). A disaccharide form of *myo*-inositol, 6-β-galactinol (6-*O*-β-D-galactopyranosyl-*myo*-inositol), comprises about one-sixth of the non-lipid *myo*-inositol in that material.

myo-Inositol is classified by the US Food and Drug Administration among the substances generally recognized as safe, and therefore can be used in the formulation of foods without the demonstrations of safety and efficacy required by the Food, Drug and Cosmetic Act. It is added to many prepared infant formulas (at about 0.1%). It is

estimated that typical American diets provide adults with about 900 mg of *myo*-inositol per day, slightly over half of which is in phospholipid form.

Absorption and Transport

Absorption

The enteric absorption of free *myo*-inositol occurs by active transport; the uptake of *myo*-inositol from the small intestine is virtually complete. The enteric absorption of phytic acid, however, depends on the ability to digest that form, and on the amounts of divalent cations in the diet/meal. Most animal species lack intestinal phytase activities, and are therefore dependent on the presence of a gut microflora that produces those enzymes. For species that harbor such microfloral populations (e.g., ruminants and long-gutted non-ruminants) phytate is digestible, thus constituting a useful dietary source of *myo*-inositol. Dietary cations (particularly Ca^{2+}) can reduce the utilization of phytate by forming insoluble (and thus non-digestible and non-absorbable) phytate **chelates**. Because a large portion of the total *myo*-inositol in mixed diets typically is in the form of phytic acid, the utilization of *myo*-inositol from high-calcium diets can be less than half of that from diets containing low to moderate amounts of the mineral.[72] Little information is available concerning the mechanism of absorption of phospholipid *myo*-inositol; it is probable that it is analogous to that of phosphatidylcholine.[73]

Transport

myo-Inositol is transported in the blood predominantly in the free form; the normal circulating concentration of *myo*-inositol in humans is about 30 μmol/l. A small but significant amount of phosphatidylinositol (PI) is found in association with the circulating lipoproteins. Free *myo*-inositol appears to be taken up by an active transport process in some tissues (kidney, brain) and by carrier-mediated diffusion in others (liver). The active process requires Na^+ and energy, and is inhibited by high levels of glucose. Apparently because of this antagonism, untreated diabetics show impaired tissue uptake and impaired urinary excretion of *myo*-inositol.

---

69. Of the total phosphorus present, phytic acid phosphorus comprises 48–73% for cereal grains (corn, barley, rye, wheat, rice, sorghum), 48–79% for brans (rice, wheat), 27–41% for legume seeds (soybeans, peas, broad beans), and 40–65% for oilseed meals (soybean meal, cottonseed meal, rapeseed meal).

70. The bioavailability of phosphorus from most plant sources is relatively good (>50%) for ruminants, which benefit from the phytase activities of their rumen microflora. Non-ruminants, however, lack their own intestinal phytase, and thus generally derive much less phosphorus from plant phytic acid, depending on the phytase contributions of their intestinal microflora. For pigs and rats such contributions appear to be significant, giving them moderate abilities (about 37% and 44%, respectively) to utilize phytic acid phosphorus. In contrast, the chick, which has a short gut and rapid intestinal transit time, and thus has only a sparse intestinal microflora, can use little (about 8%) phytic acid phosphorus.

71. Phytic acid can also form a very stable chelation complex with zinc (Zn^{2+} is held by the negative charges on adjacent pyrophosphate groups), thus reducing its nutritional availability. For this reason, the bioavailability of zinc in such plant-derived foods as soybean is very low.

---

72. For the same reason, the bioavailability of calcium is also low for high-phytate diets. This effect also occurs for the nutritionally important divalent cations Mn^{2+} and Zn^{2+}; the bioavailability of each is reduced by the presence of phytic acid in the diet.

73. This would involve hydrolysis by pancreatic phospholipase A in the intestinal lumen to produce a lysophosphatidylinositol, which, on uptake by the enterocyte, would be reacylated by an acyltransferase or hydrolyzed further to yield glycerylphosphorylinositol.

TABLE 18.5 Total *myo*-Inositol Contents of Selected Foods

Food	myo-Inositol (mg/g)	Food	myo-Inositol (mg/g)
Vegetables		**Cereals**	
Asparagus	0.29–0.68	Rice	0.15–0.30
Beans:		Wheat	1.42–11.5
Green	0.55–1.93	**Meats**	
White	2.83–4.40	Beef	0.09–0.37
Red	2.49	Chicken	0.30–0.39
Broccoli	0.11–0.30	Lamb	0.37
Cabbage	0.18–0.70	Pork	0.14–0.42
Carrots	0.52	Turkey	0.08–0.23
Cauliflower	0.15–0.18	**Liver**	
Celery	0.05	Beef	0.64
Okra	0.28–1.17	Chicken	1.31
Peas	1.16–2.35	Pork	0.17
Potatoes	0.97	**Fish**	
Spinach	0.06–0.25	Trout	0.11
Squash, yellow	0.25–0.32	Tuna	0.11–0.15
Tomatoes	0.34–0.41	**Dairy Products and Eggs**	
Fruits		Milk	0.04
Apples	0.10–0.24	Ice cream	0.09
Cantaloupe	3.55	Cheese	0.01–0.09
Grapes	0.07–0.16	**Eggs**	
Grapefruit	1.17–1.99	Whole	0.09
Oranges	3.07	Yolk	0.34
Peaches	0.19–0.58	**Nuts**	
Pears	0.46–0.73	Almonds	2.78
Strawberries	0.13	Peanuts	1.33–3.04
Watermelon	0.48		

Source: Clements, S. R. Jr and Darnell, B. (1980). *Am. J. Clin. Nutr.* 33, 1954.

Metabolism

Biosynthesis

Most, if not all, mammals appear capable of synthesizing *myo*-inositol *de novo* ultimately from glucose. Biosynthetic capacity has been found in the liver, kidney, brain, and testis of rats and rabbits, and in the kidney[74] and other tissues in humans. Biosynthesis involves the cyclization of glucose 6-phosphate to inositol 1-phosphate by inositol-1-phosphate synthase, followed by a dephosphorylation by inositol-1-phosphatase.

Fates of myo-Inositol

Free *myo*-inositol is converted to PI within cells either by *de novo* synthesis of the latter by reacting with the liponucleotide cytidine diphosphate (CDP)-diacylglycerol[75] or by an

---

74. Renal synthesis of *myo*-inositol has been found to be about 4 g/day (~2 g/kidney per day).

75. This step is catalyzed by the microsomal enzyme CDP diacylglycerol–inositol 3-phosphatidyltransferase (also called PI synthetase).

exchange with endogenous PI.[76] Phosphatidylinositol can, in turn, be sequentially phosphorylated to the monophosphate (phosphatidylinositol 4-phosphate, PIP) and diphosphate (phosphatidylinositol 4,5-diphosphate, PIP_2) forms by membrane kinases.[77] Thus, in tissues *myo*-inositol is found as the free form, as PI, PIP, and PIP_2.[78] The *myo*-inositol-containing phospholipids tend to be enriched in **stearic acid** (predominantly at the 1-position) and **arachidonic acid** (predominantly at the 2-position) in comparison with the fatty acid compositions of other phospholipids. For example, the *myo*-inositol-containing phospholipids on the plasma membrane from human platelets contain about 42 mol% stearic acid and about 44 mol% arachidonic acid. The greatest concentrations of *myo*-inositol are found in neural and renal tissues.

The turnover of the *myo*-inositol phospholipids is accomplished intracellularly. Phosphatidylinositol phosphates can be catabolized by cellular phosphomonoesterases (phosphatases), ultimately to yield PI. In the presence of cytidine monophosphate, PI synthetase functions (in the reverse direction) to break down that form to yield CDP-diacylglycerol and *myo*-inositol. The kidney appears to perform most of the further catabolism of *myo*-inositol, first clearing it from the plasma and converting it to glucose, and then oxidizing it to CO_2 via the pentose phosphate shunt. The metabolism of *myo*-inositol appears to be relatively rapid; the rat can oxidize half of an ingested dose in 48 hours.

Metabolic Functions

The metabolically active form of *myo*-inositol appears to be phosphatidylinositol, which is thought to have several physiologically important roles:

- as an affecter of membrane structure and function
- as a source of arachidonic acid for eicosanoid production
- as a mediator of cellular responses to external stimuli.

Phosphatidylinositol has been proposed to be active in the regulation of membrane-associated enzymes and transport processes. For example, phosphatidylinositol is an endogenous activator of a microsomal Na^+,K^+-ATPase, an essential constituent of acetyl-CoA carboxylase, a stimulator of tyrosine hydroxylase, a factor bound to alkaline phosphatase and 5′-nucleotidase, and a membrane anchor for acetylcholinesterase. It has been suggested that such effects involve the special membrane-active properties conferred on the phospholipid by its unique fatty acid composition. For example, its polar head group and highly non-polar fatty acyl chains may facilitate specific electrostatic interactions while providing a hydrophobic microenvironment for enzyme proteins on or in membranes. Such properties may render phosphatidylinositol an effective anchor for the hydrophobic attachment of proteins to membranes. Phosphatidylinositol also serves as a source of releasable arachidonic acid for the formation of the **eicosanoids**[79] by the cellular activities of cyclooxygenase and/or lipoxygenase. Although phosphatidylinositol is less abundant in cells than the other phospholipids (phosphatidylcholine, phosphatidylethanolamine, and phosphatidylserine), its enrichment in arachidonic acid renders it an effective source of that eicosanoid precursor.

That the metabolism of phosphatidylinositol is activated in target tissues by stimuli producing rapid (e.g., cholinergic or α-adrenergic agonists) or medium-term (e.g., mitogens) physiological responses suggests a mediating role of such responses. It is thought that this role involves the conversion of the less abundant species, phosphatidylinositol diphosphate (PIP_2), to the water-soluble metabolite **inositol 1,4,5-triphosphate (IP_3)**, which serves as a **second messenger** to activate the release of Ca^{2+} from intracellular stores.[80]

Inositol phosphate receptors on the cell surface have been shown to effect primary control over the hydrolysis of PIP_2 by regulating the activity of phospholipase C (phosphodiesterase) on the plasma membrane. Receptor occupancy thus activates the hydrolysis of PIP_2, which is favored at low intracellular concentrations of Ca^{2+}, to produce IP_3 and, perhaps, other inositol polyphosphates. The IP_3 that is produced signals the release of Ca^{2+} from discrete organelles called **calcisomes**, as well as the entry of Ca^{2+} into the cell across the plasma membrane.[81]

---

76. This reaction is stimulated by Mn^{2+}; like phosphatidylinositol synthetase, it is localized in the microsomal fraction of the cell.

77. These are ATP:phosphatidylinositol 4-phosphotransferase and ATP:phosphatidylinositol-4-phosphate 5-phosphotransferase, respectively. They are located on the cytosolic surface of the erythrocyte membrane. There is no evidence that *myo*-inositol can be isomerized or phosphorylated to the hexaphosphate level; however, such prospects would be of interest, as the isomer D-*chiro*-inositol has been shown to promote insulin function, and inositol hexaphosphate (phytic acid) has been found to be anti-carcinogenic in a variety of animal models.

78. The disaccharide 6-β-galactinol appears to be a unique mammary metabolite.

79. The eicosanoids include prostaglandins, thromboxanes, and leukotrienes. The prostaglandins are hormone-like substances secreted for short-range action on neighboring tissues; they are involved in inflammation, in the regulation of blood pressure, in headaches, and in the induction of labor. The functions of the leukotrienes and thromboxanes are less well understood; they are thought to be involved in regulation of blood pressure and in the pathogenesis of some types of disease.

80. There is some controversy concerning whether other inositol phosphates (e.g., inositol 1,3,4,5-tetraphosphate [IP_4], which is a product of a 3-kinase acting on IP_3) can also signal Ca^{2+} mobilization. Evidence suggests that, in at least some cells, IP_3 and IP_4 may have cooperative roles in Ca^{2+} signaling.

81. The Ca^{2+}-mobilizing activity of IP_3 is terminated by its dephosphorylation (via a 5-phosphatase) to the inactive inositol-1,4-bisphosphate or by its phosphorylation (via a 3-kinase) to a product of uncertain activity, IP_4.

The former process involves a specific IP_3 receptor on the calcisomal membrane; the binding of IP_3 to this receptor opens a Ca^{2+} **channel** closely associated with the receptor. The latter process is less well understood.

It is thought that the IP_3-stimulated entry of Ca^{2+} into the cell involves an increase in the permeability to Ca^{2+} of the plasma membrane that is signaled by the emptying of the IP_3-sensitive intracellular pool. The mechanism whereby the IP_3-sensitive pool and the plasma membrane communicate to effect this response is not understood. There is also some evidence to suggest that 1,2-diacylglycerol, which is formed from the receptor-stimulated metabolism of the *myo*-inositol-containing phospholipids, may also serve a second messenger function in activating protein kinase C for the phosphorylation of various proteins important to cell function. According to this hypothesis, 1,2-diacylglycerol functions with Ca^{2+} and phosphatidylserine, both of which are know to be involved in the activation of protein kinase C.

Conditions of Need

Although early reports indicated dietary needs for *myo*-inositol to prevent alopecia in rodents, fatty liver in rats, and growth retardation in chicks, guinea pigs, and hamsters, more recent studies with more complete diets have failed to confirm such needs. Hence, it has been concluded that most, if not all, of those lesions actually involved deficiencies of other nutrients (e.g., biotin, choline, and vitamin E). Because *myo*-inositol appears to be synthesized by most, if not all, species, it has been suggested that the observed responses to dietary supplements of *myo*-inositol may have involved favorable effects of the compound on the intestinal microflora. This hypothesis would suggest that the addition of *myo*-inositol to diets would be beneficial when those diets contain marginal amounts of such factors as choline and biotin, the gut synthesis of which can be important. Accordingly, it has been shown that supplements of *myo*-inositol reduced hepatic lipid accumulation in rats fed a choline-deficient diet, improved growth in rats fed a diet deficient in several vitamins, and reduced the incidence of fatty liver in and improved the growth of chicks fed a biotin-deficient diet containing an antibiotic. Thus it appears that, under certain conditions, animals can have needs for preformed *myo*-inositol. Such conditions are not well defined, but it has been suggested that they include such situations as disturbed intestinal microflora,[82] diets containing high levels of fat,[83] and a physical environment creating physiological stress.

For a few species, however, overt dietary needs for *myo*-inositol have been demonstrated. These include several fishes and the gerbil. Studies with fishes have shown dietary deprivation of *myo*-inositol to result in anorexia, fin degeneration, edema, anemia, decreased gastric emptying rate, reduced growth, and impaired efficiency of feed utilization. Studies with gerbils have shown *myo*-inositol deprivation to result in intestinal lipodystrophy, with associated hypocholesterolemia and reduced survival. It is interesting to note that these effects are observed only in female gerbils; males appear to have a sufficient testicular synthesis of *myo*-inositol. For at least these species, *myo*-inositol must be considered a dietary essential.

Health Effects

- *Pre-term infants.* Three randomized clinical trials have found inositol supplementation to improve survival, and reduce retinopathy of prematurity, bronchial dysplasia, and intraventricular hemorrhage.[84]
- *Psoriasis.* A small, randomized clinical trial with psoriasis patients found supplementation with inositol to reduce the severity of symptoms.[85]
- *Psychiatric disorders.* Because mood stabilizers such as lithium, valproate, and carbamizepine function by the stabilization of inositol signaling, it has been suggested that inositol may have value in the treatment of depression and other psychiatric disorders. While inositol supplementation has been reported to be helpful in treating bipolar depression and bulimia nervosa with binge eating, the limited findings in this area to date do not indicate clear benefits.

5. PYRROLOQUINOLINE QUINONE

In the late 1970s, studies of specialized bacteria, the methylotrophs,[86] resulted in the discovery of a new enzyme cofactor, **pyrroloquinoline quinone** (**PQQ**). That cofactor was found in several different bacterial oxidoreductases. Subsequently, PQQ has been identified in several other important enzymes (now collectively called **quinoproteins**) in yeasts, plants, and animals. In 1989, Killgore and colleagues demonstrated beneficial effects of PQQ in preventing skin lesions in mice fed a diet containing low concentrations of that factor.

---

82. For example, antibiotics can reduce the numbers of microorganisms that normally produce *myo*-inositol as well as other required nutrients.
83. High-fat diets may increase the need for *myo*-inositol for lipid transport.

84. Howlett, A. and Ohlsson, A. (2003). *Cochrane Database Syst. Rev.* CD000366.
85. Allan, S. J. R., Kavanagh, G. M., Herd, R. M. *et al.* (2004). *Br. J. Dermatol.* 150, 966.
86. The methylotrophs dissimilate single-C compounds, e.g., methane, methanol, and methylamine.

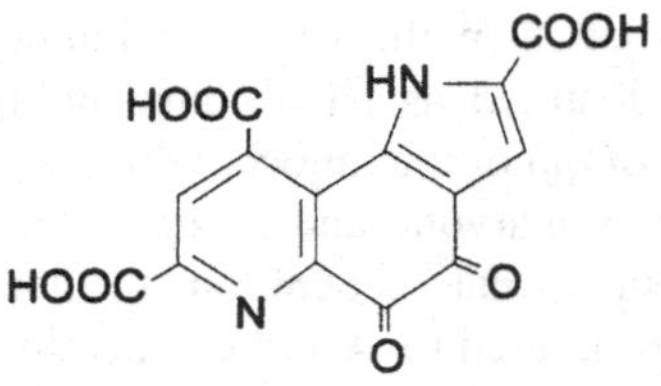

FIGURE 18.7 Chemical structure of pyrroloquinoline quinone.

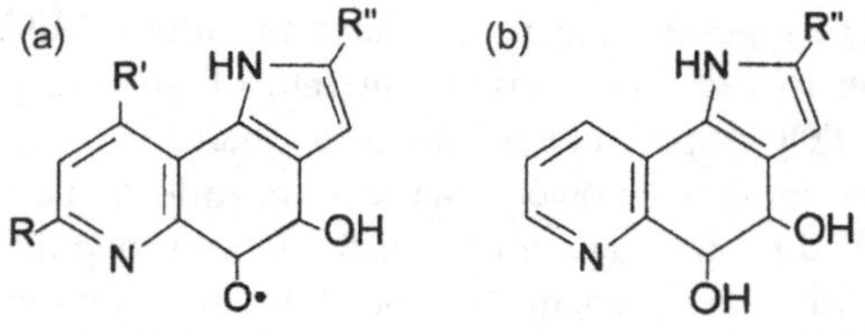

FIGURE 18.8 Chemical structures of (a) PQQH· and (b) PQQH$_2$.

Chemical Nature

PQQ, sometimes called "methoxatin," is a tricarboxylic acid with a fused heterocyclic (*o*-quinone) ring system (Fig. 18.7).[87] Its C-5 carbonyl group is very reactive toward nucleophiles,[88] leading to adduct formation.

Dietary Sources

PQQ has been found in all plants analyzed. Fruits and vegetables have been found to contain 7–34 μg/kg; legume seeds, 18.24 μg/kg; fermented products, 60–800 μg/kg; milk, 3.4 μg/kg; egg yolk, 7 μg/kg; and human milk, 140–180 μg/kg.[89] It is thought that PQQ exists in foods as imidazole and oxazole derivatives. Reports indicate PQQ to be present in casein, starch, and isolated soy protein in the range of 10–100 μg/kg. It has been estimated that most people consume 1–2 mg PQQ per day.

Absorption and Metabolism

The absorption and metabolism of PQQ have not been elucidated, although it has been shown to redox cycle and to react readily with amino acids to form imidazole and oxazole adducts. A PQQ reductase activity has been identified in bovine erythrocytes.[90] It is not clear the extent to which the hindgut microflora may contribute PQQ for enteric absorption by humans and other animals; that this source may not be substantive is suggested by the findings that some enteric species are incapable of synthesizing PQQ.

Metabolic Activities

In bacterial quinoproteins, PQQ is covalently bound to the apoprotein, probably by an amide or ester bond via its carboxylic acid group(s). The redox behavior of PQQ involves its ability to form adducts that facilitate both one- and two-electron transfers. In dehydrogenases, its function appears to involve electron transfers of both types (substrate oxidation by a two-electron transfer to PQQ, followed by single-electron transfer to such acceptors as copper-containing proteins and cytochromes). Accordingly, two forms of the cofactor, the semiquinone PQQH· and the catechol PQQH$_2$, have been found in the bacterial quinoproteins (Fig. 18.8).

The role of PQQ in eukaryotes is less clear. Although several animal quinoproteins have been proposed, more recent investigations using sensitive physical methods have failed to confirm earlier claims of covalent binding of PQQ or derivatives at the active centers of those enzymes.[91]

Evidence has been presented for PQQ having several metabolic effects:

- *Antioxidant protection.* The ability to redox cycle makes PQQ an antioxidant. Studies in animal models have shown PQQ treatment to have antioxidant effects, including protection against carbon tetrachloride hepatotoxicity, inhibition of glucocorticoid-induced lenticular glutathione depletion, and cataract.
- *Cell signaling.* PQQ has been shown to affect the activity of the oncogene ras, which is important in signal transduction pathways involved in growth and development.[92] It has also been found to affect the activity of the "Parkinson disease protein," DJ-1, a peptidase involved in androgen receptor-regulated transcription leading to mitochondrial biogenesis. PQQ has also been shown to stimulate the activation, by phosphorylation, of the promoter of peroxisome proliferator-activated receptor-γ-coactivator-1α (PGC-1α) in the stimulation of mitochondrial biogenesis.[93]

Conditions of Need

That supplemental PQQ (800 μg/kg) promoted growth in mice fed a purified diet containing low levels (<30 μg/kg)

87. 4,5-Dihydro-4,5-dioxo-1*H*-pyrrolo[2,3-*f*]quinoline-2,7,9-tricarboxylic acid.

88. For example, amino groups, thiol groups.

89. Stites, T. E., Mitchell, A. E., and Rucker, R. B. (2000). *J. Nutr.* 130, 719.

90. This activity has been called a flavin reductase or NADPH-dependent methemoglobin reductase; however, it shows a much lower K_m value for PQQ (2 μmol.l) than for flavins (about 30 μmol/l).

91. Lysyl oxidase, which plays a key role in the cross-linking of collagen and elastin, was proposed to be a quinoprotein; however, it has also been suggested that lysyl oxidase may require instead 6-hydroxydopa, which satisfies the apparent quinone requirement of plasma monoamine oxidase, after which other putative animal quinoproteins have been modeled.

92. Tchaparian, E., Marshal, L., Cutler, G., *et al.* (2010). *Biochem. J.* 429, 515.

93. Chowdnadisai, W., Bauerly, K. A., Tchaparian, E., *et al.* (2010). *J. Biol. Chem.* 285, 142.

of PQQ suggested that it may have nutritional value.[94] In addition to a clear difference in rate of growth in favor of the PQQ-supplemented mice, a quarter of the PQQ-deprived animals showed friable skin, mild alopecia, and a hunched posture, and a fifth of the latter group died by 8 weeks with aortic aneurysms or abdominal hemorrhages. The most frequent sign, friable skin, was taken to suggest an abnormality of collagen metabolism; subsequent study revealed increased collagen solubility (indicating reduced cross-linking) and abnormally low activities of lysyl oxidase in PQQ-deprived animals.[95] Attempts to breed PQQ-deprived mice were unsuccessful; mice fed the low-PQQ diet for 8–9 weeks produced either no litters or litters in which the pups were immediately cannibalized at birth. Subsequent work by the same group has shown deprivation of PQQ to have immunopathologic effects in mice: altered mitogenic responses and reduced interleukin 2 levels. More recent results from the same laboratory have shown that PQQ-deprivation of growing mice elevated the plasma levels of glucose, alanine, glycine, and serine, and reduced the amounts and function of hepatic mitochondria.[96]

Safety

High doses of PQQ have been found to cause renal toxicity and oxidative DNA damage. Such effects are thought to be due to its capacity to redox cycle, which can be pro-oxidative.

6. UBIQUINONES

The recognition of ubiquinone came as a result of efforts to understand the mechanisms of energy metabolism in the 1950s. Originally isolated as a membrane-associated lipid in beef heart mitochondria, which were being produced in huge quantities for biochemical research, the physiological significance was pursued by Folkers, who proposed as a tentative name "vitamin Q." After nearly a decade of study, interest was captured when Yamamura found ubiquinone to be useful in the treatment of heart failure.

Chemical Nature

The **ubiquinones** are a group of tetra-substituted 1,4-benzoquinone derivatives with isoprenoid side chains of variable length. Originally isolated from the unsaponifiable fractions of the hepatic lipids from vitamin A-deficient rats,

the principal species of the group (ubiquinone[50][97]) was subsequently identified as an essential component (**coenzyme Q$_{10}$ or CoQ$_{10}$**) of the mitochondrial electron transport chains of most prokaryotic and all eukaryotic cells. In the four decades since that recognition, the term "coenzyme Q" has come to be used to describe generally this family of compounds, all of which are synthesized from precursors in the inner mitochondrial membrane. The structure of the 6-chromanol portion of the CoQ group is remarkably similar to the oxidized form of vitamin E, tocopherylquinone, the difference being the two methoxyl groups on the CoQ$_{10}$ ring in place of two methyl groups on the tocopherylquinone ring.

Dietary Sources

CoQ$_{10}$ is localized in the mitochondrial electron transport chain and in other cellular membranes. Therefore, it is found in plant and animal tissues of high cellularity, and those rich in mitochrondria, such as heart and muscle (Table 18.6).

Absorption and Tissue Distribution

Ubiquinones appear to be absorbed, transported, and taken up into cells by mechanisms analogous to those of the tocopherols. In the rat, the greatest absorption of CoQ$_{10}$ has been found to occur in the duodenum, with demonstrable absorption also in the colon, ileum, and jejunum, suggesting the possibility of an enterohepatic circulation. Coenzyme Q$_{10}$ is known to be distributed in all membranes in the cell, and, particularly, the mitochondria. Relatively great concentrations of CoQ$_{10}$ are found in the liver, heart, spleen, kidney, pancreas, and adrenals.[98] The total CoQ$_{10}$ pool size in the human adult is estimated to be 0.5–1.5 g. Tissue ubiquinone levels increase under the influence of oxidative stress, cold acclimation, and thyroid hormone treatment, and appear to decrease with cardiomyopathy, other muscle diseases, and aging.

Metabolism

Biosynthesis

Coenzyme Q$_{10}$ is synthesized in most tissues. The biosynthetic process derives the isoprenyl side chain from mevalonate, the ring system from tyrosine, the hydroxyl groups

---

94. Killgore, J., Smidt, C., Duich, L., *et al.* (1989). *Science* 245, 850.
95. Steinberg, F., Stites, T.E., Anderson, P., *et al.* (2003). *Exp. Biol. Med.* 228, 160.
96. Stites, T., Storms, D., Bauerly, K., *et al.* (2006). *J. Nutr.* 136, 390.

97. The conventions of nomenclature for the ubiquinone/CoQ group are similar to those for the vitamin K group. For the ubiquinones, the number of side chain carbons is indicated parenthetically; for the CoQ designation, the number of side chain isoprenyl units is indicated in subscript.
98. The contributions of foods and feedstuffs, many of which are now known to contain appreciable concentrations of ubiquinones, to these high tissue levels are unknown.

TABLE 18.6 CoQ$_{10}$ Contents of Foods

Food	CoQ$_{10}$ (µg/g)
Vegetables	
Asparagus	2
Avocados	10
Beans	2
Broccoli	6–9
Cabbage	1–5
Cauliflower	1–7
Eggplant	1–2
Onions	1
Parsley	8–26
Pepper, sweet	3
Potatoes	1
Soybeans	7–19
Spinach	1–10
Sweet potatoes	3–4
Tomatoes	<1
Fruits	
Apples	1
Bananas	1
Oranges	1–2
Strawberries	1
Cereals	
Corn germ	7
Rice bran	5
Wheatgerm	4–7
Meats	
Beef, muscle	16–37
Beef, liver	39–51
Beef, heart	113
Pork, muscle	14–45
Pork, liver	23–54
Pork, heart	118–282
Chicken, muscle	11–25
Chicken, liver	116–132
Chicken, heart	92–192
Fish	3–130
Dairy Products and Eggs	
Milk	2
Butter	7
Eggs	1–4
Other	
Corn oil	13–139
Olive oil	4–160
Soybean oil	54–279

Source: Pravst, I., Zmitek, K., and Zmitek, J. (2010). *Crit. Rev. Food Sci. Nutr.* 50, 269.

from molecular oxygen, and the methyl groups from *S*-adenosylmethionine to produce a 50-carbon polyisoprene chain – i.e., containing 10 isoprene units. Endogenous biosynthesis appears sufficient to support membrane saturation levels; dietary supplementation increases tissue CoQ levels only in liver and spleen.

Metabolic Functions

CoQ has several essential metabolic functions.

- *Mitochondrial respiratory chain.* CoQ functions as an electron acceptor for complexes I and II of mitochondrial electron transport chains, passing electrons from flavoproteins (e.g., NADH or succinic dehydrogenases) to the cytochromes via cytochrome b_5. They perform this function by undergoing reversible reduction/oxidation to cycle between the 1,4-quinone (oxidized) and 1,4-dihydroxybenzene (reduced) species (Fig. 18.9).
- *Antioxidant.* Its redox capacity allows CoQ$_{10}$ to function as a membrane-bound antioxidant that protects and thus spares α-tocopherol in subcellular membranes, protects LDL from oxidation, and prevents the oxidation of lipids, proteins, and DNA. Administration of CoQ$_{10}$ has been found to protect against myocardial damage mediated by free-radical mechanisms (ischemia, drug toxicities) in animal models.
- *Regulation of intracellular NAD$^+$/NADH balance.* CoQ is an essential cofactor for NADH oxidase in plasma membranes; its role in regulating intracellular NAD$^+$/NADH balance is important in cell growth and development.
- *Mitochondrial membrane pore regulation.* CoQ is a factor that prevents opening of mitochondrial membrane transition pores that otherwise admit large molecules capable of antagonizing the function of that organelle. This function counters pro-apoptotic events.
- *Uncoupling of oxidative phosphorylation.* CoQ is required to deliver protons from fatty acids to uncoupling proteins, which uncouples the proton gradient from oxidative phosphorylation and results instead in the production of heat.
- *Anti-inflammatory effects.* CoQ stimulates the release by lymphocytes of factors that signal the expression of NFκB-1-dependent genes to produce anti-inflammatory factors.
- *Endothelial function.* CoQ has been shown to stimulate the release of nitric oxide by endothelia.

Conditions of Need

Metabolic needs for CoQ are linked to vitamin E status. While CoQ$_{10}$ itself does not spare vitamin E for preventing gestation-resorption syndrome in the rat, the oxidized form – that is, the 6-chromanol moiety of

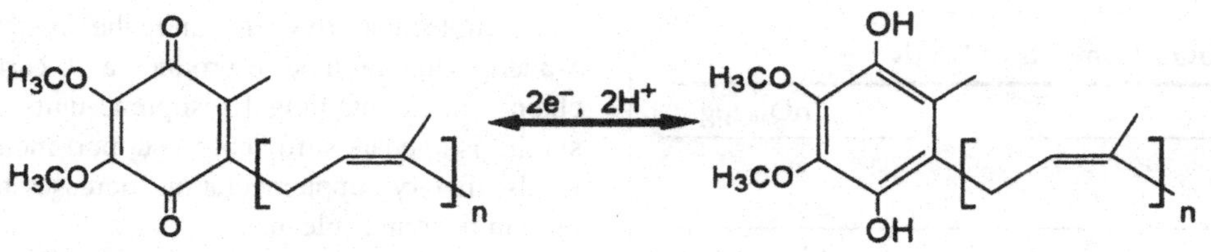

FIGURE 18.9 Structure and redox function of ubiquinones.

hexahydro-CoQ_4 – has been found to prevent the syndrome in vitamin E-deficient rats, and to produce significant reductions in both the anemia and the myopathy of vitamin E-deficient rhesus monkeys. In fact, the responses were more rapid than have been observed for α-tocopherol. Thus, it appears that dietary ubiquinones may be important as sources of antioxidant protection, but that they share that role with other antioxidant nutrients, several of which are likely to be more potent in this regard.

Deficient tissue levels of CoQ can be corrected with dietary supplements of CoQ_{10}. Such deficiencies occur in older animals, suggesting that their biosynthetic rate may decline with age. Tissue-specific deficiencies of CoQ synthesis have also been identified; these involve mutations of various genes encoding enzymes involved in that process.

Plasma and tissue CoQ levels are also lowered by statin therapy, which, by inhibing HMG-Co reductase, causes a decrease in farnasyl pyrophosphate, an intermediate in the synthesis of CoQ_{10}. Statin treatment has been associated with myopathic conditions ranging from mild myalgia to fatal rhabdomyolysis, as well as with subclinical cardiomyopathy. These conditions have been found to respond to treatment with CoQ_{10} supplementation or its synthetic analog idebenone.[99]

CoQ deficiency is manifest in humans as heterogenous diseases: encephalopathy, infantile cerebellar ataxia, myopathy, glomerulopathy; infantile multisystemic disease.[100]

Health Effects

Cardiovascular Health

Studies with animal models[101] have shown that supplemental CoQ_{10} can help maintain the integrity of cardiac muscle under cardiomyopathic conditions. Clinical trials with humans have indicated benefits of supplemental CoQ_{10} of several types.

- *Migraine*. A small, open-label trial reported CoQ_{10} to reduce headache frequency.[102]
- *Congestive heart failure*. A meta-analysis of randomized, controlled trials showed CoQ_{10} supplements to reduce dyspnea, edema, and the frequency of hospitalization.[103]
- *Hypertension*. A systematic review of eight randomized, controlled trials suggested that CoQ_{10} supplementation decreased systolic pressure by an average of 16 mmHg, and diastolic pressure by an average of 10 mmHg.[104]
- *Atherosclerosis*. A randomized controlled trial found CoQ_{10} supplementation after myocardial infarction to reduce subsequent myocardial events and cardiac deaths.[105]
- *Endothelial dysfunction*. A randomized controlled trial found CoQ_{10} supplementation to improve endothelial function of peripheral arteries of dyslipidemic patients with type 2 diabetes.[106]
- *Recovery from coronary artery bypass surgery*. In a randomized trial, patients preoperatively supplemented with CoQ_{10} were found to have significantly fewer reperfusion arrythmias and shorter hospitalizations than controls.[107]
- *Friedrich's ataxia*. High doses of CoQ_{10} administered with vitamin E have been found to improve cardiac and muscular function in patients with Friedreich's ataxia.[108,109]

99. Littarru, G. P. and Langsjoen, P. (2007). *Mitochondrion* 7S, S168–S174; Mancuso, M., Orsuci, D., Vopli, L., *et al.* (2010). *Curr. Drug Targets* 11, 111.

100. Quinzli, C. M. and Hirano, M. (2010). *Dev. Disabil. Res. Rev.* 16, 183.

101. For example, cardiomyopathy induced in the rat by feeding a fructose-based, copper-deficient diet.

102. Sándor, P. S., Di Clemente, L., Coppola, G., *et al.* (2005). *Neurology* 64, 713.

103. Soja, A. M. and Mortensen, S. A. (1997). *Mol. Aspects Med.* 18(Suppl.), S159.

104. Rosenfeldt, F., Hilton, D., Pepe, S., *et al.* (2003). *Biofactors* 18, 91.

105. Singh, R. B., Neki, N. S., Kartikey, K., *et al.* (2003). *Mol. Cell. Biochem.* 246, 75.

106. Watts, G. F., Playford, D. A., Croft, K. D., *et al.* (2002). *Diabetologia* 45, 420.

107. Makhija, N., Sendasgupta, C. Kiron, U., *et al.* (2008). *J. Cardiothorac. Vasc. Anesth.* 22, 832.

108. Cooper, J. M. and Schapira, A. H. V. (2007). *Mitochondrion* 7S, S127.

109. Friedreich's ataxia is an autosomal recessive condition involving deficient production of a mitochondrial protein, frataxin, thought to function in antioxidant regulation and/or iron metabolism. The condition manifests in adolescence as progressive limb and gait ataxias, and losses of deep tendon reflexes, and position and vibration senses due to the loss of large sensory nerones in the dorsal root ganglia and deterioration of other cerebellar–spinal tracts.

Neurologic Health

Modest improvements in symptoms were reported from a small, randomized, controlled trial with Parkinson's disease patients.[110]

Other Effects

CoQ_{10} supplementation has been reported to improve semen quality in men with idiopathic infertility, to reduce the risk of pre-eclampsia, to reduce ultraviolet light-induced skin wrinkling, and to confer protection against cardiac or hepatic toxicity associated with cancer chemotherapy.

Safety

Human studies have found CoQ to be safe and well tolerated as a supplement. A chronic, high-level treatment (0.26% of diet) has been shown to exacerbate some cognitive and sensory impairments in older mice.[111]

7. FLAVONOIDS

The group of compounds now referred to as the **flavonoids** was discovered by Szent-Györgyi as the factor in lemon juice or red peppers that potentiated the antiscorbutic activities of those foods for the guinea pig. The factor was called, by various groups "citrin," "vitamin P,"[112] and "vitamin C_2," but was ultimately found to be a mixture of phenolic derivatives of 2-phenyl-1,4-benzopyrane, the flavane nucleus.

Ubiquitous Plant Metabolites

Flavonoids are ubiquitous in foods and feedstuffs of plant origin; more than 6,000 different flavonoids have been identified, each a secondary metabolite of shikimic acid. Flavonoids have a wide variety of functions in plant tissues: natural antiobiotics,[113] predator feeding deterrents, photosensitizers, UV-screening agents, and metabolic and physiologic modulators. They represent the major sources of red, blue, and yellow pigments other than the carotenoids. Flavonoids are polyphenolic compounds containing two aromatic rings linked by an oxygen-containing, heterocyclic ring (Fig. 18.10). The hydroxyl groups of these polyphenols enable them to form glycosidic linkages with sugars, and most flavonoids occur naturally as glycosides.

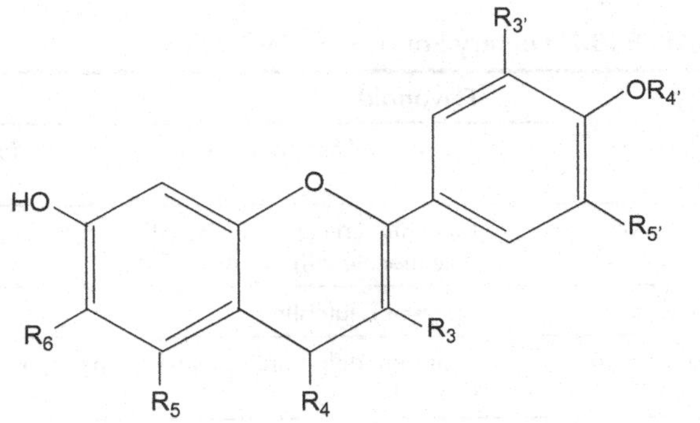

FIGURE 18.10 General structure of flavonoids.

There are six general classes of flavonoids, classified by their common ring substituents:

- *Flavonols.* These derivatives (R_3 hydroxy, R_4 keto) include quercetin, kaempferol, isorhamnetic, and myricetin, the most abundant flavonoids in human diets. Flavonols are found in a variety of fruits and vegetables, often as glycosides. Relatively high amounts (15–40 mg/100 g) are found in broccoli, kale, leeks, and onions.
- *Flavanols.* These derivatives (R_3 hydroxy), also called **catechins**, do not exist as glycosides (unlike other flavonoids). They include catechin, epicatechin, epigallocatechin, and their gallate derivatives. They are found in apples, apricots, and red grapes (2–20 mg/100 g); green tea and dark chocolate are rich in catechins (40–65 mg/100 g).
- *Flavones.* This group of some 300 compounds retains the basic flavane nucleus structure. It includes apigenin and luteolin in very high concentrations (>600 mg/100 g) in parsley, and in lower but significant amounts in cereal grains, celery, and citrus rinds (which contain polymethoxylated forms).
- *Anthocyanins.* These derivatives (R_3 and R_4 reduced) exist as glycosides, their aglycone chromophores being referred to as anthocyanidins. There are several hundred of the latter, the most common of which are cyanidin, delphinidin, malvinidin, pelargonidin, peonidin, petunidin, and malvidin. Most are red or blue pigments. The richest sources (up to 600 mg/100 g) are raspberries, blackberries, and blueberries; cherries, radishes, red cabbage, red skinned potato, red onions, and red wine are also good sources (50–150 mg/100 g). Anthocyanins have antioxidant properties. Unlike other flavonoids, anthocyanins are relatively unstable to cooking and high-temperature food processing.
- *Flavanones.* These derivatives (R_4 keto, "C" ring otherwise reduced) are found primarily in citrus pulp (15–50 mg/100 g), where they are also present as *O*- and *C*-glycosides and methoxylated derivatives. They

---

110. Muller, T., Büttner, T., Gholipour, A. F. et al. (2003). *Neurosci. Lett.* 341, 201.

111. Sumien, N., Heinrich, K. R., Shetty, R. A., et al. (2009). *J. Nutr.* 139, 1926.

112. The letter *P* indicated the permeability vitamin, because it improved capillary permeability.

113. That is, as phytoalexins.

TABLE 18.7 Dietary Sources of Flavonoids

Type	Flavonoid	Food Sources
Flavonols	Auercetin, kaemferol, myricetin, insoharmnetin	Onions (yellow), scallions, kale, broccoli, apples, berries, tea
Flavanols	Catechin, epicatechin, epigallocatechin, epicatechin gallate, epigallocatechin gallate	Tea, chocolate, grapes, berries, apples
Flavones	Apegenin, luteolin	Celery, parsley, thyme, peppers
Anthocyanins	Cyanidin, delphinidin, malvidin, pelargonidin, peonidin, petunidin	Berries, grapes, red wine
Flavanones	Hesperetin, naringenin, eriodictyol	Oranges, lemons, grapefruit
Isoflavones	Daidzein, genistein, glycitein	Soybeans and soy products, other legumes

include eriocitrin, neoericitrin, hesperetin, neohespiridin, naringin, narirutin, didymin, and poncirin.

- *Isoflavones.* These derivatives ("B" aromatic ring linked at R_3) are contained only in legumes mostly as glycosides; they include daidzein, genistein, and glycitein, which are also referred to as **phytoestrogens** due to the affinities of their 7- and 4'-hydroxyl groups to binding mammalian estrogen receptors. Soy products (soy flour, tofu, tempeh) can contain 25–200 mg/100 g).

- *Tannins.* These are polymeric flavonoids present in all plants. Those conjoined by covalent C–C bonds are not hydrozyable; they are called condensed tannins, or **proanthocyanidins**. Others, containing non-aromatic polyol carbohydrate moieties such as gallic or ellagic acids, are hydrolyzable. These polyphenols have strong antioxidant properties *in vitro*.

Dietary Sources

The consumption of flavonoids has been estimated to average 190 mg/day in the US, mostly as flavanols.[114] Similar estimates have been made for northern European diets. The greatest contributors of flavonoids in human diets are fruits and vegetables, with fruit juices, green tea, and dark chocolate being the richest sources in western diets (Table 18.7). Most flavonoids tend to be concentrated in the outer layers of fruit and vegetable tissues (e.g., skin, peel). In general, the flavonoid contents of leafy vegetables are high, whereas those of root vegetables (with the notable exception of onions with colored skins) are low. The greatest contributors to dietary flavonoid intake are tea, citrus fruits and juices, and wine.

Flavonoid aglycones are stable during food processing and cooking; however, anthocyanidins are unstable to such conditions.

114. Chun, O. K., Chung, S. J., and Song, W. O. (2007). *J. Nutr.* 137, 1244.

Absorption and Transport

Most flavonoids in foods occur as glycosides, which must be hydrolyzed by glycosidases in saliva, the brush border of the intestine, and the intestinal microflora. The efficiency of these processes appears to be low, as only 1–10% of dietary flavonoids appear to be absorbed.

Studies have shown quercitin glycosides to be absorbed intact apparently by the Na^+-dependent glucose transporter-1 (SGLT-1), to be returned to the intestinal lumen by the apical transporter multi-drug resistance-associated protein-2 (MRP-2), and to antagonize the Na^+-dependent vitamin C transporter-1 (SVCT-1). Soy isoflavones appear to be more bioavailable to children than adults.

Metabolism

Upon absorption, flavonoids are conjugated as glucuronides, sulfates, and/or methylated metabolites in the liver, and are degraded to a variety of phenolic compounds that are rapidly excreted by way of the bile and urine. Urinary flavonoids show as very highly variable, suggesting marked inter-individual variation in flavonoid metabolism.

Hindgut bacteria can degrade flavonoids by cleaving the heterocyclic ring. This results in the formation of various phenolic acids and their lactones, some of which may be absorbed from the colon. *Bacteroides* spp., which have glycases, are able to metabolize polyphenylglycones; their numbers appear to be increased by consuming flavonoid-rich foods.

Metabolic Effects

Antioxidant Activity

The ability of flavonoids to chelate divalent metal cations (e.g., Cu^{2+}, Fe^{2+}) gives them antioxidant-like activities by removing those catalysts of lipid peroxidation reactions and by scavenging radical intermediates.[115] For example, the flavonol quercetin, which has multiple phenolic hydroxyl groups (a carbonyl group at C-4, and free C-3 and C-5 hydroxyl groups), can scavenge superoxide radical ions, hydroxyl radicals, and fatty acyl peroxyl radicals. Flavonols and some proanthocyanins have been shown to inhibit macrophage-mediated LDL oxidation *in vitro*, probably by protecting LDL α-tocopherol from oxidation or by reacting with the tocopheroxyl radical to regenerate α-tocopherol. Such antioxidant potential contributes to chemical measurements of "total antioxidant capacity" in foods; however, such measures have no immediate physiologic relevance. There is no direct evidence that health effects of flavonoids involve antioxidant functions, as these compounds are extensively metabolized after ingestion, for which reason their circulating levels are low.[116]

Enzyme Modulation

Flavonoids have been found to interact with many enzymes, selectively affecting the activities of some. This includes induction of some (e.g., phase II enzymes) by binding to promoter regions of their respective genes, and inhibition of others (e.g., aldose reductase, phosphodiesterase, *O*-methyltransferase, and several serine- and threonine-kinases) by direct binding to the respective protein. For example, tea flavanols inhibit redox-sensitive transcription factors (NFκB, AP-1) and pro-oxidative enzymes (lipoxygenases, cyclooxygenases, nitric oxide synthase, xanthine oxidase), but induce phase II and antioxidant enzymes (glutathione *S*-transferases, superoxide dismutases).[117] Flavanones have been shown to induce phase II enzymes, and to exert anti-inflammatory effects; naringin in particular has been implicated in the effect of grapefruit juice in inhibiting cytochrome *P*-450-dependent drug metabolism.[118]

Some flavonoids with B-ring catechols can promote mitochondrial production of reactive oxygen species (ROS) by inhibiting succinoxidase. Others can cause uncoupling of mitochondrial oxidative phosphorylation and Ca^{2+} release by reducing membrane fluidity. Isoflavones can affect estrogen synthesis and transactivation of estrogen receptors α and β to affect signal transduction pathways.

Health Effects

Some epidemiologic studies have demonstrated associations of diets high in flavonoids with reduced risks of cardiovascular diseases and cancer. Because such diets are typically rich in fruits and vegetables, it can be difficult to determine from these results whether the protective factor(s) are flavonoids or some other phytochemicals, vitamins, or minerals (e.g., β-carotene, ascorbic acid, fiber) also provided by those foods.

Cardiovascular Health

Epidemiologic studies of the relationships of flavanoid intake and cardiovascular disease end-points have yielded inconsistent results. However, evidence points to particular types of flavanoids being protective. Meta-analyses of numerous clinical trials found evidence for chocolate consumption increasing flow-mediated dilation and reducing blood pressure, for soy protein isolate consumption being associated with diastolic blood pressure and LDL cholesterol, and for green tea consumption being associated with reduced LDL cholesterol.[119] High intakes of flavanols have been associated with modest reductions in coronary heart disease mortality.[120] Diets high in the flavonol quercetin have been associated with 21–53% reduced risks of cardiovascular disease prevalence.[121] Quercetin has been found to inhibit the activation of *c*-Jun N-terminal kinase in the modulation of angiotensin-induced hypertrophy of vascular smooth muscle cells. An analysis of data from the Nurses' Health Study found that consumption of anthocyanins, some flavones, and some flavanols was associated with prevention of hypertension.[122] Various proanthocyanins have been shown to inhibit platelet activation; to inhibit the expression of interleukin-2; and to lower serum levels of glucose, triglyceride, and cholesterol. Flavonoid-rich foods (cocoa, grape juice, red wine) have been found to be anti-thrombotic by inhibiting platelet aggregation, and to promote vascular endothelial function by stimulating nitric oxide production. Both the flavanone hesperetin and the flavanol epigallochatechin have been found

115. Galleano, M., Verstraeten, S. V., Oteiza, P. I., *et al.* (2010). *Arch. Biochem. Biophys.* 510, 23–30.
116. Hollman, P. C. H., Cassidy, A., Comte, B., *et al.* (2011). *J. Nutr.* 141, 989S.
117. Such effects have been cited as the basis of prospective anti-inflammatory roles of flavonoids (Middleton, E. Jr, Kandaswami, D., and Theoharides, T. C. [2000] *Pharmacol. Rev.* 52, 673).
118. This effect, which involves inhibition of the CYP3A4 isoform, may also involve other flavanones present in grapefruit juice. The potency of this effect is evidenced by the fact that a single glass of grapefruit juice can affect the biological activity of drugs metabolized by this enzyme system, increasing the activities of some and decreasing the activities of others.

119. Hopper, L., Kroon, P. A., Rimm, E. B., *et al.* (2008). *Am. J. Clin. Nutr.* 88, 38; Reid, K., Sullivan, T., Fakler, P., *et al.* (2010). *BMC Med.* 8, 39.
120. Huxley, R. R. and Neil, A. A. (2003). *Eur. J. Clin. Nutr.* 57, 904.
121. Arts, I. C. W. and Hollman, P. C. H. (2005). *Am. J. Clin. Nutr.* 81(Suppl.), 317S.
122. Cassidy, A., O'Reilly, E. J., Kay, C., *et al.* (2011). *Am. J. Clin. Nutr.* 93, 338.

to block oxidized LDL-induced endothelial apoptosis,[123] a key process in atherosclerosis. Intervention with quercetin was found to reduce blood pressure in hypertensive subjects.[124]

Anticarcinogenesis

Some epidemiologic studies have demonstrated associations of diets high in flavonoids with significantly reduced risks of cancers of the lung and rectum (fruit catechins), and of the lung and prostate (soy isoflavones).[125] Various flavones have been found to inhibit cell proliferation and angiogenesis *in vitro*, and to inhibit phorbol ester-induced skin cancer in the mouse model. Underlying these effects may be any of several metabolic changes shown to be evoked by various flavonoids: inhibition of protein kinase C, inhibition of nuclear poly(ADP-ribose) polymerase-1, inactivation of Akt to cause apoptosis, preservation of intracellular NAD^+, increasing expression of estrogen receptor-β, upregulation of DNA repair mechanisms, altered DNA methylation patterns, and altered cytochrome *P*-450-dependent carcinogen metabolism.

Anti-Inflammatory Effects

Epidemiologic observations have shown that diets rich in fruits and vegetables tend to be associated with relatively low levels of inflammatory markers in those people consuming them. That flavonoids may contribute to such anti-inflammatory effects was suggested by an analysis of the NHANES 1999–2002 data, which showed that intakes of the flavonols quercetin and kaempferol, the anthocyannins malvidin and peonidin, and the isoflavone genistein were each inversely related to serum concentrations of the inflammatory marker C-reactive protein.[126] Similarly, the Nurses' Health Study found flavonol intake to be inversely associated with circulating levels of soluble vascular adhesion molecule-1 (sVCAM-1), and the intakes of flavonoid-rich foods to be associated with lower circulating levels of CRP and soluble tumor necrosis factor receptor-2 (sTNF-R2).[127] Studies in animal and cell models have pointed to

several mechanisms underlying these anti-inflammatory effects, including modulation of pro-inflammatory gene expression, inhibition of NFκB activation, and inhibition of nuclear poly(ADP-ribose) polymerase-1 to reduce macrophage cytokine release.

Anti-Estrogenic Effects

That soy isoflavones can be anti-estrogenic was demonstrated by a study of Asian women in which the intake of soy products was found to be inversely associated with circulating levels of estrogen.[128] This effect involves the binding of isoflavones to intranuclear estrogen receptors α and β, thus affecting the estrogen-synthetic activity of 17β-steroid oxidoreductase and estrogen-dependent signal transduction pathways. This is believed to underlie epidemiological observations of inverse associations of soy products and symptoms of menopause or premenstrual syndrome.

Some, but not all, clinical trials have found the consumption of soy products to reduce menopausal symptoms by as much as 50–60%,[129] and a recent study found soy consumption effective in reducing premenstrual syndrome symptoms.[130] As their estrogenic character might suggest, the consumption of soy isoflavones has been associated with higher bone mineral density in a limited number of epidemiological studies. Clinical trials conducted to test the hypothesis that soy isoflavones may be useful in improving bone mineralization for the prevention of osteoporosis have yielded inconsistent results.[131]

Neurologic Health

Studies in animal models have found extracts of flavonoid-rich foods (blueberry, spinach, strawberry) to reduce age-related declines in neuronal signal transduction and cognitive function.[132] Clinical trials have found fruit flavonoids, flavanoid-rich foods (wine, tea, chocolate), and soy isoflavones to enhance cognitive function.[133] Several mechanisms have been indicated in studies with animal models: increased expression of estrogen receptor-β, and

123. Choi, J. S., Choi, Y. J., Shin, S. Y., *et al.* (2008). *J. Nutr.* 138, 983.
124. Edwards, R. L., Lyon, T., Litwan, S. E., *et al.* (2007). *J. Nutr.* 137, 2405.
125. Knekt, P., Kumpulainen, J., Jävinen, R., *et al.* (2002). *Am. J. Clin. Nutr.* 76, 560; Arts, I. C., Jacobs, D. R. Jr, and Gross, M. *et al.* (2002). *Cancer Causes Control* 13, 373; Nagata, Y., Sonoda, T., Mori, M., *et al.* (2007). *J. Nutr.* 137, 1974; Shimazu, T., Inoue, M., Sasazuki, S., *et al.* (2010). *Am. J. Clin. Nutr.* 91, 722.
126. Chun, O. K., Chung, S. J., Claycombe, K. J., *et al.* (2008). *J. Nutr.* 138, 753.
127. Landberg, R., Sun, Q., Rimm, E. B., *et al.* (2011). *J. Nutr.* 141, 618.

128. Nagata, C., Takatsuka, N., Inaba, S., *et al.* (1998). *J. Natl Cancer Inst.* 90, 1830.
129. Albertazzi, P., Pansini, F., Bonaccorsi, G., *et al.* (1998). *Obstet. Gynecol.* 91, 6; Upmalis, D. H., Lobo, R., Bradley, L., *et al.* (2000). *Menopause* 7, 236.
130. Bryant, M. (2005). *Br. J. Nutr.* 93, 731.
131. Messina, M., Ho, S., and Alekel, D. L. (2004). *Curr. Opin. Clin. Nutr. Metab. Care* 7, 649.
132. Joseph, J., Shukitt-Hale, B., Denisova, N. A., *et al.* (1999). *J. Neurosci.* 19, 8114.
133. Thorp, A. A., Sinn, N., Buckley, J. D., *et al.* (2009). *Br. J. Nutr.* 102, 1348; Nurk, E., Refsum, H., Drevon, C. A., *et al.* (2009). *J. Nutr.* 139, 120.

enhanced protein kinase and lipid kinase signaling of transcription of factors involved in synaptic plasticity and cerebrovascular blood flow.

Bone Health

Diets rich in plant-derived foods have been associated with reduced risk of osteoporosis. That flavonoids may contribute to this effect is suggested by findings from studies with animal models, including the finding that maternal consumption of soy isoflavones increased bone mineral density in the offspring. Such studies have shown flavonoids to modulate the expression of transcription factors that affect osteoblast function by affecting cellular signaling via mitogen-activated kinase (MAPK), bone morphogenic protein (BMP), estrogen receptor and osteoprotegrin/receptor activator of NFκB ligand (OPG/RANKL).[134] Due to their estrogen-mimetic effects, it has been suggested that soy isoflavones may be useful in reducing postmenopausal bone loss.

Obesity

Flavonoids have been suggested as having anti-obesity effects. A 14-year cohort study with 4,280 men and women in the Netherlands found diets rich in flavonoids were associated with lower increases in body mass index; the effect was significant only for women.[135] Other evidence comes from trials that have shown regular intake of flavonoid-rich fruits or green tea to reduce body weight. The green tea flavonol epigallocatechin-3-gallate has been shown to inhibit obesity in the mouse model, but a meta-analysis of 15 studies found that green tea catechins were effective in producing modest reductions in body mass index (BMI) and body weight, but only in the presence of caffeine.[136] Such effects are thought to be due to increased thermogenesis and appetite suppression. The consumption of flavonoids has been shown to favor *Bacteroides* spp. in the hindgut. It has been suggested that this may be a mechanism whereby flavonoids exert weight-lowering effects, as a hindgut microflora comprised of a relative abundance of *Bacteriodes* spp. over *Firmicutes* spp. is associated with a lower yield of absorbed energy from fermentation.[137]

Other Effects

It has been suggested that health benefits attributed to traditional herbal medicaments may be due to bioactive flavonoids. Evidence supporting such a hypothesis includes the findings that bilberry anthocyanins reduce retinal hemorrhage in type 2 diabetics, and that certain proanthocyanins can inhibit bacterial adherence to uroepithelial cell surfaces, suggesting a role in reducing urogenital tract infections.

8. NON-PROVITAMIN A CAROTENOIDS

The carotenoids are polyisoprenoid compounds produced by plants for the purposes of harvesting light for photosynthesis and quenching free radicals, thereby protecting plant tissues against oxidative stress. Both functions are possible due to the capabilities of the conjugated double-bond systems of these compounds, which enable them to accept unpaired electrons by delocalizing that electronegativity across multiple carbons. Accordingly, carotenoids have potent antioxidant capabilities. This property allows some carotenoids to function in vision. However most, i.e., those without the β-ionone head group necessary for that and other vitamin A functions (*see* Chapter 5), lack provitamin A activity. These include some carotenes and oxygenated analogs called xanthophylls. The most common non-provitamin A carotenoids in human diets are lycopene, lutein, zeaxanthin, and canthaxanthin (*see* Fig. 18.11).

Dietary Sources

Most non-provitamin A carotenoids are pigments, occurring in red-, yellow-, and orange-colored plant tissues (Table 18.8). The dominant form in US diets is lycopene,

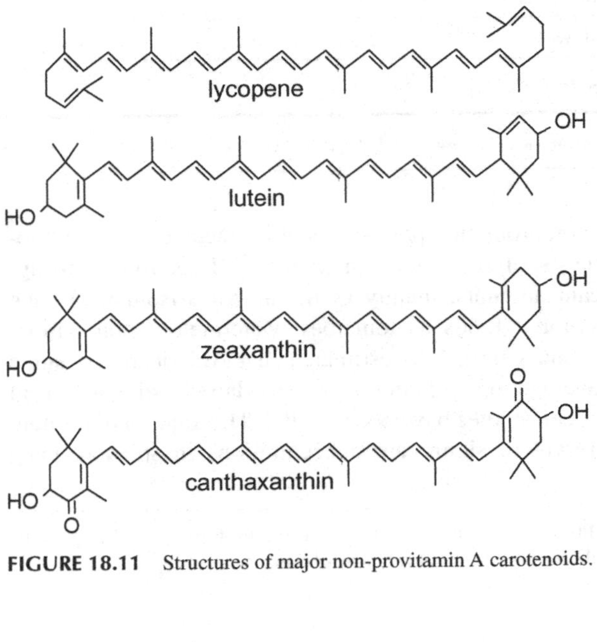

FIGURE 18.11 Structures of major non-provitamin A carotenoids.

134. Trzeciakiewicz, A., Habauzit, V., and Horcajada, M. N. (2009). *Nutr. Res. Rev.* 22, 68.
135. Hughes, L. A. E., Arts, I. C. W., Amergen, T., *et al.* (2008). *Am. J. Clin. Nutr.* 88, 1341.
136. Phung, O. J., Baker, W. L., Matthews, L. J., *et al.* (2010). *Am. J. Clin. Nutr.* 91, 73.
137. Ley, R. E., Turnbargh, P. J., Klein, S., *et al.* (2006). *Nature* 444, 1022.

TABLE 18.8 Non-Provitamin A Carotenoid Contents (μg/100 g or μg/100 ml) of Foods

Food	Lutein	Zeaxanthin	Lycopene	Food	Lutein	Zeaxanthin	Lycopene
Fruits				Endive	2,060–6,150		
Apricots	123–188	0–39	54	Kale	4,800–11,470		
Bananas	86–192		0–254	Leeks	3,680		
Figs	80		320	Lettuce	1,000–4,780		
Grapefruit, red			750	Parsley	6,400–10,650		
Guava			769–1,816	Peas, green	1,910		
Kiwi			<10	Pepper, green	92–911	0–42	
Mango			10–724	Pepper, red	248–8,506	593–1,350	
Nectarine, flesh			2–131	Pepper, yellow	419–638		
Papaya	93–318		0–7,564	Potato, sweet	50		
Pineapple			265–605	Pumpkin	630		500
Rhubarb			120	Sage	6,350		
Watermelon, red			4,770–13,523	Spinach	5,930–7,900		
Watermelon, yellow			56–287	Tomatoes	46–213		850–12,700
Vegetables				Tomato ketchup			4,710–23,400
Avocados	213–361	8–18		Tomato sauce			5,600–39,400
Basil	7,050			**Cereals**			
Beans, green	883			Corn, flakes	0–52	102–297	
Cabbage, white	450			Durum flour	164		
Carrots	254–510			Wheat flour	76–116		
Cress	6,510–7,540			**Other**			
Cucumber	459–840			Olive oil	350		
Dill weed	13,820			Butter	15–25	0–2	
Egg plant	170			Egg yolk	384–1,320		
				Milk, 4% fat	0.8–1.4	0–0.1	

From Maiani, G., Castón, M. J. P., Catasta, G., et al. (2009). Mol. Nutr. Food Res. 53, S194.

the non-aromatic, polyisoprenoid precursor to the biosynthesis of β-carotene in plants.[138] It is found in significant amounts, mainly as the all-*trans*-isomer, in such red-colored foods as tomatoes, watermelon, pink grapefruit, and guava. It is estimated that Americans consume an average of 3–11 mg lycopene daily; estimates from European studies have been similar. The xanthophyll lutein is present in significant concentrations in spinach, kale, corn, broccoli, collards, and eggs; lutein and zeaxanthin in corn represent major sources of pigmentation for poultry diets.[139] The xanthophyll astaxathin is the source of the pink coloration of salmon.

Absorption

Non-provitamin A carotenoids in foods are utilized by the same processes as those by which their provitamin

138. Plants convert lycopene to β-carotene by forming to β-rings at its ends through the action of lycopene cyclase.

139. That is, to promote coloration of egg yolks and broiler skin.

A counterparts are utilized (*see* Chapter 5). Their enteric absorption depends on their not being cleaved by the carotene oxygenases,[140] and their micelle-dependent diffusion depends on the presence of dietary fat and is subject to the antagonistic effects of binding to heat-labile food proteins. Accordingly, cooking or heat-processing improves the bioavailability of lycopene from tomato products.

Carotenoids enter the circulation as components of chylomicra, from which they are moved to lipoproteins. Most are transported by HDL; however, lutein and zeaxanthin are equally distributed among HDL and LDL. Because the transfer of lipoprotein lipid contents depends on interactions with cell surface receptors, it is thought that the distribution of lutein and zeaxanthin to peripheral tissues, and the retinal capture of lutein and zeaxanthin in particular, depends on an individual's particular lipoprotein (particularly, ApoE) profile.

While some three dozen carotenoids have been identified in human serum, lutein and zeaxanthin are preferentially captured by the retinal pigment epithelium. The human retina contains 25–200 ng of these pigments, zeaxanthin being concentrated in the center and lutein being distributed about the periphery. Greatest concentrations accumulate in the central region of the retina, known as the macula, giving that region its characteristic yellow color;[141] the greatest amounts of macular pigments occur in the central area, i.e., the fovea. The capture of the macular pigments is believed to be facilitated by a **xanthophyll-binding protein** (**XBP**). That adipose tissue may also serve as a storage site for xanthophylls is suggested by the finding that serum levels of lutein and zeaxanthin increased in response to weight loss, although macular pigment density was unaffected.[142]

Serum levels of lycopene appear to decline with age, but that affect appears to be related to the lower consumption of fat and lycopene-rich, tomato-based foods by older adults (Fig. 18.12).[143]

Metabolism

Lycopene

That *cis*-isomers of lycopene predominate (50–90%) in serum appears to be due to the combination of its superior

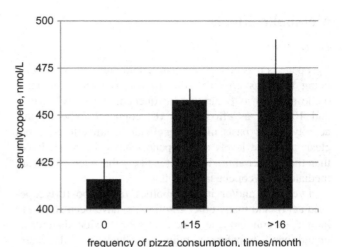

FIGURE 18.12   Relationship of serum lycopene concentration to pizza consumption in the NHANES 1988–1994 *(Gangi, V. and Kafai, M. R. [2005]* J. Nutr. *135, 567).*

absorption over the dominant all-*trans*-isomer found in plant tissues, and its continuous isomerization in the body.[144] Lycopene is turned over in the human body, with a half-life of 5 days, apparently by conversion to more polar metabolites. It has been suggested that carotene oxygenase II may be involved in this metabolism (producing acylo-retinoids), and carbonyl compounds are known to be cleaved readily by autoxidation or radical-mediated oxidation, or by singlet oxygen. Among a larger number of possible oxidative degradation products of lycopene, the following have been identified in humans or animals: 5,6-dihydroxy-5,6-dihydrolycopene; 2,6-cyclolycopene-1,5-diols; 5,6-dihydrolycopene; 5,6-dihydro-5-*cis*-lycopene; apo-8'-lycopenal; apo-10'-lycopenal; and apo-10'-lycopenoic acid.

Lutein and Zeaxanthin

While the macular pigments are exclusively of dietary orgin, they include a zeaxanthin isomer, 3R,3'S-*meso*-zeaxanthin, that is not found in the diet. This isomer appears to be formed by the oxidation–reduction and double-bond isomerization of lutein (i.e., 3R,3'R,6'R-lutein). That this isomer is not found in other tissues suggests that it may be catalyzed photochemically in the retina or by enzymes expressed only in that tissue. Studies have shown that dietary supplementation with lutein and zeaxanthin increases the macular pigments in two ways: by direct capture of the pigments, and by stimulating the migration of retinal pigment epithelial cells to the macula.[145]

140. Carotenoid cleavage activity is highly variable between individuals; this is explained in part by single nucleotide polymorphisms in the carotene oxygenase-1 (BCO1) (Leitz, G., Lange, J., and Rimbach, G. [2010] *Arch. Biochem. Biophys.* 502, 8).

141. Other mammals lack maculae; however, carotenoid-rich oil droplets have been found in the retinas of birds, reptiles, amphibians and fish.

142. Kirby, M. L., Beatty, S., Stack, J., *et al.* (2011). *Br. J. Nutr.* 105, 1036.

143. Gangi, V. and Kafai, M. R. (2005). *J. Nutr.* 135, 567.

144. Unlu, N. Z., Bohn, T., and Francis, D. M. (2007). *Br. J. Nutr.* 98, 140; Ross, A. B., Vuong, L. T., and Ruckle, J. (2011). *J. Nutr.* 93, 1263.

145. Leung, I. Y., Sandstrom, M. M., Zucker, C. L., *et al.* (2004). *Invest. Ophthalmol. Vis. Sci.* 45, 3244.

Metabolic Effects

Lycopene

Lycopene is the most potent carotenoid antioxidant *in vitro*, being twice as effective (due to its extended conjugated diene system) as β-carotene in quenching singlet oxygen, and 10 times as effective as α-tocopherol. Whether this activity is the basis of its beneficial health effects is not clear, as tissue levels of lycopene tend to be much lower than those other antioxidants. It is likely that such effects are mediated by lycopene metabolites.

Lycopene and/or its metabolites (e.g., apo-10-lycopenoid acid), and many other carotenoids, have been shown to induce several enzymes, including superoxide dismutase, glutathione-S-transferase, quinone reductase, and several phase I and II enzymes. These responses appear to involve upregulation of **antioxidant response elements** (**ARE**).

Studies have found lycopene to have a very low affinity for the RXRα (*see* Chapter 5), and thus to be a very weak transactivator of genes using the retinoic acid response element (RARE). The metabolite apo-10′-lycopenoid acid, however, has been shown to induce RARβ mRNA expression in some cell lines.[146]

Lutein and Zeaxanthin

These macular carotenoids appear to protect the primate macula from the damaging effects of blue-wavelength photons, as they absorb in the range of 420-480 nm, reducing by as much as 90% the incoming energy in this range from reaching macular photoreceptors. They accumulate selectively in the most vulnerable domains of membranes, i.e., those containing unsaturated phospholipids. Lutein and zeaxanthin can also scavenge reactive oxygen species formed in photoreceptors, thus improving visual acuity. That light-catalyzed oxidative reactions occur in the retina is indicated by the accumulation of the autofluorescent pigment **lipofuscin**[147] in that tissue.

Health Effects

Lycopene

Several health benefits have been attributed to lycopene.

- *Anticarcinogenesis.* Epidemiological investigations have shown consumption of tomato-rich diets to be associated with reduced risk of cancers of the prostate and lung.[148] A meta-analysis of 11 case–control studies and 10 cohort or nested case–control studies found prostate cancer risk to be inversely related to serum concentration of lycopene,[149] suggesting risk reduction of 25–30%. The same analysis, however, detected no significant associations of disease risk and the frequency of consumption of tomatoes, tomato products, or food lycopene, although some studies in which tomato consumption was high did report protective effects. The few intervention studies conducted to date have found lycopene supplementation to reduce cancer biomarkers and to reduce the disease progression in patients with benign prostatic hyperplasia.[150] Otherwise, few clinical studies have had sufficiently robust designs (randomized, blinded, controlled, of sufficient duration and follow-up) to yield clear determinations of anticarcinogenic efficacy.

  The relatively few studies with animal tumor models have shown lycopene treatment to affect molecular mechanisms associated with antitumorigenesis, e.g., reduced cell proliferation, increased apoptosis, reduced markers of oxidative stress. Two studies have found lycopene to reduce the incidence of spontaneous or chemically induced mammary cancers in the rat, and of chemically induced lung cancers in male (but not female) mice. Lycopene has been found to cause cell cycle arrest and apoptosis in cultured prostate cancer cells, which effect is associated with upregulation of the expression of intercellular gap junction communication associated with decreased cell proliferation.[151]

- *Skin health.* Epidemiological studies have pointed to a protective effect of tomato-rich diets against ultraviolet (UV) damage to the skin. A randomized trial found that consumption of lycopene-rich tomato paste conferred significant protection against UV damage in a small cohort of healthy women, increasing the median erythemal dose by 16%.[152]

Lutein and Zeaxanthin

The xanthophylls lutein and zeaxanthin have also been proposed as having roles in supporting visual development

146. Lian, F., Smith, D. E., Ernst, H., *et al.* (2007). *Carcinogenesis* 28, 1567.
147. This appears to result from the condensation of two molecules of *all*-trans-retinal with one of phosphatidylethanolamine, which complex is taken up by the retinal pigment epithelium and converted to a stable pyridnium bisretinoid that is cytotoxic, and causes apoptosis and, hence, macular degeneration.
148. Giovannucci, E. (2002). *Exp. Biol. Med.* 227, 852–859; Giovannucci, E. (2005). *J. Nutr.* 135(Suppl.), 2030S; Arab, L., Steck-Scott, S., Fleishaur, A. T. (2002). *Exp. Biol. Med.* 227, 894.
149. Etminan, M., Takkouche, B, and Caamano-Isorna, F. (2004). *Cancer Epidemiol. Biomarkers Prev.* 13, 340.
150. Schwarz, S., Obermüller-Jevec, U. C., HJellmis, E., *et al.* (2008). *J. Nutr.* 138, 49.
151. That is, connexin 43; Heber, D. and Lu, Q. Y. (2002). *Exp. Biol Med.* 227, 920.
152. Rizwan, M., Rodriguez-Blanco, I., Harbottle. A., *et al.* (2010). *Br. J. Dermatol.* 154, 154.

and protecting against visual fatigue, cataracts,[153] and age-related macular degeneration (a leading cause of severe vision loss in industrialized countries).[154] The Age-Related Eye Disease Study found that individuals in the highest quintile of lutein and zeaxanthin consumption had risk reductions of 27–55% for various signs of macular degeneration.[155] A randomized trial using lutein plus antioxidants as the intervention agent in patients with progressive atrophic macular degeneration reported improvements in several aspects of visual acuity;[156] however, a more recent trial detected no protective effect of a similar treatment.[157] Accordingly, data are presently considered inconclusive as to whether supplemental xanthophylls may prevent age-related macular degeneration.

Safety

Both lycopene and lutein are generally regarded as safe. A risk assessment noted Oberved Safe Levels of 75 mg/day for lycopene and 20 mg/day for lutein, but noted that higher levels have been tested without adverse effects.[158]

9. OROTIC ACID

Orotic acid was isolated in the late 1940s from distillers' dried solubles.[159] For a while, it was called "vitamin B$_{13}$." When studies failed to confirm vitamin activity, that designation was dropped. It is a normal metabolite.

Chemical Nature

Orotic acid is a substituted pyrimidine: 1,2,3,6-tetrahydro-2,6-dioxo-4-pyrimidinecarboxylic acid (Fig. 18.13).

Dietary Sources

The most important dietary sources of orotic acid are milk, milk products, and root vegetables (beets, carrots). Notably, human milk lacks orotic acid.

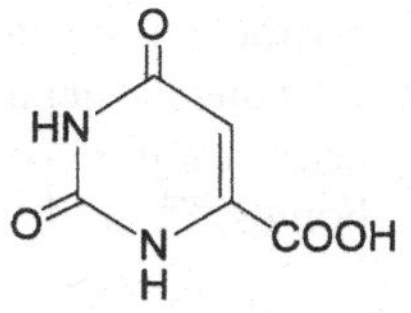

FIGURE 18.13 Chemical structure of orotic acid.

Metabolism

Orotic acid is an intermediate in the biosynthesis of **pyrimidines** (UTP, CTP, TTP). It is synthesized in the mitochondria from N-carbamylphosphate by dehydration (via dihydroorotase) and oxidation via orotate reductase to orotate, which is subsequently converted to UMP. An inborn error in UMP synthase, which catalyzes the last step, is characterized by orotic acid accumulating in the plasma and appearing in increased amounts in the urine. Orotic acid excretion is also increased in disorders of the urea cycle, and has been proposed as an indicator of arginine depletion. It has been found to be increased in cases of subclinical mastitis in dairy cows.

Health Effects

Orotic acid supplementation of experimental animals has been found to increase hepatic levels of uracil nucleotides, presumably by increasing the flux through the pyrimidine pathway. Orotate treatment has been shown to be more effective than uracil in stimulating adaptive growth of the rat jejunum after massive small bowel resection.[160]

Magnesium orotate has been used to improve ventricular function and exercise tolerance in cardiac patients;[161] however, these benefits would appear to be due to correction of magnesium depletion, and not to the orotate moiety *per se*.

Safety

Orotic acid supplements (0.1%) to the diets of rats have been found to reduce the conversion of tryptophan to niacin,[162] and to induce hepatic steatosis and hepatomegaly.[163] The latter effects were associated with increases in the sterol regulatory element binding protein-1c, the target gene for which is involved in fatty acid synthesis. Orotic

153. Olmedilla, B., Granado, F., Blanco, I., et al. (2003). J. Nutr. 19, 21.
154. With an estimate prevalence of nearly 1.5%, age-related macular degeneration affects some 1.75 million Americans.
155. AREDS (2007). Arch. Ophthamol. 125, 1225
156. Richer, S., Stiles, W., Statkute, L., et al. (2004). Optometry 75, 216.
157. Bartlett, H. E. and Eperjesi, F. (2007). Eur. J. Clin. Nutr. 61, 1121.
158. Shao, A. and Hathcock, J. N. (2006). Reg. Toxicol. Pharmacol. 45, 289.
159. This feedstuff consists of the dried aqueous residue from the distillation of fermented corn. It is used mainly as a component of mixed diets for poultry, swine, and dairy calves. It is rich in several B vitamins and has been valued as a source of unidentified growth factors, particularly for growing chicks and turkey poults.
160. Evans, M. E., Tian, J., Gu, L. H., et al. (2005). J. Parent. Nutr. 29, 315.
161. Classen, H. G. (2004). Rom. J. Intern. Med. 42, 491; Stepura, O. B. and Martynow, A. I. (2009). Intl J. Cardiol. 134, 145.
162. Fuluwatari, T., Morikawa, Y., Sugimoto, E., et al. (2002). Biosci. Biotechnol. Biochem. 66, 1196.
163. Wang, Y. M., Hu, X. Q., Xue, Y., et al. (2011). J. Nutr. 27, 571.

acid supplementation of the rat appeared to constitute an oxidative stress, as it reduced hepatic superoxide dismutase activity and increased the contents of conjugated dienes and protein carbonyls.[164]

10. p-AMINOBENZOIC ACID

Precursor to Folate in Bacteria

p-Aminobenzoic acid is an essential growth factor for many species of bacteria, which use it for the biosynthesis of folate. Some early studies showed responses of chicks (increased growth) and rats (enhanced lactation) to p-aminobenzoic acid-supplemented diets containing marginal concentrations of folate. In fact, for a time p-aminobenzoic acid was called "**vitamin B_x**." Such responses, however, were shown to be due to the use by the intestinal microflora of p-aminobenzoic acid for the synthesis of folate made available to the host either directly in the gut or indirectly via the feces. Because animals lack the enzymes of the folate synthetic pathway, they cannot convert this bacterial precursor to the actual vitamin.

Chemical Nature

p-Aminobenzoic acid (4-aminobenzoic acid or **PABA**) is comprised of a benzene ring substituted with an amino group and a carboxyl group (Fig. 18.14).

Dietary Sources

p-Aminobenzoic acid is found in significant amounts in liver, kidney, spinach, whole grains, mushrooms, yeasts, and molasses.

Metabolism

Because its metabolism is limited, p-aminobenzoic acid can be used in clinical situations as a marker of the completeness of 24-hour urine collections, 70–85% appearing in the urine after a single oral dose.[165] Its glutamyl ester, p-aminobenzoylglutamate, is the primary catabolite of folate, produced by cleavage of the C9–N10 bond (see Chapter 16). Excreted in the urine as p-acetamidobenzoylglutamate, urinary levels reflect the total body folate pool, and can thus serve as an indicator of long-term folate status.[166]

FIGURE 18.14 Chemical structure of p-aminobenzoic acid.

Health Effects

p-Aminobenzoic acid can antagonize the bacteriostatic effects of sulfonamide drugs owing to similarities in chemical structure. It also has a very high absorbance in the ultraviolet (UV) range, and is used as a UV-screening agent in sun-blocking preparations. In a randomized controlled trial,[167] it was found to inhibit fibroblast proliferation and to promote the reduction of fibrous plaques and penile deformation in patients with Peyronie's disease.[168]

11. LIPOIC ACID

Lipoic acid, also known as **thiotic acid**, was first isolated from bovine liver in 1950. It was found to be reversibly oxidized/reduced to form the redox pair, lipoic acid/dihydrolipoic acid. Unlike the other major intracellular thiol, glutathione, the reduced form of which has antioxidant activity, both oxidized and reduced forms of lipoic acid are antioxidants.

Chemistry

Lipoic acid (1,2-dithionlane-3-pentanoic acid) contains two vicinal sulfur atoms that are subject to oxidation/reduction; the reduced form, dihydrolipoic acid, has two thiols (Fig. 18.15). Lipoic acid contains one chiral center,[169] and thus two possible optical isomers.

Dietary Sources

Lipoic acid is present in a wide variety of foods, generally at low levels. Best sources are tissues rich in mitochondria (e.g., heart, kidney) or chloroplasts (i.e., dark green leafy vegetables such as spinach).[170] In these foods, most of the factor is covalently bound to lysyl residues in proteins.

164. Morifuji, M. and Aoyama, Y. (2002). J. Nutr. Biochem. 13, 403.
165. Jakobsen, J., Pedersen, A. N., and Oversen, L. (2003). Eur. J. Clin. Nutr. 57, 138.
166. Wolfe, J. M., Bailey, L. B., Herrlinger-Garcia, K., et al. (2003). Am. J. Clin. Nutr. 77, 919–923.

167. Weidner, W., Hauck, E. W., Schnitker, J., et al. (2005). Eur. Urol. 47, 530.
168. Peyronie's disease is a benign condition of the penis characterized by the formation of fibrous plaques within the tunica albuginea, usually resulting in penile deformity with some degree of erectile dysfunction.
169. That is, the lipoic acid molecule has a carbon atom to which four different moieties are bound, for which reason the molecule lacks an internal plane of symmetry.
170. Lodge, J. K., Youn, H. D., Handelman, G. J., et al. (1997). J. Appl. Nutr. 49, 3–11.

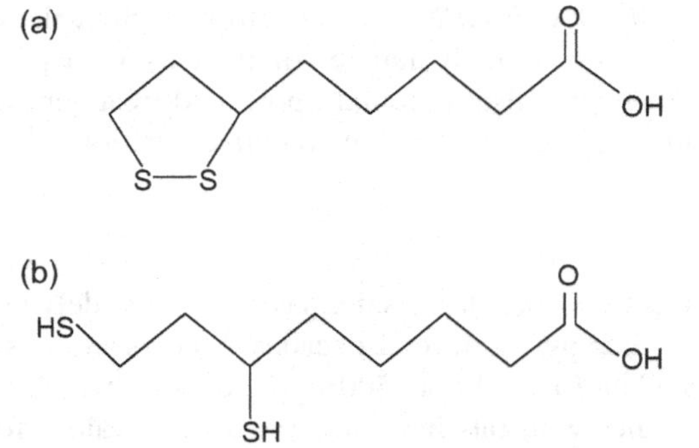

FIGURE 18.15 Chemical structures of (a) lipoic acid and (b) dihydro-lipoic acid.

Because only the R-enantiomer is produced in biological systems, that is the form found in foods; chemically synthesized sources (e.g., commercial supplements) consist of equimolar mixtures of both the R- and S-enantiomers.

Absorption and Transport

A clinical trial found R-lipoic acid to be much better (40–50%) absorbed than the *S*-enantiomer;[171] however, the bioavailability of food forms of lipoyllysine remains unexamined. It is likely to be relatively low, as studies with free R-lipoic acid have indicated efficiencies of enteric absorption of 20–38%. The enteric absorption of lipoic acid is thought to be facilitated by the monocarboxylate transporter and/or Na^+-dependent multivitamin transporter. These proteins may also be involved in facilitating the cellular uptake of the factor in tissues in which it transiently accumulates.

Metabolism

Lipoic acid is synthesized in the mitochondria from octanoic acid, and is rapidly reduced to dihydrolipoic acid. Absorbed lipoic acid appears quickly in the plasma, from which it also quickly leaves, indicating uptake by tissues, metabolism, and removal across the kidney. Lipoic acid is catabolized by extensive β-oxidation of the carbon backbone, resulting in at least a dozen apparently inactive metabolites.[172]

Metabolic Functions

Coenzyme

Lipoic acid is an essential cofactor for the oxidative decarboxylations of α-keto acids where, linked to the ε-amino

group of a lysine residue of the enzyme dihydrolipoyl transacetylase, it is one of several prosthetic groups in the multi-enzyme, lactate dehydrogenase complex. In that catalysis, the amide form, **lipoamide**, undergoes reversible acylation/deacylation to transfer acyl groups to CoA, as well as reversible redox ring opening/closing, which is coupled with the oxidation of the α-keto acid.[173] Lipoic acid thus occupies a critical position in energy metabolism, regulating the flow of carbon into the tricarboxylic acid cycle.

Antioxidant

Lipoic acid is involved in cellular antioxidant protection through direct and indirect functions.

- *Intracellular antioxidant function.* The ability to undergo interconversion between oxidized disulfide (lipoic acid) and the reduced sulfhydryl (dihydrolipoic acid) forms allows this metabolite to function as a metabolic antioxidant, quenching reactive oxygen and nitrogen species, and chelating pro-oxidant metal ions.
- *Transcriptional modulator.* Lipoic acid promotes glutathione synthesis by increasing the nuclear transcription factor Nrf2, which induces expression of catalytic and regulatory subunits of γ-glutamylcysteine ligase, the rate-limiting enzyme in glutathione synthesis.[174]

Other Modulating Effects

- *Anti-inflammatory.* Lipoic acid has been shown to impede the *in vitro* activation of the nuclear transcription factor NFκB by the pro-inflammatory cytokine TNFα, which would be expected to have an anti-inflammatory effect *in vivo*.
- *Insulin signaling.* Lipoic acid has been shown to interact with the insulin signaling pathway by stimulating the insulin receptor, and inducing the phosphorylation of Akt and the activation of phosphoinositide-3 kinase. It has been proposed that lipoic acid may also participate in the recruitment of the glucose transporter GLUT4 from its Golgi storage site in muscle.
- *Phosphatases.* Lipoic acid has been shown to inactivate protein-tyrosine phosphatases, which are otherwise stimulated by cysteine oxidation.

Conditions of Need

Some species of bacteria (e.g., *Streptococcus fecalis*, *Lactobacillus casei*) and protozoa (e.g., *Tetrahymena geleii*) have clear needs for exogenous sources of lipoic

171. Breitkamp-Grogler, K., Niebh, G., Schneider, E., *et al.* (1999). *Eur. J. Pharm. Sci.* 8, 57.
172. The dominant metabolites include bisnorlipoate, tetranorlipoate, β-hydroxy-bisnorlipoate and the corresponding bis-methylated mercapto derivatives.

173. That is, the conversion of pyruvate to acetyl-CoA, and α-ketoglutarate to succinyl-CoA.
174. Suh, J. H., Shenvi, S. V., Dixon, B. M., *et al.* (2004). *Proc. Natl Acad. Sci. USA* 101, 3381.

acid. However, no deficiency signs have been reported, and lipoic acid supplements to chicks, rats, and turkey poults fed low-lipoic acid purified diets have been without effect. While the lipoic acid, biosynthetic capacity of humans would appear adequate, there is evidence that it declines with age and in individuals with compromised health.

Health Effects

That lipoic acid can function as an antioxidant suggests that its metabolic effects, like those of other antioxidants, may be related to general antioxidant status and pro-oxidant "tone." Accordingly, it has been proposed that lipoic acid may have value in the prevention and/or treatment of conditions associated with oxidative stress.

Diabetes

Studies with experimental animals have shown that supplemental lipoic acid can reduce lipid peroxidation induced by exercise, increase the rate of glucose disposal, reduce cataract formation, and improve motor neuron conductivity in models of diabetes. Clinical trials with type 2 diabetic subjects have found lipoic acid treatment to increase glucose clearance,[175] and to reduce blood glucose levels and lipid peroxidation products.[176] A meta-analysis of four clinical trials concluded that lipoic acid treatment (600 mg/day) significantly improved diabetic polyneuropathies of the feet and lower limbs.[177]

Cardiovascular Health

A small study found oral lipoic acid (300 mg/day for 4 weeks) to improve endothelial-dependent flow-mediated vasodilation in subjects with metabolic syndrome.[178] Lipoic acid treatment has yielded inconsistent effects on blood pressure.

Other Effects

- *Multiple sclerosis.* A trial with patients indicated that lipoic acid may be useful in treatment by reducing serum matrix metalloproteinase-9, and reducing T-cell migration into the central nervous system.[179]

- *Alzheimer's disease.* An uncontrolled clinical experiment with a small number of patients with probable Alzheimer's disease found lipoic acid treatment apparently to stabilize declining cognitive function.[180]

Safety

Lipoic acid is considered safe, having been widely used in clinical therapy for several decades. Studies in dogs have indicated an LD_{50} of 400–500 mg/kg of body weight; however studies with rats have suggested LD_{50} values four to five times that range.[181] At very high doses, gastrointestinal signs (nausea, abdominal pain, vomiting, diarrhea), allergic skin reactions, and malodorous urine have been reported.

12. INEFFECTIVE FACTORS

The term "vitamin" has been used from time to time in association with factors with no apparent metabolic activity that would justify that designation. Further, opportunities afforded by patent law permitted a substance to be trade-named a vitamin in the 1940s – a situation bound to mislead the uninformed.

"Vitamin B$_{15}$"

A substance isolated from apricot kernels and other natural sources was trade-named "vitamin B$_{15}$" and was also called "pangamic acid" by its discoverers;[182] it was patented with a claim of therapeutic efficacy against a wide range of diseases of the skin, respiratory tract, nerves, and joints – despite the fact that no data were presented in the patent application. Whereas pangamic acid was originally described as *d*-gluconodimethylamino acetic acid (the ester of *d*-gluconic acid and dimethylglycine), the term now appears to be used indiscriminately; products also containing *N,N*-diisopropylammonium dichloroacetate, sodium gluconolactone, *N,N*-dimethylglycine, calcium gluconate, and/or glycine in various proportions also go by the names "pangamic acid," "pangamate," "aangamik 15," and "vitamin B$_{15}$." Thus, vitamin B$_{15}$ is not a definable chemical entity; in fact, the substance originally called pangamic acid has frequently been absent from such preparations. Of the compounds likely to be present, *N,N*-diisopropylammonium dichloroacetate is known to be hypotensive, hypothermic, and potentially toxic, and dichloroacetate is a weak mutagen.

175. Jacob, S., Henricksen, E. J., Tritschler, H. J., *et al.* (1996). *Exp. Clin. Endocrinol. Diabetes* 104, 284.
176. Ziegler, D., Hanefeld, M., Ruhnau, K. J., *et al.* (1995). *Diabetologia* 38, 1425.
177. Ziegler, D., Nowack, H., Kempler, P., *et al.* (2004). *Diabet. Med.* 21, 114.
178. Sola, S., Mir, M. Q., Cheema, F. A., *et al.* (2005). *Circulation* 111, 343.
179. Yadav, V., Marracci, G., Lovera, J., *et al.* (2005). *Mult. Scler.* 11, 159.
180. Hager, K., Marahens, A., Kenklies, M., *et al.* (2001). *Arch. Gerontol. Geriatr.* 32, 275.
181. Cremer, D. R., Rabeler, R., and Roberts, A. (2006). *Regul. Toxicol. Pharmacol.* 46, 29.
182. Ernst Krebs Sr and Ernst Krebs Jr – not to be confused with the great biochemist Sir Hans Krebs.

Despite intermittent popular interest in these preparations,[183] the only information that would appear to support positive clinical results comes from anecdotal sources and the undocumented claims by vendors of these preparations. No substantive data appear to have ever been presented to support any beneficial biological effects of the so-called vitamin B_{15}, and there is no evidence that deprivation of the factor(s) causes any physiological impairment.

Gerovital

As its name implies, Gerovital was promoted as a nutritional substance that alleviated age-related degenerative diseases. Also called "vitamin H_3" and "vitamin GH_3," it is actually a buffered solution of procaine HCl, the dental anesthetic novocaine.[184] Its health claims are unsubstantiated.

Laetrile

Sometimes called "vitamin B_{17}," laetrile was first isolated from apricot kernels. It is a discrete chemical entity, the **cyanogenic glycoside** 1-mandelonitrile-β-glucuronic acid. The term *laetrile* is often used interchangeably with the related cyanogenic glycoside amygdalin, which occurs naturally in the seeds of most fruits. These compounds have been claimed to be effective in cancer treatment, owing either to the cyanide they provide as being selectively toxic to tumor cells, or to the disease itself being a result of an unsatisfied metabolic need for laetrile. The contention that either compound provides a source of cyanide ignores the fact that animals, lacking β-glucosidases, cannot degrade the mandelonitrile moiety. In fact, laetrile and amygdalin are each non-toxic to animals and humans for that reason. Apricot kernels and other fruit seeds, however, contain β-glucosidases; if these are liberated (e.g., by crushing) before eating such seeds, then cyanide poisoning can occur.[185] Extensive animal tumor model studies and clinical trials have tested the putative anti-tumor effects of laetrile; these have yielded consistently negative results.

13. UNIDENTIFIED GROWTH FACTORS

Since the discovery of vitamin B_{12}, experimental nutritionists have observed many instances of beneficial effects, particularly stimulated growth, of natural materials added

TABLE 18.9 Sources of UGF Activity for Poultry

Condensed Fish Solubles
Fish meal
Dried whey
Brewer's dried grains
Brewer's dried yeast
Corn distillers' dried soluble
Other fermentation residues

to purified diets. Many such responses have been found to involve interrelationships of known nutrients.[186] Some have involved diet palatability, and, thus, the rate of food intake of experimental animals. Some have resulted in the discovery of new essential nutrients, e.g., selenium. Other responses remain to be understood. For young monogastrics (particularly poultry), several feedstuffs are popularly regarded as having "**unidentified growth factor**" (UGF) activity (Table 18.9).

Elucidating the nature of these UGFs has been complicated by the fact that the growth responses are small and often poorly reproducible. This suggests that other interacting effects of environment, gut microfloral, diet, etc., may be involved. As has happened in the past, perhaps the next vitamins will be discovered through studies of these or other UGFs.

14. QUESTIONS OF SEMANTICS

It has been estimated that more than 25,000 natural components of foods are biologically active. This includes the 13 families of vitamins, and some two dozen amino acids, fatty acids, and mineral elements known to be indispensible in diets. Research is making it increasingly clear that the classical vocabulary of Nutrition may not be well suited for several instances in which certain components of foods have clear health benefits and simple descriptors are lacking. There are three general cases for which present Nutrition vocabulary is lacking:

- *Need for a dietary source of a factor biosynthesized by most others.* What is the proper descriptor for the need for carnitine by an individual with an OCTN2 deficit?
- *Use of a dietary factor that renders health benefits in non-specific ways.* What are the proper descriptors for flavonoids for supporting colon health and/or reducing cardiovascular disease risk? For lycopene for reducing

---

183. For example, in its March 13, 1978, cover story, *New York Magazine* presented vitamin B_{15} as a possible cure for everything short of a transit strike.

184. Procain hydrochloride, i.e., 4-aminobenzoic acid 2-(diethylamino)-ethyl ester hydrochloride.

185. Normally, cyanide poisoning is not a problem for animals that eat apricot or peach kernels, as those seeds generally pass intact through the gut.

186. An example is the enhancement, by the natural chelating activity in corn distillers' dried solubles, of the utilization of zinc by chicks fed a soybean meal-based diet.

cancer risk? For xanthophylls for reducing risk of age-related macular degeneration?

- *Use of a non-absorbed dietary factor that renders health benefits.* What is the proper adjective for the recommended use of fiber (i.e., fermentable complex carbohydrates) or polyphenols for supporting colon health?

Study Questions and Exercises

1. List the questions that must be answered in determining the eligibility of a substance for vitamin status.
2. For each of the substances discussed in this chapter, list the available information that would support its designation as a vitamin, and that which would refute such a designation.
3. Outline the general approaches one would need to take in order to characterize the UGF activity of a natural material such as fish meal for the chick.
4. Prepare a concept map of the relationships of micronutrients and physiological function, including the specific relationships of the traditional vitamins, the quasi-vitamins, and ineffective factors.

RECOMMENDED READING

Carnitine

Arduini, A., Bonomini, M., Savica, V., et al. 2008. Carnitine in metabolic disease: Potential for pharmacological intervention. Pharmacol. Therapeut. 120, 149–156.

Calabrese, V., Stella, A.M.G., Calvani, M., Allan, D., 2005. Acetylcholine and cellular response: Roles in nutritional redox homeostasis and regulation of longevity genes. J. Nutr. Biochem. 17, 73–88.

Ferrari, R., Merli, E., Cicchitelli, G., et al. 2004. Therapeutic effects of L-carnitine and propionyl-L-carnitine on cardiovascular diseases: A review. Ann. NY Acad. Sci. 1033, 79–91.

Hathcock, J.N., Shao, A., 2006. Risk assessment for carnitine. Reg. Toxicol. Pharmacol. 46, 23–28.

Jones, L.L., McDonald, D.A., Borum, P.R., 2010. Acylcarnitines: Role in brain. Prog. Lipid Res. 49, 61–75.

Mingorance, C., Rodriguez-Rodriguez, R., Justo, M.L., et al. 2011. Pharmacological effects and clinical applications of propionyl-L-carnitine. Nutr. Rev. 69, 279–290.

Mingrove, G., 2004. Carnitine in type 2 diabetes. Ann. NY Acad. Sci. 1033, 99–107.

Natecz, K.A., Miecz, D., Berezowski, V., Cecchelli, R., 2004. Carnitine: Transport and physiological functions in the brain. Mol. Aspects Med. 25, 551–567.

Rebouche, G.J., 2006. Carnitine. In: Bowman, B.A., Russell, R.M. (Eds.), Present Knowledge in Nutrition, ninth ed. ILSI Press, Washington, DC, pp. 340–351.

Stanley, C.A., 2004. Carnitine deficiency disorders in children. Ann. NY Acad. Sci. 1033, 42–51.

Strijbis, K., Vaz, F.M., Distel, B., 2010. Enzymology of the carnitine biosynthesis pathway. IUBMB Life 62, 357–362.

Wolf, G., 2006. The discovery of a vitamin role for carnitine: The first 50 years. J. Nutr. 136, 2131–2134.

Zammit, V.A., Ramsay, R.R., Bonomini, M., Arduini, A., 2009. Carnitine, mitochondrial function and therapy. Adv. Drug Deliv. Rev. 61, 1353–1362.

Choline

Garrow, T.A., 2007. Choline. In: Zempleni, J., Rucker, R.B., McCormick, D.B., Suttie, J.W. (Eds.), Handbook of Vitamins, fourth ed. CRC Press, Washington, DC, pp. 459–487.

Glunde, K., Ackerstaff, E., Mori, N., et al. 2006. Choline phospholipid metabolism in cancer: Consequences for molecular pharmaceutical interventions. Molec. Pharmaceut. 3, 496–506.

Newborne, P.M., 2002. Choline deficiency associated with diethaniolamine carcinogenicity. Toxicol. Sci. 67, 1–3.

Ueland, P.M., 2011. Choline and betaine in health and disease. J. Inherit. Metab. Dis. 34, 3–15.

Zeisel, S.H., 2011. Nutritional genomics: Defining the dietary requirement and effects of choline. J. Nutr. 141, 531–534.

Zeisel, S.H., da Costa, K.A., 2009. Choline: An essential nutrient for public health. Nutr. Rev. 67, 615–623.

Flavonoids

Butt, M.S., Sultan, M.T., 2009. Green tea: Nature's defense against malignancies. Crit. Rev. Food Sci. Nutr. 49, 463–473.

Cederroth, C.R., Nef, S., 2009. Soy, phyoestrogens and metabolism: A review. Molec. Cell. Endocrinol. 304, 30–42.

Crozier, A., del Rio, D., Clifford, M.N., 2010. Bioavailability of dietary flavonoids and phenolic compounds. Molec. Aspects Med. 31, 446–467.

de Souza, P.L., Russel, P.J., Kearsley, J.H., Howes, L.G., 2010. Clinical pharmacology of isoflavone and its relevance for potential prevention of prostate cancer. Nutr. Rev. 68, 542–555.

del Rio, D., Borges, G., Crozier, A., 2010. Berry flavonoids and phenolics: Bioavailability and evidence of protective effects. Br. J. Nutr. 104, S67–S90.

Galleano, M., Oteiza, P.I., Fraga, C.G., 2009. Cocoa, chocolate and cardiovascular disease. J. Cardiovasc. Pharmacol. 54, 484–490.

García-Lafuente, A., Guillamón, E., Villares, A., et al. 2009. Flavonoids as anti-inflammatory agents: Implications in cancer and cardiovascular disease. Inflamm. Res. 58, 537–552.

González-Gellego, J., García-Mediavilla, M.V., Sánchez-Campos, S., Tunón, M.J., 2010. Fruit polyphenols, immunity and inflammation. Br. J. Nutr. 104, S15–S27.

Hodgson, J.M., Croft, K.D., 2010. Tea flavonoids and cardiovascular health. Molec. Aspects Med. 31, 495–502.

Huntley, A.L., 2009. The health benefits of berry flavonoids for menopausal women: Cardiovascular disease, cancer and cognition. Maturitas 63, 297–301.

Ørgaard, A., Jensen, L., 2008. The effects of soy isoflavones on obesity. Exp. Biol. Med. 233, 1066–1080.

Poulsen, R.C., Kruger, M.C., 2008. Soy phytoestrogens: Impact on postmenopausal bone loss and mechanisms of action. Nutr. Rev. 66, 359–374.

Prasain, J.K., Carlson, S.H., Wyss, J.M., 2010. Flavonoids and age related disease: Risk, benefits and critical windows. Maturitas 66, 163–171.

Rastmanesh, R., 2010. High polyphenol, low probiotic diet for weight loss because of intestinal microbiota interaction. Chem. Biol. Interact. 189, 1–8.

Rice-Evans, C.A., Packer, L. (Eds.), 2003. Flavonoids in Health and Disease. Marcel Dekker, New York, NY, 467 pp.

Romier, B., Schneider, Y.J., Larondell, Y., During, A., 2009. Dietary polyphenols can modulate the intestinal inflammatory response. Nutr. Rev. 67, 363–378.

Ross, J.A., Kasum, C.M., 2002. Dietary flavonoids: Bioavailability, metabolic effects, and safety. Annu. Rev. Nutr. 22, 19–34.

Spencer, J.P.E., 2010. The impact of fruit flavonoids on memory and cognition. Br. J. Nutr. 104 (S40–S47)

Trzeciakiewicz, A., Habauzit, V., Horcajada, M.N., 2009. When nutrition interacts with osteoblast function: Molecular mechanisms of polyphenols. Nutr. Res. Rev. 22, 68.

Vargas, A.J., Burd, R., 2010. Hormesis and synergy: Pathways and mechanisms of quercetin in cancer prevention and management. Nutr. Rev. 68, 418–428.

myo-Inositol

Bennett, M., Onnebo, S.M.N., Azevedo, C., Saiardi, A., 2006. Inosital pyrophosphates: Metabolism and signaling. Cell Mol. Life Sci. 63, 552–564.

Berridge, M.J., 2009. Inositol triphosphate and calcium signaling mechanisms. Biochim. Biophys. Acta 1793, 933–940.

Mitchell, R.H., 2008. Inositol derivatives: Evolution and functions. Nature Rev. Molec. Cell Biol. 9, 151–161.

Shears, S.B., 2009. Molecular basis for the integration of inositol phosphate signaling pathways via human ITPK1. Adv. Enzyme Reg. 49, 87–96.

York, J.D., 2006. Regulation of nuclear processes by inositol polyphosphates. Biochim. Biophys. Acta 1761, 552–559.

Lipoic Acid

Bilska, A., Wlodek, L., 2005. Lipoic acid – the drug of the future. Pharmacol. Rep. 57, 570–577.

Maczurek, A., Hager, K., Kenklies, M., et al. 2008. Lipoic acid as an anti-inflammatory and neuroprotective treatment for Alzheimer's disease. Adv. Drug Deliv. Rev. 60, 1463–1470.

Shay, K.P., Moreau, R.F., Smith, E.J., et al. 2009. Alpha-lipoic acid as a dietary supplement: Molecular mechanisms and therapeutic potential. Biochem. Biophys. Acta 1790, 1149–1160.

Singh, U., Jialal, I., 2008. Alpha-lipoic acid supplementation and diabetes. Nutr. Rev. 66, 646–657.

Non-Provitamin A Carotenoids

Carpentier, S., Knaus, M., Suh, M., 2009. Associations between lutein, zeaxanthin, and age-related macular degeneration: An overview. Crit. Rev. Food Sci. Nutr. 49, 313–326.

Giovannucci, E., 2002. A review of epidemiologic studies of tomatoes, lycopene and prostate cancer. Proc. Soc. Exp. Biol. Med. 227, 852–859.

Hammond Jr, B.R., 2008. Possible role of dietary lutein and zeaxanthin in visual development. Nutr. Rev. 66, 695–702.

Krishnadev, N., Meleth, A.D., Chew., E.Y., 2010. Nutritonal supplements for age-related macular degeneration. Curr. Opin. Ophthalmol. 21, 184–189.

Lindshield, B.L., Canene-Adams, K., Erdman Jr, J.W., 2007. Lycopenoids: Are the lycopene metabolites bioactive? Arch. Biochem. Biophys 458, 136–140.

Loane, E., Nolan, J.M., O'Donovan, O., et al. 2008. Transport and retinal capture of lutein and zeaxanthin with reference to age-related macular degeneration. Surv. Ophthalmol. 53, 68–81.

Maiani, G., Castón, M.J.P., Catasta, G., et al. 2009. Carotenoids: Actual knowledge on food sources, intakes, stability and bioavailability and their protective role in humans. Mol. Nutr. Food Res. 53, S194–S218.

Mein, J.R., Lian, F., Wang, X.D., 2008. Biological activity of lycopene metabolites: Implications for cancer prevention. Nutr. Rev. 66, 667–683.

van Breeman, R.B., Pajkovic, N., 2008. Multitargeted therapy of cancer by lycopene. Cancer Lett. 269, 339–351.

Wong, I.Y., Koo, S.C.Y., Chan, C.W.N., 2011. Prevention of age–related macular degeneration. Int. Ophthalmol. 31, 73–82.

Orotic Acid

Brosnan, M.E., Brosnan, J.T., 2007. Orotic acid excretion and arginine metabolism. J. Nutr. 137, 1656S–1661S.

Salerno, C., Crifo, C., 2002. Diagnostic value of urinary orotic acid levels: Applicable separation methods. J. Chromatog. 781, 57–71.

Wang, Y.M., Hu, X.Q., Xue, Y., et al. 2011. Study of the possible mechanism of orotic acid-induced fatty liver in rats. Nutrition 27, 571–575.

Pyrroloquinoline Quinone

Killgore, J., Smidt, C., Duich, L., et al. 2009. Potential physiological importance of pyrroloquinoline quinone. Alt. Med. Rev. 14, 268–277.

Stites, T.E., Mitchell, A.E., Rucker, R.B., 2000. Physiological importance of quinoenzymes and the o-quinone family of cofactors. J. Nutr. 130, 719–727.

Ubiquinones

Bentinger, M., Brismar, K., Dallner, G., 2007. The antioxidant role of coenzyme Q. Mitochondrion 7S, S41–S50.

Bentinger, M., Tekle, M., Dallner, G., 2007. Coenzyme Q – biosynthesis and functions. Biochem. Biophys. Res. Commun. 396, 74–79.

Crane, F.L., 2007. Discovery of ubiquinone (Q_{10}) and an overview of function. Mitochondrion 7S, S2–S7.

Gille, L., Rosenau, T., Kozlov, A.V., Gregor, W., 2008. Ubiquinone and tocopherol: Dissimilar siblings. Biochem. Pharmacol. 76, 289–302.

Lenaz, G., Fato, R., Formiggini, G., Genova, M.L., 2007. The role of coenzyme Q in mitochondrial electron transport. Mitochondrion 7S, S8–S33.

Littarru, G.P., Langsjoen, P., 2007. Coenzyme Q10 and statins: Biochemical and clinical implications. Mitochondrion 7S, S168–S174.

Littarru, G.P., Tiano, L., 2010. Clinical aspects of coenzyme Q_{10}: An update. Nutrition 26, 250–254.

López-Lluch, G., Rodríguez-Aguilera, J.C., Santos-Ocana, C., Navas, P., 2010. Is coenzyme Q a key factor in aging? Mech. Ageing Dev. 131, 225–235.

Miles, M.V., 2007. The uptake and distribution of coenzyme Q(10). Mitochondrion 7S, S72–S77.

Pepe, S., Marasco, S.F., Haas, S.J., et al. 2007. Coenzyme Q_{10} in cardiovascular disease. Mitochondrion 7S, S154–S167.

Quinzil, C.M., Hirano, M., 2010. Coenzyme Q and mitochondrial disease. Dev. Disabil. Res. Rev. 16, 183–188.

Using Current Knowledge of the Vitamins

Sources of the Vitamins

Anchoring Concepts

1. Estimates of vitamin contents of many foods and feedstuffs are available.
2. For some vitamins, only a portion of the total present in certain foods or feedstuffs is biologically available.
3. The total vitamin intake of an individual is the sum of the amounts of bioavailable vitamins in the various foods, feedstuffs, and supplements consumed.

The intakes of vitamins into the body calculated from standard tables are rarely accurate.

J. Marks

Learning Objectives

1. To understand the sources of error in estimates of vitamin contents of foods and feedstuffs.
2. To understand the concept of a core food, and to know the core foods for each of the vitamins.
3. To understand the sources of potential losses of the vitamins from foods and feedstuffs.
4. To understand which of the vitamins are most likely to be in insufficient supply in the diets of humans and livestock.
5. To understand the means available for the supplementation of individual foods and total diets with vitamins.

VOCABULARY

Bioavailability
Biofortification
Core foods
Enrichment
Fortification
Golden rice
National Nutrient Database
Nutrition Labeling and Education Act (NLEA)
Revitaminization
Supplementation
Vitamin–mineral premix
Vitaminization

1. VITAMINS IN FOODS

Vitamin Content Data

Collation of best estimates of the nutrient composition of foods and feedstuffs has been an ongoing activity by several groups in the United States since the turn of the century.[1]

---

1. The formal compilation of food composition data was initiated by the US Department of Agriculture (USDA) food chemist W. O. Atwater in 1896. Since that time, developing information on the nutrient composition of foods has been an ongoing program of the USDA. The development of nutrient composition information for feedstuffs started in the United States after the turn of the century at several land grant colleges; in more recent times those activities have passed largely into the private sector, it being in the interest of corporate feed producers to have reliable data for contents in feedstuffs of those nutrients that most directly affect the cost of their formulations (e.g., metabolizable energy, protein, indispensable amino acids, calcium, phosphorus).

The Vitamins. DOI: 10.1016/B978-0-12-381980-2.00019-0

Nutrient composition data for foods and feedstuffs are now available in many forms (e.g., books, wall charts, tables, and appendices of books, computer tapes, and diskettes). However, most compilations derive from relatively few sources. This is particularly true for the nutrient composition data for foods. For US foods, almost all current versions are renditions of the USDA **National Nutrient Database**,[2] developed through an ongoing program of the US Department of Agriculture. Similar, but less extensive, databases have been developed for other countries,[3] and efforts are being made to standardize the collection, compilation, and reporting of food nutrient composition data on a worldwide basis.[4]

The nutrient composition of feedstuffs has, with few exceptions,[5] been developed less systematically and extensively. Data sets presently in the public domain have been compiled largely from original reports in the scientific literature. Therefore, effects of uncontrolled sampling, multiple and often old analytical methods, multiple analysts, unreported analytical precision, unreported sample variance, etc., are likely to be far greater for the nutrient databases for feedstuffs than for the corresponding databases for foods.

Use of any database for estimating vitamin intake is limited for reasons concerning the accuracy and completeness of the data. Although the United States Department of Agriculture (USDA) National Nutrient Database (Table 19.1) is much more complete than most feed tables with respect to data for vitamins, it is only reasonably complete with respect to thiamin, riboflavin, and niacin; data for vitamins D and E are relatively sparse. In addition, accurate and robust[6] analytical methods are available only for vitamin E, thiamin, riboflavin, niacin, and pyridoxine. For the other vitamins, this means that the quality of available analytical data can render them unacceptable for inclusion in the database, or be low enough to raise serious questions concerning reliability.

Core Foods for Vitamins

Foods are the most important sources of vitamins in the daily diets of humans. However, the vitamins are unevenly distributed among the various foods that comprise human diets (Table 19.2). Therefore, the evaluation of the degree of vitamin adequacy of total diets is served by knowing which foods are likely to contribute significantly to the total intake of each particular vitamin, by virtue of both the frequency and the amount of the food consumed, as well as the probable concentration of the vitamin in that food. Identifying such **core foods** is difficult, because both the voluntary intakes of foods by free-living people and the concentrations of vitamins in those foods are extremely difficult to estimate quantitatively with acceptable certainty. Nevertheless, attempts to do that have indicated a manageable number of core foods for each of the vitamins; for example, for Americans it has been estimated that 80% of the total intakes of several vitamins are provided by 50–200 foods.[7]

---

2. These data constitute the databases used for US national food consumption surveys. They are obtained from scientific publications, university and government laboratories, food processors and trade groups, and USDA-funded contracts for food analysis. The data are available as the *USDA Nutrient Database for Standard Reference,* in tables and public use tapes, and can be accessed via the Internet (http://www.nal.usda.gov/fnic/foodcomp/search). Other data relevant to the vitamin status and intakes of Americans can be found in other food databases maintained by the USDA and the US Food and Drug Administration (FDA):

- *USDA US Food Supply Series* – annual estimates of amounts of about 350 foods that disappear into civilian food consumption at or before retail distribution
- *USDA Nationwide Food Consumption Surveys* (1935–1998) – food used from home supplies during 1 week by entire households and foods ingested by individual household members for 3 consecutive days
- *What We Eat in America* (WWEIA) – represents the integration (2002) of two nationwide surveys: the USDA Continuing Survey of Food Intakes by Individuals (CSFII) and the US Department of Health and Human Services National Health and Nutrition Examination Survey (NHANES)
- *FDA Total Diet Studies* – conducted yearly to assess the levels of various nutritional elements, pesticide residues, and contaminants in the US food supplement, and in representative diets of selected age–sex groups.

3. For example, the composition databases for Europe (http://www.eurofir.net/urofir_knowledge/euorpean_databases), Latin America (http://www.fao.org/infoods/tabbles_latin_en.stm), Asia (http://www.fao.org/infoods/tables_asia_en.stm), Africa (http://www.fao.org/infoods/tables_africa_en.stm), and Canada (www.hc-sc.gc.ca/fn-an/nutrition/fiche-nutri-data/index-eng.php).

4. This is the purpose of the INFOODS project of the United Nations University Food and Nutrition Program (http://www.fao.org/infoods/).

5. The notable exception in the public domain was the program at the University of Maryland, which involved the ongoing analysis of feedstuffs commonly used in feeding poultry in the United States. That program focused on macronutrients. It was discontinued in the late 1970s; the last version of the data (i.e., *1979 Maryland Feed Composition Data*) was published as a supplement in the *Proceedings of the Maryland Nutrition Conference* in that year. Other widely used feed tables were derived in part from this source – e.g., Scott, M. L., Neshiem, M. C., and Young, R. J. (1982). *Nutrition of the Chicken*. M. L. Scott Association, Ithaca, NY, p. 482.

---

6. Accuracy is minimally influenced by such factors as unusual technical skill of the analyst, particular instrumental conditions, etc.

7. Using data from the 1976 Nationwide Food Consumption Survey and the Continuing Survey of Food Intakes by Individuals, USDA nutritionists estimated that, on a national basis, Americans obtained 80% of their total intakes of the following vitamins from the following numbers of foods: vitamin A, 60; vitamin E, 100; thiamin, 168; riboflavin, 165; niacin, 159; pyridoxine, 175; folate, 129; vitamin B_{12}, 58.

TABLE 19.1 Adequacy of the USDA National Nutrient Database Vitamin Contents of Foods

Food	A	D	E	K	C	Thiamin	Riboflavin	Niacin	B6	Pantothenic Acid	Folate	B12
Baby foods	■		•	□	■	■	■	■	•	□	•	•
Baked foods:												
Breads	□		•	□		■	■	■	■	•	•	□
Sweet goods	•		•			■	■	■	•	•	•	•
Cookies/crackers	•		•			■	■	■	■	•	•	•
Beverages	•			•	•	•	•	•	•	•	•	
Breakfast cereals	•	□	•	□	•	•	•	•	•	•	•	•
Candies	•		□		□	•	•	•	•	•	•	□
Cereal grains:												
Whole	•		•	•		■	■	■	■	■	■	
Flour			•	□		■	■	■	■	■	■	
Pasta			□	□		■	■	■	■	■	■	
Dairy products	■	■	•	•	■	■	■	■	■	■	•	■
Eggs, egg products	•	•	•	•	□	■	■	■	■	■	•	■
Fast foods	■	□	•	□	•	■	■	■	•	•	□	•
Fats and oils	•	•	•	•								
Fish, shellfish:												
Raw	•	•	•	□		•	•	•	•	•	•	•
Cooked	□	•	•	□		□	□	□	□	•	•	□
Fruits:												
Raw	■		□	•	■	■	■	■	■	•	•	
Cooked	•		□	□	•	•	•	•	•	•	•	
Frozen, canned	■		□	□	•	■	□	■	•	•	•	
Legumes:												
Raw	•		■	•		■	■	■	•	■	■	
Cooked	□		■	□		■	■	■	•	■	■	
Meat:												
Beef	■	•	•	•		■	■	■	■	•	•	■
Lamb	•	•	•	□		■	■	■	•	•	•	■
Pork	•	•	•	□		■	■	■	■	•	•	■
Sausage	•	•	□	□	■	■	■	■	■	•	•	■
Veal	•	•	•	□		■	■	■	■	•	■	■
Poultry	•	•	•	□		■	■	■	•	•	•	•
Nuts, seeds	•	•	•	□	•	•	•	•	•	•	□	
Snack foods	•	•	•	□	□	•	■	■	•	•	•	□
Soups	■		□	□	■	■	■	■	•	□	□	•

(Continued)

TABLE 19.1 (Continued)

Food	A	D	E	K	C	Thiamin	Riboflavin	Niacin	B$_6$	Pantothenic Acid	Folate	B$_{12}$
Vegetables:												
Raw	■		•	•	■	■	■	■	•	□		•
Cooked	•		•	□	•	•	•	•	•	□		•
Frozen	■		□	□	■	■	■	■	□	•		•
Canned	■		□	□	■	■	■	■	□	•		•

Key: □, *few or no data;* •, *inadequate data;* ■, *substantial data.*
Source: Beecher, G. and Matthews, R. (1990). Nutrient composition of foods. In *Present Knowledge in Nutrition* (Brown, M., ed.), 6th edn. International Life Science Institute – Nutritional Foundation, Washington, DC, pp. 430–439.

TABLE 19.2 Core Foods for the Vitamins

Vitamin A	Vitamin D	Vitamin E	Vitamin K	Vitamin C	Thiamin	Riboflavin
As retinol:	Milk[a]	Vegetable oils	Broccoli	Tomatoes	Meats	Eggs
Milk (breast, animal[a])	Ghee		Asparagus	Potatoes	Potatoes	Liver
Butter, ghee, margarine[a]	Margarine[a]		Lettuce	Pumpkins	Whole grains	Meats
Liver	Cheese		Cauliflower	Citrus fruits	Some fish	Some fish
Eggs	Chicken (skin)		Cabbage	Other fruits	Legumes	Asparagus
Small fish (eaten whole)	Liver		Brussels sprouts	Yams	Oilseeds	Milk
As carotene:	Fatty fish		Turnip greens	Cassava	Milk	Whole grains
Red palm oil	Cod liver oil		Liver	Milk	Eggs	Green leaves
Dark/medium-green leaves	Egg yolk		Spinach			Legumes
Yellow/orange vegetables						
Yellow/orange fruits						
Yellow maize						

Niacin	Vitamin B$_6$	Biotin	Pantothenic Acid	Folate	Vitamin B$_{12}$
Meats	Meats	Liver	Liver	Tomatoes	Liver
Eggs	Cabbage	Egg yolk	Milk	Beets	Fish
Fish	Potatoes	Cauliflower	Meats	Potatoes	Eggs
Whole grains	Liver	Kidney	Eggs	Wheatgerm	Milk
Legumes	Beans	Peanuts	Fish	Cabbage	
Milk	Whole grains	Soybeans	Whole grains	Eggs	
Liver	Peanuts	Wheatgerm	Legumes	Meats	
Peanuts	Soybeans	Oatmeal	Spinach	Peanuts	
	Some fish	Carrots	Asparagus	Whole grains	
	Milk		Milk	Beans	

[a] *The high vitamin content is due to fortification.*

Vitamins in Human Diets

Foods of both plant and animal origin provide vitamins in mixed diets for humans (Fig. 19.1; Table 19.3):

- *Meats and meat products* are generally excellent sources of thiamin, riboflavin, niacin, pyridoxine,

and vitamin B_{12}. Liver (including that from poultry or fish) is a very good source of vitamins A, D, E, and B_{12}, as well as folacin. Eggs are good sources of biotin. Animal products, however, are generally not good sources of vitamin C, K, (except pork liver) or folate.

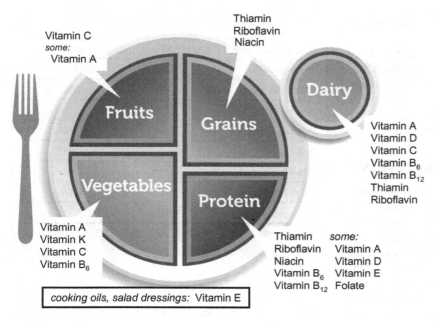

FIGURE 19.1 Vitamins provided by the major food groups, as depicted in ChooseMyPlate.gov.

TABLE 19.3 Contributions of Food Groups to the Vitamin Contents of Mixed Diets for Americans

	Vitamin A	Vitamin E	Vitamin C	Thiamin	Riboflavin	Niacin	Vitamin B_6	Folate	Vitamin B_{12}
					Per Capita Availability				
	1030 µg	21.4 mg	115 mg	2.9 mg	2.8 mg	33 mg	2.5 mg	682 µg	8.5 µg
					% Contributions by Food Group[a]				
Vegetables	**27.1**	5.9	**48.3**	8.6	5.7	9.8	20.1	12.3	0
Legumes, Nuts	0	4.8	0.1	4.5	1.6	4.0	3.6	9.3	0
Fruits	2.5	2.9	**40.5**	3.6	2.3	2.0	9.1	5.9	0
Grain products	5.3	3.9	4.7	**59.3**	**38.5**	**42.3**	18.3	**61.2**[b]	0.1
Meats, Fish	**33.2**	3.8	2.3	18.0	17.9	**37.7**	**38.1**	3.8	**77.4**
Milk products	15.7	1.7	2.5	4.3	**25.0**	1.1	6.8	3.3	18.1
Eggs	6.6	1.8	0	0.7	6.3	0.1	1.9	2.5	4.2
Fats, Oils	7.4	**74.6**	0	0	0.1	0	0	0	0.1
Other	2.2	0.6	1.7	0.9	1.9	3.2	1.9	1.7	0

[a]*Major contributing foods indicated in bold.*
[b]*Reflects mandatory fortification of wheat flour.*
Source: *From* Hiza, H. A. B., Bente, L., and Fungwe, T. (2008). *Nutrient Content of the US Food Supply*, 2008. USDA Home Economics Report No. 58, 72 pp.

- *Beans, peas and lentils* are generally good sources of thiamin, riboflavin, niacin, vitamin B_6, biotin, pantothenic acid, and folate.
- *Milk products* are important sources of vitamins A and C, thiamin, riboflavin,[8] pyridoxine, and vitamin B_{12}. Because milk is widely enriched with irradiated ergosterol (vitamin D_2), it is also an important source of vitamin D.[9]
- *Vegetables* are generally good sources of vitamins A, K, and C, and pyridoxine.
- *Fruits* are generally good sources of vitamin C; some (e.g., mangoes) are also good sources of vitamin A.
- *Grain products* are generally good sources of thiamin, riboflavin, and niacin.
- *Plant oils* are generally good sources of vitamin E. Red palm oil is a particularly good source of vitamin A.

Vitamins in Breast Milk and Formula Foods

The vitamin contents of foods that are intended for use as the main or sole components of diets (e.g., human milk, infant formulas, parenteral feeding solutions) are particularly important determinants of the vitamin status of individuals consuming them. Studies of the vitamin contents of human milk have yielded variable results, particularly for the fat-soluble vitamins. In general, it has been observed that the concentrations of all vitamins (except vitamin B_{12}) in human milk tend to increase during lactation. In comparison with cow's milk, human milk contains more of vitamins A, E, C, and niacin, but less vitamin K, thiamin, riboflavin, and pyridoxine (Table 19.4).

Because infant formulas and parenteral feeding solutions are carefully prepared and quality-controlled products, each is formulated largely from purified or partially refined ingredients to contain known amounts of the vitamins. For parenteral feeding solutions, however, some problems related to vitamin nutrition have occurred. One problem involved biotin, which was not added to such solutions before 1981 in the belief that intestinal synthesis of the vitamin was adequate for all patients except children with inborn metabolic errors or individuals ingesting large amounts of raw egg white. When it was found that children supported by total parenteral nutrition (TPN) frequently suffered from gastrointestinal abnormalities that responded to biotin,[10] the vitamin was added to TPN

TABLE 19.4 Vitamin Contents of Human and Cow's Milk

Vitamin	Human Milk	Cow's Milk
Vitamin A (retinol) (mg/l)	0.60	0.31
Vitamin D_3 (µg/l)	0.3	0.2
Vitamin E (mg/l)	3.5	0.9
Vitamin K (mg/l)	0.15	0.6
Ascorbic acid (mg/l)	38	20
Thiamin (mg/l)	0.16	0.40
Riboflavin (mg/l)	0.30	1.90
Niacin (mg/l)	2.3	0.8
Pyridoxine (mg/l)	0.06	0.40
Biotin (µg/l)	7.6	20
Pantothenic acid (mg/l)	0.26	0.36
Folate (mg/l)	0.05	0.05
Vitamin B_{12} (µg/l)	<0.1	3

Source: Porter, J. W. G. (1978). *Proc. Nutr. Soc.* 37, 225.

solutions.[11] Another problem with parenteral feeding solutions has been the loss of fat-soluble vitamins and riboflavin either by absorption to the plastic bags and tubing most frequently used, or by decomposition on exposure to light. Such effects can reduce the delivery of vitamins A, D, and E to the patient by two-thirds, and result in the loss of one-third of the riboflavin.

2. PREDICTING VITAMIN CONTENTS

Sources of Inaccuracy

The availability of data for the vitamin contents of foods and feedstuffs can be extremely useful in making judgments concerning the adequacy of food supplies and feedstuff inventories; however, estimates of the nutrient intakes of individuals as determined on the basis of these data are seldom accurate, owing to the variety of factors that may alter the nutrient composition of a food or feedstuff before it is actually ingested. The errors associated with such estimates are particularly great for vitamins.

---

8. In the United States, milk products supply an estimated 40% of the required riboflavin.

9. This practice has practically eliminated rickets in countries that use it.

10. These patients appeared to have had altered gut microflora secondary to antibiotic treatment.

11. Although there are no recommended dietary allowances (RDAs) for biotin, it has been suggested that biotin supplements be given to individuals being fed parenterally (infants, 30 µg/kg per day; adults, 5 µg/kg per day). These levels are consistent with the provisional recommendations (infants, 10 µg/day; adults, 30–100 µg/day) of the National Research Council Food Nutrition Board (1989).

Sources of error in estimating vitamins in foods and feedstuffs include:

- sampling errors
- analytical errors
- variation in the actual amount of the vitamin present
- less than full bioavailability of the specific form(s) of the vitamin present
- losses during storage
- losses during processing
- losses during cooking.

To accommodate these sources of potential error, most of which inflate estimates of nutrient intake, it is a common practice to discount by 10–25% the analytical values in the databases. It is likely that such modest discounts may still result in overestimates of intakes of at least some of the vitamins.

Analytical Errors

Errors in the sampling of and actual analysis of foods and feedstuffs for vitamins can be an important contributor to the inaccuracy of predicted values.[12] Analytical errors are less likely to be problematic for vitamin E, thiamin, riboflavin, niacin, and pyridoxine, for which robust analytical techniques (i.e., not prone to analyst effects) are available.

Natural Variation in Vitamin Contents

The concentrations of vitamins in individual foods and feedstuffs can vary widely. The vitamin contents of materials of animal origin can be affected by the conditions imposed in feeding the source animal, which can be highly variable according to country of origin, season of the year, size of the farm, age at slaughter, composition of the diets used, etc. For example, the vitamin E content of poultry meat is greater from chickens fed supplements of the vitamin than from those that are not.[13]

Genetic Sources of Variation

For foods and feedstuffs of plant origin, vitamin contents may vary among different cultivars of the same species (Fig. 19.2), and according to local agronomic factors and weather conditions that affect growth rate and yield (Tables 19.5A, 19.5B). Between-cultivar differences of as much as several orders of magnitude have been reported for most vitamins. In some cases, these differences correspond to other, readily identified characteristics of the plant. For example, the ascorbic acid contents of lettuce, cabbage, and asparagus tend to be relatively high in the colored and darker green varieties, and darker orange varieties of carrot tend to have greater provitamin A content than do lighter-colored carrots. However, vitamin contents are not necessarily related to such physical traits or to each other.

The substantial variation in reported values for vitamins in many plant foods suggests that it may be possible to breed plants for higher vitamin contents. This notion is not new; however, contents of vitamins or other micronutrients have yet to be widely included in the breeding strategies of plant breeders, whose efforts are driven primarily by agronomic issues and considerations of consumer acceptance, i.e., market demand. Unfortunately, many of the latter considerations, e.g., appearance, "freshness," have little relation to vitamin content. In a world where greater access to vitamin A would be important to some 250 million children who are at risk of that deficiency, and where greater access to ascorbic acid would reduce the anemia that affects 4 of every 10 women, the possibility of improving the vitamin and other micronutrient contents of plants (particularly the staples that feed the poor) through genetic modification cannot be ignored. Even in the industrialized world, where diet-related chronic diseases are growing problems, underexploited opportunities exist to develop vitamin/mineral-rich fruits and vegetables and to use these aspects of specific, good nutrition as marketing "hooks."

Environmental Effects

For many plants, growing conditions that favor the production of lush vegetation will result in increased

---

12. For this reason, most nutrient analytical methods have been standardized by the Association of Official Analytical Chemists.
13. A practical example of this comparison is the intensively managed commercial poultry flock fed formulated feeds in the US versus the small courtyard flock largely subsisting on table scraps, insects, and grasses in China and much of the developing world.

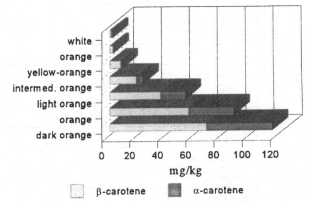

FIGURE 19.2 Variation in carotene contents among carrot cultivars. *From Leferriere, L. and Gabelman, W. H. (1968). Proc. Am. Soc. Hortic. Sci. 93, 408.*

TABLE 19.5A Fold Variations in Reported Vitamin Contents of Fruit and Vegetable Cultivars:
β-Carotene, Ascorbic Acid, α-Tocopherol, Thiamin and Riboflavin

Food	β-Carotene	Ascorbic Acid	α-Tocopherol	Thiamin	Riboflavin
Apple		29		3.0	10
Apricot	2.9			1.5	1.3
Banana		9			
Barley				2.3	
Bean	2.3	2.9		2.7	3.7
Blueberry	17	3.0		1.3	1.8
Cabbage		3.8		2.5	2.8
Carrot	80	1.4		6.9	5.5
Cassava	113	1.9			
Cauliflower		1.7		1.4	1.4
Cherry	3.5	4.2		2.0	1.5
Collard	1.4	1.6			2.1
Cowpea				2.9	3.0
Grape		3.0		7.5	3.4
Grapefruit	9.3	1.3			
Guava		11			
Lemon		1.2			
Maize	24		2.0	1.8	
Mango	3.8	91			2.0
Muskmelon		20			
Nectarine	4.8	4.7		1.0	1.3
Oat				1.8	
Orange	6.8	1.5			
Palm, oil	5.1				
Papaya	5.7	2.7		1.3	1.5
Pea	4.3	3.4		5.2	1.7
Peach	6.0	4.2		2.0	1.7
Peanut				1.4	1.9
Pear		16		7.0	5.0
Pepper, green	1.3	1.8	18		
Pepper, chili	46	10			
Plum	3.2	1.5			1.2
Potato		5.1		2.5	6.2
Rapeseed, oil			3.4		
Raspberry		2.3			
Soybean		2.4	1.2		
Spinach		1.6			

(Continued)

TABLE 19.5A (Continued)

Food	β-Carotene	Ascorbic Acid	α-Tocopherol	Thiamin	Riboflavin
Squash					
Summer		9.4			
Winter		3.5			
Strawberry		4.3			
Sunflower			2.7		
Sweet potato	89	3.1		2.9	3.1
Taro		3.2		4.9	2.5
Tomato	20	15		1.6	
Turnip, greens		1.1			
Watermelon	15				
Wheat			29	7.9	5.2
Yam		1.9		3.0	3.9

Source: Mozafar, A. (1994). *Plant Vitamins: Agronomic, Physiological and Nutritional Aspects.* CRC Press, New York, NY, p. 43.

TABLE 19.5B Fold Variations in Reported Vitamin Contents of Fruit and Vegetable Cultivars: Niacin, Pyridoxine, Biotin, Pantothenic Acid, and Folate

Food	Niacin	Vitamin B₆	Biotin	Pantothenic Acid	Folate
Apple	2.0		1.1	4.0	
Apricot	1.3				
Avocado	1.5	1.6		13	
Barley	1.1			1.2	
Bean	3.8	2.2			4.6
Blueberry	1.7				
Cherry	1.5				
Cowpea	2.2	1.5	1.5	1.3	
Grape	2.4				
Maize	5.5			1.3	
Mango	18				
Nectarine	1.3				
Oat	1.4				
Papaya	2.3				
Pea	1.2	1.3			2.2
Peach	1.2				
Peanut	1.5				

(Continued)

TABLE 19.5B (Continued)

Food	Niacin	Vitamin B$_6$	Biotin	Pantothenic Acid	Folate
Pear	4.0		1.1	2.5	
Pepper, green	1.2				
Plum	4.5				
Potato	2.7	3.2			
Rye	1.3				
Strawberry	1.3				
Sweet potato	3.4			2.2	
Taro	4.9				
Wheat	5.0	8.6		2.6	
Yam	2.7				

Source: Mozafar, A. (1994). *Plant Vitamins: Agronomic, Physiological and Nutritional Aspects.* CRC Press, New York, NY, p. 43.

concentrations of several vitamins. Such environmental factors as day length, light intensity and quality, and temperature strongly affect the concentrations of vitamins, especially carotenes and ascorbic acid. Tomatoes, for example, show very strong effects of light, with shaded fruits having less ascorbic acid than those exposed to light.[14] Low temperatures have been shown to increase the ascorbic acid contents of beans and potatoes, the thiamin contents of broccoli and cabbage, the riboflavin contents of spinach, wheat, broccoli, and cabbage, and the niacin contents of spinach and wheat. However, low temperatures decrease thiamin in beans and tomatoes, riboflavin in beans, niacin in tomatoes, and carotenoids in carrots, sweet potatoes, and papayas. Conditions at harvest can also affect the vitamin content of some crops – for example, the vitamin E content of fungus-infected corn grain can be less than half that of non-blighted corn. Legumes such as alfalfa and soybeans contain the enzyme lipoxygenase, which, if not inactivated (by drying) soon after harvest, catalyzes lipid oxidation reactions resulting in massive destruction of carotenoids and vitamin E. Accordingly, the vitamin contents of plant foods can be markedly different in different parts of the world, and can show marked seasonal and annual changes. These fluctuations can be as great as 8-fold for the α-tocopherol content of alfalfa hay within a single season, and 11-fold for the ascorbic acid content of apples produced in different years.

Effects of Agronomic Practices

Soil and crop management practices can affect the vitamin contents of edible plant tissues. These relationships are extremely complex, varying according to the soil type, plant species, and vitamin in question. In general, mineral fertilization can increase plant contents of ascorbic acid (P, K, Mn, B, Mo, Cu, Zn, Co), carotenes (N, Mg, Mn, Cu, Zn, B), thiamin (N, P, B), and riboflavin (N); however, nitrogen fertilization tends to decrease ascorbic acid concentrations despite increased yields. Organic fertilizers appear to increase the concentrations of some vitamins, in particular thiamin and vitamin B$_{12}$. Some of these effects may be due to the lower nitrate contents in organic fertilizers compared with inorganic ones, but organic fertilizers tend also to contain those vitamins that plant roots have recently been found able to absorb.

Practices that affect light exposure and plant growth rate can also affect plant vitamin contents. For example, ascorbic acid, the biosynthesis of which is related to plant carbohydrate metabolism, has been shown to be greater in field-grown versus greenhouse-grown tomatoes;[15] in peas, grapes, and tomatoes grown at lower planting densities; in lower-yielding or smaller apples; and in field-ripened versus artificially ripened apples.[16]

Tissue Variation

Another source of variation in food vitamin contents comes from the fact that most vitamins are not distributed uniformly among the various edible tissues of plants

---

14. This effect has been documented in response to a cloudy day, and to the difference between exposed and shaded sides of the same fruit.

15. This difference can be as great as two-fold.
16. By exposure to ethylene either in storage or in transit to market.

(Figs 19.3–19.5). In fact, gradients of several nutrients are found in plants, corresponding to the anatomical distribution of the phloem and xylem vascular network. Thus, relatively higher concentrations have been observed for ascorbic acid in the stem end of oranges and pears, the top end of pineapples, the apical ends of potatoes, the tuber end of sweet potatoes, both the lower and upper portions of carrots, the top end of turnips, and the stem tips of asparagus, bamboo shoots, and cucumbers. In general, exposed tissues (i.e., skin/peel and outer leaves) tend to contain greater concentrations of vitamins, particularly ascorbic acid, which is largely distributed in chloroplasts in tissues exposed to light.[17] In cereal grains, thiamin and niacin tend to be concentrated in tissues[18] that are removed in milling; therefore, breads made from refined wheat flour tend to be much lower in those vitamins than are products made from maize, which is not milled. Mobilization of seed vitamin stores and, in some cases, biosynthesis of vitamins occurs during germination such that young seedlings (sprouts) tend to have relatively high vitamin contents.

Accommodating Variation in Vitamin Contents of Foods and Feedstuffs

Natural variation in the nutritional composition of foods and feedstuffs has generally been accommodated by the analysis of multiple representative samples of each material of interest. Nevertheless, most databases include only a single value, the mean of all analyses, and fail to indicate the variance around that mean. The practical necessity of using databases so constructed means that the nutritionist is faced with the dilemma of estimating vitamin intake through the use of data that are likely to be inaccurate, but to an uncertain and unascertainable degree. Thus, if an average value of 150 mg/kg is used to represent the ascorbic acid concentration of potatoes, as is frequently the case, then it must be recognized that half of all samples will exceed that value (thus yielding an underestimate) while half will contain less than that value (thus yielding an overestimate). In constructing databases for use in meal planning or feed formulation, a better way to accommodate such natural variation is to enter into the database values discounted by a multiple of the standard deviation that would yield an acceptably low probability of

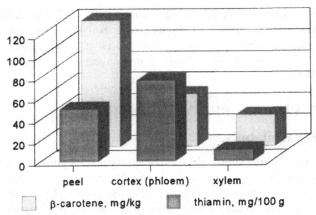

FIGURE 19.3 Tissue distribution of vitamins in carrots. *From Yamaguchi, Y. (1952). Proc. Am. Soc. Hortic. Sci. 60, 351.*

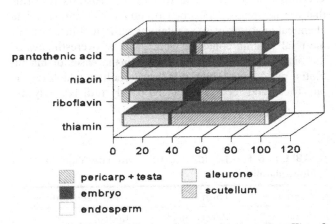

FIGURE 19.4 Tissue distribution of vitamins in wheat. *From Hinto, I. (1953). Nature (Lond.) 173, 993.*

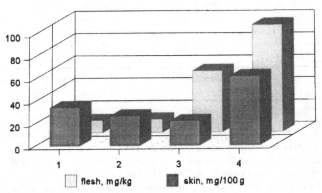

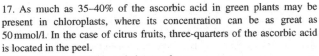

FIGURE 19.5 Tissue distribution of ascorbic acid in four apple cultivars. *From Gross, E. (1943). Gartenbauwissenschaften 17, 500.*

17. As much as 35–40% of the ascorbic acid in green plants may be present in chloroplasts, where its concentration can be as great as 50 mmol/l. In the case of citrus fruits, three-quarters of the ascorbic acid is located in the peel.
18. That is, the scutellum and aleurone layer.

overestimating actual nutrient amounts.[19] That approach, however, requires a fairly extensive body of data from which to generate meaningful estimates of variance. Few, if any, sets of food/feedstuff vitamin composition data are that extensive.

3. VITAMIN BIOAVAILABILITY

Bioavailability

Apart from considerations of analytical accuracy and natural variation associated with estimates of the vitamin contents of foods and feedstuffs, chemical analyses of vitamin contents may not provide useful information regarding the amounts of vitamin that are biologically available (Table 19.6). The concept of **bioavailability** (*see* Chapter 3) relates to the proportion of an ingested nutrient that is absorbed, retained, and metabolized through normal pathways to exert normal physiological function. Many vitamins can be present in foods and feedstuffs in forms that are not readily absorbed by humans or animals. The chemical analyses of such vitamins will yield measures of the total vitamin contents, which will be overestimates of the biologically relevant amounts. In the cases of niacin, biotin, pyridoxine, vitamin B_{12}, and choline, which in certain foods and feedstuffs can be poorly utilized, only the

biologically available amounts have nutritional relevance. For those, in particular, it would be useful to have methods for assessing vitamin bioavailability. An *in vitro* method has been developed to measure niacin bioavailability,[20] but for most vitamins bioavailability must be determined experimentally using appropriate animal models.

Vitamin bioavailabilities can be affected by several extrinsic and intrinsic factors.

Extrinsic factors:
- Concentration – effects on solubility and absorption kinetics
- Physical form – effects of physical interactions with other food components and/or of coatings, emulsifiers, etc., with vitamin supplements.
- Food/diet composition – effects on intestinal transit time, digestion, vitamin emulsification, vitamin absorption, and/or intestinal microflora
- Non-food agonists – cholestyramine, alcohol, and other drugs that may impair vitamin absorption or metabolism.

Intrinsic factors:
- Age – age-related differences in gastrointestinal function
- Health status – effects on gastrointestinal function.

4. VITAMIN LOSSES

Storage Losses

The storage of untreated foods can result in considerable losses due to post-harvest oxidation and enzymatic decomposition. The ascorbic acid contents of cold-stored apples and potatoes can drop by two-thirds and one-third, respectively, within 1–2 months. Those of some green vegetables can drop to 20–78% of original levels after a few days of storage at room temperature. Such losses can vary according to specific techniques of food processing and preservation.

Milling Losses

The milling of grain to produce flour involves the removal of large amounts of the bran and germ portions of the native product. Because those portions are typically rich in vitamin E and many of the water-soluble

TABLE 19.6 Foods and Feedstuffs with Low Vitamin Bioavailabilities

Vitamin	Form	Food/Feedstuff
Vitamin A	Provitamins A	Corn
Vitamin E	Non-tocopherols	Corn oil, soybean oil
Ascorbic acid	Ascorbinogen	Cabbage
Niacin	Niacytin	Corn, potatoes, rice, sorghum grain, wheat
Pyridoxine	Pyridoxine 5'-β-glucoside	Corn, rice bran, unpolished rice, peanuts, soybeans, soybean meal, wheat bran, whole wheat
Biotin	Biocytin	Barley, fishmeal, oats, sorghum grain, wheat

19. This approach was originated in the 1950s by G. F. Combs, Sr (the author's father) at the University of Maryland, when he developed the Maryland Feed Composition Table. The data included in that table were based on replicate analyses from multiple samples of each feedstuff, and were expressed as the mean −0.9 SD units. That adjustment was selected to allow a likelihood of overestimating actual nutrient concentration of $P = 0.20$, which level was acceptable in his judgment.

20. This method, developed by Carpenter, involves the comparison of the amounts of niacin determined chemically before (free niacin) and after (total niacin) alkaline hydrolysis. The free niacin thus determined correlates with the available niacin determined using the growth response of niacin-deficient rats fed a low-tyrosine diet.

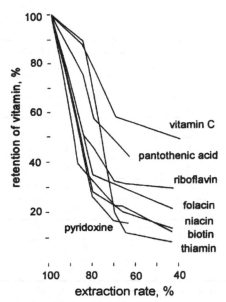

FIGURE 19.6 Loss of vitamins in the making of flour. *Source: Moran, T. (1959). Nutr. Abstr. Rev. 29, 1.*

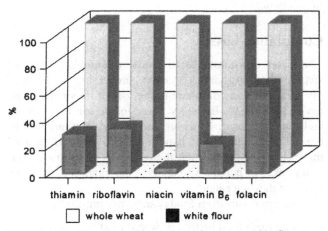

FIGURE 19.7 Vitamin contents of whole wheat versus white flour.

TABLE 19.7 Effects of Food Processing Techniques on Vitamin Contents of Foods

Technique	Main Effects	Vitamins Destroyed[a]
Blanching	Partial removal of oxygen	Vitamin C (10–60%)[b,c]
	Partial heat inactivation of enzymes	Thiamin (2–30%), riboflavin (5–40%), niacin (15–50%), carotene (<5%)[c]
Pasteurization	Removal of oxygen[d]	Thiamin (10–15%)
	Inactivation of enzymes	Minor losses (1–5%) of niacin, vitamin B_6, riboflavin, and pantothenic acid
Canning	Exclusion of oxygen	Highly variable losses[e,f]
Freezing[g]	Inhibition of enzyme activity[h]	Very slight losses of most vitamins
Frozen storage[g]		Substantial losses of vitamin C and pantothenic acid; moderate losses of thiamin and riboflavin
Freeze drying	Removal of water	Very slight losses of most vitamins
Hot air drying	Removal of water	10–15% losses of vitamin C and thiamin
γ-Irradiation	Inactivation of enzymes	Some losses (about 10%) of vitamins C, E, K and thiamin

[a]Actual losses are variable, depending on exact conditions of time, temperature, etc.
[b]Loss of vitamin C is due to both oxidation and leaching.
[c]Losses of oxidizable vitamins can be reduced by rapid cooling after blanching.
[d]Vitamin losses are usually small, owing to the exclusion of oxygen during this process.
[e]Losses in addition to those associated with heat sterilization before canning.
[f]For example, 15% loss of vitamin C after 2 years at 10°C.
[g]Thawing losses are associated with vitamin leaching into the syrup.
[h]While enzymatic decomposition is completely inhibited in frozen vegetables, reactivation occurs during thawing such that significant vitamin losses can occur. This is avoided by rapidly blanching before freezing.

vitamins, highly refined flours[21] are low in these vitamins (Figs 19.6, 19.7).

Processing Losses

Vitamins are subjected to destructive forces during thermal processing in the preservation of foods. Blanching, which is a mild heat treatment used to inactivate potentially deleterious enzymes, reduce microbial numbers, and decrease interstitial gases, usually is minimally destructive, although it can result in the leaching of water-soluble vitamins from foods blanched in hot water. Otherwise, blanching usually improves vitamin stability. In contrast, canning and other forms of high-temperature treatment can accelerate reactions of vitamin degradation, depending on the chemical nature of the food (i.e., its pH, dissolved oxygen and moisture contents, presence of transition metals and/or other reactive compounds) (Tables 19.7, 19.8).

---

21. Such flours consist mainly of the starchy endosperm.

TABLE 19.8 Typical Losses of Vitamins through Canning

Food	Vitamin A	Vitamin C	Thiamin	Riboflavin	Niacin	Vitamin B$_6$	Biotin	Pantothenic Acid	Folate
Asparagus	43	54	67	55	47	64	0		75
Lima beans	55	76	83	67	64	47		72	62
Green beans	52	79	62	64	40	50		60	57
Beet	50	70	67	60	75	9		33	80
Carrot	9	75	67	60	33	80	40	54	59
Corn	32	58	80	58	47	0	63	59	72
Mushroom		33	80	46	52		54	54	84
Green pea	30	67	74	64	69	69	78	80	59
Spinach	32	72	80	50	50	75	67	78	35
Tomato	0	26	17	25	0		55	30	54

Source: Lund, D. (1988). In *Nutritional Evaluation of Food Processing*, third ed. (Karmas, E. and Harris, R. S., eds). Van Nostrand Reinhold, New York, NY, p. 319.

Vitamins A, E, C, thiamin, and folate are sensitive to moist, high-temperature conditions such as occur during extrusion. Losses as great as 80% can occur, depending on the amount of water added to the food mixture and the temperature of the system.

Freezing and drying usually result in only minor losses of most vitamins. Losses associated with ionizing (γ) irradiation vary according to the energy dose, but are generally low (less than 10%).

Cooking Losses

Cooking can introduce further losses of vitamins from native food materials. However, methods used for cooking vary widely between different cultures and among different individuals, making vitamin losses associated with cooking highly variable. The washing of vegetables in water before cooking can result in the extraction of water-soluble vitamins, particularly if they are soaked for long periods of time. The peeling of vegetables can remove vitamins associated with the outer tissues.[22] Vitamin losses associated with cooking processes are also highly variable, but generally amount to about 50% for the less stable vitamins. The greatest losses are associated with long cooking times under conditions of exposure to air. Vitamin losses are less when food is cooked rapidly, as in a pressure cooker or a microwave oven, or by high-temperature stir frying. The baking of bread can reduce the thiamin

content of flour by about 25% without affecting its content of niacin or riboflavin.

Cumulative Losses

The losses of vitamins from foods are cumulative. Every step in the post-harvest storage, processing, and cooking of a food can contribute to the loss of its vitamin contents (Table 19.9). In theory, these losses can be modeled and thus predicted. However, in practice, the variation in the actual conditions of handling foods through each of these steps is so great that the only way to estimate vitamin intakes of people is to analyze the vitamin contents of foods as they are eaten.

Vitamin losses from foods can be minimized by:

- using fresh food instead of stored food
- using minimum amounts of water in food preparation and cooking
- using minimum cooking (using high temperatures for short periods of time)
- avoiding the storage of cooked food before it is eaten.

5. VITAMIN FORTIFICATION

Availability of Purified Vitamins

All of the vitamins are produced commercially in pure forms. Most are produced by chemical synthesis, but some are also isolated from natural sources (e.g., vitamin A from fish liver, vitamin D$_3$ from liver oil or irradiated yeast, vitamin E from soybean or corn oils, and vitamin K from fish meal) and some are produced microbiologically (e.g., thiamin, riboflavin, folate, pyridoxine, biotin, pantothenic

---

22. For example, the peeling of potatoes can substantially reduce the ascorbic acid content of that food.

TABLE 19.9 General Stabilities of Vitamins to Food Processing and Cooking

Vitamin	Conditions that Enhance Loss
Vitamin A	Highly variable but significant losses during storage and preparation
Vitamin D	(Stable to normal household procedures)
Vitamin E	Frying can result in losses of 70–90%; bleaching of flour destroys 100%; other losses in preparation or baking are small
Vitamin K	(Losses not significant due to synthesis by intestinal microflora)
Ascorbic acid	Readily lost by oxidation and/or extraction in many steps of food preparation, heat sterilization, drying, and cooking
Thiamin	Readily lost by leaching, by removal of thiamin-rich fractions from native foods (e.g., flour milling) and by heating; losses as great as 75% may occur in meats, and 25–33% in breads
Riboflavin	Readily lost on exposure to light (90% in milk exposed to sunlight for 2 hours, 30% from milk exposed to room light for 1 day), but very stable when stored in dark; small losses (12–25%) on heating during cooking
Niacin	Leached during blanching of vegetables ($\leq$40%), but very stable to cooking
Vitamin B_6	Leached during food preparation; pasteurization causes losses of 67%; roasting of beef causes losses of about 50%
Biotin	(Apparently very stable; limited data)
Pantothenic Acid	Losses of 60% by milling of flour and of about 30% by cooking of meat; small losses in vegetable preparation
Folate	(Data not available)
Vitamin B_{12}	Only small losses on irradiation of milk by visible or ultraviolet light

With the notable exception of the tocopherols,[25] there is no basis to the notion that biopotencies of vitamins prepared by chemical/microbiological synthesis are not at least as great as those of vitamins isolated from natural sources. In some cases, synthetic vitamins may be appreciably more bioavailable than the vitamin from natural sources (e.g., purified niacin versus protein-bound niacytin; purified biotin versus protein-bound biocytin).

The use of purified vitamins offers obvious advantages for purposes of ensuring vitamin potency in a wide variety of formulated products, including fortificants for foods, premixed supplements for feeds, nutritional supplements, pharmaceuticals, and ingredients in cosmetics. Commercial vitamin production has grown steadily since the discovery of vitamin B_{12}. At that time, annual world vitamin production was estimated to be only <1,500 metric tons; by 2005 it exceeded 20,000 metric tons. Vitamins are produced by at least 30 firms in some 17 countries, but a half-dozen companies presently dominate the world market.

Addition of Vitamins to Foods

The addition of vitamins to certain foods is a common practice in most countries. In some cases, vitamins are added to selected, widely used foods (e.g., bread) for several purposes:

- *Fortification* – ensuring vitamin adequacy of populations; e.g., white flour (folate), milk (vitamins A and D), margarine (vitamin A), formula foods (multiple vitamins in infant formulas, liquid nutrient supplements, enteral formulas used for tube feeding, parenteral formulas used for intravenous feeding).
- *Vitaminization* – to make foods carriers of vitamins not normally present; e.g., many breakfast cereals (multiple vitamins), orange juice (vitamin D), wheat (vitamin A), table salt (multiple vitamins).
- *Revitaminization* – to restore the vitamin content to that originally present before processing; e.g., white flour (thiamin, riboflavin, niacin).
- *Enrichment* – to increase the amounts of vitamins already present.

These processes are subject to regulation by national food authorities. In the United States, the addition of nutrients to foods is regulated by the Food and Drug Administration (FDA), which has identified as candidates for addition to foods 22 nutrients, including 12 vitamins

acid, and vitamin B_{12}[23]). Before their commercial synthesis became feasible, which began only in the 1940s, vitamins were extracted from such natural sources as fish oils and rose-hip syrup. Today, the production of vitamins is based predominantly on their chemical synthesis and/or microbiological production, the latter having been greatly impacted by the emergence of new techniques in biotechnology.[24]

---

23. The commercial production of vitamin B_{12} is strictly from microorganisms.
24. The industrial production of the vitamins has been nicely reviewed: O'Leary, M. J. (1993). Industrial production. In *The Technology of Vitamins in Food* (Ottaway, P. B., ed.). Chapman & Hall, London, p. 63.

---

25. There is evidence that several vitamers E produced by chemical synthesis vary in biopotency (see Chapters 3 and 7). This is due to both the positions and numbers of their ring-methyl groups, the most biopotent being the trimethylated (α) form, as well as the stereochemical form of the isoprenoid side chain. The most potent vitamer is the one that occurs naturally, *RRR*-α-tocopherol.

TABLE 19.10 Vitamins Approved by the FDA for Addition to Foods

Vitamin	Recommended Level of Addition (per 100 kcal)
Vitamin A	250 IU
Vitamin D	20 IU
Vitamin E	1.5 IU
Vitamin C	3 mg
Thiamin	75 μg
Riboflavin	85 μg
Niacin	1.0 mg
Vitamin B$_6$	0.1 mg
Biotin	15 μg
Pantothenic acid	0.5 mg
Folate	20 μg; wheat flour products:[a] 140 μg/100 g
Vitamin B$_{12}$	0.3 μg

[a]Mandated for most enriched flour, breads, corn meals, rice, noodles, macaroni, and other grain products.

(Table 19.10).[26] Fortification of wheat flour with folate has been mandatory in the US since 1998 and in Ireland since 2006. Since 1966, the USDA and USAID[27] have also routinely fortified or enriched foods provided as foreign aid under Public Law 480 (Table 19.11).[28] In addition, many antixerophthalmia programs have used vitamin A fortification of such foods as dried milk, wheat flour, sugar, tea, margarine, and monosodium glutamate (MSG).[29]

Stabilities of Vitamins Added to Foods

The stabilities and bioavailabilities of vitamins added to foods depend on the form of vitamin used, the composition of the food to which it is added, and the absorption status of the individual ingesting that food. The less stable vitamins can be lost from foods during storage, depending on the conditions (time, temperature, and moisture) of that storage (Fig. 19.8).

Vitamin Intakes from Foods

Food intake patterns are determined by many factors (e.g., tradition, taste, access, cost, ease of preparation), but seldom nutrient content. In addition, patterns of food intake

TABLE 19.11 Vitamins Added to P.L. 480 Title II Commodities

Vitamin	Amount Added per 100 g[a]				
	Wheat–Soy Blend	Corn–Soy Blend	Soy-Fortified Cereals	Non-Fat Dry Milk	Others
Vitamin A (IU)	2314	2314	2204–2645	5000–7000	2204–2645
Vitamin D (IU)	198	198			
Vitamin E[b] (IU)	7.5	7.5			
Vitamin C (mg)	40.1	40.1			
Thiamin[c] (mg)	0.28	0.28	0.44–0.66		0.44–0.66
Riboflavin (mg)	0.39	0.39	0.26–0.40		0.26–0.40
Niacin (mg)	5.9	5.9	3.5–5.3		3.5–5.3
Vitamin B$_6$[d] (mg)	0.165	0.165			
Pantothenic Acid (mg)	2.75	2.75			
Folate (μg)	198	198			
Vitamin B$_{12}$ (μg)	3.97	3.97			

[a]Processed blended foods are also fortified with Ca, P, Fe, Zn, I, and Na; soy-fortified cereals and other processed foods are fortified with Ca and Fe.
[b]As all-rac-α-tocopheryl acetate.
[c]As thiamin mononitrate.
[d]As pyridoxine hydrochloride.

---

26. In addition to these vitamins, the following nutrients are approved: protein, calcium, phosphorus, magnesium, potassium, manganese, iron, copper, zinc, and iodine.
27. United States Agency for International Development.

28. The cost of this fortification is very low relative to the total value of the commodities. The ingredients (vitamins and minerals) used to enrich the processed and soy-fortified commodities cost less than 2.5% of the value of the product; those used to enrich the more expensive blended food supplements cost less than 5% of the product value.
29. That is, monosodium glutamate, used as a seasoning.

change.[30] The most current estimates indicate that the majority of Americans obtain most of their vitamins from their foods (Table 19.12); however, the intakes of vitamins A, D, E, K, B_6, B_{12}, thiamin, riboflavin, niacin, and folate were found to be generally lower among individuals living under 131% of poverty compared to other economic groups.[31] Individuals consuming strict vegetarian diets will not obtain vitamins D and B_{12} from those foods.

Dinner, typically the largest meal of the day, tends to be the most important for providing vitamins, but breakfast tends to be the most important in providing vitamin D, likely due to the consumption of vitamin D-fortified milk. Snacking, now practiced by 97% of Americans,[32] provided nearly a fifth of vitamin intake. An even greater amount, some 30%, was obtained from meals consumed away from the home – i.e., prepared by others.

6. BIOFORTIFICATION

Agricultural technologies can increase the bioavailable vitamin contents of foods. Increasing vitamin content has not been an explicit goal of crop improvement, which has centered on economically important traits such as those directly related to yield and disease resistance. However, that narrow attitude has changed with the recognition of the "hidden hunger" of micronutrient malnutrition – i.e., the persistent, debilitating shortages of vitamins and essential minerals in the face of remarkable gains in the global production of total staple foods and total calories. That one-sixth of the world's population does not have access to the foods necessary for a nutritionally balanced diet has made it impossible to overlook shortages of vitamin A, folate, iron, iodine, and zinc, particularly among the

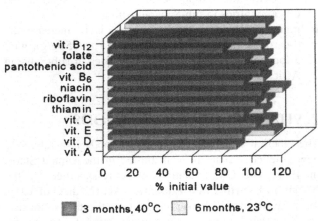

FIGURE 19.8 Stabilities of vitamins added to a breakfast cereal. *From Anderson, A. K. (1976). Food Technol. 30, 110.*

TABLE 19.12 Average Daily Intakes of Vitamins from Foods by Americans, by Percentile of Usual Intake

Vitamin	Average Intake from Foods	% Consumed in				% Consumed Away from Home
		Breakfast		Lunch	Dinner	
Vitamin A, µg	607 ± 15	29	21	32	18	27
Vitamin D, µg	4.6 ± 0.1	36	18	28	18	23
Vitamin E, mg	7.2 ± 0.2	17	24	34	25	35
Vitamin K, mg	88.9 ± 4.2	8	29	52	11	38
Vitamin C, mg	84.2 ± 3.5	22	20	30	28	27
Thiamin, mg	1.59 ± 0.03	25	24	34	17	31
Riboflavin, mg	2.16 ± 0.04	30	20	28	22	30
Niacin, mg	23.9 ± 0.34	19	25	39	17	34
Vitamin B_6, mg	1.91 ± 0.04	24	22	35	19	31
Pantothenic acid, mg	—	—	—	—	—	—
Folate, µg	527 ± 10	29	22	42	17	29
Vitamin B_{12}, µg	5.19 ± 0.12	27	23	34	16	31

Source: From US Department of Agriculture, Agricultural Research Service (2010). *What We Eat in American, NHANES 2007–2008* (www.ars. usda.gov/ba/bhnrc/fsrg).

---

30. How else would a sushi chef have a chance in North Dakota?
31. US Department of Agriculture, Agricultural Research Service (2010). *What We Eat in American, NHANES 2007–2008* (www.ars.usda.gov/ba/bhnrc/fsrg).

32. Piernas, C. and Popkin, B. M. (2010). *J. Nutr.* 140, 325.

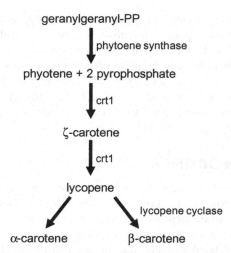

FIGURE 19.9 Pathway of β-carotene synthesis in plants.

world's poor. The international agricultural community has responded with a number of coordinated efforts to use modern breeding techniques to enhance the micronutrient contents of selected staple foods. This effort has been called *field fortification* and *biofortification*.[33]

Biofortification is a new approach to improving the nutritional value of crops. It involves breeding for increased levels of key vitamins and essential minerals in several staple crops relied on by the world's poor. Proof of this principle was accomplished by Potrykus, Beyer and colleagues,[34] who used molecular biological techniques to produce Golden Rice with as much as 35 μg/g β-carotene in the endosperm. This pioneering work involved the insertion into the rice genome of two of the three[35] genes needed for β-carotene synthesis (Fig. 19.9): PSY (phytoene synthase) from daffodil (*Narcissus pseudonarcissus*), and CRT1 from a soil bacterium (*Erwinia uredovora*). Simon and colleagues used visual color-scoring in breeding a biofortified carrot particularly high in β-carotene.[36]

Subsequent efforts have used conventional breeding with such modern methods as marker-assisted selection to produce germplasm with enhanced micronutrient content to be used by national agricultural research systems in their programs of breeding high-yielding cultivars. These efforts have targeted the following crops, the available cultivars of which show heterogeneity with respect to β-carotene content:[37]

- orange-fleshed sweet potato – the goal is to increase the β-carotene content 16-fold (from 2 to 32 μg/g), with a focus on Mozambique, Uganda, and other countries in sub-Saharan Africa.
- yellow cassava – the goal is to increase the β-carotene content 30-fold (from 0.5 to 15.5 μg/g), with a focus on the Democratic Republic of the Congo, Nigeria, and other countries in sub-Saharan Africa.
- high-β-carotene maize – the goal is to increase the β-carotene content 30-fold (from 0.5 to 15 μg/g), with a focus on Zambia and other countries in sub-Saharan Africa.[38]

7. VITAMIN LABELING OF FOODS

The labeling of nutrient contents of foods is a relatively new practice, having been instituted in the United States in 1972. The US regulations were re-specified by the **Nutrition Labeling and Education Act (NLEA)** of 1990, the purpose of which was to provide, through a consistent food label format, useful information to consumers about the foods they eat in the context of their daily diets. The nutrition labeling of foods has the potential to influence consumer food use choices to the extent that the label information is accessible and can be acquired, processed, and used (Fig. 19.10).

This US program involves compulsory labeling for most prepared and packaged foods.[39] It encourages voluntary labeling, either for individual products or at the point of purchase, for the most frequently consumed fresh fruits,[40] vegetables[41] or seafood;[42] and at the point of purchase for fresh poultry and meats, and for prepared foods served in

33. The motive force for this effort has been the Consultative Group for International Agricultural Research (CGIAR) led by the International Food Policy Research Institute (IFPRI). With support from the Bill and Melinda Gates Foundation, IFPRI put together a global alliance called Harvest Plus (HarvestPlus@cgiar.org and www.HarvestPlus.org).
34. Ye, X. Al-Babili, and S. Klöti, A. (2000). *Science* 287, 303.
35. The LYC (lycopene cyclase) gene is expressed in rice endosperm.
36. Mills, J. P., Simon, P. W., and Tanumihardjo, S. A. (2008). *J. Nutr.* 138, 1692.

37. Other targets include high-iron pearl millet (India), bean (D.R. Congo, Rwanda), high-zinc rice (Bangladesh, India), and wheat (India, Pakistan).
38. This has been found to be achievable by taking advantage of rare genetic variations in the β-carotene hydroxylase-1 (crtRB1) gene (Yan, J., Kandiani, C. B., Harjes, C. E., *et al* (2010). *Nature Genetics* 42, 322) or by upregulating the expression of PSY1 (Toledo-Ortiz, G., Huq, E., and Rodriguz-Concepcíon, M. (2010). *Proc. Natl Acad. Sci.* 107, 116 126).
39. The act excludes foods containing few nutrients, e.g., plain coffee, tea, spices; foods produced by small businesses; and foods prepared and served by the same establishment.
40. Bananas, apples, watermelons, oranges, cantaloupe, grapes, grapefruit, strawberries, peaches, pears, nectarines, honeydew melons, plums, avocados, lemons, pineapples, tangerines, cherries, kiwi fruits, and limes.
41. Potatoes, iceberg lettuce, tomatoes, onions, carrots, celery, corn, broccoli, cucumbers, bell peppers, leaf lettuce, sweet potatoes, mushrooms, green onions, green beans, summer squash, and asparagus.
42. Shrimp, cod, pollack, catfish, scallops, salmon, flounder, sole, oysters, orange roughy, mackerel, ocean perch, rockfish, whiting, clams, haddock, blue crabs, rainbow trout, halibut, and lobster.

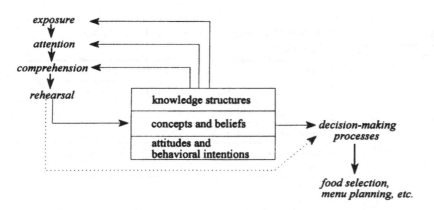

FIGURE 19.10 Model for stages of information processing in decision making. After Olson, J. H. and Sims, C. A. [1980]. *J. Nutr. 12, 157.*

Nutrition Facts

Serv. Size 1/2 cup (122 g)
Servings about 3.5

Amount Per Serving	
Calories 50	Fat Cal. 10

	% Daily Value*
Total Fat 1 g	1%
Sodium 260 mg	11%
Total Carb. 8 g	3%
Fiber 1 g	5%
Sugars 7 g	
Protein 1 g	

Vitamin A 6%	•	Vitamin C 4%
Calcium 4%	•	Iron 4%

Not a significant source of saturated fat and cholesterol.
* Percent Daily Values are based on a 2,000 calorie diet.

FIGURE 19.11 Nutrition information food label, United States.

INGREDIENTS: WHOLE WHEAT, WHEAT BRAN, SUGAR, SALT, MALT, THIAMIN HYDROCHLORIDE, PYRIDOXINE HYDROCHLORIDE, FOLIC ACID, REDUCED IRON, BHT.

NUTRITION INFORMATION
PER 30 g
SERVING CEREAL
(175 mL, ¾ CUP)

ENERGY	Cal	100
	kJ	420
PROTEIN	g	3.0
FAT	g	0.6
CARBOHYDRATE		24.0
SUGARS	g	4.4
STARCH	g	16.6
FIBRE	g	3.0
SODIUM	mg	265
POTASSIUM	mg	168

PERCENTAGE OF
RECOMMENDED
DAILY INTAKE

THIAMIN	%	46
NIACIN	%	6
VITAMIN B$_6$	%	10
FOLACIN	%	8
IRON	%	28

FIGURE 19.12 Nutrition information food label, Canada.

restaurants. In addition to information about the name of the product and its manufacturer, and the measure/count of food contents, the act requires the food label to carry information about the ingredients, serving size, and number of servings, and quantities of specified food components and nutrients (Figs 19.11, 19.12). Vitamin and mineral content information must be presented in comparison with a standard, the Reference Daily Intakes (RDIs) (Table 19.13).[43] For the information they present, nutrition labels draw on the USDA National Nutrient Data Bank or an alternative data bank developed by the Produce Marketing Association.

43. RDI values are based on the respective Recommended Dietary Allowances (RDAs), and are compared to Daily Reference Values (DRVs). The RDIs replaced (largely in name only) the US RDA values, which are based on the 1968 RDAs, used before 1990. Most labels use the RDIs developed for adults and children 4 years of age or older; foods targeted to a certain age group must use the RDI developed for that group.

The NLEA requires that information about vitamin A and vitamin C be carried on all food labels. It makes optional the disclosure of contents of other nutrients including vitamins for which RDAs have been established. In all cases, information must be presented according to the specified format.

TABLE 19.13 US RDAs Used in Food Labeling

Nutrient	Amount
Protein[a]	50 g
Minerals	
Calcium	1,000 mg
Iron	18 mg
Iodine	150 µg
Copper	2 mg
Vitamins	
Vitamin A	5,000 IU
Vitamin D	400 IU
Vitamin E	30 IU
Vitamin C	60 mg
Thiamin	1.5 mg
Riboflavin	1.7 mg
Niacin	20 mg
Vitamin B_6	2 mg
Pantothenic acid	10 mg
Biotin	300 µg
Folate	400 µg

TABLE 19.14 Dietary Supplement Use (%) by Americans

Supplement	NHANES I 1992[a] Men	NHANES I 1992[a] Women	CSFII 1994–1996[b] Men	CSFII 1994–1996[b] Women
Any supplement	20.2	26.8	41.9	55.8
Multivitamin	16.8	21.5	20.2	22.1
Multivitamin/ mineral			13.1	21.5
Single vitamin/ mineral			14.5	22.4
Vitamin A	1.4	1.6		
Vitamin C	7.2	7.8		
Vitamin E	4.0	4.7		

[a]National Health and Nutrition Examination Survey I.
[b]Continuing Survey of Food Intakes of Individuals.
Source: From Radimer, K. L. (2003). *J. Nutr.* 133, 2003S.

8. VITAMIN SUPPLEMENTATION

Vitamin Supplements

More than half of the US population takes dietary supplements,[44] the most popular being multivitamin/mineral supplements used by 40% of American adults, particularly women (Table 19.14),[45] and more than 30% of children.[46] Studies have shown that nutritional supplement users tend to be more health conscious, and have better diets, more education, and higher incomes than the general population. Supplement use is greater among health professionals, vegetarians, the elderly, and readers of health-focused magazines. Vitamin supplement use is most frequent among individuals who believe that diet affects disease; non-drinkers and lighter drinkers of alcohol; former smokers and individuals who never smoked; and individuals in

the lowest three quartiles of BMI. Users of vitamin supplements have markedly greater vitamin intakes than non-users (Table 19.15). The median levels of supplement use by Americans are one to two times the RDAs for vitamins A, D, and B_6, niacin, pantothenic acid, and folate; and greater than twice the RDAs for vitamins E, C, and B_{12}, thiamin, and riboflavin.

Studies in the US have shown that regular users of multivitamin/mineral supplements are more likely to have adequate vitamin intakes (Table 19.16), less likely to have suboptimal blood nutrient concentrations, and more likely to have optimal levels of biomarkers of chronic disease.[47] However, it is not clear whether those users actually are at reduced chronic disease risk as a result of that practice. This is, in part, because users have generally healthy lifestyles, which independently reduce their low risk. Systematic reviews of the relevant clinical literature are limited by the paucity of rigorous studies conducted to date; most available results do not provide strong evidence for beneficial health effects of multivitamin/mineral supplements for most people.[48] However, there have been indications of some benefits, including reduced fracture risks in postmenopausal women, and marginal increases in cognitive performance in children.[49]

44. That is, 53.4% of respondents in the NHANES 2003–2006 (Bailey, R. L., Gahche, J. J., Lentino, C. V., *et al.* [2011] *J. Nutr.* 141, 261).
45. Rock, C. L. (2007). *Am. J. Clin. Nutr.* 85, 277S; Gahche, J., Bailey, R., Burt, V., *et al.* (2011). NCHS Data Brief No. 61.
46. Picciano, M. F., Dwyer, J. T., Radimer, K. L., *et al.* (2007). *Arch. Pediatr. Adolesc. Med.* 161, 978.

47. Block, G., Jensen, C. D., Norkus, E. P., *et al.* (2007). *Nutr. J.* 6, 30.
48. NIH State-of-the-Science Panel (2007). *Am. J. Clin. Nutr.* 85(Suppl.), 257S; McCormick, D. B. (2010). *Nutr. Rev.* 68, 207.
49. Eilander, A., Gera, T., Sachdev, H. S., *et al.* (2010). *Am. J. Clin. Nutr.* 91, 115.

TABLE 19.15 Contributions of Commonly Used Dietary Supplements to Vitamin Intakes of Americans

Vitamin	Sex	EAR[a]	Intake from food	Intake from Supplement[a]	Total Intake
Vitamin A	Male	625 µg	656 µg	1,050 µg	1,706 µg
	Female	500 µg	564 µg	1,050 µg	1,614 µg
Vitamin E	Male	12 µg	8.2 µg	13.5 µg	21.7 µg
	Female	12 µg	6.3 µg	13.5 µg	19.8 µg
Vitamin C	Male	75 mg	105 mg	60 mg	165 mg
	Female	60 mg	84 mg	60 mg	144 mg

[a]Vitamin content based on that of the most commonly consumed multivitamin/mineral supplement in the NHANES 2001–2002.
Source: From Dwyer, J. T., Holden, J., Andrews, K., et al. (2007). *Anal. Bioanal. Chem.* 389, 37.

TABLE 19.16 Effect of Multivitamin (MV) Use on Prevalence of Adequate Vitamin Intakes of Subjects in the Hawaii–Los Angeles Multiethnic Cohort

Vitamin	Men			Women		
	Non-users	Users		Non-users	Users	
	From Food	From Food	Total (Food + MV)	From Food	From Food	Total (Food ± + ± MV)
Vitamin A	59 ± 42[a]	61 ± 41	87 ± 29	69 ± 39	73 ± 39	89 ± 28
Vitamin E	27 ± 41	28 ± 41	68 ± 43	22 ± 38	23 ± 39	60 ± 45
Vitamin C	72 ± 42	76 ± 40	89 ± 29	82 ± 37	86 ± 33	93 ± 25
Thiamin	82 ± 35	84 ± 33	94 ± 23	79 ± 37	82 ± 35	92 ± 25
Riboflavin	87 ± 30	89 ± 28	95 ± 19	77 ± 20	79 ± 19	90 ± 16
Niacin	89 ± 27	90 ± 25	96 ± 18	84 ± 32	86 ± 30	94 ± 22
Vitamin B$_6$	79 ± 37	81 ± 36	93 ± 24	73 ± 41	77 ± 39	90 ± 28
Folate	93 ± 23	94 ± 21	97 ± 15	88 ± 30	90 ± 27	95 ± 20
Vitamin B$_{12}$	90 ± 28	90 ± 27	96 ± 18	83 ± 18	85 ± 34	94 ± 23

[a]Mean ± SE.
Source: From Murphy, S. P., White, K. K., Park, S. Y., et al. (2007). *Am. J. Clin. Nutr.* 85(Suppl.), 280S.

Interventions with multivitamin supplements have been useful in undernourished populations. For example, antenatal multiple-micronutrient supplementation reduced combined fetal loss and neonatal deaths by 11% and increased birth weight by 14% in Indonesia (particularly in undernourished and/or anemic mothers), and increased birth weight in Nepal.[50] The use of vitamin supplements among peoples in developed countries has become great enough to make this means a significant contributor to the vitamin nutriture of those populations. Still, for most of the poor in those countries, access to multivitamin supplements remains limited by cost, making food-based approaches more sustainable for addressing prevalent multi-micronutrient shortages.

Global sales of vitamins are expected to reach $3.3 billion by 2015. The drivers of this increasing trend have been identified as increasing awareness of consumers, increasing demand by the food industry facilitated by the GRAS[51] status of vitamins, increasing consumer demand

50. The Supplementation with Multiple Micronutrients Intervention Trial (SUMMIT) Study Group (2008). *Lancet* 371, 215; Vaidya, A., Saville, N., Shresthra, B. P., et al. (2008). *Lancet* 371, 492.

51. Generally Recognized As Safe.

for health-beneficial products, and increasing demand by livestock producers for performance-enhancers. The US is the largest consumer of vitamins, 40% of which are from multivitamin supplements, most (>70%) of which are sold in drug stores, supermarkets, and health food stores. In the US, dietary supplements are regulated by the FDA under the Dietary Supplement Health and Education Act of 1994.[52]

Guidelines for Supplement Use

Healthy individuals can and should obtain adequate amounts of all nutrients, including vitamins, from a well-balanced diet based on a variety of foods of good quality. Such an approach minimizes the risks of deficiencies as well as excesses of all nutrients. It also acknowledges that foods can provide health benefits that have yet to be fully elucidated. Nevertheless, certain circumstances may warrant the use of vitamin supplements, including:

- folate for women who may conceive, are pregnant or are lactating;[53]
- multivitamins for individuals with very low caloric intakes (such that their consumption of total food is insufficient to provide all nutrients)
- vitamin B_{12} for strict vegetarians, individuals with gastric achlorhydria or gastric resection, and most people over 50 years of age
- vitamin D for most people living in northern latitudes
- vitamin K (single dose) for newborn infants to prevent abnormal bleeding
- when recommended by a health professional for patients with diseases or medications that interfere with vitamin utilization.

For other persons with varied, balanced diets, the actual benefit of taking vitamin supplements is doubtful.

9. VITAMINS IN LIVESTOCK FEEDING

Vitamins in Animal Feeds

The economic considerations in feeding livestock generally dictate the use of a relatively small number of feedstuffs with few (if any) day-to-day changes in diet

formulation.[54] In livestock production, the continued use of the same or very similar diets has resulted in the empirical development of knowledge about the vitamin contents of feedstuffs (Table 19.17).

Unlike human diets, formulated diets for livestock generally do not provide adequate amounts of vitamins unless they are supplemented with either certain vitamin-rich feedstuffs or purified vitamins. In general, the relative vitamin adequacy of unsupplemented animal feeds depends on the relative complexity (the number of feedstuffs used in the mixture) of the diet. The vitamin contents of simple rations tend to be less than those of complex ones (Table 19.18). For example, complex diets such as those used for feeding poultry or swine in the 1950s[55] would be expected to contain in their constituent feedstuffs more than an adequate amount of vitamin B_{12}, and adequate (or nearly so) amounts of vitamin K, vitamin E, thiamin, riboflavin, niacin, pyridoxine, pantothenic acid, folate, and choline. In contrast, the simpler rations (based almost exclusively on corn and soybean meal) that are used today contain lower amounts, if any, of the more costly vitamin-rich feedstuffs previously used. Such simple rations can be expected to contain in constituent feedstuffs adequate levels only of vitamin E, thiamin, pyridoxine, and biotin (Tables 19.19, 19.20). The availability of stable, biologically available, and economical vitamins facilitated this change in complexity of animal feeds by replacing the more costly vitamin-rich feedstuffs used previously with inexpensive mixtures of vitamins.

Losses of Vitamins from Feedstuffs and Finished Feeds

The vitamin contents of feedstuffs and finished feeds[56] are subject to destruction in ways very similar to those of foods. The storage losses that can occur in particular feedstuffs are dependent on the conditions of temperature and moisture during storage; heat and humidity enhance oxidation reactions of several of the vitamins (vitamins A and E, thiamin, riboflavin, and biotin). Vitamin losses are therefore minimized by drying feedstuffs quickly, and storing them dry in weather-proof bins. Where the drying of a feedstuff is slow[57] or incomplete,[58] or where leaky bins are used for its storage, vitamin losses are greatest.

---

52. This legislation charges the FDA to establish a framework for assuring safety, outline guidelines for literature displayed where supplements are sold, provide for use of claims and nutritional support statements, require ingredient and nutritional labeling, and establish good manufacturing practice regulations. The law changes previous legislation in that dietary supplements are no longer subject to the pre-market safety evaluations required of other food ingredients.
53. Pregnant and lactating women may also need supplements of iron and calcium.

54. For example, starting broiler chicks are typically fed the same diet from hatching to 3 weeks of age, and laying hens in some management systems may be fed the same diet for 20 weeks before the formula is changed.
55. Those complex diets contained, in addition to a major grain and soybean meal, small amounts of the following: alfalfa meal, corn distillers' dried solubles, fish meal, meat, and bone meal.
56. Complete, blended, ready-to-feed rations.
57. For example, sun-drying enhances the destruction of vitamin E in corn (although sun-curing of cut hay is essential to provide vitamin D activity).
58. Where moisture is not reduced to less than about 15%.

TABLE 19.17 Feedstuffs Containing Significant Amounts of Vitamins

Vitamin A	Vitamin D	Vitamin E	Vitamin K	Vitamin C	Thiamin	Riboflavin
None[a]	None[a]	Alfalfa, dehydrated	Alfalfa, dehydrated	none[a,b]	none[a]	Dried skim milk
	Alfalfa, sun-cured	Alfalfa, sun-cured				Peanut meal
	Wheatgerm meal					Brewers' yeast
	Corn germ meal					Dried buttermilk
	Stabilized vegetable oils					Dried whey
						Torula yeast
						Corn distillers' solubles
						Liver/glandular meal

Niacin	Vitamin B₆	Biotin	Pantothenic Acid	Folate	Vitamin B₁₂	Choline
Barley	Sunflower seed meal	Corn germ meal	Molasses	Dried brewers' grains	Dried fish solubles	Liver/glandular meal
Cottonseed meal	Sesame meal	Brewers' yeast	Rice polishings	Alfalfa, dehydrated	Liver/glandular meal	Dried fish solubles
Dried fish solubles	Meat/bone meal	Molasses	Sunflower seed meal	Brewers' yeast	Hydrolyzed feathers	Soybean meal
Rice bran, polishings	Torula yeast	Torula yeast	Peanut meal	Soybean meal	Fish meals	Corn distillers' solubles
Wheat bran	Hydrolyzed feathers	Liver/glandular meal	Torula yeast	Torula yeast	Crab meal	
Corn gluten feed	Safflower meal		Liver/glandular meal	Meat/bone meal	Dried skim milk	
Fish meals			Brewers' yeast	Corn distill. solubles	Dried butter milk	
Peanut meal				Alfalfa, sun-cured	Meat/bone meal	
Torula yeast						
Corn gluten meal						
Corn distillers' solubles						
Liver/glandular meal						
Sunflower seed meal						
Brewers' yeast						

[a]Instability of the vitamin in most feedstuffs renders few, if any, predictable sources of appreciable amounts of it.
[b]Not required by livestock species.

TABLE 19.18 Vitamins Provided by Constituent Feedstuffs in Older (Complex) and Modern (Simple) Chick Starter Diets

	1936 Diet[a] (%)	Modern Diet (%)
Ingredients		
Cornmeal	27.5	65.61
Oats	10.0	
Wheat bran	20.0	
Wheat middlings	10.0	
Soybean meal, 49% protein	10.0	19.08
Meat and bone meal	10.0	4.78
Poultry by-product meal		7.00
Dried whey	5.0	
Dehydrated alfalfa meal	5.0	
Blended fat		3.18
Limestone	2.0	
Salt	0.5	0.25
D,L-methionine (98%)		0.10
Trace minerals	+[b]	+[c]
Vitamins	+[d]	+[e]
Vitamins Provided by Feedstuffs		
Vitamin A (IU)	6,000 (400)[f]	1,360 (91)[f,g]
Vitamin E (IU)	27 (270)	20.5 (205)[g]
Vitamin K (mg)	0.73 (146)	0 (0)[g]
Thiamin (mg)	4.7 (261)	3.1 (172)
Riboflavin (mg)	5.4 (150)	2.3 (64)[g]
Niacin (mg)	69.9 (259)	28.3 (105)[g]
Pyridoxine (mg)	5.7 (190)	5.3 (177)
Biotin (g)	208 (139)	141 (94)
Pantothenic acid (mg)	15.7 (157)	7.4 (74)[g]
Folate (mg)	0.81 (145)	0.32 (58)
Vitamin B_{12} (g)	6.5 (72)	21.0 (233)[g]
Choline (mg)	1,115 (86)	1,395 (107)[g]

[a]This was a state-of-the-art diet for starting chicks at Cornell University in 1942.
[b]$MnSO_4$, 125 mg/kg.
[c]Provides, per kilogram of diet: ZnO, 66 mg; $MnSO_4$, 220 mg; Na_2SeO_3, 220 g.
[d]Vitamin D_3, 790 IU/kg.
[e]Provides, per kilogram of diet: vitamin A, 4,400 IU; vitamin D_3, 2,200 IU; vitamin E, 5.5 IU; vitamin K_3, 2 mg; riboflavin, 4 mg; nicotinic acid, 33 mg; pantothenic acid, 11 mg; vitamin B_{12}, 1 g; choline, 220 mg.
[f]Numbers in parentheses give amounts of each vitamin as a percentage of current (1984) recommendations of the US National Research Council.
[g]Included in the vitamin–mineral premix.

TABLE 19.19 Vitamins Most Likely to be Limiting in Non-Ruminant Livestock Feeds

Vitamin A
Vitamin E
Niacin
Pantothenic acid
Vitamin D, if raised indoors
Vitamin K, if raised on slatted or raised wire floors
Vitamin B_{12}, if raised on slatted or raised wire floors
Choline, chicks only

TABLE 19.20 Insufficient Amounts of Vitamins in Turkey Starter Diet Feedstuffs

Vitamin	Level from Feedstuffs, % NRC Requirement	
	Simple Feed[a]	Complex Feed[b]
Vitamin A	20	40
Vitamin E	130	130
Thiamin	170	160
Riboflavin	60	90
Niacin	30	60
Vitamin B_6	130	100
Pantothenic acid	90	110
Biotin	120	140
Folate	50	50
Vitamin B_{12}	0	74
Choline	90	100

[a]Corn, 40.5%; soybean meal, 51.2%; animal fat, 4%; $CaHPO_4$, 3%; limestone, 0.8%; salt, 0.3%; methionine, 0.15%; trace minerals, 0.05%.
[b]Milo, 20.5%; wheat, 20%; soybean meal, 33.9%; poultry meal, 6%; animal fat, 5%; meat and bone meal, 5%; fish meal, 4%; alfalfa meal, 2%; distillers' grains and solubles, 2%; limestone, 0.7%; $CaHPO_4$, 0.5%; salt, 0.3%; methionine, 0.13%; trace minerals, 0.05%.
Source: Anonymous (1989). *Vitamin Nutrition for Poultry*. Hoffman-La Roche, Inc., Nutley, NJ, pp. 13–14.

Vitamin losses from finished feeds are usually greater than those of individual feedstuffs. Finished feeds are supplemented with essential trace elements, some of which (Cu^{2+}, Fe^{3+}) can act as catalytic centers of oxidation reactions leading to vitamin destruction. Such effects are particularly important in high-energy feeds (e.g., broiler diets), which generally contain significant amounts of polyunsaturated fats. It is a common practice in many countries to compress many of these (and other) feeds into pellet form[59] by processes involving steam, heat, and pressure. Evidence suggests that pelleting can enhance the bioavailability of niacin and biotin, which occur in feedstuffs in bound forms, but it generally results in the destruction of vitamins A, D, E, K_3, C, and thiamin.

Vitamin Premixes for Animal Feeds

As purified sources of the vitamins have become available at low cost, it has become possible to use fewer feedstuffs in less complicated blends to produce diets of high quality that will support efficient and predictable animal performance. Thus, many feedstuffs formerly valuable as sources of vitamins (e.g., brewers' yeast, dried buttermilk, green feeds[60]) are no longer economical to use in intensive animal management systems.[61]

The use of purified vitamins as supplements to animal feeds has increased the economy of animal feeding by obviating the need to include relatively expensive vitamin-rich feedstuffs in favor of lower-priced feedstuffs that are lower in vitamin content but provide useful energy and protein. In modern practice, the addition of vitamins to animal feeds is accomplished by preparing a mixture of the specific vitamins required with a suitable carrier[62] to ensure homogeneous distribution in the feed as it is mixed. Such a preparation is referred to as **vitamin premix** (Tables 19.21, 19.22) and is handled in much the same way as other feedstuffs in the blending of animal feeds. Typically, vitamin premixes are formulated to be blended into diets at rates of 0.5–1.0%.[63]

TABLE 19.21 Vitamins Generally Included in Vitamin Premixes for Livestock Diets

Vitamin	Poultry	Piglets	Hogs	Calves	Cattle
Vitamin A	+	+	+	+	+
Vitamin D₃	+	+	+	+	+
Vitamin E	+	+	+	+	+[a]
Vitamin K	+	+	+[a]		
Ascorbic acid	+[b] +	+			
Thiamin	+[a]	+[a]			
Riboflavin	+	+	+	+[a]	
Niacin	+	+			
Vitamin B₆	+[a]	+		+[a]	+[a]
Pantothenic acid	+	+	+	+[a]	
Biotin	+[a]	+[a]	+[a]		
Folate					
Vitamin B₁₂	+	+	+[a]		
Choline	+	+			

[a]Sometimes added.
[b]Added in situations of stress.

Premixes generally also contain synthetic antioxidants (e.g., ethoxyquin, butylated hydroxytoluene [BHT]) to enhance vitamin stability during storage.[64] In many cases, trace minerals are included in **vitamin–mineral premixes**.[65] It is standard practice in the formulation of vitamin premixes to use amounts of vitamins that, when added to the expected amounts intrinsic to the component feedstuffs, will provide a comfortable excess above those levels found experimentally to be required to prevent overt deficiency signs. This is done in view of the many potential causes of increased vitamin needs; owing to the low cost of vitamin supplementation, this approach is considered a kind of low-cost nutrition insurance, as vitamin

---

59. There are many reasons for pelleting finished feeds. Pelleting prevents demixing of the feed during handling and transit. It can improve the economy of feed handling owing to the associated increase in bulk density. For the same reason, it can improve the consumption of bulky, low-density feeds. It can also improve the efficiency of feed utilization by reducing wastage at the feeder. It is thought that the metabolizable energy values of some feedstuffs may be improved by the steam treatment used in pelleting (e.g., soybean meal with significant residual antitryptic activity). Pelleting also improves the handling of feeds that are very dusty.

60. For example, fresh cabbage, grass.

61. This phenomenon is most true in the economically developed parts of the world. In the developing world, such factors as the shortage of hard currency may make purified sources of vitamins too expensive to use in animal diets, thus making natural sources of the vitamins more valuable. Under such circumstances, it is prudent to exploit a wide variety of local feedstuffs, food wastes, and food by-products in the formulation of animal feeds that are adequate in terms of vitamins as well as all other known nutrients.

62. Examples include soybean meal, finely ground corn or wheat, corn gluten meal, wheat middlings.

63. The cost of the vitamin premix typically represents less than 2% of the total cost of most finished feeds. Of that amount, approximately two-thirds of the vitamin cost is accounted for by vitamin E, niacin, vitamin A, and riboflavin (roughly in that order).

64. Studies have shown that the loss of vitamin A from poultry feeds stored at moderate temperatures (about 15% in 30 days) was slightly reduced (to about 10%) by the addition of any of several synthetic antioxidants. Under conditions of high temperature and high humidity, vitamin A losses from finished feeds can be much greater (e.g., 80–95%). Maximal protection by antioxidants is expected under conditions in which vitamin oxidation is moderate (e.g., short-term feed storage in hot, humid environments).

65. Owing to the presence of mineral catalysts in oxidative reactions, the stabilities of oxidant-sensitive vitamins in compound premixes can be expected to be less than in premixes of the vitamins alone.

TABLE 19.22 Examples of Vitamin Premixes for Animal Feeds

Vitamin	Units/1,000 kg Diet		
	Practical Diet[a] for Chicks[c]	Semipurified Diet[b] for Chicks[d]	Semipurified Diet[b] for Rats[e]
Vitamin A[f] (IU)	8,800,000	50,000	40,000,000
Vitamin D$_3$ (IU)	2,200,000	4,500,000	1,000,000
Vitamin E (IU)[g]	5,500	50,000	50,000
Menadione NaHSO$_3$ (g)	2.2	1.5	50
Thiamin HCl (g)	15	6	
Riboflavin (g)	4.4	15	6
Niacin (g)	33	50	30
Pyridoxine HCl (g)	6	7	
d-Calcium pantothenate (g)	11	20	16
Biotin (mg)	0.6	0.2	
Folic acid (g)	6	2	
Vitamin B$_{12}$ (mg)	10	20	10
Choline chloride (g)	220	2,000	+[h]
Minerals	+[i]	+[j]	+[j]
Other ingredients:			
Antioxidant[k] (g)	125	100	100
Carrier (g)	to weight[l]	to weight[m]	to weight[m]

[a]Composed of nonpurified natural feedstuffs (e.g., corn, soybean meal).
[b]Composed of purified/partially purified ingredients (e.g., isolated soy protein, casein, sucrose, starch).
[c]From Scott, M. L., Nesheim, M. C., and Young, R. J. (1982). Nutrition of the Chicken, 3rd edn. M. L. Scott & Assoc., Ithaca, NY, p. 494.
[d]From Scott, M. L., Nesheim, M. C., Young, R. J. (1982). Nutrition of the Chicken, 3rd edn. M. L. Scott & Assoc., Ithaca, NY, p. 546.
[e]AIN-76 diet.
[f]As all-trans-retinyl palmitate.
[g]As all-rac-tocopheryl acetate.
[h]Added as 0.2% choline bitartrate.
[i]Includes 66g of ZnO, 220g of MnSO$_4$, and 220mg of Na$_2$SeO$_3$.
[j]Includes CaHPO$_4$·2H$_2$O, CaCO$_3$, KH$_2$PO$_4$, NaHCO$_3$, KHCO$_3$, KCl, NaCl, MnSO$_4$·H$_2$O, FeSO$_4$·7H$_2$O, MgCO$_3$, MgSO$_4$, KIO$_3$, CuO$_4$·5H$_2$O, ZnCO$_3$, CoCl$_2$, NaMoO$_4$·2H$_2$O, and/or Na$_2$SeO$_3$ in amounts appropriate for the composition of the particular diet.
[k]For example, ethoxyquin, BHT.
[l]Corn meal.
[m]Sucrose.

premixes usually account for only 1–2% of the total cost of feeds for non-ruminant livestock. It should be remembered, however, that purified vitamins may not always be cheap, particularly in developing countries. Under those circumstances, the appropriate way to assess the value of using vitamin supplements is to compare their market prices with the estimated loss of production realized by not supplementing feeds that can be economically produced using locally available feedstuffs.

Stabilities of Vitamins in Feeds

Vitamins tend to be less stable in vitamin–mineral premixes used for livestock feeds, owing to the redox reactions catalyzed by trace elements and physical abrasion of protective coatings (Table 19.23). Vitamin premixes that contain choline chloride typically show accelerated losses of vitamin B$_6$, which reacts with choline. During the storage of finished feeds, the migration of moisture to the shady,

TABLE 19.23 Typical Stabilities of Vitamins in a Broiler Feed

Vitamin	% Retained Activity			
	Premix Storage (2 Months)	Pelleting/ Conditioning (93°C, 1 min)	Feed Storage (2 Weeks)	Cumulative
Vitamin A[a]	98	90	92	81
Vitamin D$_3$	98	93	93	85
Vitamin E[b]	99	97	98	94
Vitamin K[c]	92	65	85	51
Thiamin[d]	99	89	98	86
Riboflavin	99	89	97	85
Niacin	99	90	93	83
Vitamin B$_6$[e]	99	87	95	82
Biotin	99	89	95	84
Pantothenic acid[f]	99	89	98	86
Folate	99	89	98	86
Vitamin B$_{12}$	100	96	98	86

[a]all-trans-Retinyl acetate.
[b]all-rac-α-Tocopheryl acetate.
[c]Menadione sodium bisulfite complex.
[d]Thiamin mononitrate.
[e]Pyridoxine hydrochloride.
[f]Calcium pantothenate.
Source: BASF Keeping Current, No. 9138, 1992.

relatively cool side of a feed bin can result in the development of pockets of relatively high moisture, which can enhance the chemical degradation of vitamins as well as support the growth of vitamin-consuming fungi. Feeds that are pelleted or extruded are also exposed to friction, pressure, heat, and humidity, all of which enhance vitamin loss.

The chemical stabilities of some vitamins can be improved by using a more stable chemical form or formulation. For example, the calcium salt of pantothenic acid is more stable than the free acid form, and esters of vitamins A and E (retinyl acetate, tocopheryl acetate) are much more resistant to oxidation than the free alcohol forms. Vitamin preparations can also be coated or encapsulated[66] in ways that exclude oxygen and/or moisture, thus rendering them more stable. They are often spray-dried, spray-congealed or prepared as adsorbates to improve their handling characteristics. Owing to such approaches, purified vitamins added to foods have been found to be as stable and bioavailable, if not more so, than the forms of the vitamins intrinsic to foods.

66. Gelatin, edible fats, starches, and sugars are used for this purpose.

Study Questions and Exercises

1. For a core food for any particular vitamin, construct a flow diagram showing all of the processes, from the growing of the food to the eating of it by a human, that might reduce the useful amount of that vitamin in the food.

2. In consideration of the core foods for the vitamins and your personal food habits, which vitamin(s) might you expect to have the lowest intakes from your diet? Which might you expect to be low in the typical American diet? Which might you expect to be low in vegetarian and low-meat diets?

3. Use a concept map to show the relationships of vitamin supplementation of animal feeds to the concepts of chemical stability, bioavailability, and physiological utilization.

4. Prepare a flow diagram to show the means by which you might first evaluate the dietary vitamin status of a specific population (e.g., in an institutional setting), and then improve it, if necessary.

5. What principles should be used in planning diets to ensure adequacy with respect to the vitamins (and other nutrients)?

RECOMMENDED READING

Allen, L.H., 2006. New approaches for designing and evaluating food fortification programs. J. Nutr. 136, 1055–1058.

Allen, L., de Benoist, F., Dary, O., Hurrell, R. (Eds.),, 2006. Guidelines on Food Fortification with Micronutrients. FAO/WHO, Geneva, p. 376.

Backstrand, J.R., 2002. The history and future of food fortification in the United States: A public health perspective. Nutr. Rev. 60, 15–26.

Food and Nutrition Board, 2003. Dietary Reference Intakes: Applications in Dietary Planning. National Academy Press, Washington, DC, p. 237.

Food and Nutrition Board, 2003. Dietary Reference Intakes: Guiding Principles for Nutrition Labeling and Fortification. National Academy Press, Washington, DC, p. 205.

Hiza, H.A.B., Bente, L., Fungwe, T., 2008. Nutrient Content of the US Food Supply, 2008. USDA Home Economics Report No. 58, p. 72.

Holden, J.M., Harnly, J.M., Beecher, G.R., 2006. Food composition. In: Bowman, B.A., Russell, R.M. (Eds.), Present Knowledge in Nutrition. International Life Science Institute – Nutritional Foundation, Washington, DC, pp. 781–794.

Mozafar, A., 1994. Plant Vitamins: Agronomic, Physiological and Nutritional Aspects. CRC Press, New York, NY, p. 43.

Ottaway, P.B., 1993. The Technology of Vitamins in Foods. Chapman & Hall, London.

Park, Y.K., Sempos, C.T., Barton, C.N., et al. 2000. Effectiveness of food fortification in the United States: The case of pellagra. Am. J. Pub. Health 90, 727–738.

Porter, D.V., Earle, R.O. (Eds.),, 1990. Nutrition Labeling: Issues and Directions for the 1990s. National Academy Press, Washington, DC, p. 325.

Yates, A.A., 2006. Which dietary reference intake is best suited to serve as the basis for nutrition labeling for daily values? J. Nutr. 136, 2457–2462.

Assessing Vitamin Status

Anchoring Concepts

1. Detection of suboptimal vitamin status at early stages (before manifestation of overt deficiency disease) is desirable for the reason that vitamin deficiencies are most easily correctable in their early stages.
2. Vitamin status can be estimated by evaluating diets and food habits, but these methods are not precise.
3. Vitamin status can be determined by measuring the concentrations of vitamins and metabolites, and the activities of vitamin-dependent enzymes, in samples of tissues and urine.
4. Suboptimal status is more probable for some vitamins than for others.

... the old idea, that the state of nutrition of a child could be at once established by mere cursory inspection by the doctor, has to be abandoned ... [Such methods] gave us very little information about the occurrence of the milder degrees of deficiency, or of the earlier stages of their development.

L. J. Harris

Learning Objectives

1. To understand the requirements of valid methods for assessing vitamin status.
2. To understand the methods available for assessing the vitamin status of humans and animals.
3. To be familiar with available information regarding the vitamin status of human populations.

VOCABULARY

Anthropometric assessment
Biochemical assessment
Biomarker
Clinical assessment
Dietary assessment
Food frequency questionnaires (FFQ)
Hidden hunger
Nutrient loading
Nutritional assessment
Nutritional status
Sociologic assessment

1. GENERAL ASPECTS OF NUTRITIONAL ASSESSMENT

Nutritional assessment, in any application, has three general purposes:

- Detection of deficiency states
- Evaluation of nutritional qualities of diets, food habits, and/or food supplies
- Prediction of health effects.

The need to understand and describe the health status of individuals, a basic tenet of medicine, spawned the development of methods to assess nutrition status as appreciation grew for the important relationship between nutrition and health. The first applications of nutritional assessment were in investigations of feed-related health and production problems of livestock, and, later, in examinations of human populations in developing countries. Activities of the latter type, consisting mainly of organized nutrition surveys, resulted in the first

The Vitamins. DOI: 10.1016/B978-0-12-381980-2.00020-7

efforts to standardize both the methods employed to collect such data and the ways in which the results are interpreted.[1] More recently, nutritional assessment has also become an essential part of the nutritional care of hospitalized patients, and has become increasingly important as a means of evaluating the impact of public nutrition intervention programs.

Systems of Nutritional Assessment

Three types of nutritional assessment systems have been employed both in population-based studies and in the care of hospitalized patients:

- *Nutrition surveys* – cross-sectional evaluations of selected population groups; conducted to generate baseline nutritional data, to learn overall nutrition status, and to identify subgroups at nutritional risk
- *Nutrition surveillance* – continuous monitoring of the nutritional status of selected population groups (e.g., at-risk groups) for an extended period of time; conducted to identify possible causes of malnutrition
- *Nutrition screening* – comparison of individuals' parameters of nutritional status with predetermined standards; conducted to identify malnourished individuals requiring nutritional intervention.

Methods of Nutritional Assessment

Systems of nutritional assessment can employ a wide variety of specific methods. In general, however, these methods fall into five categories:

- *Dietary assessment* – estimation of nutrient intakes from evaluations of diets, food availability, and food habits (using such instruments as food frequency questionnaires, food recall procedures, diet histories, food records)
- *Anthropometric assessment* – estimation of nutritional status on the basis of measurements of the physical dimensions and gross composition of an individual's body
- *Clinical assessment* – estimation of nutritional status on the basis of recording a medical history and conducting a physical examination to detect signs (observations made by a qualified observer) and symptoms (manifestations reported by the patient) associated with malnutrition

- *Biochemical assessment* – estimation of nutritional status on the basis of measurements of nutrient stores, functional forms, excreted forms, and/or metabolic functions
- *Sociologic assessment* – collection of information on non-nutrient-related variables known to affect or be related to nutritional status (e.g., socioeconomic status, food habits and beliefs, food prices and availability, food storage and cooking practices, drinking water quality, immunization records, incidence of low birth-weight infants, breastfeeding and weaning practices, age- and cause-specific mortality rates, birth order, family structure).

Typically, nutritional assessment systems employ several of these methods for the complete evaluation of nutritional status. Some of these approaches, however, are more informative than others with respect to specific nutrients, and, particularly, to early stages of vitamin deficiencies (Table 20.1).

2. ASSESSMENT OF VITAMIN STATUS

Risk of suboptimal vitamin status is determined largely by factors that limit access to a diet that provides adequate amounts of these and other essential nutrients, as well as factors that limit the body's ability to utilize them after ingestion. These factors can be best ascertained by assessing dietary practices, clinical status, and biochemical indicators (biomarkers) of vitamin status.

Dietary Assessment

Diets and food habits that are likely to provide insufficient amounts of available vitamins can be identified by dietary evaluation.

- *Monotonous diet* – a diet with little food variety,[2] particularly one based primarily on milled cereal grains
- *Low caloric intake* – low intake of total food
- *Enteric malabsorption* – due to deficiencies and/or imbalances in other dietary components, e.g., fat.

There is no universal method for dietary assessment. Instead, several methods, each with certain advantages and limitations, are used:

- *24-hour recalls.* Interviews have been used. The USDA has improved this methodology with the development of a five-step computerized dietary recall instrument, the USDA Automated Multiple-Pass Method (AMPM); a web-based, self-administered instrument has been developed.[3]

1. In 1955, the US government organized the Interdepartmental Committee on Nutrition for National Defense (ICNND) to assist developing countries in assessing the nutritional status of their peoples, identifying problems of malnutrition, and developing practical ways of solving their nutrition-related problems. The ICNND teams conducted nutrition surveys in 24 countries. In 1963, the ICNND published the first comprehensive manual (ICNND [1963] *Manual for Nutrition Surveys*, 2nd edn. US Government Printing Office, Washington, DC) in which analytical methods were described and interpretive guidelines were presented.

2. This may include a strict vegetarian diet that does not include some source of vitamin B_{12}.
3. Thompson, F. E., Subar, A. F., Loria, C. M., *et al.* (2010). *J. Am. Diet. Assoc.* 110, 48.

TABLE 20.1 Relevance of Assessment Methods to the Stages of Vitamin Deficiency

Stage of Deficiency	Most Informative Methods				
	Dietary	Biochemical	Anthropometric	Clinical	Sociologic
1. Depletion of vitamin stores	+	+			
2. Cellular metabolic changes		+	+	+	
3. Clinical defects		+	+	+	
4. Morphological changes			+	+	
5. Behavioral signs					+

- *Dietary records*. Written records (food diaries) have been used; cell phone-based methods are under development to use video (with food image processing), voice and/or text input to capture eating episodes. Some studies have indicated that these methods can yield useful results for vitamins if conducted for as many as 6 days.[4]
- *Food frequency questionnaires (FFQ)*. This approach depends on respondent memory, and uses fixed lists of foods. FFQs yield information about the usual diet of the past and are relatively low cost,[5] for which reasons they are the methods of choice for epidemiological studies. FFQs have been found to yield useful estimates of biomarkers of some vitamins.[6] A systematic review[7] found that vitamin intake estimated from FFQs and 24-hour recall methods showed correlation coefficients of 0.26–0.38, and that those from FFQs and dietary record methods showed correlation coefficients of 0.41–0.53.

For the purpose of assessing vitamin intake/status, these methods almost always yield imprecise estimates, with greater inherent variability for vitamins than for macronutrients. This occurs for reasons of predictive uncertainty, as discussed in Chapter 19.

Clinical Assessment

Pathophysiological factors that may limit vitamin utilization can be ascertained by clinical evaluation. These factors include:

- *Enteric malabsorption* – acquired or innate problems affecting the absorptive surface of the gut; e.g., enteritis, gastrointestinal surgery

- *Impaired vitamin retention/utilization* – acquired or innate problems of hepatic or renal vitamin metabolism; e.g., hepatitis, nephritis.

Diagnoses of vitamin deficiencies are generally most possible in the later stages, when physiologic dysfunction and/or morphological changes can be detected. However, overt vitamin deficiency signs and symptoms (see Chapter 4) are relatively rare compared with the incidence of suboptimal vitamin status.

Biochemical Assessment

Nutritional status with respect to the vitamins refers to the functional reserve capacity provided by the amounts of vitamins in tissue. Therefore, the ideal biomarker of vitamin status would be a measure of stored metabolic function of the vitamin, which would offer the best chance for detecting early-stage (i.e., subclinical) vitamin deficiency.

A useful biomarker of vitamin status must:

- correlate with the rate of vitamin intake, at least within the nutritionally significant range, and respond to deprivation of the vitamin
- relate to a meaningful period of time
- relate to normal physiologic function
- be measurable in an accessible specimen
- be technically feasible, reproducible, and affordable
- have an available base of normative data.

Available Biomarkers of Vitamin Status

In some cases it is possible to assess the stored metabolic function of a vitamin.[8] In most cases, however, direct measurement of vitamin function is not possible owing to the

4. Presse, N., Payettr, H., Shatenstein, B., *et al.* (2011). *J. Nutr.* 141, 341.
5. Block, G., Thompson, F. E., Hartman, A. M., *et al.* (1990). *J. Clin. Epidemiol.* 43, 1327.
6. Tangney, C. C., Bienias, J. L., Evans, D. A., *et al.* (2004). *J. Nutr.* 134, 927.
7. Henríque-Sánchez, P., Sánchez-Villegas, A., Doreste-Alonso, J., *et al.* (2009). *Br. J. Nutr.* 102, S10.

8. Examples include the measurement of prothrombin time to assess vitamin K status, and the measurement of stimulation coefficients of erythrocyte transketolase and glutathione reductase to assess thiamin and riboflavin status, respectively.

TABLE 20.2 Tissues Accessible for Assessing Biomarkers of Vitamin Status

Tissue or Cell Type	Relevance
Blood	
Plasma/serum	Contains newly absorbed vitamins being transported to other tissues; therefore, tends to reflect recent vitamin intake; this effect can be reduced by collecting fasting blood
Erythrocytes	With a half-life of about 120 days, they tend to reflect chronic nutrient status; analyses can be technically difficult
Leukocytes	Have relatively short half-lives and, therefore, can be used to monitor short-term changes in nutrient status
Tissues	
Liver, adipose, muscle, marrow	Sampling is invasive, requiring research or clinical settings
Hair, nails	Easily collected and stored specimens offer advantages for studies of trace element status; not useful for assessing vitamin status
Skin, macula	Can be scanned non-invasively by resonance Raman spectroscopy for assessing carotenoids

effort, Biomarkers of Nutrition in Development (BOND), has undertaken to do this. BOND will establish a process to identify the best available biomarkers for given uses (assessment of nutrient intake, status, function, and effects), and provide the evidence needed by researchers and clinicians to make informed diagnoses, and by policy-makers to make informed decisions.

Interpreting Biomarker Data

The relevance of biomarkers of vitamin status of individuals is not straightforward, owing to issues of intra-individual variation and confounding effects, which may be quantitatively more significant for individuals than for populations. For example, intra-individual (within-person) variation is frequently noted in serum analytes. Therefore, a measurement of a single blood sample may not be appropriate for estimating the usual circulating level of the analyte of an individual, even though it may be useful in estimating the mean level of a population. Several factors can confound the interpretation of parameters of vitamin status: those affecting the response parameters directly, drugs that can increase vitamin needs, seasonal effects related to the physical environment[12] or food availability,[13] use of parenteral feeding solutions,[14] use of vitamin supplements,[15] smoking,[16] etc. (Table 20.4).

The guidelines originally developed by the ICNND remain useful for the interpretation of the results of biomarkers of

absence of a functional biomarker,[9] the existence of more than one metabolic function with different sensitivities to vitamin supply,[10] and/or the function of the vitamin in a loosely bound fashion unstable to methods of tissue preparation.[11] Therefore, other biomarkers can be useful: measurements of the vitamin, particular metabolites, and other enzymes in accessible tissues or urine (Tables 20.2, 20.3).

Currently, there are no formal conventions regarding the optimal biomarkers for nutritional uses. An international

---

9. For example, vitamin E appears to function as a biological lipid antioxidant, but measuring that function is not possible with any physiological relevance because all of the known products of lipid peroxidation (e.g., malonaldehyde, alkanes) are known to be metabolized. This makes such measurements difficult to interpret with respect simply to vitamin E status.

10. For example, pyridoxal phosphate is a cofactor for each of two enzymes involved in the metabolic conversion of tryptophan to niacin: kynureninase and a transaminase. Although the cofactor is essential for the activity of each enzyme, kynureninase has a much greater affinity for pyridoxal phosphate ($K_m = 10^{-3}$ mol/l) than does the transaminase ($K_m = 10^{-8}$ mol/l). Therefore, under conditions of pyridoxine deprivation, the transaminase activity can be reduced even though kynureninase activity is unaffected.

11. For example, the metabolically active forms of niacin, NAD(P)H, function as the co-substrates of many redox enzymes. These enzyme-co-substrate complexes are only transiently associated; therefore, dilution of biological specimens results in their dissociation and, usually, in the oxidation of the co-substrate.

12. For example, individuals living in northern latitudes typically show peak plasma levels of 25-hydroxyvitamin D_3 (25-OH-D_3) around September and low levels around February, with inverse patterns of plasma parathyroid hormone (PTH) concentrations, owing to the seasonal variation in exposure to ultraviolet light.

13. For example, residents of Finland showed peak plasma ascorbic acid levels in August–September and lowest levels in November–January, owing to seasonal differences in the availability of vitamin C-rich fruits and vegetables.

14. Individuals supported by total parenteral nutrition (TPN) have frequently been found to be of low status with respect to biotin (owing to their abnormal intestinal microflora) and the fat-soluble vitamins (owing to absorption by the plastic bags and tubing, and to destruction by UV light used to sterilize TPN solutions).

15. The NHANESI survey showed that more than 51% of Americans over 18 years of age used vitamin/mineral supplements, with 23.1% doing so on a daily basis. Further, the National Ambulatory Medical Care Survey (1981) showed that 1% of office visits to physicians (in particular, general and family practitioners) involved a prescription or recommendation for multivitamins. Multivitamins appear to be the most commonly used supplements, followed by vitamin C, calcium, vitamin E, and vitamin A. The use of vitamin supplements has been found to have greater impact than that of vitamin-fortified food on both the mean and coefficient of variation (CV) of estimates of vitamin intake in free-living populations.

16. Smokers have been found to have abnormally low plasma levels of ascorbic acid (with a corresponding increase in dehydroascorbic acid), pyridoxal, and pyridoxal phosphate.

TABLE 20.3 Biomarkers of Vitamin Status

Vitamin	Functional Parameters	Tissue Levels	Urinary Excretion
Vitamin A		Serum retinol[a] Change in serum retinol after oral load[b] Liver retinyl esters	
Vitamin D		Serum 25-(OH)$_2$-vitamin D$_3$[a] Serum vitamin D$_3$ Serum 1,25-(OH)$_2$-D$_3$ Serum alkaline phosphatase	
Vitamin E	Erythrocyte hemolysis	Serum tocopherols[a] Serum malondialdehyde Serum 1,4-isoprostanes Breath alkanes	
Vitamin K	Clotting time Prothrombin time[a]		
Vitamin C		Serum ascorbic acid Leukocyte ascorbic acid[a]	Ascorbic acid Ascorbic acid after load[c]
Thiamin	Erythrocyte transketolase stimulation[a]	Blood thiamin Blood pyruvate	Thiamin (thiochrome) Thiamin after load[c]
Niacin		RBC NAD[a] RBC NAD:NADP ratio Plasma tryptophan	1-methylnicotinamide 1-methyl-6-pyridone-3-carboxamide
Riboflavin	RBC glutathione reductase stimulation[a]	Blood riboflavin	Riboflavin Riboflavin after load[c]
Vitamin B$_6$	RBC transaminase	Plasma pyridoxal phosphate RBC transaminase stimulation RBC pyridoxal phosphate Plasma pyridoxal	Xanthurenic acid after tryptophan load[a,c] Quinolinic acid 4-Pyridoxic acid
Biotin		Blood biotin[a]	Biotin
Pantothenic acid	RBC sulfanilamide acetylase	Serum pantothenic acid RBC pantothenic acid Blood pantothenic acid[a]	Pantothenic acid
Folate		Serum folates[a] RBC folates[a] Leukocyte folates Liver folates	FIGLU[c] after histidine load[a,c] Urocanic acid after histidine load[c]
Vitamin B$_{12}$		Serum vitamin B$_{12}$[a] RBC vitamin B$_{12}$	FIGLU[d] Methylmalonic acid[a]

[a]Most useful parameter.
[b]Relative dose–response test.
[c]Single large oral dose.
[d]FIGLU, formiminoglutamic acid.

TABLE 20.4 Limitations of Some Biomarkers of Vitamin Status

Vitamin	Biomarker	Limitations
Vitamin A	Plasma[a] retinol	Reflects body vitamin A stores only at severely depleted or excessive levels; confounding effects of protein and zinc deficiencies and renal dysfunction
Vitamin D	Plasma[a] alkaline phosphatase	Affected by other disease states
Vitamin E	Plasma[a] tocopherol	Affected by blood lipid transport capacity
Thiamin	Plasma[a] thiamin	Low sensitivity to changes in thiamin intake
Riboflavin	Plasma[a] riboflavin	Low sensitivity to changes in riboflavin intake
Vitamin B_6	RBC glutamic-pyruvic	Genetic polymorphism transaminase
Folate	RBC folates	Also reduced in vitamin B_{12} deficiency
	Urinary FIGLU[b]	Also increased in vitamin B_{12} deficiency
Vitamin B_{12}	Urinary FIGLU[b]	Also increased in folate deficiency

[a]For serum.
[b]FIGLU, formiminoglutamic acid.

vitamin status (Table 20.5). It is important to note, however, that those interpretive guidelines were developed for use in surveys of populations.

3. VITAMIN STATUS OF HUMAN POPULATIONS

Reserve Capacities of Vitamins

The reserve capacities of the vitamins vary; each is affected by the history of vitamin intake, the metabolic needs for the vitamin, and the general health status of the individual. Typical reserve capacities of a healthy, adequately nourished human adult are as follows; these are expressed in terms of time-equivalents based on abilities to meet normal metabolic needs.

4–10 days	Thiamin, biotin and pantothenic acid
2–6 weeks	Vitamins D, E, K, and C; riboflavin, niacin, and vitamin B_6
3–4 months	Folate
1–2 years	Vitamin A
3–5 years	Vitamin B_{12}

Differences in reserve capacities reflect differential abilities to retain and store the vitamins, and lead, therefore, to differential sensitivities to vitamin deprivation. For example, individuals with histories of generally adequate vitamin nutriture can be expected to sustain longer periods of deprivation of vitamins A or B_{12} than they could of thiamin, biotin or pantothenic acid. Similarly, metabolic

and physiologic lesions caused by deficiencies of thiamin, biotin or pantothenic acid can be expected to appear much sooner than those of vitamins A or B_{12}, which may remain occult. Because nutritional intervention is typically most efficacious and cost-effective in earlier stages of vitamin deficiencies, the early detection of occult deficiencies is important for designing effective therapy and prophylaxis programs.

National Nutrition Surveillance

The United States has had a series of programs to track the nutritional adequacy of the food supply and/or the nutritional status of people (Table 20.6). These have included efforts to obtain information on most of the high-risk vitamins, e.g., vitamins A, E, C, B_6, and B_{12}, thiamin, riboflavin, and niacin. Other countries have conducted similar studies,[17] as well as regular food reporting.

Vitamins in US Food Supply

The desire to reduce the prevalence of obesity and preventable deaths, particularly due to heart disease and cancer, is driving changes in American diets, which are seen as contributing to these diseases. It is estimated that at least a fifth of heart disease and a third of all cancers could be prevented by improving the American diet; specifically, by increasing

17. For example, the New Zealand National Nutrition Survey; Luxembourg Nutritional Surveillance System; United Kingdom Expenditure and Food Survey, National Food Survey and School Nutrition Dietary Assessment.

TABLE 20.5 Interpreting Biomarkers of Vitamin Status

Vitamin	Parameter	Age Group	Deficient (High Risk)	Low (Moderate Risk)	Acceptable (Low Risk)
			Values, by Category of Status[a]		
Vitamin A	Plasma[b] retinol (μg/dl)	<5 months	<10	10–19	>20
		0.5–17 years	<20	20–29	>30
		Adult	<10	10–19	>20
Vitamin D	Plasma[b] 25-(OH)-D$_3$[c] (ng/ml)	All ages	<20[v]	20–29[v]	≥30[v]
	Plasma[b] alkaline phosphatase[c] (U/ml)	Infants	>390	298–390	99–298
		Adults	<40	40–56	57–99
Vitamin E	Plasma[b] α-tocopherol (mg/dl)	All ages	<0.35	0.35–0.80	>0.80
Vitamin K	Clotting time (min)	All ages	>10	About 10	
	Prothrombin time (min)	__[d]	__[d]	__[d]	
Vitamin C	Plasma[b] ascorbic acid (mg/dl)	All ages	<0.20	0.20–0.30	>0.30
	Leukocyte ascorbic acid (mg/dl)	All ages	<8	8–15	>15
	Whole blood ascorbic acid (mg/dl)	All ages	<0.30	0.30–0.50	>0.50
Thiamin	Urinary thiamin (μg/g creatinine)	1–3 years	<120	120–175	>175
		4–6 years	<85	85–120	>120
		7–9 years	<70	70–180	>180
		10–12 years	<60	60–180	>180
		13–15 years	<50	50–150	>150
		Adults	<27	27–65	>65
		Pregnant, 2nd trim.	<23	23–55	>55
		Pregnant, 3rd trim.	<21	21–50	>50
	Urinary thiamin:				
	μg/24 hr	Adults	<40	40–100	>100
	μg/6 hr	Adults	<10	10–25	>25
	Urinary thiamin after load[e] (μg/4 hr)	Adults	<20	20–80	>80
	RBC transketolase stimulated by TPP[f,g] (%)	Adults	>25	15–25	<15
Riboflavin	Urinary riboflavin (μg/g creatinine)	1–3 years	<150	150–500	>500
		4–6 years	<100	100–300	>300
		7–9 years	<85	85–270	>270
		10–15 years	<70	70–200	>200
		Adults	<27	27–80	>80
		Pregnant, 2nd trim.	<39	39–120	>120
		Pregnant, 3rd trim.	<30	30–90	>90
	Urinary riboflavin (μg/24 hr)	Adults	<40	40–120	>120
	Urinary riboflavin (μg/6 hr)	Adults	<10	10–30	>30
	Urinary riboflavin load[h] (μg/4 hr)	Adults	<1,000	1,000–1,400	>1,400
	RBC riboflavin (μg/day)	Adults	<10.0	10.0–14.9	>14.9
	RBC glutathione reductase FAD[i] stimulation (%)	Adults	>40	20–40	<20
Niacin	Urinary N′-methylnicotinamide (μg/g creatinine)	Adults	<0.5	0.5–1.6	>1.6
		Pregnant, 2nd trim.	<0.6	0.6–2.0	>2.0
		Pregnant, 3rd trim.	<0.8	0.8–2.5	>2.5
	Urinary N′-methylnicotinamide (μg/6 hr)	Adults	<0.2	0.2–0.6	>0.6
	Urinary 2-pyridone[j]:N′-methylnicotinamide	All ages	__[k]	<1.0	≥1.0
Vitamin B$_6$	Plasma PalP[l] (nmol/l)	All ages	__[k]	<60[m]	≥60[m]
	Urinary vitamin B$_6$ (μg/g creatinine)	1–3 years	__[k]	<90[m]	≥90[m]
		4–6 years	__[k]	<75[m]	≥75[m]
		7–9 years	__[k]	<50[m]	≥50[m]

(Continued)

TABLE 20.5 (Continued)

Vitamin	Parameter	Age Group	Values, by Category of Status[a]		
			Deficient (High Risk)	Low (Moderate Risk)	Acceptable (Low Risk)
		10–12 years	—[k]	<40[m]	≥40[m]
		13–15 years	—[k]	<30[m]	≥30[m]
		Adults	—[k]	<20[m]	≥20[m]
	Urinary 4-pyridoxic acid (mg/24 hr)	Adults	<0.5[m]	0.5–0.8[m]	>0.8[m]
	Urinary xanthurenic acid after tryptophan load[h] (mg/24 hr)	Adults	>50[m]	25–50[m]	<25[m]
	Urinary 3-OH-kynurenine after tryptophan load[h] (mg/24 hr)	Adults	>50[m]	25–50[m]	<25[m]
	Urinary kynurenine after tryptophan load[h] (mg/24 hr)	Adults	>50[m]	10–50[m]	<10[m]
	Quinolinic acid after tryptophan load[h] (mg/24 hr)	Adults	>50[m]	25–50[m]	<25[m]
	Erythrocyte alanine aminotransferase stimulation by PalP[l] (%)	Adults	—[k]	>25[m]	≤25[m]
	Erythrocyte aspartate aminotransferase stimulation by PalP[l] (%)	Adults	—[k]	>50[m]	≤50[m]
Biotin	Urinary biotin (μg/24 hr)	Adults	<10[m]	10–25[m]	>25[m]
	Whole blood biotin (ng/ml)	Adults	<0.4[m]	0.4–0.8[m]	>0.8[m]
Pantothenic acid	Plasma[b] pantothenic acid (μg/dl)	Adults	—[k]	<6[m]	≥6[m]
	Blood pantothenic acid (μg/dl)	Adults	—[k]	<80[m]	≥80[m,n]
	Urinary pantothenic acid (mg/24 hr)	Adults	—[k]	<1[m]	≥1[m,o]
Folate	Plasma[b] folates (ng/ml)	All ages	<3	3–6	>6
	RBC folates (ng/ml)	All ages	140	140–160	>160
	Leukocyte folates (ng/ml)	All ages	—[k]	<60	>60
	Urinary FIGLU[p] after histidine load[q] (mg/8 hr)	Adults	>50[m]	5–50	<5[r]
Vitamin B_{12}	Plasma[b] vitamin B_{12} (pg/ml)	All ages	100	100–150	>150[s]
	Urinary methylmalonic acid after valine load[t] (mg/24 hr)	Adults	≥300	2–300	≤2
	Urinary excretion of labeled B_{12} after a flushing dose[u] (%)	Adults	<3	3–8	>8

[a]ICNND (1963). *Manual for Nutrition Surveys*, 2nd edn. US Government Printing Office, Washington, DC; Sauberlich, H. E., Skala, J. H., and Dowdy, R. P. (1974). *Laboratory Tests for the Assessment of Nutritional Status*. CRC Press, Cleveland, OH; Gibson, R. S. (1990). *Principles of Nutritional Assessment*. Oxford University Press, New York, NY.
[b]Or serum.
[c]Subject to effects of season and sex.
[d]Results vary according to assay conditions; most assays are designed such that normal prothrombin times are 12–13 seconds, with greater values indicating suboptimal vitamin K status.
[e]Single oral 2 mg dose.
[f]TPP, thiamin pyrophosphate.
[g]The TPP effect.
[h]Single oral 2 g dose.
[i]FAD, flavin adenine dinucleotide, reduced form, 1–3 μmol/l.
[j]N′-methyl-2 pyridone-5-carboxamide.
[k]Database is insufficient to support a guideline.
[l]PalP, pyridoxal phosphate.
[m]These values have only a small database, and therefore are considered as tentative.
[n]Normal values are about 100 μg/dl.
[o]Normal values are 2–4 mg/24 hours.
[p]FIGLU, formiminoglutamic acid.
[q]single oral 2 to 20 mg dose.
[r]Normal adults excrete 5–20 mg/8 hours.
[s]Most healthy individuals show 200–900 pg/ml.
[t]Single oral 5 to 10 g dose.
[u]This is the Schilling test; it involves measurement of labeled vitamin B_{12} excreted from a 0.5 to 2 μg tracer dose after a large flushing dose (e.g., 1 mg) given 1 hour after the tracer.
[v]Modified according to discussion in Chapter 6.

TABLE 20.6 National Surveys of Dietary Intake and Nutritional Status in North America

Survey	Description
USDA historical data of US food supply	Tracking since 1909 of foods available to the American public by disappearance to wholesale and retail markets
US Nationwide Food Consumption Surveys	USDA studies (conducted at about 10-year intervals since 1935) of dietary intakes and food use patterns of American households and individuals
Ten-State Nutrition Survey	NIH study (1968–1970) of nutritional status of >60,000 individuals in 10 US states (California, Kentucky, Louisiana, Maine, Mississippi, South Carolina, Texas, Wisconsin, New York, and West Virginia) selected to include low-income groups
Total Diet Study	FDA study of average intakes of certain essential mineral elements (I, Fe, Na, K, Cu, Mn, Zn), pesticides, toxicants, and radionuclides, based on analyses of foods purchased in grocery stores across the United States
Nutrition Canada	Canadian study conducted in the early 1970s of the nutritional status of >19,000 individuals
National Health and Nutrition Examination (NHANES)	USDA-CDC studies conducted to monitor the overall nutritional status of the US population; NHANESI (1971–1974); NHANESII, (1976–1980); Hispanic HANES (HHANES, 1982–1984); NHANESIII (1988–1994); continuing NHANES
Consumer Survey of Food Intakes of Individuals (CSFII)	USDA studies conducted to monitor food intake patterns of the US population; CSFII 1994–1996, CSFII 1998
What We Eat in America	The integration (in 2002) of NHANES and CSFII; conducted as part of NHANES

the consumption of fruits and vegetables, and reducing intakes of saturated and total fat. The most visible effort of this type has been the 5-A-Day for Better Health Program, initiated in 1991 by the US National Cancer Institute with joint support from American food industry groups, and implemented at the state level under various names – for example, "5-A-Day," "High Five," "Gimme 5."

Despite an emerging picture of health benefits of diets richer in fruits and vegetables, surveys have shown that the regular intakes of fruits and vegetables of many Americans continue to fall short of the 5-A-Day goal. During the past decade these intakes have increased by nearly 29% for vegetables and 38% for non-citrus fruits; however, the list of most frequently consumed fruits and vegetables continues to be short, with lowest consumption observed among lower socioeconomic groups and among individuals unaware of the health benefits attached to fruits and vegetables.

While it is generally accepted that vegetarian, but not fruitarian, diets can be nutritionally adequate if sensibly selected, it is clear that problems can arise in any type of diet if the variety of food is restricted, and particularly, if the consumption of dairy products is low. Therefore, important questions must be raised concerning the impacts on vitamin and overall nutrient intakes of an emphasis on fruits and vegetables, particularly in the context of reduced intakes of meats (vitamins A, B_6, and B_{12}, thiamin, and niacin), and replacement of vegetable oils (important sources of vitamin

E) with reduced- and no-fat substitutes. Such diet changes may increase intakes of vitamins A and C, but they may also reduce intakes of several of the B vitamins.

Historical records of the American food supply would indicate general increases in the amounts of most of the vitamins available for consumption (Table 20.7).[18] Whether such increases have been reflected in the actual intakes of vitamins, or whether they have been distributed democratically across the American population, is not indicated by such gross evaluations of the food supply. It is clear that people with low incomes tend to consume less food, although their food tends to have greater nutritional value per calorie than that consumed by people with greater incomes. While differences in diet quality due to income status appear to be small on average, variation in nutrient intake within groups of individuals appears to be very large. On average, at least, the vitamin intakes of Americans would appear to be generally adequate (Table 20.8). However, studies indicate that the vitamin intakes of many Americans may not meet the Recommended Dietary Allowances (RDAs).

18. These result from increases in the availability of vegetables (+26%), fruits (+22%), grains (+44%), added fats and oils (+56%), and meats (+13%); and decreases in milk (−34%), eggs (−19%) over that same period of time (1970–2007) (Barnard, N. D. [2010] *Am. J. Clin. Nutr.* 91(Suppl.), 1530S).

Nutritional Surveillance Reveals Vitamin Deficiencies

Data on food-nutrient supplies and apparent nutrient consumption are necessary for national food and health policy planning, but they yield no information useful in addressing questions of nutritional status of individuals within populations. To produce such data, nutrition surveys were initiated.

NHANES surveys have revealed:

- *Folate deficiency.* The prevalence of low folate status (erythrocyte folate <140 ng/ml) dropped from 30.4% in 1988–1994 to 2.8% (including 4.5% of women of child-bearing age) in 1999–2000.[19]
- *Vitamin B_{12} deficiency.* Results in 2003–2004 showed <1–3% of children and 3–6% of adults to be of deficient vitamin B_{12} status (serum B_{12} levels <148 pmol/l), with 20% showing marginal status.[20] The prevalence of low status increases with age and with vegetarian (particularly vegan) dietary practice.[21]
- *Vitamin C deficiency.* The prevalence of deficiency (serum vitamin C <11.4 µmol/l) was 7.7% in 2003–2004, with smokers and individuals of low socioeconomic status at elevated risk to being deficient.[22]
- *Vitamin B_6 deficiency.* The prevalence of low vitamin B_6 status (plasma pyridoxal 5′-phosphate <20 nmol/l) was 19–27% across age groups in 2003–2004, with women of child-bearing age or using oral contraceptives, and smokers, being at greatest risk of being deficient.

It is also likely that suboptimal status with respect to other vitamins (e.g., vitamins C, D, E, and B_6) also occur among obese Americans, as has been observed in Norway.[23]

Global Malnutrition

Under the auspices of national programs, bilateral programs, and international agencies, many nutrition surveys have been conducted in developing countries where malnutrition continues to be a problem. These have shown that more than 925 million people (13.1% of the world population) are estimated not to have access to enough food to

19. McDowell, M. A., Lacher, D. A., Pfeiffer, C. M., *et al.* (2008). NCHS Data Brief No. 6.
20. Allen, L. H. (2009). *J. Nutr.* 89(Suppl.), 693S.
21. Elmadfa, I. and Singer, I. (2009). *Am. J. Clin. Nutr.* 89(Suppl.), 1693S.
22. Schleicher, R. L., Carroll, M. D., Ford, E. S., *et al.* (2009). *Am. J. Clin. Nutr.* 90, 1252.
23. Aasheim, E. T., Hofso, D., Hjelmesoeth, J., *et al.* (2008). *J. Nutr.* 87, 362.

TABLE 20.7 Vitamins Available for Consumption[a] by Americans

Vitamin	1909–19	1920–29	1930–39	1940–49	1950–59	1960–69	1970–79	1980–89	1990–99	2000	2005
Vitamin A (IU)	1,040	1,090	1,070	1,210	1,140	1,150	1,260	1,230	1,270	1,260	1,030
Vitamin E (mg)[b]	7.7	8.5	9.2	10.3	10.6	11.7	13.9	15.6	16.8	20.0	21.4
Vitamin C (mg)	95	100	104	112	98	93	112	119	127	130	115
Thiamin (mg)	1.5	1.5	1.4	1.9	1.8	1.9	2.3	2.6	3.0	3.0	2.9
Riboflavin (mg)	1.8	1.8	1.8	2.3	2.3	2.2	2.5	2.8	2.9	2.9	2.8
Niacin (mg)	18	17	16	20	20	20	25	29	32	33	33
Vitamin B_6 (mg)	2.1	2.0	1.9	2.0	1.8	1.8	2.0	2.2	2.4	2.5	2.5
Folate (µg)	309	305	309	325	297	284	326	356	449	706	682
Vitamin B_{12} (µg)	7.8	7.6	7.2	8.6	8.6	8.9	8.9	8.1	7.9	8.2	8.5

[a] Per person per day.
[b] α-Tocopherol equivalents.
Source: Hiza, H. A. B., Bente, L., and Fungwe, T. (2008). *Nutrient Content of the US Food Supply.* Home Economics Report 58, USDA, Washington, DC, p. 72.

TABLE 20.8 Vitamin Intakes from Foods by Americans

Vitamin	Males, by Age Group (years)								
	2–5	6–11	12–19	20–29	30–39	40–49	50–59	60–69	≥70
Vitamin A, µg	619 ± 32[a]	614 ± 29	680 ± 47	597 ± 29	637 ± 27	669 ± 36	660 ± 26	650 ± 30	706 ± 25
Vitamin D, µg	6.5 ± 0.3	5.5 ± 0.3	5.9 ± 0.4	4.9 ± 0.2	4.9 ± 0.4	5.2 ± 0.5	5.3 ± 0.5	4.4 ± 0.2	4.9 ± 0.2
Vitamin E, mg	4.6 ± 0.1	6.0 ± 0.2	7.7 ± 0.5	7.9 ± 0.3	9.1 ± 0.7	8.4 ± 0.3	8.8 ± 0.3	7.6 ± 0.4	7.1 ± 0.3
Vitamin C, mg	4.5 ± 5.0	87.0 ± 4.5	86.6 ± 5.7	93.1 ± 7.1	102.4 ± 8.9	87.0 ± 6.1	91.1 ± 8.7	83.2 ± 4.4	86.1 ± 3.7
Thiamin, mg	1.27 ± 0.04	1.58 ± 0.05	1.88 ± 0.06	2.18 ± 0.19	1.85 ± 0.05	1.97 ± 0.08	1.86 ± 0.06	1.69 ± 0.06	1.59 ± 0.05
Riboflavin, mg	1.93 ± 0.06	2.15 ± 0.07	2.58 ± 0.11	2.60 ± 0.14	2.53 ± 0.09	2.76 ± 0.12	2.55 ± 0.11	2.33 ± 0.08	2.20 ± 0.08
Niacin, mg	15.8 ± 0.4	21.7 ± 0.9	28.9 ± 1.3	34.2 ± 1.6	30.5 ± 1.2	31.7 ± 0.6	29.9 ± 1.0	25.7 ± 0.8	21.9 ± 0.8
Vitamin B$_6$, mg	1.46 ± 0.03	1.74 ± 0.07	2.29 ± 0.13	2.57 ± 0.14	2.39 ± 0.09	2.43 ± 0.05	2.25 ± 0.09	2.06 ± 0.09	1.97 ± 0.06
Folate, µg	427 ± 22	530 ± 14	610 ± 16	692 ± 40	625 ± 34	633 ± 40	586 ± 23	546 ± 19	521 ± 15
Vitamin B$_{12}$, µg	4.50 ± 0.19	5.21 ± 0.21	6.68 ± 0.28	6.95 ± 0.38	6.39 ± 0.36	6.46 ± 0.46	6.13 ± 0.49	6.01 ± 0.46	5.40 ± 0.32

Vitamin	Females, by Age Group (years)								
	2–5	6–11	12–19	20–29	30–39	40–49	50–59	60–69	≥70
Vitamin A, µg	556 ± 27	523 ± 22	528 ± 34	532 ± 33	553 ± 32	555 ± 55	614 ± 26	651 ± 36	616 ± 28
Vitamin D, µg	6.1 ± 0.3	4.6 ± 0.2	3.8 ± 0.2	3.6 ± 0.3	3.6 ± 0.3	3.6 ± 0.2	4.4 ± 0.4	3.9 ± 0.2	3.8 ± 0.2
Vitamin E, mg	4.4 ± 0.2	6.0 ± 0.3	6.0 ± 0.4	6.5 ± 0.5	7.3 ± 0.4	6.6 ± 0.5	7.8 ± 0.5	6.5 ± 0.3	6.2 ± 0.2
Vitamin C, mg	86.7 ± 4.8	75.4 ± 6.1	73.8 ± 5.6	80.8 ± 8.5	77.6 ± 6.2	68.6 ± 6.5	87.3 ± 9.4	75.6 ± 4.7	76.9 ± 3.6
Thiamin, mg	1.19 ± 0.03	1.39 ± 0.06	1.45 ± 0.09	1.38 ± 0.04	1.37 ± 0.03	1.40 ± 0.07	1.43 ± 0.09	1.29 ± 0.05	1.33 ± 0.04
Riboflavin, mg	1.81 ± 0.05	1.81 ± 0.07	1.78 ± 0.07	1.81 ± 0.10	1.87 ± 0.08	1.94 ± 0.09	1.97 ± 0.07	1.90 ± 0.07	1.80 ± 0.04
Niacin, mg	14.2 ± 0.5	18.9 ± 0.6	20.8 ± 0.8	21.0 ± 0.7	21.2 ± 1.0	21.0 ± 1.0	21.2 ± 0.8	18.5 ± 0.4	17.8 ± 0.4
Vitamin B$_6$, mg	1.31 ± 0.05	1.57 ± 0.05	1.63 ± 0.06	1.66 ± 0.09	1.69 ± 0.05	1.63 ± 0.09	1.78 ± 0.13	1.60 ± 0.04	1.54 ± 0.05
Folate, µg	401 ± 16	470 ± 19	509 ± 33	460 ± 19	471 ± 14	470 ± 28	470 ± 28	446 ± 21	451 ± 12
Vitamin B$_{12}$, µg	4.12 ± 0.19	4.52 ± 0.28	4.14 ± 0.23	4.17 ± 0.23	4.38 ± 0.22	4.39 ± 0.29	4.32 ± 0.26	4.31 ± 0.35	4.37 ± 0.26

[a]Mean ± SE
Source: What We Eat In America 2007–2008.

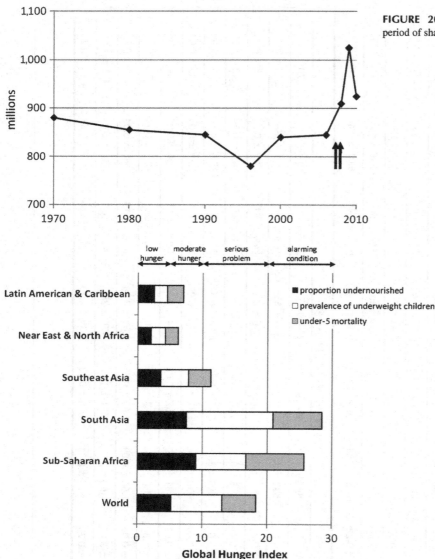

FIGURE 20.1 Trends in global hunger. Arrows indicate period of sharply rising global food prices (FAO estimates).

FIGURE 20.2 Regional rankings according to the 2010 Global Hunger Index. From: *von Grebmer, K., Ruel, M. T., Menon, P., et al. (2010). 2010* Global Hunger Index, The Challenge of Hunger: Focus on the Crisis of Child Undernutrition. *International Food Policy Institute, Washington, DC.*

meet their basic daily needs.[24] Malnutrition and underlying food insecurity have as their root causes poverty. Accordingly, the rise (by some 36%) in global food prices in 2007 was estimated to have increased the number of extremely poor by more than 100 million people by 2008, increasing the numbers of the world's hungry (Fig. 20.1).

Undernutrition has its most visible effects on children (Fig. 20.2). With diets inadequate in quantity and quality, a third of the children in developing countries are stunted.[25] Undernourished children can experience as much as 160 days of illness in a year. At least half of the 10.9 million child deaths that occur each year are because of malnutrition and its potentiating effects on infectious disease[26] – the rate in developing countries is 10 times that in developed countries. The increases in global food prices in 2007 have been estimated to have increased by 44 million in 2008 the number of children experiencing permanent physical and cognitive setbacks due to malnutrition.

---

24. Such widespread malnutrition exists despite impressive gains in global agricultural production. In the past three decades cereal yields have more than doubled, and per capita supplies of food energy are at all-time high levels – exceeding present global needs. However, the newly developed, high-yielding, *green revolution* varieties of major staple grains, being much more profitable than traditional crops (including pulses), have displaced the latter and led to substantial reductions in the diversity of cropping systems. This appears to have contributed to micronutrient malnutrition while increasing caloric output.

---

25. That is, below the third height-for-age percentile. In 2008, at least 90% of the children were stunted in 36 countries (Bhutta, Z. A., Ahmed, T., Black, R. E., *et al.* [2008] *Lancet* 371, 417).
26. That is, diarrheal diseases (61% of deaths), malaria (57%), pneumonia (52%), and measles (45%) (Bryce, J., Boschi-Pinto, C., Shibuya, K., *et al.* [2005] *Lancet* 365, 1147).

"Hidden Hunger"

It is clear that the view of malnutrition resulting mainly from insufficient supplies of macronutrients (i.e., energy and protein) has led to gross underestimates of problems caused by deficiencies of critical micronutrients (i.e., vitamins and trace elements) – problems now being referred to as "hidden hunger." Two billion people live at risk of diseases resulting from deficiencies of vitamin A, iodine, and iron; most of them are women and children living in the less-developed countries of sub-Saharan Africa, the eastern Mediterranean, southern and Southeast Asia, Latin America, the Caribbean, and the western Pacific.[27]

Vitamin A Deficiency

It is estimated that more than 140 million children worldwide are deficient in vitamin A, some 90% of whom live in south Asia and Africa. In recent years, substantial progress has been made in reducing the magnitude of this problem. In 1994, nearly 14 million pre-school children (three-quarters from south Asia) were estimated to have xerophthalmia. By 2005 that prevalence had declined, yet some 5.2 million children remain affected by night blindness. If untreated, two-thirds of those children die within months of going blind owing to their increased susceptibility to infections enhanced by the deficiency. Vitamin A deficiency remains the single most important cause of childhood blindness in developing countries.

A third of the world's pre-school children appear to be growing up with insufficient vitamin A (see Chapter 5, Table 5.1). Subclinical vitamin A deficiency is associated with increased child mortality, having public health significance in at least 122 countries in Africa, Asia, and some parts of Latin America. Studies have shown that providing vitamin A can reduce child mortality by about 25%, and birth-related, maternal mortality by 40%. More than 19 million pregnant women in developing countries are also vitamin A-deficient; a third are affected by night blindness. Most studies show that children with histories of xerophthalmia consume fewer dark green leafy vegetables than their healthier counterparts.

Anemia

An estimated 42% of the world's women are anemic.[28] While anemia can have multiple causes (including malaria and intestinal parasitism), it is thought that at least half of the anemia worldwide is due to nutritional iron deficiency. The prevalence of anemia may underestimate that of low iron status, which affects more than 2.1 billion people, particularly women of reproductive age and pre-school children living in tropical and subtropical zones, but also school-aged children and working men in many areas. Iron deficiency can reduce work capacity, impair learning ability, increase susceptibility to infections, and increase risk of death associated with pregnancy and childbirth.[29]

It is bitterly ironic that iron, the fourth most abundant element in the Earth's crust, is so widely deficient in many diets. This results from the very low bioavailability of inorganic and most plant forms of the element,[30] which comprise the major sources of iron in the diets of the world's poor. For this reason it has been said that anemia may better be described as a vitamin C deficiency disease, as the presence of ascorbic acid in the lumen of the gut is known to markedly improve the bioavailability of dietary iron. Vitamin A has also been shown to affect iron utilization, and deficiencies of vitamin E, vitamin B_6, folate, and vitamin B_{12} also cause anemias.[31]

Other Vitamin Shortages

It is difficult to determine the degree of adequacy of intakes of other micronutrients including vitamins. This is due to limitations in the scope of nutritional surveillance programs, which naturally focus on nutrients associated with prevalent health issues. Still, it is very likely that many groups are at risk of deficiencies of vitamin B_{12} and folate.

Study Questions and Exercises

1. List biomarkers that might be useful in assessing the vitamin status of (i) a food, (ii) a meal, (iii) a national food supply, (iv) an individual, and (v) a population.
2. Devise a system of biochemical measurements that could be performed on a 7 ml sample of fresh blood to yield as much information as possible about the

---

27. An estimated 1,600 million people live in iodine-deficient areas. The most prevalent outcome of iodine deficiency is goiter, affecting some 200 million people. In addition, some 6 million infants born annually to iodine-deficient mothers develop severe mental and neurological impairment known as *cretinism* (half of this number is in southern Asia). The deficiency also increases the rates of stillbirths, abortions, and infant deaths.
28. This ranges from a high in southern Asia (64%) to the lowest, but still surprisingly high, rates (>20%) in industrialized countries.
29. It is estimated that a fifth of maternal mortality is due to the direct (heart failure) or indirect (inability to tolerate hemorrhage) effects of anemia; severe anemia is responsible for nearly one-third of fatalities among children who are not given immediate transfusions.
30. These non-heme sources of iron are typically absorbed at less than 5–15% efficiency. In plants, much of this can be bound to phytic acid, which, being indigestible by most monogastric animals, further reduces iron bioavailability. In contrast, some 25–30% of heme iron in animal products is absorbed by monogastrics.
31. The anemia produced by iron deficiency is easily distinguishable from anemia produced by other vitamin deficiencies. Iron deficiency anemia is normocytic and hypochromic; vitamin E deficiency is characterized by a hemolytic anemia with reticulocytosis; and folate or vitamin B_{12} deficiency anemias are macrocytic and normochromic.

vitamin status of the donor. (Assume enzyme activities can be assayed using no more than 20 μl of plasma or erythrocyte lysate, and other biochemical measurements can be made using no more than 100 μl each.)

3. Give an example of a situation wherein a particular biochemical test may be necessary for the diagnosis of a vitamin-related disorder detected by clinical examination.

4. In general, what are the relationships of biomarkers and clinical findings in the assessment of vitamin status?

5. In general terms, discuss the advantages and disadvantages of the various biomarkers for assessing vitamin status (e.g., functional tests, load tests, urinary excretion tests, circulating metabolite tests).

RECOMMENDED READING

Briefel, R.R., 2006. Nutrition monitoring in the United States. In: Bowman, B.A., Russell, R.M. (Eds.), Present Knowledge in Nutrition, ninth ed. ILSI Press, Washington, DC, pp. 838–858.

Combs Jr, G.F., Welch, R.M., Duxbury, J.M., et al. 1996. Food-Based Approaches to Preventing Micronutrient Malnutrition: An International Research Agenda. Cornell University, Ithaca, NY, p. 68.

Dwyer, J.T., 1991. Nutritional consequences of vegetarianism. Annu. Rev. Nutr. 11, 61–91.

Gary, P.J., Koehler, K.M., 1990. Problems in interpretation of dietary and biochemical data from population studies. In: Brown, M. (Ed.), Present Knowledge in Nutrition, sixth ed. International Life Science Institute – Nutritional Foundation, Washington, DC, pp. 407–414.

Gibson, R.S., 1990. Principles of Nutritional Assessment. Oxford University Press, New York, NY, p. 691.

Henríque-Sánchez, P., Sánchez-Villegas, A., Doreste-Alonso, J., et al. 2009. Dietary assessment methods for micronutrient intake: A systematic review on vitamins. Br. J. Nutr. 102, S10–S37.

Interdepartmental Committee on Nutrition for National Defense, 1963. Manual for Nutrition Surveys, second ed. US Government Printing Office, Washington, DC, p. 327.

Pelletier, D.L., Olson, C.L., Frongillo, E.A., 2006. Food insecurity, hunger and undernutrition. In: Bowman, B.A., Russell, R.M. (Eds.), Present Knowledge in Nutrition, ninth ed. ILSI Press, Washington, DC, p. 906–922.

Román-Vinas, B., Serra-Majem, L., Ribas-Barba, L., et al. 2009. Overview of methods used to evaluate the adequacy of nutrient intakes for individuals and populations. Br. J. Nutr 101 (Suppl. 2), S6–S11.

Sauberlich, H.E., 1999. Laboratory Tests for the Assessment of Nutritional Status, second ed. CRC Press, Cleveland, OH, p. 486.

Quantifying Vitamin Needs

Chapter Outline

Anchoring Concepts

1. The vitamins have many metabolic role(s) essential to normal physiological function; these roles can be compromised by quantitatively insufficient or temporarily irregular vitamin intakes.
2. Vitamin needs can be determined by monitoring responses of parameters related to the metabolic functions and/or body reserves of the vitamins.
3. Quantitative information is available concerning the vitamin contents of many common foods and feedstuffs.

There is not always agreement on the criteria for deciding when a requirement has been met. If the requirement is considered to be the minimal amount that will maintain normal physiological function and reduce the risk of impairment of health from nutritional inadequacy to essentially zero, we are left with questions such as: "What is normal physiological function?" "What is health?" and "What degree of reserve or stores of the nutrient is adequate?" Differences in judgment on such issues are to be expected.

A. E. Harper

Learning Objectives

1. To understand the concepts of minimum requirement, optimal requirement, and allowance as used with respect to vitamins.
2. To understand the methods available for estimating minimal and optimal vitamin requirements of animals and humans.
3. To understand the basis for establishing allowances for vitamins in human and animal feeding.
4. To be familiar with the sources of information concerning vitamin requirements and allowances.

VOCABULARY

Adequate intake (AI)
Daily values (DVs)
Dietary recommendations
Dietary reference intake (DRI)
Dietary standards
Estimated average requirement (EAR)
Estimated safe and adequate daily dietary intake (ESADDI)
Food and Agricultural Organization (FAO)
Food and Nutrition Board
Institute of Medicine (IOM)
Margin of safety
Metabolic profiling
Minimum requirement
National Academy of Sciences
National Research Council
Nutrient allowances
Nutritional essentiality
Optimal requirement
Protective Nutrient Intake
Recommended Daily Allowance (RDA)
Recommended Dietary Intake (RDI)
Recommended Nutrient Intake (RNI)
Upper limit (UL)
World Health Organization (WHO)

The Vitamins. DOI: 10.1016/B978-0-12-381980-2.00021-9

1. DIETARY STANDARDS

Purposes of Dietary Standards

The need to formulate healthy diets for both humans and animals has stimulated the translation of current nutrition knowledge into a variety of **dietary standards** for the intakes of specific nutrients. As these are typically developed by committees of experts reviewing the pertinent scientific literature, they are frequently referred to as **dietary recommendations**. Formally, they may be called **Recommended Dietary Allowances (RDAs)** or **Recommended Dietary Intakes (RDIs)**.

Allowances and Requirements

Regardless of the ways they may be named, dietary standards differ from nutrient requirements, although they are derived from the latter. Dietary standards are relevant to populations; they describe the average amounts of particular nutrients that should satisfy the needs of almost all healthy individuals in defined groups. In contrast, **nutrient requirements** are relevant to individuals; they describe amounts of particular nutrients that satisfy certain criteria related to the metabolic activity of those nutrients or to general physiological function. Because recommended allowances and intakes are designed to satisfy the needs of groups of individuals whose nutrient requirements vary, they, by definition, exceed the average requirement.

2. DETERMINING DIETARY STANDARDS FOR VITAMINS

Determining Nutrient Requirements

The nutrient requirement is a theoretical construct that describes the intake of a particular nutrient that supports a body pool of the nutrient and/or its metabolically active forms adequate to maintain normal physiological function. In practice, it is generally used in reference to the lowest intake that supports normal function; that is, the **minimum requirement**. Minimum requirements, while seemingly physiologically relevant, are difficult to define and impossible to measure with any reasonable precision. They can vary according to the criteria by which they are defined. This problem is illustrated by the widely varying estimates of the vitamin A requirement for calves; various estimates may be derived by different criteria (Table 21.1).

Therefore, in order to be relevant to the overall health of the typical individual, estimates of minimum nutrient requirements must be based on responses of obvious physiological importance. For many nutrients (e.g., the indispensable amino acids) it may thus be appropriate to define minimum requirements on the basis of a fairly non-specific parameter such as growth. For vitamins, however, it is appropriate to define minimum requirements on the basis of biomarkers more specifically related to their respective metabolic functions, such as enzyme activities and tissue concentrations, as these can reflect changes at the earlier stages of vitamin deficiencies. The most useful biomarkers of vitamin status are those that respond early to deprivation of the vitamin, as such changes can be used to detect suboptimal vitamin status at the early and most readily corrected stages.

Quantifying the minimum requirement, even with the use of an appropriate biomarker, is not straightforward. It generally requires an experimental approach in which the test animals are fed a basal diet constructed to be deficient in the nutrient of interest but otherwise adequate with respect to all known nutrients, with this diet being supplemented with known amounts of the nutrient of interest. This may necessitate the use of uncommon feedstuffs such that the diet bears little similarity to those used in practice, and it usually means that the test nutrient is provided largely in free form, which may not resemble its form in practical foods and feedstuffs. Even with these caveats in mind, the level of nutrient intake to be identified as the minimum requirement is not always clear, as the optimal value for that biomarker is usually a matter of judgment.

Most responses of specific nutrient-depleted animals to input of the relevant nutrient appear to be curvilinear (Fig. 21.1, right panel). However, in most nutrient requirement experiments both rectilinear and curvilinear models usually fit equally well. For this reason, many investigators have used rectilinear models to impute requirements (e.g., the x value of the intercept of the two linear regressions of the observed data in broken-line regression analyses;[1] Fig. 21.1, left panel). Others, however, have

TABLE 21.1 Vitamin A Requirements of Calves, Based on Different Criteria

Criteria	Estimated Requirement, IU/Day
Prevention of nyctalopia	20
Normal growth	32
Normal serum retinol levels	40
Moderate hepatic retinyl ester reserves	250
Substantial hepatic retinyl ester reserves	1,024

Source: Marks, J. (1968). The Vitamins in Health and Disease: A Modern Reappraisal. Churchill, London, p. 32.

---

1. This approach offers the advantage of rendering a requirement value that is derived mathematically from the observed data; however, that value tends to be in the region of greatest variation in the input–response curve.

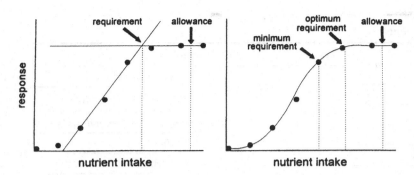

FIGURE 21.1 Requirements and allowances for nutrients are determined from the responses of physiologically meaningful parameters to the level of nutrient intake.

used curvilinear models, which consider the variations in the experimental population of both the measured response and the nutrient need for maintenance (usually related to body size). From the proposition of curvilinearity, it follows that no value can be properly described as the "requirement" of the test populations. Nevertheless, the approach can be used to determine the risk of not fully satisfying the requirement for given proportions of the experimental population. The level of intake associated with acceptable risk of deficiency (a matter of judgment) is frequently called the "optimum requirement" or "level of optimum intake." In public health, such levels are determined on the basis of assumptions regarding putative health risks and in consideration of inter-individual variation. In livestock production, where the cost of feeding accounts for as much as three-quarters of the total cost of production, it is necessary to determine intakes of the more costly nutrients (protein, limiting amino acids, energy) that optimize economic efficiency.

3. FACTORS AFFECTING VITAMIN REQUIREMENTS

Many factors can affect nutrient requirements (Tables 21.2, 21.3), such that those of individuals with the same general characteristics can vary substantially. For most nutrients, the requirements of individuals in given populations appear to be normally distributed. For this reason, in the absence of clear information, it is reasonable to assume that the variations in vitamin requirements are similar to those typically observed in biological systems – that is, normally distributed with a coefficient of variation of 10–15%.

Developing Vitamin Allowances

Because nutrient requirements, even for the best cases, are quantitative estimates based on data of uncertain precision derived from a limited number of subjects, those values have limited practical usefulness. In practice, **nutrient allowances**, or recommended intakes, are far more useful. They are selected to meet the needs of those individuals with the greatest requirements. That is, an allowance is set at the right-hand tail of the natural distribution of requirements. An allowance exceeds the **estimated average requirement (EAR)** for the population by an increment sometimes referred to as a **margin of safety**. Allowances for vitamins, particularly in livestock feeding, have often been set on the basis of practical experience; however, rational approaches are available and have been used for establishing official nutrient allowances for both animals and humans. The latter are generally described in statistical terms relating to the proportion of the target population, the requirements of which would be met by the recommended level of intake. For example, committees of the Food and Nutrition Board and WHO/FAO[2] have set allowances at 2 standard deviations (SD) above the EAR, a decision appropriate to the needs of approximately 97.5% of the population (Fig. 21.2).[3] This method has yielded satisfactory results, likely due in part to the generous estimates of EARs generally made by expert committees.

Allowances, therefore, are derived from estimates of EARs (of typical individuals) made from actual biological data, usually from nutritional experiments. Because they are used as standards for populations, allowances are developed in consideration of risk of nutrient deficiency. Therefore, allowances are relevant to specified populations, with their characteristic food habits and inherent variations in nutrient requirements. For example, the Recommended Dietary Allowances established by the US Food and Nutrition Board are implicitly intended to relate to the US population. These recommendations were originally developed to facilitate the wartime planning of food supplies, but have become a key source of information for making food and health policy in the US and elsewhere (Table 21.4).

---

2. World Health Organization and Food and Agricultural Organization, respectively, of the United Nations.
3. A notable exception is in the setting of allowances for energy; these are typically set at the estimated average requirements of classes of individuals, for the reason that, unlike other nutrients, both intake and expenditure of energy appear to be regulated such that free-living individuals with free access to food maintain (at least very nearly) energy balance.

TABLE 21.2 Factors Affecting Vitamin Requirements

Factor	Examples	Factor	Examples
Physiological determinants	Active growth		Gluten enteropathy
	Pregnancy		Intestinal parasitism (e.g., hookworm, *Strongyloides*, *Giardia lamblia*, *Dibothriocephalus latus*)
	Lactation		
	Aging		Enteritis
	Intra-individual variation		Cystic fibrosis
	Level of physical activity		Certain drug treatments
	Obesity	Hypermetabolic states	Thyrotoxicosis
Hereditary conditions	Inborn vitamin-dependent diseases; polymorphisms of vitamin transporters, receptors, vitamin-dependent enzymes, and enzymes of vitamin metabolism		Pyrexial disease
			Infections
		Conditions causing decreased nutrient utilization	Chronic liver disease
			Chronic renal disease
Conditions causing maldigestion/malabsorption	Pancreatitis	Conditions involving increased cell turnover	Congenital and acquired hemolytic anemias
	Gastrointestinal surgery		Sickle cell disease
	Endocrine disorders (e.g.,diabetes, hypoparathyroidism, congenital or acquired hemolytic anemias, Addison's disease)	Conditions increasing nutrient turnover/loss	Extensive burns
			Bullous dermatoses
	Hepatobiliary disease		Enteropathy
	Intestinal resection/bypass		Nephrosis
	Pernicious anemia		Surgery
	Regional ileitis		Hemodialysis
	Radiation injury		Smoking
	Kwashiorkor		
	Pellagra		

Differences between Requirements and Allowances

Confusion surrounds the allowances for the vitamins (and other nutrients) that have been developed by various expert committees. Some questions arise, particularly concerning dietary recommendations for livestock, because the rationales for such values are frequently not presented. A fairly common example is the mistaken impression, on the part of formulators of animal feeds, that vitamin allowances are requirements; this mistake can lead to vitamin over-fortification of those feed vitamins. Other questions arise over the publication of differing recommendations by different committees of experts, all of whom consider the same basic data in their respective reviews of the pertinent literature. This situation results from the paucity of clear and compelling data on nutrient requirements; differences in

environmental conditions and food supplies; and the lack of consensus on such issues as criteria for defining requirements, appropriate margins of safety, and whether standards should be based on intakes of food as consumed or as purchased. These considerations make the variable factor of scientific judgment important in estimating the nature of nutrient requirements. Thus, dietary recommendations are revised periodically[4] as new information becomes available.

---

4. As available information bases improve, expert committees typically have reduced the levels of their recommendations for nutrient allowances. This likely reflects the basically conservative nature of the committee system used for these purposes, whereby the paucity of data tends to be handled by generously estimating quantitative needs.

TABLE 21.3 Physiologically Significant Drug–Vitamin Interactions

Vitamin	Drugs	Vitamin	Drugs
Vitamin A	Diuretic: spironolactone		Anticholinergic, anti-Parkinsonian: L-dopa
	Bile acid sequestrant: cholestyramine, colestipol		Antihypertensive: hydralazine
	Laxative: phenolphthalein, mineral oil		Chelating agent, antiarthritic: penicillamine
Vitamin D	Antibacterial: isoniazid		Ethanol
	Anticonvulsant: phenytoin, diphenylhydantoin, primidone		Oral contraceptives
	Bile acid sequestrant: colestipol		Smoking
	Laxative: phenolphthalein, mineral oil	Biotin	None reported
Vitamin E	Smoking	Pantothenic acid	None reported
Vitamin K	Anticoagulant: warfarin	Folate	Antacid: sodium bicarbonate, aluminum hydroxide
	Anticonvulsant: phenytoin, diphenylhydantoin, primidone		Anti-bacterial: sulfasalazine, trimethoprim
	Bile acid sequestrant: colestipol		Anticonvulsant: phenytoin
	Immunosuppressant: cyclosporins		Anti-inflammatant: sulfasalazine, aspirin
	Laxative: mineral oil, phenolphthalein		Anti-malarial: pyrimethamine
Vitamin C	Anti-inflammatory: aspirin		Antineoplastic: methotrexate
	Oral contraceptives		Bile acid sequestrant: cholestyramine, colestipol
	Smoking		Diuretic: triamterene
Thiamin	Ethanol		Ethanol
Riboflavin	Antibacterial: boric acid		Oral contraceptives
	Tranquilizer: chlorpromazine	Vitamin B$_{12}$	Analytical reagent: biquanides
Niacin	Antibacterial: isoniazid		Antibacterials: p-aminosalicylic acid, neomycin
	Anti-inflammatant: phenylbutazone		Antihistaminic: cimetidine, ranitidine
Vitamin B$_6$	Analytical reagent: thiosemicarbazide		Anti-inflammatant, gout suppressant: colchicine
	Antibacterial: isoniazid		Bile acid sequestrant: cholestyramine, colestipol

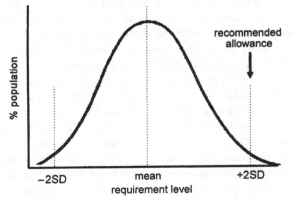

FIGURE 21.2 The "mean plus 2 SD" conversion algorithm for determining recommended dietary allowances.

Applications of RDAs

Questions about nutrient allowances for humans arise owing to the application of those values to purposes for which they were not intended. Although the RDAs were developed originally for use in planning food supplies for groups of people,[5] they are in fact today used for many other purposes: evaluating the nutritional adequacy of diets; evaluating results of dietary surveys; setting standards for food assistance programs, institutional feeding programs, and food and nutrition regulations; developing

5. The RDAs were originated in 1941 for use in planning US food policy during World War II.

TABLE 21.4 History of the Recommended Daily Allowances (RDAs) for Vitamins[a]

	1941	1948	1957	1968	1976	1980	1989	1997–2001	2010
Vitamin A (mg RE)	1,000	1,000	1,000	1,000	1,000	1,000	1,000	900	*[b]
Vitamin D (IU/μg)	—	—	—	400 IU	400 IU	5 μg	5 μg	[10][c]	*[b]
Vitamin E (IU/ mg)	—	—	—	30 IU	15 IU	10 IU	10 IU	15 mg	15 mg
Vitamin K (μg)	—	—	—	—	—	—	80	[120][c]	*[b]
Vitamin C (mg)	75	75	70	60	45	60	60	90	*[b]
Thiamin (mg)	2.3	1.5	0.9	1.3	1.4	1.4	1.5	1.2	*[b]
Riboflavin (mg)	3.3	1.8	1.3	1.7	1.6	1.6	1.7	1.3	*[b]
Niacin (mg)	23	15	15	17	18	18	19	16	*[b]
Vitamin B_6 (mg)	—	—	—	2.0	2.0	2.2	2.0	1.7	*[b]
Pantothenic acid (mg)	—	—	—	—	—	—	—	[5][c]	*[b]
Biotin (mg)	—	—	—	—	—	—	—	[30][c]	*[b]
Folate (μg)	—	—	—	400	400	400	200	400	*[b]
Vitamin B_{12} (μg)	—	—	—	3.0	3.0	5.0	2.0	2.4	*[b]

[a]Values shown are for males, 25–50 years of age or (for vitamin D in 2010) 31–50 years of age.
[b]Values set in 1997–2001 remain in use.
[c]RDAs not available for these vitamins; values shown are Adequate Intakes (AIs).

food and nutrition education programs; and formulating new food products and special dietary foods. Many of these uses have fostered criticism of the RDAs for not dealing with associations of diet and chronic and degenerative diseases, for not including guidelines for appropriate intakes of fat, cholesterol, and fiber, and for not providing guidance for food selection and prevention of obesity. These problems stem from fundamental misunderstandings that, while the RDAs may be used in certain programs to implement sound public health policy, they are not intended to be policy recommendations *per se*. In fact, RDAs cannot serve as general dietary guidelines; by definition, they are reference standards dealing with nutrients, whereas dietary guidelines deal primarily with foods. The RDAs are, in fact, standards on which sound dietary guidelines are to be based.

It should be kept in mind that the RDAs, like other nutrient allowances, are intended to relate to intakes of nutrients as part of normal diets[6] of specified populations. They are intended to be average daily intakes based on periods as short as 3 days (for nutrients with fast turnover rates) to several weeks or months (for nutrients with slower turnover rates).

The RDA Concept

In one sense, the RDA construct is somewhat archaic in that it fails to pertain to biological functions of nutrients that may be non-specific or non-traditional, in the context of being outside the known functions of nutrients. The conceptual framework upon which the RDA was derived is being replaced by a new, more individualistic view of nutrition that relates more broadly to health. This view is the basis of problems that have become apparent concerning the RDA. To retain the practical utility of the RDA, it will be necessary to reconstruct it; such reconstruction must be based on new paradigms for nutritional science that, informed by the "genomics revolution," explicitly consider individual metabolic characteristics.

The RDA was developed to facilitate food planning for the US population. It is a child of the central concept of the field of nutrition, **nutritional essentiality**, which has been used to describe those factors in the external chemical environment that are specifically required for normal metabolic functions and, accordingly, those exogenous sources on which organisms depend for normal physiologic functions (e.g., growth, reproductive success, survival, and freedom from certain clinical/metabolic disorders). The

---

6. The RDA subcommittee emphasized that the RDAs can typically be met or closely approximated by diets that are based on the consumption of a variety of foods from diverse food groups that contain adequate energy.

vitamins are among the more than 40 such factors generally considered to be nutritionally essential – i.e., indispensable in the diets of animals and humans. Deprivation of any one of these is made manifest by clinical signs that are usually specific in nature. Nutritional essentiality has been based on empirical findings that nutrients function to prevent ill health in very specific ways. Under this paradigm, nutrient deficiency diseases have played important roles in the development of our knowledge of nutrition: their specific prevention has been used both to define nutrient essentiality and to quantify nutrient needs. Indeed, a nutrient has not been considered essential unless a clinical disease has been related specifically to its deprivation. Therefore, as the term has been used, nutritional essentiality clearly connotes the specific prevention of deficiency disease. This connotation is expressed in the quantitative estimation of population-based nutrient needs, the RDAs; but it now serves to limit the essentiality paradigm as a conceptual framework in modern nutrition, which is cast in a different context – one in which optimum health is more broadly conceptualized.

These limitations have produced a number of major questions concerning RDAs:

- Which level of nutrient need should define a requirement? Should this be the level that supports all/some dependent enzymes at 50, 80 or 100% of maximal activity?
- Can a nutrient be conditionally essential (e.g., choline for individuals with low methionine intakes; glutamine for surgical patients)?
- How can an individual's varying nutrient requirements be described (e.g., effects of infection, oxidative stress)?
- Can nutrients be said to be required for their non-specific effects (e.g., antioxidants)?
- Can a non-nutrient be said to be required (e.g., dietary fiber)?

The specific deficiency disease connotation of the RDA is, perhaps, most troublesome in dealing with issues of diet and health, for the essentiality paradigm does not pertain to functions of nutrients that are either non-specific or non-traditional, i.e., outside the known functions of nutrients.

Considering Non-Classical Functions of Nutrients

For some dietary factors, functions influencing the risk of chronic disease have been suggested by epidemiological and experimental animal model studies and, to a lesser extent, clinical trials. Reduced risks of several chronic diseases have been associated with increased intakes

and/or status of several vitamins. The metabolic bases of these linkages remain to be elucidated; indeed, these areas are among the most active in contemporary nutritional science:

- *Cancer* and foods containing vitamin A or vitamin C, intakes of riboflavin, and plasma levels or intakes of α-tocopherol, carotenoids, and 25-OH-vitamin D.
- *Cardiovascular disease* and intakes of vitamin C, vitamin E, and β-carotene.
- *Neural tube defects* and periconceptual folate intake.
- *Diabetes* and *multiple sclerosis* and plasma 25-OH-vitamin D.

The case of the apparent effects of antioxidants illustrates the limitation of the RDA construct. Current thinking is that antioxidant nutrients (vitamins E and C, selenium, and, perhaps, β-carotene) participate in a system of protection against the deleterious metabolic effects of free radicals. Because many diseases are thought to involve enhanced free-radical production, protection from oxidative stress is thought to be critical to normal physiologic function. According to this hypothesis, antioxidants would be expected to suppress radical-induced DNA damage involved in the initiation of carcinogenesis, to inhibit the oxidation of cholesterol in low-density lipoproteins (LDLs) in atherosclerosis, and to inhibit the oxidation of lens proteins in cataracts. Antioxidant nutrients have been shown to enhance immune functions, which may also contribute to reduced risks of cancer as well as infectious disease. Many of these antioxidant effects do not appear to have the specificity connoted by the essentiality paradigm. For example, the complementary natures of the antioxidant functions of vitamins E and C and selenium suggest that any one may spare needs for the others in protecting against subcellular free-radical damage and LDL oxidation. It is likely that, through such biochemical mechanisms, the antioxidant nutrients may be modifiers of disease risk rather than primary agents in disease etiologies. However non-specific they may be, such effects raise legitimate questions concerning nutrient need – questions not easily addressed under the essentiality paradigm or translated into RDAs.

New Paradigms for Nutrition

The term "essentiality" has become rather elastic in its application. Nutrients have come to be described as being "required" or "essential" for particular functions. Some are called "dispensable" or "indispensable" under specific conditions; several are recognized as "beneficial" at levels greater than those that are considered to be "required." Indeed, the translation of nutritional knowledge into dietary guidance requires such language. However, the emergence of this sort of terminology indicates that the essentiality paradigm is, in fact, being displaced by a new conceptualization of nutrition.

It is likely that new paradigms of nutrition will encompass an individualized view of organisms that recognizes both endogenous and exogenous conditions as determinants of the nature and amounts of factors available from the external chemical environment that must be obtained to support definable health outcomes. Accordingly, such factors will be considered as nutrients *if* and *when* their activities, in the metabolism of the host and/or the associated microflora, are beneficial to those outcomes. This view will recognize a variety of outcomes as being appropriate for various individuals, both within and between a species/ population. Thus, freedom from overt physiological dysfunction as well as reduced risk of chronic diseases will be important outcomes in human nutrition, whereas such outcomes as maximal growth rate, optimal efficiency of feed utilization, and minimal susceptibility to infection will be priorities in livestock nutrition.

The old paradigm is being outgrown at an increasing pace with progress in the modern field of molecular biology. It has now become clear that some nutrients function as gene regulators, and that predisposition to disease can have genetic bases. The mapping of the human genome and human microbiome has led to the development of powerful tools to study individual metabolic characteristics. As **metabolic profiling** becomes more feasible, it will become possible to address individuals' nutritional needs on the basis of their peculiar metabolic characteristics. Not only clinicians, but also dieticians, will be able to ask such questions as whether an individual has sodium-sensitive hypertension, a cystathionine β-synthase mutation or the methylenetetrahydrofolate reductase (MTHFR) C677T/C genotype. The time is quickly approaching when it will be possible to identify disease predisposition, metabolic characteristics, and specific dietary needs of individuals based on rapid, genomic/metabolomic analyses. As that becomes practicable, the population-based paradigm will lose much of its value.

Reconstructing the RDA

This crisis in conceptualization became manifest in the lively discussion concerning the need for new approaches to the development of dietary recommendations that preceded the development of the **Dietary Reference Intakes (DRIs)** in the 1990s. The challenge was to recreate the RDA as a useful construct under this emerging paradigm that addressed both the prevention of overt nutritional deficiencies as well as the maintenance of health. It became clear that DRIs must accommodate the possibility that a nutrient can have beneficial action at levels above those previously thought to be "required" for normal physiologic

function. To do this, it is necessary to convey information concerning three levels of nutrient activity:[7]

- the amount of nutrient required to prevent overt deficiency disease
- if applicable, the amount of nutrient that may provide other health benefits
- the amount of nutrient that may carry specific health hazards.

4. VITAMIN ALLOWANCES FOR HUMANS

Several Standards

The first nutrient allowances were published 50 years ago by the US National Academy of Sciences. Based on available information, those Recommended Dietary Allowances (RDAs) have since been revised periodically. Since the first publication of RDAs, similar dietary standards have been produced by several countries and international organizations. For the reasons mentioned previously, the various recommendations tend to be similar but not always identical. For example, most are based on food as consumed; however, some are based on food as purchased, making them appear higher than those of other countries.

RDAs

The RDAs are probably the most widely referenced of the dietary standards, and the set that is most comprehensive with respect to the vitamins. It is worth noting that the RDAs for vitamins are still not complete; that is, quantitative recommendations on some (e.g., vitamin D, vitamin K, biotin, pantothenic acid) have not been made owing to a still-insufficient information base. In 1980, the problem of dealing with nutrients known to be essential for humans but for which insufficient data are available was handled

---

7. The need for a tri-level allowance is perhaps best illustrated by the impending situation concerning the essential trace element selenium (Se). The RDA for Se was established in 1989, largely on the basis of the amount judged sufficient to support maximal activities of the Se-dependent glutathione peroxidase in the plasma of young men: adult women, 50 μg; adult men, 70 μg. These levels may even be high, a point of little importance to the US population, which probably has regular Se intakes in the range of 80–200 μg/day. What is particularly important, however, is that clinical intervention trials have confirmed a large body of animal tumor model studies in showing cancer-chemopreventive activity of Se at intakes substantially greater than those sufficient to support maximal activities of selenoenzymes. In a decade-long, double-blind trial involving a cohort of older Americans, our group (Clark, L. C., Combs, G. F. Jr, Turnbull, B. W., *et al.* [1996] *J. Am. Med. Assoc.* 276, 1957) found that the use of a daily oral supplement of Se (200 μg, i.e., about twice the normal dietary intake) was associated with significant reductions in risks of total cancer and all leading cancers (lung, colorectal, prostate) in that population.

by including provisional recommendations. The ninth and tenth editions of the RDAs included **Estimated Safe and Adequate Daily Dietary Intake (ESADDI)** ranges of daily dietary intakes for such nutrients. This terminology is no longer used; instead, estimates of **Adequate Intakes (AIs)** are now used in cases where available data are judged to be insufficient for developing RDAs.

The setting of dietary allowances is an exercise of experts who evaluate published scientific literature. That different expert panels can reach different conclusions from the same body of published data is evidenced by the differences in national dietary allowances. That the growing body of relevant data also changes over time is evidenced by the changes in RDAs over the history of that institution (Table 21.4). For example, only in 1968 were RDAs established for vitamins D, E, C, and B$_{12}$, and folate. In 1989, an RDA for vitamin K was first set; however, that value, as well as that for vitamin D, was replaced with AI values in 2000. Only in 2010 were RDAs for vitamin D established.[8]

Accordingly, the setting of dietary allowances is a continuing process. There is growing appreciation of the need and advantages of harmonizing these processes as carried out in different countries.[9]

The DRIs

The most recent (2010) edition of dietary allowances for vitamin D and calcium builds on the broad, previous (1997–2001) edition of dietary allowances (Table 21.5) produced by the Food and Nutrition Board. These were preceded in the 1990s by a series of workshops that addressed the conceptual framework upon which that work was based. This resulted in an expansion of the former RDAs with a system of Dietary Reference Intakes (DRIs).[10] This system involved four types of reference values (Fig. 21.3):

- *Estimated Average Requirements (EARs)* – intakes to meet the requirements of half the healthy individuals in each age–sex specific demographic sub-group of the American population.
- *Recommended Daily Allowances (RDAs)* – average daily intake levels sufficient to meet the requirements of nearly all (97–97%) of the healthy individuals in each age–sex specific demographic subgroup. The RDA is calculated from the EAR:

$$RDA = EAR + 2SD_{EAR}$$

where SD_{EAR} is the standard deviation of the EAR. While the RDA resembles that construct used previously, in fact it is different in that it assumes the SD_{EAR} to be 10% of the EAR, whereas a value to 15% had been used previously. For this reason, many of the new RDAs are lower than earlier ones.

- *Adequate Intakes (AIs)* – observed and/or experimentally determined approximations of nutrient intakes of groups of healthy individuals extrapolated to each age–sex demographic subgroup; used when data are judged to be insufficient for the estimation of an EAR and subsequent calculation of an RDA.
- *Tolerable Upper Intake Limits (ULs)* – the highest level of daily intake that is likely to pose no risks of adverse health effects to almost all healthy individuals in each age–sex specific demographic subgroup. The use of ULs will facilitate the development of recommendations of nutrient intakes at what might be called "supra-nutritional" levels when such intakes have been shown to have health benefits. Pertinent to this consideration is emerging understanding of the roles of at least several vitamins in reducing chronic disease risk.

This approach to the development of dietary allowances presumes the availability of empirical data for the distribution of individual nutrient requirements, necessary for calculating both the EAR and the SD_{EAR}. However, such data are available for very few nutrients. Thus, the DRI process involved a consensus opinion to assume that

8. These were met with instant dispute for having been set too low to support optimal vitamin D status: Hall, L. M., Kimlin, M. G., Aronov, P. A., *et al.* (2010). *J. Nutr.* 140, 542; Heaney, R. P. and Holick, M. F. (2011). *J. Bone Mineral Res.* 26, 455.

9. Fairweather-Tait, S., Gurinović, M., van Ommen, B., *et al.* (2010). *Eur. J. Clin. Nutr.* 64, S26.

10. Questions concerning the means of developing consistent and reliable standards led the Tenth RDA Committee to review the scientific basis of the entire RDA table. The Committee recommended a lower RDA for vitamin A (reducing the RDA for men 1,000 to 700 IU, and that for women from 800 to 600 IU) and vitamin C (reducing the RDA for men from 60 to 40 mg, that for women from 60 to 30 mg, and that for infants from 35 to 25 mg). It was reported that these reductions were resisted by the Food and Nutrition Board, which had been advised by another subcommittee to increase the intakes of these nutrients based on cancer risk reduction potential. In an unexpected move, the Board elected not to accept the recommendations of the RDA Committee. This move prompted lively discussion in the Nutrition community, ultimately resulting in a re-thinking of the RDA construct and the development of the DRIs.

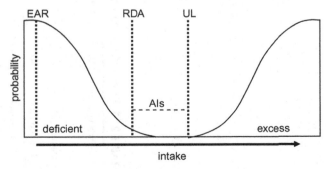

FIGURE 21.3 Conceptual basis for DRIs.

TABLE 21.5 Food and Nutrition Board Recommended Daily Allowances (RDAs) for Vitamins

Age–Sex Group	Vitamin A (μg[a])	Vitamin D (μg)	Vitamin E (mg[b])	Vitamin K (μg)	Vitamin C (mg)	Thiamin (mg)	Riboflavin (mg)	Niacin (mg[c])	Vitamin B$_6$ (μg)	Pantothenic Acid (μg)	Biotin (μg)	Folate (μg[d])	Vitamin B$_{12}$ (μg)
Infants:													
0–6 months	[400][e]	10	4	[2.0][e]	[40][e]	[0.2][e]	[0.3][e]	[2][e]	[0.1][e]	[1.7][e]	[5][e]	[65][e]	[0.4][e]
7–11 months	[500][e]	10	5	[2.5][e]	[50][e]	[0.3][e]	[0.4][e]	[4][e]	[0.3][e]	[1.8][e]	[6][e]	[80][e]	[0.5][e]
Children:													
1–3 years	300	15	6	[30][e]	15	0.5	0.5	6	0.5	2	[8][e]	[150][e]	0.9
4–8 years	400	15	7	[55][e]	25	0.6	0.6	8	0.6	3	[12][e]	[200][e]	1.2
Males:													
9–13 years	600	15	11	[60][e]	45	0.9	0.9	12	1.0	4	[20][e]	[300][e]	1.8
14–18 years	900	15	15	[75][e]	75	1.2	1.3	16	1.3	5	[25][e]	[400][e]	2.4
19–30 years	900	15	15	[120][e]	90	1.2	1.3	16	1.3	5	[30][e]	[400][e]	2.4
31–50 years	900	15	15	[120][e]	90	1.2	1.3	16	1.3	5	[30][e]	[400][e]	2.4
51–70 years	900	15	15	[120][e]	90	1.2	1.3	16	1.7	5	[30][e]	[400][e]	2.4
>70 years	900	20	15	[120][e]	90	1.2	1.3	16	1.7	5	[30][e]	[400][e]	2.4
Females:													
9–13 years	600	15	11	[60][e]	45	0.9	0.9	12	1.0	4	[20][e]	[300][e]	1.8
14–18 years	700	15	15	[75][e]	65	1.0	1.0	14	1.2	5	[25][e]	[400][e]	2.4

19–30 years	700	15	15	[90]e	75	1.1	1.1	14	1.3	5	[30]e	[400]e	2.4
31–50 years	700	15	15	[90]e	75	1.1	1.1	14	1.3	5	[30]e	[400]e	2.4
51–70 years	700	15	15	[90]e	75	1.1	1.1	14	1.5	5	[30]e	[400]e	2.4
>70 years	700	20	15	[90]e	75	1.1	1.1	14	1.5	5	[30]e	[400]e	2.4
Pregnancy:													
≤18 years	750	15	15	[75]e	80	1.4	1.4	18	1.9	6	[30]e	[600]e	2.6
19–30 years	770	15	15	[90]e	85	1.4	1.4	18	1.9	6	[30]e	[600]e	2.6
31–50 years	770	15	15	[90]e	85	1.4	1.4	18	1.9	6	[30]e	[600]e	2.6
Lactation:													
≤18 years	1,200	15	19	[75]e	115	1.6	1.4	17	2.0	7	[35]e	[550]e	2.8
19–30 years	1,300	15	19	[90]e	120	1.6	1.4	17	2.0	7	[35]e	[550]e	2.8
31–50 years	1,300	15	19	[90]e	120	1.6	1.4	17	2.0	7	[35]e	[550]e	2.8

aRetinol equivalents.
bα-Tocopherol.
cNiacin equivalents.
dFolate equivalents.
eRDA has not been set; AI is given instead.
Sources: Food and Nutrition Board (1997). *Dietary Reference Intakes for Calcium, Phosphorus, Magnesium, Vitamin D and Fluoride.* National Academy Press, Washington, DC, 432 pp.; Food and Nutrition Board (2000). *Dietary Reference Intakes for Thiamin, Riboflavin, Niacin, Vitamin B6, Folate, Vitamin B12, Pantothenic Acid, Biotin and Choline.* National Academy Press, Washington, DC, 564 pp.; Food and Nutrition Board (2000). *Dietary Reference Intakes for Vitamin C, Vitamin E, Selenium and Carotenoids.* National Academy Press, Washington, DC, 506 pp.; Food and Nutrition Board (2001). *Dietary Reference Intakes for Vitamin A, Vitamin K, Arsenic, Boron, Chromium, Copper, Iodine, Iron, Manganese, Molybdenum, Nickel, Silicon, Vanadium and Zinc.* National Academy Press, Washington, DC, 773 pp.; Food and Nutrition Board (2010). *Dietary Reference Intakes: Calcium, Vitamin D.* National Academy Press, Washington, DC, p. 1115.

the distributions of individual nutrient requirements are each normal with a coefficient of variation (CV) of 10%, with only two exceptions: for vitamin A, CV = 20%; for niacin, CV = 15%.

International Standards

The FAO and WHO have established standards for energy, protein, calcium, iron, and eight of the vitamins (Table 21.6). This system of recommendations is intended for international use, and thus to be relevant to varied population groups. It includes reference values similar to those used by the Food and Nutrition Board: requirements similar to EARs, **Recommended Nutrient Intakes (RNIs)**, similar to RDA; and Upper Tolerable Nutrient Intake Levels, similar to the ULs. In addition, the FAO/WHO system provides a value applicable for nutrients that may be protective against a specified nutritional or health risk of published health relevance, **Protective Nutrient Intakes**.

5. VITAMIN ALLOWANCES FOR ANIMALS

Public and Private Information

The development of livestock production enterprises for the economical production of human food and fiber has superimposed practical needs on the formulation of animal feeds that do not exist in the area of human nutrition. Most notably, this involves access to current information on nutrient requirements and feedstuff nutrient composition. Often, it is the availability of such accurate data that enables commercial animal nutritionists to formulate nutritionally balanced feeds, using computerized linear programming techniques, that maintain cost competitiveness in a context wherein the cost of feeding can be the largest cost of production.[11] Thus, while recent expansion of understanding in human nutrition has come from discoveries made in public-sponsored research, research in food animal nutrition has, over the past few decades, moved progressively out of the public sector and into the research divisions of agribusinesses with immediate interests in generating such data. The result is that a diminishing proportion of practical animal nutrition data

(particularly in the area of amino acid nutrition) remains in the public sector and is thus available to the scrutiny of experts. As a consequence, two types of dietary standards are in use: the first is the standard developed by review of open data available in the scientific literature; the second is the standard developed through in-house testing and/or practical experience by animal producers. Whereas the former data are in the public domain, the latter usually are not.

Public information on nutrient allowances is reviewed by expert committees in the United States, the United Kingdom, and several other countries under programs charged with the responsibility of establishing nutrient recommendations on the basis of the best available data. Perhaps the most widely used source of such recommendations is the Committee on Animal Nutrition of the US National Research Council (Table 21.7). Through expert subcommittees, each dedicated to a particular species, the NRC maintains the periodic review of nutrient standards, many of which serve as the bases of recommendations for animal feed formulation throughout the world. Currently, many gaps remain in our knowledge of vitamin requirements. This is particularly true for ruminant species, for which the substantial rumenal destruction of vitamins appears to be compensated by adequate microbial synthesis, and for several non-ruminant species that are not widely used for commercial purposes. Therefore, many of the standards for vitamins and other nutrients are imputed from available data on related species; in part for this reason, the requirements for some nutrients (e.g., selenium) appear to be very similar among many species.

Study Questions and Exercises

1. Prepare a concept map illustrating the relationships of the concepts of minimal and optimal nutrient requirements and nutrient allowances to the concepts of physiological function and health.

2. What issues relate to the application of dietary allowances, as they are currently defined, to individuals?

3. What issues relate to the consideration of nutritional status in such areas as immune function or chronic and degenerative diseases in the development of dietary standards?

---

11. For example, the feed costs for broiler chickens can account for 60–70% of the total cost of producing poultry meat.

TABLE 21.6 FAO/WHO Recommended Nutrient Intakes (RNIs) for Vitamins

Age–sex Group	Vitamin A (μg)[a]	Vitamin D (μg)	Vitamin E (mg)	Vitamin K (μg)	Vitamin C (mg)	Thiamin (mg)	Riboflavin (mg)	Niacin (mg)[b]	Vitamin B₆ (μg)	Pantothenic Acid (μg)	Biotin (μg)	Folate (μg)	Vitamin B₁₂ (μg)
Infants:													
0–6 months	375	5	2.7	5	25	0.2	0.3	2[c]	0.1	1.7	5	80	0.4
7–11 months	400	5	2.7	10	30	0.3	0.4	4	0.3	1.8	6	80	0.5
Children:													
1–3 years	400	5	5	15	30	0.5	0.5	6	0.5	2	8	160	0.9
4–6 years	450	5	5	20	30	0.6	0.6	8	0.6	3	12	200	1.2
7–9 years	500	5	7	25	35	0.9	0.9	12	1.0	4	20	300	1.8
Adolescents, 10–18 Years:													
Males	600	5	10	35–65	40	1.2	1.3	16	1.3	5	25	400	2.4
Females	600	5	7.5	35–55	40	1.1	1.0	16	1.2	5	25	400	2.4
Adults:													
Males 19–65 years	600	5[d],10[e]	10	65	45	1.2	1.3	16	1.3[d],1.7[e]	5	30	400	2.4
Females 19–50 years	500	5	7.5	55	45	1.1	1.1	14	1.3	5	30	400	2.4
Females 51–65 years	500	10	7.5	55	45	1.1	1.1	14	1.5	5	30	400	2.4
Older adults, >65 Years:													
Men[c]	600	15	10	65	45	1.2	1.3	16	1.7	5	—	400	2.4
Women[c]	600	15	7.5	55	45	1.1	1.1	14	1.5	5	—	400	2.4
Pregnancy	800	5	—	55	55	1.4	1.4	18	1.9	6	30	600	2.6
Lactation	850	5	—	55	70	1.5	1.6	17	2.0	7	35	500	2.8

[a]Retinol equivalents.
[b]Niacin equivalents.
[c]Pre-formed niacin.
[d]19–50 years.
[e]>50 years.

Source: Joint WHO/FAO Expert Consultation (2001). *Human Vitamin and Mineral Requirements*. Food and Agricultural Organisation, Rome, p. 286.

TABLE 21.7 Estimated Vitamin Requirements (units/kg diet) of Domestic and Laboratory Animals

Species	Vitamin A (IU)	Vitamin D (IU)	Vitamin E (mg)[a]	Vitamin K (µg)[b]	Vitamin C (mg)	Thiamin (mg)	Riboflavin (mg)	Niacin (mg)	Vitamin B6 (mg)	Folate (mg)	Pantothenite (mg)	Biotin (µg)	Vitamin B12 (µg)	Choline (g)
Birds:														
Chickens														
Growing chicks	1,500	200	10	0.5		1.8	3.6	27	2.5–3	0.55	10	0.1–0.15	3–9	0.5–1.3
Laying hens	4,000	500	5	0.5	0.8		2.2	10	3	0.25	2.2	0.1	4	
Breeding hens	4,000	500	10	0.5	0.8		3.8	10	4.5	0.25	10	0.15	4	
Ducks														
Growing	4,000	220		0.4			4	55	2.6		11			
Breeding	4,000	500		0.4			4	40	3		11			
Geese														
Growing	1,500	200				2.5–4		35–55			15			
Breeding	4,000	200				4	20							
Pheasants							3.5	40–60			10			1–1.5
Quail														
Growing bobwhite							3.8	30			13			1.5
Breeding bobwhite							4	20			15			1.0
Growing coturnix	5,000	1,200	12	1		2	4	40	3	1	10	0.3	3	2.0
Breeding coturnix	5,000	1,200	25	1		2	4	20	3	1	15	0.15	3	1.5
Turkeys														
Growing poults	4,000	900	12	0.8–1		2	3.6	40–70	3–4.5	0.7–1	9–11	0.1–0.2	3	0.8–1.9
Breeding hens	4,000	900	25	1		2	4	30	4	1	16	0.15	3	1.0
Cats	10,000	1,000	80			5	5	45	4	1	10	0.5	20	2.0
Cattle:														
Dry heifers	2,200	300												
Dairy bulls	2,200	300												
Lactating cows	3,200	300												
Beef cattle	2,200	300												
Dogs	5,000	275	50			1	2.2	11.4	1	0.18	10	0.1	22	1.25

Fishes:											
Bream	10,000	300				5–6	7	30–50		1	4.0
Carp	2,000	1,000			28	5–6	9	10–20	20	1	3.0
Catfish	2,500	30	60		3		14	40		5	
Coldwater spp.	2,400	30	100	10	10		150	20		10	
Foxes	2,440		1	1	5.5	1.8	9.6	7.4		0.2	
Goats	60[c]		200								
Guinea pigs	12.9[c]	50	5	2	3		10	20	10	0.3	1.0
Hamsters	23,333	1,000	4	20	6	15	90	40		0.6	2.0
	3,636	2,484	3								
Horses:											
Ponies	25[c]										
Pregnant mares	50[c]										
Lactating mares	55–65[c]										
Yearlings	40[c]										
2-year olds	30[c]										
Mice	500	150	3	5	1		10	10		0.2	0.6
Mink	5,930	27		1.3	1.6	1.6	20	8		0.12	32.6
Primates[d]	15,000	50	0.1	5	2.5		50	10		0.2	0.1
Rabbits:											
Growing	580	40		3	39		180				1.2
Pregnant	>1,160	40	0.2								
Lactating		40									
Rats	4,000	1,000	0.5	4	6		20	8		1	1.0
Sheep:											
Ewes											
Early pregnancy	26[c]	5.6[c]									
Late pregnancy/											
Lactating	35[c]	5.6[c]									
Rams	43[c]	5.6[c]									

(Continued)

TABLE 21.7 (Continued)

Species	Vitamin A (IU)	Vitamin D (IU)	Vitamin E (mg)[a]	Vitamin K (µg)[b]	Vitamin C (mg)	Thiamin (mg)	Riboflavin (mg)	Niacin (mg)	Vitamin B6 (mg)	Folate (mg)	Pantothenite (mg)	Biotin (µg)	Vitamin B12 (µg)	Choline (g)
Lambs														
Early weaned	35[c]	6.6[c]												
Finishing	26[c]	5.5[c]												
Shrimps					10		120			120	120			0.6
Swine:														
Growing	2,200	200	11	2		1.3	2.2–3	10–22	1.5	0.6	11–13	0.1	22	0.4–1.1
Bred gilt/sow	4,000	200	10	2			3	10	1	0.6	12	0.1	15	1.25
Lactating gilt/sow	2,000	200	10	2			3	10	1	0.6	12	0.1	15	1.25
Boars	4,000	200	10	2			3	10	1	0.6	12	0.1	15	1.25

[a]α-Tocopherol.
[b]Menadione.
[c]Unlike almost all of the other values in this table, this requirement is expressed in international units (IU) per kilogram body weight.
[d]Non-human species.

Sources: National Research Council (1984). *Nutrient Requirements of Poultry*, 8th edn (rev.). National Academy Press, Washington, DC; National Research Council (1978). *Nutrient Requirements of Cats* (rev.). National Academy Press, Washington, DC; National Research Council (1978). *Nutrient Requirements of Dairy Cattle*, 5th edn (rev.). National Academy Press, Washington, DC; National Research Council (1984). *Nutrient Requirements of Beef Cattle*, 6th edn (rev.). National Academy Press, Washington, DC; National Research Council (1985). *Nutrient Requirements of Dogs* (rev.). National Academy Press, Washington, DC; National Research Council (1983). *Nutrient Requirements of Warmwater Fishes and Shellfishes* (rev.). National Academy Press, Washington, DC; National Research Council (1981). *Nutrient Requirements of Coldwater Fishes*. National Academy Press, Washington, DC; National Research Council (1982). *Nutrient Requirements of Mink and Foxes*, 2nd edn (rev.). National Academy Press, Washington, DC; National Research Council (1981). *Nutrient Requirements of Goats: Angora, Dairy and Meat Goats in Temperate and Tropical Countries*. National Academy Press, Washington, DC; National Research Council (1978). *Nutrient Requirements of Laboratory Animals*, 3rd edn (rev.). National Academy Press, Washington, DC; National Research Council (1982). *Nutrient Requirements of Mink and Foxes*, 2nd edn (rev.). National Academy Press, Washington, DC; National Research Council (1978). *Nutrient Requirements of Nonhuman Primates*. National Academy Press, Washington, DC; National Research Council (1978). *Nutrient Requirements of Horses*, 4th edn (rev.). National Academy Press, Washington, DC; National Research Council (1977). *Nutrient Requirements of Rabbits*, 2nd edn (rev.). National Academy Press, Washington, DC; National Research Council (1975). *Nutrient Requirements of Sheep*, 5th edn (rev.). National Research Council (1979). *Nutrient Requirements of Swine*, 8th edn (rev.). National Academy Press, Washington, DC.

RECOMMENDED READING

Beaton, G.H., 2005. When is an individual an individual vs. a member of a group? An issue in application of the dietary reference intakes. Nutr. Rev. 64, 211–225.

Chan, L.N., 2006. Drug–nutrient interactions. In: Shils, M.E., Shike, M., Caballero, B., Cousins, R. (Eds.), Modern Nutrition in Health and Disease, tenth ed. Lippincott, New York, NY, pp. 1539–1553.

Dwyer, J.T., 2010. Dietary guidelines 2010. Nutr. Today 45, 144–153.

Food and Nutrition Board, 1997. Dietary Reference Intakes for Calcium, Phosphorus, Magnesium, Vitamin D and Fluoride. National Academy Press, Washington, DC, p. 432.

Food and Nutrition Board, 2000. Dietary Reference Intakes for Thiamin, Riboflavin, Niacin, Vitamin B_6, Folate, Vitamin B_{12}, Pantothenic Acid, Biotin and Choline. National Academy Press, Washington, DC, p. 564.

Food and Nutrition Board, 2000. Dietary Reference Intakes for Vitamin C, Vitamin E, Selenium and Carotenoids. National Academy Press, Washington, DC, p. 506.

Food and Nutrition Board, 2001. Dietary Reference Intakes for Vitamin A, Vitamin K, Arsenic, Boron, Chromium, Copper, Iodine, Iron, Manganese, Molybdenum, Nickel, Silicon, Vanadium and Zinc. National Academy Press, Washington, DC, p. 773.

Food and Nutrition Board, 2003. Dietary Reference Intakes: Applications in Dietary Planning. National Academy Press, Washington, DC, p. 237.

Food and Nutrition Board, 2003. Dietary Reference Intakes: Guiding Principles for Nutrition Labeling and Fortification. National Academy Press, Washington, DC, p. 205.

Food and Nutrition Board, 2010. Dietary Reference Intakes, Calcium, Vitamin D. National Academy Press, Washington, DC, p. 1115.

Joint WHO/FAO Expert Consultation, 2001. Human Vitamin and Mineral Requirements. Food and Agricultural Organisation, Rome, p. 286.

Joint WHO/FAO Consultation, 2002. Diet, Nutrition and the Prevention of Chronic Diseases. World Health Organisation, Geneva, p. 149.

Joost, H.G., Gibney, M.J., Cashman, K.D., et al. 2007. Personalized nutrition: Status and perspectives. Br. J. Nutr. 98, 26–31.

King, J.C., 2007. An evidence-based approach for establishing dietary guidelines. J. Nutr. 137, 480–483.

Mattys, C., Bucchini, L., Busstra, M.C., et al. 2006. Dietary standards in the United States. In: Bowman, B.A,. Russell, R.M. Present Knowledge in Nutrition, ninth ed. vol. II. ILSI, Washington, DC, pp. 859–875.

Murphy, S., 2006. The recommended dietary allowance (RDA) should not be abandoned: An individual is *both* an individual and a member of a group. Nutr. Rev. 64, 313–318.

Rosenberg, I.H., 2007. Challenges and opportunities in the translation of the science of vitamins. Am. J. Clin. Nutr. 85 (Suppl), 325S–327S.

Russell, R.M., 2008. Current framework for DRI development: What are the pros and cons? Nutr. Rev. 66, 455–458.

Taylor, C.L., Albert, J., Weisell, R., Nishida, C., 2006. International dietary standards: FAO and WHO. In: Bowman, B.A., Russell, R.M. Present Knowledge in Nutrition, ninth ed. vol. II. ILSI, Washington, DC, pp. 876–887.

Trumbo, P., 2008. Challenges with using chronic disease endpoints in setting dietary reference intakes. Nutr. Rev. 66, 459–464.

Walter, P., Hornig, D.H., Moser, U. (Eds.), 2001. Functions of Vitamins Beyond Recommended Daily Allowances. Karger, Basel, p. 214.

Yates, A.A., 2006. Dietary reference intakes: Rationale and applications. In: Shils, M.E., Shike, M., Caballero, B., Cousins, R. (Eds.), Modern Nutrition in Health and Disease, tenth ed. Lippincott, New York, NY, pp. 1655–1672.

Vitamin Safety

Chapter Outline

Anchoring Concepts

1. Vitamins are typically used in human feeding, in animal diets, and in treating certain clinical conditions, at levels in excess of their requirements.
2. Several of the vitamins, most notably vitamins A and D, can produce adverse physiological effects when consumed in excessive amounts.

Nutriment is both food and poison. The dosage makes it either poison or remedy.

Paracelsus

Learning Objectives

1. To understand the concept of upper safe use level as used with respect to the vitamins.
2. To understand the margins of safety above their respective requirements for intakes of each of the vitamins.
3. To understand the factors affecting vitamin toxicities.
4. To understand the signs/symptoms of vitamin toxicities in humans and animals.

VOCABULARY

Carotenodermia
Hypervitaminosis
Lowest observed adverse effect level (LOAEL)
Margin of safety
No observed adverse effect level (NOAEL)
Range of safe intake
Safety index (SI)
Tolerable Upper Intake Limit (UL)
Toxic threshold

1. USES OF VITAMINS ABOVE REQUIRED LEVELS

Typical Uses Exceed Requirements

Most normal diets that include varieties of foods can be expected to provide supplies of vitamins that meet those levels required to prevent clinical signs of deficiencies. In addition, most intentional uses of vitamins are designed to exceed those requirements for most individuals. Indeed, that is the principle by which vitamin allowances are set. Thus, the formulation of diets, the planning of meals, the vitamin fortification of foods, and the designing of vitamin supplements are all done to provide vitamins at levels contributing to total intakes that exceed the requirements of most individuals by some **margin of safety**. This approach minimizes the probability of producing vitamin deficiencies in populations.

Clinical Conditions Requiring Elevated Doses

Some clinical conditions require the use of vitamin supplements at levels greater than those normally used to accommodate the usual margins of safety. These include specific vitamin deficiency disorders (e.g., xerophthalmia, rickets, and polyneuritis encephalopathy related to alcohol abuse) and certain rare inherited metabolic defects (e.g., vitamin B_6-responsive cystathionase deficiency, vitamin B_{12}-responsive transcobalamin II deficiency, biotin-responsive

The Vitamins. DOI: 10.1016/B978-0-12-381980-2.00022-0

biotinidase deficiency).[1] In such cases, vitamins are pre-scribed at doses that far exceed requirement levels.

Other Putative Benefits of Elevated Doses

Elevated doses of vitamins are also frequently prescribed by physicians or taken as over-the-counter supplements by affected individuals in the treatment of certain other pathological states, including neurological pains, psycho-sis, alopecia, anemia, asthenia, premenstrual tension, car-pal tunnel syndrome, and prevention of the common cold. Although the efficacies of vitamin supplementation in most of these conditions remain untested in double-blind clinical trials, vitamin prophylaxis and/or therapy for at least some conditions is perceived as effective by many people in the medical community as well as in the general public. This view supports the widespread use of oral vita-min supplements at dosages greater than 50–100 times the Recommended Daily Allowances (RDAs).[2]

2. HAZARDS OF EXCESSIVE VITAMIN INTAKES

Non-Linear Risk Responses to Vitamin Dosages

The risks of adverse effects (toxicity) of the vitamins, like those of any other potentially toxic compounds, are func-tions of dose level. In general, the risk–dosage function is curvilinear, indicating a **hazard threshold** for vitamin dosage at some level greater than the requirement for that vitamin. Thus, a dosage increment exists between the level required to prevent deficiency and that sufficient to pro-duce adverse effects. That increment, the **range of safe intake**, is bounded on the low-dosage side by the allow-ance, and on the high-dosage side by the upper safe limit, each of which is set on the basis of similar considerations of risk of adverse effects within the population (Fig. 22.1).

Factors Affecting Vitamin Toxicity

Several factors can affect the toxicity of any vitamin. These include the route of exposure, the dose regimen (number of doses and intervals between doses), the general

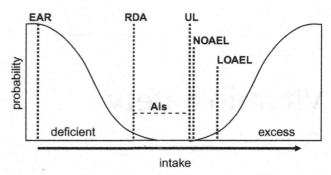

FIGURE 22.1 Vitamin safety follows a biphasic dose–response cure: just as very low intakes of vitamins can produce deficiency disorders, very high intakes can produce adverse effects.

health of the subject, and potential effects of food and drugs. For example, parenteral routes of vitamin admin-istration may increase the toxic potential of high vitamin doses, as the normal routes of controlled absorption and hepatic first-pass metabolism may be circumvented. Large single doses of the water-soluble vitamins are rarely toxic, as they are generally rapidly excreted, thus minimally affecting tissue reserves. However, repeated multiple doses of these compounds can produce adverse effects. In con-trast, single large doses of the fat-soluble vitamins can produce large tissue stores that can steadily release toxic amounts of the vitamin thereafter. Some disease states, such as those involving malabsorption, can reduce the potential for vitamin toxicity; however, most increase that potential by compromising the subject's ability to metabo-lize and excrete the vitamin,[3] or by rendering the subject particularly susceptible to **hypervitaminosis**.[4] Foods and some drugs can reduce the absorption of certain vitamins, thus reducing their toxicities.

3. SIGNS OF HYPERVITAMINOSES

The signs of intoxication for each vitamin vary with the species affected and the time-course of overexposure (see Tables 22.1 and 22.2, pages 497–500). Nevertheless, cer-tain signs or syndromes are characteristic for each vitamin:

Vitamin A

The potential for vitamin A intoxication is greater than that for other hypervitaminoses, as its range of safe intakes is relatively small. For humans, acute exposures as low as

---

1. Other examples are given in Chapter 4.
2. For example, several studies have shown that athletes and their coaches generally believe that athletes require higher levels of vitamins than non-athletes. This attitude appears to affect their behavior, as athletes use vita-min (and mineral) supplements with greater frequencies than the general public. One study found that 84% of international Olympic competitors used vitamin supplements. Despite this widespread belief, it remains unclear whether any of the vitamins at levels of intake greater than RDAs can affect athletic performance.

---

3. For example, individuals with liver damage (e.g., alcoholic cirrhosis, viral hepatitis) have increased plasma levels of free (unbound) retinol and a higher incidence of adverse reactions to large doses of vitamin A.
4. For example, patients with nephrocalcinosis are particularly suscepti-ble to hypervitaminosis D.

TABLE 22.1 Signs and Symptoms of Vitamin Toxicities in Humans

Vitamin	Children	Adults
Vitamin A	*Acute:* Anorexia, bulging fontanelles, lethargy, high intracranial fluid pressure, irritability, nausea, vomiting	*Acute:* Abdominal pain, anorexia, blurred vision, lethargy, headache, hypercalcemia, irritability, muscular weakness, nausea, vomiting, peripheral neuritis, skin desquamation
	Chronic: Alopecia, anorexia, bone pain, bulging fontanelles, cheilitis, cranio tabes, hepatomegaly, hyperostosis, photophobia, premature epiphyseal closure, pruritus, skin desquamation, erythema	*Chronic:* Alopecia, anorexia, ataxia, bone pain, cheilitis, conjunctivitis, diarrhea, diplopia, dry mucous membranes, dysuria, edema, high CSF pressure, fever, headache, hepatomegaly, hyperostosis, insomnia, irritability, lethargy, menstrual abnormalities, muscular pain and weakness, nausea, vomiting, polydypsia, pruritus, skin desquamation, erythema, splenomegaly, weight loss
Vitamin D	Anorexia, diarrhea, hypercalcemia, irritability, lassitude, muscular weakness, neurological abnormalities, pain, polydypsia, polyuria, poor weight gain, renal impairment	Anorexia, bone demineralization, constipation, hypercalcemia, muscular weakness and pain, nausea, vomiting, polyuria, renal calculi
Vitamin E	No adverse effects reported	Mild gastrointestinal distress, some nausea, coagulopathies in patients receiving anticonvulsants
Vitamin K[a]	No adverse effects reported	No adverse effects reported
Vitamin C	No adverse effects reported	Gastrointestinal distress, diarrhea, oxaluria
Thiamin[b]	No adverse effects reported	Headache, muscular weakness, paralysis, cardiac arrhythmia, convulsions, allergic reactions
Riboflavin	No adverse effects reported	No adverse effects reported
Niacin	No adverse effects reported	Vessel dilation, itching, headache, anorexia, liver damage, jaundice, cardiac arrhythmia
Vitamin B_6	No adverse effects reported	Neuropathy, skin lesions
Pantothenic acid	No adverse effects reported	Diarrhea[c]
Biotin	No adverse effects reported	No adverse effects reported
Folate	No adverse effects reported	Allergic reactions[c]
Vitamin B_{12}	No adverse effects reported	Allergic reactions[c]

[a]Adverse effects observed only for menadione; phylloquinone and the menaquinones appear to have negligible toxicities.
[b]Adverse effects have been observed only when the vitamin was administered parenterally; none have been observed when it has been given orally.
[c]This sign has been observed in only a few cases.

25 times the RDA are thought to be potentially intoxicating, although actual cases of hypervitaminosis A have been very rare[5] at chronic doses less than about 9,000 μg of retinol equivalents (RE) per day. Hypervitaminosis A occurs when plasma retinol levels exceed 3 μmol/l (caused by increases in retinyl esters), which in humans can occur in response to single large doses (>660,000 IU for adults, >330,000 IU for children) or after doses >100,000 IU/day have been taken for several months.

Acute Toxicity

Children with hypervitaminosis A develop transient (1–2 days) signs: nausea, vomiting, signs due to increased cerebrospinal fluid pressure (headache, vertigo, blurred or double vision), and muscular incoordination. Studies have found that 3–9% of children given high, single therapeutic doses (200,000 IU) show transient nausea, vomiting, headache, and general irritability; a similar percentage of younger children may show fontanelle bulging, which subsides in 48–96 hours.

5. According to Bendich ([1989] *Am. J. Clin. Nutr.*, 49, 358), fewer than 10 cases per year were reported in 1976–1987. Several of those occurred in individuals with concurrent hepatic damage due to drug exposure, viral hepatitis or protein-energy malnutrition.

TABLE 22.2 Signs of Vitamin Toxicities in Animals

Vitamin	Sign	Species
Vitamin A	Alopecia	Rat, mouse
	Anorexia	Cat, cattle, chicken, turkey
	Cartilage abnormalities	Rabbit
	Convulsions	Monkey
	Elevated heart rate	Cattle
	Fetal malformations	Hamster, monkey, mouse, rat
	Hepatomegaly	Rat
	Gingivitis	Cat
	Irritability	Cat
	Lethargy	Cat, monkey
	Reduced CSF pressure	Cattle, goat, pig
	Poor growth	Chicken, pig, turkey
	Skeletal abnormalities	Cat, cattle, chicken, dog, duck, mouse, pig, rabbit, rat, turkey, horse
Vitamin D[a]	Anorexia	Cattle, chicken, fox, pig, rat
	Bone abnormalities	Pig, sheep
	Cardiovascular calcinosis	Cattle, dog, fox, horse, monkey, mouse, pig, rat, sheep, rabbit
	Renal calcinosis	Cattle, chicken, dog, fox, horse, monkey, mouse, pig, rat, sheep, turkey
	Cardiac dysfunction	Cattle, pig
	Hypercalcemia	Cattle, chicken, dog, fox, horse, monkey, mouse, pig, rat, sheep, trout
	Hyperphosphatemia	Horse, pig
	Hypertension	Dog
	Myopathy	Fox, pig
	Poor growth, weight loss	Catfish, chicken, horse, mouse, pig, rat
	Lethality	Cattle
Vitamin E	Atherosclerotic lesions	Rabbit
	Bone demineralization	Chicken, rat
	Cardiomegaly	Rat
	Hepatomegaly	Chicken
	Hyperalbuminemia	Rat
	Hypertriglyceridemia	Rat
	Hypocholesterolemia	Rat
	Impaired muscular function	Chicken
	Increased hepatic vitamin A	Chicken, rat
	Increased prothrombin time	Chicken
	Reduced adrenal weight	Rat
	Poor growth	Chicken
	Increased hematocrit	Rat

TABLE 22.2 (Continued)

Vitamin	Sign	Species
	Reticulocytosis	Chicken
	Splenomegaly	Rat
Vitamin K[b,c]	Anemia[c]	Dog
	Renal failure[c]	Horse
	Lethality	Chicken,[c,d] mouse,[c,d] rat[c]
Vitamin C	Anemia	Mink
	Bone demineralization	Guinea pig
	Decreased circulating thyroid hormone	Rat
	Liver congestion	Guinea pig
	Oxaluria	Rat
Thiamin	Respiratory distress (i.p. dose)	Rat
	Cyanosis (i.p. dose)	Rat
	Epileptiform convulsions (i.p. dose)	Rat
Riboflavin	No adverse effects reported for oral doses	
	Lethality (parental dose)	Rat
Niacin	Impaired growth	Chicken (embryo)
	Developmental abnormalities	Chicken (embryo), mouse (fetus)
	Liver damage	Mouse
	Mucocutaneous lesions	Chicken
	Myocardial damage	Mouse
	Decreased weight gain	Chicken
	Lethality (i.p. dose)	Chicken (embryo), mouse[d]
Vitamin B$_6$	Anorexia	Dog
	Ataxia	Dog, rat
	Convulsions	Rat
	Lassitude	Dog
	Muscular weakness	Dog
	Neurologic impairment	Dog
	Vomiting	Dog
	Lethality	Mouse, rat
Pantothenic acid	No adverse effects reported for oral doses	
	Lethality (i.p. dose)	Rat
Biotin	No adverse effects reported for oral doses	
	Irregular estrus (i.p. dose)	Rat
	Fetal resorption (i.p. dose)	Rat

(Continued)

TABLE 22.2 (Continued)

Vitamin	Sign	Species
Folate	No adverse effects reported for oral doses	
	Epileptiform convulsions (i.p. dose)	Rat
	Renal hypertrophy (i.p. dose)	Rat
Vitamin B_{12}	No adverse effects reported for oral doses	
	Irregular estrus (i.p. dose)	Rat
	Fetal resorption (i.p. dose)	Rat
	Reduced fetal weights (i.p. dose)	Rat

[a]Vitamin D_3 is much more toxic than vitamin D_2.
[b]Only menadione produces adverse effects; phylloquinone and the menaquinones have negligible toxicities.
[c]These effects observed after parenteral administration of the vitamin.
[d]Nicotinamide is more toxic than nicotinic acid.

Chronic Toxicity

Chronic hypervitaminosis A occurs with recurrent exposures exceeding 12,500 IU (infants) to 33,000 IU (adult). The early sign is commonly dry lips (cheilitis), which is often followed by dryness and fragility of the nasal mucosa, dry eyes, and conjunctivitis. Skin lesions include dryness, pruritis, erythema, scaling, peeling of the palms and soles, hair loss (alopecia), and nail fragility. Headache, nausea, and vomiting (signs of increased intracranial pressure) can also occur. Infants and young children can show painful periostitis.

In animals, adverse effects have been reported at intakes as low as 10 times the RDA, but intoxication typically follows chronic intakes of 100- to 1,000-fold RDA levels. The most frequently observed signs are loss of appetite, loss of weight or reduced growth, skeletal malformations, spontaneous fractures, and internal hemorrhages. Most signs can be reversed by discontinuing excessive exposure to the vitamin. Ruminants appear to tolerate high intakes of vitamin A better than non-ruminants, apparently due to destruction of the vitamin by the rumen microflora.

That retinoids can be embryotoxic raises concerns about the safety of high-level vitamin A supplementation for pregnant animals and humans. High doses of retinol, all-*trans*-retinoic acid, or 13-*cis*-retinoic acid can disrupt cephalic neural crest cell activity, producing craniofacial, central nervous system, cardiovascular, and thymus malformations. Fetal malformations have been reported in cases of oral use of 20,000–25,000 IU/day all-*trans*-retinoic acid in treating *acne vulgaris*. Regular intakes exceeding 10,000 IU/day (of preformed vitamin A) have been associated with increased risk of birth defects in a small cohort of women with very high vitamin A intakes (mean >21,000 IU/day). Rare cases of premature closure of lower limb epiphyses have been reported in animals, e.g., "hyena disease" in calves.

The toxicities of carotenoids appear to be low. Regular intakes as great as 30 mg β-carotene per day are without side effects other than carotenodermia.

Vitamin D

Intakes as low as 50 times the RDA have been reported to be toxic to humans, particularly in children. Excessive intakes of vitamin D increase circulating levels of 25-OH-D_3, which at high levels appears to bind VDR, thus bypassing the regulation of the 25-OH-D_3-1-hydroxylase to induce transcriptional responses normally signaled only by 1,25-$(OH)_2$-D_3. Risk of hypervitaminosis D is increased under conditions, such as chronic inflammation, in which the normal feedback regulation of the renal 25-OH-D_3-1-hydroxylase is compromised. Studies with animals indicate that vitamin D_3 is 10–20 times more toxic than vitamin D_2,[6] apparently because it is more readily metabolized than the latter vitamer to the 25-hydroxy metabolites.

Hypervitaminosis D is characterized by increases in both the enteric absorption and bone resorption of calcium. This produces hypercalcemia and, ultimately, calcinosis – i.e., deposition of calcium and phosphate in soft tissues (heart, kidney, and vascular and respiratory systems). Thus, the risk of hypervitaminosis D depends on concomitant

---

6. That is, vitamin D_3 can produce effects comparable to those of vitamin D_2 at doses representing only 5–10% of the latter.

intakes of calcium and phosphorus. Vitamin D-intoxicated individuals show anorexia, vomiting, headache, drowsiness, diarrhea, and polyuria.

Vitamin D_3 has been found safe for pregnant and lactating women and their children at oral doses of 100,000 IU/day. There have been no documented cases of hypervitaminosis D due to excessive sunlight exposure.

Vitamin E

Vitamin E is one of the *least toxic* of the vitamins. Both animals and humans appear to be able to tolerate rather high levels. For animals, doses at least two orders of magnitude above nutritional requirements, e.g., 1,000–2,000 IU/kg, are without untoward effects. For humans, daily doses as high as 400 IU have been be considered harmless, and large oral doses as great as 3,200 IU have not been found to have consistent ill effects.

Studies with animals indicate that excessive dosages of tocopherols exert most, if not all, of their adverse effects by antagonizing the utilization of the other fat-soluble vitamins, thus reducing hepatic vitamin A storage, impairing bone mineralization, and producing coagulopathies. In each case, these signs could be corrected with supplements of the appropriate vitamin (A, D, and K, respectively). These effects appear to involve impaired absorption, and inhibition of retinyl ester hydrolase and vitamin K-dependent carboxylations.

Isolated reports of signs in humans consuming up to 1,000 IU per day included headache, fatigue, nausea, double vision, muscular weakness, mild creatinuria, and gastrointestinal distress.

Vitamin K

Phylloquinone and Menaquinones

The toxic potentials of the naturally occurring forms of vitamin K are negligible. Phylloquinone exhibits no adverse effects when administered to animals in massive doses by any route. The menaquinones are similarly thought to have little, if any, toxicity.

Menadione

The synthetic vitamer menadione, when administered parenterally, can at high doses produce fatal anemia, hyperbilirubinemia, and severe jaundice. However, its toxic threshold appears to be at least three orders of magnitude greater than nutritionally required levels. At such doses menadione appears to cause oxidative stress by reduction to the semiquinone radical, which, in the presence of O_2, is reoxidized to the quinone, resulting in the formation of the superoxide radical anion. Menadione can also react with free sulfhydryl groups; thus, high levels

may deplete reduced glutathione (GSH) levels. The horse appears to be particularly vulnerable to menadione toxicity. Parenteral doses of 2–8 mg/kg have been found to be lethal in that species, whereas the parenteral LD_{50}[7] values for most other species are an order of magnitude greater than that.

Vitamin C

The only adverse effects of large doses of vitamin C that have been consistently observed in humans are gastrointestinal disturbances and diarrhea occurring at levels of intake nearly 20–80 times the RDA. Concern has also been expressed that excess ascorbic acid may be pro-oxidative, may competitively inhibit the renal reabsorption of uric acid, may enhance the enteric destruction of vitamin B_{12}, may enhance the enteric absorption of non-heme iron (thus leading to iron overload), may produce mutagenic effects, and may increase ascorbate catabolism that would persist after returning to lower intakes of the vitamin. Present knowledge indicates that most, if not all, of these concerns are not warranted.

That ascorbic acid can enhance the enteric absorption of dietary iron has led to concern that megadoses may lead to progressive iron accumulation in iron-replete individuals (iron storage disease). This hypothesis has not been supported by results of studies with animal models. Nevertheless, patients with hemochromatosis or other forms of excess iron accumulation should avoid taking vitamin C supplements with their meals.

Perhaps the greatest concern associated with high intakes of vitamin C regards increased oxalate production. In humans, unlike other animals, oxalate is a major metabolite of ascorbic acid, accounting for 35–50% of the 35–40 mg of oxalate excreted in the urine each day.[8] The health concern is that high vitamin C intake may lead to increased oxalate production, and thus to increased risk of urinary calculi.[9] Metabolic studies have indicated that the turnover of ascorbic acid is limited, for which reason high intakes of vitamin C would not be expected to greatly affect oxalate production. Clinical studies have revealed slight oxaluria in patients given daily multiple-gram doses of vitamin C. It is not clear whether this effect is clinically significant, as its magnitude is low and within normal

---

7. The LD_{50} value is a useful parameter indicative of the degree of toxicity of a compound. It is defined as the lethal dose for 50% of a reference population, and is calculated from experimental dose–survival data using the probit analysis.

8. The balance of urinary oxalate comes mainly from the degradation of glycine (about 40% of the total); but some also can come from the diet (5–10%).

9. There is also some question as to whether oxalate may have been produced as an artifact of the analytical procedure.

variation.[10] Nevertheless, prudence dictates the avoidance of doses greater than 1,000 mg of vitamin C for individuals with a history of forming renal stones.

Little information is available on vitamin C toxicity in animals, although acute LD_{50} (50% lethal dose) values for most species and routes of administration appear to be at least several grams per kilogram of body weight. Dietary vitamin C intakes 100–1,000 times the allowance levels appear safe for most species.

Thiamin

The toxic potential of thiamin appears to be low, particularly when administered orally. Parenteral doses of the vitamin at 100–200 times the RDA have been reported to cause intoxication in humans, characterized by headache, convulsions, muscular weakness, paralysis, cardiac arrhythmia, and allergic reactions.

Most of the available information pertinent to its toxic potential is for thiamin hydrochloride. At very high doses (1,000-fold the levels required to prevent deficiency signs), that form can be fatal by suppressing the respiratory center. Such doses of the vitamin to animals produce curare-like signs, suggestive of blocked nerve transmission: restlessness, epileptiform convulsions, cyanosis, and dyspnea. Lower levels, ≤300 mg/day, are used therapeutically in humans without adverse reactions.

Riboflavin

High oral doses of riboflavin are very safe, probably owing to the relatively poor absorption of the vitamin at high levels. Oral riboflavin doses as great as 2–10 g/kg body weight produce no adverse effects in dogs and rats. The vitamin is somewhat more toxic when administered parenterally. The LD_{50} (50% lethal dose) values for the rat given riboflavin by the intraperitoneal, subcutaneous, and oral routes have been estimated to be 0.6, 5, and >10 g/kg, respectively. No adverse effects in humans have been reported.

Niacin

Acute Toxicity

In humans, small doses (10 mg) of nicotinic acid can cause flushing, although this effect is not associated with other seriously adverse reactions. At high dosages (2–4 mg/day) in humans, nicotinic acid can cause vasodilation, itching, urticaria (hives), gastrointestinal discomfort (heartburn, nausea, vomiting, rarely diarrhea), and headaches. These responses appear to be mediated by the niacin receptor, which is expressed by macrophages and bone marrow-derived cells of the skin. They can be minimized by using a slow-release formulation of nicotinic acid or by using a cyclooxygenase inhibitor (e.g., aspirin, indomethacin).

Nicotinamide only rarely produces these reactions. Many patients have taken daily oral doses of 200–1,000 mg for periods of years, with only occasional side effects (skin rashes, hyperpigmentation, reduced glucose tolerance in diabetics, some liver dysfunction) at the higher dosages. Doses 50–100 times the RDA are considered safe.

Chronic Toxicity

The longer-term effects of high nicotinic acid doses include cases of insulin resistance, which may involve a rebound in lipolysis that results in increased free fatty acid levels. Transient hepatic dysfunction has also been reported. Chronic, high intakes of nicotinamide may deplete methyl groups (to excrete the vitamin), which would be exacerbated by low intakes of methyl donors methionine and choline, and suboptimal status with respect to folate and/or vitamin B_{12}.

Available information on the niacin tolerances of animals is scant, but suggests that toxicity requires daily doses greater than 350–500 mg of nicotinic acid equivalents per kilogram body weight.

Vitamin B_6

The toxicity of vitamin B_6 appears to be relatively low, with intakes as great as 100 times the RDA having been used safely by most people. Very high doses of the vitamin (several grams per day) have been shown to induce reversible sensory neuropathies marked by changes in gait and peripheral sensation. The primary target appears to be the peripheral nervous system; although massive doses of the vitamin have produced convulsions in rats, central nervous system abnormalities have not been reported frequently in humans. Reports of individuals taking massive doses of the vitamin (>2 g/day) indicate that the earliest detectable signs are ataxia and loss of small motor control. Doses up to 750 mg/day for extended periods of time (years) have been found safe.

The vitamin can increase the conversion of L-dopa to dopamine, which can interfere with the action of the former when used in the management of Parkinson's disease in patients not taking a decarboxylase inhibitor.

Substantial information concerning the safety of large doses of vitamin B_6 in animals is available only for the dog and the rat. That information indicates that doses less than 1,000 times the allowance levels are safe for those species, and, by inference, for other animal species.

---

10. Forty percent of subjects given 2 g of ascorbic acid daily showed increases in urinary oxalate excretion by more than 10% (Chai, W., Liebman, M., Kynast-Gales, S., *et al.* [2004] *Am. J. Kidney Dis.* 44, 1060).

Biotin

Biotin is generally regarded as non-toxic. Adverse effects of large doses of biotin have not been reported in humans or animals given the vitamin in doses as high as 200 mg orally or 20 mg intravenously. Limited data suggest that biotin is safe for most people at doses as great as 500 times the RDA, and for animals at probably more than 1,000 times allowance levels.

Animal studies have revealed few, if any, indications of toxicity, and it is probable that animals can tolerate the vitamin at doses at least an order of magnitude greater than nutritional levels.

Pantothenic Acid

Pantothenic acid is generally regarded as non-toxic. A few reports indicate diarrhea occurring in humans consuming 10–20 g of the vitamin per day. Thus, pantothenic acid is thought to be safe for humans at doses at least 100 times the RDA.

No adverse reactions have been reported in any species following the ingestion of large doses of the vitamin. It has been estimated that animals can tolerate doses of pantothenic acid as great as at least 1,000 times their respective nutritional requirements for the vitamin. Parenteral administration of very large amounts (e.g., 1 g per kg body weight) of the calcium salt have been shown to be lethal to rats.

Folate

Folate is generally regarded as being non-toxic. Other than a few cases of apparent allergic reactions, the only proposed adverse effect in humans (interference with the enteric absorption of zinc) is not supported with adequate data. Intakes of 400 mg of folate per day for several months have been tolerated without side effects in humans, indicating that levels at least as great as 2,000 times the RDA are safe.

No adverse effects of high oral doses of folate have been reported in animals, although parenteral administration of pharmacologic amounts (e.g., 250 mg/kg, which is about 1,000 times the dietary requirement) has produced epileptic responses and renal hypertrophy in rats. High-folate treatment has been found to exacerbate teratogenic effects of nutritional Zn deficiency.

Vitamin B$_{12}$

Vitamin B$_{12}$ has no appreciable toxicity. No adverse reactions have been reported for humans or animals given high levels of the vitamin. Upper safe limits of vitamin B$_{12}$ use are therefore highly speculative; it appears that doses at least as great as 1,000 times the RDA/allowances are safe for humans and animals.

4. SAFE INTAKES OF VITAMINS

Ranges of Safe Intakes

The vitamins fall into four categories of relative toxicity at levels of exposure above typical allowances:

Greatest toxic potential	Vitamin A, vitamin D
Moderate toxic potential	Niacin
Low toxic potential	Vitamin E, vitamin C, thiamin, riboflavin, vitamin B$_6$
Negligible toxic potential	Vitamin K, pantothenic acid, biotin, folate, vitamin B$_{12}$

Under circumstances of vitamin use at levels appreciably greater than the standard allowances (RDAs for humans or recommended use levels for animals), prudence dictates giving special consideration to those vitamins with greatest potentials for toxicity (those in the first two or three categories). In practice, it may only be necessary to consider the most potentially toxic vitamins of the first category (vitamins A and D).

Quantifying Safe Intakes

There is no standard algorithm for quantifying the ranges of safe intakes of vitamins, but an approach developed for environmental substances that cause systemic toxicities has recently been employed for this purpose with nutritionally essential inorganic elements. This approach involves the imputation of an acceptable daily intake (ADI)[11] based on the application of a safety factor (SF)[12] to an experimentally determined highest **no observed adverse effect level (NOAEL)** of exposure to the substance. In the absence of sufficient data to ascertain an NOAEL, an experimentally determined **lowest observed adverse effect level (LOAEL)** is used:

$$ADI = LOAEL \div SF$$

An extension of this approach is to express the comparative safety of nutrients using a **safety index (SI)**. This index is analogous to the therapeutic index (TI) used for drugs; it is the ratio of the minimum toxic dose and the recommended intake (RI) derived from the RDA:

$$SI = LOAEL \div RI$$

11. The US Environmental Protection Agency has replaced the ADI with the **reference dose (R$_f$D)**, a name the agency considers to be more value neutral, i.e., avoiding any implication that the exposure is completely safe or acceptable.
12. SF values are selected according to the quality and generalizability of the reported data in the case selected as the reference standard. Higher values (e.g., 100) may be used if animal data are extrapolated to humans, whereas lower values (e.g., 1 or 3) may be used if solid clinical data are available.

TABLE 22.3 Use of a Safety Index to Quantitate the Toxic Potentials of Selected Vitamins for Humans

Parameter	Vitamin A	Vitamin D	Vitamin C	Niacin	Vitamin B_6
RDI[a]	3,300 IU	20 μg	60 mg	20 mg	2 mg
LOAEL	25,000 IU	250 μg	2,000 mg	500 mg	50 mg
Safety index (SI)	7.6	12.5	33	25	25

[a]The greatest RDA for persons ≥4 years of age, excluding pregnant and lactating women.
Sources: Hathcock, J. N. (1993). *J. Nutr. Rev.* 51, 278; Hathcock, J. N., Shao, A., Vieth, R., et al. (2007). *Am. J. Clin. Nutr.* 85, 6.

This approach was used by Hathcock and colleagues[13] to express quantitatively the safety limits of several vitamins for humans (Table 22.3).

The Dietary Reference Intakes (DRIs) of the Food and Nutrition Board (1997–2001, 2010) addressed the safety of high doses of essential nutrients with the Tolerable Upper Intake Limit (UL). The UL is defined as the highest level of daily intake that is likely to pose *no* risks of adverse health effects to almost all healthy individuals in each age–sex specific demographic subgroup. In this context, "adverse effect" is defined as any significant alteration in structure or function. It should be noted that the Food and Nutrition Board chose to use the term "tolerable intake" to avoid implying possible beneficial effects of intakes greater than the RDA.[14] The ULs are based on chronic intakes.

The ULs are derived through a multi-step process of hazard identification and dose–response evaluation:

1 *Hazard identification*, involving the systematic evaluation of all information pertaining to adverse effects of the nutrient;
2 *Dose–response assessment*, involving the determination of the relationship between level of nutrient intake and incidence/severity of adverse effects;
3 *Intake assessment*, involving the evaluation of the distribution of nutrient intakes in the general population; and
4 *Risk characterization*, involving the expression of conclusions from the previous steps in terms of the fraction of the exposed population having nutrients in excess of the estimated UL.

In practice, ULs are set at less than the respective LOAELs and no greater than the NOAELs (Fig. 22.2) from

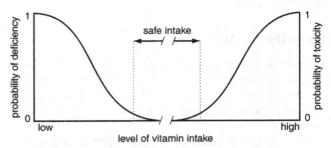

FIGURE 22.2   Conceptual basis for Tolerable Upper Intake Limit (UL) in DRI system.

which they are derived, subject to uncertainty factors (UFs) used to characterize the level of uncertainty associated with extrapolating from observed data to the general population.[15] The ULs for the vitamins are presented in Table 22.4 and 22.5.

Study Questions and Exercises

1. Which vitamins are most likely to present potential for hazards for humans?
2. For each vitamin, determine the range of safe intakes (which satisfy needs without risking toxicity) for humans or animals, and identify the most useful parameters by which that safety can be ascertained.
3. Identify treatments that can reduce the manifestations of vitamin toxicities.
4. Use specific examples to discuss the relationship of the toxic potential of vitamins to their absorption and metabolic disposition.

---

13. Hathcock, J. (1993). *Nutr. Rev.* 51, 278–285.
14. *See* Food and Nutrition Board (1998). *Dietary Reference Intakes: a Risk Assessment Model for Establishing Upper Intake Levels for Nutrients.* National Academy Press, Washington, DC, 71 pp.

15. Small UFs (close to 1) are used in cases where little population variability is expected for the adverse effects, where extrapolation from primary data is not believed to under-predict the average human response, and where a LOAEL is available. Larger UFs (as high as 10) are used in cases where the expected variability is great, where extrapolation is necessary from primary animal data, and where a LOAEL is not available and a NOAEL value must be used.

TABLE 22.4 Food and Nutrition Board Tolerable Upper Intake Limits (ULs) for Vitamins

Age-Sex Group	Vitamin A (µg[a])	Vitamin D (µg)	Vitamin E (mg[b])	Vitamin K (µg)	vitamin C (mg)	Thiamin (mg)	Riboflavin (mg)	Niacin (mg[c])	Vitamin B6 (µg)	Pantothenic Acid (µg)	Biotin (µg)	Folate (µg[d])	Vitamin B12 (µg)
Infants													
0–11 months	600	25–38	—	—	—	—	—	—	—	—	—	—	—
Children													
1–3 years	600	63	200	—	400	—	—	10	—	—	—	300	—
4–8 years	600	75	300	—	650	—	—	15	—	—	—	400	—
Males													
9–13 years	1,700	100	600	—	1,200	—	—	20	—	—	—	600	—
14–18 years	2,800	100	800	—	1,800	—	—	30	—	—	—	800	—
19+ years	3,000	100	1,000	—	2,000	—	—	35	—	—	—	1,000	—
Females													
9–13 years	1,700	100	600	—	1,200	—	—	20	—	—	—	600	—
14–18 years	2,800	100	800	—	1,800	—	—	30	—	—	—	800	—
>18 years	3,000	100	1,000	—	2,000	—	—	35	—	—	—	1,000	—
Pregnancy													
≤18 years	2,800	100	800	—	1,800	—	—	30	—	—	—	800	—
>18 years	2,800	100	1,000	—	2,000	—	—	35	—	—	—	1,000	—
Lactation													
≤18 years	2,800	100	800	—	1,800	—	—	30	—	—	—	800	—
>18 years	3,000	100	1,000	—	2,000	—	—	35	—	—	—	1,000	—

[a]Retinol equivalents.
[b]α-Tocopherol.
[c]Niacin equivalents.
[d]Folate equivalents.
—, UL not established.

Sources: Food and Nutrition Board (1997). *Dietary Reference Intakes for Calcium, Phosphorus, Magnesium, Vitamin D and Fluoride.* National Academy Press, Washington, DC, 432 pp.; Food and Nutrition Board (2000). *Dietary Reference Intakes for Thiamin, Riboflavin, Niacin, Vitamin B6, Folate, Vitamin B12, Pantothenic Acid, Biotin and Choline.* National Academy Press, Washington, DC, 564 pp.; Food and Nutrition Board (2000). *Dietary Reference Intakes for Vitamin C, Vitamin E, Selenium and Carotenoids.* National Academy Press, Washington, DC, 506 pp.; Food and Nutrition Board (2001). *Dietary Reference Intakes for Vitamin A, Vitamin K, Arsenic, Boron, Chromium, Copper, Iodine, Iron, Manganese, Molybdenum, Nickel, Silicon, Vanadium and Zinc.* National Academy Press, Washington, DC, 773 pp; Food and Nutrition Board (2010). *Dietary Reference Intakes: Calcium, Vitamin D.* National Academy Press, Washington, DC, p. 1115.

TABLE 22.5 Recommended Upper Safe Intakes of the Vitamins for Animals

Vitamin	× Allowance[a]
High Toxic Potential	
Vitamin A	10[b]–30[c]
Vitamin D	10–20[d]
Moderate Toxic Potential	
Niacin[e]	50–100
Low Toxic Potential	
Vitamin E	100
Vitamin C	100–1,000
Thiamin	500
Riboflavin	100–500
Vitamin B$_6$	100–1,000
Negligible Toxic Potential	
Vitamin K[f]	1,000
Pantothenic acid	1,000
Biotin	1,000
Folate	1,000
Vitamin B$_{12}$	1,000

[a]From *Committee on Animal Nutrition (1987). Vitamin Tolerance in Animals. National Academy Press, Washington, DC.*
[b]*For non-ruminant species.*
[c]*For ruminant species.*
[d]*Vitamin D$_3$ is more toxic than vitamin D$_2$.*
[e]*Nicotinamide is more toxic than nicotinic acid.*
[f]*Only menadione has significant (low) toxicity.*

RECOMMENDED READING

Food and Nutrition Board, 1997. Dietary Reference Intakes for Calcium, Phosphorus, Magnesium, Vitamin D and Fluoride. National Academy Press, Washington, DC, p. 432.

Food and Nutrition Board, 1998. Dietary Reference Intakes: A Risk Assessment Model for Establishing Upper Intake Levels for Nutrients. National Academy Press, Washington, DC, p. 71.

Food and Nutrition Board, 2000. Dietary Reference Intakes for Thiamin, Riboflavin, Niacin, Vitamin B$_6$, Folate, Vitamin B$_{12}$, Pantothenic Acid, Biotin and Choline. National Academy Press, Washington, DC, p. 564.

Food and Nutrition Board, 2000. Dietary Reference Intakes for Vitamin C, Vitamin E, Selenium and Carotenoids. National Academy Press, Washington, DC, p. 506.

Food and Nutrition Board, 2001. Dietary Reference Intakes for Vitamin A, Vitamin K, Arsenic, Boron, Chromium, Copper, Iodine, Iron, Manganese, Molybdenum, Nickel, Silicon, Vanadium and Zinc. National Academy Press, Washington, DC, p. 773.

Food and Nutrition Board, 2010. Dietary Reference Intakes, Calcium, Vitamin D. National Academy Press, Washington, DC, p. 1115.

Hathcock, J.N., 1997. Vitamins and minerals: Efficacy and safety. Am. J. Clin. Nutr. 66, 427–437.

Joint WHO/FAO Consultation, 2002. Diet, Nutrition and the Prevention of Chronic Diseases. World Health Organisation, Geneva, p. 149.

Joint WHO/FAO Expert Consultation, 2001. Human Vitamin and Mineral Requirements. Food and Agricultural Organisation, Rome, p. 286.

Mason, P., 2007. One is okay, more is better? Pharmacological aspects and safe limits of nutritional supplements. Proc. Nutr. Soc 66, 493–507.

Subcommittee on Vitamin Tolerance, Committee on Animal Nutrition, National Research Council, 1987. Vitamin Tolerance in Animals. National Academy Press, Washington, DC.

Current and Obsolete Designations of Vitamins (**Bolded**) and Other Vitamin-Like Factors

Name	Explanation
aneurin	infrequently used synonym for thiamin
A-N factor	obsolete term for the "*anti-neuritic factor*" (thiamin)
bios factors	obsolete terms for yeast growth factors now known to include biotin
citrovorum factor	infrequently used term for a naturally occurring form of folic acid (N^5-formyl-5,6,7,8-tetrahydropteroylmonoglutamic acid) which is required for the growth of *Leuconostoc citrovorum*
extrinsic factor	obsolete term for the anti-anemic activity in liver, now called vitamin B_{12}
factor U	obsolete term for chick anti-anemic factor now known as a form of folate
factor R	obsolete term for chick anti-anemic factor now known as a form of folate
factor X	obsolete term used at various times to designate the rat fertility factor now called vitamin E and the rat growth factor now called vitamin B_{12}
filtrate factor	obsolete term for the anti-black tongue disease activity, now known to be niacin, that could be isolated from the "*B_2 complex*" by filtration through fuller's earth; also used to describe the chick anti-dermatitis factor, now known to be pantothenic acid, isolated from acid solutions of the "*B_2 complex*" by filtration through fuller's earth
flavin	term originally used to describe the water-soluble fluorescent rat growth factors isolated from yeast and animal tissues; now, a general term for isoalloxazine derivatives including riboflavin and its active forms, FMN and FAD
hepatoflavin	obsolete term for the water-soluble rat growth factor, now known to be riboflavin, isolated from liver
intrinsic factor	accepted designation for the vitamin B_{12}-binding protein produced by gastric parietal cells and necessary for the enteric absorption of the cobalamins
lactoflavin	obsolete term for the water-soluble rat growth factor, now known to be riboflavin, isolated from whey
LLD factor	obsolete term for the activity in liver that promoted the growth of *Lactobacillus lactis* Dorner, now known to be vitamin B_{12}
norit eluate	obsolete term for *Lactobacillus casei* growth-promotant, factor now known as folic acid, that could be isolated from liver and yeasts by adsorption on Norit
ovoflavin	obsolete term for the water-soluble rat growth factor, now known to be riboflavin, isolated from egg white
P-P factor	obsolete term for the thermostable "*pellagra-preventive*" component, now known as niacin, of the "*water-soluble B*" activity of yeast

Name	Explanation
rhizopterin	obsolete synonym for the *SLR factor*," i.e., a factor from Rhizobius sp. fermentation that stimulated the growth of *Streptococcus lactis* R. (now called *S. fecalis*), which is now known to be a folate activity
SLR factor	obsolete term for the *Streptococcus lactis* R. (now called *S. faecalis*) growth promotant later called "*rhizopterin*" and now known to be a folic acid activity
streptogenin	a peptide present in liver and in enzymatic hydrolysates of casein and other proteins which promotes growth of mice and certain micro-organisms (hemolytic streptococci and lactobacilli); not considered a vitamin
vitamin A	accepted designation of retinoids that prevent xerophthalmia and nyctalopia, and are essential for epithelial maintenance
vitamin B	original anti-beriberi factor; now known to be a mixture of factors and designated as the vitamin B complex
vitamin B complex	term introduced when it became clear that "*water-soluble B*" contained more than one biologically active substance (such preparations were subsequently found to be mixtures of thiamin, niacin, riboflavin, pyridoxine, and pantothenic acid); the term has contemporary lay use as a non-specific name for all of the B-designated vitamins
vitamin B$_1$	synonym for thiamin
vitamin B$_2$	synonym for riboflavin
vitamin B$_2$ complex	obsolete term for the thermostable "second nutritional factor" in yeast, which was found to be a mixture of niacin, riboflavin, pyridoxine, and pantothenic acid
vitamin B$_3$	infrequently used synonym for pantothenic acid; was also used for nicotinic acid
vitamin B$_4$	unconfirmed activity preventing muscular weakness in rats and chicks; believed to be a mixture of arginine, glycine, riboflavin, and pyridoxine
vitamin B$_5$	unconfirmed growth promotant for pigeons; probably niacin
vitamin B$_6$	synonym for pyridoxine
vitamin B$_7$	unconfirmed digestive promoter for pigeons; may be a mixture; also "*vitamin I*"
vitamin B$_8$	adenylic acid; no longer classified as a vitamin
vitamin B$_9$	unused designation
vitamin B$_{10}$	growth promotant for chicks; likely a mixture of folic acid and vitamin B$_{12}$
vitamin B$_{11}$	apparently the same as "*vitamin B$_{10}$*"
vitamin B$_{12}$	accepted designation of the cobalamins (cyano- and aquo-cobalamins) that prevent pernicious anemia and promote growth in animals
vitamin B$_{12a}$	synonym for aquacobalamin
vitamin B$_{12b}$	synonym for hydroxocobalamin
vitamin B$_{12c}$	synonym for nitritocobalamin
vitamin B$_{13}$	synonym for orotic acid, an intermediate of pyrimidine metabolism; not considered a vitamin
vitamin B$_{14}$	unconfirmed
vitamin B$_{15}$	synonym for "*pangamic acid*;" no proven biological value
vitamin B$_{17}$	synonym for laetrile, a cyanogenic glycoside with unsubstantiated claims of anticarcinogenic activity; not considered a vitamin
vitamin B$_c$	obsolete term for pteroylglutamic acid
vitamin B$_p$	activity preventing perosis in chicks; replaceable by choline and Mn
vitamin B$_t$	activity promoting insect growth; identified as carnitine
vitamin B$_x$	activity associated with pantothenic acid and *p*-aminobenzoic acid
vitamin C	accepted designation of the anti-scorbutic factor, ascorbic acid

Name	Explanation
vitamin C_2	unconfirmed anti-pneumonia activity; also called "*vitamin J*"
vitamin D	accepted designation of the anti-rachitic factor (the calciferols)
vitamin D_2	accepted designation for ergocalciferol (a vitamin D-active substance derived from plant sterols)
vitamin D_3	accepted designation for cholecalciferol (a vitamin D-active substance derived from animal sterols)
vitamin E	accepted designation for tocopherols active in preventing myopathies and certain types of infertility in animals
vitamin F	obsolete term for essential fatty acids; also an abandoned term for thiamin activity
vitamin G	obsolete term for riboflavin activity; also an abandoned term for the "*pellagra-preventive factor*" (niacin)
vitamin H	obsolete term for biotin activity
vitamin I	mixture also formerly called "*vitamin B_7*"
vitamin J	postulated anti-pneumonia factor also formerly called "*vitamin C_2*"
vitamin K	accepted designation for activity preventing hypoprothrombinemic hemorrhage shared by related napthoquinones
vitamin K_1	accepted designation for phylloquinones (vitamin K-active substances produced by plants)
vitamin K_2	accepted designation for prenylmenaquinones (vitamin K-active substances synthesized by microorganisms and produced from other vitamers K by animals)
vitamin K_3	accepted designation for menadione (synthetic vitamin K-active substance not found in nature)
vitamin L_1	unconfirmed liver filtrate activity, probably related to anthranilic acid, proposed as necessary for lactation
vitamin L_2	unconfirmed yeast filtrate activity, probably related to adenosine, proposed as necessary for lactation
vitamin M	obsolete term for anti-anemic factor in yeast now known to be pteroylglutamic acid
vitamin N	obsolete term for a mixture proposed to inhibit cancer
vitamin O	unused designation
vitamin P	activity reducing capillary fragility related to citrin, which is no longer classified as a vitamin
vitamin Q	unused designation (the letter was used to designate coenzyme Q)
vitamin R	obsolete term for folic acid; from Norris' chick anti-anemic "*factor R*"
vitamin S	chick growth activity related to the peptide "*streptogenin;*" the term was also applied to a bacterial growth activity probably related to biotin
vitamin T	unconfirmed group of activities isolated from termites, yeasts or molds and reported to improve protein utilization in rats
vitamin U	unconfirmed activity from cabbage proposed to cure ulcers and promote bacterial growth; may have folic acid activity
vitamin V	tissue-derived activity promoting bacterial growth; probably related to NAD
Wills' factor	obsolete term for the anti-anemic factor in yeast now known to be a form of folate
zoopherin	obsolete term for a rat growth factor now known as vitamin B_{12}

Original Reports for Case Studies

Chapter 5 *Case 1*
McLaren, D. S., Ahirajian, E., Tchalian, M. G., and Koury, G. (1965). Xerophthalmia in Jordan. *Am. J. Clin. Nutr.* 17, 117–130.

Case 2
Wechsler, H. L. (1979). Vitamin A deficiency following small-bowel bypass surgery for obesity. *Arch. Dermatol.* 115, 73–75.

Case 3
Sauberlich, H. E., Hodges, R. E., Wallace, D. L., Kolder, H., Canham, J. E., Hood, H., Racia, N. Jr, and Lowry, L. K. (1974). Vitamin A metabolism and requirements in the human studied with the use of labeled retinol. *Vit. Horm.* 32, 251–275.

Chapter 6 *Case*
Marx, S. J., Spiegel, A. M., Brown, E. M., Gardner, D.G ., Downs, R.W. Jr, Attie, M., Hamstra , A. J., and DeLuca, H. F. (1978). A familial syndrome of decrease in sensitivity to 1,25-dihydroxyvitamin D. *J. Clin. Endocrinol. Metab.* 47, 1303–1310.

Chapter 7 *Case 1*
Boxer, L. A., Oliver, J. M., Spielberg, S. P., Allen, J. M., and Schulman, J. D. (1979). Protection of granulocytes by vitamin E in glutathione synthetase deficiency. *N. Engl. J. Med.* 301, 901–905.

Case 2
Harding, A. E., Matthews, S., Jones, S., Ellis, C. J. K., Booth, I. W., and Muller, D. P. R. (1985). Spinocerebellar degeneration associated with a selective defect in vitamin E absorption. *N. Engl. J. Med.* 313, 32–35.

Chapter 8 *Case 1*
Colvin, B. T. and Lloyd, M. J. (1977). Severe coagulation defect due to a dietary deficiency of vitamin K. *J. Clin. Pathol.* 30, 1147–1148.

Case 2
Corrigan, J. and Marcus, F. I. (1974). Coagulopathy associated with vitamin E ingestion. *J. Am. Med. Assoc.* 230, 1300–1301.

Chapter 9 *Case 1*
Hodges, R. E., Hood, J., Canham, J. E., Sauberlich, H. E., and Baker, E. M. (1971). Clinical manifestations of ascorbic acid deficiency in man. *Am. J. Clin. Nutr.* 24, 432–443.

Case 2
Dewhurst, K. (1954). A case of scurvy simulating a gastric neoplasm. *Br. Med. J.* 2, 1148–1150.

Chapter 10 *Case 1*
Burwell, C. S. and Dexter, L. (1947). Beriberi heart disease. *Trans. Assoc. Am. Physiol.* 60, 59–64.

Case 2
Blass, J. P. and Gibson, G. E. (1977). Abnormality of a thiamin-requiring enzyme in patients with Wernicke-Korsakoff syndrome. *New Engl. J. Med.* 297, 1367–1370.

Chapter 11	*Case* Dutta, P., Gee, M., Rivlin, R. S., and Pinto, J. (1988). Riboflavin deficiency and glutathione metabolism in rats: Possible mechanisms underlying altered responses to hemolytic stimuli. *J. Nutr.* 118, 1149–1157.
Chapter 12	*Case* Vannucchi, H. and Moreno, F. S. (1989). Interaction of niacin and zinc metabolism in patients with alcoholic pellagra. *Am. J. Clin. Nutr.* 50, 364–69.
Chapter 13	*Case 1* Barber, G. W. and Spaeth, G. L. (1969). The successful treatment of homocystinuria with pyridoxine. *J. Pediatr.* 75, 463–478. *Case 2* Schaumberg, H., Kaplan, J., Windebank, A., Vick, N., Rasmus, S., Pleasure, D., and Brown, M. J. (1983). Sensory neuropathy from pyridoxine abuse. *New Engl. J. Med.* 309, 445–448.
Chapter 14	*Case* Mock, D. M., DeLorimer, A. A., Liebman, W. M., Sweetman, L., and Baker, H. (1981). Biotin deficiency: An unusual complication of parenteral alimentation. *New Engl. J. Med.* 304, 820–823.
Chapter 15	*Case* Lacroix, B., Didier, E., and Grenier, J. F. (1988). Role of pantothenic and ascorbic acid in wound healing processes: *In vitro* study on fibroblasts. *Intl J. Vit. Nutr. Res.* 58, 407–413.
Chapter 16	*Case* Freeman, J. M., Finkelstein, J. D., and Mudd, S. H. (1975). Folate-responsive homocystinuria and schizophrenia. A defect in methylation due to deficient 5,10-methylenetetrahydrofolic acid reductase activity. *New Engl. J. Med.* 292, 491–496.
Chapter 17	*Case* Higginbottom, M. C., Sweetman, L., and Nyhan, W. L. (1978). A syndrome of methylmalonic aciduria, homocystinuria, megaloblastic anemia and neurological abnormalities of a vitamin B_{12}-deficient breast-fed infant of a strict vegetarian. *New Engl. J. Med.* 299, 317–323.

A Core of Current Vitamin Literature

The following tables list journals publishing original research and reviews about the vitamins, as well as website and reference texts useful to students, researchers, and clinicians.

TABLE C.1 Journals Presenting Original Research and Reviews

Title	Publisher/URL[a,b]
American Journal of Clinical Nutrition	American Society for Nutrition http://www.ajcn.org
American Journal of Epidemiology	Oxford Journals http://aje.oxfordjournals.org/
American Journal of Public Health	American Public Health Association http://www.ajph.aphypublications.org/
Annals of Internal Medicine	American College of Physicians http://www.annals.org/
Annals of Nutrition and Metabolism	S. Karger, Basel http://content.karger.com/ProdukteDB/produkte.asp?Aktion=JournalHome&ProduktNr=223977
Annual Review of Biochemistry	Annual Reviews http://www.annualreviews.org/journal/biochem
Annual Review of Nutrition	Annual Reviews http://www.annualreviews.org/journal/nutr
Archives of Internal Medicine	American Medical Association http://archinte.ama-assn.org/
Australian Journal of Nutrition and Dietetics	Dietetics Association of Australia http://wwsw.ajnd.org.au/
BBA (Biochemica et Biophysical Acta)	Elsevier http://www.elsevier.com/wps/find/journaldescription.cws_home/506062/description
Biochemical and Biophysical Research Communications	Elsevier http://www.elsevier.com/wps/find/journaldescription.cws_home/622790/description
Biochemistry	American Chemical Society http://pubs.acs.org/journal/bichaw
Biofactors	ILR Press, Oxford http://onlinelibrary.wiley.com/journal/10.1002/(ISSN)1872-8081

(Continued)

TABLE C.1 (Continued)

Title	Publisher/URL[a,b]
British Journal of Nutrition	Nutrition Society http://journals.cambridge.org/action/displayJournal?jid=BJN
British Medical Journal	BMJ Publishing Group, London http://www.bmj.com/
Cancer Epidemiology, Biomarkers and Prevention	American Association for Cancer Research http://cebp.aacrjournals.org/
Cell	Cell Press http://www.cell.com/
Clinical Biochemistry	Canadian Society of Clinical Chemists http://www.elsevier.com/wps/find/journaldescription.cws_home/525463/description
Clinical Chemistry	American Association of Clinical Chemists http://www.clinchem.org/
Critical Reviews in Biochemistry and Molecular Biology	Informa Healthcare http://informahealthcare.com/bmg
Critical Reviews in Food Science and Nutrition	Taylor & Francis Group http://www.tandf.co.uk/journals/bfsn
Current Nutrition and Food Science	Bentham Science Publishers http://www.benthamscience.com/cnf/index.htm
Current Opinion in Clinical Nutrition and Metabolic Care	Lippincott, Williams & Wilkins http://journals.lww.com/co-clinicalnutrition/pages/default.aspx
Epidemiology	Lippincott, Williams & Wilkins http://journals.lww.com/epidem/pages/default.aspx
European Journal of Biochemistry (FEBS Journal)	Federation of European Biochemical Societies http://onlinelibrary.wiley.com/journal/10.1111/(ISSN)1742-4658
European Journal of Clinical Nutrition	Stockton Press http://www.nature.com/ejcn/index.html
FASEB Journal	Federation of American Societies for Experimental Biology http://fasebj.org/
FEBS Letters	Federation of European Biochemical Societies http://www.febsletters.org
Gastroenterology	American Gastroenterology Association Institute http://www.gastrojournal.org/
International Journal of Epidemiology	Oxford Journals http://ide.oxforedjournjals.org
International Journal of Food Sciences and Nutrition	Informa Healthcare http://informahealthcare.com/ijf
International Journal for Vitamin and Nutrition Research	Hogrefe & Huber Publishers http://www.verlag.hanshuber.com/zetschriften/journal.php?abbrev=VIT
International Journal of Nutrition and Metabolism	Academic Journals http://www.academicjournals.org/IJNAM/index.htm
International Journal of Obesity	Nature Publishing Group http://www,nature.com/ijo/index.html
Journal of Agricultural and Food Chemistry	American Chemical Society http://pubs.acs.org/journal/jafcau

(Continued)

TABLE C.1 (Continued)

Title	Publisher/URL[a,b]
Journal of the American Dietetic Association	American Dietetic Association http://www.adajournal.org/
JAMA (Journal of the American Medical Association)	American Medical Association http://www.jama.ama-assn.org/
Journal of Biological Chemistry	American Society of Biological Chemists http://www.jbc.org/
Journal of Clinical Biochemistry and Nutrition	Institute of Applied Biochemistry http://www.jstage.jst.go.jp/browse/jcbn/
Journal of Food Composition and Analysis	Elsevier, Amsterdam http://www.elsevier.com/wps/find/journaldescription.cws_home/622878/description
Journal of Immunology	American Association of Immunologists http://www.jimmunol.org/
Journal of Lipid Research	American Society for Biochemistry and Molecular Biology http://www.jlr.org/
Journal of Nutrition	American Society for Nutrition http://www.nutrition.org/publications/the-journal-of-nutrition
Journal of Nutritional Biochemistry	Elsevier http://www.elsevier.com/wps/find/journaldescription.cws_home/525013/description#description
Journal of Nutritional Sciences and Vitaminology	The Vitamin Society of Japan, and Japanese Society of Nutrition and Food Science http://www.jstage.jst.go.jp/browse.jnsv
Journal of Parenteral and Enteral Nutrition	American Society of Parenteral and Enteral Nutrition http://pen.sagepub.com/
Journal of Pediatric Gastroenterology and Nutrition	Lippicott Williams & Wilkens http://journals.lww.com/jpgn/pages.default.aspx
Lipids	American Oil Chemists Society http://www.springer.com/life+sciences/journal/11745
New England Journal of Medicine	Massachusetts Medical Society http://www.nejm.org/
Nutrition Abstracts and Reviews Series A (human, experimental)	CABI http://www.cabi.org/default.aspx?site=170&page=1016&pid=79
Nutrition Abstracts and Reviews Series B (feeds, feeding)	CABI http://www.cabi.org/default.aspx?site=170&page=1016&pid=2181
Nutrition and Food Science	MICS Publishing Group http://omisconline.org/infshome.php
Nutrition in Clinical Care	Wiley http://onlinelibrary.wiley.com/journal/10.1111/(ISSN)1523-5408
Nutrition Journal	Cell & Bioscience http://www.nutritionj.com/
Nutrition Research	Elsevier http://www.elsevier.com/wps/find/journaldescription.cws_home/525483/description
Nutrition Research Reviews	The Nutrition Society http://www.nutritionsociety.org/node/237

(Continued)

TABLE C.1 (Continued)

Title	Publisher/URL[a,b]
Nutrition Reviews	Wiley-Blackwell http://www.wiley.com/bw/journal.asp?ref=0029-6643
Nutrition Today	Lippincott Williams & Wilkins http://journals.lww.com/nutritiontodayonline/pages/default.aspx
Nutritional Biochemistry	Elsevier http://www.elsevier.com/wps/find/journaldescription.cws_home/525013/description#description
Obesity	The Obesity Society http://www.nature.com/oby/index.html
PNAS (Proceedings of the National Academy of Sciences)	National Academy of Sciences (US) http://www.pnas.org/
Proceedings of the Nutrition Society	Nutrition Society http://www.nutritionsociety.org/node/238
Proceedings of the Society for Experimental Biology and Medicine	Society for Experimental Biology and Medicine http://www.sebm.org/journal.php

[a]URL, uniform resource locator.
[b]Sites accessed July 14, 2011.

TABLE C.2 Some Useful Websites

Programs/Information	URL[a,b]
United Nations	
Food and Agricultural Organization (FAO)	http://www.fao.org/
Agriculture and Consumer Protection	http://www.fao.org/ag/portal/index_en/en/
Codex Alimentarius[c]	http://www.codexalimentarius.net/web/index_en.jsp
Committee of World Food Security	http://www.fao.org/cfs/en/
Food Composition - INFOODS[d] project	http://www.fao.org/infoods/index_en.stm
Hunger	http://www.fao.org/hunger/en/
Millenium Development Goals	http://www.fao.org/mdg/en/
National Policies and Strategies	http://www.fao.org/ag/agn/nutrition/nationalpolicies_en.stm
Nutritional Assessment	http://www.fao.org/ag/agn/nutrition/assessment_en.stm
Nutrition Country Profiles	http://www.fao.org/ag/agn/nutrition/profiles_en.stm
Nutrition Education and Consumer Awareness	http://www.fao.org/ag/humannutrition/nutritioneducation/en/
Nutrition Requirements	http://www.fao.org/ag/humannutrition/nutrition/en/
Statistics	http://www.fao.org/corp/statistics/en/
World Food Situation	http://www.fao.org/worldfoodsituation/wfs-home/en/
UN University	http://www.unu.edu/
World Health Organization (WHO)	http://www.who.int/en/
Child Growth Standards	http://www.who.int/childgrowth/en/

(Continued)

TABLE C.2 (Continued)

Programs/Information	URL[a,b]
Data and Statistics	http://www.who.int/research/en/
Global Database on Child Growth and Malnutrition	http://www.who.int/nutgrowthdb/en/
Global Health Library	http://www.who.int/ghl/en/
Growth Refence Data 5-19 years	http://www.who.int/growthref/en/
Nutrition for Health and Development	http://www.who.int/nutrition/en/
Vitamin and Mineral Information Systems (VMNIS)	http://www.who.int/vmnis/en/
World Health Report	http://www.who.int/whr/en/index.html
World Health Statistics	http://www.who.int/whosis/whostat/en/index.html
United States Government	
Department of Agriculture (USDA)	http://www.usda.gov/wps/portal/usdahome
Center for Nutrition Policy and Promotion	http://www.cnpp.usda.gov/
Diet and Health	http://www.ers.usda.gov/Emphases/Healthy/
Economic Data on Food, Agriculture and the Rural Economy	http://www.ers.usda.gov/Data/
Food Availability (per capita) Data System	http://www.ers.usda.gov/Data/foodconsumption/
Food and Nutrition Information Center	http://fnic.nal.usda.gov/nal_display/index.php?info_center=4&tax_level=1
Child Nutrition Programs	http://www.fns.usda.gov/cnd/
ChooseMyPlate.gov	http://www.choosemyplate.gov/
Dietary Assessment Tools	http://fnic.nal.usda.gov/nal_display/index.php?info_center=4&tax_level=2&tax_subject=256&topic_id=1325
Dietary Guidelines for Americans, 2010	http://www.cnpp.usda.gov/DGAs2010-PolicyDocument.htm
National Agricultural Library	http://www.nalusda.gov/
National Institute for Food and Agriculture	http://www.csrees.usda.gov/
Food, Nutrition and Health Programs	http://www.csrees.usda.gov/nea/food/food.cfm
National Nutrient Database for Standard Reference	http://www.nal.usda.gov/fnic/foodcomp/search/
National Program in Human Nutrition Research Department of Health and Human Services	http://www.hhs.gov/
Center for Disease Control and Prevention	http://www.cdc.gov/
National Center for Health Statistics	http://www.cdc.gov/nchs/
National Health and Examination Survey (NHANES)	http://www.cdc.gov/nchs/nhanes.htm
US National Physical Plan	http://www.physicalactivityplan.org/
Food and Drug Administration (FDA)	http://www.fda.gov/
Center for Food Safety and Applied Nutrition (CSFAN)	http://www.fda.gov/AboutFDA/CentersOffices/CFSAN/default.htm
Dietary Supplements	http://www.fda.gov/Food/DietarySupplements/default.htm
National Institutes of Health (NIH)	http://www.nih.gov/
Eunice Kennedy Shriver National Institute of Child and Human Development	http://www.nichd.nih.gov/
National Cancer Institute	http://www.cancer.gov/

(Continued)

TABLE C.2 (Continued)

Programs/Information	URL[a,b]
National Human Genome Institute	http://www.genome.gov/
National Heart, Lung and Blood Institute	http://www.nhlbi.nih.gov/
National Institute of Diabetes and Digestive and Kidney Diseases	http://www2.niddk.nih.gov/
National Institute on Aging	http://www.nia.nih.gov/
National Library of Medicine	http://www.nlm.nih.gov/
Office of Dietary Supplements	http://ods.od.nih.gov/
Let's Move	http://www.letsmove.gov/
Nutrition.gov	http://www.nutrition.gov/nal_display/index.php?info_center=11&tax_level=1
US Global Health Initiative	http://www.ghi.gov/
Professional Societies	
American Dietetic Association	http://www.eatright.org
American Society of Nutrition	http://www.cnpp.usda.gov/
University On-Line Resources	
Cornell University: "*Cornell NutritionWorks*"	http://www.nutritionworks.cornell.edu/home/index.cfm
Harvard University : "*The Nutrition Source*"	http://www.nutritionworks.cornell.edu/home/index.cfm
Johns Hopkins School of Public Health: "*Johns Hopkins Public Health*"	http://www.nutritionworks.cornell.edu/home/index.cfm
Tufts University: "Health and Nutrition Newsletter"	http://www.tuftshealthletter.com/

[a]URL, uniform resource locator.
[b]Sites accessed July 14, 2011.
[c]Joint program of FAO and WHO.
[d]International Network of Food Data Systems.

TABLE C.3 A Bookshelf of Useful References

Ball, G. F. M. (2004). *Vitamins: Their Role in the Human Body*. Wiley-Blackwell, New York, NY, p. 448

Ball, G. F. M. (2005). *Vitamins in Foods: Analysis, Bioavailability and Stability*. CRC Press, New York, NY p. 824

Bender, D. A. (2009). *Nutritional Biochemistry of the Vitamins*, 2nd edn. Cambridge University Press, Cambridge, p. 516

Berdanier, C. D. and Moutaid-Moussa, N., eds (2004). *Genomics and Proteomics in Nutrition*. Marcel Dekker, New York, NY, p. 507

Bowman, B.A. Russell, R. M., eds. (2006). *Present Knowledge in Nutrition*, 9th edn, Vols I and II. ILSI Press, Washington, DC, p. 967

Cheeke, P. R. and Dierenfeld, E. A. (2010). *Comparative Animal Nutrition and Metabolism*. CABI, New York, NY, p. 336

Eitenmiller, R. R., Landen, W. O. Jr, and Ye, L. (2007). *Vitamin Analyses for the Health and Food Sciences*, 2nd edn. CRC Press, New York, NY, p. 664

Food and Nutrition Board (1997). *Dietary Reference Intakes for Calcium, Phosphorus, Magnesium, Vitamin D and Fluoride*. National Academy Press, Washington, DC, p. 432

Food and Nutrition Board (2000). *Dietary Reference Intakes for Thiamin, Riboflavin, Niacin, Vitamin B_6, Folate, Vitamin B_{12}, Pantothenic Acid, Biotin and Choline*. National Academy Press, Washington, DC, p. 564

Food and Nutrition Board (2000). *Dietary Reference Intakes for Vitamin C, Vitamin E, Selenium and Carotenoids*. National Academy Press, Washington, DC, p. 506

(Continued)

TABLE C.3 (Continued)

Food and Nutrition Board (2001). *Dietary Reference Intakes for Vitamin A, Vitamin K, Arsenic, Boron, Chromium, Copper, Iodine, Iron, Manganese, Molybdenum, Nickel, Silicon, Vanadium and Zinc.* National Academy Press, Washington, DC, p. 773

Food and Nutrition Board (2003). *Dietary Reference Intakes: Applications in Dietary Planning.* National Academy Press, Washington, DC, p. 237

Food and Nutrition Board (2003). *Dietary Reference Intakes: Guiding Principles for Nutrition Labeling and Fortification.* National Academy Press, Washington, DC, p. 205

Food and Nutrition Board (2010). *Dietary Reference Intakes: Calcium, Vitamin D.* National Academy Press, Washington, DC, p. 1105

Gauch, H. G. Jr (2003). *Scientific Method in Practice.* Cambridge University Press, Cambridge, p. 435

Gibson, R. S. (1990). *Principles of Nutritional Assessment.* Oxford University Press, New York, NY, p. 525

Health and Human Services (2001). *Healthy People 2010*, Vols 1–2. International Medical Publishing, McLean, VA, p. 1226

Higdon, J. (2003). *An Evidence-Based Approach to Vitamins and Minerals.* Thieme, New York, NY, p. 253

Insel, P., Turner, R. E., and Ross, D. (2009). *Discovering Nutrition*, 3rd edn. Jones and Bartlett, Boston, MA, p. 7264

Leeson, S. and Summers, J. D. (2001). *Scott's Nutrition of the Chicken*, 4th edn. University Press, Toronto, p. 535

Mahan, L. K. and Escott-Stump, S. (2007). *Krause's Food, Nutrition, & Diet Therapy*, 12th edn. W.B. Saunders, Philadelphia, PA, p. 1376

Maulik, N. and Maulik, G. (2011). *Nutrition, Epigenetic Mechanisms, and Human Disease.* CRC Press, New York, NY, p. 426

McDonald, P., Edwards, R. A., Greenhalgh, J. F. D., Morgan, C. A., Sinclair, L. A., and Wilkinson, R. G. (2010). *Animal Nutrition*, 7th edn. Benjamen-Cummings, New York, NY, p. 692

Ottaway, P. B., ed. (1999). *The Technology of Vitamins in Food.* Aspen Publishers, Gaithersburg, MD, p. 270

Pond, W. G., Church, D. C., Pond, K. R., and Schokneckt, P. A. (2005). *Basic Animal Nutrition and Feeding*, 5th edn. John Wiley & Sons, New York, NY, p. 580

Sauberlich, H. E. (1999). *Laboratory Tests for the Assessment of Nutritional Status*, 2nd edn, CRC Press, New York, NY, p. 486

Shils, M. E., Shike, M., Ross, A. C., Caballero, B., and Cousins, R. J., eds (2006). *Modern Nutrition in Health and Disease*, 10th edn. Lippincott Williams & Wilkins, New York, NY, p. 2069

Stipanuk, M. H. and Caudill, M. A. (2012). *Biochemical and Physiological Aspects of Human Nutrition*, 3rd edn. W.B. Saunders, New York, NY.

Whitney, E. and Rolfes, S. R. (2007). *Understanding Nutrition*, 11th edn. Wadsworth Publishing, New York, NY, p. 720

Zemplini, J., Rucker, R. B., McCormick, D. B., and Suttie, J. W., eds (2007). *Handbook of Vitamins*, 4th edn. CRC Press, New York, NY, p. 593

Vitamin Contents of Foods
(units per 100 g edible portion)

FOOD* by Major Food Group	Vit A (IU)	Vit D (IU)	Vit E (mg TE)	Vit K (mg)	Vit C (mg)	Thiamin (mg)	Riboflavin (mg)	Niacin (mg)	Vit. B_6 (mg)	Pantothenic (mg)	Folate (µg)	Vit. B_{12} (µg)
CEREALS												
BARLEY, PEARLED, CKD	7		0.05		0	0.083	0.062	2.06	0.115	0.135	16	0
BUCKWHEAT GROATS, RSTD, CKD	0		0.236		0	0.04	0.039	0.94	0.077	0.359	14	0
BULGUR, CKD	0		0.029		0	0.057	0.028	1	0.083	0.344	18	0
CORNFLOUR, WHOLEGRAIN, YELLOW	469		0.25		0	0.246	0.08	1.9	0.37	0.658	25	0
CORNMEAL, WHOLEGRAIN, YELLOW	469		0.67		0	0.385	0.201	3.63	0.304	0.425	25.4	0
COUSCOUS, CKD	0		0.013		0	0.063	0.027	0.98	0.051	0.371	15	0
HOMINY, CANNED, WHITE	0		0.05		0	0.003	0.006	0.03	0.005	0.154	1	0
HOMINY, CANNED, YELLOW	110		0		0	0.003	0.006	0.03	0.005	0.154	1	0
MACARONI, COOKED, ENR	0		0.03		0	0.204	0.098	1.67	0.035	0.112	7	0
MILLET, CKD	0		0.18		0	0.106	0.082	1.33	0.108	0.171	19	0
NOODLES, CHINESE, CHOW MEIN	85		0.16		0	0.578	0.421	5.95	0.11	0.533	22	0
NOODLES, EGG, CKD, ENR	20		0.05		0	0.186	0.083	1.49	0.036	0.145	7	0.09
NOODLES, EGG, SPINACH, CKD, ENR	103		0.05		0	0.245	0.123	1.47	0.114	0.233	21	0.14
NOODLES, JAPANESE, SOBA, CKD	0		0		0	0.094	0.026	0.51	0.04	0.235	7	0
NOODLES, JAPANESE, SOMEN, CKD	0		0		0	0.02	0.033	0.1	0.013	0.172	2	0
OAT BRAN, CKD	0		0		0	0.16	0.034	0.14	0.025	0.217	6	0
OATS	0		0.7		0	0.763	0.139	0.96	0.119	1.349	56	0
RICE, BROWN, LONG-GRAIN, CKD	0		0.72		0	0.096	0.025	1.53	0.145	0.285	4	0
RICE, WHITE, GLUTINOUS, CKD	0		0.03		0	0.02	0.013	0.29	0.026	0.215	1	0
RICE, WHITE, LONG-GRAIN, PARBOILED, CKD, ENR	0		0.05		0	0.25	0.018	1.4	0.019	0.324	4	0
RICE, WHITE, LONG-GRAIN, REG, CKD	0		0.05		0	0.163	0.013	1.48	0.093	0.39	3	0
RYE FLOUR, MEDIUM	0		1.33		0	0.287	0.114	1.73	0.268	0.492	19	0
SEMOLINA, ENR	0		0.06		0	0.811	0.571	5.99	0.103	0.58	72	0
SORGHUM	0		0		0	0.237	0.142	2.93	0	0	0	0
SPAGHETTI, CKD, ENR, CKD	0		0.06		0	0.204	0.098	1.67	0.035	0.112	7	0
SPAGHETTI, SPINACH, CKD	152		0		0	0.097	0.103	1.53	0.096	0.183	12	0
SPAGHETTI, WHOLEWHEAT, CKD	0		0.05		0	0.108	0.045	0.71	0.079	0.419	5	0

Food											
TAPIOCA, PEARL, DRY	0	0		0	0.004	0	0	0.008	0.135	4	0
WHEAT BRAN, CRUDE	0	2.32		0	0.523	0.577	13.6	1.303	2.181	79	0
WHEAT FLOUR, WHOLE-GRAIN	0	1.23		0	0.447	0.215	6.37	0.341	1.008	44	0
WHEAT FLOUR, WHITE, ALL-PURPOSE, ENR, BLEACHED	0	0.06	0.6	0	0.785	0.494	5.9	0.044	0.438	26	0
WHEATGERM, CRUDE	0	0		0	1.882	0.499	6.81	1.3	2.257	281	0
WILD RICE, COOKED	0	0.23		0	0.052	0.087	1.29	0.135	0.154	26	0
BREADS, CAKES AND PASTRIES											
BAGELS, PLAIN, ENR	0	0.033		0	0.463	0.305	4.42	0.049	0.254	17	0
BISCUITS, PLAIN/BUTTERMLK	2	2.875		0	0.427	0.292	3.35	0.047	0.3	7	0.14
BREAD, BANANA, W/VEG SHORTENING	92	0		1.7	0.173	0.198	1.46	0.151	0.261	11	0.09
BREAD, CORNBREAD, W/2% MILK	277	0		0.3	0.291	0.294	2.25	0.113	0.339	19	0.15
BREAD, CRACKED-WHEAT	0	0.564		0	0.358	0.24	3.67	0.304	0.512	39	0.03
BREAD, FRENCH/VIENNA/SOURDOUGH	0	0.236		0	0.52	0.329	4.75	0.043	0.387	31	0
BREAD, IRISH SODA	194	1.057		0.8	0.298	0.269	2.41	0.083	0.25	10	0.05
BREAD, ITALIAN	0	0.277		0	0.473	0.292	4.38	0.048	0.378	30	0
BREAD, MIXED-GRAIN	0	0.615		0.3	0.407	0.342	4.37	0.333	0.512	48	0.07
BREAD, OAT BRAN	5	0.395		0	0.504	0.346	4.83	0.04	0.159	25	0
BREAD, OATMEAL	16	0.343		0.4	0.399	0.24	3.14	0.068	0.341	27	0.02
BREAD, PITA, WHITE, ENR	0	0.038		0	0.599	0.327	4.63	0.034	0.397	24	0
BREAD, PITA, WHOLEWHEAT	0	0.934		0	0.339	0.08	2.84	0.231	0.548	35	0
BREAD, PUMPERNICKEL	0	0.507		0	0.327	0.305	3.09	0.126	0.404	34	0
BREAD, RAISIN, ENR	2	0.758		0.5	0.339	0.398	3.47	0.069	0.387	34	0
BREAD, RYE	4	0.552		0.2	0.434	0.335	3.81	0.075	0.44	51	0
BREAD, WHEAT BRAN	0	0.674		0	0.397	0.287	4.4	0.176	0.536	25	0
BREAD, WHEATGERM	1	0.87		0.3	0.369	0.375	4.5	0.098	0.313	55	0.07
BREAD, WHEAT	0	0.546		0	0.418	0.28	4.12	0.097	0.436	41	0
BREAD, WHITE	0	0.286		0	0.472	0.341	3.97	0.064	0.39	34	0.02
BREAD, WHOLEWHEAT	0	1.036		0	0.351	0.205	3.84	0.179	0.552	50	0.01
CAKE, ANGELFOOD	0	0		0	0.102	0.491	0.88	0.031	0.198	3	0.07

(Continued)

FOOD*, by Major Food Group	Vit. A (IU)	Vit. D (IU)	Vit. E (mg TE)	Vit. K (mg)	Vit. C (mg)	Thiamin (mg)	Riboflavin (mg)	Niacin (mg)	Vit. B6 (mg)	Pantothenic (mg)	Folate (µg)	Vit. B12 (µg)
CAKE, BOSTON CREAM PIE	80		1.064		0.1	0.408	0.27	0.19	0.026	0.301	8	0.16
CAKE, FRUITCAKE	78		3.12		0.4	0.05	0.099	0.79	0.046	0.226	3	0.06
CAKE, GINGERBREAD	55		1.372		0.1	0.189	0.186	1.56	0.038	0.224	10	0.07
CAKE, POUND	606		0		0.1	0.137	0.229	1.31	0.035	0.341	11	0.18
CAKE, SHORTCAKE, BISCUIT-TYPE	72		0		0.2	0.311	0.272	2.57	0.03	0.248	10	0.07
CAKE, SPONGE	154		0.45		0	0.243	0.269	1.93	0.052	0.478	13	0.24
CAKE, WHITE, WO/FRSTNG	56		1.825		0.2	0.186	0.242	1.53	0.021	0.184	7	0.08
CAKE, YELLOW, WO/FRSTNG	139		2.056		0.2	0.183	0.233	1.46	0.036	0.31	10	0.16
CHEESECAKE	552		1.05		0.6	0.028	0.193	0.2	0.052	0.571	15	0.17
COOKIES, ANIMAL CRACKERS	0		1.827		0	0.35	0.326	3.47	0.022	0.376	14	0.05
COOKIES, BROWNIES	69		2.134		0.1	0.255	0.21	1.72	0.035	0.547	12	0.15
COOKIES, BUTTER, ENR	600		0.464		0	0.37	0.335	3.19	0.036	0.488	6	0.25
COOKIES, CHOC CHIP, LOW FAT	3		0		0.3	0.289	0.266	2.77	0.262	0.146	6	0
COOKIES, CHOC SANDWICH, W/CREME FILLING	1		3.03		0	0.079	0.179	2.07	0.019	0.172	5	0.02
COOKIES, FIG BARS	44		0.702		0.2	0.158	0.217	1.87	0.075	0.364	10	0.02
COOKIES, FORTUNE	10		0.344		0	0.182	0.13	1.84	0.013	0.297	10	0.05
COOKIES, GINGERSNAPS	1		1.488		0	0.2	0.293	3.24	0.183	0.339	6	0
COOKIES, GRAHAM CRACKERS	0		1.907		0	0.222	0.314	4.12	0.065	0.537	17	0
COOKIES, MOLASSES	1		2.08		0	0.355	0.264	3.03	0.241	0.425	7	0
COOKIES, OATMEAL	16		2.822		0.4	0.267	0.23	2.23	0.054	0.207	7	0
COOKIES, PEANUT BUTTER	29		3.516		0	0.17	0.18	4.27	0.07	0.437	32	0.05
COOKIES, RAISIN, SOFT-TYPE	41		1.543		0.3	0.216	0.206	1.97	0.054	0.31	9	0.1
COOKIES, VANILLA WAFERS	1		0		0	0.361	0.209	2.98	0.026	0.31	8	0.05
CRACKERS, CHEESE	162		1.012		0	0.57	0.428	4.67	0.553	0.526	25	0.46
CRACKERS, CHEESE, W/PEANUT BUTTER FILLING	319		4.419		0	0.403	0.344	6.52	1.492	0.51	25	0.01
CRACKERS, MATZO	0		0.402		0	0.387	0.291	3.89	0.115	0.443	14	0
CRACKERS, MELBA TOAST	0		0.234		0	0.413	0.273	4.11	0.098	0.693	26	0

Food										
CRACKERS, RUSK TOAST	46	0	0	0.404	0.399	4.63	0.038	0.406	64	0.07
CRACKERS, RYE	0	1.362	0	0.243	0.145	1.04	0.21	0.676	22	0
CRACKERS, RYE, WAFERS	23	1.999	0.1	0.427	0.289	1.58	0.271	0.569	45	0
CRACKERS, SALTINES	0	1.653 2	0	0.565	0.462	5.25	0.038	0.456	31	0
CRACKERS, WHEAT	0	4.012	0	0.505	0.327	4.96	0.136	0.522	18	0
CROISSANTS, BUTTER	539	0.43	0.2	0.388	0.241	2.19	0.058	0.861	28	0.3
CROUTONS, PLAIN	0	0	0	0.623	0.272	5.44	0.026	0.429	22	0
DANISH PASTRY, CHEESE	203	2.583	0.1	0.19	0.26	2	0.044	0.27	25	0.24
DANISH PASTRY, FRUIT, ENR	52	1.759	3.9	0.263	0.22	1.99	0.028	0.634	16	0.09
DOUGHNUTS, CAKE-TYPE, PLAIN	57	3.457	0.2	0.222	0.24	1.85	0.056	0.276	8	0.23
DOUGHNUTS, CAKE-TYPE, PLAIN, SUGARED/GLAZED	10	0	0.1	0.233	0.198	1.51	0.027	0.353	12	0.2
ENGLISH MUFFINS, PLAIN, ENR	0	0.126	0.1	0.384	0.275	3.8	0.042	0.315	29	0.04
ENGLISH MUFFINS, WHEAT	2	0.31	0	0.431	0.292	3.36	0.09	0.302	39	0
FRENCH TOAST, W/LOW FAT (2%) MILK	484	0	0.3	0.204	0.321	1.63	0.074	0.549	23	0.31
HUSH PUPPIES	142	1.05	0.2	0.352	0.332	2.78	0.102	0.357	20	0.19
MUFFINS, BLUEBERRY	34	1.049	1.1	0.14	0.12	1.1	0.022	0.335	16	0.58
MUFFINS, CORN	208	1.224	0.1	0.273	0.326	2.04	0.084	0.444	34	0.19
MUFFINS, PLAIN, W/LOW FAT (2%) MILK	140	0	0.3	0.284	0.301	2.31	0.042	0.351	13	0.15
PANCAKES, BLUEBERRY	199	0	2.2	0.195	0.272	1.52	0.049	0.395	12	0.2
PANCAKES, PLAIN	196	0	0.3	0.201	0.281	1.57	0.046	0.405	12	0.22
PIE, APPLE, ENR, FLR	124	1.649	3.2	0.028	0.027	0.26	0.038	0.119	4	0
PIE, BANANA CREME	261	1.472	1.6	0.139	0.207	1.05	0.133	0.388	11	0.25
PIE, BLUEBERRY	140	1.479	2.7	0.01	0.03	0.3	0.037	0.089	4	0
PIE, CHERRY	237	1.408	0.7	0.023	0.029	0.2	0.041	0.319	8	0
PIE, CHOCOLATE CREME	2	2.387	0.3	0.036	0.107	0.68	0.02	0.393	7	0.04
PIE, COCONUT CREME	90	1.607	0	0.05	0.08	0.2	0.068	0.315	5	0.19
PIE, LEMON MERINGUE	175	1.429	3.2	0.062	0.209	0.65	0.03	0.793	8	0.15
PIE, PEACH	105	1.722	1	0.061	0.033	0.2	0.023	0.094	4	0
PIE, PECAN	175	2.53	1.1	0.091	0.122	0.25	0.021	0.424	6	0.08
PIE, PUMPKIN	4515	1.608	1.5	0.055	0.153	0.19	0.057	0.507	15	0.39

(Continued)

FOOD* by Major Food Group	Vit. A (IU)	Vit. D (IU)	Vit. E (mg TE)	Vit. K (mg)	Vit. C (mg)	Thiamin (mg)	Riboflavin (mg)	Niacin (mg)	Vit. B6 (mg)	Pantothenic (mg)	Folate (µg)	Vit. B$_{12}$ (µg)
ROLLS, DINNER, PLAIN	0		0.776		0.1	0.493	0.319	4.03	0.054	0.505	30	0.03
ROLLS, DINNER, WHEAT	0		1.041		0	0.433	0.273	4.07	0.085	0.154	15	0
ROLLS, FRENCH	4		0.361		0	0.523	0.3	4.35	0.041	0.222	33	0
ROLLS, HAMBURGER/HOTDOG, PLAIN	0		0.462		0	0.484	0.312	3.93	0.047	0.529	27	0.02
ROLLS, HARD (INCL KAISER)	0		0.181		0	0.478	0.336	4.24	0.055	0.222	15	0
STRUDEL, APPLE	30		1.78		1.7	0.04	0.025	0.33	0.043	0.181	6	0.15
SWEET ROLLS, CINNAMON W/RAISINS	215		2.857		2	0.324	0.265	2.38	0.107	0.338	24	0.12
TACO SHELLS, BAKED	350		3.033		0	0.228	0.053	1.35	0.368	0.47	6	0
WAFFLES, PLAIN	228		0		0.4	0.263	0.347	2.07	0.056	0.485	15	0.25
WONTON WRAPPERS	14		0.082		0	0.519	0.378	5.42	0.03	0.025	17	0.02
BREAKFAST CEREALS												
ALL-BRAN	2500		1.843	2	50	1.3	1.4	16.7	1.7	1.734	300	5
CORN CEREAL, EXTRUDED CIRCLES	4167		0.687		50	1.25	1.42	16.7	1.67	0.077	333	0
CORN CEREAL, EXTRUDED WAFFLE-TYPE	504		0.25		53	1.3	0.24	17.6	1.8	0.163	353	5.3
CORNFLAKES	2500		0.125	0.04	50	1.3	1.4	16.7	1.7	0.329	353	0
CORN GRITS, CKD W/H$_2$O	0		0.05		0	0.1	0.06	0.81	0.024	0.064	1	0
CREAM OF RICE, CKD W/H$_2$O	0		0.02		0	0	0	0.4	0.027	0.076	3	0
CREAM OF WHEAT, CKD W/H$_2$O	0		0.013		0	0.1	0	0.6	0.014	0.075	4	0
FARINA, CKD W/H$_2$O	0		0.013		0	0.08	0.05	0.55	0.01	0.056	2	0
GRANOLA (OATS, WHEAT GERM)	37		12.88		1.4	0.74	0.28	2.05	0.32	0.603	86	0
MIXED BRAN (WHEAT, BARLEY)	0		2.32		95	2.4	2.7	31.7	3.2	1.93	71	9.5
OAT BRAN	1531		0.662		30.6	0.765	0.867	10.2	1.02	0.747	278	0
OATMEAL, WO/FORT, CKD W/H$_2$O	16	3	0.1		0	0.11	0.02	0.13	0.02	0.2	4	0
PUFFED RICE	0		0.101	0.08	0	0.41	0.05	6.25	0	0.34	10	0
PUFFED WHEAT	0		0	2	0	2.6	1.8	35.3	0.17	0.518	32	0
RAISIN BRAN	1364		0.912		0	0.7	0.8	9.1	0.9	0.66	200	2.7
RICE CEREAL, CRISPY STYLE	2500		0.125		50	1.3	1.4	16.7	1.7	0.976	353	0
RICE CEREAL, EXTRUDED, CHECK-STYLE	60		0.13		53	1.3	0.03	17.6	1.8	0.353	353	5.3

Food											
SHREDDED WHEAT	0	0.53	0.7	0	0.28	0.28	4.57	0.253	0.814	50	0
WHEAT FLAKES	2500	1.231		50	1.25	1.42	16.7	1.67	0.796	333	0
WHEATGERM, TOASTED	0	18.14		6	1.67	0.82	5.59	0.978	1.387	352	0
VEGETABLES											
ALFALFA SEEDS, SPROUTED, RAW	155	0.02		8.2	0.076	0.126	0.48	0.034	0.563	36	0
AMARANTH LEAVES, CKD, BLD, DRND	2770	0		41.1	0.02	0.134	0.56	0.177	0.062	56.8	0
ARTICHOKES, CKD, BLD, DRND	177	0.19	800	10	0.065	0.066	1	0.111	0.342	51	0
ASPARAGUS, CKD, BLD, DRND	539	0.38		10.8	0.123	0.126	1.08	0.122	0.161	146	0
BALSAM-PEAR (BITTER GOURD), TIPS, BLD, DRND	1733	0.5		55.6	0.147	0.282	1	0.76	0.06	87.6	0
BAMBOO SHOOTS, CKD, BLD, DRND	0	0		0	0.02	0.05	0.3	0.098	0.066	2.3	0
BEANS, NAVY, SPROUTED, CKD, BLD, DRND	4	0		17.3	0.381	0.235	1.26	0.198	0.854	106	0
BEANS, PINTO, IMMAT SEEDS, FRZ, CKD, BLD, DRND	0	0		0.7	0.274	0.108	0.63	0.194	0.258	33.5	0
BEANS, SNAP, GREEN, CKD, BLD, DRND	666	0.14	38	9.7	0.074	0.097	0.61	0.056	0.074	33.3	0
BEANS, SNAP, YELLOW, CKD, BLD, DRND	81	0.29		9.7	0.074	0.097	0.61	0.056	0.074	33.3	0
BEET GRNS, CKD, BLD, DRND	5100	0.3		24.9	0.117	0.289	0.5	0.132	0.329	14.3	0
BEETS, CKD, BLD, DRND	35	0.3		3.6	0.027	0.04	0.33	0.067	0.145	80	0
BEETS, PICKLED, CND, SOL & LIQUIDS	11	0		2.3	0.01	0.048	0.25	0.05	0.137	26.5	0
BROADBEANS, IMMAT SEEDS, CKD, BLD, DRND	270	0		19.8	0.128	0.09	1.2	0.029	0.066	57.8	0
BROCCOLI, CKD, BLD, DRND	1388	1.69	205	74.6	0.055	0.113	0.57	0.143	0.508	50	0
BROCCOLI, RAW	1542	1.66	270	93.2	0.065	0.119	0.64	0.159	0.535	71	0
CABBAGE, CHINESE (PAK-CHOI), CKD, BLD, DRND	2568	0.12		26	0.032	0.063	0.43	0.166	0.079	40.6	0
CABBAGE, CKD, BLD, DRND	132	0.105		20.1	0.057	0.055	0.28	0.113	0.139	20	0
CABBAGE, RAW	133	0.105	145	32.2	0.05	0.04	0.3	0.096	0.14	43	0
CABBAGE, RED, CKD, BLD, DRND	27	0.12		34.4	0.034	0.02	0.2	0.14	0.22	12.6	0
CABBAGE, SAVOY, CKD, BLD, DRND	889	0		17	0.051	0.02	0.02	0.152	0.159	46.3	0
CARROTS, BABY, RAW	15010	0		8.4	0.031	0.05	0.89	0.077	0.229	33	0
CARROTS, CKD, BLD, DRND	24554	0.42	18	2.3	0.034	0.056	0.51	0.246	0.304	13.9	0

(Continued)

FOOD*, by Major Food Group	Vit. A (IU)	Vit. D (IU)	Vit. E (mg TE)	Vit. K (mg)	Vit. C (mg)	Thiamin (mg)	Riboflavin (mg)	Niacin (mg)	Vit. B6 (mg)	Pantothenic (mg)	Folate (µg)	Vit. B12 (µg)
CARROTS, FRZ, CKD, BLD, DRND	17702	0	0.42		2.8	0.027	0.037	0.44	0.129	0.161	10.8	0
CARROTS, RAW	28129	0	0.46	5	9.3	0.097	0.059	0.93	0.147	0.197	14	0
CASSAVA, RAW	25	0	0.19		20.6	0.087	0.048	0.85	0.088	0.107	27	0
CATSUP	1016	0	1.465		15.1	0.089	0.073	1.37	0.175	0.143	15	0
CAULIFLOWER, CKD, BLD, DRND	17	0	0.04	10	44.3	0.042	0.052	0.41	0.173	0.508	44	0
CAULIFLOWER, RAW	19	0	0.04	10	46.4	0.057	0.063	0.53	0.222	0.652	57	0
CELERY, RAW	134	0	0.36	12	7	0.046	0.045	0.32	0.087	0.186	28	0
CHARD, SWISS, CKD, BLD, DRND	3139	0	1.89	660	18	0.034	0.086	0.36	0.085	0.163	8.6	0
CHIVES, RAW	4353	0	0.21	190	58.1	0.078	0.115	0.65	0.138	0.324	105	0
CILANTRO, RAW	6130	0	2.041		35.3	0.063	0.182	1.31	0.132	0.57	62	0
COLLARDS, CKD, BLD, DRND	3129	0	0.88		18.2	0.04	0.106	0.58	0.128	0.218	93	0
CORIANDER, RAW	2767	0	2.5	310	10.5	0.074	0.12	0.73	0.105	0.185	10.3	0
CORN, SWEET, YELLOW, CKD, BLD, DRND	217	0	0.09		6.2	0.215	0.072	1.61	0.06	0.878	46.4	0
CORN, SWEET, YELLOW, RAW	281	0	0.09	0.5	6.8	0.2	0.06	1.7	0.055	0.76	45.8	0
CORN, SWEET, YELLOW, CND, SOL & LIQUIDS	152	0	0		5.5	0.026	0.061	0.94	0.037	0.522	38.1	0
COWPEAS (BLACKEYES), CKD, BLD, DRND	791	0	0.22		2.2	0.101	0.148	1.4	0.065	0.154	127	0
CUCUMBER, WITH PEEL, RAW	215	0	0.079	19	5.3	0.024	0.022	0.22	0.042	0.178	13	0
DANDELION GREENS, RAW	14000	0	2.5		35	0.19	0.26	0.81	0.251	0.084	27.2	0
EGGPLANT, CKD, BLD, DRND	64	0	0.03		1.3	0.076	0.02	0.6	0.086	0.075	14.4	0
ENDIVE, RAW	2050	0	0.44		6.5	0.08	0.075	0.4	0.02	0.9	142	0
GARLIC, RAW	0	0	0.01		31.2	0.2	0.11	0.7	1.235	0.596	3.1	0
GINGER ROOT, RAW	0	0	0.26		5	0.023	0.029	0.7	0.16	0.203	11.2	0
GOURD, CALABASH, CKD, BLD, DRND	0	0	0		8.5	0.029	0.022	0.39	0.038	0.144	4.3	0
HEARTS OF PALM, CANNED	0	0	0		7.9	0.011	0.057	0.44	0.022	0.126	39	0
KALE, CKD, BLD, DRND	7400	0	0.85	650	41	0.053	0.07	0.5	0.138	0.049	13.3	0
KOHLRABI, CKD, BLD, DRND	35	0	1.67		54	0.04	0.02	0.39	0.154	0.16	12.1	0
LEEKS, CKD, BLD, DRND	46	0	0	11	4.2	0.026	0.02	0.2	0.113	0.072	24.3	0
LEMON GRASS (CITRONELLA), RAW	11	0	0		2.6	0.065	0.135	1.1	0.08	0.05	75	0

Food												
LETTUCE, BUTTERHEAD, RAW	970	0	0.44	122	8	0.06	0.06	0.3	0.05	0.18	73.3	0
LETTUCE, COS OR ROMAINE, RAW	2600	0	0.44		24	0.1	0.1	0.5	0.047	0.17	136	0
LETTUCE, ICEBERG, RAW	330	0	0.28		3.9	0.046	0.03	0.19	0.04	0.046	56	0
LETTUCE, LOOSELEAF, RAW	1900	0	0.44	210	18	0.05	0.08	0.4	0.055	0.2	49.8	0
LIMA BEANS, CKD, BLD, DRND,	370	0	0.14		10.1	0.14	0.096	1.04	0.193	0.257	26.3	0
LOTUS ROOT, CKD, BLD, DRND	0	0	0.01		27.4	0.127	0.01	0.3	0.218	0.302	7.9	0
MUNG BEANS, SPROUTED, CKD, STIR-FRIED	31	0	0		16	0.14	0.18	1.2	0.13	0.559	69.6	0
MUSHROOM, CLOUD FUNGUS, DRIED	0	0	0		0	0.015	0.844	6.27	0.112	0.481	38	0
MUSHROOM, OYSTER, RAW	48	0			0	0.055	0.36	3.58	0.122	1.291	47	0
MUSHROOMS, CND, DRND SOL	0	0	0.12		0	0.085	0.021	1.59	0.061	0.811	12.3	0
MUSHROOMS, RAW	0	76	0.12	0.02	3.5	0.102	0.449	4.12	0.097	2.2	21.1	0
MUSHROOMS, SHIITAKE, DRIED	0	1660	0.12		3.5	0.3	1.27	14.1	0.965	21.879	163	0
MUSHROOMS, STRAW, CND, DRND SOL	0	0	0		0	0.013	0.07	0.22	0.014	0.412	38	0
MUSTARD GRNS, CKD, BLD, DRND	3031	0	2.01	130	25.3	0.041	0.063	0.43	0.098	0.12	73.4	0
NEW ZEALAND SPINACH, CKD, BLD, DRND	3622	0	0		16	0.03	0.107	0.39	0.237	0.256	8.3	0
OKRA, CKD, BLD, DRND	575	0	0.69		16.3	0.132	0.055	0.87	0.187	0.213	45.7	0
ONIONS, CKD, BLD, DRND	0	0	0.13	2	5.2	0.042	0.023	0.17	0.129	0.113	15	0
ONIONS, RAW	0	0	0.13	540	6.4	0.042	0.02	0.15	0.116	0.106	19	0
PARSLEY, RAW	5200	0	1.79		133	0.086	0.098	1.31	0.09	0.4	152	0
PARSNIPS, CKD, BLD, DRND	0	0	1		13	0.083	0.051	0.72	0.093	0.588	58.2	0
PEAS, EDIBLE-PODDED, CKD, BLD, DRND	131	0	0.39	20	47.9	0.128	0.076	0.54	0.144	0.673	29.1	0
PEAS, EDIBLE-PODDED, RAW	145	0	0.39	25	60	0.15	0.08	0.6	0.16	0.75	41.7	0
PEAS, GREEN, RAW	640	0	0.39	36	40	0.266	0.132	2.09	0.169	0.104	65	0
PEAS, GREEN, CKD, BLD, DRND	597	0	0.39		14.2	0.259	0.149	2.02	0.216	0.153	63.3	0
PEPPER, BANANA, RAW	340	0	0.69		82.7	0.081	0.054	1.24	0.357	0.265	29	0
PEPPERS, CHILI, GREEN, CND	126	0	0		34.2	0.01	0.03	0.63	0.12	0.084	54	0
PEPPERS, HUNGARIAN, RAW	140	0	0		92.9	0.079	0.055	1.09	0.517	0.205	53	0
PEPPERS, JALAPENO, RAW	215	0	0.473		44.3	0.144	0.057	1.12	0.508	0.228	47	0

(Continued)

FOOD*, by Major Food Group	Vit. A (IU)	Vit. D (IU)	Vit. E (mg TE)	Vit. K (mg)	Vit. C (mg)	Thiamin (mg)	Riboflavin (mg)	Niacin (mg)	Vit. B6 (mg)	Pantothenic (mg)	Folate (µg)	Vit. B12 (µg)
PEPPERS, SWEET, GREEN, RAW	632	0	0.69	17	89.3	0.066	0.03	0.51	0.248	0.08	22	0
PEPPERS, SWEET, RED, RAW	5700	0	0.69		190	0.066	0.03	0.51	0.248	0.08	22	0
PICKLES, CUCUMBER, SWEET	126	0	0.16		1.2	0.009	0.032	0.17	0.015	0.12	1	0
PICKLES, CUCUMBER, DILL	329	0	0.16	26	1.9	0.014	0.029	0.06	0.013	0.054	1	0
PIGEONPEAS, CKD, BLD, DRND	130	0	0.17		28.1	0.35	0.166	2.15	0.053	0.63	100	0
PIMENTO, CANNED	2655	0	0.69		84.9	0.017	0.06	0.62	0.215	0.01	6	0
POTATOES, AU GRATIN, PREPD W/BUTTER	264	0	0		9.9	0.064	0.116	0.99	0.174	0.387	8.1	0
POTATOES, BKD, FLESH	0	0	0.04		12.8	0.105	0.021	1.4	0.301	0.555	9.1	0
POTATOES, CND, DRND	0	0	0.05		5.1	0.068	0.013	0.92	0.188	0.354	6.2	0
POTATOES, FRENCH FRIES, FRZ, OVEN HEATED	0	0	0.19	5	10.1	0.113	0.028	2.09	0.308	0.337	12	0
POTATOES, HASHED BROWN	0	0	0.19		5.7	0.074	0.02	2	0.278	0.499	7.7	0
POTATOES, MASHED, PREPD W/WHOLE MILK & MARG	169	0	0.3		6.1	0.084	0.04	1.08	0.224	0.57	7.9	0
POTATOES, MICROWAVED, CKD IN SKIN, FLESH	0	0	0		15.1	0.129	0.025	1.63	0.319	0.597	12.4	0
POTATOES, SCALLOPED, PREPD W/BUTTER	135	0	0		10.6	0.069	0.092	1.05	0.178	0.514	8.7	0
PUMPKIN, CND	22056	0	1.06	16	4.2	0.024	0.054	0.37	0.056	0.4	12.3	0
RADISHES, RAW	8	0	0.001	0.1	22.8	0.005	0.045	0.3	0.071	0.088	27	0
RUTABAGAS, CKD, BLD, DRND	561	0	0.15		18.8	0.082	0.041	0.72	0.102	0.155	15	0
SAUERKRAUT, CND, SOL & LIQUIDS	18	0	0.1	25	14.7	0.021	0.022	0.14	0.13	0.093	23.7	0
SHALLOTS, RAW	1190	0	0		8	0.06	0.02	0.2	0.345	0.29	34.2	0
SOYBEANS, GRN, CKD, BLD, DRND	156	0	0.01		17	0.26	0.155	1.25	0.06	0.128	111	0
SPINACH, CKD, BLD, DRND	8190	0	0.955		9.8	0.095	0.236	0.49	0.242	0.145	146	0
SPINACH, RAW	6715	0	1.89	400	28.1	0.078	0.189	0.72	0.195	0.065	194	0
SQUASH, ACORN, CKD, BKD	428	0	0		10.8	0.167	0.013	0.88	0.194	0.504	18.7	0
SQUASH, BUTTERNUT, CKD, BKD	7001	0	0		15.1	0.072	0.017	0.97	0.124	0.359	19.2	0
SQUASH, HUBBARD, CKD, BKD	6035	0	0		9.5	0.074	0.047	0.56	0.172	0.447	16.2	0

Food												
SQUASH, SPAGHETTI, CKD, BLD, DRND/BKD	110	0	0.12		3.5	0.038	0.022	0.81	0.099	0.355	8	0
SQUASH, SUMMER, CKD, BLD, DRND	287	0	0.12		5.5	0.049	0.049	0.51	0.094	0.137	20.1	0
SQUASH, ZUCCHINI, CKD, BLD, DRND	240	0	0.12		4.6	0.041	0.041	0.43	0.078	0.114	16.8	0
SUCCOTASH (CORN & LIMAS), CKD, BLD, DRND	294	0	0		8.2	0.168	0.096	1.33	0.116	0.567	32.8	0
SWEET POTATO LEAVES, CKD, STEAMED	916	0	0.96		1.5	0.112	0.267	1	0.16	0.2	48.8	0
SWEET POTATO, CKD, BKD IN SKIN	21822	0	0.28	4	24.6	0.073	0.127	0.6	0.241	0.646	22.6	0
TARO LEAVES, CKD, STEAMED	4238	0	0		35.5	0.139	0.38	1.27	0.072	0.044	48.3	0
TARO SHOOTS, CKD	51	0	0		18.9	0.038	0.053	0.81	0.112	0.076	2.6	0
TARO, CKD	0	0	0.44		5	0.107	0.028	0.51	0.331	0.336	19.2	0
TOMATO JUICE, CND	556	0	0.91	4	18.3	0.047	0.031	0.67	0.111	0.25	19.9	0
TOMATO PASTE, CND	2445	0	4.3		42.4	0.155	0.19	3.22	0.38	0.753	22.4	0
TOMATO SAUCE, CND	979	0	1.4	7	13.1	0.066	0.058	1.15	0.155	0.309	9.4	0
TOMATOES, GREEN, RAW	642	0	0.38		23.4	0.06	0.04	0.5	0.081	0.5	8.8	0
TOMATOES, RED, RIPE, CND, STEWED	541	0	0.38		11.4	0.046	0.035	0.71	0.017	0.114	5.4	0
TOMATOES, RED, RIPE, RAW	623	0	0.38	6	19.1	0.059	0.048	0.63	0.08	0.247	15	0
TURNIP GRNS, CKD, BLD, DRND	5498	0	1.721	200	27.4	0.045	0.072	0.41	0.18	0.274	118	0
TURNIPS, CKD, BLD, DRND	0	0	0.03	0.06	11.6	0.027	0.023	0.3	0.067	0.142	9.2	0
WATERCHESTNUTS, CHINESE, CND	4	0	0.5		1.3	0.011	0.024	0.36	0.159	0.221	5.8	0
WATERCRESS, RAW	4700	0	1	250	43	0.09	0.12	0.2	0.129	0.31	9.2	0
WINGED BNS, CKD, BLD, DRND	88	0	0		9.8	0.086	0.072	0.65	0.082	0.041	35.1	0
YAM, CKD, BLD, DRND, BKD	0	0	0.16		12.1	0.095	0.028	0.55	0.228	0.311	16	0
YARDLONG BEAN, CKD, BLD, DRND	450	0	0		16.2	0.085	0.099	0.63	0.024	0.051	44.5	0
FRUITS AND FRUIT JUICES												
APPLE JUICE, CND/BOTTLED, WO/VIT C	1	0	0.01	0.1	0.9	0.021	0.017	0.1	0.03	0.063	0.1	0
APPLES, RAW, W/SKIN	53	0	0.32		5.7	0.017	0.014	0.08	0.048	0.061	2.8	0
APPLESAUCE, CND, WO/VIT C	29	0	0.01	0.5	1.2	0.013	0.025	0.19	0.026	0.095	0.6	0
APRICOT NECTAR, CND, WO/VIT C	1316	0	0.08	5	0.6	0.009	0.014	0.26	0.022	0.096	1.3	0
APRICOTS, DEHYDRATED	12669	0	0		9.5	0.043	0.148	3.58	0.52	1.067	4.4	0
APRICOTS, RAW	2612	0	0.89		10	0.03	0.04	0.6	0.054	0.24	8.6	0

(Continued)

FOOD* by Major Food Group	Vit. A (IU)	Vit. D (IU)	Vit. E (mg TE)	Vit. K (mg)	Vit. C (mg)	Thiamin (mg)	Riboflavin (mg)	Niacin (mg)	Vit. B6 (mg)	Pantothenic (mg)	Folate (µg)	Vit. B12 (µg)
AVOCADOS, RAW	612	0	1.34	40	7.9	0.108	0.122	1.92	0.28	0.971	61.9	0
BANANAS, RAW	81	0	0.27	0.5	9.1	0.045	0.1	0.54	0.578	0.26	19.1	0
BLACKBERRIES, RAW	165	0	0.71		21	0.03	0.04	0.4	0.058	0.24	34	0
BLUEBERRIES, RAW	100	0	1	6	13	0.048	0.05	0.36	0.036	0.093	6.4	0
CANTALOUPES, RAW	3224	0	0.15	1	42.2	0.036	0.021	0.57	0.115	0.128	17	0
CASABA MELONS, RAW	30	0	0.15		16	0.06	0.02	0.4	0.12	0	17	0
CHERRIES, SOUR, RED, RAW	1283	0	0.13		10	0.03	0.04	0.4	0.044	0.143	7.5	0
CHERRIES, SWEET, RAW	214	0	0.13		7	0.05	0.06	0.4	0.036	0.127	4.2	0
CRABAPPLES, RAW	40	0	0	0.01	8	0.03	0.02	0.1	0	0	0	0
CRANBERRIES, RAW	46	0	0.1		13.5	0.03	0.02	0.1	0.065	0.219	1.7	0
CRANBERRY SAUCE, CND	20	0	0.1		2	0.015	0.021	0.1	0.014	0	1	0
CURRANTS, EUROPEAN BLACK, RAW	230	0	0.1		181	0.05	0.05	0.3	0.066	0.398	0	0
CUSTARD-APPLE, (BULLOCK'S-HEART), RAW	33	0	0		19.2	0.08	0.1	0.5	0.221	0.135	0	0
DATES, DOMESTIC, DRIED	50	0	0.1		0	0.09	0.1	2.2	0.192	0.78	12.6	0
ELDERBERRIES, RAW	600	0	1		36	0.07	0.06	0.5	0.23	0.14	6	0
FIGS, DRIED, UNCOOKED	133	0	0		0.8	0.071	0.088	0.69	0.224	0.435	7.5	0
FIGS, RAW	142	0	0.89		2	0.06	0.05	0.4	0.113	0.3	6	0
FRUIT COCKTAIL, CND, H$_2$O PK, SOL & LIQUIDS	250	0	0.29	0.8	2.1	0.016	0.011	0.36	0.052	0.062	2.7	0
GOOSEBERRIES, RAW	290	0	0.37		27.7	0.04	0.03	0.3	0.08	0.286	6	0
GRAPEFRUIT JUICE, CND	7	0	0.05	0.2	29.2	0.042	0.02	0.23	0.02	0.13	10.4	0
GRAPEFRUIT, RAW, PINK/RED/WHITE	124	0	0.25	0.02	34.4	0.036	0.02	0.25	0.042	0.283	10.2	0
GRAPE JUICE, CND/BOTTLED, WO/VIT C	8	0	0	0.2	0.1	0.026	0.037	0.26	0.065	0.041	2.6	0
GRAPES, ADHERENT SKIN TYPE, RAW	73	0	0.7	3	10.8	0.092	0.057	0.3	0.11	0.024	3.9	0
GUAVAS, RAW	792	0	1.12		183.5	0.05	0.05	1.2	0.143	0.15	14	0
HONEYDEW MELONS RAW	40	0	0.15		24.8	0.077	0.018	0.6	0.059	0.207	6	0
JACKFRUIT, RAW	297	0	0.15		6.7	0.03	0.11	0.4	0.108	0	14	0
KIWI FRUIT, RAW	175	0	1.12	25	98	0.02	0.05	0.5	0.09	0	38	0
KUMQUATS, RAW	302	0	0.24		37.4	0.08	0.1	0.5	0.06	0	16	0

Food												
LEMON JUICE, CND/BOTTLED	15	0	0.09		24.8	0.041	0.009	0.2	0.043	0.091	10.1	0
LEMONS, RAW, WO/PEEL	29	0	0.24		53	0.04	0.02	0.1	0.08	0.19	10.6	0
LIME JUICE, CND/BOTTLED	16	0	0.07		6.4	0.033	0.003	0.16	0.027	0.066	7.9	0
LITCHIS, RAW	0	0	0.7		71.5	0.011	0.065	0.6	0.1	0	14	0
MANGOS, RAW	3894	0	1.12		27.7	0.058	0.057	0.58	0.134	0.16	14	0
NECTARINES, RAW	736	0	0.89		5.4	0.017	0.041	0.99	0.025	0.158	3.7	0
OLIVES, RIPE, CND	403	0	3		0.9	0.003	0	0.04	0.009	0.015	0	0
ORANGE JUICE, CHILLED, INCL FROM CONC	78	0	0.19	0.1	32.9	0.111	0.021	0.28	0.054	0.191	18.1	0
ORANGES, RAW	205	0	0.24	0.1	53.2	0.087	0.04	0.28	0.06	0.25	30.3	0
PAPAYAS, RAW	284	0	1.12		61.8	0.027	0.032	0.34	0.019	0.218	38	0
PASSION-FRUIT, PURPLE, RAW	700	0	1.12		30	0	0.13	1.5	0.1	0	14	0
PEACHES, CND, H2O PK, SOL & LIQUIDS	532	0	0.89		2.9	0.009	0.019	0.52	0.019	0.05	3.4	0
PEACHES, RAW	535	0	0.7	3	6.6	0.017	0.041	0.99	0.018	0.17	3.4	0
PEARS, CND, H2O PK, SOL & LIQUIDS	0	0	0.5	0.5	1	0.008	0.01	0.05	0.014	0.022	1.2	0
PEARS, RAW	20	0	0.5		4	0.02	0.04	0.1	0.018	0.07	7.3	0
PINEAPPLE, CND, H2O PK, SOL & LIQUIDS	15	0	0.1		7.7	0.093	0.026	0.3	0.074	0.1	4.8	0
PINEAPPLE, RAW	23	0	0.1	0.1	15.4	0.092	0.036	0.42	0.087	0.16	10.6	0
PLANTAINS, COOKED	909	0	0.14		10.9	0.046	0.052	0.76	0.24	0.233	26	0
PLUMS, RAW	323	0	0.6	12	9.5	0.043	0.096	0.5	0.081	0.182	2.2	0
POMEGRANATES, RAW	0	0	0.55		6.1	0.03	0.03	0.3	0.105	0.596	6	0
PRICKLY PEARS, RAW	51	0	0.01		14	0.014	0.06	0.46	0.06	0	6	0
PRUNES, CND, HVY SYRUP PK, SOL & LIQUIDS	797	0	0		2.8	0.034	0.122	0.87	0.203	0.1	0.1	0
PRUNES, DEHYDRATED, STEWED	523	0	0		0	0.046	0.03	0.99	0.191	0.108	0.2	0
PRUNES, DEHYDRATED, UNCOOKED	1762	0	0		0	0.118	0.165	3	0.745	0.418	1.9	0
QUINCES, RAW	40	0	0.55		15	0.02	0.03	0.2	0.04	0.081	3	0
RAISINS, GOLDEN SEEDLESS	44	0	0.7		3.2	0.008	0.191	1.14	0.323	0.14	3.3	0
RAISINS, SEEDLESS	8	0	0.7		3.3	0.156	0.088	0.82	0.249	0.045	3.3	0
RASPBERRIES, RAW	130	0	0.45		25	0.03	0.09	0.9	0.057	0.24	26	0

(Continued)

FOOD*, by Major Food Group	Vit. A (IU)	Vit. D (IU)	Vit. E (mg TE)	Vit. K (mg)	Vit. C (mg)	Thiamin (mg)	Riboflavin (mg)	Niacin (mg)	Vit. B6 (mg)	Pantothenic (mg)	Folate (µg)	Vit. B12 (µg)
RHUBARB, FRZ, CKD	69	0	0.2		3.3	0.018	0.023	0.2	0.02	0.05	5.3	0
STRAWBERRIES, CND, HVY SYRUP, SOL & LIQUIDS	26	0	0.14		31.7	0.021	0.034	0.06	0.049	0.179	28	0
STRAWBERRIES, RAW	27	0	0.14		56.7	0.02	0.066	0.23	0.059	0.34	17.7	0
TANGERINES, RAW	920	0	0.24		30.8	0.105	0.022	0.16	0.067	0.2	20.4	0
WATERMELON, RAW	366	0	0.15		9.6	0.08	0.02	0.2	0.144	0.212	2.2	0
BEANS AND PEAS												
BLACK BEANS, CKD, BLD	6	0	0		0	0.244	0.059	0.51	0.069	0.242	149	0
BROAD BEANS (FAVA), CKD, BLD	15	0	0.09		0.3	0.097	0.089	0.71	0.072	0.157	104	0
CHICKPEAS CKD, BLD	27	0	0.35		1.3	0.116	0.063	0.53	0.139	0.286	172	0
COWPEAS (BLACKEYES), CKD, BLD	15	0	0.28		0.4	0.202	0.055	0.5	0.1	0.411	208	0
FALAFEL	13	0	0		1.6	0.146	0.166	1.04	0.125	0.292	77.6	0
FRENCH BEANS, CKD, BLD	3	0	0		1.2	0.13	0.062	0.55	0.105	0.222	74.7	0
GREAT NORTHERN BEANS, CKD, BLD	1	0	0		1.3	0.158	0.059	0.68	0.117	0.266	102	0
HUMUS, RAW	25	0	1		7.9	0.092	0.053	0.41	0.398	0.288	59.4	0
KIDNEY BEANS, CKD, BLD	0	0	0.21		1.2	0.16	0.058	0.58	0.12	0.22	130	0
LENTILS, CKD, BLD	8	0	0.11		1.5	0.169	0.073	1.06	0.178	0.638	181	0
LIMA BEANS, LRG, CKD, BLD	0	0	0.18		0	0.161	0.055	0.42	0.161	0.422	83.1	0
LUPINS, CKD, BLD	7	0	0		1.1	0.134	0.053	0.5	0.009	0.188	59.3	0
MUNG BEANS, CKD, BLD	24	0	0.51		1	0.164	0.061	0.58	0.067	0.41	159	0
NAVEY BEANS, CKD, BLD	2	0	0		0.9	0.202	0.061	0.53	0.164	0.255	140	0
PEANUT BUTTER, SMOOTH STYLE	0	0	10		0	0.083	0.105	13.4	0.454	0.806	74	0
PEANUTS, CKD, BLD	0	0	3.17		0	0.259	0.063	5.26	0.152	0.825	74.6	0
PEANUTS, DRY-ROASTED	0	0	7.41		0	0.438	0.098	13.5	0.256	1.395	145	0
PEANUTS, OIL-ROASTED	0	0	7.41		0	0.253	0.108	14.3	0.255	1.39	126	0
PEAS, SPLIT, CKD, BLD	7	0	0.39		0.4	0.19	0.056	0.89	0.048	0.595	64.9	0
PIGEON PEAS (RED GM), CKD, BLD	3	0	0		0	0.146	0.059	0.78	0.05	0.319	111	0
PINTO BEANS, CKD, BLD	2	0	0.94		2.1	0.186	0.091	0.4	0.155	0.285	172	0

Food											
REFRIED BEANS, CANNED	0	0	0		6	0.027	0.016	0.32	0.143	0.097	11	0
SOY FLOUR, FULL-FAT, RSTD	110	0	0		0	0.412	0.941	3.29	0.351	1.209	227	0
SOY MILK	32	0	0.01	3	0	0.161	0.07	0.15	0.041	0.048	1.5	0
SOYBEANS, RSTD	200	0	1.95	37	2.2	0.1	0.145	1.41	0.208	0.453	211	0
TEMPEH	686	0	0		0	0.131	0.111	4.63	0.299	0.355	52	1
TOFU, RAW	85	0	0.01	2	0.1	0.081	0.052	0.2	0.047	0.068	15	0
WINGED BEANS, CKD, BLD	0	0	0		0	0.295	0.129	0.83	0.047	0.156	10.4	0
YARDLONG BEANS, CKD, BLD	16	0	0		0.4	0.212	0.064	0.55	0.095	0.398	146	0
NUTS												
ACORNS, DRIED	0	0	0		0	0.149	0.154	2.41	0.695	0.94	115	0
ALMONDS, DRY RSTD, UNBLANCHED	0	0	5.55		0.7	0.13	0.599	2.82	0.074	0.254	63.8	0
BRAZILNUTS, DRIED, UNBLANCHED	0	0	7.6		0.7	1	0.122	1.62	0.251	0.236	4	0
BUTTERNUTS, DRIED	124	0	3.5		3.2	0.383	0.148	1.05	0.56	0.633	66.2	0
CASHEW NUTS, DRY RSTD	0	0	0.57		0	0.2	0.2	1.4	0.256	1.217	69.2	0
CHESTNUTS, EUROPEAN, RSTD	24	0	1.2		26	0.243	0.175	1.34	0.497	0.554	70	0
COCONUT MEAT, DRIED, FLAKED	0	0	0.73		0	0.03	0.02	0.3	0.261	0.696	7.8	0
COCONUT MEAT, RAW	0	0	0.73		3.3	0.066	0.02	0.54	0.054	0.3	26.4	0
COCONUT MILK, CND	0	0	0		1	0.022		0.64	0.028	0.153	13.5	0
COCONUT WATER	0	0	0		2.4	0.03	0.057	0.08	0.032	0.043	2.5	0
FILBERTS (HAZELNUTS), DRY RSTD	69	0	23.9		1	0.213	0.213	2.77	0.635	1.19	74.5	0
HICKORY NUTS, DRIED	131	0	5.21		2	0.867	0.131	0.91	0.192	1.746	40	0
MACADAMIA NUTS, OIL RSTD	9	0	0.41		0	0.213	0.109	2.02	0.198	0.442	15.9	0
PECANS, DRY RSTD	133	0	3.1		2	0.317	0.106	0.92	0.195	1.774	40.7	0
PINE NUTS, PINYON, DRIED	29	0	0		2	1.243	0.223	4.37	0.111	0.21	57.8	0
PISTACHIO NUTS, DRY RSTD	238	0	5.21	70	7.3	0.423	0.246	1.41	0.255	1.212	59.1	0
PUMPKIN & SQUASH SEEDS, WHOLE, RSTD	62	0	0		0.3	0.034	0.052	0.29	0.037	0.056	9	0
SUNFLOWER KERNELS, DRIED	50	0	50.27		1.4	2.29	0.25	4.5	0.77	6.745	227	0
TAHINI, FROM RSTD & TSTD SESAME KRNLS	67	0	2.27		0	1.22	0.473	5.45	0.149	0.693	97.7	0
WALNUTS, BLACK, DRIED	296	0	2.62		3.2	0.217	0.109	0.69	0.554	0.626	65.5	0

(Continued)

FOOD* by Major Food Group	Vit. A (IU)	Vit. D (IU)	Vit. E (mg TE)	Vit. K (mg)	Vit. C (mg)	Thiamin (mg)	Riboflavin (mg)	Niacin (mg)	Vit. B6 (mg)	Pantothenic (mg)	Folate (µg)	Vit. B$_{12}$ (µg)
POULTRY												
CHICKEN, DARK MEAT, MEAT & SKIN, FRIED W/BATTER	103		0		0	0.117	0.218	5.61	0.25	0.953	9	0.27
CHICKEN, DARK MEAT, MEAT & SKIN, RSTD	201		0		0	0.066	0.207	6.36	0.31	1.111	7	0.29
CHICKEN, GIBLETS, SIMMERED	7431		1.302		8	0.087	0.953	4.1	0.34	2.959	376	10.14
CHICKEN, LIGHT MEAT, MEAT & SKIN, FRIED W/BATTER	79		0		0	0.113	0.147	9.16	0.39	0.794	6	0.28
CHICKEN, LIGHT MEAT, MEAT & SKIN, RSTD	110		0		0	0.06	0.118	11.1	0.52	0.926	3	0.32
CHICKEN, LIVER, SIMMERED	16375		1.44		15.8	0.153	1.747	4.45	0.58	5.411	770	19.39
DUCK, MEAT ONLY, RSTD	77		0.7		0	0.26	0.47	5.1	0.25	1.5	10	0.4
DUCK, MEAT & SKIN, RSTD	210		0.7		0	0.174	0.269	4.83	0.18	1.098	6	0.3
GOOSE, MEAT ONLY, RSTD	40		0		0	0.092	0.39	4.08	0.47	1.834	12	0.49
TURKEY, BREAST, MEAT & SKIN, RSTD	0		0		0	0.057	0.131	6.37	0.48	0.634	6	0.36
TURKEY, DARK MEAT, RSTD	0		0.64		0	0.063	0.248	3.65	0.36	1.286	9	0.37
TURKEY, MEAT ONLY, RSTD	0		0.329		0	0.062	0.182	5.44	0.46	0.943	7	0.37
TURKEY, MEAT & SKIN, RSTD	0		0.339		0	0.057	0.177	5.09	0.41	0.858	7	0.35
BEEF												
BRISKET, WHOLE, LEAN, 1/4" FAT, BRSD	0		0.14		0	0.07	0.22	3.71	0.29	0.36	8	2.6
CHUCK, ARM POT RST, 1/4" FAT, BRSD	0		0.22		0	0.07	0.24	3.18	0.28	0.33	9	2.95
CHUCK, BLADE RST, 1/4" FAT, BRSD	0		0.2		0	0.07	0.24	2.42	0.26	0.31	5	2.28
CORNED BEEF, CND	0		0.15		0	0.02	0.147	2.43	0.13	0.626	9	1.62
DRIED BEEF, CURED	0		0.14		0	0.083	0.2	5.45	0.35	0.611	11	2.66
FLANK, BRSD	0		0		0	0.14	0.18	4.42	0.35	0.37	9	3.3
GROUND, LEAN, BKD, MED	0		0		0	0.05	0.19	4.28	0.2	0.27	9	1.77
GROUND, REG, BKD, MED	0		0		0	0.03	0.16	4.75	0.23	0.22	9	2.34
RIB, EYE, SMALL END (10–12), 1/4" FAT, BROILED	0		0		0	0.09	0.19	4.22	0.35	0.31	7	3.01
RIB, WHOLE (RIBS 6–12), 1/4" FAT, BROILED	0		0.236		0	0.08	0.17	3.25	0.27	0.32	6	2.87

ROUND, BOTTOM ROUND, 1/4" FAT, BRSD	0	0.19	0	0.07	0.24	3.73	0.33	0.38	10	2.35
ROUND, EYE OF ROUND, 1/4" FAT, RSTD	0	0.179	0	0.08	0.16	3.49	0.35	0.42	7	2.1
ROUND, FULL CUT, 1/4" FAT, BROILED	0	0.19	0	0.09	0.21	3.99	0.38	0.38	9	3.01
ROUND, TOP ROUND, 1/4" FAT, BROILED	0	0.15	0	0.11	0.26	5.71	0.53	0.46	11	2.42
SHORT LOIN, PORTERHOUSE STEAK, 1/4" FAT, BROILED	0	0.217	0	0.09	0.21	3.89	0.34	0.3	7	2.11
SHORT LOIN, TOP LOIN, 1/4" FAT, BROILED	0	0.021	0	0.08	0.18	4.7	0.37	0.33	7	1.94
TENDERLOIN, 1/4" FAT, BROILED	0	0.19	0	0.11	0.26	3.52	0.39	0.34	6	2.41
TOP SIRLOIN, 1/4" FAT, BROILED	0	0.18	0	0.11	0.27	3.93	0.41	0.36	9	2.69
PORK										
BACON, CANADIAN-STYLE, GRILLED	0	0.26	0	0.824	0.197	6.92	0.45	0.52	4	0.78
BACON, CKD, BROILED/PAN-FRIED/RSTD	0	0.54	0	0.692	0.285	7.32	0.27	1.055	5	1.75
CURED HAM, BONELESS, EXTRA LEAN (5% FAT), RSTD	0	0.26	0	0.754	0.202	4.02	0.4	0.403	3	0.65
CURED HAM, BONELESS, REG (11% FAT), RSTD	0	0.26	0	0.73	0.33	6.15	0.31	0.72	3	0.7
CURED HAM, EXTRA LEAN (4% FAT), CND	0	0.26	0	0.836	0.23	5.3	0.45	0.492	6	0.82
CURED HAM, REG (13% FAT), CND	0	0.26	0	0.963	0.231	3.22	0.48	0.394	5	0.78
LEG (HAM), WHOLE, LEAN, RSTD	9	0.26	0.4	0.69	0.349	4.94	0.45	0.67	12	0.72
LOIN, BLADE (CHOPS), BONE-IN, LEAN & FAT, BRSD	8	0.26	0.6	0.476	0.232	3.59	0.297	0.562	2	0.65
LOIN, TENDERLOIN, LEAN, RSTD	7	0.26	0.4	0.94	0.39	4.71	0.42	0.687	6	0.55
LOIN, TOP LOIN (CHOP), BONELESS, LEAN & FAT, BRSD	7	0	0.3	0.552	0.256	4.54	0.329	0.638	4	0.46
LOIN, TOP LOIN (ROAST), BONELESS, LEAN & FAT, RSTD	8	0	0.4	0.614	0.296	5.13	0.373	0.545	8	0.55
LOIN, WHOLE, LEAN, BROILED	7	0.26	0.7	0.923	0.338	5.24	0.492	0.729	6	0.72
SHOULDER, ARM PICNIC, LEAN, RSTD	7	0	0.3	0.578	0.357	4.31	0.41	0.592	5	0.78
SHOULDER, WHOLE, LEAN, RSTD	7	0.26	0.6	0.628	0.37	4.26	0.317	0.651	5	0.86
SPARE RIBS, LEAN & FAT, BRSD	10	0.26	0	0.408	0.382	5.48	0.35	0.75	4	1.08

(Continued)

FOOD*, by Major Food Group	Vit. A (IU)	Vit. D (IU)	Vit. E (mg TE)	Vit. K (mg)	Vit. C (mg)	Thiamin (mg)	Riboflavin (mg)	Niacin (mg)	Vit. B6 (mg)	Pantothenic (mg)	Folate (µg)	Vit. B12 (µg)
SAUSAGES AND LUNCHEON MEATS												
BOLOGNA, BEEF	0	28	0.19		0	0.05	0.109	2.41	0.15	0.28	5	1.42
BOLOGNA, PORK	0	56	0.26		0	0.523	0.157	3.9	0.27	0.72	5	0.93
BOLOGNA, TURKEY	0		0.53		0	0.055	0.165	3.53	0.22	0.7	7	0.27
BRATWURST, COOKED, PORK	0	44	0.25		1	0.505	0.183	3.2	0.21	0.32	2	0.95
CHICKEN ROLL, LIGHT MEAT	82		0.265		0	0.065	0.13	5.29	0.21	0.39	2	0.15
FRANKFURTER, BEEF	0	36	0.19		0	0.051	0.102	2.42	0.12	0.29	4	1.54
FRANKFURTER, BEEF & PORK	0	36	0.25		0	0.199	0.12	2.63	0.13	0.35	4	1.3
FRANKFURTER, CHICKEN	130		0.215		0	0.066	0.115	3.09	0.32	0.83	4	0.24
FRANKFURTER, TURKEY	0		0.617		0	0.041	0.179	4.13	0.23	0.72	8	0.28
HAM, CHOPPED, CANNED	0		0.25		2	0.535	0.165	3.2	0.32	0.28	1	0.7
HAM, SLICED, EXRA LEAN, (5% FAT)	0		0.29		0	0.932	0.223	4.84	0.46	0.47	4	0.75
HAM, SLICED, REG (11% FAT)	0		0.29		0	0.863	0.252	5.25	0.34	0.45	3	0.83
ITALIAN SAUSAGE, CKD, PORK	0		0.25		2	0.623	0.233	4.17	0.33	0.45	5	1.3
KIELBASA, PORK, BEEF & NONFAT DRY MILK	0		0.22		0	0.228	0.214	2.88	0.18	0.82	5	1.61
KNOCKWURST, PORK & BEEF	0		0.57		0	0.342	0.14	2.73	0.17	0.32	2	1.18
OLIVE LOAF, PORK	200	44	0.25		0	0.295	0.26	1.84	0.23	0.77	2	1.26
PASTRAMI, TURKEY	0		0.216		0	0.055	0.25	3.53	0.27	0.58	5	0.24
POLISH SAUSAGE, PORK	0		0		1	0.502	0.148	3.44	0.19	0.45	2	0.98
SALAMI, BEEF	0	36	0.19		0	0.103	0.189	3.24	0.18	0.95	2	3.06
SALAMI, BEEF & PORK	0		0.22		0	0.239	0.376	3.55	0.21	0.85	2	3.65
SMOKED LINK SAUSAGE, PORK	0	52	0.25		2	0.7	0.257	4.53	0.35	0.78	5	1.63
SMOKED LINK SAUSAGE, PORK & BEEF	0	28	0.22		0	0.26	0.17	3.23	0.17	0.44	2	1.51
TURKEY BREAST MEAT	0		0.09		0	0.04	0.107	8.32	0.36	0.59	4	2.02
TURKEY HAM	0		0.64		0	0.052	0.247	3.53	0.24	0.85	6	0.24
TURKEY ROLL, LIGHT MEAT	0		0.134		0	0.089	0.226	7	0.32	0.42	4	0.24
VIENNA SAUSAGE, CND, BEEF & PORK	0		0.22		0	0.087	0.107	1.61	0.12	0.35	4	1.02

FISH AND SEAFOOD													
ABALONE, MIXED SPECIES, FRIED	5		0	18	1.8	0.22	0.13	1.9	0.15	2.87	5.4	0.69
ANCHOVY, CND IN OIL, DRND SOL	70		5		0	0.078	0.363	19.9	0.203	0.909	12.5	0.88
CARP, COOKED, DRY HEAT	32		0		1.6	0.14	0.07	2.1	0.219	0.87	17.3	1.471
CATFISH, CHANNEL, BREADED & FRIED	28		0		0	0.073	0.133	2.28	0.19	0.73	16.5	1.9
CAVIAR, BLACK & RED	1868	232	7		0	0.19	0.62	0.12	0.32	3.5	50	20
CLAM, MIXED SPECIES, BREADED & FRIED	302	4	0		10	0.1	0.244	2.06	0.06	0.43	18.2	40.269
COD, ATLANTIC, CKD, DRY HEAT	46		0.3		1	0.088	0.079	2.51	0.283	0.18	8.1	1.048
COD, ATLANTIC, DRIED & SALTED	141		0.6		3.5	0.268	0.24	7.5	0.864	1.675	24.7	10
CRAB, ALASKA KING, CKD, MOIST HEAT	29		0		7.6	0.053	0.055	1.34	0.18	0.4	51	11.5
CRAB, BLUE, CKD, MOIST HEAT	6		1		3.3	0.1	0.05	3.3	0.18	0.43	50.8	7.3
CRAYFISH, MIXED SPECIES, WILD, CKD, MOIST HEAT	50		1.5		0.9	0.05	0.085	2.28	0.076	0.58	44	2.15
EEL, MIXED SPECIES, CKD, DRY HEAT	3787		5.1		1.8	0.183	0.051	4.49	0.077	0.28	17.3	2.885
FLATFISH (FLOUNDER/SOLE), CKD, DRY HEAT	38		1.89		0	0.08	0.114	2.18	0.24	0.58	9.2	2.509
GEFILTEFISH	89		0		0.8	0.065	0.059	1	0.08	0.2	2.8	0.844
HADDOCK, CKD, DRY HEAT	63		0		0	0.04	0.045	4.63	0.346	0.15	13.3	1.387
HALIBUT, CKD, DRY HEAT	179		1.09		0	0.069	0.091	7.12	0.397	0.38	13.8	1.366
HERRING, CKD, DRY HEAT	102		1.34		0.7	0.112	0.299	4.12	0.348	0.74	11.5	13.141
HERRING, KIPPERED	128		1		1	0.126	0.319	4.4	0.413	0.88	13.7	18.701
HERRING, PICKLED	861	680	1		0	0.036	0.139	3.3	0.17	0.081	2.4	4.27
LOBSTER, NORTHERN, CKD, MOIST HEAT	87		1		0	0.007	0.066	1.07	0.077	0.285	11.1	3.11
MACKEREL, CKD, DRY HEAT	180		0		0.4	0.159	0.412	6.85	0.46	0.99	1.5	19
MACKEREL, JACK, CND, DRND SOL	434	228	1.4		0.9	0.04	0.212	6.18	0.21	0.305	5	6.94
MUSSEL, BLUE, CKD, MOIST HEAT	304		0		13.6	0.3	0.42	3	0.1	0.95	75.6	24
OCEAN PERCH, CKD, DRY HEAT	46		0		0.8	0.13	0.134	2.44	0.27	0.42	10.4	1.154
OYSTER, BREADED & FRIED	302		0		3.8	0.15	0.202	1.65	0.064	0.27	13.6	15.629
OYSTER, CANNED	300		0.85		5	0.15	0.166	1.24	0.095	0.18	8.9	19.133
OYSTER, RAW	100		0.85	0.1	3.7	0.1	0.095	1.38	0.062	0.185	10	19.46
PERCH, CKD, DRY HEAT	32		0		1.7	0.08	0.12	1.9	0.14	0.87	5.8	2.2

(Continued)

FOOD*, by Major Food Group	Vit. A (IU)	Vit. D (IU)	Vit. E (mg TE)	Vit. K (mg)	Vit. C (mg)	Thiamin (mg)	Riboflavin (mg)	Niacin (mg)	Vit. B6 (mg)	Pantothenic (mg)	Folate (µg)	Vit. B$_{12}$ (µg)
PIKE, NORTHERN, CKD, DRY HEAT	81		0		3.8	0.067	0.077	2.8	0.135	0.87	17.3	2.3
POLLOCK, WALLEYE, CKD, DRY HEAT	76		0.2		0	0.074	0.076	1.65	0.069	0.16	3.6	4.2
ROE, MIXED SPECIES, RAW	263		7		16	0.24	0.74	1.8	0.16	1	80	10
SALMON, CHINOOK, SMOKED	88		1.35		0	0.023	0.101	4.72	0.278	0.87	1.9	3.26
SALMON, COHO, WILD, CKD, MOIST HEAT	108		0		1	0.115	0.159	7.78	0.556	0.834	9	4.48
SALMON, PINK, CND, SOL W/BONE & LIQ	55	624	1.35		0	0.023	0.186	6.54	0.3	0.55	15.4	4.4
SALMON, SOCKEYE, CKD, DRY HEAT	209		0		0	0.215	0.171	6.67	0.219	0.7	5	5.8
SARDINE, CND IN OIL, DRND SOL	224	272	0.3		0	0.08	0.227	5.25	0.167	0.642	11.8	8.94
SCALLOP, MIXED SPECIES, BREADED & FRIED	75		0		2.3	0.042	0.11	1.51	0.14	0.2	18.2	1.318
SEA BASS, MIXED SPECIES, CKD, DRY HEAT	213				0	0.13	0.15	1.9	0.46	0.87	5.8	0.3
SHARK, MIXED SPECIES, BATTERED & FRIED	180		0		0	0.072	0.097	2.78	0.3	0.62	5.2	1.211
SHRIMP, MIXED SPECIES, BREADED & FRIED	189		0		1.5	0.129	0.136	3.07	0.098	0.35	8.1	1.87
SHRIMP, MIXED SPECIES, CKD, MOIST HEAT	219	152	0.51		2.2	0.031	0.032	2.59	0.127	0.34	3.5	1.488
SMELT, RAINBOW, CKD, DRY HEAT	58		0		0	0.01	0.146	1.77	0.17	0.74	4.6	3.969
SNAPPER, MIXED SPECIES, CKD, DRY HEAT	115		0		1.6	0.053	0.004	0.35	0.46	0.87	5.8	3.5
SQUID, MIXED SPECIES, FRIED	35		0		4.2	0.056	0.458	2.6	0.058	0.51	5.3	1.228
SURIMI	66		0		0	0.02	0.021	0.22	0.03	0.07	1.6	1.6
SWORDFISH, CKD, DRY HEAT	137		0		1.1	0.043	0.116	11.8	0.381	0.38	2.3	2.019
TROUT, RAINBOW, WILD, CKD, DRY HEAT	50		0		2	0.152	0.097	5.77	0.346	1.065	19	6.3
TUNA, FRSH, BLUEFIN, CKD, DRY HEAT	2520		0		0	0.278	0.306	10.5	0.525	1.37	2.2	10.878
TUNA, LT, CND IN H$_2$O, DRND SOL	56	236	0.53		0	0.032	0.074	13.3	0.35	0.214	4	2.99
WHITING, MIXED SPECIES, CKD, DRY HEAT	114		0.3		0	0.068	0.06	1.67	0.18	0.25	15	2.6

DAIRY PRODUCTS AND EGGS												
BUTTER	3058	56	1.58	7	0	0.005	0.034	0.04	0.003	0.11	3	0.125
CHEESE, AMERICAN	1209.6		0.46		0	0.027	0.353	0.07	0.071	0.482	7.8	0.696
CHEESE, BLUE	721		0.64		0	0.029	0.382	1.02	0.166	1.729	36.4	1.217
CHEESE, BRIE	667		0.655		0	0.07	0.52	0.38	0.235	0.69	65	1.65
CHEESE, CAMEMBERT	923	12	0.655		0	0.028	0.488	0.63	0.227	1.364	62.2	1.296
CHEESE, CHEDDAR	1059	12	0.36	3	0	0.027	0.375	0.08	0.074	0.413	18.2	0.827
CHEESE, COLBY	1034		0.35		0	0.015	0.375	0.09	0.079	0.21	18.2	0.826
CHEESE, COTTAGE, 1% FAT	37		0.11		0	0.021	0.165	0.13	0.068	0.215	12.4	0.633
CHEESE, COTTAGE, CREAMED	163		0.122		0	0.021	0.163	0.13	0.067	0.213	12.2	0.623
CHEESE, CREAM	1427		0.941		0	0.017	0.197	0.1	0.047	0.271	13.2	0.424
CHEESE, CREAM, FAT FREE	930		0.03		0	0.05	0.172	0.16	0.05	0.194	37	0.55
CHEESE, EDAM	916	36	0.751		0	0.037	0.389	0.08	0.076	0.281	16.2	1.535
CHEESE, FETA	447		0.03		0	0.154	0.844	0.99	0.424	0.967	32	1.69
CHEESE, GOUDA	644		0.35		0	0.03	0.334	0.06	0.08	0.34	20.9	1.535
CHEESE, GRUYERE	1219		0.35		0	0.06	0.279	0.11	0.081	0.562	10.4	1.6
CHEESE, MONTEREY	950		0.34		0	0.015	0.39	0.09	0.079	0.21	18.2	0.826
CHEESE, MOZZARELLA, SKIM MILK	584		0.43		0	0.018	0.303	0.11	0.07	0.079	8.8	0.817
CHEESE, MOZZARELLA, WHOLE MILK	792		0.35		0	0.015	0.243	0.08	0.056	0.064	7	0.654
CHEESE, MUENSTER	1120		0.465		0	0.013	0.32	0.1	0.056	0.19	12.1	1.473
CHEESE, PARMESAN	701	28	0.8		0	0.045	0.386	0.32	0.105	0.527	8	1.4
CHEESE, PROVOLONE	815		0.35		0	0.019	0.321	0.16	0.073	0.476	10.4	1.463
CHEESE, RICOTTA, SKIM MILK	432		0.214		0	0.021	0.185	0.08	0.02	0.242	13.1	0.291
CHEESE, RICOTTA, WHOLE MILK	490		0.35		0	0.013	0.195	0.1	0.043	0.213	12.2	0.338
CHEESE, SWISS	845	44	0.5		0	0.022	0.365	0.09	0.083	0.429	6.4	1.676
CREAM, HALF & HALF	434		0.11		0.86	0.035	0.149	0.08	0.039	0.289	2.5	0.329
CREAM, LT, COFFEE/TABLE	633		0.15		0.76	0.032	0.148	0.06	0.032	0.276	2.3	0.22
CREAM, SOUR	790		0.566	1	0.86	0.035	0.149	0.07	0.016	0.36	10.8	0.3
EGG, WHITE, DRIED	0	0	0		0	0.005	2.53	0.87	0.036	0.775	18	0.18
EGG, WHITE, RAW, FRESH	0	0	0	0.01	0	0.006	0.452	0.09	0.004	0.119	3	0.2

(Continued)

FOOD*, by Major Food Group	Vit. A (IU)	Vit. D (IU)	Vit. E (mg TE)	Vit. K (mg)	Vit. C (mg)	Thiamin (mg)	Riboflavin (mg)	Niacin (mg)	Vit. B6 (mg)	Pantothenic (mg)	Folate (µg)	Vit. B12 (µg)
EGG, WHOLE, FRIED	857	32	1.64		0	0.057	0.523	0.08	0.143	1.224	38	0.92
EGG, WHOLE, HARD-BOILED	560		1.05		0	0.066	0.513	0.06	0.121	1.398	44	1.11
EGG, WHOLE, RAW	635	52	1.05	2	0	0.062	0.508	0.07	0.139	1.255	47	1
EGG, YOLK, RAW	1945		3.16		0	0.17	0.639	0.02	0.392	3.807	146	3.11
MILK, BUTTRRMILK, FROM SKIM MILK	33		0.06		0.98	0.034	0.154	0.06	0.034	0.275	5	0.219
MILK, CND, EVAPORATED, WHOLE, WO/VIT A	243		0.18		1.88	0.047	0.316	0.19	0.05	0.638	7.9	0.163
MILK, DRY, SKIM, NON-FAT SOL, REG, WO/VIT A	36		0.021		6.76	0.415	1.55	0.95	0.361	3.568	50.1	4.033
MILK, GOAT	185	12	0.09		1.29	0.048	0.138	0.28	0.046	0.31	0.6	0.065
MILK, HUMAN	241	4	0.9		5	0.014	0.036	0.18	0.011	0.223	5.2	0.045
MILK, LOW FAT, 1% FAT, W/VIT A	205	40	0.04		0.97	0.039	0.167	0.09	0.043	0.323	5.1	0.368
MILK, LOW FAT, 2% FAT, W/VIT A	205	40	0.07		0.95	0.039	0.165	0.09	0.043	0.32	5.1	0.364
MILK, SKIM, W/VIT A	204	40	0.04	0.02	0.98	0.036	0.14	0.09	0.04	0.329	5.2	0.378
MILK, WHOLE, 3.3% FAT	126	40	0.1	0.3	0.94	0.038	0.162	0.08	0.042	0.314	5	0.357
YOGURT, PLAIN, WHOLE MILK	123		0.088		0.53	0.029	0.142	0.08	0.032	0.389	7.4	0.372
FATS AND OILS												
FAT, CHICKEN	0		2.7		0	0	0	0	0	0	0	0
LARD	0		1.2		0	0	0	0	0	0	0	0
MARGARINE, HARD, CORN (HYDR & REG)	3571	0	14.94		0.16	0.01	0.037	0.02	0.009	0.084	1.18	0.095
MARGARINE, HARD, CORN, SOY & COTTONSEED (HYDR)	3571	0	0	51	0.16	0.01	0.037	0.02	0.009	0.084	1.18	0.095
MARGARINE, SOFT, CORN (HYDR & REG)	3571	0	0		0.141	0.009	0.032	0.02	0.008	0.075	1.05	0.084
MARGARINE, SOFT, SAFFLOWER, COTTONSEED & PNUT (HYDR)	3571	0	0		0.141	0.009	0.032	0.02	0.008	0.075	1.05	0.084
MARGARINE, SOFT, SOYBEAN (HYDR & REG)	3571	0	0		0.141	0.009	0.032	0.02	0.008	0.075	1.05	0.084
MAYONNAISE	220		4	81	0	0.013	0.024	0	0.017	0.243	6.28	0.208
OIL, CANOLA	0	0	20.95	141	0	0	0	0	0	0	0	0

	C1	C2	C3	C4	C5	C6	C7	C8	C9	C10	C11
OIL, COCOA BUTTER	0	0	0	0	0	0	0	0	0	0	0
OIL, COCONUT	0	0	0.28	0	0	0	0	0	0	0	0
OIL, COD LIVER	100000	16700	0	0	0	0	0	0	0	0	0
OIL, CORN				3	0	0	0	0	0	0	0
OIL, MUSTARD	0	0	0	0	0	0	0	0	0	0	0
OIL, OLIVE	0	0	12.4	49	0	0	0	0	0	0	0
OIL, PALM	0	0	21.76	0	0	0	0	0	0	0	0
OIL, PEANUT	0	0	12.92	0.7	0	0	0	0	0	0	0
OIL, RICE BRAN	0	0	0	0	0	0	0	0	0	0	0
OIL, SESAME	0	0	4.09	10	0	0	0	0	0	0	0
OIL, SOYBEAN	0	0	18.19	193	0	0	0	0	0	0	0
OIL, SUNFLOWER	0	0	0	9	0	0	0	0	0	0	0
OIL, WHEATGERM	0	0	192.4	0	0	0	0	0	0	0	0
SALAD DRESSING, 1000 ISLAND, LOW FAT	320	0	1.19	0	0.011	0.021	0	0.015	0.216	0.185	5.58
SALAD DRESSING, 1000 ISLAND	320	0	1.14	0	0.013	0.024	0	0.017	0.243	0.208	6.28
SALAD DRESSING, FRENCH, LOW FAT	1300	0	1.19	0							
SALAD DRESSING, ITALIAN, LOW FAT	0	0	1.5	0							
SALAD DRESSING, RUSSIAN	690	0	10.2	6	0.05	0.05	0.6	0.03	0.4	0.3	10.4
SALAD DRESSING, RUSSIAN, LOW FAT	56	0	0.76	6	0.007	0.013	0	0.009	0.135	0.116	3.49
SHORTENING, SOYBEAN & COTTONSEED (HYDR)	0	0	8.28	0	0	0	0	0	0	0	0
TALLOW	0	0	2.7	0	0	0	0	0	0	0	0
SPICES											
ALLSPICE	540	0	1.03	39	0.101	0.063	2.86	0.34	0	0	36
ANISE SEED	311	0	1.03	21	0.34	0.29	3.06	0.34	0.797	0	10
BASIL	9375	0	1.69	61	0.148	0.316	6.95	1.21	0	0	274
BAY LEAF	6185	0	1.786	47	0.009	0.421	2.01	1	0	0	180
CARAWAY SEED	363	0	2.5	21	0.383	0.379	3.61	0.34	0	0	10
CARDAMON	0	0	0	21	0.198	0.182	1.1	0	0	0	0
CELERY SEED	52	0	1.03	17.1	0.34	0.29	3.06	0.34	0	0	10
CHILI POWDER	34927	0	1.03	64	0.349	0.794	7.89	1.87	0	0	100

(Continued)

FOOD*, by Major Food Group	Vit. A (IU)	Vit. D (IU)	Vit. E (mg TE)	Vit. K (mg)	Vit. C (mg)	Thiamin (mg)	Riboflavin (mg)	Niacin (mg)	Vit. B6 (mg)	Pantothenic (mg)	Folate (µg)	Vit. B12 (µg)
CINNAMON	260	0	0.01		28	0.077	0.14	1.3	0.25	0	29	0
CLOVES	530	0	1.69		81	0.115	0.267	1.46	1.29	0	93	0
CORIANDER LEAF, DRIED	5850	0	1.03		567	1.252	1.5	10.7	1.21	0	274	0
CORIANDER SEED	0	0	0		21	0.239	0.29	2.13	0	0	0.03	0
CUMIN SEED	1270	0	1.03		7.7	0.628	0.327	4.58	0.34	0	10	0
CURRY POWDER	986	0	0.3		11.4	0.253	0.281	3.47	0.68	0	154	0
DILL SEED	53	0	1.03		21	0.418	0.284	2.81	0.34	0	10	0
DILL WEED, DRIED	5850	0			50	0.418	0.284	2.81	1.461	0	0	0
FENNEL SEED	135	0	0		21	0.408	0.353	6.05	0	0	0	0
GARLIC POWDER	0	0	0.01		18	0.466	0.152	0.69	2.7	0	2	0
GINGER	147	0	0.28		7	0.046	0.185	5.16	1.12	0	39	0
MACE	800	0	2.5		21	0.312	0.448	1.35	0.3	0	76	0
MARJORAM, DRIED	8068	0	1.69		51	0.289	0.316	4.12	1.21	0	274	0
MUSTARD SEED, YELLOW	62	0	2.5		3	0.543	0.381	7.89	0.3	0	76	0
NUTMEG	102	0	2.5		3	0.346	0.057	1.3	0.3	0	76	0
OREGANO	6903	0	1.69		50	0.341	0.32	6.22	1.21	0	274	0
PAPRIKA	60604	0	0.69		71	0.645	1.743	15.3	2.06	1.78	106	0
PEPPER, BLACK	190	0	1.03		21	0.109	0.24	1.14	0.34	0	10	0
PEPPER, RED/CAYENNE	41610	0	4.8		76	0.328	0.919	8.7	2.06	0	106	0
PEPPER, WHITE	0	0	2.5		21	0.022	0.126	0.21	0.34	0	10	0
PEPPERMINT, FRESH	4248	0	0.34		31.8	0.082	0.266	1.71	0.129	0.338	114	0
POPPY SEED	0	0	2.72		3	0.849	0.173	0.98	0.444	0	58	0
ROSEMARY, DRIED	3128	0	0		61	0.514	0	1	0	0	0	0
SAFFRON	530	0	1.69		81	0.115	0.267	1.46	1.3	0	93	0
SAGE	5900	0	1.69		32.3	0.754	0.336	5.72	1.21	0	274	0
SAVORY	5130	0	0		50	0.366	0	4.08	0	0	0	0
SPEARMINT, FRESH	4054	0	0.34		13	0.078	0.175	0.95	0.158	0.25	105	0
TARRAGON	4200	0	1.69		50	0.251	1.339	8.95	1.21	0	274	0
THYME	3800	0	1.69		50	0.513	0.399	4.94	1.21	0	274	0

TURMERIC	0	0.07	26	0.152	0.233	5.14	1.8	0	39	0
VANILLA EXTRACT	0	0	0	0.011	0.095	0.43	0.026	0.035	0	0
SOUPS										
BEAN W/PORK, CND	662	0.285	1.2	0.065	0.025	0.42	0.031	0.07	23.7	0.03
BLACK BEAN, CND	445	0.058	0.2	0.042	0.039	0.41	0.07	0.16	20	0
CHICKEN BROTH, CND	0	0.028	0	0.006	0.046	2.23	0.02	0.04	4	0.2
CHICKEN GUMBO, CND	108	0.031	4	0.2	0.3	0.53	0.05	0.16	5	0.02
CHICKEN NOODLE, CND	532	0.051	0	0.052	0.054	1.23	0.021	0.14	1.8	0.13
CHICKEN W/RICE, CND	539	0.043	0.1	0.014	0.02	0.92	0.02	0.14	0.9	0.13
CLAM CHOWDER, MANHATTAN, CND	767	0.581	3.2	0.024	0.032	0.65	0.08	0.15	8	3.23
CLAM CHOWDER, NEW ENGLAND, CND	8	0.066	1.9	0.016	0.03	0.74	0.06	0.26	2.9	7.82
CREAM OF ASPARAGUS, CND	355	0.5	2.2	0.043	0.062	0.62	0.01	0.11	19	0.04
CREAM OF CELERY, CND	244	0.15	0.2	0.023	0.039	0.27	0.01	0.92	1.9	0.04
CREAM OF CHICKEN, CND	446	0.13	0.1	0.023	0.048	0.65	0.013	0.17	1.3	0.07
CREAM OF MUSHROOM, CND	0	1.04	0.9	0.024	0.066	0.65	0.01	0.2	3	0.1
CREAM OF ONION, CND	236	0.68	1	0.04	0.06	0.4	0.02	0.24	5.7	0.04
CREAM OF POTATO, CND	230	0.06	0	0.028	0.029	0.43	0.03	0.7	2.4	0.04
GAZPACHO, CND, READY TO SERVE	1067	0.19	2.9	0.02	0.01	0.38	0.06	0.07	4	0
LENTIL W/HAM, CND, READY TO SERVE	145	0	1.7	0.07	0.045	0.55	0.09	0.14	20	0.12
MINESTRONE, CND	1908	0.6	0.9	0.044	0.036	0.77	0.08	0.28	13.1	0
OYSTER STEW, CND	58	0	2.6	0.017	0.029	0.19	0.01	0.1	2	1.79
PEA, GREEN, CND	153	0.099	1.3	0.082	0.052	0.94	0.04	0.1	1.4	0
PEA, SPLIT W/HAM, CND	331	0	1.1	0.11	0.056	1.1	0.05	0.2	1.9	0.2
TOMATO RICE, CND	588	0.6	11.5	0.048	0.039	0.82	0.06	0.1	11	0
TOMATO, CND	555	2.02	53	0.07	0.04	1.13	0.09	0.12	11.7	0
TURKEY NOODLE, CND	233	0.047	0.1	0.059	0.051	1.11	0.03	0.14	1.8	0.13
VEGETABLE BEEF, CND	1508	0	1.9	0.029	0.039	0.82	0.06	0.28	8.4	0.25
VEGETARIAN VEGETABLE, CND	2453	0.26	1.2	0.044	0.037	0.75	0.045	0.28	8.6	0
BEVERAGES										
BEER, REG	0	0	0	0.006	0.026	0.45	0.05	0.058	6	0.02
CLAM & TOMATO JUICE, CND	215	0	4.1	0.04	0.03	0.19	0.084	0.251	15.9	30.6

(Continued)

FOOD*, by Major Food Group	Vit. A (IU)	Vit. D (IU)	Vit. E (mg TE)	Vit. K (mg)	Vit. C (mg)	Thiamin (mg)	Riboflavin (mg)	Niacin (mg)	Vit. B6 (mg)	Pantothenic (mg)	Folate (µg)	Vit. B12 (µg)
COCOA MIX, WO/ADDED NUTR, PDR	14	0	0.15		1.8	0.096	0.565	0.59	0.114	0.893	0	1.32
COFFEE, BREWED, ESPRESSO	0	0	0		0.2	0.001	0.177	5.21	0.002	0.028	1	0
COFFEE, BREWED, REGULAR	0	0	0	10	0	0	0	0.22	0	0.001	0.1	0
DISTILLED (GIN/RUM/VODKA/WHISKEY), 80 PROOF	0	0	0		0	0.006	0.004	0.01	0.001	0	0	0
SODAS, GINGER ALE/GRAPE/ORANGE	0	0	0		0	0	0	0	0	0	0	0
SODAS, LEMON-LIME	0	0	0		0	0	0	0.02	0	0	0	0
TEA, BREWED	0	0	0	0.05	0	0	0.014	0	0	0.011	5.2	0
WINE, TABLE	0	0	0	0.01	0	0.004	0.016	0.07	0.024	0.028	1.1	0.01
SNACK FOODS AND DESSERTS												
BANANA CHIPS	83	0	5.4		6.3	0.085	0.017	0.71	0.26	0.62	14	0
CANDIES, CARAMELS	32	0	0.463	0	0.5	0.01	0.181	0.25	0.035	0.592	5	0
CANDIES, GUMDROPS	0	0	0	0	0	0	0.002	0	0	0.006	0	0
CANDIES, HARD	0	0	0	0	0	0.004	0.003	0.01	0.003	0.008	0	0
CANDIES, JELLYBEANS	0	0	0	0	0	0	0	0	0	0	0	0
CANDIES, MARSHMALLOWS	1	0	0	0	0	0.001	0.001	0.08	0.002	0.005	1	0
CANDIES, MILK CHOCOLATE	185	0	1.24		0.4	0.079	0.301	0.32	0.042	0.424	8	0.39
CANDIES, SEMISWEET CHOC	21	0	1.19		0	0.055	0.09	0.43	0.035	0.105	3	0
FROSTING, VANILLA, CREAMY, READY TO EAT	746	0	2.016		0	0	0.006	0.01	0	0	0	0
GELATINS, PREPD W/H2O	0	0	0	0	0	0	0.003	0	0.002	0.002	0	0
ICE MILK, VANILLA	165	0	0		0.8	0.058	0.265	0.09	0.065	0.505	6	0.67
JAMS AND PRESERVES	12	0	0		8.8	0	0.022	0.04	0.02	0.02	33	0
JELLIES	17	0	0		0.9	0.001	0.026	0.04	0.02	0.197	1	0
MARMALADE, ORANGE	47	0	0		4.8	0.005	0.006	0.05	0.014	0.015	36	0
MOLASSES	0	0	0		0	0.041	0.002	0.93	0.67	0.804	0	0
POPCORN, AIR-POPPED	196	0	0.12		0	0.203	0.283	1.94	0.245	0.42	23	0
POPCORN, CAKES	72	0	0.12		0	0.075	0.178	6.01	0.181	0.434	18	0

Food											
POPCORN, CARAMEL-COATED, W/PNUTS	64	1.5	0	0	0.051	0.126	1.99	0.185	0.23	16	0
POPCORN, OIL-POPPED	154	0.12	0	0.3	0.134	0.136	1.55	0.209	0.305	17	0
POTATO CHIPS, PLAIN	0	4.88	10	31.1	0.167	0.197	3.83	0.66	0.402	45	0
POTATO CHIPS, BARBECUE	217	5	10	33.9	0.215	0.215	4.69	0.622	0.617	83	0
PRETZELS, HARD, SALTED	0	0.21	1	0	0.461	0.623	5.25	0.116	0.288	83	0
PUDDING, CHOC, PREPD W/2% MILK	185	0		0.6	0.03	0.139	0.13	0.034	0.26	4	0.23
PUDDING, CHOC, W/WHOLE MILK	107	0.06		0.9	0.033	0.141	0.1	0.038	0.269	4	0.3
PUDDING, LEMON, PREPD W/2% MILK	170	0		0.8	0.033	0.137	0.07	0.036	0.267	4	0.3
PUDDING, RICE, PREPD W/WHOLE MILK	107	0		0.7	0.075	0.139	0.45	0.034	0.285	4	0.24
PUDDING, TAPIOCA, PREPD W/WHOLE MILK	109	0		0.7	0.03	0.141	0.07	0.038	0.274	4	0.25
RICE CAKES, BROWN RICE, PLAIN	46	0.72	0	0	0.061	0.165	7.81	0.15	1	21	0
SUGAR, BROWN	0	0	0	0	0.008	0.007	0.08	0.026	0.111	1	0
SUGAR, GRANULATED	0	0	0	0	0	0.019	0	0	0	0	0
SYRUP, CHOC, FUDGE-TYPE	90	0	0	0.5	0.03	0.22	0.2	0.035	0.306	4	0.3
SYRUP, CORN	0	0	,	0	0.011	0.009	0.02	0.009	0.023	0	0
SYRUP, MAPLE	0	0	0	0	0.006	0.01	0.03	0.002	0.036	0	0
SYRUP, SORGHUM	0	0	0	0	0.1	0.155	0.1	0.67	0.804	0	0
TORTILLA CHIPS, PLAIN	196	1.36	0	0	0.075	0.184	1.28	0.286	0.788	10	0
TORTILLA CHIPS, TACO-FLAVOR	905	0	0	0.9	0.242	0.204	2	0.297	0.29	21	0
AMERICAN FAST FOODS											
BISCUIT, W/EGG, CHS, & BACON	450	0		1.1	0.21	0.3	1.6	0.07	0.82	26	0.73
BISCUIT, W/HAM	118	1.984		0.1	0.45	0.28	3.08	0.12	0.36	7	0.03
BURRITO, W/BNS & CHS	672	0		0.9	0.12	0.38	1.92	0.13	0.86	44	0.48
BURRITO, W/BNS & MEAT	275	0		0.8	0.23	0.36	2.34	0.16	0.97	32	0.75
CHEESEBURGER, SINGLE PATTY, W/CONDMNT	409	0.47	12	1.7	0.22	0.2	3.29	0.1	0.28	16	0.83
CHICKEN FILLET SNDWCH, W/CHS	272	0		1.3	0.18	0.2	3.98	0.18	0.59	20	0.2
CHICKEN, BREADED & FRIED (BREAST/WING)	118	0		0	0.09	0.18	7.35	0.35	1.59	5	0.41
CHICKEN, BREADED & FRIED (DRUMSTK/THIGH)	150	0		0	0.09	0.29	4.87	0.22	1.66	6	0.56

(Continued)

FOOD*, by Major Food Group	Vit. A (IU)	Vit. D (IU)	Vit. E (mg TE)	Vit. K (mg)	Vit. C (mg)	Thiamin (mg)	Riboflavin (mg)	Niacin (mg)	Vit. B6 (mg)	Pantothenic (mg)	Folate (µg)	Vit. B12 (µg)
CHILI CON CARNE	657		0		0.6	0.05	0.45	0.98	0.13	1.42	12	0.45
CROISSANT, W/EGG, CHS, & HAM	297		0		7.5	0.34	0.2	2.1	0.15	0.82	24	0.66
ENCHILADA, W/CHS & BF	591				0.7	0.05	0.21	1.31	0.14	0.75	100	0.53
ENGLISH MUFFIN, W/EGG, CHS, & CANADIAN BACON	428		0.621		1.3	0.361	0.327	2.43	0.107	0.653	32	0.49
FRENCH TOAST STKS	32		2.81		0	0.16	0.18	2.1	0.18	0.4	95	0.05
HAMBURGER, SINGLE PATTY, W/CONDMNT	70	12	0.012		2.1	0.274	0.222	3.69	0.111	0.263	49	1.03
HOTDOG, PLAIN	0		0		0.1	0.24	0.28	3.72	0.05	0.52	30	0.52
HOTDOG, W/CHILI	51		0		2.4	0.19	0.35	3.28	0.04	0.48	44	0.26
HOTDOG, W/CORN FLR COATING (CORNDOG)	118		0		0	0.16	0.4	2.38	0.05	0.77	34	0.25
HUSH PUPPIES	120		0		0	0	0.03	2.6	0.13	0.28	27	0.22
NACHOS, W/CHS	495		0		1.1	0.17	0.33	1.36	0.18	1.16	9	0.73
ONION RINGS, BREADED & FRIED	10		0.4		0.7	0.1	0.12	1.11	0.07	0.24	14	0.15
PIZZA W/CHEESE, MEAT, & VEG	663		0		2	0.27	0.22	2.48	0.12	1.05	34	0.46
PIZZA W/CHEESE	607		0		2	0.29	0.26	3.94	0.07	0.35	93	0.53
PIZZA W/PEPPERONI	397		0		2.3	0.19	0.33	4.29	0.08	0.35	74	0.26
POTATO SALAD	100		0		1.1	0.07	0.11	0.27	0.15	0.37	25	0.12
POTATO, BKD & TOPPED W/CHS SAU & BACON	210		0		9.6	0.09	0.08	1.33	0.25	0.43	10	0.11
POTATO, BKD & TOPPED W/CHS SAU & BROCCOLI	500		0		14.3	0.08	0.08	1.06	0.23	0.42	18	0.1
POTATO, MASHED	41		0		0.4	0.09	0.05	1.2	0.23	0.48	8	0.05
POTATOES, HASHED BROWN	25		0.17		7.6	0.11	0.02	1.49	0.23	0.47	11	0.02
ROAST BEEF SANDWICH, PLAIN	151		0		1.5	0.27	0.22	4.22	0.19	0.6	29	0.88
SALAD, TACO	297		0		1.8	0.05	0.18	1.24	0.11	0.68	20	0.32

Food										
SALAD, TOSSED, WO/DRSNG, W/TURKEY, HAM & CHS	323	0	5	0.12	0.12	1.83	0.13	0.28	31	0.26
SALAD, TOSSED, WO/DRSNG, W/CHICK	429	0	8	0.05	0.06	2.7	0.2	0.27	31	0.09
SALAD, TOSSED, WO/DRSNG, W/CHS & EGG	379	0	4.5	0.04	0.08	0.45	0.05	0.24	39	0.14
SALAD, TOSSED, WO/DRSNG	1136	0	23.2	0.03	0.05	0.55	0.08	0.12	37	0
SUBMARINE SANDWICH, W/COLD CUTS	186	0	5.4	0.44	0.35	2.41	0.06	0.39	24	0.48
SUBMARINE SANDWICH, W/RST BF	191	0	2.6	0.19	0.19	2.76	0.15	0.36	21	0.84
SUBMARINE SANDWICH, W/TUNA SALAD	73	0	1.4	0.18	0.13	4.43	0.09	0.73	22	0.63
TACO	500	0	1.3	0.09	0.26	1.88	0.14	0.99	14	0.61
TOSTADA, W/BNS, BF, & CHS	567	0	1.8	0.04	0.22	1.27	0.11	0.83	43	0.5

*Abbreviations: BKD, baked; BLD, boiled; BRSD, braised; CHS, cheese; CKD, cooked; CND, canned; DRND, drained; ENR, enriched; FRZ, frozen; HYDR, hydrogenated; PREPD, prepared; RST, roasted; SAU, sausage; STWD, stewed.
Sources: Nutrient Data Laboratory, Agricultural Research Service, US Department of Agriculture.

Vitamin Content of Feedstuffs (units per kg)

Feedstuff	Vit. E (IU)	Riboflavin (mg)	Niacin (mg)	Vit. B_6 (mg)	Biotin (mg)	Pantothenic Acid (mg)	Folate (mg)	Vit. B_{12} (µg)	Choline (mg)
Alfalfa leaf meal, dehydrated	140	15	55	11	0.35	33	4		1,600
Alfalfa meal, dehydrated	120	13	46	10	0.33	27	3.5		1,600
Alfalfa meal, sun-cured	66	11	40	9	0.3	20	3.3		1,500
Bakery product, dehydrated	25	0.8	50	4.4	0.07	9	0.15		660
Barley	36	2	57	2.9	15	6.6	0.5		1,100
Beans, field	1	1.8	24	0.3	0.11	3.1	1.3		
Blood meal	0	4.2	29	0		5.3			280
Brewers' dried grains	26	1.5	44	0.66		8.8	9.7		1,600
Buckwheat		11	18			5.9			13,000
Buttermilk, dried	6.3	30	9	2.4	0.3	30	0.4	20	1,800
Casein, purified	1.5	1.3	0.4		2.6	0.4		200	
Citrus pulp, dried	2.2	22			13			900	
Coconut oil	35	0	0	0	0	0	0	0	0
Coconut oil meal (copra meal)	3.5	24	4.4		6.6	0.3		1,100	
Corn and cob meal	20	1.1	20	5	0.05	5	0.3		550
Corn germ meal	87	3.7	42		3	3.3	0.7		1,540
Corn gluten feed	24	2.2	66	5.3	0.3	0.5	0.2		1,100
Corn gluten meal	42	1.5	50	8	0.15	10	0.7		330
Corn gluten meal, 60% protein	50	1.8	60	9.6	0.2	12	0.84		400
Corn oil	280	0	0	0	0	0	0	0	0
Corn, dent, no. 2, yellow	22	1.3	22	7	0.06	5.7	0.36		620
Cottonseed meal, dehulled	5.7	51	7	0.1	15	1.1		3,300	
Cottonseed meal, hyradraulic/expeller	40	4	5	5.3	0.1	11	1		2,800
Cottonseed meal, solvent	15	5	44	6.4	0.1	13	1		2,900
Crab meal		5.9	44			6.6		330	2,000
Distillers' dried grains (corn)	30	3.1	42		0.7	5.9			1,900
Distillers' dried grains w/sol's (corn)	40	8.6	66		1.1	11	0.9		2,500
Distillers' dried solubles (corn)	55	17	115	10	1.5	22	2.2		4,800
Feathers, poultry, hydrolyzed	2	24	44	44	11	0.22	70	900	

Fish meal, anchovetta	3.4	6.6	64	3.5	8.8	0.26	0.2	100	3,700
Fish meal, herring	27	9	89	3.7	11	0.42	0.24	240	4,000
Fish meal, menhaden	9	4.8	55	3.5	8.8	0.26	0.2	88	3,500
Fish meal, pilchard	9	9.5	55	3.5	9	0.26	0.2	100	2,200
Fish meal, redfish waste	6			3.3		0.08	0.2	100	3,500
Fish meal, whitefish waste	9	9	70	3.3	8.8	0.08	0.2	100	2,200
Fish oils, stabilized	70	0	0	0	0	0	0	0	0
Fish solubles, dried	6	7.7	230	0.13	45	0.26		400	5,300
Hominy feed, yellow	2.2	44	11	0.13	7.7	0.28		1,000	
Lard, stabilized	23	0	0	0	0	0	0	0	0
Liver and glandular meal	40	160	5	0.8	4	105	440	10,500	
Meat and bone meal, 45% protein	1	5.3	38	2.3	2.4	0.1	0.05	44	2,000
Meat and bone meal, 50% protein	1	4.4	49	2.5	3.7	0.14	0.05	44	2,200
Meat meal, 55% protein	1	5.3	57	3	4.8	0.26	0.05	44	2,200
Milk, dried skim	20	11	4.9	0.33	0.02	33	60	1,400	
Milo (grain sorghum)	12	1.2	40	4	11	0.18	0.24		680
Molasses, beet	0.4	40	5.4	88	0.2	66		880	
Molasses, cane	5	2.5	100	4.4	58	100	0.04		880
Oat mill by-product	1.5	10				3.3		440	
Oatmeal feed	24	1.8	13	2.2	15	0.22	0.35		1,200
Oats, heavy	20	1.1	18	1.3	13	0.11	0.3		1,100
Olive oil	125								
Peanut meal, dehulled, solvent	3	12	180	10	60	0.39	0.36		2,100
Peanut meal, solvent	3	11	170	10	53	0.39	0.36		2,000
Peanut oil	280	0	0	0	0	0	0	0	0
Peas, field dry	1.8	37	1	0.18	0.36	10	0.36		
Potato meal, white, dried	0.7	33	14	0.1	0.6	20		2,600	
Poultry by-product meal	2	11	40		8.8	0.3	1	6,000	2,000
Poultry offal fat, stabilized	30	0	0	0	0	0	0	0	0
Rapeseed meal	19	3.7	155	9		0.3		6,600	
Rice bran oil	420								

(Continued)

Feedstuff	Vit. E (IU)	Riboflavin (mg)	Niacin (mg)	Vit. B$_6$ (mg)	Biotin (mg)	Pantothenic Acid (mg)	Folate (mg)	Vit. B$_{12}$ (µg)	Choline (mg)
Rice bran	60	2.6	300		0.42	23			1,300
Rice polishings	90	1.8	530		0.62	57			1,300
Rice, rough	14	0.5	37			3.3	0.25		800
Rice, white, polished	3.6	0.6	14	0.4		3.3	0.15		900
Safflower meal	1	4	6		1.4	4	0.44		2,600
Safflower oil	500	0	0	0	0	0	0	0	0
Sesame meal	3.3	30	12.5		6			1,500	
Sesame oil	250	0	0	0	0	0	0	0	0
Soybean meal	2	3.3	27	8	0.32	14.5	3.6		2,700
Soybean meal, dehulled	3.3	3.1	22	8	0.32	14.5	3.6		2,700
Soybean oil	280	0	0	0	0	0	0	0	0
Soybean, isolated protein	0	1.25	4.9	1.3		0.63			0
Soybeans, full-fat, processed	50	2.6	22	11	0.37	15	2.2		2,800
Sunflower oil	350	0	0	0	0	0	0	0	0
Sunflower seed meal, dehulled, solvent	20	7.2	106	16		40			4,200
Sunflower seed meal, solvent	11	6.4	91	16	0	10			2,900
Tallow, stabilized	13	0	0	0	0	0	0	0	0
Tomato pomace, dried	6.2								
Wheat bran	17	3.1	200	10	0.48	29	0.78		1,000
Wheatgerm meal	130	5	50	13	0.22	12	2.4		3,300
Wheat middlings	44	2	100	11	0.37	20	1.1		1,100
Wheat shorts	57	2	95	11	0.37	8	1.1		930
Wheat, hard, northern US & Canada	11	1.1	60	4	0.11	13	0.4		1,000
Wheat, hard, south-central US	11	2	60	4	0.11	13	0.35		1,000
Wheat, soft	11	1.1	60	4	0.11	13	0.3		1,000
Whey, dried		30	11	2.5	0.25	47	0.58	0.3	200
Whey product, dried		40	15	3.2	0.28	60	0.8	40	260
Yeast, brewers', dried	0	35	450	3.3	1.3	110	12		3,900
Yeast, torula, dried	0	44	500		2	83	21		2,900

Source: *From Scott, M. L., Nesheim, M. C., and Young, R. J. (1982). The Nutrition of the Chicken, 3rd edn. M.L. Scott & Associates, Ithaca, NY, pp.490–493; with permission.*

Note: Page numbers followed by f, t, or n indicate figures, tables, and footnotes respectively

Printed in the United States
By Bookmasters